Prokaryotic Metabolism and Physiology

Determination of the genome sequences for a wide range of bacteria and archaea has made an in-depth knowledge of prokaryotic metabolic function even more essential in order to give biochemical, physiological and ecological meaning to the genomic information. Clearly describing the important metabolic processes that occur under different conditions and in different environments, this advanced text provides an overview of the key cellular processes that determine prokaryotic roles in the environment, biotechnology and human health. Structure and composition are described as well as the means by which nutrients are transported into cells across membranes. Discussion of biosynthesis and growth is followed by detailed accounts of glucose metabolism through glycolysis, the TCA cycle, electron transport and oxidative phosphorylation, as well as other trophic variations found in prokaryotes including the use of organic compounds other than glucose, anaerobic fermentation, anaerobic respiration, chemolithotrophy and photosynthesis. The regulation of metabolism through control of gene expression and enzyme activity is also covered, as well as the survival mechanisms used under starvation conditions.

Professor Byung Hong Kim is an expert on anaerobic metabolism, organic degradation and bioelectrochemistry. He graduated from Kyungpook National University, Korea and obtained a PhD from University College Cardiff. He has carried out research at several universities around the world, with an established career in the Korea Institute of Science and Technology. Currently he is teaching at the National University of Malaysia. He has been honoured by the Korean Government, which designated his research group a National Research Laboratory, the Bioelectricity Laboratory, and has served as President of the Korean Society for Microbiology and Biotechnology. Professor Kim wrote the classic Korean microbiology text on *Microbial Physiology* and has published over 200 refereed papers and reviews, and holds over 20 patents relating to applications of his research in environmental and microbial biotechnology.

Professor Geoffrey Michael Gadd is an authority on microbial interactions with metals and minerals, their geomicrobial significance and applications in environmental biotechnology. He holds the Boyd Baxter Chair of Biology and leads the Geomicrobiology Group at the University of Dundee and was founding Head of the Division of Molecular Microbiology in the School of Life Sciences. He has published over 300 refereed scientific papers, books, chapters and reviews and has received invitations to speak at international conferences in over 30 countries. Professor Gadd has served as President of the British Mycological Society and is an elected Fellow of the Royal Society of Biology, the American Academy of Microbiology, the Linnean Society, the Learned Society of Wales, the Royal Society of Edinburgh and elected Member of the European Academy of Microbiology. He has received the Berkeley Prize and President's Award from the British Mycological Society, the Charles Thom Award from the Society for Industrial Microbiology and the Colworth Prize from the Microbiology Society for his research contributions to the microbiological sciences.

Prokaryotic Metabolism and Physiology

SECOND EDITION

Byung Hong Kim

Korea Institute of Science and Technology
National University of Malaysia

Geoffrey Michael Gadd

University of Dundee

Shaftesbury Road, Cambridge CB2 8EA, United Kingdom

One Liberty Plaza, 20th Floor, New York, NY 10006, USA

477 Williamstown Road, Port Melbourne, VIC 3207, Australia

314–321, 3rd Floor, Plot 3, Splendor Forum, Jasola District Centre, New Delhi – 110025, India

103 Penang Road, #05–06/07, Visioncrest Commercial, Singapore 238467

Cambridge University Press is part of Cambridge University Press & Assessment, a department of the University of Cambridge.

We share the University's mission to contribute to society through the pursuit of education, learning and research at the highest international levels of excellence.

www.cambridge.org
Information on this title: www.cambridge.org/9781316622919

DOI: 10.1017/9781316761625

First published 2008
3rd printing 2013
Second edition 2019

A catalogue record for this publication is available from the British Library

ISBN 978-1-107-17173-2 Hardback
ISBN 978-1-316-62291-9 Paperback

Additional resources for this publication are at www.cambridge.org/ProkaryoticMetabolism.

To our families
Hyungock Hong, Kyoungha Kim and Youngha Kim
and
Julia, Katie and Richard Gadd

Contents in brief

Contents

Preface for the second edition

Since the first edition of *Bacterial Physiology and Metabolism* was published in 2008, significant progress in many areas has been made, requiring extensive revision of the first edition. Furthermore, some important topics were not adequately covered in the first edition. These include the modified TCA cycles in cyanobacteria and obligately fermentative bacteria (Section 5.2.2), novel TCA cycle intermediate replenishment mechanisms (ethylmalonyl-CoA pathway, Section 5.3.3 and methylaspartate cycle, Section 5.3.4), archaeal pentose metabolism (Section 7.2.3), methane oxidation in anaerobic environments (Section 9.9.2), elucidation of novel CO_2 fixation cycles (the 4-hydroxybutyrate cycles, Section 10.8.5), bacterial immune systems (Section 13.5), toxin/antitoxin systems (Section 13.4.2) and competence (Section 13.6). Also included in this edition are accounts of the synthesis of the non-canonical amino acids, pyrrolysine and selenocysteine, and their codon usage. Analysis of bacterial genomes has led to the identification of many novel mechanisms of metabolic regulation, including two-component systems and small non-coding RNAs (discussed in Chapter 12). Another intriguing discovery is the use of certain rare earth elements by methylotrophs (Section 2.1). The book title has also been amended to *Prokaryotic Metabolism and Physiology* to reflect the increasing content of archaeal processes. We hope this second edition is received as well as the first edition.

The authors would like to express their appreciation to Professors K. S. Kim, S. H. Bang, J. H. Shun, J. K. Lee and I. S. Chang for reading parts of the manuscript, Ms. Y. J. Kim for preparing the figures and the staff of Cambridge University Press involved in various stages of the publication process, including Katrina Halliday, Jenny van der Meijden and Lindsey Tate.

Byung Hong Kim

Geoffrey Michael Gadd

Preface for the first edition

Knowledge of the physiology and metabolism of prokaryotes underpins our understanding of the roles and activities of these organisms in the environment, including pathogenic and symbiotic relationships, as well as their exploitation in biotechnology. Prokaryotic organisms include bacteria and archaea and, although remaining relatively small and simple in structure throughout their evolutionary history, exhibit incredible diversity regarding their metabolism and physiology. Such metabolic diversity is reflective of the wide range of habitats where prokaryotes can thrive and in many cases dominate the biota, and is a distinguishing contrast with eukaryotes that exhibit a more restricted metabolic versatility. Thus, prokaryotes can be found almost everywhere under a wide range of physical and chemical conditions, including aerobic to anaerobic, light and dark, low to high pressure, low to high salt concentrations, extremes of acidity and alkalinity, and extremes of nutrient availability. Some physiologies, e.g. chemolithotrophy and nitrogen fixation, are only found in certain groups of prokaryotes, while the use of inorganic compounds, such as nitrate and sulfate, as electron acceptors in respiration is another prokaryotic ability. The explosion of knowledge resulting from the development and application of molecular biology to microbial systems has perhaps led to a reduced emphasis on their physiology and biochemistry, yet paradoxically has enabled further detailed analysis and understanding of metabolic processes. Almost in a reflection of the bacterial growth pattern, the number of scientific papers has grown at an exponential rate, while the number of prokaryotic genome sequences determined is also increasing rapidly. This production of genome sequences for a wide range of organisms has made an in-depth knowledge of prokaryotic metabolic function even more essential in order to give biochemical, physiological and ecological meaning to the genomic information. Our objective in writing this new textbook was to provide a thorough survey of the prokaryotic metabolic diversity that occurs under different conditions and in different environments, emphasizing the key biochemical mechanisms involved. We believe that this approach provides a useful overview of the key cellular processes that determine bacterial and archaeal roles in the environment, biotechnology and human health. We concentrate on bacteria and archaea but, where appropriate, also provide comparisons with eukaryotic organisms. It should be noted that many important metabolic pathways found in prokaryotes also occur in eukaryotes further emphasizing prokaryotic importance as research models in providing knowledge of relevance to eukaryotic processes.

This book can be considered in three main parts. In the first part, prokaryotic structure and composition is described as well as the means by which nutrients are transported into cells across membranes. Discussion of biosynthesis and growth is followed by detailed accounts of glucose metabolism through glycolysis, the TCA cycle, electron transport and oxidative phosphorylation, largely based on the model bacterium *Escherichia coli*. In the second part, the trophic variations found in prokaryotes are described, including the use of organic compounds other than glucose, anaerobic fermentation, anaerobic respiration, chemolithotrophy and photosynthesis. In the third part, the regulation of metabolism through control of gene expression and enzyme activity is covered, as well as the survival mechanisms used by prokaryotes under starvation conditions. This text is relevant to advanced undergraduate and postgraduate courses, as well as being of use to teachers and researchers in microbiology, molecular biology, biotechnology, biochemistry and related disciplines.

We would like to express our thanks to all those who helped and made this book possible. We appreciate the staff of Academy Publisher (Seoul, Korea) who redrew the figures for the book, and those at Cambridge University Press involved at various stages of the publication process, including Katrina Halliday, Clare

Georgy, Dawn Preston, Alison Evans and Janice Robertson. Special thanks also go to Diane Purves in Dundee, who greatly assisted correction, collation, editing and formatting of chapters, and production of the index, and Dr Nicola Stanley-Wall, also in Dundee, for the cover illustration images. Thanks also to all those teachers and researchers in microbiology around the world who have helped and stimulated us throughout our careers. Our families deserve special thanks for their support and patience.

Byung Hong Kim
Geoffrey Michael Gadd

Chapter 1

Introduction to prokaryotic metabolism and physiology

The biosphere has been shaped both by physical events and by interactions with the organisms that occupy it. Among living organisms, prokaryotes are much more metabolically diverse than eukaryotes and can also thrive under a variety of extreme conditions where eukaryotes cannot. This is possible because of the wealth of genes, metabolic pathways and molecular processes that are unique to prokaryotic cells. For this reason, prokaryotes are very important in the cycling of elements, including carbon, nitrogen, sulfur and phosphorus, as well as metals and metalloids such as copper, mercury, selenium, arsenic and chromium. Prokaryotes are important not only for shaping the biosphere, but are also involved in the health of plants and animals including humans. Disease-causing bacteria have been a major concern in microbiology from the dawn of the science, while recent developments in 'microbiome' research reveal paramount roles in the well-being of higher plants and animals. A full understanding of the complex biological phenomena that occur in the biosphere therefore requires a deep knowledge of the unique biological processes that occur in this vast prokaryotic world.

After publication in 1995 of the first full DNA sequence of a free-living bacterium, *Haemophilus influenzae*, whole genome sequences of thousands of prokaryotes have now been determined and many others are currently being sequenced (see https://gold.jgi.doe.gov/ and https://www.ncbi.nlm.nih.gov/guide/genomes-maps/). Our knowledge of the whole genome profoundly influences all aspects of microbiology. Determination of entire genome sequences, however, is only a first step in fully understanding the properties of an organism and its interactions with the environment in which it lives. The functions encoded by these sequences need to be elucidated to give biochemical, physiological and ecological meaning to the information. Furthermore, sequence analysis indicates that the biological functions of substantial portions of complete genomes are so far unknown. Defining the role of each gene in the complex cellular metabolic network is a formidable task. In addition, genomes contain hundreds to thousands of genes, many of which encode multiple proteins that interact and function together as multicomponent systems for accomplishing specific cellular processes. The products of many genes are often co-regulated in complex signal transduction networks, and understanding how the genome functions as a whole presents an even greater challenge. It is also known that for a significant proportion of metabolic activities, no representative genes have been identified across all organisms, such activities being termed as 'orphan' to indicate they are not currently assigned to any gene. This also represents a major future challenge and will require both computational and experimental approaches.

It is widely accepted that less than 1 per cent of prokaryotes have been cultivated in pure culture under laboratory conditions. This is also true of the majority of species associated with higher organisms, including humans, which have not been isolated in pure culture but play important roles in the well-being of the host. Development of new sequencing techniques has allowed us to obtain genomic information from the multitudes of unculturable prokaryotic species and complex microbial populations that exist in nature. Such information might provide a basis for the development of new cultivation techniques. Elucidation of the function of unknown genes through a better understanding of biochemistry and physiology could ultimately result in a fuller understanding of the complex biological phenomena occurring in the biosphere.

Unlike multicellular eukaryotes, individual cells of unicellular prokaryotes are more exposed to the continuously changing environment, and have evolved unique structures and metabolic processes to survive under such conditions. Chapter 2 describes the main aspects of the composition and structure of prokaryotic cells.

Life can be defined as a reproduction process using materials available from the environment according to the genetic information possessed by the organism. Utilization of the materials available in the environment necessitates transport into cells that are separated from the environment by a membrane. Chapter 3 outlines transport mechanisms, not only for intracellular entry of nutrients, but also for excretion of materials including extracellular enzymes and materials that form cell surface structures.

Many prokaryotes, including *Escherichia coli*, can grow in a simple mineral salts medium containing glucose as the sole organic compound. Glucose is metabolized through glycolytic pathways and the tricarboxylic acid (TCA) cycle, supplying all carbon skeletons, energy in the form of ATP and reducing equivalents in the form of NADPH for growth and reproduction. Glycolysis is described in Chapter 4 with emphasis on the reverse reactions of the EMP pathway and on prokaryote-specific metabolic pathways. When substrates other than glucose are used, parts of the metabolic pathways are employed in either forward or reverse directions. Chapter 5 describes the TCA cycle and related metabolic pathways, and energy transduction mechanisms. Chapter 6 describes the biosynthetic metabolic processes that utilize carbon skeletons, ATP and NADPH, the production of which is discussed in the previous chapters. These chapters summarize the biochemistry of central metabolism that is employed by prokaryotes to enable growth on a glucose–mineral salts medium.

The next five chapters describe metabolism in some of the various trophic variations found in prokaryotes. These are the use of organic compounds other than glucose as carbon and energy sources (Chapter 7), anaerobic fermentation (Chapter 8), anaerobic respiratory processes (Chapter 9), chemolithotrophy (Chapter 10) and photosynthesis (Chapter 11). Some of these metabolic processes are prokaryote specific, while others are found in both prokaryotes and eukaryotes.

Prokaryotes only express a proportion of their genes at any given time, just like eukaryotes. This enables them to grow in the most efficient way under any given conditions. Metabolism is regulated not only through control of gene expression but also by controlling the activity of enzymes. These regulatory mechanisms are discussed in Chapter 12. Finally, the survival of prokaryotic organisms under starvation conditions is discussed in terms of storage materials, resting cell structures and population survival in Chapter 13.

This book has been written as a text for senior students at undergraduate level and postgraduates in microbiology and related subjects. A major proportion of the book has been based on review papers published in various scientific journals including those listed below:

Annual Review of Microbiology
Annual Review of Biochemistry
Current Opinion in Microbiology
FEMS Microbiology Reviews
Journal of Bacteriology
Microbiology and Molecular Biology Reviews (formerly *Microbiology Reviews*)
Nature Reviews Microbiology
Trends in Microbiology

The authors would also like to acknowledge the authors of the books listed below that have been consulted during the preparation of this book.

Caldwell, D. R. (2000). *Microbial Physiology and Metabolism*, 2nd edn. Belm, CA: Star Publishing Co.

Dawes, D. A. (1986). *Microbial Energetics*. Glasgow: Blackie.

Dawes, I. W. & Sutherland, I. W. (1992). *Microbial Physiology*, 2nd edn. Basic Microbiology Series, 4. Oxford: Blackwell.

Gottschalk, G. (1986). *Bacterial Metabolism*, 2nd edn. New York: Springer-Verlag.

Ingraham, J. L., Maaloe, O. & Neidhardt, F. C. (1983). *Growth of the Bacterial Cell*. Sunderland, MA: Sinauer Associates Inc.

Mandelstam, J., McQuillin, K. & Dawes, I. (1982). *Biochemistry of Bacterial Growth*, 3rd edn. Oxford: Blackwell.

Moat, A. G., Foster, J. W. & Spector, M. P. (2002). *Microbial Physiology*, 4th edn. New York: Wiley.

Neidhardt, F. C. & Curtiss, R. (eds.) (1996). *Escherichia coli and Salmonella: Cellular and Molecular Biology*, 2nd edn. Washington, DC: ASM Press.

Neidhardt, F. C., Ingraham, J. L. & Schaechter, M. (1990). *Physiology of the Bacterial Cell: A Molecular Approach*. Sunderland, MA: Sinauer Associates Inc.

Stanier, R. J., Ingraham, J. L., Wheelis, M. K. & Painter, P. R. (1986). *The Microbial World*, 5th edn. Upper Saddle River, NJ: Prentice-Hall.

White, D. (2000). *The Physiology and Biochemistry of Prokaryotes*, 2nd edn. Oxford: Oxford University Press.

Further Reading

Note this section contains key references only. Additional recommended references are available at www.cambridge.org/ProkaryoticMetabolism.

General

Downs, D. M. (2006). Understanding microbial metabolism. *Annual Review of Microbiology* **60**, 533–559.

Solden, L., Lloyd, K. & Wrighton, K. (2016). The bright side of microbial dark matter: lessons learned from the uncultivated majority. *Current Opinion in Microbiology* **31**, 217–226.

Diversity

Achtman, M. & Wagner, M. (2008). Microbial diversity and the genetic nature of microbial species. *Nature Reviews Microbiology* **6**, 431–440.

Bertin, P. N., Medigue, C. & Normand, P. (2008). Advances in environmental genomics: towards an integrated view of micro-organisms and ecosystems. *Microbiology* **154**, 347–359.

Fernandez, L. A. (2005). Exploring prokaryotic diversity: there are other molecular worlds. *Molecular Microbiology* **55**, 5–15.

Ecology and Geomicrobiology

Ehrlich, H. L., Newman, D. K. & Kappler, A. (2015). *Ehrlich's Geomicrobiology*, 6th edn. Boca Raton, FL, USA: CRC Press.

Gadd, G. M., Semple, K. T. & Lappin-Scott, H. M. (2005). *Micro-organisms and Earth Systems: Advances in Geomicrobiology*. Cambridge: Cambridge University Press.

Konhauser, K. O. (2007). *Introduction to Geomicrobiology*. Malden, MA, USA: Blackwell Science Ltd.

Madsen, E. L. (2015). *Environmental Microbiology: From Genomes to Biogeochemistry*, 2nd edn. New York: Wiley.

Shively, J. M., English, R. S., Baker, S. H. & Cannon, G. C. (2001). Carbon cycling: the prokaryotic contribution. *Current Opinion in Microbiology* **4**, 301–306.

Vorholt, J. A. (2012). Microbial life in the phyllosphere. *Nature Reviews Microbiology* **10**, 828–840.

Evolution

Boucher, Y., Douady, C. J., Papke, R. T., Walsh, D. A., Boudreau, M. E., Nesbo, C. L., Case, R. J. & Doolittle, W. F. (2003). Lateral gene transfer and the origins of prokaryotic groups. *Annual Review of Genetics* **37**, 283–328.

Boyd, E. F., Almagro-Moreno, S. & Parent, M. A. (2009). Genomic islands are dynamic, ancient integrative elements in bacterial evolution. *Trends in Microbiology* **17**, 47–53.

Koch, A. L. & Silver, S. (2005). The first cell. *Advances in Microbial Physiology* **50**, 227–259.

van der Meer, J. R. & Sentchilo, V. (2003). Genomic islands and the evolution of catabolic pathways in bacteria. *Current Opinion in Biotechnology* **14**, 248–254.

Genomics

Chun, J. & Rainey, F. A. (2014). Integrating genomics into the taxonomy and systematics of the Bacteria and Archaea. *International Journal of Systematic and Evolutionary Microbiology* **64**, 316–324.

Francke, C., Siezen, R. J. & Teusink, B. (2005). Reconstructing the metabolic network of a bacterium from its genome. *Trends in Microbiology* **13**, 550–558.

Loman, N. J. & Pallen, M. J. (2015). Twenty years of bacterial genome sequencing. *Nature Reviews Microbiology* **13**, 787–794.

Medini, D., Serruto, D., Parkhill, J., Relman, D. A., Donati, C., Moxon, R., Falkow, S. & Rappuoli, R. (2008). Microbiology in the post-genomic era. *Nature Reviews Microbiology* **6**, 419–430.

Ward, N. & Fraser, C. M. (2005). How genomics has affected the concept of microbiology. *Current Opinion in Microbiology* **8**, 564–571.

Extreme environments

Bowers, K. & Wiegel, J. (2011). Temperature and pH optima of extremely halophilic archaea: a mini-review. *Extremophiles* **15**, 119–128.

Cowan, D. A. (2004). The upper temperature for life – where do we draw the line? *Trends in Microbiology* **12**, 58–60.

Javaux, E. J. (2006). Extreme life on Earth: past, present and possibly beyond. *Research in Microbiology* **157**, 37–48.

Human microbiome

Bashan, A., Gibson, T. E., Friedman, J., Carey, V. J., Weiss, S. T., Hohmann, E. L. & Liu, Y.-Y. (2016). Universality of human microbial dynamics. *Nature* **534**, 259–262.

Bauer, K. C., Huus, K. E. & Finlay, B. B. (2016). Microbes and the mind: emerging hallmarks of the gut microbiota–brain axis. *Cellular Microbiology* **18**, 632–644.

Borody, T. J. & Khoruts, A. (2012). Fecal microbiota transplantation and emerging applications. *Nature Reviews Gastroenterology and Hepatology* **9**, 88–96.

Consortium, T. H. M. P. (2012). Structure, function and diversity of the healthy human microbiome. *Nature* **486**, 207–214.

Garrett, W. S. (2015). Cancer and the microbiota. *Science* **348**, 80–86.

Grice, E. A. & Segre, J. A. (2011). The skin microbiome. *Nature Reviews Microbiology* **9**, 244–253.

Hooper, L. V., Littman, D. R. & Macpherson, A. J. (2012). Interactions between the microbiota and the immune system. *Science* **336**, 1268–1273.

O'Toole, P. W. & Jeffery, I. B. (2015). Gut microbiota and aging. *Science* **350**, 1214–1215.

Sharon, G., Sampson, T. R., Geschwind, D. H. & Mazmanian, S. K. (2016). The central nervous system and the gut microbiome. *Cell* **167**, 915–932.

Sommer, F., Anderson, J. M., Bharti, R., Raes, J. & Rosenstiel, P. (2017). The resilience of the intestinal microbiota influences health and disease. *Nature Reviews Microbiology* **15**, 630–638.

Chapter 2

Composition and structure of prokaryotic cells

Like all organisms, microorganisms grow, metabolize and replicate utilizing materials available from the environment. Such materials include those chemical elements required for structural aspects of cellular composition and metabolic activities such as enzyme regulation and redox processes. To understand bacterial metabolism, it is therefore helpful to know the chemical composition of the cell and component structures. This chapter describes the elemental composition and structure of prokaryotic cells, and the kinds of nutrients needed for biosynthesis and energy-yielding metabolism.

2.1 Elemental composition

From over 100 naturally occurring elements, microbial cells generally only contain 12 in significant quantities. These are known as major elements, and are listed in Table 2.1 together with some of their major functions and predominant chemical forms used by microorganisms.

They include elements such as carbon (C), oxygen (O) and hydrogen (H) constituting organic compounds like carbohydrates. Nitrogen (N) is found in microbial cells in proteins, nucleic acids and coenzymes. Sulfur (S) is needed for S-containing amino acids, such as methionine and cysteine, and for various coenzymes. Phosphorus (P) is present in nucleic acids, phospholipids, teichoic acid and nucleotides including NAD(P) and ATP. Potassium is the major inorganic cation (K^+), while chloride (Cl^-) is the major inorganic anion. K^+ is required as a cofactor for certain enzymes, e.g. pyruvate kinase. Chloride is involved in the energy conservation process utilized by halophilic archaea (Section 11.6). Sodium (Na^+) participates in several transport and energy transduction processes, and plays a crucial role in microbial growth under alkaline conditions (Section 5.7.4). Magnesium (Mg^{2+}) forms complexes with phosphate groups including those found in nucleic acids, ATP, phospholipids and lipopolysaccharides. Several microbial intracellular enzymes, e.g. monomeric alkaline phosphatase, are calcium dependent. Ferrous and ferric ions play a crucial role in oxidation–reduction reactions as components of electron carriers such as Fe-S proteins and cytochromes.

In addition to these 12 major elements, others are also found in microbial cells as minor elements (Table 2.2). All the metals listed in Table 2.2 are required for specific enzymes. It is interesting to note that the atomic number of tungsten is far higher than that of the other elements, and that this element is only required in rare cases.

In addition to those listed in Table 2.2, some unusual elements are used by microorganisms. Under molybdenum-limited conditions, a vanadium-containing nitrogenase is synthesized in nitrogen-fixing organisms (Section 6.2.1.2). A claim that arsenate can substitute for phosphate in the synthesis of certain macromolecules,

Table 2.1 | Major elements found in microbial cells with their functions and predominant chemical forms used by microorganisms.

Element	Atomic number	Chemical forms used by microbes	Function
C	6	organic compounds, CO, CO_2	major constituents of cell material in proteins, nucleic acids, lipids, carbohydrates and others
O	8	organic compounds, CO_2, H_2O, O_2	
H	1	organic compounds, H_2O, H_2	
N	7	organic compounds, NH_4^+, NO_3^-, N_2	
S	16	organic sulfur compounds, SO_4^{2-}, HS^-, S^0, $S_2O_3^{2-}$	proteins, coenzymes
P	15	HPO_4^{2-}, organophosphate organic phosphonates, phosphite	nucleic acids, phospholipids, teichoic acid, coenzymes
K	19	K^+	major inorganic cation, compatible solute, enzyme cofactor
Mg	12	Mg^{2+}	enzyme cofactor, bound to cell wall, membrane and phosphate esters including nucleic acids and ATP
Ca	20	Ca^{2+}	enzyme cofactor, bound to cell wall
Fe	26	Fe^{2+}, Fe^{3+}	cytochromes, ferredoxin Fe-S proteins, enzyme cofactor
Na	11	Na^+	involved in transport and energy transduction
Cl	17	Cl^-	major inorganic anion

including DNA, in a bacterium isolated from an arsenic-rich lake has been extensively debated. A similarly toxic metal, cadmium, is contained in the carbonic anhydrase of a marine diatom. Growth of extremely acidophilic methanotrophic *Methylacidiphilum fumariolicum* on methanol is strictly dependent on the presence of lanthanides, a group of the rare earth elements (REEs) that includes lanthanum (La), cerium (Ce), praseodymium (Pr) and neodymium (Nd). Lanthanides are an essential cofactor in its methanol dehydrogenase (MDH). This enzyme (XoxF) is different from the MxaF-type MDH, containing calcium as a catalytic cofactor of *Methylobacterium extorquens* (Section 7.10.2). The XoxF-type MDH is induced by La and Ce in other methylotrophs, *Bradyrhizobium* sp. and *Methylobacterium radiotolerans*. Metagenomic studies have shown that XoxF-type MDH is much more prominent in nature than the MxaF-type enzymes. Silicate can be solubilized by a group of bacteria known as silicate bacteria, including *Bacillus circulans*, but silicon does not appear to have any essential roles in prokaryotic biology.

2.2 Importance of chemical form

2.2.1 Five major elements

The elements listed in Tables 2.1 and 2.2 need to be supplied or be present in the chemical forms that the organisms can use. Carbon is the most abundant element in all living organisms. Prokaryotes are broadly classified according to the carbon sources they use: organotrophs

Table 2.2 Minor elements found in microbial cells with their functions and predominant chemical form used by microorganisms.

Element	Atomic number	Chemical form used by microbes	Function
Mn	25	Mn^{2+}	superoxide dismutase, photosystem II
Co	27	Co^{2+}	coenzyme B_{12}
Ni	28	Ni^{2+}	hydrogenase, urease
Cu	29	Cu^{2+}	cytochrome oxidase, oxygenase
Zn	30	Zn^{2+}	alcohol dehydrogenase, aldolase, alkaline phosphatase, RNA and DNA polymerase, arsenate reductase
Se	34	SeO_3^{2-}	formate dehydrogenase, glycine reductase
Mo	42	MoO_4^{2-}	nitrogenase, nitrate reductase, formate dehydrogenase, arsenate reductase
W	74	WO_4^{2-}	formate dehydrogenase, aldehyde oxidoreductase

(heterotrophs) use organic compounds as their carbon source while CO_2 is used by lithotrophs (autotrophs). These groups are divided further according to the form of energy they use: chemotrophs (chemoorganotrophs and chemolithotrophs) depend on chemical forms for energy, while phototrophs (photoorganotrophs and photolithotrophs) utilize light energy ('organo' refers to an organic substance while 'litho' refers to an inorganic substance).

Nitrogen sources commonly used by microbes include organic nitrogenous compounds such as amino acids, and inorganic forms such as ammonium and nitrate. Gaseous N_2 can serve as a nitrogen source for nitrogen-fixing prokaryotes. Nitrogen fixation is not known in eukaryotic microorganisms. Some chemolithotrophs can use ammonium as their energy source (electron donor, Section 10.2) while nitrate can be used as an electron acceptor by denitrifiers (Section 9.1).

Sulfate is the most commonly used sulfur source, while other sulfur sources used include organic sulfur compounds, sulfide, elemental sulfur and thiosulfate. Sulfide and sulfur can serve as electron donors in certain chemolithotrophs (Section 10.3), and sulfate and elemental sulfur are used as electron acceptors and are reduced to sulfide by sulfidogens (Section 9.3).

Phosphate is the most common phosphorus (P) source used by microorganisms, and many microorganisms use organophosphate. When the phosphate supply is limited, various organophosphonates are used as a P source in microorganisms. Unlike organophosphates that possess C–O–P linkages, phosphonates have direct C–P linkages. These are not only produced in nature as antibiotics (e.g. fosfomycin) and herbicides (e.g. phosphinothricin), but are also chemically synthesized for various commercial applications such as the broad-spectrum herbicide glyphosate. Phosphite is also used as a P source by many marine microorganisms.

2.2.2 Oxygen

Oxygen in cells originates mainly from organic compounds, water or CO_2. Molecular oxygen (O_2) is seldom used in biosynthetic processes. Some prokaryotes use O_2 as the electron acceptor, but some cannot grow in its presence. Thus, organisms can be grouped according to their reaction with O_2, into aerobes that require O_2, facultative anaerobes that use O_2 when it is available but can also grow in its absence, and obligate anaerobes that do not use O_2. Some obligate anaerobes cannot grow, and die in the presence of O_2, while others can tolerate it. The former are termed strict anaerobes and the latter aerotolerant anaerobes.

2.2.3 Growth factors

Some organotrophs, such as *Escherichia coli*, can grow in simple media containing glucose and mineral salts, while others, like lactic acid bacteria, require complex media containing various compounds such as vitamins, amino acids and nucleic acid bases. This is because the latter organisms cannot synthesize certain essential cellular materials from only glucose and mineral salts. These required compounds should therefore be supplied in the growth media: such compounds are known as growth factors. Growth factor requirements differ between organisms, with vitamins being the most commonly required growth factors (Table 2.3).

Table 2.3 Common growth factors required by prokaryotes and their major functions.

Growth factor	Function
p-aminobenzoate	component of tetrahydrofolate, a one-carbon unit carrier
Biotin	prosthetic group of carboxylase and mutase
Coenzyme M	methyl carrier in methanogenic archaea
Folate	component of tetrahydrofolate
Haemin (Hemin)	precursor of cytochromes and haemoproteins
Lipoate	prosthetic group of 2-keto acid decarboxylase
Nicotinate	precursor of pyridine nucleotides (NAD^+, $NADP^+$)
Pantothenate	precursor of coenzyme A and acyl carrier protein
Pyridoxine	precursor of pyridoxal phosphate
Riboflavin	precursor of flavins (FAD, FMN)
Thiamine	precursor of thiamine pyrophosphate
Vitamin B_{12}	precursor of coenzyme B_{12}
Vitamin K	precursor of menaquinone

2.3 Structure of microbial cells

Microorganisms are grouped into either prokaryotes or eukaryotes according to their cellular structure. With only a few exceptions, prokaryotic cells do not have subcellular organelles separated from the cytoplasm by phospholipid membranes, such as the eukaryotic nuclear and mitochondrial membranes. Organelles, like the nucleus, mitochondria and endoplasmic reticulum, are only found in eukaryotic cells. The detailed structure of prokaryotic cells is described below.

2.3.1 Flagella and pili

Motile prokaryotic cells have an appendage called a flagellum (plural, flagella), involved in motility, and a similar but smaller structure, the fimbria (plural, fimbriae). Fimbriae are not involved in motility and are composed of proteins.

The bacterial flagellum consists of three parts. These are a basal body, a hook and a filament (Figure 2.1). The basal body is embedded in the cytoplasmic membrane and cell surface structure and connected to the filament through the hook. In Gram-negative bacteria the basal body consists of a cytoplasmic membrane ring, a periplasmic ring and an outer-membrane ring through which the central rod passes. The diameter of the rings can be 20–50 nm depending on the species. The cytoplasmic ring of the basal body is associated with additional proteins known as the Mot complex. The Mot complex rotates the basal body with the entire flagellum consuming a proton motive force (or sodium motive force). The cytoplasmic membrane ring is therefore believed to function as a motor with the Mot complex. A more detailed description of motility is given in Section 12.2.11. In addition to the Mot complex, the basal body is associated with an export apparatus through which the building blocks of the filament are transported.

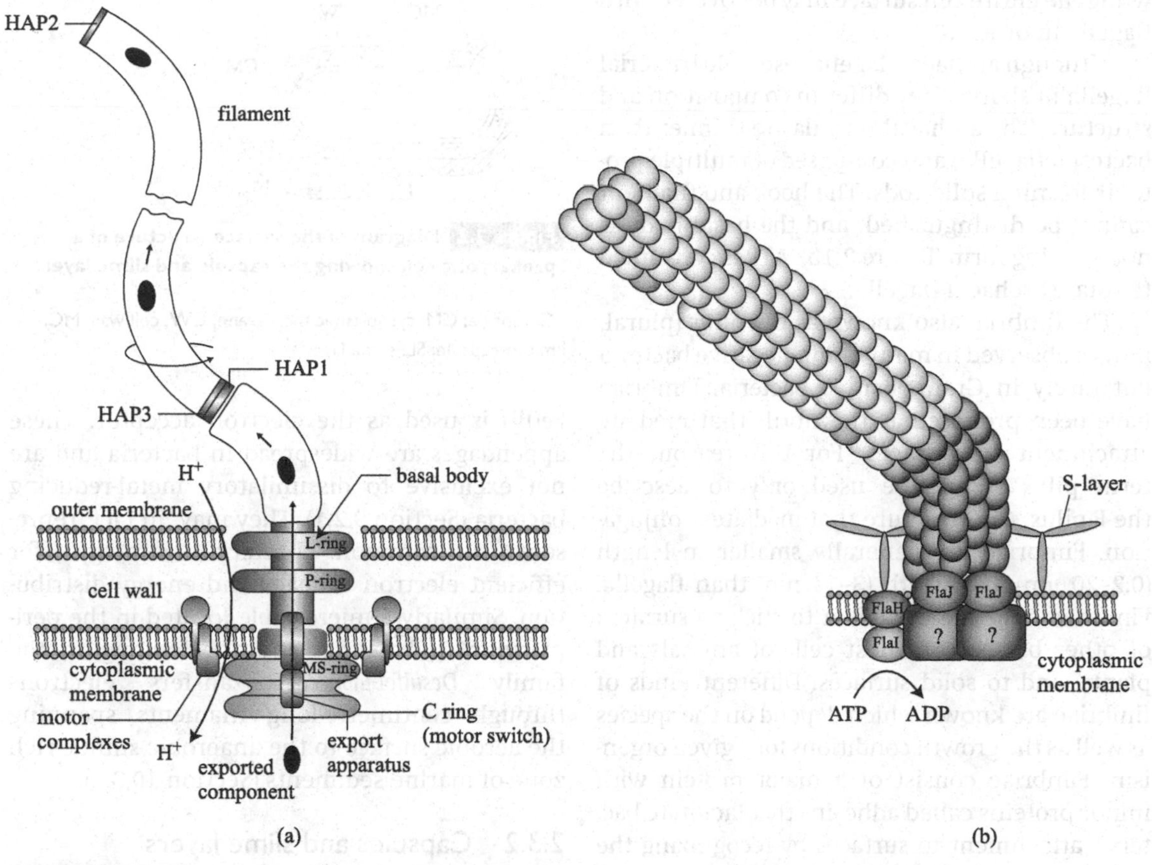

Figure 2.1 **The structure of the flagellum in (a) Gram-negative bacteria and (b) archaea.**

(Modified from *Nature Rev. Microbiol.* **6**: 466–476, 2008)

(a) Gram-negative bacterial flagellum. Three rings of the basal body are embedded in the cytoplasmic membrane, peptidoglycan layer and outer membrane. The outer filament is connected to the basal body. The cytoplasmic membrane ring of the basal body is associated with the Mot complex. This complex functions like a motor, rotating the flagellum thus rendering the cell motile. Energy for this rotation is provided from the proton (or sodium) motive force.

(b) The archaeal flagellum. The filament is composed of multiple proteins and the hook cannot be distinguished. ATP provides energy for flagellar rotation.

HAP1 and HAP3, hook-associated proteins; HAP2, filament cap.

Springer, with permission.

The hook connects the central rod of the basal body to the filament and is composed of a single protein called the hook protein. The filament, with a diameter of 10–20 nm, can be dissolved at pH 3–4 with surfactants to a single protein solution of flagellin. The molecular weight of flagellin varies from 20 to 65 kD depending on the bacterial species. The hook and the filament are tube-shaped and the flagellin moves through the tube to the growing tip of the filament. The tip of the filament is covered with filament cap protein. Flagellin can be exported to the medium in mutants defective in expression of this protein.

The number and location of flagella vary depending on the bacterial species. In some prokaryotes they are located at one or both poles, while the entire cell surface may be covered with flagella in others.

Although archaeal flagella resemble bacterial flagella in shape, they differ in composition and structure. The archaeal flagella are thinner than bacterial flagella, and composed of multiple proteins forming solid rods. The hook and filament cannot be distinguished, and the basal body is not in a ring form (Figure 2.1b). ATP is consumed to rotate archaeal flagella.

The fimbria, also known as the pilus (plural, pili), is observed in many Gram-negative bacteria but rarely in Gram-positive bacteria. Fimbriae have been proposed as the fibrils that mediate attachment to surfaces. For this reason, the term pilus should be used only to describe the F-pilus, the structure that mediates conjugation. Fimbriae are generally smaller in length (0.2–20 μm) and width (3–14 nm) than flagella. Fimbriae help the organism to stick to surfaces of other bacteria, to host cells of animals and plants, and to solid surfaces. Different kinds of fimbriae are known which depend on the species as well as the growth conditions for a given organism. Fimbriae consist of a major protein with minor proteins called adhesins that facilitate bacterial attachment to surfaces by recognizing the appropriate receptor molecules. They are classified as type I through to type IV according to these receptor recognition properties. Adhesive properties are inhibited by sugars such as mannose, galactose and their oligomers, suggesting that the receptors are carbohydrate in nature.

A fibril bigger in size than fimbriae occurs in many Gram-negative bacteria that harbour the conjugative F-plasmid. This is called the F-pilus or sex pilus and mediates attachment between mating cells for the purpose of transmitting DNA from the donor cell by means of the F-pilus to a recipient cell. The F-pilus recognizes a receptor molecule on the surface of the recipient cell and after attachment, the F-pilus is depolymerized so that there is direct contact between the cells for DNA transmission.

Shewanella oneidensis and *Geobacter sulfurreducens* produce electrically conductive pilus-like appendages called bacterial nanowires, when Fe(III) is used as the electron acceptor. These appendages are widespread in bacteria and are not exclusive to dissimilatory metal-reducing bacteria (Section 9.2.1). They may, in fact, represent a common bacterial strategy for efficient electron transfer and energy distribution. Similarly a micro-cable located in the periplasm of a bacterial strain belonging to the family *Desulfobulbaceae* transfers electrons through centimetre-long filaments spanning the aerobic surface to the anaerobic sulfide-rich zone of marine sediments (Section 10.3.1).

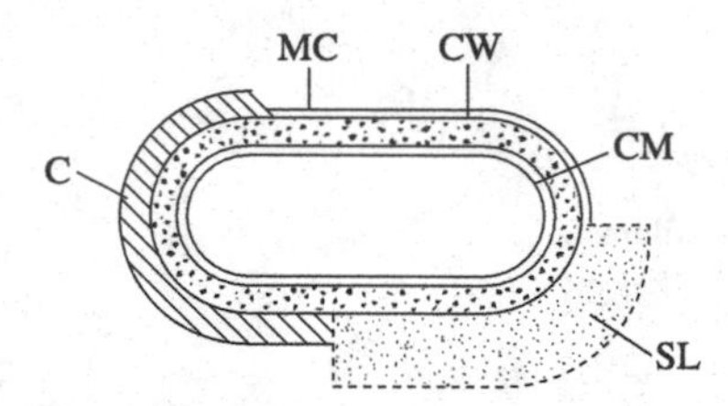

Figure 2.2 Diagram of the surface structure of a prokaryotic cell showing the capsule and slime layer.

C, capsule; CM, cytoplasmic membrane; CW, cell wall; MC, micro-capsule; SL, slime layer.

2.3.2 Capsules and slime layers

Many prokaryotic cells are covered with polysaccharides known as extracellular polymeric substances (EPS). In some cases the polymers are tightly integrated with the cell, while in others they are loosely associated. The former is called a capsule, and the latter a slime layer (Figure 2.2). Slime layer materials can diffuse into the medium, their structure and composition being dependent on growth conditions. An important role for these structures is adhesion to host cells for invasion or to a solid surface to initiate and stabilize biofilm formation. These structures are also responsible for resistance to phagocytosis, thereby increasing virulence. In some bacteria the capsule functions as a receptor for phage. Since the polysaccharides are hydrophilic, they can also protect cells from desiccation.

The term glycocalyx can be used to describe extracellular structures including the capsule and S-layer, the latter being described below.

2.3.3 S-layer, outer membrane and cell wall

Unicellular prokaryotes have elaborate surface structures. These include the S-layer, outer membrane and cell wall. The cell wall determines the physical shape of prokaryotic cells in most cases. Prokaryotes can be classified into four groups according to their cell wall structure. These are mycoplasmas, Gram-negative bacteria, Gram-positive bacteria and archaea. Cell walls are not found in mycoplasmas which are obligate intracellular pathogens. Peptidoglycan (murein) is the sole building block of the cell wall in Gram-negative bacteria, which also have an outer membrane. Peptidoglycan and teichoic acid constitute the cell wall in Gram-positive bacteria, which do not possess an outer membrane. Some archaea have cell walls that do not contain peptidoglycan, while others are devoid of cell walls.

2.3.3.1 S-layer

A protein or glycoprotein layer is found on the surface of all prokaryotic cells except mycoplasmas. This is called an S-layer (Figure 2.3). All prokaryotic cells studied (except certain intracellular pathogens of animal cells) are surrounded by this layer, but this property can be lost in some laboratory strains. This suggests that the layer is indispensable in natural environments. The proposed functions of the S-layer are (1) protection from toxic compounds, (2) adhesion to solid surfaces, (3) a phage receptor, (4) a physical structure to maintain cell morphology and (5) a binding site for certain extracellular enzymes.

Strains devoid of an S-layer are less resistant to peptidoglycan- and protein-hydrolysing enzymes, and are more prone to release from biofilms with environmental changes, such as pH. The S-layer serves as a cell wall in some archaea. Amylase is bound to the S-layer in *Bacillus stearothermophilus*.

Lysinibacillus sphaericus isolated from a toxic metal contaminated environment possesses an S-layer with outstanding metal-binding and recrystallization properties that enable the bacteria to survive by binding metals with high affinity. Conversely, essential trace elements are able to cross the S-layer and reach the interior of the cell. The genome of this bacterium encodes at least eight putative S-layer protein genes with distinct differences. The presence of multiple S-layer gene copies may enable bacterial strains to quickly adapt to changing environments.

The protein forming the S-layer is one of the most abundant proteins in bacterial cells, comprising 5–10 per cent of total cellular protein. In a fast growing organism, this protein needs to be synthesized and exported very efficiently. The promoter and signal sequence of this protein has therefore been studied for use in foreign protein production using bacteria.

2.3.3.2 Outer membrane

Gram-negative bacteria are more resistant to lysozyme, hydrolytic enzymes, surfactants, bile salts and hydrophobic antibiotics than Gram-positive bacteria. These properties are due to the presence of the outer membrane in Gram-negative bacteria (Figure 2.3). The outer membrane (OM) is different in structure from the cytoplasmic membrane (CM). The CM consists of phospholipids while lipopolysaccharide (LPS) forms the outer leaflet of the OM, with the inner leaflet composed of phospholipids. Lipopolysaccharide provides a permeability barrier against the hydrophobic compounds listed above. In addition to these lipids, the OM contains proteins and lipoproteins (Table 2.4).

Lipopolysaccharide consists of three components: lipid A, core polysaccharide and repeating polysaccharide. The repeating polysaccharide is referred to as O-antigen. Lipid A is embedded in the membrane to form the lipid layer, and the sugar moieties extend into the cell surface. The sugar moieties of LPS consist of hexoses, hexosamines, deoxyhexoses and keto-sugars with different structures depending on the species and on the culture conditions. Figure 2.4 shows the LPS structures found in *Salmonella typhimurium* and *Escherichia coli*.

O-antigen mutants of pathogens are less virulent and show a different phage susceptibility, suggesting that the O-antigen is involved in pathogenesis and phage recognition processes.

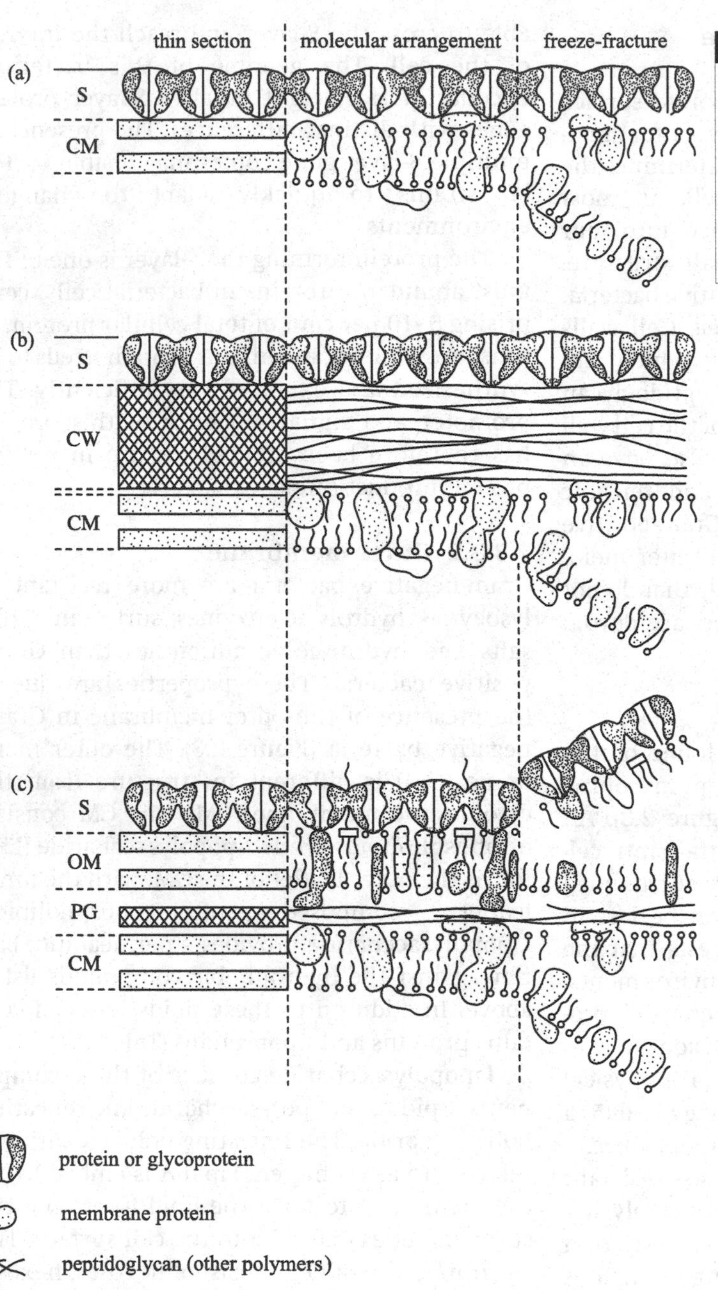

Figure 2.3 Cell surface structures of prokaryotic cells.

(a) Archaea, (b) Gram-positive bacteria, (c) Gram-negative bacteria.

S, S-layer; CM, cytoplasmic membrane; CW, cell wall; OM, outer membrane; PG, peptidoglycan.

Table 2.4 Outer membrane (OM) components and their functions in *Escherichia coli*.

Component	Function
Phospholipid	inner leaflet
Lipopolysaccharide	outer leaflet, hydrophilic in nature providing a barrier against hydrophobic compounds. Stabilization of the surface structure by bonding with metal ions such as Mg^{2+}
Lipoprotein	lipid part is embedded in the OM hydrophobic region, and the sugar part is covalently bound to peptidoglycan, which stabilizes the OM
OM protein A	maintenance of OM stability, receptor for amino acids and peptides, F-pilus in recipient cell for conjugation
Porin	three different porins, OmpC, OmpF and PhoE, each consisting of three peptides, act as specific and non-specific channels for hydrophilic solutes
Receptor proteins	for sugars, amino acids, vitamins, etc.
Other proteins	enzymes such as phospholipase, protease, etc., extracellular protein export machinery

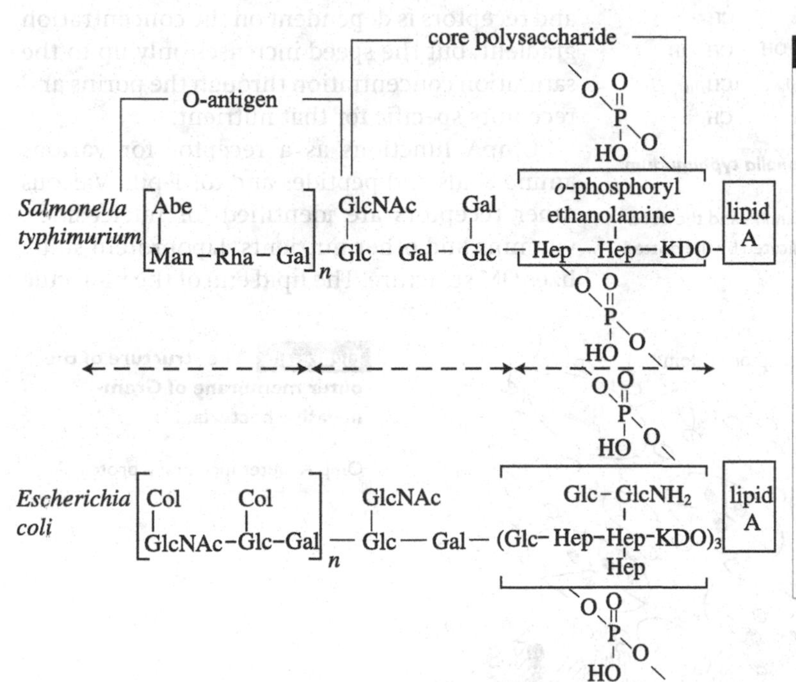

Figure 2.4 **Lipopolysaccharide structure in *Salmonella typhimurium* and *Escherichia coli*.**

Lipopolysaccharide (LPS) consists of lipid A, core polysaccharide and repeating polysaccharide. Lipid A forms a bilayer membrane with the lipid part of phospholipid and the sugar part extending into the surface. The repeating polysaccharide is involved in pathogenesis and is called O-antigen. LPS contains the unique sugars L-glycero-D-mannoheptose (Hep) and 2-keto-3-deoxyoctonate (KDO), and rare sugars such as abequose (Abe) and colitose (Col). Galactose (Gal), glucose (Glc), mannose (Man) and rhamnose (Rha) can also be present.

The structure of O-antigen varies with changes in growth conditions such as temperature, and this polysaccharide may therefore protect the cell during such changes. Lipid A consists of 6–7 molecules of 3-hydroxy fatty acids bound to phosphorylated glucosamine (Figure 2.5).

Because of the phosphate and carboxylic moieties in LPS, the Gram-negative bacterial cell surface is negatively charged. Divalent cations such as Mg^{2+} cross-link LPS, thus stabilizing the OM structure. The OM becomes more permeable to hydrophobic compounds when the cations are removed using chelating agents such as EDTA. In addition to LPS and phospholipids, the OM contains a variety of proteins, including lipoprotein, which have various functions (Figure 2.6).

Active transport of nutrients takes place across the cytoplasmic membrane. Nutrients therefore need to cross the OM, and for this the OM has porins and receptors. Porins are hydrophilic channels formed by three peptides spanning the OM, allowing the diffusion of solutes of molecular weight less than 600 Da. *Escherichia coli* has three porins, OmpC, OmpF and PhoE – OmpC and OmpF being non-specific. OmpF is slightly bigger than OmpC, the synthesis of which is stimulated under high osmotic pressure when the synthesis of the former is repressed (Section 12.2.9). PhoE is synthesized under phosphate-limited conditions and is specific for anions. The pore size of porins varies depending on the species, and is regulated by the molecular arrangement around it. Nutrients with a higher molecular weight cross the OM through receptor proteins. The speed of diffusion across the OM through non-specific porins and receptors is dependent on the concentration gradient, but the speed increases only up to the saturation concentration through the porins and receptors specific for that nutrient.

OmpA functions as a receptor for various amino acids and peptides and for F-pili. Various other receptors are identified for nucleosides, vitamins and other nutrients. Lipoprotein stabilizes OM structure. The lipid end of the molecule

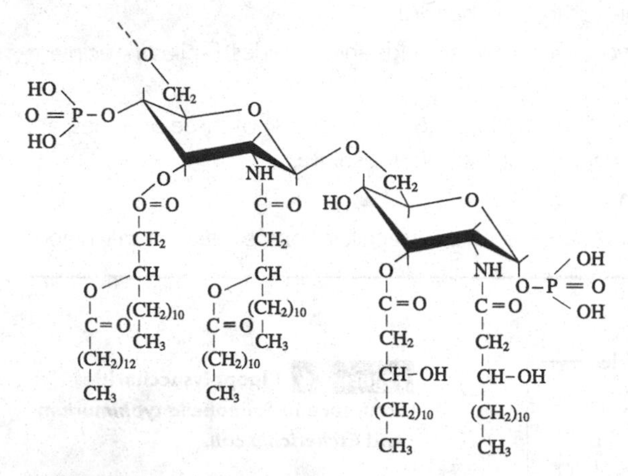

Figure 2.5 Lipid A structure in *Salmonella typhimurium*.

3-hydroxy fatty acids are bound to glucosamine and the core polysaccharide is linked to carbon 6, as indicated by the dotted line.

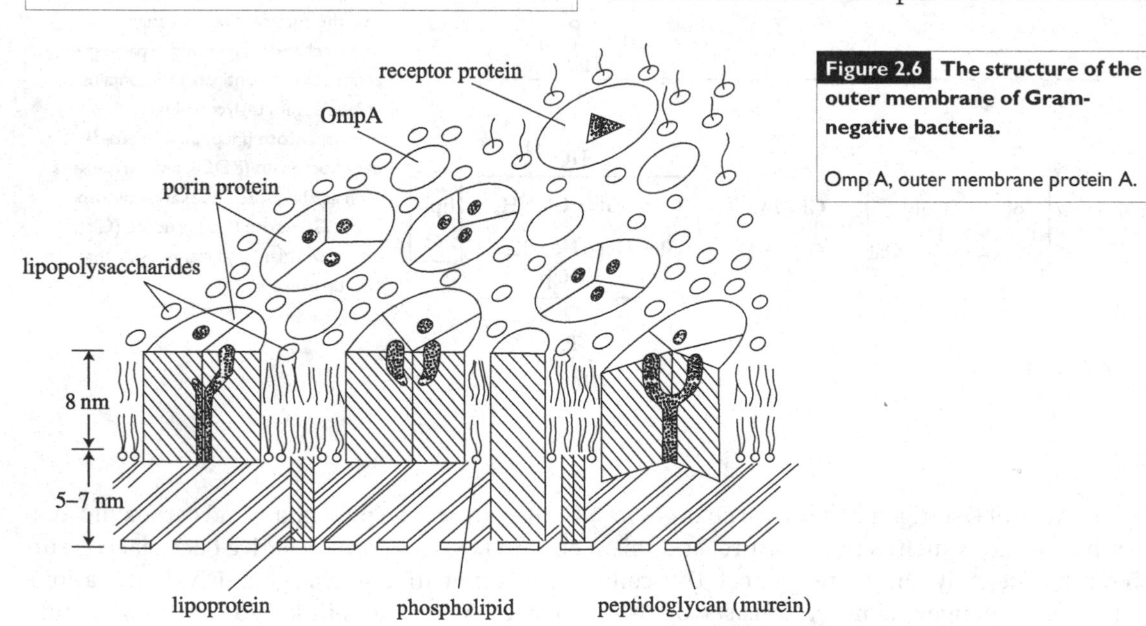

Figure 2.6 The structure of the outer membrane of Gram-negative bacteria.

Omp A, outer membrane protein A.

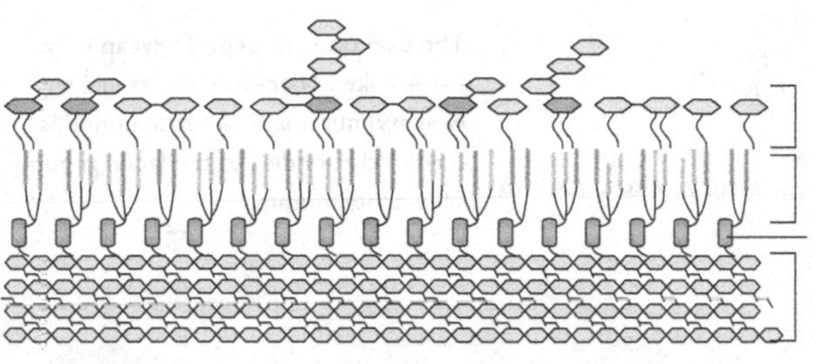

Figure 2.7 Outer membrane-like mycomembrane in species of *Mycobacterium* and *Corynebacterium*.

The mycomembrane consists of arabinogalactan, mycolic acids and lipids. Arabinogalactan esterified by the mycolic acids is covalently linked to the cell wall peptidoglycan.

is embedded in the lipid area of the OM, while the protein end is covalently bound to peptidoglycan. A lipoprotein in a plant pathogenic bacterium, *Erwinia chrysanthemi*, has pectin methylesterase activity.

Like other microbes, many Gram-negative bacteria can use polymers as their carbon and energy sources (Section 7.1) and excrete hydrolysing enzymes into their external environment. The OM has proteins responsible for the translocation of cell surface structure components (Section 3.10). Cytochromes and ferric reductase are located in the OM in Fe(III)-reducing bacteria such as *Shewanella putrefaciens* and *Geobacter sulfurreducens* (Section 9.2.1).

Many Gram-negative bacteria release outer membrane vesicles (OMVs) under conditions of stress. Their functions include protein secretion, immune activation and suppression, modulation of stress, attachment and virulence. OMVs released from *Pseudomonas aeruginosa* contain lytic enzymes that cause the lysis of other bacteria.

An anaerobic hyperthermophilic archaeon, *Ignococcus hospitalis*, is unique in having an inner and an outer membrane (OM) forming two cell compartments, a tightly packed cytoplasm surrounded by a weakly staining intermembrane compartment (IMC). This chemolithotroph uses hydrogen as the electron donor and sulfur as the electron acceptor (Section 10.5.4). The OM is energized with the energy transducing ATPase and H_2:sulfur oxidoreductase localized on it. Therefore, this membrane is referred to as the outer cellular membrane (OCM).

Some Gram-positive bacteria of the actinobacteria group, including species of *Mycobacterium* and *Corynebacterium*, have a unique outer membrane-like cell surface structure called a mycomembrane, consisting of arabinogalactan, mycolic acids and lipids. Their cell wall peptidoglycan is covalently linked to arabinogalactan which, in turn, is esterified by the mycolic acids (Figure 2.7). Mycolic acids are very long hydroxylated fatty acids with a carbon number ranging between 22 and 90, depending on the strain. In addition, lipids are found in non-covalent hydrophobic association with the mycolic acids. These lipids include trehalose dimycolate, known as the cord factor, and a wide variety of phospholipids and glycolipids in mycobacteria, and trehalose mono- and dimycolate and some phospholipids in corynebacteria. A specific mechanism is involved in protein excretion through the mycomembrane (Section 3.10.2.8). The mycolic acids of the mycomembrane render these bacteria a distinctive staining property called 'acid' or 'acid–alcohol' fastness.

2.3.3.3 Cell wall and periplasm

With a few exceptions, prokaryotic cells have a cell wall that provides the physical strength to maintain their shape. Peptidoglycan is the main component of the cell wall of bacteria. The cell wall in Gram-negative bacteria is much thinner than in Gram-positive bacteria, which have a complex cell wall with other polymers and do not possess an outer membrane (Figure 2.3).

Peptidoglycan is a polymer with a backbone of β-1,4-linked *N*-acetylglucosamine and *N*-acetylmuramate (Figure 2.8), the lactyl group of which is cross-linked through tetrapeptides (Figure 2.9). Some Gram-positive bacteria, including *Lactobacillus acidophilus*, are resistant to lysozyme since they have peptidoglycan *o*-acetylated at the 6-OH of *N*-acetylmuramate.

D-glutamate and D-alanine occupy the second and fourth positions of the tetrapeptide in most

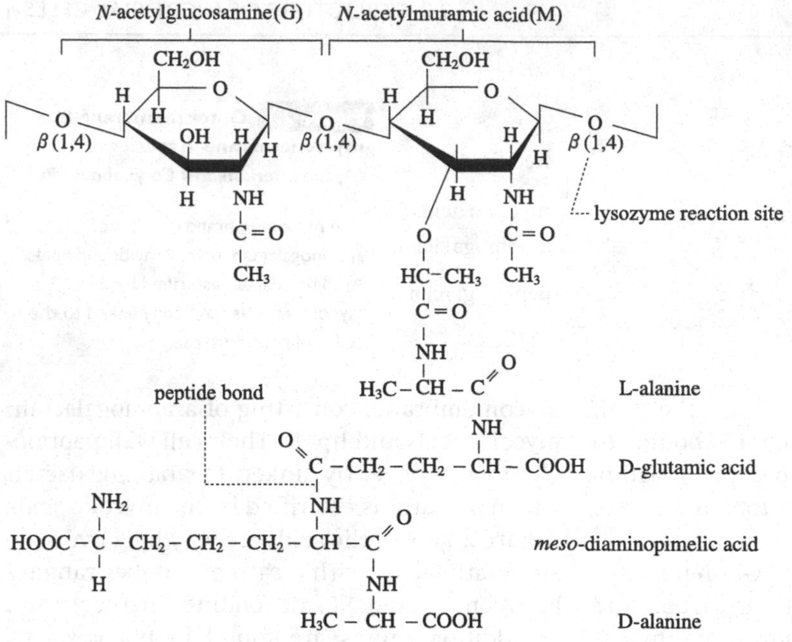

Figure 2.8 The basic unit of peptidoglycan in *Escherichia coli*.

The backbone of peptidoglycan is β-1,4-linked *N*-acetylglucosamine and *N*-acetylmuramate. A tetrapeptide is covalently bonded to the lactyl group of *N*-acetylmuramate.

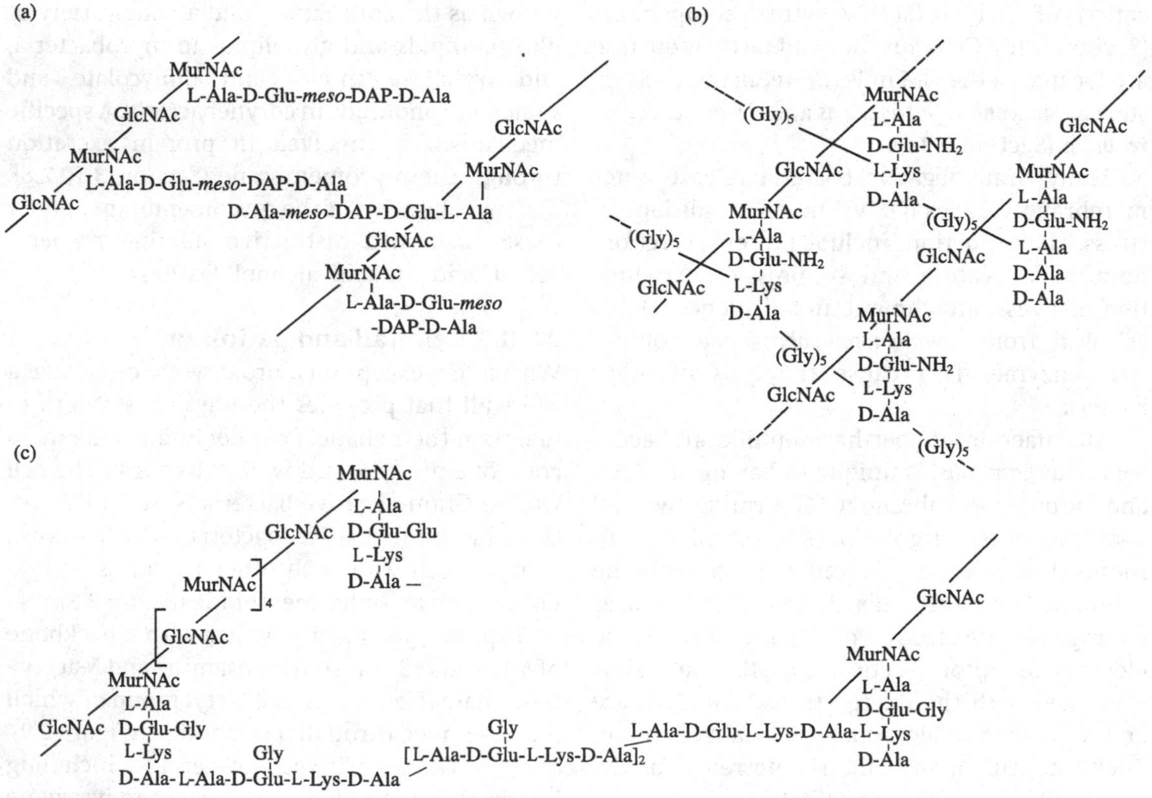

Figure 2.9 The structure of peptidoglycan: the main component of the eubacterial cell wall.

The backbone of peptidoglycan consists of a repeat of β-1,4-linked *N*-acetylglucosamine and *N*-acetylmuramate that is bonded with a tetrapeptide. The tetrapeptides of neighbouring backbones are cross-linked to make a net-like structure.

(a) *Escherichia coli*; (b) *Staphylococcus aureus*; (c) *Micrococcus luteus* (*Micrococcus lysodeikticus*).

cases, and L-serine or glycine the first position, depending on the species. Various amino acids are found in the third position, including L-ornithine, L-lysine, L,L-diaminopimelate, *meso*-diaminopimelate and rarely L-homoserine. The third amino acid in the peptidoglycan tetrapeptide is a key criterion used for bacterial classification. The degree of cross-linking differs depending on the species, and is low at the growing tip of a single bacterial cell. Peptidoglycan with reduced cross-linking is more susceptible to hydrolyzing enzymes.

Gram-positive bacteria do not have an outer membrane but have a much thicker cell wall containing teichoic acid, lipoteichoic acid and lipoglycan in addition to peptidoglycan. The term secondary cell wall polymer (SCWP) is used for Gram-positive bacterial cell wall polysaccharides other than peptidoglycan including teichoic acid.

The structure of teichoic acid differs depending on the bacterial species. Teichoic acids are polymers of ribitol phosphate, glycerol phosphate and their derivatives. In some cases, the hydroxyl groups of the poly alcohols are bonded with amino acids or sugars. Teichoic acid is linked to peptidoglycan through a linkage unit (Figure 2.10). While peptidoglycan is a structure rendering physical strength, teichoic acid has some physiological functions. Because of the phosphate in teichoic acids, the surface of Gram-positive bacteria is negatively charged, attracting cations such as Mg^{2+} which stabilize cell wall structure. An autolytic enzyme is necessary to hydrolyse peptidoglycan at the growing tips to enable cell growth. This enzyme is attached to teichoic acid, ensuring it is not released into the medium and enabling control of activity. Teichoic acid is also a receptor for certain phages. Gram-positive bacteria synthesize teichuronic acid in place of teichoic acid under phosphate-limited conditions, suggesting that this cell wall constituent is essential. It has been found that teichoic acid is dispensable in *Bacillus subtilis* and *Staphylococcus aureus*, but not lipoteichoic acid.

Lipoteichoic acid has a similar structure to teichoic acid. A diacylglycerol-containing sugar is attached to the terminal polyalcohol phosphate. This lipid part of lipoteichoic acid is embedded in the cytoplasmic membrane, while the polyalcohol phosphate partly constitutes the cell wall.

The Gram-positive bacterial cell wall contains various proteins, including the autolytic enzymes mentioned above. They are attached to the cell wall through the action of an enzyme called sortase (Section 6.9.2.4) after they are excreted through the cytoplasmic membrane. Their functions include (1) metabolism of cell surface structures, (2) invasion into the host, (3) hydrolysis of polymers such as proteins and polysaccharides and (4) adhesion to solid surfaces.

Members of the family *Mycoplasmataceae* and the class *Chlamydiae* are obligate intracellular pathogens of humans and other animals and do not have cell wall structures. However, the genome of *Chlamydiae* includes a complete set of genes for the synthesis of peptidoglycan. They are also susceptible to antibiotics such as β-lactams and D-cycloserine, which inhibit peptidoglycan synthesis. This is because an intermediate of peptidoglycan biosynthesis, bactoprenyl-P-*N*-acetylmuramate-*N*-acetylglucosamine (lipid II, Section 6.9.2.2) is essential for the formation of the multiprotein machinery responsible for cell division, the divisome complex. Cell division and cell wall biosynthesis in all prokaryotes are driven by partially overlapping multiprotein machineries whose activities are tightly controlled and coordinated. These organisms have cysteine-rich proteins in their outer membrane. Disulfide bonds in the proteins render the cells resistant to osmotic pressure.

The phylum *Planctomycetes* is one of the major divisions of the domain bacteria and is phylogenetically deep-rooted. These bacteria lack peptidoglycan in their cell wall, which mainly comprises protein, and is probably an S-layer.

The archaeal cell wall is different from the bacterial cell wall. The main components of the archaeal cell wall include pseudopeptidoglycan (pseudomurein), sulfonated polysaccharide and glycoprotein. Pseudopeptidoglycan is found in the cell wall of *Methanobacterium* species. The structure of pseudopeptidoglycan is different from that of peptidoglycan. *N*-acetyltalosaminuronate is linked to

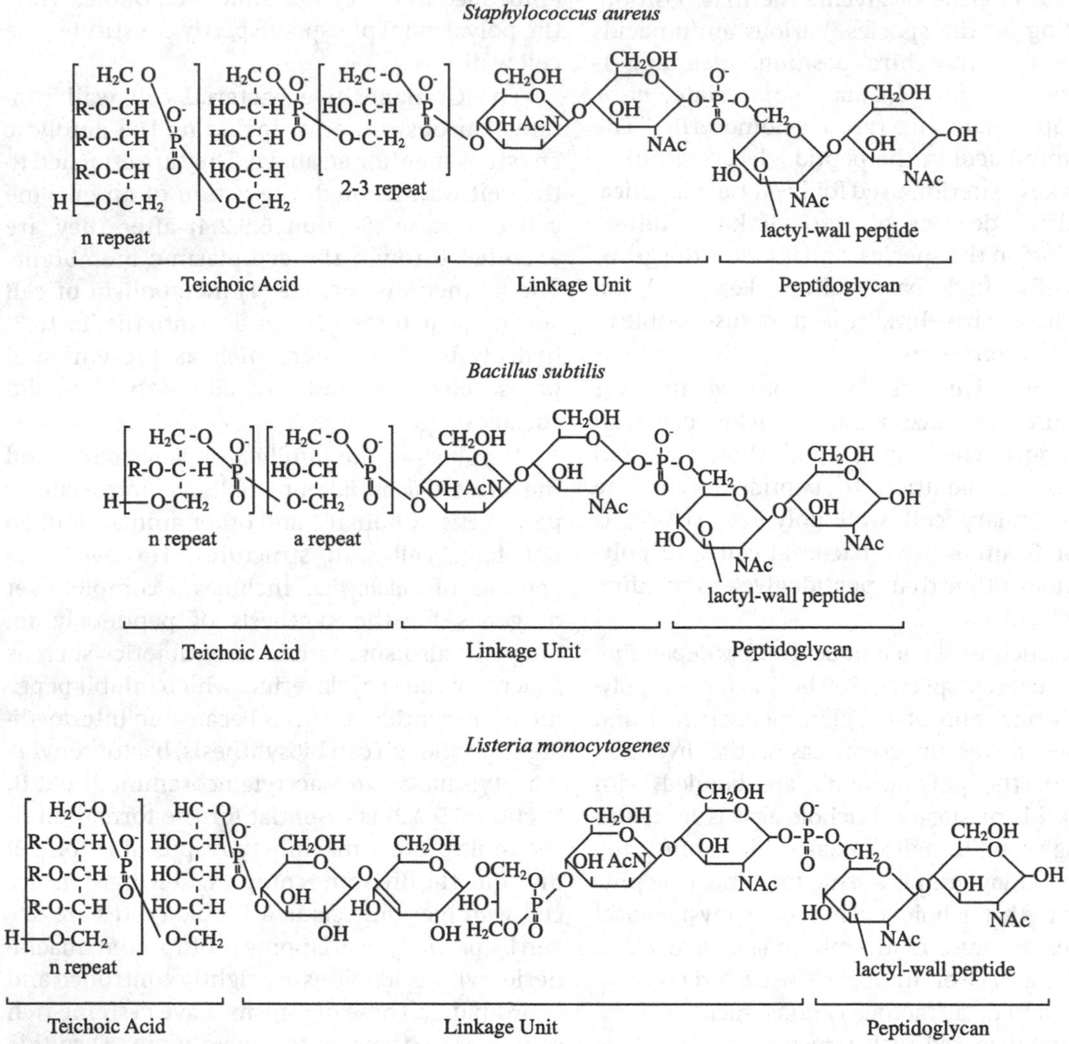

Staphylococcus aureus

Teichoic Acid | Linkage Unit | Peptidoglycan

Bacillus subtilis

Teichoic Acid | Linkage Unit | Peptidoglycan

Listeria monocytogenes

Teichoic Acid | Linkage Unit | Peptidoglycan

Figure 2.10 Structure of teichoic acid.

(*Microbiol. Mol. Biol. Rev.* 63:174–229, 1999)

Teichoic acid is a polymer of ribitol phosphate or glycerol phosphate. In some Gram-positive bacteria, the poly-alcohols are bonded with alanine, galactose and *N*-acetylglucosamine. Linkage units link teichoic acid to peptidoglycan.

N-acetylglucosamine through a β-1,3-linkage in place of the β-1,4-linked *N*-acetylmuramate. Consequently lysozyme does not hydrolyse pseudopeptidoglycan. D-amino acids are not found in pseudopeptidoglycan. Members of the genus *Methanosarcina* have a cell wall consisting of polysaccharide, and the halophilic archaeon *Halococcus* has a cell wall of sulfonated polysaccharide. Members of the genus *Halobacterium* and hyperthermophilic archaea have glycoprotein as their cell wall, probably as an S-layer.

The term periplasm is used to describe a separate compartment between the outer membrane and the cytoplasmic membrane in Gram-negative bacteria. Peptidoglycan (cell wall) is contained within this compartment. The periplasm is in a gel state and contains proteins and oligosaccharides. Under high osmotic pressure conditions, Gram-negative bacteria accumulate an oligosaccharide known as osmoregulated periplasmic glucan, which buffers the osmolarity.

A variety of proteins are found in the periplasm, including sensor proteins (Section 12.2.10), enzymes for the synthesis of cell surface components (Section 6.9), transporters (Chapter 3), solute-binding proteins, part of the electron transport system (Section 5.8), and hydrolytic enzymes such as β-lactamase, amylase and alkaline phosphatase. *Pseudomonas aeruginosa* releases outer membrane vesicles containing autolytic enzymes into the medium, causing the lysis of other bacteria.

Some filamentous bacteria have micro-cables in their periplasm that transfer electrons centimetre-distances, enabling sulfide oxidation in the sulfide-rich anaerobic zone coupled with oxygen reduction in the aerobic zone (Section 10.3.1). These are called cable bacteria and belong to the family *Desulfobulbaceae*.

2.3.4 Cytoplasmic membrane

2.3.4.1 Properties and functions

The cell contents (cytoplasm) needs to be isolated from the external environment, but at the same time nutrients must be transported into the cell. The cytoplasmic membrane mediates not only these functions but also other important physiological activities. These include solute transport (Chapter 3), oxidative phosphorylation through electron transport (Section 5.8), photosynthetic electron transport in photosynthetic prokaryotes (Section 11.4.3), maintenance of electrochemical gradients and ATP synthesis (Section 5.8.4), motility (Section 12.2.11), signal transduction (Section 12.2.10), synthesis of cell surface structures and protein secretion (Section 3.10). The cytoplasmic membrane consists of phospholipid (35–50 per cent) and protein (50–65 per cent). The phospholipid is responsible for the isolating property of the membrane, with the various proteins being involved in the rest of the membrane functions.

Phospholipids have unique properties which determine isolation of cytoplasm from the cell exterior. They are amphipathic compounds, having hydrophobic hydrocarbon chains (tails) and a hydrophilic head group in a single molecule. When an amphipathic compound is mixed with water, the hydrophilic heads and water interact with each other and the hydrophobic tails do so in such a way that the molecules form micelles or liposomes (Figure 2.11). Because of this

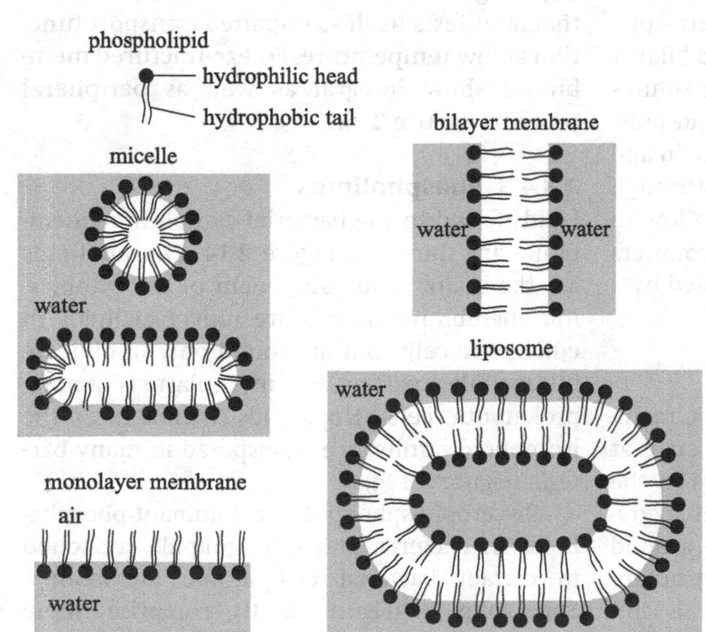

Figure 2.11 Diagram of a micelle and liposome formed when an amphipathic phospholipid is mixed with water.

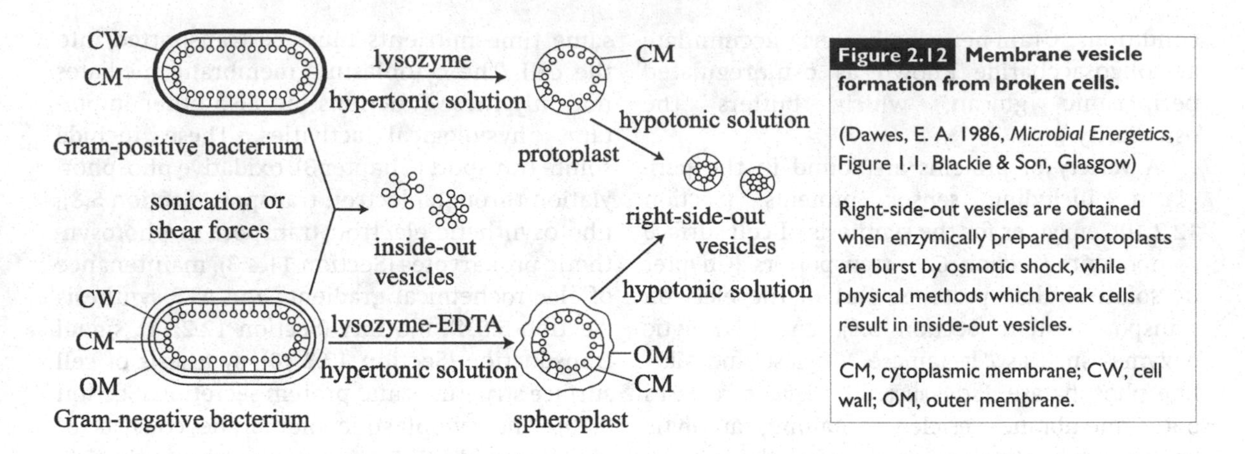

Figure 2.12 Membrane vesicle formation from broken cells.

(Dawes, E. A. 1986, *Microbial Energetics*, Figure 1.1. Blackie & Son, Glasgow)

Right-side-out vesicles are obtained when enzymically prepared protoplasts are burst by osmotic shock, while physical methods which break cells result in inside-out vesicles.

CM, cytoplasmic membrane; CW, cell wall; OM, outer membrane.

property, a damaged cytoplasmic membrane can restore its structure spontaneously, and when the cells are broken the membrane forms smaller vesicles (Figure 2.12).

When cells are broken using physical methods such as sonication or the French press, inside-out vesicles are formed, while right-side-out vesicles are formed when the protoplast is osmotically lysed after the cell wall is removed using cell wall hydrolysing enzymes. Vesicles are a useful tool to study membranes without the interference of cytosolic activities.

Phospholipid forms both inner and outer leaflets of the cytoplasmic membrane, but the membrane is asymmetrical due to proteins present in the membrane. The phospholipid bilayer membrane is permeable to hydrophobic solutes and water but not to charged solutes and polymers. Membrane proteins transport these in and out of the cell. Though water can diffuse through the membrane, the diffusion rate is too low to explain the rapid water flux which counters osmotic shock. This rapid flux is mediated by a protein, aquaporin.

2.3.4.2 Membrane structure

The hydrocarbon part of the cytoplasmic membrane phospholipids is in a semi-solid state at physiological temperature. The proteins in the membrane are clustered in functional aggregates, with the aggregates floating around within the membrane. This membrane structure is called a fluid mosaic model (Figure 2.13). Thin sections of the cytoplasmic membrane stained with metal compounds show two lines of strong electron absorption separated by a less electron dense zone (Figure 2.3). The electron dense area is the metal-binding head group separated by the hydrocarbon part which binds less metal. This structure is consistent with the behaviour of an amphipathic phospholipid.

The fluidity of the membrane is determined by the melting point of the hydrocarbon part of the phospholipid. Phospholipid containing more unsaturated fatty acid has a low melting point and is found in the cytoplasmic membrane of psychrophiles. Thermophiles have a high degree of saturation in their membrane that manifests itself as impaired transport function at low temperature. Freeze-fractured membranes show integral as well as peripheral proteins (Figure 2.3).

2.3.4.3 Phospholipids

Lipids found in the bacterial cytoplasmic membrane are shown in Figure 2.14. Phospholipids are the major lipid component of the cytoplasmic membrane. Sterols are universal lipids in eukaryotic cells, but are found only in obligate intracellular pathogenic mycoplasmas among prokaryotic cells. Hopanoids, members of the isoprenoid family, are widespread in many bacteria (Figure 2.14g).

Glycerophospholipids are dominant phospholipids in bacteria, and sphingolipids are found in certain bacterial cytoplasmic membranes. Sphingolipids (Figure 2.14f) comprise more than 50 per cent of cytoplasmic membrane

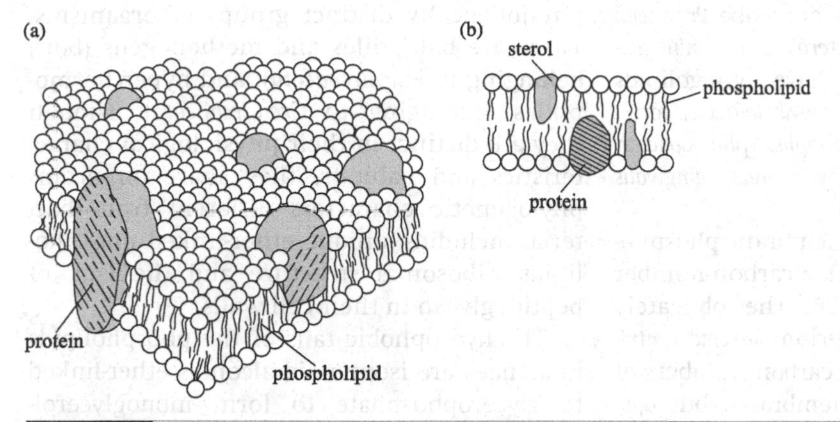

Figure 2.13 Schematic diagram of the cytoplasmic membrane.

Protein comprises about 50–65 per cent of the membrane. The proteins in the membrane are clustered in functional aggregates, and the aggregates float around within the membrane. This membrane structure is called a fluid mosaic model. Sterols are found only in obligate intracellular pathogenic mycoplasmas among prokaryotes, as in eukaryotic cells (b), but not in most prokaryotes (a). Hopanoids, members of the isoprenoid family, are widespread in many bacteria.

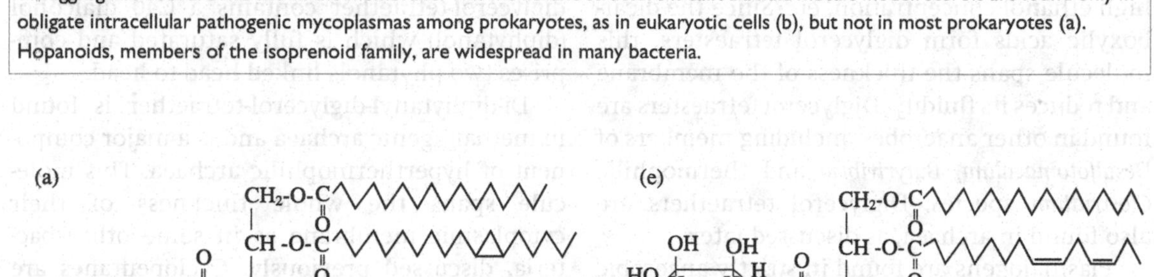

Figure 2.14 Phospholipids in bacteria.

Hydrophobic tails of phospholipids in the bacterial cytoplasmic membrane include saturated, unsaturated, branched and cyclopropane fatty acids. Phosphatidyls are the most common lipid in the bacterial cytoplasmic membrane. In most phospholipids, fatty acids form an ester linkage with glycerol, and in *Clostridium butyricum* ether forms are found as plasmalogens. Sphingolipids are the main phospholipids in certain anaerobes and hopanoids are found in many bacteria.

(a) Phosphatidylethanolamine; (b) phosphatidylglycerol; (c) phosphatidylethanolamine plasmalogen; (d) phosphatidylcholine; (e) phosphatidylinositol; (f) sphingomyelin; (g) hopanoids.

phospholipids in the obligate anaerobe *Prevotella melaninogenicus* (formerly *Bacteroides melaninogenicus*). Other bacteria known to have sphingolipids include *Prevotella ruminicola, Bdellovibrio bacterivorus, Flectobacillus major, Bacteroides fragilis, Sphingobacterium spiritivorum* and *Porphyromonas gingivalis* (formerly *Bacteroides gingivalis*).

As shown in Figure 2.14, membrane phospholipids contain fatty acids with a carbon number of 10–20, but mainly 16–20. The obligately anaerobic fermentative bacterium *Sarcina ventriculi* contains fatty acids with carbon numbers of 14–18 in its cytoplasmic membrane, but α,ω-dicarboxylic acids with carbon numbers of 28–36 become the major fatty acids in its membrane in hostile environments, such as those of low pH, high ethanol concentration, etc. Since the dicarboxylic acids form diglycerol tetraesters, this molecule spans the thickness of the membrane and reduces its fluidity. Diglycerol tetraesters are found in other anaerobes, including members of *Desulfotomaculum, Butyrivibrio* and thermophilic *Clostridium* species. Diglycerol tetraethers are also found in archaea, as discussed later.

Plasmalogens are found in strictly anaerobic bacteria (Figure 2.14 c). The C-1 of glycerol in this phospholipid has an ether linkage to a long chain alcohol. Plasmalogens are not found in aerobic, facultative anaerobic and microaerophilic bacteria. They are known in species of *Bacteroides, Clostridium, Desulfovibrio, Peptostreptococcus, Propionibacterium, Ruminococcus, Selenomonas* and *Veillonella* among others.

To maintain membrane fluidity, more saturated fatty acids are found in cells growing at higher temperature. In addition to temperature, fatty acid composition is influenced by other environmental factors, such as pH, osmotic pressure and organic solvents. Fatty acids containing a cyclopropane ring are found in many bacteria. These are known as lactobacillic acids and were first identified in *Lactobacillus arabinosus*. They are formed under certain growth conditions or at a certain growth phase. They are not present in vegetative cells of *Azotobacter vinelandii*, but cysts of this organism possess them as major fatty acids.

Archaea possess different phospholipids from bacteria. Archaea comprise three physiologically distinct groups of organisms. These are halophiles and methanogens (both belonging to *Euryarchaeota*), and hyperthermophiles (belonging to *Crenarchaeota*). Though they are distinct in their physiological characteristics and habitats, they share important phylogenetic characters different from bacteria, including unique ether-linked phospholipids, ribosomal structure and the lack of peptidoglycan in their cell walls.

The hydrophobic tails of the phospholipids in archaea are isoprenoid alcohols ether-linked to glycerophosphate to form monoglycerol-diether or diglycerol-tetraether (Figure 2.15). The alcohols are either 20 or 25 carbons long in monoglycerol-diethers. The di-diphytanyl-diglycerol-tetraether contains a C40 dialcohol (diphytanol) which is fully saturated and comprises two phytanols linked head-to-head.

Di-diphytanyl-diglycerol-tetraether is found in methanogenic archaea and is a major component of hyperthermophilic archaea. This molecule spans the whole thickness of their cytoplasmic membrane as in some other bacteria, discussed previously. Cyclopentanes are found in the hydrophobic tails of hyperthermophilic archaeal membranes. In addition to glycerol, tetritol and nonitol constitute methanogenic archaeal phospholipids. Unlike bacteria, the hydrophilic head groups in archaea are oligosaccharides or sulfonated oligosaccharides in addition to phosphate compounds.

In bacteria and eukaryotes, membrane fluidity is determined by the degree of saturation of the phospholipid fatty acid according to the temperature at which they grow. Little is known about how membrane fluidity is controlled in archaea, which can have optimum temperatures that span from ambient to over 100°C. Studies using hyperthermophiles, including *Archeaoglubus fulgidus, Acidilobus sulfurireducens, Thermophilus acidophilum, Methanocaldococcus jannaschii* and *Ignisphaera aggregans*, have shown that the cytoplasmic membrane of cells cultivated at higher temperature contains more membrane-spanning glycerol dialkyl glycerol tetraether, cyclopentane-containing biphytanyl and carbohydrate content in the hydrophilic head. The carbohydrate stabilizes the membrane through hydrogen bonding, and the

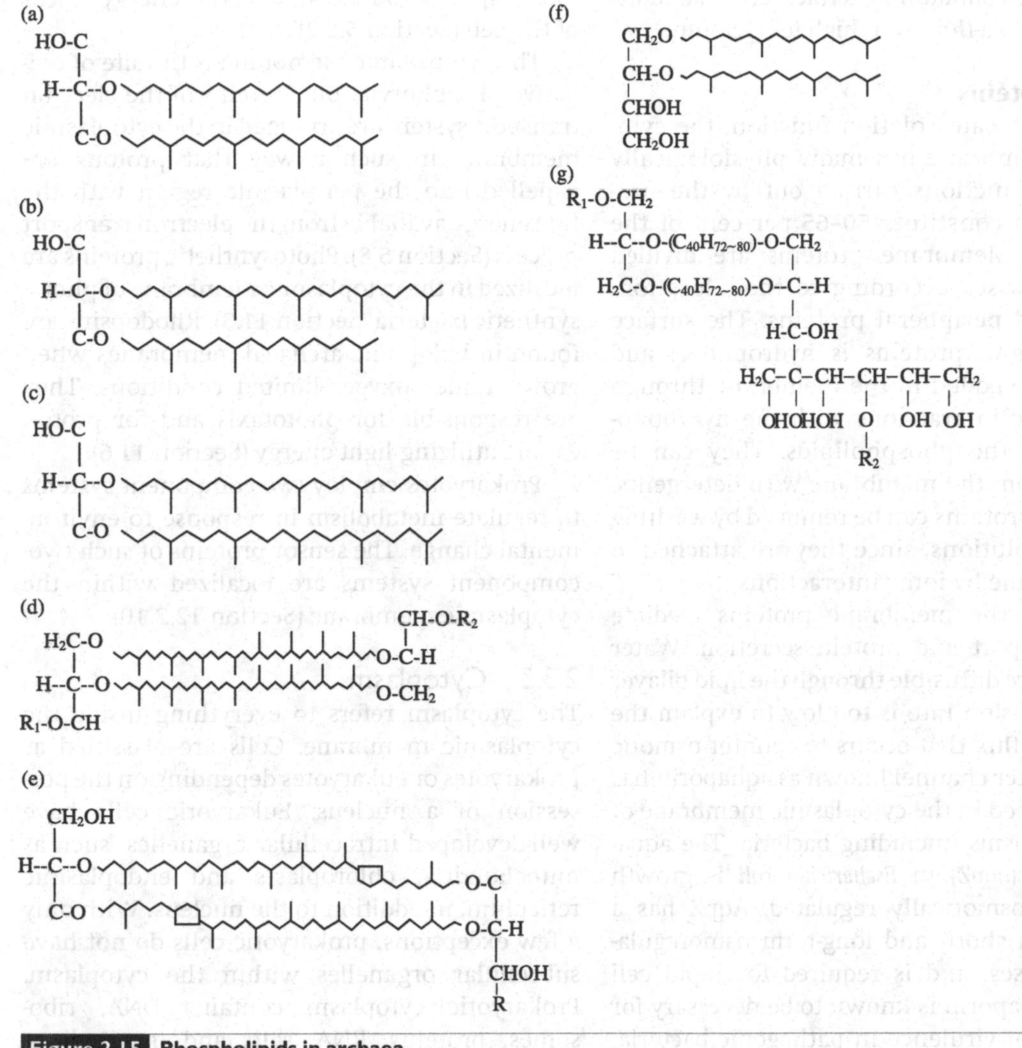

Figure 2.15 Phospholipids in archaea.

The hydrophobic tails of the phospholipids in archaea are isoprenoid alcohols, either 20 (phytanyl) or 25 (sesterterpanyl) carbons long and fully saturated. They are ether-linked to polyalcohols, mainly glycerol (a, b, c), and tetritol (f) and nonitol (g) are also found in archaeal phospholipids.

Phospholipids in halophilic archaea: complex lipids consisting of (a) 2,3-di-*O*-phytanyl-*sn*-glycerol (b) 2-*O*-sesterterpanyl-3-*O*-phytanyl-*sn*-glycerol, and (c) 2,3-di-*O*-sesterterpanyl-*sn*-glycerol, the CI of which is linked to phosphate compounds, oligosaccharides or sulfonated oligosaccharides.

The common phospholipids in methanogenic archaea are (a) 2,3-di-*O*-phytanyl-*sn*-glycerol and (d) di-diphytanyl-diglycerol-tetraether. The C40 molecule (diphytanol) comprises two phytanols linked head-to-head as in (d). Oligosaccharides and phosphate compounds are linked to the CI of glycerol. In some methanogenic archaeal phospholipids, tetritol replaces glycerol.

Di-diphytanyl-diglycerol-tetraether is the major phospholipid in hyperthermophilic archaea. Cyclopentanes are found in their hydrophobic tails (e). Glycerol is the common polyalcohol in the tetraether (R = H in (e)), and rarely nonitol occupies one end of the tetraether compounds (g and R = $C_6H_{13}O_6$ in (e)).

cyclopentane-containing tetraether maintains membrane semi-fluidity at high temperature.

2.3.4.4 Proteins

In addition to an isolation function, the cytoplasmic membrane has many physiologically important functions carried out by the proteins, which constitute 50–65 per cent of the membrane. Membrane proteins are divided into two classes according to their location: integral and peripheral proteins. The surface of the integral proteins is hydrophobic and they are embedded in the membrane through hydrophobic interactions with the hydrophobic tails of the phospholipids. They can be removed from the membrane with detergents. Peripheral proteins can be removed by washing using salt solutions, since they are attached to the membrane by ionic interactions.

Many of the membrane proteins mediate solute transport and protein secretion. Water molecules are diffusible through the lipid bilayer but the diffusion rate is too low to explain the rapid water flux that occurs to counter osmotic shock. A water channel known as aquaporin has been identified in the cytoplasmic membrane of many organisms, including bacteria. The aquaporin gene (*aqpZ*) in *Escherichia coli* is growth phase and osmotically regulated. AqpZ has a role in both short- and long-term osmoregulatory responses, and is required for rapid cell growth. Aquaporin is known to be necessary for expression of virulence in pathogenic bacteria, and for sporulation and germination.

Enzymes responsible for synthesis and turnover of surface structures, such as phospholipase, protease and peptidase, are also associated with the cytoplasmic membrane. The bacterial chromosome is attached to the membrane, and also some of the enzymes involved in DNA replication. Bacterial DNA is attached to the cytoplasmic membrane via a protein that is believed to be responsible for chromosome segregation into daughter cells during cell division.

The ATP synthase (ATPase) enzyme complex is a membrane protein that couples ATP synthesis and hydrolysis to transmembrane proton transfer, thereby governing the proton motive force, Δp (Section 5.6.4), and the energy status of the cell (Section 5.6.2).

The cytoplasmic membrane is the site of oxidative phosphorylation. Proteins of the electron transport system are arranged in the cytoplasmic membrane in such a way that protons are expelled into the periplasmic region with the free energy available from the electron transport process (Section 5.8). Photosynthetic proteins are localized in the cytoplasmic membrane of photosynthetic bacteria (Section 11.3). Rhodopsins are found in halophilic archaeal membranes when grown under oxygen-limited conditions. They are responsible for phototaxis and for proton export utilizing light energy (Section 11.6).

Prokaryotes employ two-component systems to regulate metabolism in response to environmental change. The sensor proteins of such two-component systems are localized within the cytoplasmic membrane (Section 12.2.10).

2.3.5 Cytoplasm

The cytoplasm refers to everything inside the cytoplasmic membrane. Cells are classified as prokaryotes or eukaryotes depending on the possession of a nucleus. Eukaryotic cells have well-developed intracellular organelles, such as mitochondria, chloroplasts and endoplasmic reticulum, in addition to the nucleus. With only a few exceptions, prokaryotic cells do not have subcellular organelles within the cytoplasm. Prokaryotic cytoplasm contains DNA, ribosomes, proteins, RNA, salts and metabolites, and is viscous due to the high concentration of macromolecules. Some of these macromolecules form aggregates, while others are soluble. The soluble part is called the cytosol. Though the cytoplasm is not compartmentalized, the cytoplasm is not a random mixture of its components. Proteins in the cytoplasm are in high concentration and can interact with each other to form a kind of network. The enzymes involved in a particular metabolic process are adjacently located for their required interaction. The term 'metabolon' has been proposed to describe such a set of enzymes and their cofactors involved in such a fashion. A bacterium from the surgeonfish, *Epulopiscium fishelsoni*, has a cell size (80×600 μm) that can be seen with the naked eye. It was

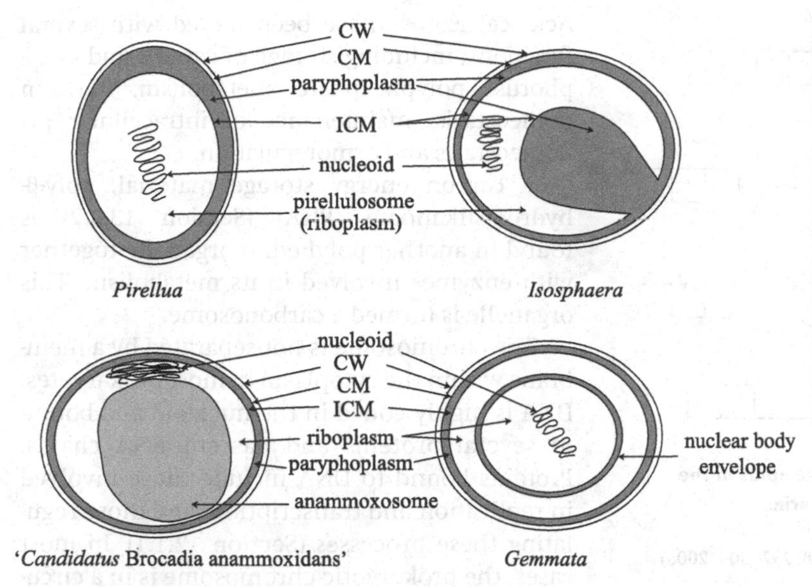

Figure 2.16 Cell organization and compartmentalization in (upper) *Pirellula marina* and *Isosphaera pallida* (also applies to *Planctomyces maris*), and (lower) '*Candidatus* Brocadia anammoxidans' and *Gemmata obscuriglobus*.

(*Arch. Microbiol.* **175**: 413–429, 2001)

CW, cell wall; CM, cytoplasmic membrane; ICM, intracytoplasmic membrane.

predicted that a cell of this size could not function properly without possessing a metabolic network such as the metabolon. Polyhedral organelles have been recognized as extremely large macromolecular complexes in bacteria, consisting of metabolic enzymes encased within a multiprotein membrane. They act as microcompartments that enhance metabolic processes by selectively concentrating specific metabolites. Polyhedral organelles are widely used by bacteria for optimizing metabolic processes. A typical example of these organelles is the carboxysome (see below and Section 10.8.1).

There are a few examples of intracytoplasmic membrane structures in bacteria. Some of these are continuous with the cytoplasmic membrane and are therefore believed to be derived from invaginations of chemically modified areas of the cytoplasmic membrane. Others are independent of the cytoplasmic membrane.

Thylakoids of cyanobacteria are the best known example of an intracellular organelle among prokaryotes. This organelle contains photosynthetic pigments (Section 11.3.1). Other examples are in the members of the phylogenetically deep-rooted bacterial family, the *Planctomycetaceae*, which have a compartmentalized cytoplasm (Figure 2.16). They have a bilayered intracytoplasmic membrane (ICM) that is discontinuous with the cytoplasmic membrane. The compartment enclosed by the ICM is called the riboplasm and contains the nucleoid and ribosomes. Little is known of the function of the paryphoplasm that occupies the region between the ICM and the cytoplasmic membrane. *Gemmata obscuriglobus*, a planctomycete originally isolated from fresh water, has been found to possess another unique feature not found in other bacteria. This organism has a membrane-bound nuclear body consisting of two membranes surrounding the fibrillar DNA-containing nucleoid. Another planctomycete not isolated in pure culture, *Candidatus* Brocadia anammoxidans, has yet another subcellular organelle, the anammoxosome (Figure 2.16), that performs anaerobic ammonia oxidation using nitrite as the electron donor (Section 9.1.4). The anammoxosome membrane consists of ladderane lipids containing three or four cyclobutane rings (Figure 2.17).

Many bacteria have intracytoplasmic membrane structures continuous with the cytoplasmic membrane. These include nitrogen fixers (Section 6.2.1), methanotrophs (Section 7.10), nitrifiers (Section 10.2) and photosynthetic bacteria (Section 11.3). Magnetotactic bacteria have an intracytoplasmic membrane-bound structure called the magnetosome which

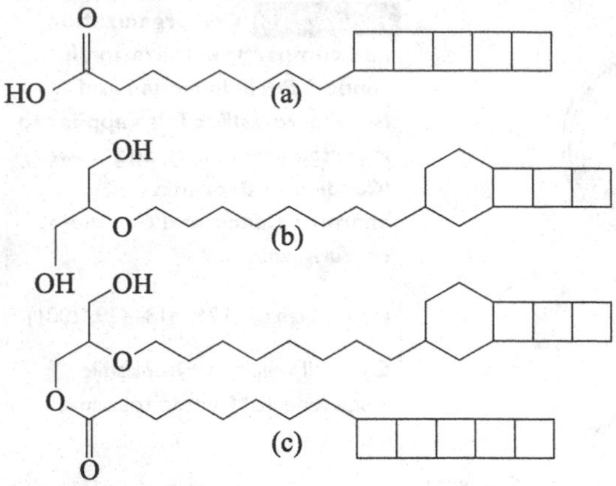

Figure 2.17 Structure of ladderane lipids in the annamoxosome of anammox bacteria.

(Modified from *FEMS Microbiol. Lett.* **258**: 297–304, 2006)

Three classes of ladderane lipids are known, (a) free fatty acid with cyclobutane rings, (b) glycerol ether with a cyclohexane and cyclobutane rings and (c) glyceride with ether and ester links.

contains magnetite crystals. The intracytoplasmic membrane structures in these bacteria contain proteins that determine the specific physiological properties of the bacteria.

As stated above, a large number of polyhedral organelles have been identified in bacteria, such as the carboxysome, acidocalcisome and carbonosome. Obligate lithotrophs use CO_2 as their carbon source. Ribulose-bisphosphate carboxylase fixes CO_2 and this enzyme is included in an enzyme complex called a carboxysome (Section 10.8.1). Similar polyhedral organelles have been identified in many bacteria. In enteric bacteria, including *Salmonella enterica*, enzymes needed for the utilization of 1,2-propanediol (Pdu) and ethanolamine (Eut) are encapsulated in separate polyhedral organelles. It is interesting to note that some cellulolytic bacteria produce a similar but extracellular complex known as a cellulosome.

An energy storage material, polyphosphate (also known as metachrome or volutin) (Section 13.2.4), is accumulated in a subcellular organelle called the acidocalcisome. This polyphosphate granule is rich in calcium and polyphosphate, and is the only organelle that has been conserved during evolution from prokaryotes to eukaryotes. Acidocalcisomes have been linked with several functions, including storage of cations and phosphorus, polyphosphate metabolism, calcium homeostasis, maintenance of intracellular pH homeostasis and osmoregulation.

A carbon energy storage material, poly-β-hydroxyalkanoate (PHA) (Section 13.2.2) is found in another polyhedral organelle together with enzymes involved in its metabolism. This organelle is termed a carbonosome.

The chromosome is not separated by a membrane within the cytoplasm in most prokaryotes. DNA is highly coiled in the nucleoid and bound to several proteins and nascent RNA chains. Proteins bound to DNA include those involved in replication and transcription, and those regulating these processes (Section 12.1.1). In most cases, the prokaryotic chromosome is in a circular form. Some bacteria, e.g. *Streptomyces lividans*, have linear DNA. Since transcription and translation takes place in the same compartment and an intervening sequence is absent in prokaryotes, ribosomes bind to the nascent mRNA for translation (Section 6.12). For this reason, certain genes in prokaryotes can be controlled by attenuation (Section 12.1.4) and this regulatory mechanism is not known in eukaryotes. Intervening sequences have, however, been identified in some prokaryotes, including archaea. In addition to the chromosome, smaller DNA molecules are common in prokaryotes. They are called plasmids. Cells harbouring plasmids show different phenotypic traits, including resistance to particular antibiotics and toxic metals, production of bacteriocins and toxins, virulence, conjugation and others. Some large-sized plasmids carry essential genes. These are called megaplasmids or second chromosomes.

A major proportion of the RNA in the cytoplasm is present as ribosomes. The ribosome is a complex consisting of 50 different proteins and three different kinds of RNA (i.e. 23S, 16S and 5S rRNA). Ribosomes are the site of protein synthesis. Multiple ribosomes bind mRNA, and this structure is called a polysome.

The liquid part of the cytoplasm, the cytosol, contains many enzymes, including those mediating central metabolism, metabolites and salts. The cytosol contains small molecular weight organic compounds such as betaine, amino acids and sugars to balance the external osmotic pressure, in addition to inorganic salts. These are called compatible solutes. A hyperthermophilic methanogenic archaeon, *Methanopyrus kandleri*, produces 2,3-diphosphoglycerate (cDPG) as its compatible solute. Di-myo-inositol-1,1′-phosphate is used for this purpose in the hyperthermophilic archaea *Pyrococcus woesei* and *Pyrococcus furiosus*.

2.3.6 Resting cells

Many bacteria differentiate into resting cells when the growth environment becomes unfavourable, such as by depletion of nutrients (Section 13.3). The best known resting cells are spores, as found in the Gram-positive aerobic *Bacillus* and anaerobic *Clostridium* genera. Cysts are another form of resting cells. These are resistant to physical and chemical stresses such as desiccation and ionizing radiation. Spores are resistant to high temperatures, but cysts are not. Spores can remain viable for several decades under dry conditions.

Spore-specific structures include the outer coat, inner coat and cortex. The coats mainly comprise protein and the cortex peptidoglycan. The cortex, occupying the region between the spore wall and the coats, confers the resistance properties of the spores. The structure of cysts is similar to vegetative cells, except for the exine and intine. The exine is the outer cyst wall, mainly consisting of alginate, protein and lipid, and the intine is the inner cyst wall comprising polymannuronic acid. The exine is stabilized with divalent metal ions such as Ca^{2+} which bridge the carboxyl groups of alginate. A cyst central body can be prepared by treating cysts with chelating agents such as EDTA and citrate. The cyst central body transforms to a vegetative cell under favourable growth conditions, but is not resistant to external stresses, showing that the exine is responsible for resistance. Resting cells are discussed in detail later (Section 13.3).

Further Reading

Note this section contains key references only. Additional recommended references are available at www.cambridge.org/ProkaryoticMetabolism.

Cellular elements

Beinert, H. (2000). A tribute to sulfur. *European Journal of Biochemistry* **267**, 5657–5664.

Dosanjh, N. S. & Michel, S. L. J. (2006). Microbial nickel metalloregulation: NikRs for nickel ions. *Current Opinion in Chemical Biology* **10**, 123–130.

Hille, R. (2002). Molybdenum and tungsten in biology. *Trends in Biochemical Sciences* **27**, 360–367.

Jakubovics, N. S. & Jenkinson, H. F. (2001). Out of the Iron Age: new insights into the critical role of manganese homeostasis in bacteria. *Microbiology* **147**, 1709–1718.

Kobayashi, M. & Shimizu, S. (1999). Cobalt proteins. *European Journal of Biochemistry* **261**, 1–9.

Lane, T. W., Saito, M. A., George, G. N., Pickering, I. J., Prince, R. C. & Morel, F. M. M. (2005). A cadmium enzyme from a marine diatom. *Nature* **435**, 42.

Pol, A., Barends, T. R. M., Dietl, A., Khadem, A. F., Eygensteyn, J., Jetten, M. S. M. & Op den Camp, H. J. M. (2014). Rare earth metals are essential for methanotrophic life in volcanic mudpots. *Environmental Microbiology* **16**, 255–264.

Stadtman, T. C. (2002). Discoveries of vitamin B_{12} and selenium enzymes. *Annual Review of Biochemistry* **71**, 1–16.

Wolfe-Simon, F., Blum, J. S., Kulp, T. R., Gordon, G. W., Hoeft, S. E., Pett-Ridge, J., Stolz, J. F., Webb, S. M., Weber, P. K., Davies, P. C. W., Anbar, A. D. & Oremland, R. S. (2011). A bacterium that can grow by using arsenic instead of phosphorus. *Science* **332**, 1163–1166.

Cell surface appendages

Albers, S.-V. & Jarrell, K. F. (2015). The archaellum: how archaea swim. *Frontiers in Microbiology* **6**, 23.

Beatson, S. A., Minamino, T. & Pallen, M. J. (2006). Variation in bacterial flagellins: from sequence to structure. *Trends in Microbiology* **14**, 151–155.

Gorby, Y. A., Yanina, S., McLean, J. S., Rosso, K. M., Moyles, D., Dohnalkova, A., Beveridge, T. J., Chang, I. S., Kim, B. H., Kim, K. S., Culley, D. E., Reed, S. B., Romine, M. F., Saffarini, D. A., Hill, E. A., Shi, L., Elias, D. A., Kennedy, D. W., Pinchuk, G.,

Watanabe, K., Logan, B., Nealson, K. H. & Fredrickson, J. K. (2006). Electrically conductive bacterial nanowires produced by *Shewanella oneidensis* strain MR-1 and other microorganisms. *Proceedings of the National Academy of Sciences of the USA* **103**, 11358–11363.

Persat, A., Inclan, Y. F., Engel, J. N., Stone, H. A. & Gitai, Z. (2015). Type IV pili mechanochemically regulate virulence factors *in Pseudomonas aeruginosa*. *Proceedings of the National Academy of Sciences of the USA* **112**, 7563–7568.

Pfeffer, C., Larsen, S., Song, J., Dong, M., Besenbacher, F., Meyer, R. L., Kjeldsen, K. U., Schreiber, L., Gorby, Y. A., El-Naggar, M. Y., Leung, K. M., Schramm, A., Risgaard-Petersen, N. and Nielsen, L. P. (2012). Filamentous bacteria transport electrons over centimetre distances. *Nature* **491**, 218–221.

Scott, J. R. & Zahner, D. (2006). Pili with strong attachments: Gram-positive bacteria do it differently. *Molecular Microbiology* **62**, 320–330.

S-layer and other surface structures

Ahn, J. S., Chandramohan, L., Liou, L. E. & Bayles, K. W. (2006). Characterization of CidR-mediated regulation in *Bacillus anthracis* reveals a previously undetected role of S-layer proteins as murein hydrolases. *Molecular Microbiology* **62**, 1158–1169.

Albers, S.-V. & Meyer, B. H. (2011). The archaeal cell envelope. *Nature Reviews Microbiology* **9**, 414–426.

Bush, C. A., Yang, J., Yu, B. & Cisar, J. O. (2014). Chemical structures of *Streptococcus pneumoniae* capsular polysaccharide type 39 (CPS39), CPS47F, and CPS34 characterized by nuclear magnetic resonance spectroscopy and their relation to CPS10A. *Journal of Bacteriology* **196**, 3271–3278.

Johnson, B., Selle, K., O'Flaherty, S., Goh, Y. J. & Klaenhammer, T. (2013). Identification of extracellular surface-layer associated proteins in *Lactobacillus acidophilus* NCFM. *Microbiology* **159**, 2269–2282.

Park, S., Kelley, K. A., Vinogradov, E., Solinga, R., Weidenmaier, C., Misawa, Y. & Lee, J. C. (2010). Characterization of the structure and biological functions of a capsular polysaccharide produced by *Staphylococcus saprophyticus*. *Journal of Bacteriology* **192**, 4618–4626.

Pohlschroder, M. & Albers, S.-V. (2016). Archaeal cell envelope and surface structures. *Frontiers in Microbiology* **6**, 1515.

Rothfuss, H., Lara, J. C., Schmid, A. K. & Lidstrom, M. E. (2006). Involvement of the S-layer proteins Hpi and SlpA in the maintenance of cell envelope integrity in *Deinococcus radiodurans* R1. *Microbiology* **152**, 2779–2787.

Outer membrane in Gram-negative bacteria

Biller, S. J., Schubotz, F., Roggensack, S. E., Thompson, A. W., Summons, R. E. & Chisholm, S. W. (2014). Bacterial vesicles in marine ecosystems. *Science* **343**, 183–186.

Bishop, R. E. (2014). Emerging roles for anionic non-bilayer phospholipids in fortifying the outer membrane permeability barrier. *Journal of Bacteriology* **196**, 3209–3213.

Burghardt, T., Nather, D. J., Junglas, B., Huber, H. & Rachel, R. (2007). The dominating outer membrane protein of the hyperthermophilic Archaeum *Ignicoccus hospitalis*: a novel pore-forming complex. *Molecular Microbiology* **63**, 166–176.

Koebnik, R., Locher, K. P. & Van Gelder, P. (2000). Structure and function of bacterial outer membrane proteins: barrels in a nutshell. *Molecular Microbiology* **37**, 239–253.

Küper, U., Meyer, C., Müller, V., Rachel, R. & Huber, H. (2010). Energized outer membrane and spatial separation of metabolic processes in the hyperthermophilic Archaeon *Ignicoccus hospitalis*. *Proceedings of the National Academy of Sciences of the USA* **107**, 3152–3156.

Nikaido, H. (2003). Molecular basis of bacterial outer membrane permeability revisited. *Microbiology and Molecular Biology Reviews* **67**, 593–656.

Schulz, G. E. (2002). The structure of bacterial outer membrane proteins. *Biochimica et Biophysica Acta* **1565**, 308–317.

Cell wall

Brown, S., Santa Maria, J. P. & Walker, S. (2013). Wall teichoic acids of Gram-positive bacteria. *Annual Review of Microbiology* **67**, 313–336.

Cabeen, M. T. & Jacobs-Wagner, C. (2005). Bacterial cell shape. *Nature Reviews Microbiology* **3**, 601–610.

Henrichfreise, B., Schiefer, A., Schneider, T., Nzukou, E., Poellinger, C., Hoffmann, T.-J., Johnston, K. L., Moelleken, K., Wiedemann, I., Pfarr, K., Hoerauf, A. & Sahl, H. G. (2009). Functional conservation of the lipid II biosynthesis pathway in the cell wall-less bacteria *Chlamydia* and *Wolbachia*: why is lipid II needed? *Molecular Microbiology* **73**, 913–923.

Patin, D., Bostock, J., Chopra, I., Mengin-Lecreulx, D. & Blanot, D. (2012). Biochemical characterisation of the chlamydial MurF ligase, and possible sequence of the chlamydial peptidoglycan pentapeptide stem. *Archives of Microbiology* **194**, 505–512.

Schneewind, O. & Missiakas, D. (2014). Lipoteichoic acids, phosphate-containing polymers in the envelope of Gram-positive bacteria. *Journal of Bacteriology* **196**, 1133–1142.

Turner, R. D., Vollmer, W. & Foster, S. J. (2014). Different walls for rods and balls: the diversity of peptidoglycan. *Molecular Microbiology* **91**, 862–874.

Wanner, S., Schade, J., Keinhörster, D., Weller, N., George, S. E., Kull, L,. Bauer, J., Grau, T., Winstel, V., Stoy, H., Kretschmer, D., Kolata, J., Wolz, C., Bröker, B. M. & Weidenmaier, C. (2017). Wall teichoic acids mediate increased virulence in *Staphylococcus aureus*. *Nature Microbiology* **2**, 16257.

Periplasm

Bohin, J. P. (2000). Osmoregulated periplasmic glucans in Proteobacteria. *FEMS Microbiology Letters* **186**, 11–19.

Flores, E., Herrero, A., Wolk, C. P. & Maldener, I. (2006). Is the periplasm continuous in filamentous multicellular cyanobacteria? *Trends in Microbiology* **14**, 439–443.

Matias, V. R. F. and Beveridge, T. J. (2008). Lipoteichoic acid is a major component of the *Bacillus subtilis* periplasm. *Journal of Bacteriology* **190**, 7414–7418.

Cytoplasmic membrane

Bernstein, H. D. (2000). The biogenesis and assembly of bacterial membrane proteins. *Current Opinion in Microbiology* **3**, 203–209.

Boyd, E., Hamilton, T., Wang, J., He, L. & Zhang, C. (2013). The role of tetraether lipid composition in the adaptation of thermophilic archaea to acidity. *Frontiers in Microbiology* **4**:00063.

Cavicchioli, R. (2011). Archaea – timeline of the third domain. *Nature Reviews Microbiology* **9**, 51–61.

Cronan, J. E. (2006). A bacterium that has three pathways to regulate membrane lipid fluidity. *Molecular Microbiology* **60**, 256–259.

Engelman, D. M. (2005). Membranes are more mosaic than fluid. *Nature* **438**, 578–580.

Gumbart, J., Wang, Y., Aksimentiev, A., Tajkhorshid, E. & Schulten, K. (2005). Molecular dynamics simulations of proteins in lipid bilayers. *Current Opinion in Structural Biology* **15**, 423–431.

Kung, C. & Blount, P. (2004). Channels in microbes: so many holes to fill. *Molecular Microbiology* **53**, 373–380.

Mansilla, M. C., Cybulski, L. E., Albanesi, D. & de Mendoza, D. (2004). Control of membrane lipid fluidity by molecular thermosensors. *Journal of Bacteriology* **186**, 6681–6688.

Schmerk, C. L., Bernards, M. A. & Valvano, M. A. (2011). Hopanoid production is required for low-pH tolerance, antimicrobial resistance, and motility in *Burkholderia cenocepacia*. *Journal of Bacteriology* **193**, 6712–6723.

Cytoplasm

Borrero-de Acuña, J. M., Rohde, M., Wissing, J., Jänsch, L., Schobert, M., Molinari, G., Timmis, K. N., Jahn, M. & Jahn, D. (2016). Protein network of the *Pseudomonas aeruginosa* denitrification apparatus. *Journal of Bacteriology* **198**, 1401–1413.

Bowman, G. R., Lyuksyutova, A. I. & Shapiro, L. (2011). Bacterial polarity. *Current Opinion in Cell Biology* **23**, 71–77.

Cabeen, M. T. & Jacobs-Wagner, C. (2010). The bacterial cytoskeleton. *Annual Review of Genetics* **44**, 365–392.

Mathews, C. K. (1993). The cell – bag of enzymes or network of channels? *Journal of Bacteriology* **175**, 6377–6381.

Matturro, B., Cruz Viggi, C., Aulenta, F. & Rossetti, S. (2017). Cable bacteria and the bioelectrochemical snorkel: the natural and engineered facets playing a role in hydrocarbons degradation in marine sediments. *Frontiers in Microbiology* **8**, 952.

Noirot, P. & Noirot-Gros, M. F. (2004). Protein interaction networks in bacteria. *Current Opinion in Microbiology* **7**, 505–512.

Sleator, R. D. & Hill, C. (2002). Bacterial osmoadaptation: the role of osmolytes in bacterial stress and virulence. *FEMS Microbiology Reviews* **26**, 49–71.

Spitzer, J. J. & Poolman, B. (2005). Electrochemical structure of the crowded cytoplasm. *Trends in Biochemical Sciences* **30**, 536–541.

Prokaryotic intracellular organelles

Chowdhury, C., Sinha, S., Chun, S., Yeates, T. O. & Bobik, T. A. (2014). Diverse bacterial microcompartment organelles. *Microbiology and Molecular Biology Reviews* **78**, 438–468.

Cornejo, E., Abreu, N. & Komeili, A. (2014). Compartmentalization and organelle formation in bacteria. *Current Opinion in Cell Biology* **26**, 132–138.

Kerfeld, C. A. and Erbilgin, O. (2015). Bacterial microcompartments and the modular construction of microbial metabolism. *Trends in Microbiology* **23**, 22–34.

Kerfeld, C. A., Sawaya, M. R., Tanaka, S., Nguyen, C. V., Phillips, M., Beeby, M. & Yeates, T. O. (2005). Protein structures forming the shell of primitive bacterial organelles. *Science* **309**, 936–938.

Lewis, P. J. (2004). Bacterial subcellular architecture: recent advances and future prospects. *Molecular Microbiology* **54**, 1135–1150.

Martin, T. (2011). Good things come in small packages: subcellular organization and development in bacteria. *Current Opinion in Microbiology* **14**, 687–690.

Niftrik, L. (2013). Cell biology of unique anammox bacteria that contain an energy conserving prokaryotic organelle. *Antonie van Leeuwenhoek* **104**, 489–497.

Saier Jr, M. H. & Bogdanov, M. V. (2013). Membranous organelles in bacteria. *Journal of Molecular Microbiology and Biotechnology* **23**, 5–12.

Chapter 3

Membrane transport – nutrient uptake and protein excretion

Microbes import the materials needed for growth and survival from their environment and export metabolites. As described in the previous chapter, the cytoplasm is separated from the environment by the hydrophobic cytoplasmic membrane, which is impermeable to hydrophilic solutes. Because of this permeability barrier exerted by the phospholipid component, almost all hydrophilic compounds can only pass through the membrane by means of integral membrane proteins. These are called carrier proteins, transporters or permeases.

Solute transport can be classified as diffusion, active transport or group translocation according to the mechanisms involved. Diffusion does not require energy; energy is required for active transport; and solutes transported by group translocation are chemically modified during this process. Some solutes are accumulated in the cell against a concentration gradient of several orders of magnitude, and energy needs to be invested for such accumulation.

3.1 Ionophores: models of carrier proteins

There are two models that explain solute transport mediated by carrier proteins: the mobile carrier model and the pore model. The solute binds the carrier at one side of the membrane and dissociates at the other side according to the mobile carrier model, while the pore model proposes that the carrier protein forms a pore across the membrane through which the solute passes. A certain group of antibiotics can make the membrane permeable to ions. These are called ionophores and are useful compounds to assist the study of membrane transport.

One ionophore, valinomycin, transports ions according to the carrier model, while gramicidin A, another ionophore, makes a pore across the membrane. Valinomycin is a circular molecule consisting of valine, lactate and hydroxy isovalerate (Figure 3.1). Hydrophobic methyl and isopropyl groups form the surface of the circular molecule while hydrophilic carbonyl groups are arranged inside. Cations such as K^+ bind to the hydrophilic interior of the molecule. The complex moves through the hydrophilic membrane to the other side of the membrane where the cation dissociates. The efficiency of a mobile carrier is temperature dependent and becomes less efficient at low temperature due to decreased membrane fluidity. Uncouplers (Section 5.8.5) are mobile carriers of H^+.

Gramicidin A is a peptide consisting of 15 amino acid residues (Figure 3.2). This linear peptide has one hydrophilic side and one hydrophobic side. Two molecules of this compound form a hydrophilic pore across the membrane with the interaction between the hydrophobic side of the compound and the membrane lipid. Various cations can move through the pore thus created, the efficiency being temperature independent. It is believed that some carrier proteins are mobile carriers while others form pores.

3.2 Diffusion

Some solutes enter cells by diffusion, without energy expenditure, according to the concentration gradient. Hydrophobic solutes diffuse

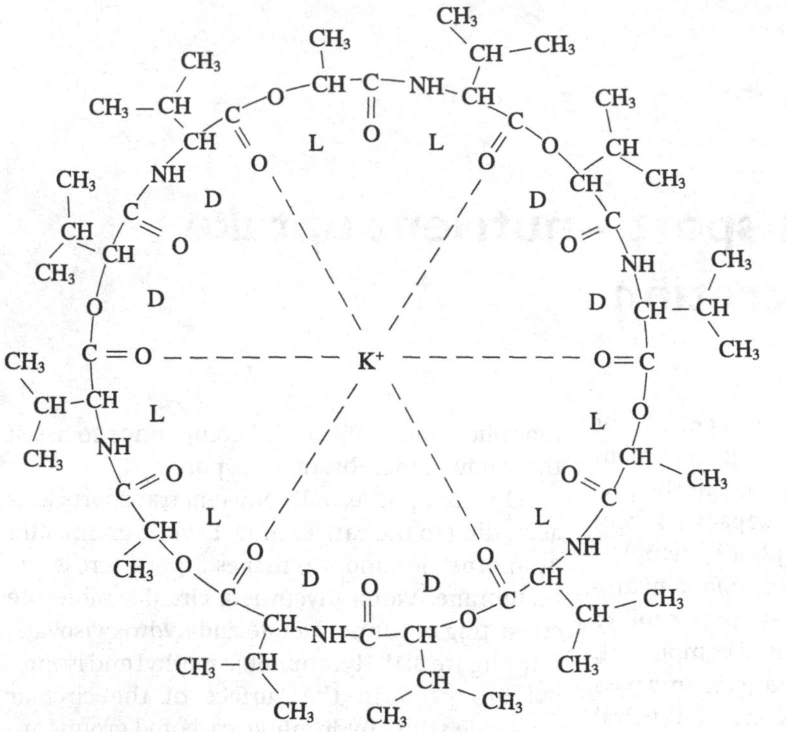

Figure 3.1 Structure of valinomycin, a model mobile carrier.

Cations bind to the hydrophilic interior of the molecule at the side of the membrane with the higher concentration of the cation, and the complex moves through the hydrophilic membrane to the other side where the cation concentration is lower. The cations dissociate to equilibrate with the low concentration.

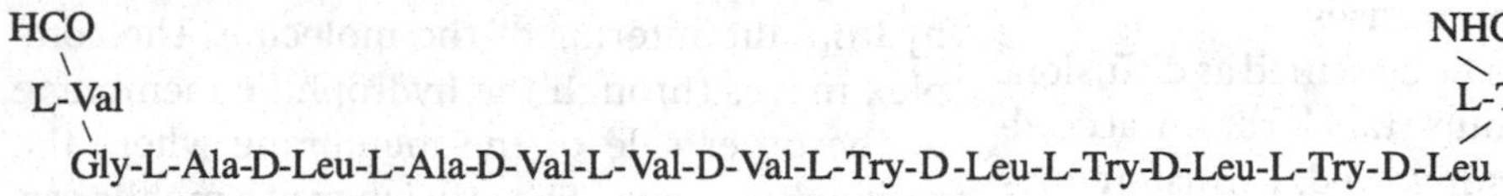

HCO–L-Val–Gly-L-Ala-D-Leu-L-Ala-D-Val-L-Val-D-Val-L-Try-D-Leu-L-Try-D-Leu-L-Try-D-Leu–L-Try–NHCH$_2$CH$_2$OH

Figure 3.2 Gramicidin A, a model of a pore-forming carrier protein.

Two molecules of this peptide with 15 amino acid residues form a pore across the membrane through a hydrophobic interaction between the side chains of the amino acids and the membrane lipid. Solutes can move through the pore.

through the lipid part of the membrane, and others diffuse through carrier proteins. The former is called simple diffusion and the latter, facilitated diffusion. Facilitated diffusion shows different kinetics from that of simple diffusion. The initial diffusion rate is proportional to the concentration gradient in simple diffusion, while facilitated diffusion shows a relationship between the diffusion rate and the concentration of the solute similar to the Michaelis–Menten kinetics known in enzyme catalysis. Solutes transported through facilitated diffusion do not passively leak into the cell to any significant extent, and the rate of transport is directly proportional to the fraction of carrier proteins associated with them. When the carrier protein is fully saturated with the solute, the rate of transport reaches a maximum, and the rate does not increase further with any further increase in solute concentration. By definition, charged solutes are not transported through diffusion, since the transport of charged solutes changes the membrane potential (Section 5.7.1).

3.3 Active transport and role of electrochemical gradients

Solute transport coupled to energy transduction is divided into primary and secondary transport according to the energy source. Primary transport systems are driven by energy-generating metabolism. Primary transport includes proton

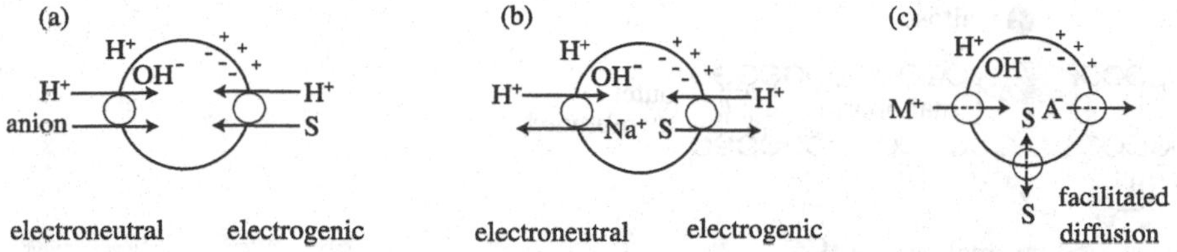

(a) H^+ OH^- H^+ anion H^+ S electroneutral electrogenic

(b) H^+ OH^- H^+ H^+ Na^+ S electroneutral electrogenic

(c) H^+ OH^- M^+ S A^- S facilitated diffusion

Figure 3.3 Electrochemical gradient-dependent active transport.

(Dawes, E. A. 1986, *Microbial Energetics*, Figure 6.1. Blackie & Son, Glasgow)

Some solutes move across the membrane together with protons (or sodium ions) in the same direction (a, symport), while others move in opposite directions (b, antiport). Some ions cross the membrane along the electrochemical gradient (c, uniport). Transport involving the membrane potential is called electrogenic transport. Uniport and symport of uncharged solutes are electrogenic. When anions with a total negative charge of *n* are symported with *m* protons, the transport becomes electroneutral in the case of $n = m$ and electrogenic in the case of $n \neq m$. In antiport, where cations with a total positive charge of *n* are exchanged with *m* protons it becomes electroneutral in the case of $n = m$ and electrogenic in the case of $n \neq m$.

export driven by electron transport in respiration (Section 5.8) and photosynthesis (Section 11.4.3), by ATP hydrolysis (Section 5.8.4) and by light in halophilic archaea (Section 11.6). Also included in primary transport are chloride ion import in halophilic archaea driven by light (Section 11.6), sodium ion export coupled to decarboxylation reactions, proton export coupled to fumarate reduction and fermentation product excretion (Section 5.8.6), and the import of sugars through group translocation (Section 3.5). Protons (or Na^+) are exported by energy-converting hydrogenase (Section 5.8.6.5) and H^+ (Na^+)-translocating ferredoxin:NAD^+ oxidoreductase (Section 5.8.6.6) in anaerobes. These ion transport mechanisms are energy-conserving processes, except for sugar transport by group translocation, and will be discussed in the appropriate sections. This section is devoted to energy-dependent transport of materials needed for growth and survival and includes the secondary transport and group translocation of sugars.

Energy needed for secondary transport is supplied as an electrochemical gradient (a proton motive or sodium motive force) or from high energy phosphate bonds. The proton gradient (internal alkaline) and membrane potential are established during primary transport (Section 5.8). Since sodium ions are exported due to their coupling with some energy-yielding reactions such as sodium pump (Section 5.8.3.3), decarboxylation and anaerobic respiration (Chapter 9), a sodium gradient across the prokaryotic membrane is also established. Proton and sodium gradients are collectively termed electrochemical gradients, and they are used as energy for many secondary transport processes. All the carrier proteins studied have 12 helices spanning the membrane, some of which function as binding sites for solutes and others for protons (or sodium ions).

The proton motive force consists of a proton gradient (ΔpH) and membrane potential ($\Delta\Psi$) (Section 5.7). Depending on the nature of the solute, transport requires energy in the form of either ΔpH or $\Delta\Psi$, or both. According to the carrier proteins involved, electrochemical gradient-dependent solute transport mechanisms can be classed as symport, antiport and uniport (Figure 3.3). A solute crosses the membrane in the same direction with protons (or sodium ions) in the symport mechanism but in the opposite direction in an antiport system. Uniporters transport ions along the electrochemical gradient without involving protons or sodium ions. Uniporters consume only the $\Delta\Psi$ part of the proton (or sodium) motive force. This is called electrogenic transport. When a monovalent anion is symported (cotransported) with a proton, the ΔpH is reduced without any change in the $\Delta\Psi$. This is called electroneutral transport. When an uncharged solute is symported with protons, the ΔpH as well as the $\Delta\Psi$ supply the energy needed for the accumulation of the solute. Since $\Delta\Psi$ is reduced this becomes an electrogenic transport system.

Many bacteria have five to nine distinct genes that encode cytoplasmic Na^+/H^+ or K^+/H^+

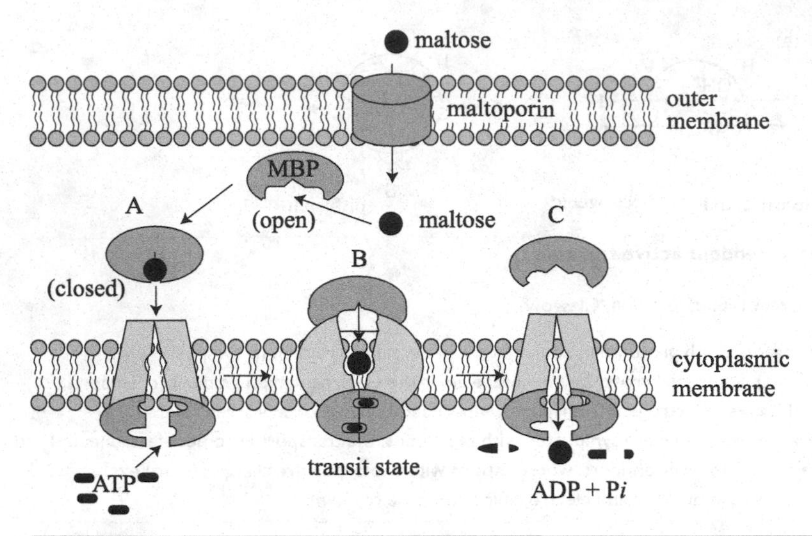

Figure 3.4 Maltose transport through the ATP-binding cassette (ABC) transporter.

(Modified from *Mol. Microbiol.* **77**: 1354–1366, 2010)

(A) Maltose-binding protein (MBP) binds maltose, undergoing a change from an open to a closed conformation, generating a high-affinity sugar-binding site. In the closed conformation, MBP binds the maltose ATP-binding cassette (ABC) to initiate transport and ATP hydrolysis. (B) In the transition state for ATP hydrolysis, the MBP becomes tightly bound to the maltose ABC to transfer the sugar. (C) Maltose is transported, and MBP is released.

antiporters. In contrast, pathogens that live primarily inside host cells usually possess none or one such antiporter, while some other stress-exposed bacteria exhibit even higher numbers. These monovalent cation/proton antiporters enable the bacteria to meet the challenge of high or fluctuating pH (Section 5.7.2), salt, temperature or osmolarity.

3.4 ATP-dependent transport: the ATP-binding cassette (ABC) pathway

Solute transport can be driven not only by the electrochemical gradient but also by ATP hydrolysis. An *unc* (ATPase) mutant of *Escherichia coli* was unable to take up maltose under conditions where a large proton motive force was established, but the disaccharide was transported when the mutant was supplied with substrates that can produce ATP through substrate-level phosphorylation (Section 5.6.5), such as 1,3-diphosphoglycerate and phosphoenolpyruvate. Maltose transport requires not only ATP but also a binding protein in the periplasm.

Gram-negative bacteria have solute-binding proteins in the periplasm that are released when cells are subjected to osmotic shock with EDTA and $MgCl_2$ (cold osmotic shock). For this reason, solute-binding protein-dependent transport is called a shock-sensitive transport system. A variety of nutrients including sugars, amino acids and ions can be transported through the shock-sensitive transport system. Solutes cross the outer membrane through porins and bind specific binding proteins before being transported through the cytoplasmic membrane by a membrane-bound protein complex. This protein complex is a member of a large super family of proteins that import nutrients or export cell surface constituents and extracellular proteins (Section 3.10). They have an ATP-binding motif and hydrolyse ATP to supply energy for transport. They are called ATP-binding cassette (ABC) transporters and are known to occur in all organisms (Figure 3.4).

The facultative human pathogen *Vibrio cholerae* has an inducible ABC transport system for

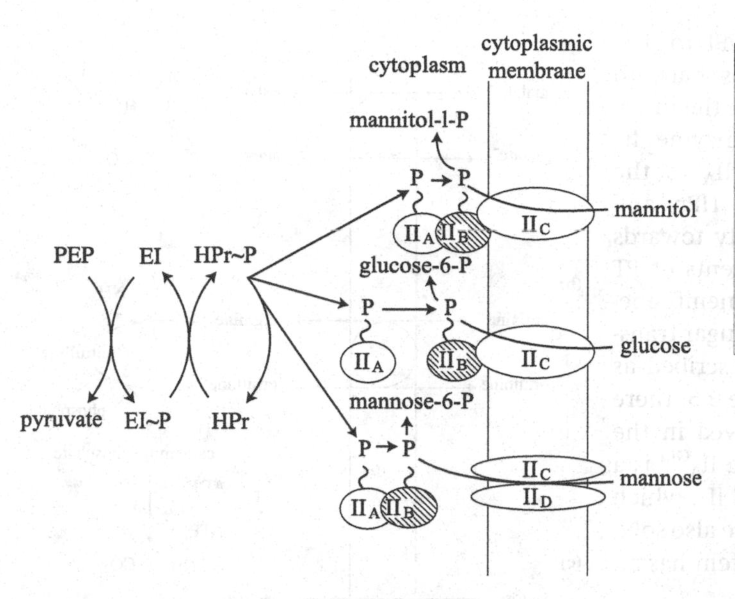

Figure 3.5 Group translocation of sugars mediated by the phosphotransferase system in anaerobic bacteria.

(*Microbiol. Rev.* **57**: 543–594, 1993)

Sugars are transported into the cell as phosphorylated forms mediated by the phosphotransferase (PT) system. Phosphoenolpyruvate (PEP) serves as the phosphate donor.

hexose-6-phosphates under phosphate-limited conditions.

Prokaryotes transport micronutrients such as vitamins through ABC transporters that employ substrate binding integral membrane proteins called S-components to replace the water-soluble substrate binding proteins or domains. This transport mechanism is known as energy coupling factor (ECF) transport. In some cases, the ECF module is dedicated to a single S-component, but in many cases, the ECF module can interact with several different S-components that are unrelated in sequence and bind diverse substrates. The modular organization with exchangeable S-components on a single ECF module allows the transport of chemically different substrates via a common route.

A glutamate-binding protein is known in *Rhodobacter sphaeroides*, but glutamate transport is driven by the electrochemical gradient in this bacterium. This system is known as the TRAP (tripartite ATP-independent periplasmic) transporter. TRAP transporters are widespread in prokaryotes, and are shock-tolerant high-affinity Na^+-dependent unidirectional secondary transporters.

Through active transport, nutrients are accumulated in the cell with the expenditure of a large amount of energy in the form of the electrochemical gradient or ATP. Microbes can grow efficiently in environments with low nutrient concentrations due to active transport systems. Active transport can be summarized as follows:

(1) Carrier proteins have solute specificity as in the enzyme–substrate relationship.
(2) Energy is needed to change the affinity of the transporter for the transported solute at the other side of the membrane.
(3) The transported solute can be accumulated against a concentration gradient.
(4) The structure of the solute does not change during active transport.

3.5 Group translocation

Sugars are phosphorylated during their transport in many bacteria, especially in anaerobes. The phosphate donor in these transport systems is phosphoenolpyruvate (PEP), a glycolytic intermediate. Since the solute is phosphorylated, this transport is referred to as group translocation. This system is not known in eukaryotes.

A group of proteins known as the phosphotransferase (PT) system transports and phosphorylates sugars (Figure 3.5). They are cytoplasmic enzyme I and HPr (histidine-containing protein), and membrane-bound enzymes II_A, II_B and II_C.

Enzyme I transfers phosphate from PEP to HPr. Phosphorylated HPr transfers phosphate to enzyme II_A. The solute passes through the membrane-embedded enzyme II_C and enzyme II_B transfers phosphate from enzyme II_A to the solute. The cytoplasmic proteins HPr and enzyme I do not have any specificity towards the sugar and are common components of PT systems. On the other hand, the membrane-bound proteins are specific for each sugar transported. The specific enzymes are described as enzyme $II_A{}^{Glu}$ etc. As shown in Figure 3.5, there are variations in the proteins involved in the transport of different sugars. Enzyme $II_A{}^{Glu}$ is a soluble protein, and enzymes II_A and II_B, which are involved in mannose transport, are also soluble. In addition, the mannose PT system has an extra membrane protein, enzyme II_D.

The bacterial PT system not only transports sugars into the cell, but also plays important roles in metabolic regulation including carbon catabolite repression (Section 12.1.3). In addition to the PT system catalysing group translocation of sugars, many Gram-negative bacteria possess the analogous nitrogen PT system (PTS^{Ntr}). In analogy to the sugar PTS, EI^{Ntr} and NPr catalyse the PEP-dependent phosphorylation of protein $EIIA^{Ntr}$. EIIB- and EIIC-like domains are absent in PTS^{Ntr}. PTS^{Ntr} does not participate in sugar transport, but exclusively serves regulatory functions using $EIIA^{Ntr}$ as an output domain in the regulation of nitrogen and carbon metabolism (Section 12.2.2), virulence and K^+ homeostasis. Unusually, the phosphorylated $EIIA^{Ntr}$ leads to an increase in succinate-mediated catabolite repression in *Sinorhizobium meliloti*.

3.6 Precursor/product antiport

Some lactic acid bacteria utilize the potential energy developed by the accumulation of fermentation products in the cell to drive nutrient transport in a similar manner to antiport. Instead of H^+ or Na^+, a fermentation product is exchanged with its precursor and this system is therefore referred to as the precursor/product antiport (Figure 3.6).

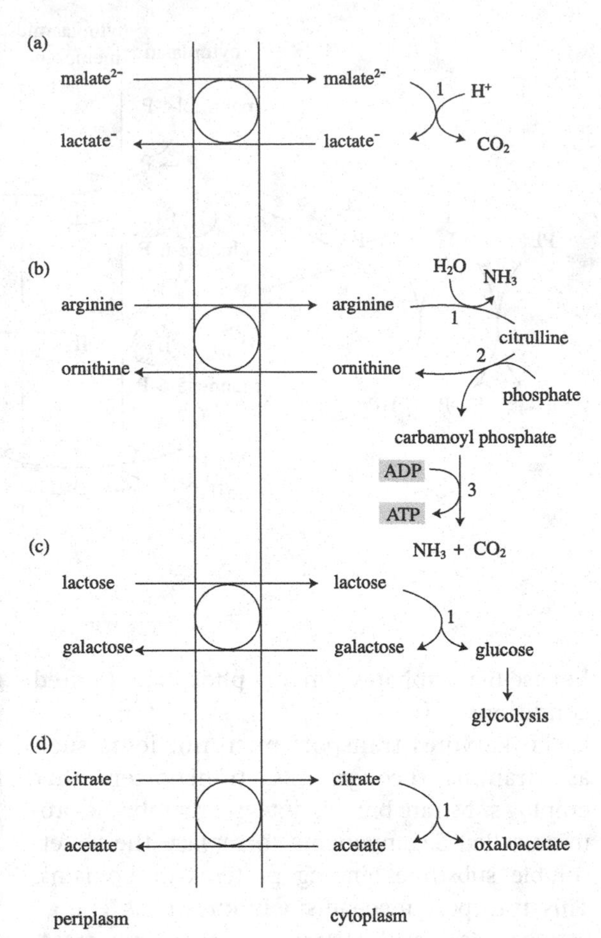

Figure 3.6 Precursor/product antiport systems of lactic acid bacteria.

(Modified from *Mol. Microbiol.* **4**: 1629–1636, 1990)

Some anaerobic fermentative bacteria, including lactic acid bacteria, utilize the potential energy of a high concentration of fermentation product inside the cell to import the precursor.

(a) Malate/lactate antiport. 1, malolactic fermentation.
(b) Arginine/ornithine antiport. 1, arginine deiminase; 2, ornithine carbamoyltransferase; 3, carbamate kinase.
(c) Lactose/galactose antiport. 1, β-galactosidase. (d) Citrate/acetate (acetoin, pyruvate) antiporter. 1, citrate lyase.

A malate/lactate antiport system is known in species of *Lactobacillus, Streptococcus, Leuconostoc* and *Pediococcus*. These organisms generate a proton motive force fermenting malate to lactate through the well-documented malolactic fermentation pathway (Section 8.4.6). Species of *Streptococcus* ferment arginine to ornithine through citrulline to produce ATP. Ornithine is exchanged with arginine in a 1:1 ratio.

Lactose is imported through a H^+-symport system in *Streptococcus thermophilus* and *Lactobacillus bulgaricus*. Lactose can be transported by a lactose/galactose antiport system in these bacteria when the lactose concentration is high. Lactose is hydrolysed to glucose and galactose by β-galactosidase. Glucose is metabolized through glycolysis and galactose is exchanged with lactose. Excreted galactose is then utilized after all the lactose is consumed.

Citrate is metabolized to pyruvate through oxaloacetate in *Lactococcus lactis*. Pyruvate in turn is converted to acetate and acetoin (Section 8.4.6). Citrate is imported by the action of an antiporter with acetate, acetoin or pyruvate.

In addition to lactic acid bacteria, precursor/product antiport systems are known in other anaerobic fermentative bacteria, such as the oxalate/formate antiporter in *Oxalobacter formigenes* and the betaine/*N,N*-dimethyl glycine antiporter in *Eubacterium limosum*. Figure 3.7 summarizes some of the nutrient uptake pathways known in prokaryotes.

3.7 Ferric ion (Fe(III)) uptake

In natural aerobic ecosystems, almost all iron is present as the ferric ion (Fe(III)) since ferrous iron (Fe(II)) is auto-oxidized with molecular oxygen at neutral pH. Ferric iron is virtually insoluble in water, with a solubility of around 10^{-20} M and this is much lower than the 10^{-6} M necessary to supply adequate iron for most microbes. To overcome this problem, many microbes synthesize and excrete low molecular weight ferric iron chelating compounds known as siderophores, for the sequestration and uptake of iron. Siderophores form complexes with Fe(III) that are imported into the cell by an ABC transport system. The complex is transported across the outer membrane by TonB-dependent active transport (Section 3.8) powered by the proton motive force through a cytoplasmic membrane protein complex in Gram-negative bacteria. Siderophores are a collection of compounds with a variety of chemical structures. Two main structural classes of siderophores are the catecholamides and the hydroxamates (Figure 3.8). A given organism may produce siderophores of one or both classes.

In spite of the preference of siderophores for Fe(III), they can also chelate other metals with variable affinities. Metals other than iron can activate the production of siderophores by bacteria, thereby implicating siderophores in the homeostasis of metals other than iron and tolerance to toxic metals.

3.8 TonB-dependent active transport across the outer membrane in Gram-negative bacteria

Gram-negative bacteria invest energy in the form of a proton motive force to transport certain nutrients such as ferric–siderophore complexes, vitamin B_{12} and nickel complexes across the outer membrane. An outer membrane protein complex, the TonB-dependent transporter (TBDT), transports them into the periplasmic region, utilizing energy supplied by a cytoplasmic membrane protein complex, TonB–ExbB–ExbD, and therefore consuming the proton motive force (Figure 3.9). TBDTs are specific for the nutrient transported, and the TonB–ExbB–ExbD complex is a universal energy transducing apparatus for TBDTs. The synthesis of the TBDT system is regulated by the availability of the particular nutrient. The ABC transporters transport the nutrients into the cytoplasm.

3.9 Multidrug efflux pump

Many bacteria are resistant to antimicrobial agents including antibiotics. There are different resistance mechanisms including active efflux. Based on the source of energy and amino acid sequence, the bacterial efflux transporters are classified into five different superfamilies: (1) the major facilitator superfamily (MFS), (2) the small multidrug resistant superfamily (SMR), (3)

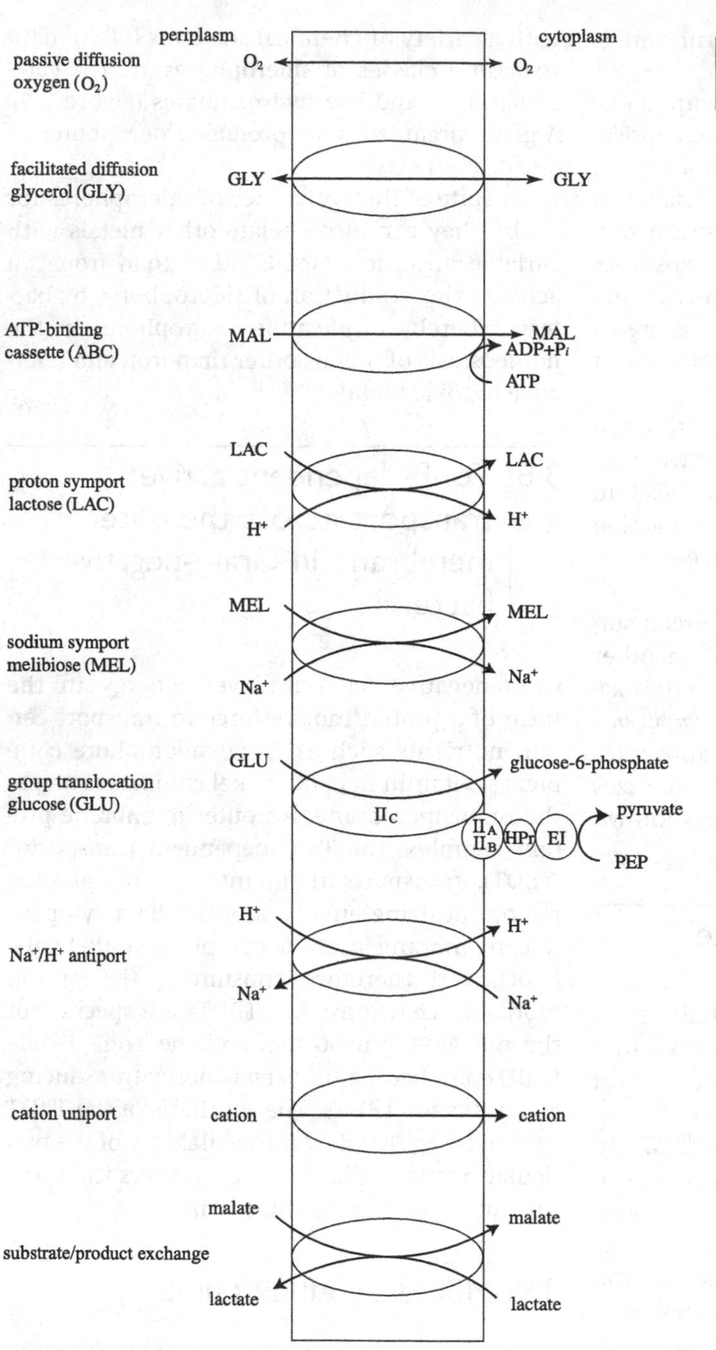

Figure 3.7 Nutrient import systems in prokaryotes.

the multi antimicrobial extrusion protein superfamily (MATE), (4) the ATP-binding cassette (ABC) superfamily and (5) the resistance-nodulation-cell division superfamily (RND). ATP hydrolysis provides energy for the ABC superfamily that is known in all domains of life. The other four superfamilies known in bacteria are antiporters, consuming the proton (sodium) motive force to export antimicrobial agents. Some of them can extrude a wide range of chemically diverse compounds, including antibiotics, antiseptics, dyes, detergents and solvents.

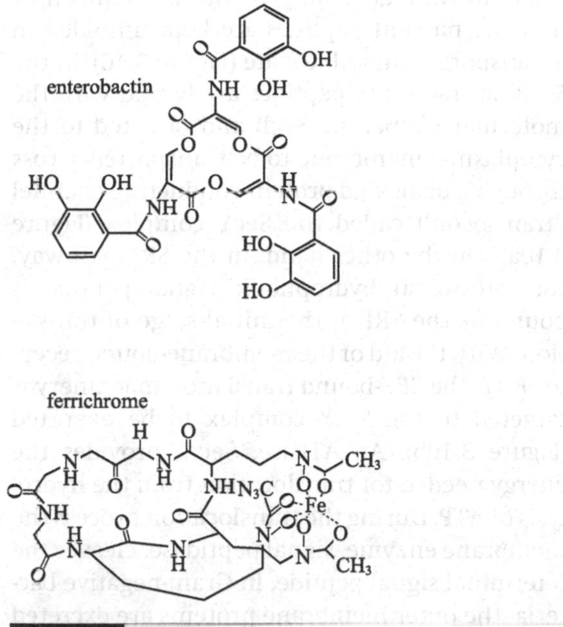

In Gram-negative bacteria these pumps span the cytoplasmic membrane, cell wall and outer membrane, consisting of outer membrane proteins, cytoplasmic membrane proteins and periplasmic membrane fusion proteins. They are referred to as tripartite efflux pumps. They are structurally similar to the type I protein exporter (Section 3.10.2.2, Figure 3.13).

3.10 Export of cell surface structural components

Large numbers of proteins and polysaccharides are present in the cytoplasmic membrane, periplasm (including the cell wall) and outer membrane. Many prokaryotes also secrete extracellular enzymes and toxins. These must be translocated through the cytoplasmic membrane after their synthesis in the cytoplasm. Translocation into and through the cytoplasmic

Figure 3.8 Structures of the siderophores enterobactin (a catecholamide) and ferrichrome (a hydroxamate).

(*FEMS Microbiol. Rev.* **27**: 215–237, 2003)

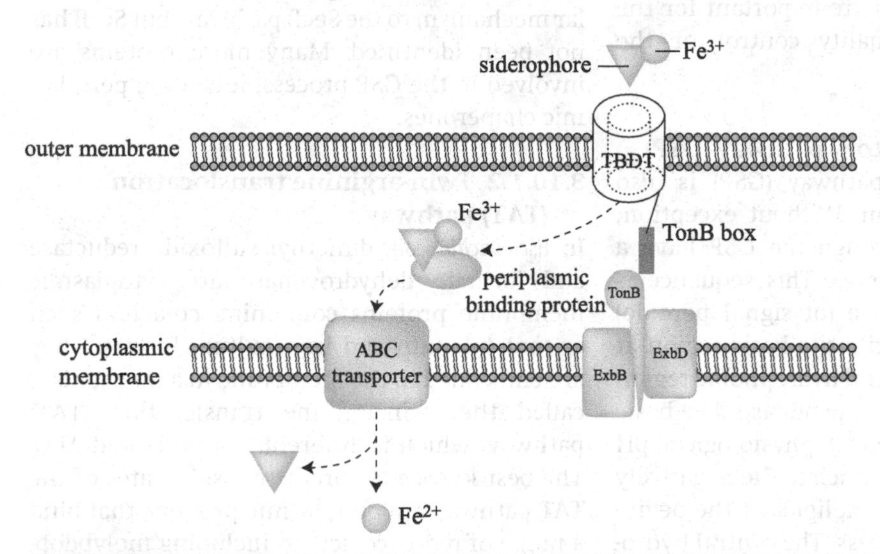

Figure 3.9 Ferric–siderophore transport mediated by the TonB-dependent active transporter and ABC transporter.

(Modified from *Ann. Rev. Microbiol.* **64**: 43–60, 2010)

The ferric–siderophore complex is transported into the periplasm through an outer membrane protein complex, the TonB-dependent outer membrane transporter (TBDT), powered by energy supplied by a cytoplasmic membrane-bound protein complex (TonB–ExbB–ExbD), thereby consuming the proton motive force. A periplasmic binding protein directs the ferric–siderophore complex to an ABC transporter through which the complex is transported into the cytoplasm.

Annual Reviews, with permission.

membrane is referred to as protein transport. The term 'secretion' is used to refer to protein translocation away from the cytoplasmic membrane to the cell surface and to the extracellular medium. Proteins are transported through one of three mechanisms. These are the general secretory pathway (GSP), the ABC pathway and the twin-arginine translocation (TAT) pathway. In addition to these pathways some of the proteins exported through the outer membrane (Section 3.10.2) pass through the cytoplasmic membrane by other specific mechanisms in Gram-negative bacteria. It should be noted that the proteins forming flagella (Section 6.13.1) and fimbriae are transported across the cytoplasmic membrane through specific mechanisms, the flagella-export apparatus (FEA) and the fimbrilin-protein exporter (FPE), respectively.

3.10.1 Protein transport

Proteins exported through the cytoplasmic membrane have *N*- or *C*-terminal sequences that are cleaved during transport by proteolytic enzymes. These enzymes are important for the exact targeting and quality control of the exported proteins.

3.10.1.1 General secretory pathway (GSP)

The general secretory pathway (GSP) is also known as the Sec system. Without exception, proteins transported through the GSP have a unique *N*-terminal sequence. This sequence is called the signal sequence (or signal peptide) and consists of three regions: a basic region at the *N*-terminal end, a central hydrophobic region and a recognition site for a peptidase. The basic end is positively charged at physiological pH values and is believed to attach to the negatively charged membrane phospholipids at the beginning of the transport process. The central hydrophobic region inserts itself into the cytoplasmic membrane, facilitating the transport of the main peptide, and the recognition site is cleaved by a signal peptidase at the membrane during or after transport.

Folded proteins cannot be transported through the GSP. The GSP can be classified into a SRP (signal recognition particle) pathway and a SecB pathway according to the mechanism of how the nascent peptides are kept unfolded in a transport compatible state (Figure 3.10). In the SecB pathway, the peptides are bound with the molecular chaperone SecB and targeted to the cytoplasmic membrane to be transported across through a dedicated protein-conducting channel ('translocon') called the SecY complex (Figure 3.10a). On the other hand, in the SRP pathway, the *N*-terminal hydrophobic signal peptide is bound by the SRP at the initial stage of translation. With the aid of the membrane-bound receptor FtsY, the SRP-bound translation machinery is targeted to the SecY complex to be exported (Figure 3.10b). An ATPase, SecA, provides the energy needed for translocation from the hydrolysis of ATP. During the translocation process the membrane enzyme, signal peptidase, cleaves the *N*-terminal signal peptide. In Gram-negative bacteria, the outer membrane proteins are excreted exclusively through the SecB pathway, and the cytoplasmic membrane proteins are embedded into the membrane by the SRP pathway. Proteins are excreted in Gram-positive bacteria by a similar mechanism to the SecB pathway, but SecB has not been identified. Many more proteins are involved in the GSP process, including periplasmic chaperones.

3.10.1.2 Twin-arginine translocation (TAT) pathway

In *Escherichia coli*, dimethyl sulfoxide reductase and formate dehydrogenase are cytoplasmic membrane proteins containing cofactors such as molybdenum and iron–sulfur clusters. They are embedded in the membrane in a mechanism called the twin-arginine translocation (TAT) pathway, which is different from GSP and ABC. The best known (or predicted) substrates of the TAT pathway are periplasmic proteins that bind a range of redox cofactors, including molybdopterin, Fe–S, Ni–Fe centres and others. These cofactors can be inserted into the peptide in the cytoplasm only and this requires substantial or complete folding of the mature protein, so it is believed that fully or largely folded structures are transported across the cytoplasmic membrane through this pathway. This system is also used for the translocation of other proteins

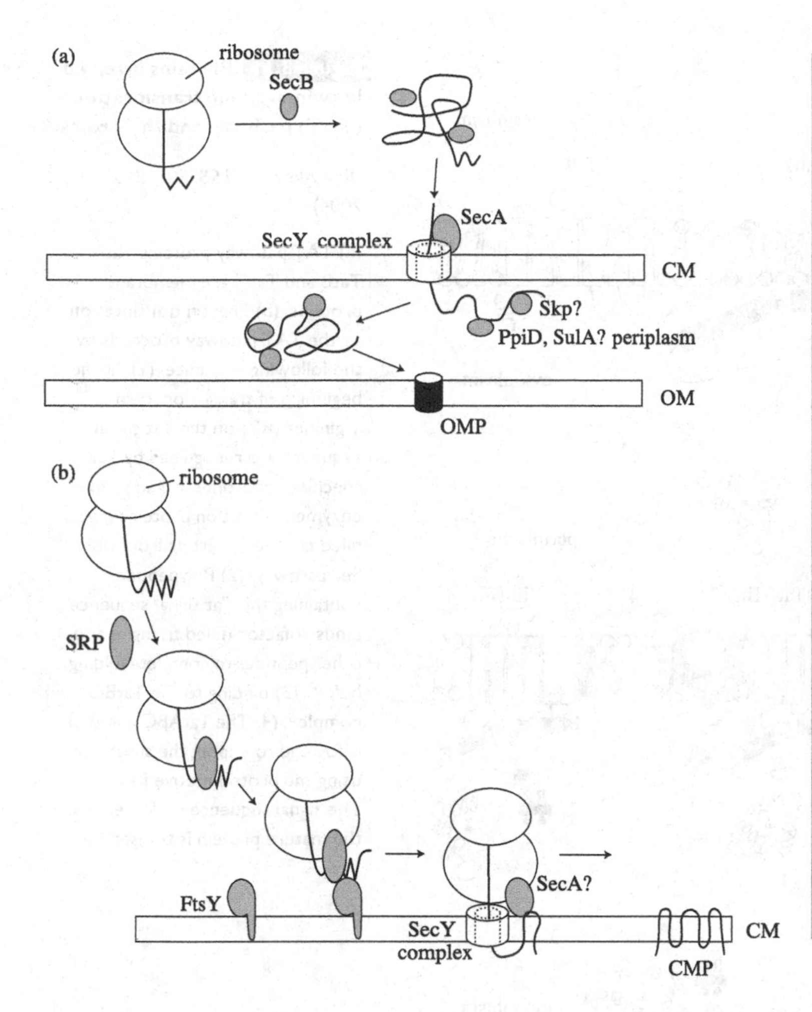

Figure 3.10 Protein translocation through the general secretory pathway (GSP).

(*Curr. Opin. Microbiol.* **3**: 203–209, 2000)

The GSP is classified according to the mechanisms by which the nascent peptides are prevented from folding before excretion. The molecular chaperone SecB binds peptides in the SecB pathway (a) and a signal recognition particle (SRP) plays a similar role in the SRP pathway. (b) An ATPase, SecA, provides the energy needed for translocation with the hydrolysis of ATP. During the translocation process the membrane enzyme, signal peptidase, cleaves the *N*-terminal signal peptide. In Gram-negative bacteria the outer membrane proteins (OMP) are excreted exclusively through the SecB pathway, and the cytoplasmic membrane proteins (CMP) are embedded in the membrane by the SRP pathway. Proteins are excreted by Gram-positive bacteria by a similar mechanism, but SecB has not been identified.

CM, cytoplasmic membrane; OM, outer membrane.

that do not contain any cofactors, as these might fold too rapidly for the Sec system to handle. The TAT pathway has a function of protein quality control, translocating only correctly folded substrate.

Proteins transported through the TAT pathway contain an *N*-terminal signal sequence. The TAT signal sequence is longer and less hydrophobic than the Sec signal sequence. In all TAT signal sequences a twin-arginine motif is present and this motif is an absolute requirement to route a protein through this pathway.

The TAT apparatus consists of cytoplasmic TatD and membrane-bound TatA, TatB, Tat C and Tat E. TatA and TatE seem to have overlapping functions, and TatE is dispensable. TatA and TatB have a single transmembrane helix while TatC has six helices (Figure 3.11a). Though the precise mechanism for protein translocation by the TAT pathway is not known, it is believed that TatB and TatC form a TatBC complex before binding the substrate, by recognizing the Tat signal sequence, to bring the substrate to TatA. The TatABC complex transports the substrate (Figure 3.11b). ATP is not required at any stage of this pathway, and the proton motive force (Δp) is consumed to provide the energy needed for protein translocation.

The TAT pathway was discovered in plants as a protein translocation mechanism into the thylakoids. A wide variety of prokaryotes have the unique ability to transport folded proteins through tightly sealed membranes. These include tetrachloroethene dehalogenase

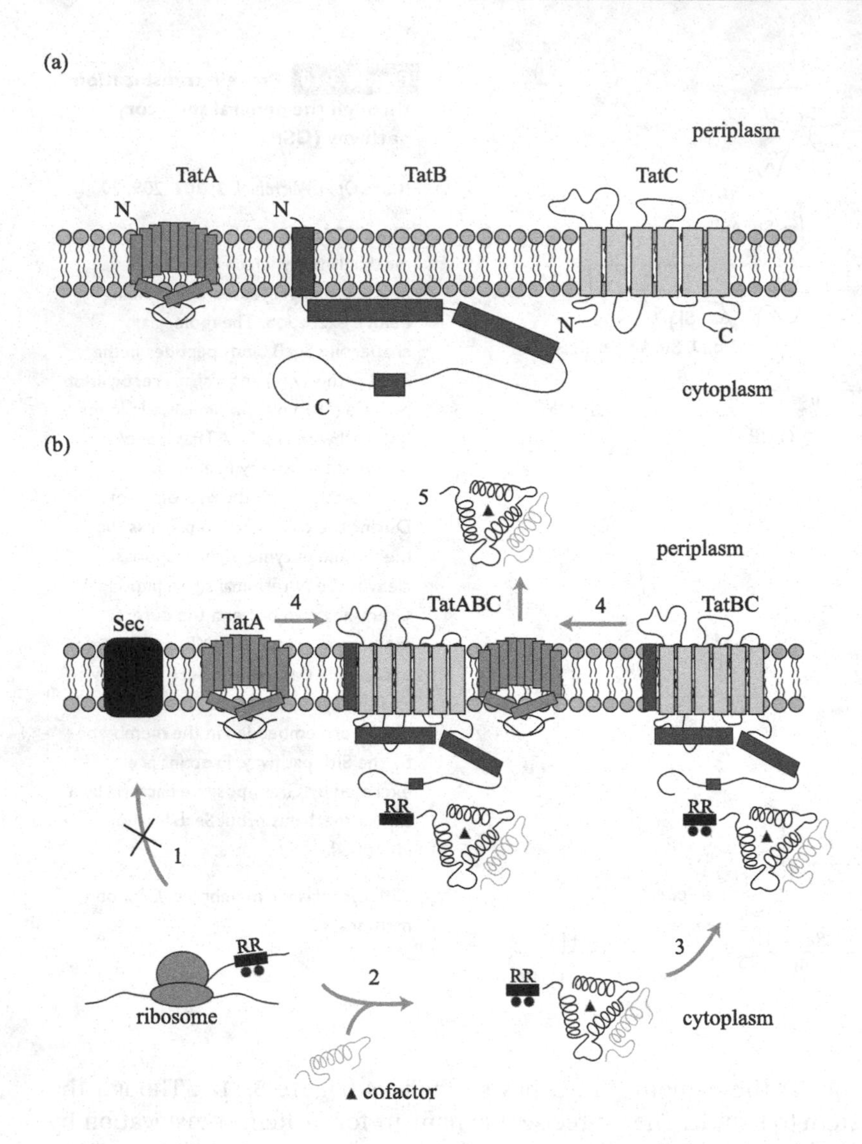

Figure 3.11 Proteins involved in twin-arginine translocation (TAT) pathway and their roles.

(*Res. Microbiol.* **155**: 803–810, 2004)

(a) TAT pathway proteins TatA, TatB and TatC are membrane proteins. (b) Protein translocation by the TAT pathway proceeds by the following sequence: (1) At the beginning of translation, twin arginines (RR) on the Tat signal sequence are recognized by Tat specific chaperones, called redox enzyme maturation proteins (small filled circles), diverting from the Sec pathway. (2) Polypeptide containing the Tat signal sequence binds cofactor (filled triangle) and other peptides to complete folding, before (3) binding to the TatBC complex. (4) The TatABC complex is formed to export the substrate using the proton motive force. (5) The signal sequence is cleaved and the mature protein is transported.

in *Dehalospirillum multivorans*, hydrogenase in *Desulfovibrio vulgaris, Escherichia coli* and *Wolinella succinogenes*, nitrous oxide reductase in *Pseudomonas stutzeri* and glucose-fructose oxidoreductase in *Zymomonas mobilis*. The *Tat* genes have not been found in the complete genome sequences of mycoplasmas and certain methanogens.

Some complex proteins are exported through the TAT pathway. A TAT signal sequence in one of the subunits facilitates the transport of the entire complex. The periplasmic nickel-containing hydrogenase in *Escherichia coli* is composed of a small subunit containing a TAT signal sequence and a large subunit devoid of an export signal. This hydrogenase complex is formed in the cytoplasm before being transported.

Peroxidase (YwbN) and phosphodiesterase in *Bacillus subtilis* and staphylococcal iron-dependent peroxidase (FepB) are translocated through the TAT pathway. In Gram-positive bacteria, TatA and TatC perform the translocation without TatB.

3.10.1.3 ATP-binding cassette (ABC) pathway

ABC transporters are evolutionarily related and have various functions. This pathway is known not only in bacteria but also in eukaryotes and archaea. As described previously (Section 3.4), ABC facilitates the transport of various nutrients, including sugars, amino acids and ions in prokaryotes. Lipopolysaccharide is assembled in the periplasm and embedded in the outer membrane (Section 6.9.4) through the ABC pathway. Capsular polysaccharides are exported to the cell surface through a similar mechanism. The ABC pathway is also involved in the extrusion of noxious substances such as antibiotics, extracellular toxins and proteins, and the targeting of membrane and surface structures. ABC transporters have been most studied in Gram-negative bacteria, and are discussed in Section 3.10.2.2.

3.10.1.4 Protein translocation through the cell wall in Gram-positive bacteria

Autolysin-defective mutants of *Bacillus subtilis* and *B. licheniformis* are non-motile under autolysin deficient conditions while they produce flagella with motility under autolysin producing conditions. Proteins larger than the pore size (about 2 nm) of the normal peptidoglycan layer need peptidoglycan hydrolysing enzyme activities to cross the cell wall. Some proteins require dedicated peptidoglycan hydrolysing enzymes, while others take advantage of gaps in the peptidoglycan layer that are created during normal metabolism by peptidoglycan-degrading enzymes.

Many pathogenic Gram-positive bacteria excrete virulence related peptides and proteins in the precise time governed by quorum sensing (Section 12.2.8). The regulation of their secretion is controlled not only by gene expression, but also by export through the cell envelope.

3.10.2 Protein translocation across the outer membrane in Gram-negative bacteria

In bacteria with two membranes (diderm bacteria), including Gram-negative bacteria, proteins that are targeted for the cell surface or extracellular environment must also cross the additional barrier to secretion formed by the outer membrane. Monoderm bacteria, such as Gram-positive bacteria, do not have this problem. Nine dedicated protein secretion systems have been identified so far, named the chaperone/usher system and type I to type VIII; differing greatly in their composition and mechanism of action. The two membranes present in Gram-negative bacteria are negotiated either by one-step transport mechanisms (types I, III and VI), where the unfolded substrate is translocated directly into the extracellular space, without any periplasmic intermediates, or by two-step mechanisms (chaperone/usher system and, types II, IV, V, VII and VIII), where the substrate is first transported into the periplasm to allow folding before a second transport step across the outer membrane.

The type I protein translocation machinery is classed within the family of ATP-binding cassette (ABC) transporters described previously. Proteins translocated by the type II systems are transported to the periplasm by the GSP before crossing the outer membrane using energy provided by ATP hydrolysis on the inner face of the cytoplasmic membrane. Many Gram-negative plant and animal pathogenic bacteria utilize a specialized type III secretion system as a molecular syringe to inject effector proteins directly into host cells. Flagellin (Section 2.3.1) is exported in a similar way to the type III secretion system. The type IV secretion system is used to release proteins through the outer membrane in some pathogenic bacteria after they are exported through the cytoplasmic membrane by the GSP. DNA injection into a plant cell by *Agrobacterium tumefaciens* is another form of the type IV system and bacterial conjugation with the transport of DNA is also related to this system. The type V secretion system is the simplest protein secretion pathway and is known as the autotransporter pathway. The type VI secretion system is a cell envelope-spanning machine that translocates toxic effector proteins into eukaryotic and prokaryotic cells in a similar way to the type III system, and which has important roles in pathogenesis and bacterial competition.

Another type VII secretion system has been proposed to describe protein translocation through the mycomembrane of Gram-positive bacteria belonging to genera of *Mycobacterium* and *Corynebacterium* (Section 2.3.3.2).

The type VIII secretion system, which is known as the curli biogenesis system, is

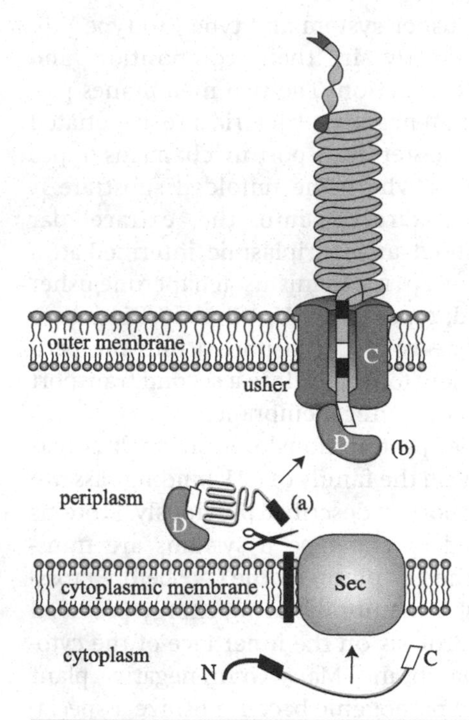

Figure 3.12 Model for the biogenesis of pili by the chaperone/usher system.

(*Curr. Opin. Cell Biol.* **12**: 420–430, 2000)

(a) Pilus subunits cross the cytoplasmic membrane via the GSP (Sec) system, followed by cleavage of their *N*-terminal signal peptide. The periplasmic chaperone PapD binds to each subunit via a conserved carboxy-terminal subunit motif (white box), allowing proper subunit folding and preventing premature subunit–subunit binding. (b) The PapD–pilin complex is targeted to the outer membrane protein, PapC, the usher protein for assembly into pili and secretion across the outer membrane. *C*-terminal (white box) motifs of the secreted pilin interact with the *N*-terminal (black box) motif of the subunit at the tip of the pilus.

D, chaperone PapD; C, usher PapC; Sec, translocon.

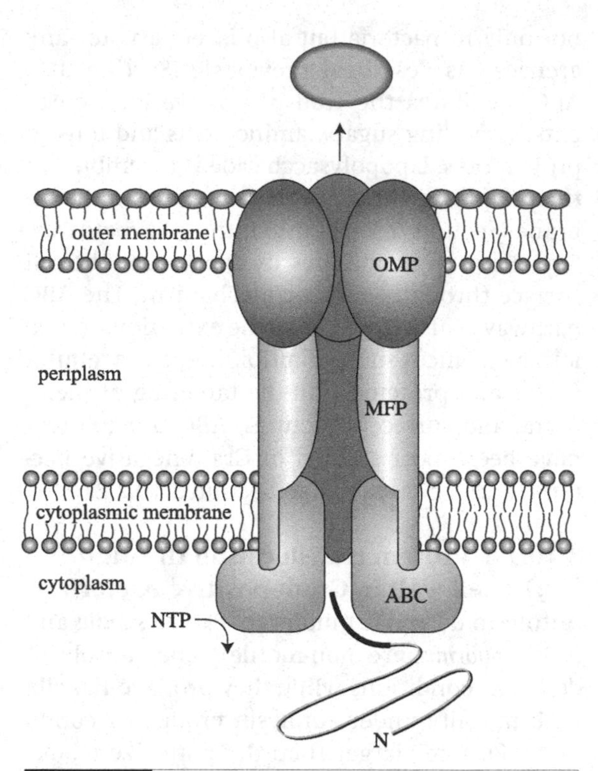

Figure 3.13 Protein secretion in Gram-negative bacteria through the ABC (type I) pathway.

(*Curr. Opin. Cell Biol.* **12**: 420–430, 2000)

Proteins exported by the ABC pathway do not have an *N*-terminal signal sequence but have a *C*-terminal secretion sequence. An ABC exporter consists of cytoplasmic membrane ABC, periplasmic membrane fusion protein (MFP) and an outer membrane protein (OMP). ABC hydrolyses ATP to export proteins through the MFP and OMP.

ABC, ATP-binding cassette; MFP, membrane fusion protein; NTP, nucleotide triphosphate (ATP).

responsible for outer membrane secretion and assembly of thin and aggregative pili called curli. Curli are functional amyloid fibres that are involved in cell aggregation, bacterial adhesion and the formation of mature biofilms.

3.10.2.1 Chaperone/usher system

Certain cell surface proteins associated with virulence are exported through the chaperone/usher system. They are translocated across the cytoplasmic membrane by the GSP system before binding to a periplasmic chaperone, PapD, which prevents premature folding of the peptides. Periplasmic chaperone-peptide complexes are targeted to an outer membrane protein called the usher, through which the peptide is secreted. Some proteins cross the cytoplasmic membrane by the TAT pathway before being secreted by this system. Energy is not needed in this process. Pili are formed in this system (Figure 3.12).

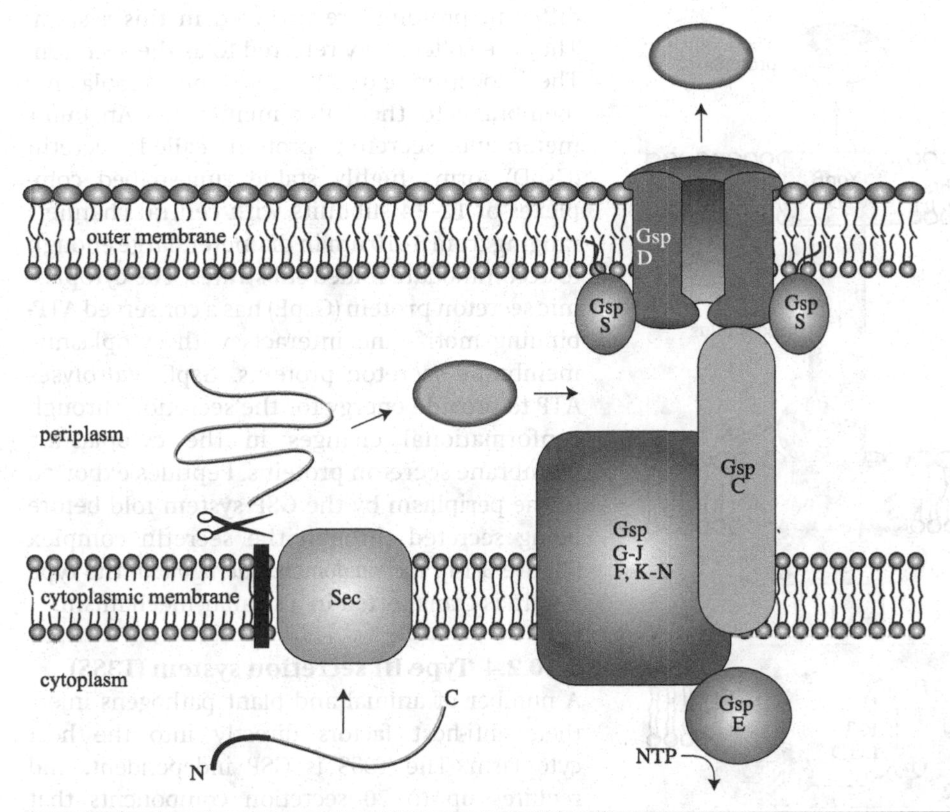

Figure 3.14 The type II system for secretion of proteins through the outer membrane in Gram-negative bacteria.

(*Curr. Opin. Cell Biol.* **12**: 420–430, 2000)

Type II substrates cross the cytoplasmic membrane via the Sec system followed by signal-sequence cleavage and protein folding in the periplasm. The GspD secretin, indicated here as a complex with the GspS lipoprotein, serves as a gated channel for secretion of substrates to the cell surface. GspC may transmit energy from the cytoplasmic membrane, presumably generated by the cytoplasmic GspE nucleotide-binding protein, to the outer membrane complex.

Gsp, general secretion pathway protein; NTP, nucleotide triphosphate, such as ATP.

3.10.2.2 Type I secretion system (T1SS): ATP-binding cassette (ABC) pathway

The type I (ABC) system is employed to secrete extracellular proteins such as hydrolytic enzymes (proteases and lipases) and toxins in Gram-negative bacteria. The T1SS is GSP independent, and the secretion process begins when a secretion signal located within the *C*-terminal end of the secreted effector molecule interacts with the ABC transporter protein. In general, this signal sequence is specific and is recognized only by the dedicated ABC transporter. The cytoplasmic membrane protein complex ABC pushes the peptide through a membrane fusion protein (MFP) and an outer membrane protein (OMP), with the hydrolysis of ATP providing energy needed for the translocation. The MFP anchors within the cytoplasmic membrane and spans the periplasm (Figure 3.13).

3.10.2.3 Type II secretion system (T2SS)

This system represents a third branch of the GSP system. The T2SS is employed for the secretion of a wide variety of proteins in Gram-negative bacteria, including the cholera toxin of *Vibrio cholerae*, exotoxin A of *P. aeruginosa* and several polysaccharide hydrolysing enzymes such as pullulanase of *Klebsiella oxytoca*. More than 12

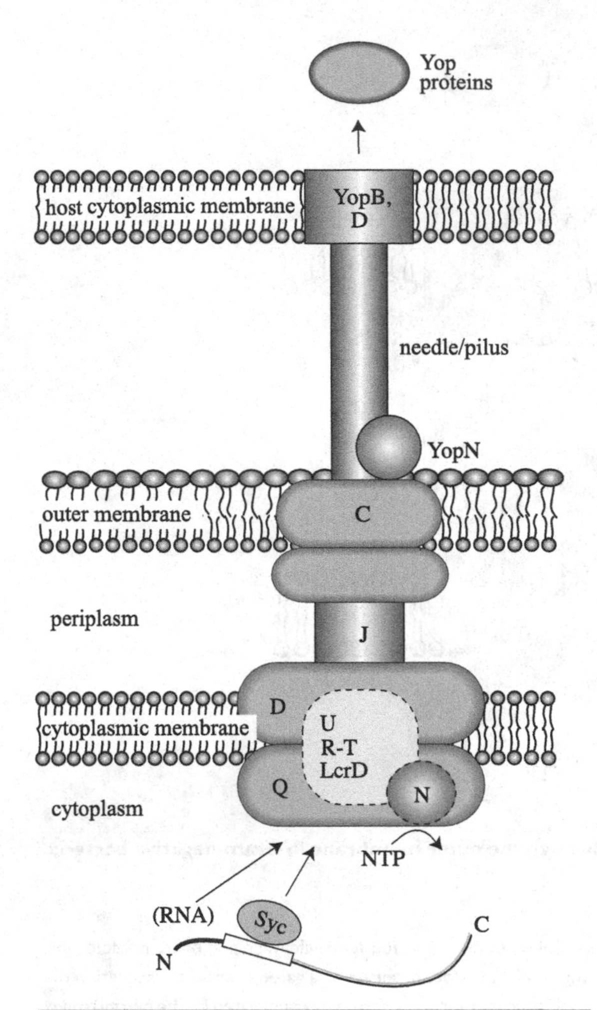

Figure 3.15 Protein injection into a host cell through the type III excretion system.

(*Curr. Opin. Cell Biol.* **12**: 420–430, 2000)

The figure shows the type III apparatus in species of *Yersinia*. Some type III secretion substrates contain two *N*-terminal secretion sequences that direct the substrate to the secretion machinery, one encoded by the mRNA and the second serving as a binding site for cytoplasmic Syc chaperones. By homology with flagellar proteins, YscR-U and LcrD may form a central secretion apparatus, energized by YscN. The YscC secretin presumably provides a secretion channel across the outer membrane and the surface-localized YopN protein is thought to serve as a channel gate. By analogy with flagella, type III substrates (Yop proteins) may travel through a central channel in the type III needle. Translocation of Yop proteins into the target eukaryotic cell may take place via a channel formed in the plasma membrane by YopB and YopD.

C, YscC secretin; D, YscD; J, YscJ; N, YscN; Q, YscQ; R-T, YscR-T.

different proteins are involved in this system. They are collectively referred to as the secreton. Their location extends from the cytoplasmic membrane to the outer membrane. An outer membrane secreton protein called secretin (GspD) forms highly stable ring-shaped complexes of 12–14 subunits with central channels ranging from 5–10 nm in diameter, large enough to accommodate folded substrates. The cytoplasmic secreton protein (GspE) has a conserved ATP-binding motif, and interacts with cytoplasmic membrane secreton proteins. GspE hydrolyses ATP to provide energy for the secretion through conformational changes in the cytoplasmic membrane secreton proteins. Peptides exported to the periplasm by the GSP system fold before being secreted through the secretin complex (Figure 3.14). In *Pseudomonas aeruginosa*, the type 4 pilin is transported in a similar mechanism.

3.10.2.4 Type III secretion system (T3SS)

A number of animal and plant pathogens inject their anti-host factors directly into the host cytoplasm. The T3SS is GSP independent, and requires up to 20 secretion components that assemble into a large structure, called the injectisome, which spans both bacterial membranes and the host membrane. The channel-forming protein on the outer membrane shares homology with the secretin of the type II system. Proteins are excreted through a tube-like structure called the needle. ATP hydrolysis provides the energy needed for the excretion (Figure 3.15). Flagella are very similar in structure to the type III excretion machinery and flagellin is excreted in a similar mechanism to that of the type III system (Section 6.13.1). Proton motive force is consumed to export flagellin.

Many bacteria contain more than one T3SS that might be of importance at different stages of the infection process. Their expression is highly regulated and two *N*-terminal secretion signals have been identified for export. The first signal appears to reside in the mRNA and may target the RNA–ribosome complex to the type III machinery for coupled translation and secretion. The second secretion signal serves as the binding site for cytoplasmic chaperones, termed Syc proteins, and may specifically target protein to the type III machinery for translocation into host cells.

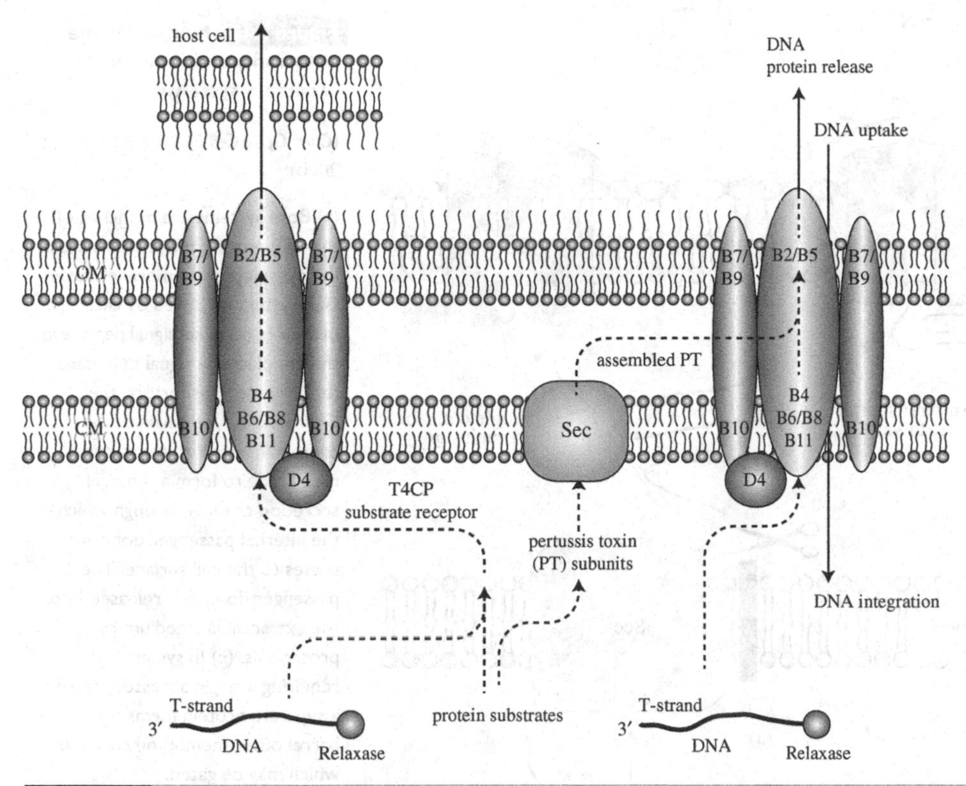

Figure 3.16 Model of type IV secretion for injection of toxin proteins into the eukaryotic host cell.

(Modified from *Microbiol. Mol. Biol. Rev.* **73**: 775–808, 2009)

This is a protein secretion system identified in pathogenic *Legionella pneumophila* and *Helicobacter pylori*. DNA is transported from *Agrobacterium tumefaciens* into plant host cells by a similar mechanism. Protein secretion by the type IV system may take place via a periplasmic intermediate, with substrates first transported through the GSP (Sec) system or a GSP-independent mechanism into the periplasm (left). DNA secretion probably takes place from the cytoplasm in a single step without a periplasmic intermediate. The cytoplasmic membrane components contain nucleotide-binding activity and probably energize aspects of the secretion process. DNA is transported from the donor cell to the recipient cell through the conjugation system which is a subfamily of T4SS. In the donor cell, DNA transfer and replication (Dtr) proteins, including relaxase, bind the origin-of-transfer and process DNA to a single DNA strand destined for translocation (T-strand) that is transferred to the recipient cell.

3.10.2.5 Type IV secretion system (T4SS)

This system has been identified in *Legionella pneumophila* and *Helicobacter pylori* and secretes protein toxins into host cells. Peptides secreted through this system are exported through the cytoplasmic membrane either by the GSP system or by a GSP-independent mechanism. The periplasmic intermediate is secreted into the host cell by a similar mechanism to that of the type III system. However, the outer membrane channel of the type IV system is not homologous with that of the type III system (Figure 3.16). Most proteins secreted by this system are GSP independent. They have a *C*-terminal positive charged tail or *C*-terminal hydrophobic tail recognized by a chaperone that directs the protein to the type IV secretion machinery. Other proteins such as the pertussis toxin of *Bordetella pertussis* are translocated across the cytoplasmic membrane through GSP.

This type IV system is similar to the mechanism that facilitates the translocation of DNA from *Agrobacterium tumefaciens* into the plant host cells. DNA is transported directly from the

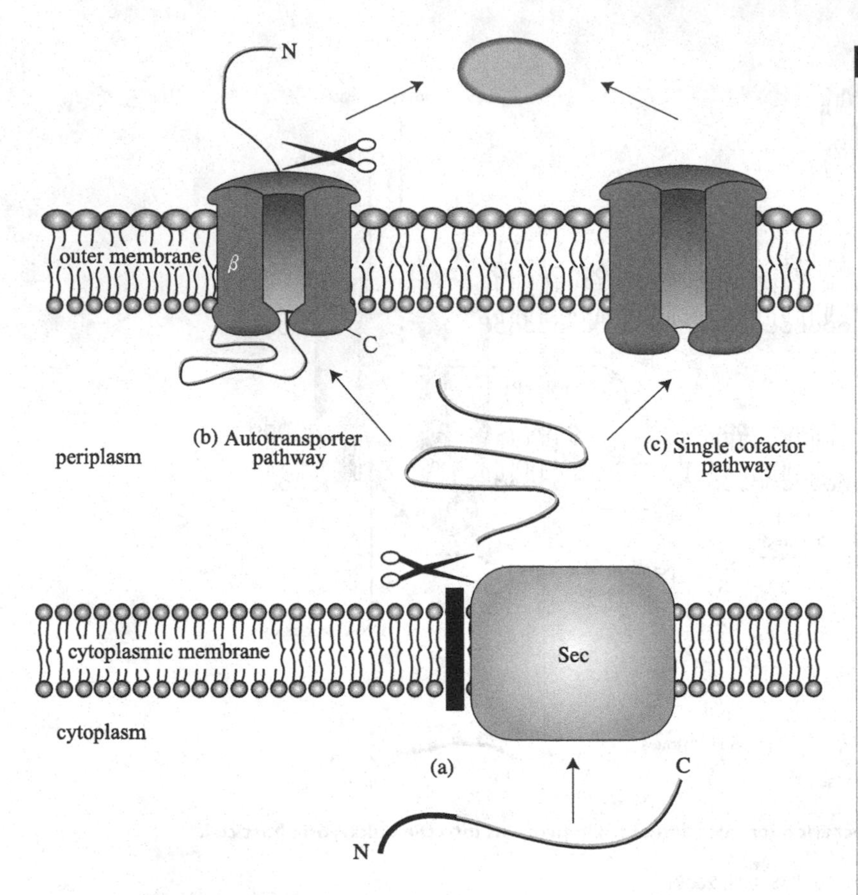

Figure 3.17 Models for the autotransporter and single accessory systems.

(*Curr. Opin. Cell Biol.* **12**: 420–430, 2000)

(a) Both systems are branches of the GSP system, and proteins cross the cytoplasmic membrane via the GSP system, followed by cleavage of their *N*-terminal signal peptide in the periplasm by signal peptidase. (b) For autotransporters, the *C*-terminal or β-domain of the protein inserts into the outer membrane to form a β-barrel secretion channel through which the internal passenger domain passes to the cell surface. The passenger domain is released into the extracellular medium by proteolysis. (c) In systems requiring a single accessory factor, a separate protein forms the β-barrel outer membrane channel, which may be gated.

C, *C*-terminal; N, *N*-terminal; OM, outer membrane; Sec, translocon.

bacterial cytoplasm. Similarity is also found with bacterial conjugation. The cytoplasmic membrane type IV components hydrolyse ATP to provide energy for the secretion.

This system is found in both Gram-negative and Gram-positive bacteria and also in some archaea.

3.10.2.6 Type V secretion system (T5SS): autotransporter and proteins requiring single accessory factors

Proteases, toxins and cell surface proteins such as those used for adhesion to solid surfaces (adhesins) and for invasion into the host cell (invasins) are excreted through the outer membrane by the T5SS. This system is subdivided into five sub-types. These are (i) the classical autotransporter (T5aSS), (ii) the two-partner secretion system (T5bSS), (iii) the trimeric autotransporter (T5cSS), (iv) the hybrid autotransporter between T5aSS and T5bSS (T5dSS) and (v) the inverted autotransporter (T5eSS).

The proteins translocated through the classical autotransporter (T5aSS) possess three domains: an *N*-terminal signal sequence for secretion across the cytoplasmic membrane by the GSP system, an internal passenger (functional) domain and a *C*-terminal β-domain. This β-domain forms a porin-like structure at the outer membrane through which the passenger domain passes to the cell surface. The passenger domain is released into the extracellular medium by proteolysis (Figure 3.17). In some cases, proteins lacking a β-domain are secreted through a β-domain formed by a separate protein (a single accessory factor). Foreign proteins were released into the environment when a segment of DNA containing genes for the protein, signal peptide and *C*-terminal β-domain was introduced into *Escherichia coli*, showing that the autotransporter is not selective for the passenger proteins.

In the two-partner secretion (T5bSS) system, the passenger proteins are translocated

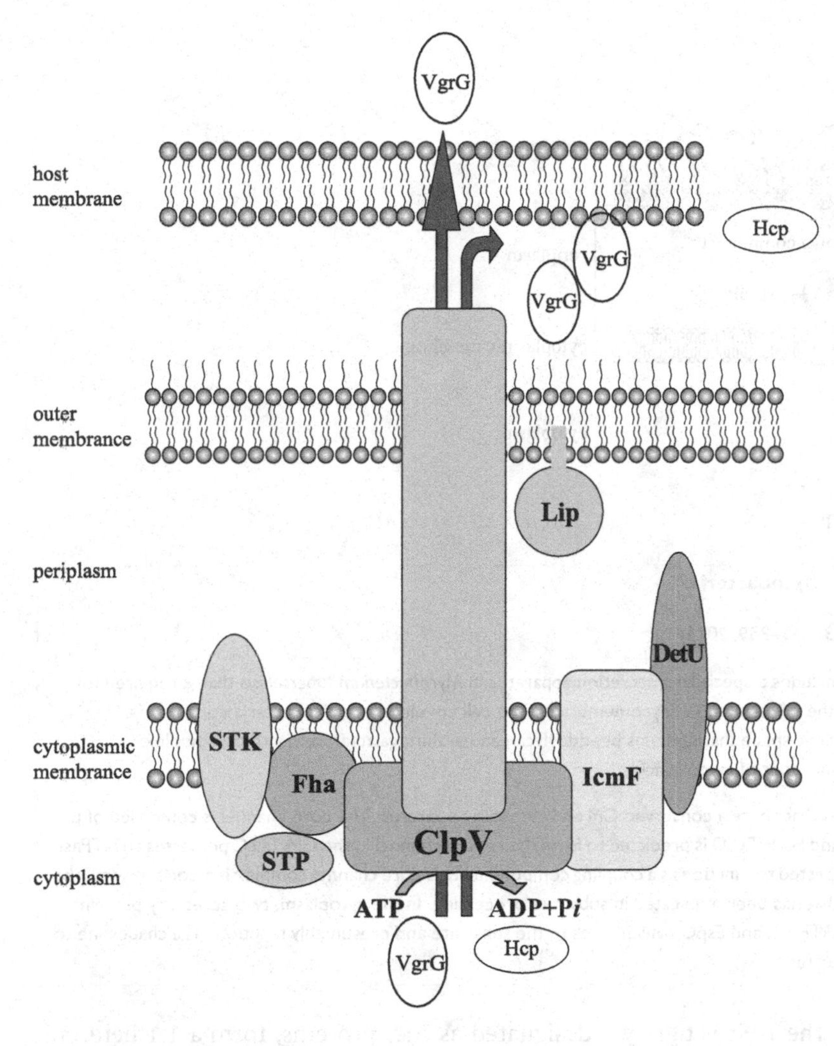

Figure 3.18 Model for type VI secretion system for injection of effector protein into a eukaryotic or prokaryotic cell for virulence or symbiosis.

(Modified from *Microbiol.* **154**: 1570–1583, 2008)

The ClpV ATPase may help to transport the channel-forming Hcp and VgrG across the cell envelope. The VgrG puncturing device is involved either in injecting the *C*-terminus of evolved VgrG into the eukaryotic cell or in releasing VgrG into the extracellular milieu or at the bacterial cell surface. Lip is a putative outer-membrane lipoprotein. IcmF and DotU are inner-membrane proteins. The level of phosphorylation of the Fha protein regulates T6SS activity. STK is the Ser/Thr kinase, whereas STP is the Ser/Thr phosphatase.

separately from those forming β-domain pores on the outer membrane as a single accessory factor. In the T5cSS, a trimer in the *C*-terminal β-domain serves as an outer membrane anchor for cell-surface exposure of the passenger domains. Proteins translated with the passenger domain separately form the hybrid autotransporter (T5dSS). Invasins of some pathogens are targeted to the cell surface through the inverted autotransporter (T5eSS), in that the translocation unit is located at the *N*-terminal instead of the *C*-terminal end of the classical autotransporter.

3.10.2.7 Type VI secretion system (T6SS)

The bacterial T6SS functions as a virulence factor, capable of attacking both eukaryotic and prokaryotic target cells by a process that involves protein transport through a contractile bacteriophage tail-like structure. The T6SS is a cell envelope-spanning machine that has a pivotal role in pathogenesis, symbiosis and bacterial competition. The T6SS is composed of 13 conserved, essential core components and several accessory components (Figure 3.18). Components that have been characterized include an IcmF-associated homologue, the ATPase ClpV, a regulatory FHA domain protein and the secreted VgrG and Hcp (haemolysin co-regulated protein) proteins. Hcp is a specific chaperone binding effectors to stabilize them in the bacterial cytoplasm. Type VI secretion is clearly a key virulence factor for some important pathogenic bacteria, including *Vibrio cholerae* and *Salmonella typhimurium*. However, type VI secretion systems (T6SSs) are widespread in nature and not

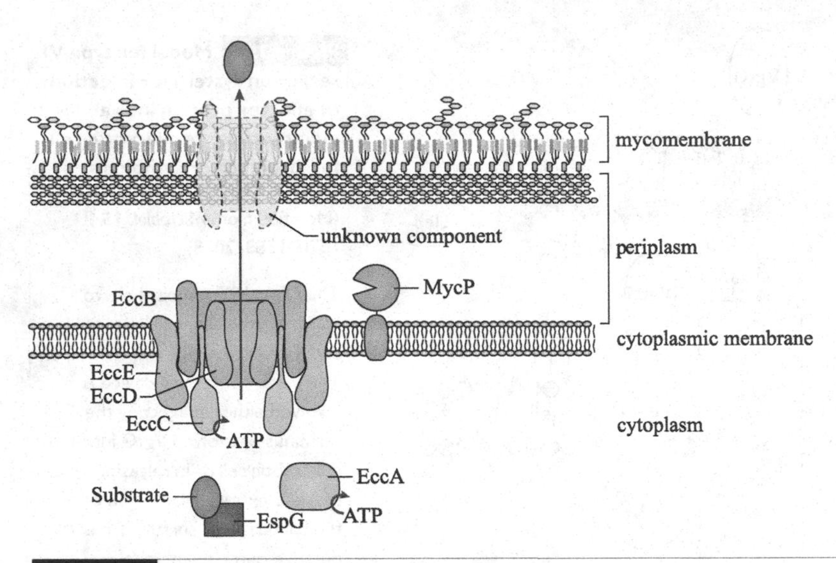

Figure 3.19 Type VII secretion in mycobacteria.

(Modified from *Nature Rev. Microbiol.* **13**: 343–359, 2015)

The type VII secretion system (T7SS) includes a specialized secretion apparatus in *Mycobacterium tuberculosis* that is required for secretion of virulence factors through the mycobacterial mycomembrane. The cell envelope of mycobacteria consists of a cytoplasmic membrane (CM), a periplasmic space that contains peptidoglycan and arabinogalactan, and a thick, complex mycomembrane that contains a waxy lipid coat of mycolic acids.

T7SSs are ~1.5 MDa protein complexes that share a conserved CM and cytosolic apparatus. The core channel is composed of the membrane proteins EccB, EccC, EccD and EccE. EccD is predicted to form the central channel in the CM. EccC possesses an ATPase cytoplasmic domain which has been suggested to function as a coupling component. This core channel complex is associated with the mycosin MycP, a membrane protease that has been implicated in substrate processing. In the cytoplasm, two accessory proteins facilitate substrate secretion: EccA, an ATPase, and EspG, which binds to the substrate and presumably functions as a chaperone to guide the substrate to the secretion apparatus.

confined to known pathogens. The T6SS is tightly regulated and controlled at both transcriptional and post-transcriptional levels to avoid host defence systems.

3.10.2.8 Type VII secretion system (T7SS)

Some Gram-positive bacteria of the actinobacteria group, including species of *Mycobacterium* and *Corynebacterium*, have a unique outer membrane-like cell surface structure called the mycomembrane (Section 2.3.3.2). These bacteria possess a specialized secretion apparatus that is required to transport proteins across the mycomembrane. These proteins are related to the virulence of mycobacteria, such as *Mycobacterium tuberculosis*. They include the 6 kDa early secreted antigenic target (ESAT-6) and the 10 kDa culture filtrate protein (CFP-10). These two proteins, designated as ESX proteins, form a 1:1 heterodimeric complex, and are among the most important proteins of *M. tuberculosis* that are involved in host–pathogen interactions. This bacterium has up to five of these secretion systems, named ESX-1 to ESX-5. At least three of these secretion systems are essential for mycobacterial virulence and/or viability. They are known as type VII secretion systems (T7SS). T7SSs are known in many other actinobacteria and Gram-positive bacteria, including *Bacillus anthracis* and *Staphylococcus aureus*. The Gram-positive bacterial T7SSs differ from the actinobacterial systems and are named T7bSS.

Detailed composition of T7SS machineries and their functions are not fully elucidated, but believed to be similar to those of T4SS and T6SS (Figure 3.19). Substrates for the T7SS have a unique

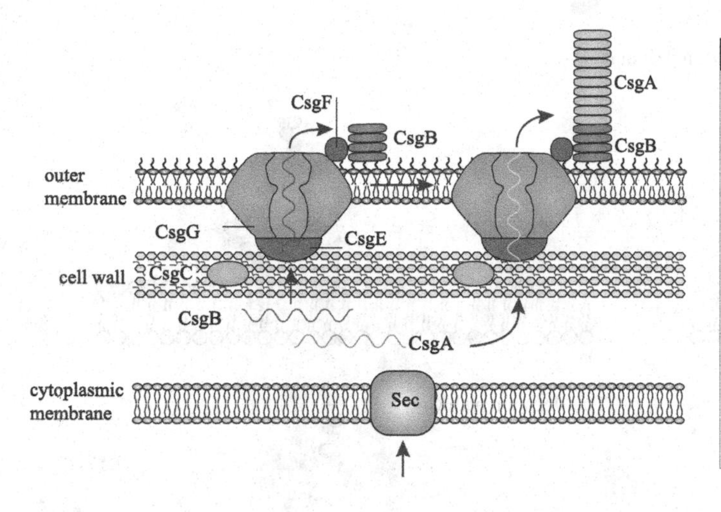

Figure 3.20 Subunit secretion through the type VIII secretion system and biosynthesis of curli.

(Modified from *Nature Rev. Microbiol.* **13**: 343–359, 2015)

The major curli subunit protein CsgA and the minor nucleator molecule CsgB are translocated by the general secretory pathway to the periplasm, where a chaperone, CsgE, prevents them from prematurely aggregating. After secretion across the outer membrane, CsgB nucleates the polymerization of CsgA subunits into curli fibres with assistance from CsgF.

signal sequence on the *C*-terminal, different from other protein secretion mechanisms.

3.10.2.9 Type VIII secretion (curli biogenesis) system (T8SS)

Curli are functional amyloid fibres attached to the Gram-negative bacterial cell surface which are involved in cell aggregation, bacterial adhesion and the formation of mature biofilms. They are secreted across the outer membrane through an extracellular nucleation-precipitation (ENP) pathway known as the type VIII secretion system (T8SS) after being translocated across the cytoplasmic membrane by GSP. Since the major curli subunit protein CsgA is highly aggregative, and curli assembly intermediates are toxic, their synthesis and secretion are highly coordinated with temporal and spatial control of their assembly.

The major curli subunit protein CsgA is translocated through the outer membrane in an unfolded state and is then assembled into fibres with the help of the minor curli subunit CsgB. CsgA contains a SecYEG secretion signal sequence and a *C*-terminal amyloid core domain. The nucleator molecule CsgB contains residues that are necessary for the attachment of CsgB to the cell wall. After secretion, CsgB nucleates the polymerization of CsgA subunits into curli fibres. Both subunits are transported across the cell wall to the outer membrane via a specialized pore-forming lipoprotein, CsgG, which acts in concert with the periplasmic and extracellular accessory proteins, CsgE and CsgF, respectively. The periplasmic factor CsgE binds directly to CsgA, thereby preventing the latter from prematurely aggregating during transit through the periplasm. CsgF, conversely, appears to be exposed to the surface and is critical for CsgB-mediated nucleation of CsgA fibres (Figure 3.20). The curli biogenesis pathway represents a structurally and mechanistically unique class among bacterial protein secretion systems.

3.10.3 Export of polysaccharides

Many bacteria export polysaccharides that constitute extracellular and capsular polysaccharides (Section 2.3.2) and lipopolysaccharides (LPS, Section 2.3.3). These polymers exhibit remarkably diverse structures and play important roles in the biology of free-living, commensal and pathogenic bacteria. These high molecular weight hydrophilic polymers must be assembled and exported in a process spanning the cell envelope. They are exported across the cytoplasmic membrane through one of three pathways: Wzx/Wzy-dependent, ABC transporter-dependent and synthase-dependent, and across the outer membrane by the ABC transporter family.

The repeating units of the lipopolysaccharide (LPS) are synthesized on the inner face of the cytoplasmic membrane and attached to membrane-bound undecaprenyl phosphate (Section 6.9.2) by the action of glycosyl transferases. The subunits are flipped over by the action of Wzx translocase to the outer face of the membrane

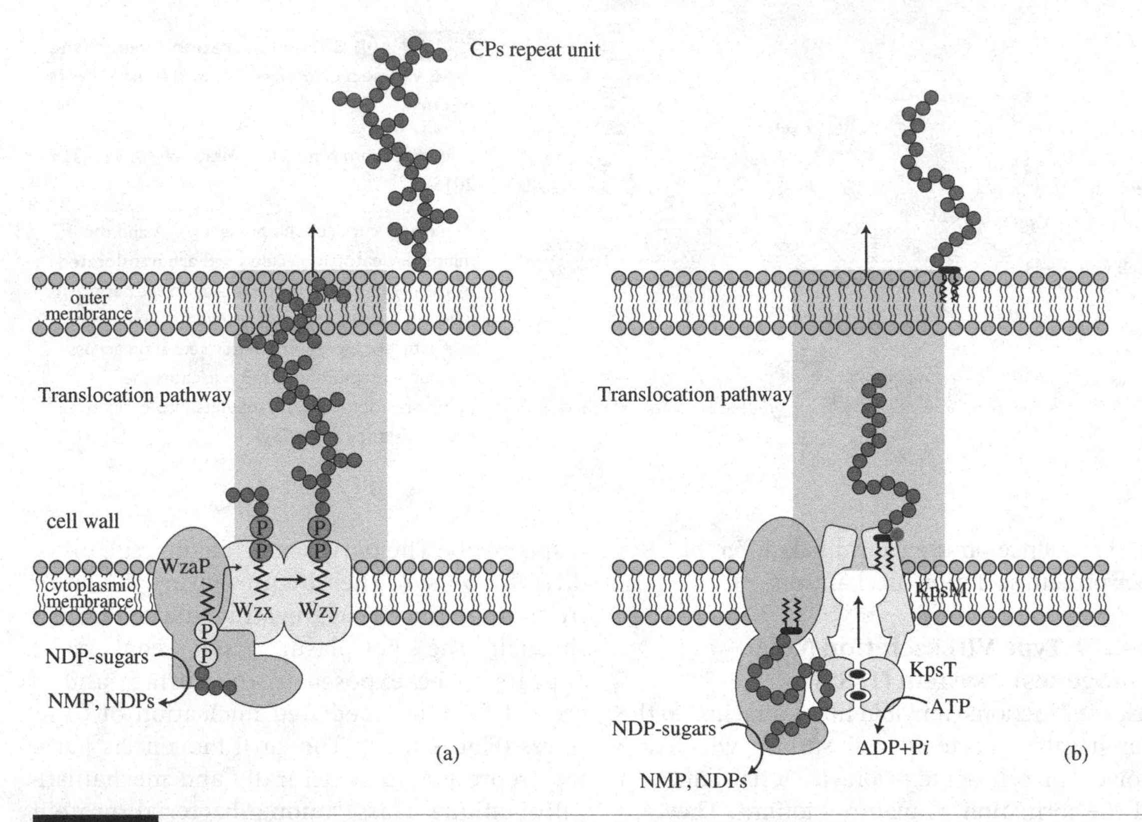

Figure 3.21 Polysaccharide export through (a) the Wzy-dependent and (b) ABC transporter-dependent pathways.

(Modified from *Microbiol. Mol. Biol. Rev.* **73**: 155–177, 2009)

In the Wzx/Wzy-dependent pathway, the repeating units of the lipopolysaccharide (LPS) are synthesized and bound to undecaprenyl phosphate (WzaP) on the inner face of the cytoplasmic membrane by glycosyl transferases, before being flipped over by the action of Wzx translocase to the outer face of the membrane, where the subunits are polymerized by a polymerase, Wzy, and transferred to core polysaccharide attached to lipid A to form LPS. The Wzx/Wzy-dependent pathway is involved in the synthesis of many bacterial surface polysaccharides.

Polysaccharides transported by the ABC transporter-dependent pathway are fully polymerized by sequential glycosyl transfer at the cytoplasmic face of the inner membrane using undecaprenyl phosphate or diacylglycerol phosphate, depending on the individual polysaccharide. Less is known about the third synthase-dependent pathway.

American Society for Microbiology, with permission.

where the subunits are polymerized by a polymerase, Wzy, before being transferred to core polysaccharide attached to lipid A to form LPS (Section 6.8.4). This transport of LPS is referred to as the Wzx/Wzy-dependent pathway (Figure 3.21a). The Wzx/Wzy-dependent pathway is involved in the synthesis of many bacterial surface polysaccharides.

Polysaccharides assembled by ABC transporters are fully polymerized by sequential glycosyl transfer at the cytoplasmic face of the inner membrane (Figure 3.21b). The glycan can be assembled using undecaprenyl phosphate or diacylglycerol phosphate, depending on the individual polysaccharide.

Less is known about the synthase-dependent pathway. Synthases are glycosyltransferases and are involved in the formation of some important biological molecules, including bacterial cellulose, hyaluronan, chondroitin, alginate and poly-β-D-N-acetylglucosamine. In the synthase pathway, the nature of the acceptor on which the polymer grows is not always certain, nor is the process by which the polymer is exported. This pathway is catalysed by three components, including an inner-membrane-embedded polysaccharide synthase, a periplasmic tetratricopeptide repeat (TPR)-containing scaffold protein and an outer-membrane β-barrel porin, but details of the process require further studies to be performed.

3.10.4 Protein secretion in archaea

Protein translocation in the archaea has not been studied as much as in bacteria. Based on comparative genome analysis with well-studied bacteria and eukarya and biochemical and genetic studies on representative archaea, it is known that their protein-translocation machineries are similar to those of bacteria, though proteins involved in transport have less homology. Proteins are translocated across the cytoplasmic membrane in archaea through the general secretion pathway (GSP), ATP-binding cassette (ABC) pathway and twin-arginine translocation (TAT) pathway, as in bacteria. Some archaeal proteins involved in GSP are similar to eukaryotic counterparts, while others are similar to bacterial ones. These secretion machineries seem to have a mixture of eukaryotic and bacterial characteristics, but also possess unique archaeal traits.

Secreted proteins of halophilic archaea are rich in surface-exposed negatively charged amino acids that improve the solubility and flexibility of these proteins at high salt concentrations, preventing them from precipitation (salting out). These proteins can fold quickly and they have to be translocated through the TAT pathway. In the extreme halophile *Halobacterium salinarum*, most proteins seem to be secreted through the TAT pathway. In *Haloarcula marismortui*, *Haloarcula volcanii* and the haloalkaliphile *Natronomonas pharaonis*, there is a strong preference for the TAT pathway, which indicates that the use of the TAT pathway may be a specific adaptation to the high-saline environment. Another unique feature of the TAT pathway in haloarchaea is that translocation is driven by the sodium motive force, whereas in most other microorganisms, the proton motive force is the driving force.

Ignicoccus hospitalis is the only known archaeon with an outer membrane like Gram-negative bacteria, and most archaea usually have a layer of crystalline protein, the S-layer. Archaea have some homologous proteins with some found in Gram-negative bacteria that are involved in protein translocation across the outer membrane (Section 3.10.2). These are types I, II and IV secretion systems, that are probably involved in protein translocation across the S-layer and other surface structures, but much less is known about these than in bacteria.

3.11 Metallochaperones

Metals required for growth of microorganisms include cobalt, copper, nickel and zinc which can be toxic above certain threshold concentrations (Section 2.1). They are transported into the cell by the proton (sodium) antiporter or ABC transporter, and must reach vital destinations within cells without inflicting damage or becoming trapped by adventitious binding sites. Their concentration in the cytoplasm is tightly controlled through import, export (Section 3.9) and binding proteins. After import into the cell, they bind specific proteins such as CopA for copper. These proteins are referred to as metallochaperones. They assist metals in reaching vital cellular destinations without damage to the cell. Metal ions are specifically released from metallochaperones on contact with their cognate apoproteins and metal transfer is thought to proceed by ligand substitution. When the concentration of metals in the cell is in excess, they are exported through an efflux pump (Section 3.9) with the assistance of metallochaperones. Metalloregulatory proteins, with the aid of metalloregulatory riboswitches (Section 12.1.6), sense the metal concentration and regulate the transcription of metal-responsive genes.

Further Reading

Note this section contains key references only. Additional recommended references are available at www.cambridge.org/ProkaryoticMetabolism.

General

Busch, W. & Saier, M. H., Jr. (2002). The transporter classification (TC) system, 2002. *Critical Reviews in Biochemistry and Molecular Biology* **37**, 287–337.

Eggeling, L. & Sahm, H. (2003). New ubiquitous translocators: amino acid export by *Corynebacterium glutamicum* and *Escherichia coli*. *Archives of Microbiology* **180**, 155–160.

Harold, F. M. (2005). Molecules into cells: specifying spatial architecture. *Microbiology and Molecular Biology Reviews* **69**, 544–564.

Hedfalk, K., Tornroth-Horsefield, S., Nyblom, M., Johanson, U., Kjellbom, P. & Neutze, R. (2006). Aquaporin gating. *Current Opinion in Structural Biology* **16**, 447–456.

Lolkema, J. S., Poolman, B. & Konings, W. N. (1998). Bacterial solute uptake and efflux systems. *Current Opinion in Microbiology* **1**, 248–253.

Pudlik, A. M. & Lolkema, J. S. (2011). Citrate uptake in exchange with intermediates in the citrate metabolic pathway in *Lactococcus lactis* IL1403. *Journal of Bacteriology* **193**, 706–714.

Active transport

Krulwich, T. A., Hicks, D. B. & Ito, M. (2009). Cation/proton antiporter complements of bacteria: why so large and diverse? *Molecular Microbiology* **74**, 257–260.

Mesbah, N. M., Cook, G. M. & Wiegel, J. (2009). The halophilic alkalithermophile *Natranaerobius thermophilus* adapts to multiple environmental extremes using a large repertoire of $Na^+(K^+)/H^+$ antiporters. *Molecular Microbiology* **74**, 270–281.

Psakis, G., Saidijam, M., Shibayama, K., Polaczek, J., Bettaney, K. E., Baldwin, J. M., Baldwin, S. A., Hope, R., Essen, L.-O., Essenberg, R. C. & Henderson, P. J. F. (2009). The sodium-dependent D-glucose transport protein of *Helicobacter pylori*. *Molecular Microbiology* **71**: 391–403.

Sobczak, I. & Lolkema, J. S. (2005). The 2-hydroxycarboxylate transporter family: physiology, structure, and mechanism. *Microbiology and Molecular Biology Reviews* **69**, 665–695.

ATP-binding cassette (ABC) pathway

Albers, S. V., Koning, S. M., Konings, W. N. & Driessen, A. J. (2004). Insights into ABC transport in archaea. *Journal of Bioenergetics and Biomembranes* **36**, 5–15.

Cabezon, E. & de la Cruz, F. (2006). TrwB: an F1-ATPase-like molecular motor involved in DNA transport during bacterial conjugation. *Research in Microbiology* **157**, 299–305.

Cheng, J., Poduska, B., Morton, R. A. & Finan, T. M. (2011). An ABC-type cobalt transport system is essential for growth of *Sinorhizobium melilotiat* trace metal concentrations. *Journal of Bacteriology* **193**, 4405–4416.

Davidson, A. L., Dassa, E., Orelle, C. & Chen, J. (2008). Structure, function, and evolution of bacterial ATP-binding cassette systems. *Microbiology and Molecular Biology Reviews* **72**, 317–364.

Erkens, G. B., Majsnerowska, M., ter Beek, J. & Slotboom, D. J. (2012). Energy coupling factor-type ABC transporters for vitamin uptake in prokaryotes. *Biochemistry* **51**, 4390–4396.

Pohl, A., Devaux, P. F. & Herrmann, A. (2005). Function of prokaryotic and eukaryotic ABC proteins in lipid transport. *Biochimica et Biophysica Acta* **1733**, 29–52.

Tripartite ATP-independent periplasmic (TRAP) transporters

Fischer, M., Zhang, Q. Y., Hubbard, R. E. & Thomas, G. H. (2010). Caught in a TRAP: substrate-binding proteins in secondary transport. *Trends in Microbiology* **18**, 471–478.

Mulligan, C., Fischer, M. & Thomas, G. H. (2011). Tripartite ATP-independent periplasmic (TRAP) transporters in bacteria and archaea. *FEMS Microbiology Reviews* **35**: 68–86.

Winnen, B., Hvorup, R. N. & Saier, M. H., Jr. (2003). The tripartite tricarboxylate transporter (TTT) family. *Research in Microbiology* **154**, 457–465.

Group translocation

Barabote, R. D. & Saier, M. H., Jr. (2005). Comparative genomic analyses of the bacterial phosphotransferase system. *Microbiology and Molecular Biology Reviews* **69**, 608–634.

Deutscher, J., Aké, F. M. D., Derkaoui, M., Zébré, A. C., Cao, T. N., Bouraoui, H., Kentache, T., Mokhtari, A., Milohanic, E. & Joyet, P. (2014). The bacterial phosphoenolpyruvate:carbohydrate phosphotransferase system: regulation by protein phosphorylation and phosphorylation-dependent protein-protein interactions. *Microbiology and Molecular Biology Reviews* **78**, 231–256.

Goodwin, R. A. & Gage, D. J. (2014). Biochemical characterization of a nitrogen-type phosphotransferase system reveals that enzyme EI^{Ntr} integrates carbon and nitrogen signaling in *Sinorhizobium meliloti*. *Journal of Bacteriology* **196**, 1901–1907.

Pflüger-Grau, K. & Görke, B. (2010). Regulatory roles of the bacterial nitrogen-related phosphotransferase system. *Trends in Microbiology* **18**, 205–214.

Iron uptake and siderophores

Braun, V. & Braun, M. (2002). Active transport of iron and siderophore antibiotics. *Current Opinion in Microbiology* **5**, 194–201.

Llamas, M. A. & Bitter, W. (2006). Iron gate: the translocation system. *Journal of Bacteriology* **188**, 3172–3174.

Schalk, I. J., Hannauer, M. & Braud, A. (2011). New roles for bacterial siderophores in metal transport and tolerance. *Environmental Microbiology* **13**, 2844–2854.

Wandersman, C. & Delepelaire, P. (2004). Bacterial iron sources: from siderophores to hemophores. *Annual Review of Microbiology* **58**, 611–647.

TonB-dependent active transport across the outer membrane in Gram-negative bacteria

Balhesteros, H., Shipelskiy, Y., Long, N. J., Majumdar, A., Katz, B. B., Santos, N. M., Leaden, L., Newton, S. M., Marques, M. V. & Klebba, P. E. (2017). TonB-dependent heme/hemoglobin utilization by *Caulobacter crescentus* HutA. *Journal of Bacteriology* **199**, e00723-16.

Celia, H., Noinaj, N., Zakharov, S. D., Bordignon, E., Botos, I., Santamaria, M., Barnard, T. J., Cramer, W. A., Lloubes, R. & Buchanan, S. K. (2016). Structural insight into the role of the Ton complex in energy transduction. *Nature* **538**, 60–65.

Noinaj, N., Guillier, M., Barnard, T. J. & Buchanan, S. K. (2010). TonB-dependent transporters: regulation, structure, and function. *Annual Review of Microbiology* **64**, 43–60.

Postle, K. & Kadner, R. J. (2003). Touch and go: tying TonB to transport. *Molecular Microbiology* **49**, 869–882.

Multidrug efflux pump

Du, D., van Veen, H. W. & Luisi, B. F. (2015). Assembly and operation of bacterial tripartite multidrug efflux pumps. *Trends in Microbiology* **23**, 311–319.

Hinchliffe, P., Symmons, M. F., Hughes, C. & Koronakis, V. (2013). Structure and operation of bacterial tripartite pumps. *Annual Review of Microbiology* **67**, 221–242.

Paulsen, I. T. (2003). Multidrug efflux pumps and resistance: regulation and evolution. *Current Opinion in Microbiology* **6**, 446–451.

Protein translocation

Dalbey, R. E., Wang, P. & van Dijl, J. M. (2012). Membrane proteases in the bacterial protein secretion and quality control pathway. *Microbiology and Molecular Biology Reviews* **76**, 311–330.

Holland, I. B. (2004). Translocation of bacterial proteins – an overview. *Biochimica et Biophysica Acta* **1694**, 5–16.

Pohlschroeder, M., Hartmann, E., Hand, N. J., Dilks, K. & Haddad, A. (2005). Diversity and evolution of protein translocation. *Annual Review of Microbiology* **59**, 91–111.

Pugsley, A. P., Francetic, O., Driessen, A. J. & de Lorenzo, V. (2004). Getting out: protein traffic in prokaryotes. *Molecular Microbiology* 52, 3–11.

General secretion pathway (GSP)

Desvaux, M., Parham, N. J., Scott-Tucker, A. & Henderson, I. R. (2004). The general secretory pathway: a general misnomer? *Trends in Microbiology* **12**, 306–309.

Tjalsma, H., Antelmann, H., Jongbloed, J. D. H., Braun, P. G., Darmon, E., Dorenbos, R., Dubois, J. -Y. F., Westers, H., Zanen, G., Quax, W. J., Kuipers, O. P., Bron, S., Hecker, M. & van Dijl, J. M. (2004). Proteomics of protein secretion by *Bacillus subtilis*: separating the 'secrets' of the secretome. *Microbiology and Molecular Biology Reviews* **68**, 207–233.

Tsirigotaki, A., De Geyter, J., Sostaric, N., Economou, A. & Karamanou, S. (2017). Protein export through the bacterial Sec pathway. *Nature Reviews Microbiology* **15**(1), 21–36.

van der Sluis, E. O. & Driessen, A. J. M. (2006). Stepwise evolution of the Sec machinery in Proteobacteria. *Trends in Microbiology* **14**, 105–108.

Twin-arginine translocation (TAT) pathway

Berks, B. C. (2015). The twin-arginine protein translocation pathway. *Annual Review of Biochemistry* **84**, 843–864.

Palmer, T. & Berks, B. C. (2012). The twin-arginine translocation (Tat) protein export pathway. *Nature Reviews Microbiology* **10**, 483–496.

Robinson, C. & Bolhuis, A. (2004). Tat-dependent protein targeting in prokaryotes and chloroplasts. *Biochimica et Biophysica Acta* **1694**, 135–147.

Protein translocation through the ABC pathway

Gebhard, S. (2012). ABC transporters of antimicrobial peptides in Firmicutes bacteria – phylogeny, function and regulation. *Molecular Microbiology* **86**, 1295–1317.

Protein translocation through the cell wall in Gram-positive bacteria

Buist, G., Ridder, A. N. J. A., Kok, J. & Kuipers, O. P. (2006). Different subcellular locations of secretome components of Gram-positive bacteria. *Microbiology* **152**, 2867–2874.

Forster, B. M. & Marquis, H. (2012). Protein transport across the cell wall of monoderm Gram-positive bacteria. *Molecular Microbiology* **84**, 405–413.

Scheurwater, E. M. & Burrows, L. L. (2011). Maintaining network security: how macromolecular structures cross the peptidoglycan layer. *FEMS Microbiology Letters* **318**, 1–9.

Protein translocation in Gram-negative bacteria

Blanco, L. P., Evans, M. L., Smith, D. R., Badtke, M. P. & Chapman, M. R. (2012). Diversity, biogenesis and function of microbial amyloids. *Trends in Microbiology* **20**, 66–73.

Büttner, D. (2012). Protein export according to schedule: architecture, assembly, and regulation of type III secretion systems from plant- and animal-pathogenic bacteria. *Microbiology and Molecular Biology Reviews* **76**, 262–310.

Christie, P. J., Atmakuri, K., Krishnamoorthy, V., Jakubowski, S. & Cascales, E. (2005). Biogenesis, architecture, and function of bacterial type IV secretion systems. *Annual Review of Microbiology* **59**, 451–485.

Costa, T. R. D., Felisberto-Rodrigues, C., Meir, A., Prevost, M. S., Redzej, A., Trokter, M. & Waksman, G. (2015). Secretion systems in Gram-negative bacteria: structural and mechanistic insights. *Nature Reviews Microbiology* **13**, 343–359.

Dalbey, R. E. & Kuhn, A. (2012). Protein traffic in Gram-negative bacteria – how exported and secreted proteins find their way. *FEMS Microbiology Reviews* **36**, 1023–1045.

Evans, L. D. B., Hughes, C. & Fraser, G. M. (2014). Building a flagellum outside the bacterial cell. *Trends in Microbiology* **22**, 566–572.

Ghosh, P. (2004). Process of protein transport by the type III secretion system. *Microbiology and Molecular Biology Reviews* **68**, 771–795.

Hachani, A., Wood, T. E. & Filloux, A. (2016). Type VI secretion and anti-host effectors. *Current Opinion in Microbiology* **29**: 81–93.

Henderson, I. R., Navarro-Garcia, F., Desvaux, M., Fernandez, R. C. & Ala'Aldeen, D. (2004). Type V protein secretion pathway: the autotransporter story. *Microbiology and Molecular Biology Reviews* **68**, 692–744.

Kanonenberg, K., Schwarz, C. K. W. & Schmitt, L. (2013). Type I secretion systems – a story of appendices. *Research in Microbiology* **164**, 596–604.

Kim, D. S. H., Chao, Y. & Saier, M. H., Jr. (2006). Protein-translocating trimeric autotransporters of Gram-negative bacteria. *Journal of Bacteriology* **188**, 5655–5667.

Lara-Tejero, M., Kato, J., Wagner, S., Liu, X. & Galán, J. E. (2011). A sorting platform determines the order of protein secretion in bacterial type III systems. *Science* **331**, 1188–1191.

Lasica, A. M., Ksiazek, M., Madej, M. & Potempa, J. (2017). The type IX secretion system (T9SS): highlights and recent insights into its structure and function. *Frontiers in Cellular & Infection Microbiology* **7**: 215.

Stoop, E. J. M., Bitter, W. & van der Sar, A. M. (2012). Tubercle bacilli rely on a type VII army for pathogenicity. *Trends in Microbiology* **20**, 477–484.

Export of polysaccharides

Cuthbertson, L., Kos, V. & Whitfield, C. (2010). ABC Transporters involved in export of cell surface glycoconjugates. *Microbiology and Molecular Biology Reviews* **74**, 341–362.

Cuthbertson, L., Mainprize, I. L., Naismith, J. H. & Whitfield, C. (2009). Pivotal roles of the outer membrane polysaccharide export and polysaccharide copolymerase protein families in export of extracellular polysaccharides in Gram-negative bacteria. *Microbiology and Molecular Biology Reviews* **73**, 155–177.

Okuda, S., Freinkman, E. & Kahne, D. (2012). Cytoplasmic ATP hydrolysis powers transport of lipopolysaccharide across the periplasm in *E. coli*. *Science* **338**, 1214–1217.

Putker, F., Bos, M. P. & Tommassen, J. (2015). Transport of lipopolysaccharide to the Gram-negative bacterial cell surface. *FEMS Microbiology Reviews* **39**, 985–1002.

Qiao, S., Luo, Q., Zhao, Y., Zhang, X. C. & Huang, Y. (2014). Structural basis for lipopolysaccharide insertion in the bacterial outer membrane. *Nature* **511**, 108–111.

Whitney, J. C. & Howell, P. L. (2013). Synthase-dependent exopolysaccharide secretion in Gram-negative bacteria. *Trends in Microbiology* **21**, 63–72.

Protein secretion in Archaea

Albers, S. V., Szabo, Z. & Driessen, A. J. M. (2006). Protein secretion in the Archaea: multiple paths towards a unique cell surface. *Nature Reviews Microbiology* **4**, 537–547.

Gehring, A. M., Walker, J. E. & Santangelo, T. J. (2016). Transcription regulation in archaea. *Journal of Bacteriology* **198**, 1906–1917.

Gimenez, M. I., Dilks, K. & Pohlschroder, M. (2007). *Haloferax volcanii* twin-arginine translocation substates include secreted soluble, C-terminally anchored and lipoproteins. *Molecular Microbiology* **66**, 1597–1606.

Karr, E. A. (2014). Transcription regulation in the third domain. *Advances in Applied Microbiology* **89**, 101–133.

Kwan, D. C., Thomas, J. R. & Bolhuis, A. (2008). Bioenergetic requirements of a Tat-dependent substrate in the halophilic archaeon *Haloarcula hispanica*. *FEBS Journal* **275**(24), 6159–6167.

Pohlschroder, M., Gimenez, M. I. & Jarrell, K. F. (2005). Protein transport in Archaea: Sec and twin arginine translocation pathways. *Current Opinion in Microbiology* **8**, 713–719.

Saleh, M., Song, C., Nasserulla, S. & Leduc, L. G. (2010). Indicators from archaeal secretomes. *Microbiological Research* **165**, 1–10.

Metallochaperones

Abdul Ajees, A., Yang, J. & Rosen, B. (2011). The ArsD As (III) metallochaperone. *BioMetals* **24**, 391–399.

Bleriot, C., Effantin, G., Lagarde, F., Mandrand-Berthelot, M.-A. & Rodrigue, A. (2011). RcnB is a periplasmic protein essential for maintaining intracellular Ni and Co concentrations in *Escherichia coli*. *Journal of Bacteriology* **193**, 3785–3793.

Braymer, J. J. & Giedroc, D. P. (2014). Recent developments in copper and zinc homeostasis in bacterial pathogens. *Current Opinion in Chemical Biology* **19**, 59–66.

Chacon, K. N., Mealman, T. D., McEvoy, M. M. & Blackburn, N. J. (2014). Tracking metal ions through a Cu/Ag efflux pump assigns the functional roles of the periplasmic proteins. *Proceedings of the National Academy of Sciences of the USA* **111**, 15373–15378.

Chandrangsu, P., Rensing, C. & Helmann, J. D. (2017). Metal homeostasis and resistance in bacteria. *Nature Reviews Microbiology* **15**, 338–350.

Furukawa, K., Ramesh, A., Zhou, Z., Weinberg, Z., Vallery, T., Winkler, W. C. & Breaker, R. R. (2015). Bacterial riboswitches cooperatively bind Ni^{2+} or Co^{2+} ions and control expression of heavy metal transporters. *Molecular Cell* **57**, 1088–1098.

Osman, D., Patterson, C. J., Bailey, K., Fisher, K., Robinson, N. J., Rigby, S. E. J. & Cavet, J. S. (2013). The copper supply pathway to a *Salmonella* Cu, Zn-superoxide dismutase (SodCII) involves P1B-type ATPase copper efflux and periplasmic CueP. *Molecular Microbiology* **87**, 466–477.

Chapter 4

Glycolysis

Escherichia coli can grow on a simple medium containing glucose and mineral salts and this bacterium can synthesize all cell constituents using materials provided in this medium. Glucose is metabolized through the Embden–Meyerhof–Parnas (EMP) pathway and hexose monophosphate (HMP) pathway and the metabolic product, pyruvate, is decarboxylated oxidatively to acetyl-CoA, to be oxidized through the tricarboxylic acid (TCA) cycle. Twelve intermediates of these pathways are used as carbon skeletons for biosynthesis (Table 4.1). Heterotrophs that utilize organic compounds other than carbohydrates convert their substrates into one or more of these intermediates. For this reason, glucose metabolism through glycolysis and the TCA cycle is called central metabolism. All reactions in central metabolism are indispensable, but deletion mutants of many genes of enzymes catalysing these reactions can grow. Some of the mutants have genes for isoenzymes, and multifunctional enzymes catalyse the reactions catalysed by knocked-out genes in the others. One example of the latter is the transaldolase of the hexose monophosphate pathway that converts sedoheptulose-7-phosphate and glyceraldehyde-3-phosphate to erythrose-4-phosphate and fructose-6-phosphate. In the absence of this enzyme, sedoheptulose-7-phosphate is phosphorylated by phosphofructokinase to sedoheptulose-1,7-diphosphate that is cleaved to erythrose-4-phosphate and glyceraldehyde-3-phosphate by fructose diphosphate aldolase (Section 4.3.1). Another example is xylonate dehydratase which replaces gluconate dehydratase catalysing the modified ED pathway (Section 4.4.3.2) in the halophilic archaeon *Haloferax volcanii*.

Eukaryotes metabolize glucose through the EMP pathway to generate ATP, pyruvate and NADH, and the HMP pathway is needed to supply the metabolic intermediates not available from the EMP pathway, such as pentose-5-phosphate and erythrose-4-phosphate, and NADPH. Most prokaryotes employ similar mechanisms, but some prokaryotes metabolize glucose through unique pathways known only in prokaryotes, e.g. the Entner–Doudoroff (ED) pathway and the phosphoketolase (PK) pathway. Some prokaryotes have genes for the ED pathway in addition to the EMP pathway; genes for these pathways are expressed at the same time in several prokaryotes including *Escherichia coli*, a thermophilic bacterium (*Thermotoga maritima*), a thermophilic archaeon (*Thermoproteus tenax*) and a halophilic archaeon (*Halococcus saccharolyticus*). *Escherichia coli* metabolizes glucose via the EMP pathway, but gluconate is oxidized through the ED pathway. The PK pathway known in heterofermentative lactic acid bacteria and bifidobacteria is found in other organisms including some fungi (Section 4.5.2).

Glycolytic pathways in archaea are substantially different from those in bacteria and eukarya. Most archaea metabolize glucose either through modified EMP pathways (Section 4.1.4) or modified ED pathways (Section 4.4.3.2), and the hexose monophosphate pathway (HMP) is

Table 4.1 Metabolic intermediates used as carbon skeletons for biosynthesis.

Carbon skeleton	From	Precursor for
Glucose-6-phosphate	EMP	polysaccharides
Fructose-6-phosphate	EMP	peptidoglycan
Ribose-5-phosphate	HMP	nucleic acids
Erythrose-4-phosphate	HMP	amino acids
Triose-phosphate	EMP	lipids
3-phosphoglycerate	EMP	amino acids
Phosphoenolpyruvate	EMP	amino acids
Pyruvate	EMP	amino acids
Acetyl-CoA	Pyruvate	fatty acids
2-ketoglutarate	TCA	amino acids
Succinyl-CoA	TCA	amino acids
Oxaloacetate	TCA	amino acids

absent. They obtain pentose-5-phosphate in a different way (Section 4.3.5), and pentose degradation (Section 7.2.3) significantly differs from that known for bacterial model organisms such as *Escherichia coli*. Erythrose-4-phosphate is not the precursor for aromatic amino acid synthesis in archaea (Sections 4.3.5 and 6.4.4).

Carbohydrates are phosphorylated before they are metabolized in most cases. It is believed that phosphorylated intermediates are less likely to diffuse away through the cytoplasmic membrane. Some bacteria and archaea also phosphorylate intermediates of glucose metabolism in modified glycolytic pathways.

This chapter describes glucose oxidation to pyruvate and related metabolic pathways. Pyruvate metabolism will be further discussed in Chapters 5, 8 and 9.

4.1 EMP pathway

Many anaerobic and enteric bacteria transport glucose via group translocation (phosphotransferase system, PTS, Section 3.5) in the form of glucose-6-phosphate. Glucose transported through active transport is phosphorylated by hexokinase:

$$\text{glucose} + \text{ATP} \xrightarrow{\text{hexokinase}} \text{glucose-6-phosphate} + \text{ADP}$$

Hexokinase can phosphorylate other hexoses, such as mannose, and requires Mg^{2+} for activity. The enzyme cannot catalyse the reverse reaction.

Glucose-6-phosphate can also be obtained from glycogen:

$$\underset{\text{(glycogen)}}{[\text{glucose}]_n} + P_i \xrightarrow{\text{phosphorylase}} \underset{\text{(glycogen)}}{[\text{glucose}]_{n-1}} + \text{glucose-1-phosphate}$$

$$\text{glucose-1-phosphate} \xrightleftharpoons{\text{phosphoglucomutase}} \text{glucose-6-phosphate}$$

Glucose-6-phosphate is a precursor for the biosynthesis of polysaccharides, as well as a substrate of the EMP pathway (Figure 4.1), which is the commonest glycolytic pathway in all kinds of organisms.

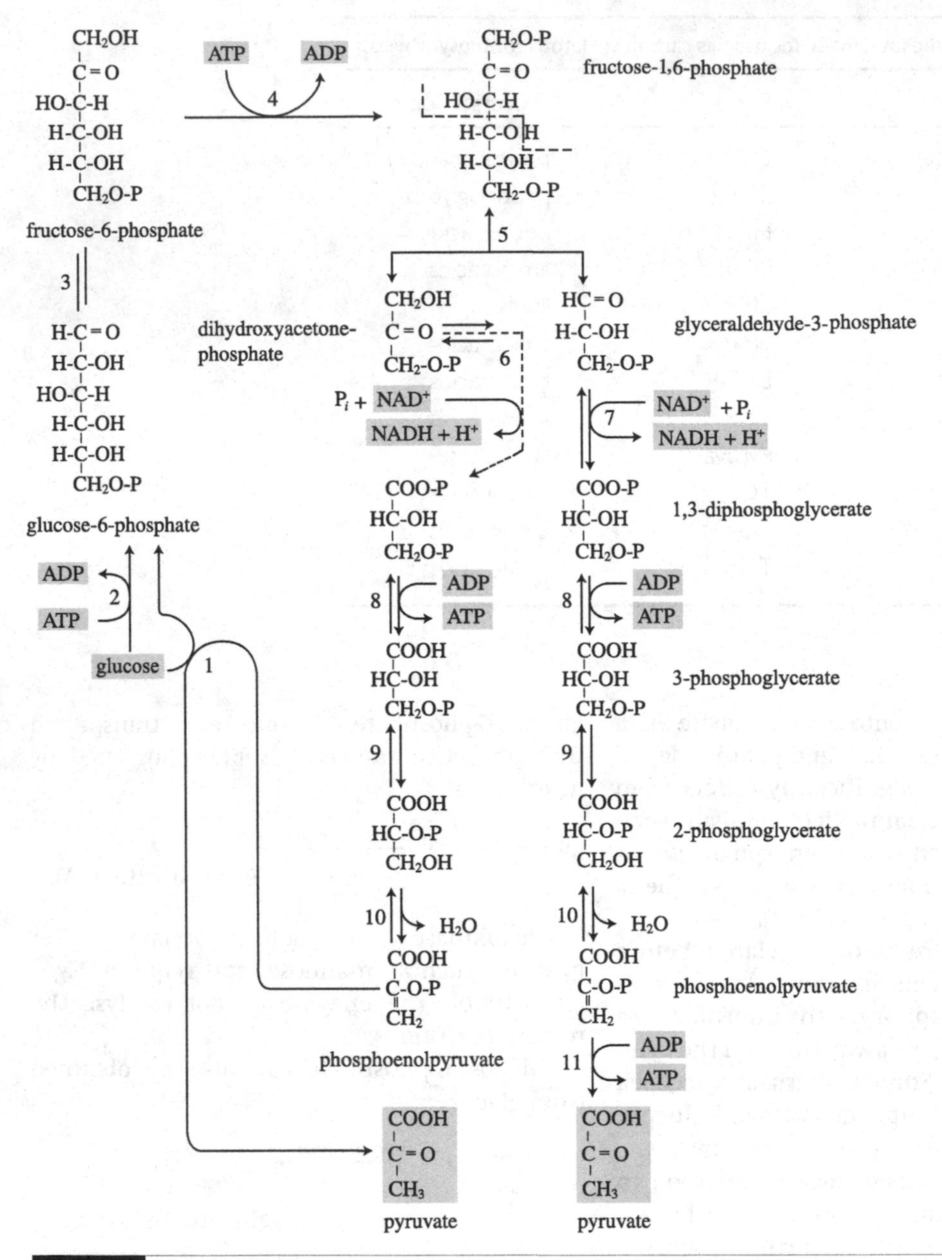

Figure 4.1 **Glucose oxidation via the Embden–Meyerhof–Parnas pathway.**

Glucose is phosphorylated to glucose-6-phosphate by PEP: glucose phosphotransferase (1) during group translocation (phosphotransferase system, PTS), or by hexokinase (2) after uptake via active transport.

3, glucose-6-phosphate isomerase; 4, phosphofructokinase; 5, fructose diphosphate aldolase; 6, triose-phosphate isomerase; 7, glyceraldehyde-3-phosphate dehydrogenase; 8, 3-phosphoglycerate kinase; 9, phosphoglycerate mutase; 10, enolase; 11, pyruvate kinase.

4.1.1 Phosphofructokinase (PFK): key enzyme of the EMP pathway

Glucose-6-phosphate is isomerized to fructose-6-phosphate before being phosphorylated to fructose-1,6-diphosphate by the action of phosphofructokinase (PFK). These two reactions require Mg^{2+}. Glucose-6-phosphate isomerase catalyses the reverse reaction, but phosphofructokinase does not. The irreversibility of an enzyme is due to thermodynamic reasons, and many enzymes that do not catalyse the reversible reaction are regulated. PFK is the key enzyme of the EMP pathway. If this enzyme is present in a given prokaryote, it can be assumed that this organism catabolizes glucose through the EMP pathway. Fructose-6-phosphate is a precursor of amino sugars and their polymers, such as peptidoglycan (Section 6.8.2).

Fructose-1,6-diphosphate aldolase cleaves fructose-1,6-diphosphate to two molecules of triose-phosphate. This aldolase catalyses the reverse reaction, and participates in gluconeogenesis (Section 4.2), producing hexose phosphate when non-carbohydrate substrates are used as carbon sources.

4.1.2 ATP synthesis and production of pyruvate

Triose-phosphate isomerase equilibrates dihydroxyacetone phosphate and glyceraldehyde-3-phosphate produced from fructose-1,6-diphosphate. Under standard conditions the equilibrium shifts to the formation of dihydroxyacetone phosphate ($\Delta G^{0\prime}$= −7.7 kJ/mol glyceraldehyde-3-phosphate), but the reverse reaction is favoured because glyceraldehyde-3-phosphate is continuously consumed in subsequent reactions. Phospholipids are synthesized from glyceraldehyde-3-phosphate (Section 6.6.2).

Glyceraldehyde-3-phosphate is oxidized to 1,3-diphosphoglycerate by glyceraldehyde-3-phosphate dehydrogenase. This endergonic reaction ($\Delta G^{0\prime}$ = +6.3 kJ/mol glyceraldehyde-3-phosphate) is efficiently pulled by the following exergonic reaction catalysed by 3-phosphoglycerate kinase ($\Delta G^{0\prime}$ = – 12.5 kJ/mol 1,3-phosphoglycerate). This enzyme requires Mg^{2+}, as do most kinases, and ATP generation in this reaction is an example of substrate-level-phosphorylation. 3-phosphoglycerate is the starting material for the synthesis of amino acids, serine, glycine and cysteine (Section 6.4.2).

$$\text{1,3-diphosphoglycerate} + \text{ADP} \xrightleftharpoons{\text{3-phosphoglycerate kinase}} \text{3-phosphoglycerate} + \text{ADP}$$
$$(\Delta G^{0\prime} = -12.5\ \text{kJ/mol 1,3-diphosphoglycerate})$$

3-phosphoglycerate is converted to 2-phosphoglycerate by phosphoglycerate mutase, which requires 2,3-diphosphoglycerate as a coenzyme. 2-phosphoglycerate is dehydrated to phosphoenolpyruvate (PEP) by an enolase in the presence of divalent cations, such as Mg^{2+} and Mn^{2+}. PEP is used to generate ATP with the reaction of the last enzyme in the EMP pathway in the presence of Mg^{2+} and K^+. PEP supplies energy in group translocation, and is used to synthesize aromatic amino acids (Section 6.4.4). Glyceraldehyde-3-phosphate is an intermediate in the HMP and ED pathways and the reactions from this triose-phosphate are shared with both these pathways.

Four ATPs are synthesized and two high energy phosphate bonds are consumed in this pathway, resulting in a net gain of two ATPs per glucose oxidized. The NADH reduced in the glycolytic pathway is reoxidized in aerobic (Section 5.8) and anaerobic respiration (Chapter 9), and in fermentation (Chapter 8), reducing various electron acceptors, depending on the organism and on their availability.

4.1.3 Modified EMP pathways in bacteria

Some prokaryotes, unlike eukaryotes, metabolize glucose through modified EMP pathways, depending on the growth conditions.

4.1.3.1 Use of atypical cofactors

Some organisms use atypical cofactors while others metabolize glucose through different intermediates. Polyphosphate (PP) is used as the phosphate donor to phosphorylate glucose in some bacteria. Polyphosphate–glucose phosphotransferases (PPGPT) of bacteria belonging to the order *Actinomycetales* use ATP as well as PP as a phosphate donor while those of nitrogen-fixing cyanobacteria, *Corynebacterium glutamicum* and

PP-accumulating *Microlunatus phosphovorus*, are only active on PP. PPGPT can phosphorylate other hexoses such as mannose, and requires Mg^{2+} for activity. PP is a phosphate and energy reserve material (Section 13.2.4).

A ruminal bacterium, *Streptococcus bovis*, has a $NADP^+$-dependent glyceraldehyde-3-phosphate dehydrogenase (GAPDH) in addition to the usual NAD^+-dependent enzyme. The $NADP^+$-dependent GAPDH plays an important role in NADPH production in microorganisms that do not have other NADPH-producing systems, such as the hexose monophosphate pathway and NADPH:NAD^+ oxidoreductase.

A cellulolytic anaerobic bacterium, *Clostridium thermocellum*, does not have the pyruvate kinase gene. Phosphoenolpyruvate (PEP) is converted to pyruvate through oxaloacetate and malate in $NADP^+$-dependent reactions phosphorylating GDP to GTP. When this bacterium is cultivated on cellobiose, the substrate is imported by the ABC transporter (Section 3.4) and hydrolysed by cellobiose phosphorylase (Section 7.1.4). GTP is preferred over ATP in the hexokinase reaction to phosphorylate glucose, and phosphofructokinase uses pyrophosphate instead of ATP. GDP is phosphorylated to GTP by the action of phosphoglycerate kinase:

PEP
GDP → GTP
Oxaloacetate
NADH → NAD^+
Malate
NADP+ → NADPH
Pyruvate

The phosphofructokinase of the extremely thermophilic bacterium *Caldicellulosiruptor saccharolyticus* is pyrophosphate (PP_i)-dependent as in archaea (Section 4.1.4). This enzyme can catalyse the reverse reaction. In this bacterium, pyruvate phosphate dikinase catalyses PEP conversion to pyruvate (Section 4.2.1). This organism has a membrane-bound pyrophosphatase but not cytosolic pyrophosphatase.

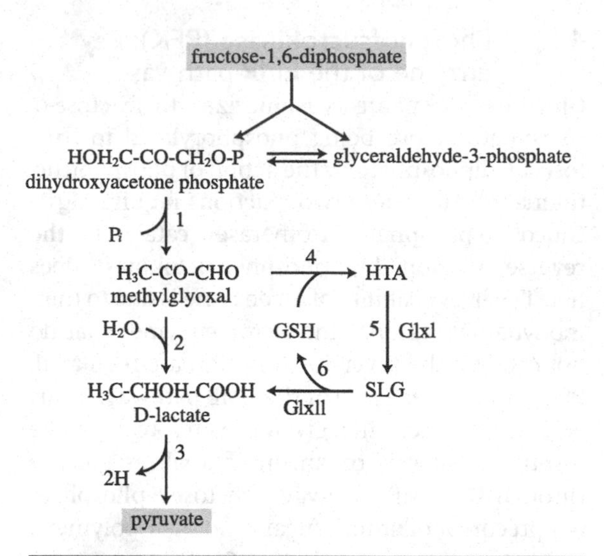

Figure 4.2 The methylglyoxal bypass, a modified EMP pathway under phosphate-limited conditions.

Under phosphate-limited conditions, bacteria such as *Escherichia coli* metabolize dihydroxyacetone phosphate to pyruvate to supply precursors for biosynthesis with a reduced ATP yield. The intermediate, methylglyoxal, is toxic and glyoxalase I, II and III detoxify it.

1, methylglyoxal synthase; 2, glyoxalase III; 3, lactate oxidase; 4, spontaneous chemical reaction; 5, glyoxalase I; 6, glyoxlase II.

HTA, hemithioacetal, SLG, S-lactoylglutathione; GSH, glutathione.

4.1.3.2 Methylglyoxal bypass

Under phosphate-limited conditions with accumulation of dihydroxyacetone phosphate, *Escherichia coli* and *Pseudomonas saccharophila* oxidize dihydroxyacetone phosphate to pyruvate through methylglyoxal (MG, Figure 4.2). This diversion enables acetyl-CoA synthesis through pyruvate when glyceraldehyde-3-phosphate dehydrogenase cannot function due to a low concentration of inorganic phosphate (P_i), one of its substrates, with a reduced ATP yield.

Methylgloxal is very reactive, destroying nucleic acids and proteins by reacting with guanine and adenine, and with amino acids such as arginine, lysine and cysteine. When this toxic compound accumulates in the cell, various proteins are synthesized, including membrane proteins to excrete it, and glyoxalase (Glx) I, II and III to increase the rate of MG conversion to lactate. GlxIII, formerly known as heat shock protein Hsp31, converts MG directly to lactate, but the GlxI–GlxII system is the more important detoxification route. Through a spontaneous chemical reaction with glutathione, MG is converted to hemithioacetal that is isomerized to *S*-lactoylglutathione (SLG) by GlxI. SLG is hydrolysed to lactate and glutathione. SLG activates the K^+ efflux system coupled to influx of H^+, lowering the cytoplasmic pH. MG is less reactive at low pH. There are reductase and dehydrogenase enzymes catalysing MG conversion, but their contribution is minor. Stationary-phase *Escherichia coli* cells became more susceptible to MG when the GlxIII gene was deleted.

Methylgloxal synthase activity is regulated by P_i through feedback inhibition, and is activated by its substrate, dihydroxyacetone phosphate. Because the concentration of dihydroxyacetone phosphate in the cell is around 0.5 mM, lower than the K_m value for methylglyoxal synthase even under phosphate-limited conditions, the MG concentration does not reach toxic levels. However, toxic levels of MG accumulate with a limited supply of phosphate when other nutrients such as nitrogen are limited. For example, a rumen bacterium, *Prevotella ruminicola*, loses viability when excess carbohydrate is supplied under nitrogen- and phosphate-limited conditions, due to MG accumulation. The non-phosphorylated form of Crh (catabolite repression HPr, Section 12.1.3.3) inhibits MG synthase in *Bacillus subtilis*, preventing glucose metabolism through this bypass under glucose-limited conditions.

A thermophilic anaerobic bacterium, *Clostridium thermosaccharolyticum*, reduces methylglyoxal to 1,2-propanediol via acetol (CH_3COCH_2OH).

4.1.4 Modified EMP pathways in archaea

Most archaea utilizing carbohydrates employ modified ED pathways (Section 4.4.3.2), but a few metabolize sugars in modified EMP pathways. The halophilic archaeon *Haloarcula vallismortis* transports fructose through an active transport mechanism and a ketohexokinase phosphorylates the free sugar to fructose-1-phosphate, which is transformed to fructose-1,6-diphosphate before being oxidized through the normal EMP pathway (Section 7.2.1):

$$\text{fructose} + \text{ATP} \xrightarrow{\text{ketohexokinase}} \text{fructose-1-phosphate} + \text{ADP}$$

$$\text{fructose-1-phosphate} + \text{ATP} \xrightarrow{\text{1-phosphofructokinase}} \text{fructose-1,6-diphosphate} + \text{ADP}$$

Hyperthermophilic archaea such as *Pyrococcus furiosus*, *Thermococcus celer* and *Desulfurococcus amylolyticus* employ yet another modified EMP pathway to ferment sugars (Figure 4.3). *Pyrococcus furiosus* does not use glucose, but ferments starch. This organism transports oligosaccharides produced from starch by α-amylase. Oligosaccharides are hydrolysed to glucose by α-glucosidase. The differences from glycolysis in this organism are (1) phosphorylation of glucose and fructose-6-phosphate is catalysed by ADP-dependent kinases, and (2) glyceraldehyde-3-phosphate is reduced to 3-phosphoglycerate by glyceraldehyde-3-phosphate:ferredoxin oxidoreductase (GAPOR), without generating ATP. It has been hypothesized that this organism has evolved in high temperature environments to use ADP in place of ATP, since ADP is more stable than ATP under such conditions. The reason why energy is not conserved during the oxidation of glyceraldehyde-3-phosphate to 1,3-diphosphoglycerate might be due to the use of ferredoxin as the electron acceptor, which has a lower redox potential than NAD^+.

Another hyperthermophilic archaeon, *Thermococcus kodakarensis*, oxidizes glyceraldehyde-3-phosphate (GAP) to 3-phosphoglycerate by GAPOR, as above, and has two other enzymes that react with GAP. These are non-phosphorylating $NADP^+$-dependent GAP dehydrogenase (GAPN) and the classical phosphorylating GAP dehydrogenase (GAPDH). GAPOR is the main enzyme for glycolysis, and GAPN is responsible for regeneration of NADPH in archaea (Section 4.3.5). The functional hexose monophosphate pathway is not found in many archaea (Section 4.3.5). GAPOR and GAPN cannot catalyse the reverse reactions. GAPDH is involved in gluconeogenesis (Section 4.2) with 3-phosphoglycerate kinase.

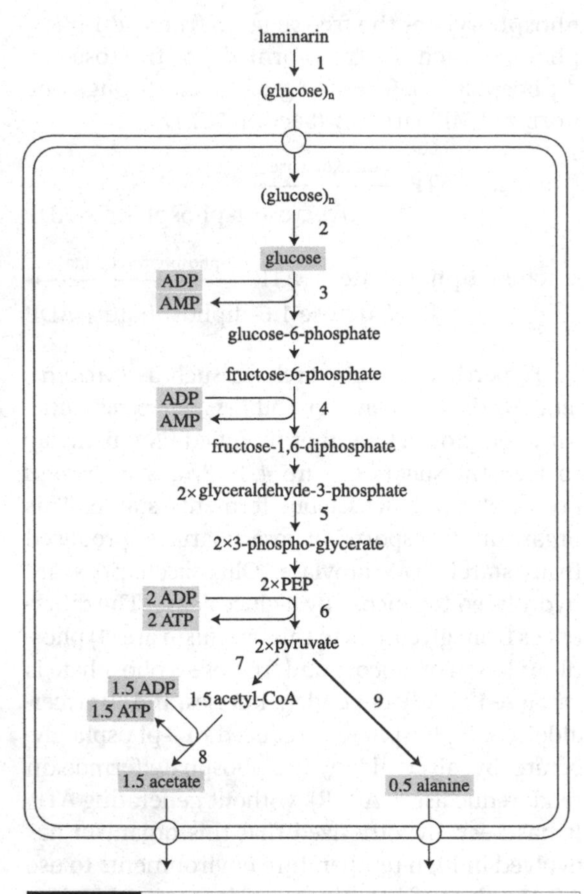

Figure 4.3 **Modified EMP pathway for metabolism of oligosaccharides produced by α-amylase in *Pyrococcus furiosus*.**

(*Extremophiles* **2**: 201–205, 1998)

Oligosaccharides produced by the extracellular α-amylase (1) are transported into the cell to be hydrolysed to glucose by α-glucosidase (2). Glucose and fructose-6-phosphate are phosphorylated by ADP-dependent glucokinase (3) and ADP-dependent phosphofructokinase (4). Glyceraldehyde-3-phosphate is oxidized to 3-phosphoglycerate by glyceraldehyde-3-phosphate; ferredoxin oxidoreductase (5) without ATP generation.

6, pyruvate kinase; 7, pyruvate:ferredoxin oxidoreductase; 8, acetyl-CoA synthetase; 9, alanine aminotransferase.

In addition to these ADP-dependent kinases, a pyrophosphate (PP_i)-dependent phosphofructokinase is known in *Thermoproteus tenax* and in the extremely thermophilic bacterium *Caldicellulosiruptor saccharolyticus* (Section 4.1.3.1). The PP_i-dependent phosphofructokinase catalyses the reverse reaction and is not regulated by AMP and PEP, unlike the ATP (ADP)-dependent enzymes. *Thermoproteus tenax* metabolizes carbohydrates in both modified EMP (85%) and modified ED pathways (15%) in parallel (Section 4.4.3.2).

Pyruvate is reduced as an electron acceptor to alanine in *Pyrococcus furiosus* (Section 8.10).

4.1.5 Regulation of the EMP pathway

The EMP pathway serves not only to generate ATP but also to provide precursors for biosynthesis (Table 4.1). The EMP pathway is regulated by the energy status of the cell as well as by the concentration of certain metabolic intermediates. A parameter, the adenylate energy charge (EC), was devised to describe the energy status of a microbial culture (Section 5.6.2). Generally, the glycolytic pathway is activated at low EC values and repressed at high EC values. As stated earlier, the enzymes that do not catalyse the reverse reaction are regulated. In the EMP pathway, these enzymes are phosphofructokinase and pyruvate kinase. In organisms where glucose is transported by active transport, hexokinase is repressed by its product, glucose-6-phosphate.

4.1.5.1 Regulation of phosphofructokinase

The EMP pathway is controlled mainly by regulating phosphofructokinase activity. This enzyme is activated by ADP and GDP, and repressed by PEP in bacteria. On the other hand, AMP activates, and ATP and citrate repress this enzyme in yeast. In general, the enzyme is repressed when the EC value (Section 5.6.2) and the concentration of intermediates used as precursors for biosynthesis are high, and activated when the organism needs ATP and biosynthetic precursors.

Fructose-1,6-diphosphate is dephosphorylated to fructose-6-phosphate and P_i in gluconeogenesis by fructose-1,6-diphosphatase. If fructose-1,6-diphosphatase and phosphofructokinase are both active in a cell, ATP is wasted. This is called a futile cycle and is avoided through an elaborate regulatory mechanism (Section 4.2.4).

4.1.5.2 Regulation of pyruvate kinase

Pyruvate kinase is activated when fructose-1,6-diphosphate (FDP) is accumulated in the cell. This kind of regulation is termed feedforward activation or precursor activation. The term feedforward activation is used to describe the activation of the last enzyme of a metabolic pathway by the substrate of the first enzyme of the same pathway. Feedback inhibition describes the inhibition of the first enzyme by the final product in an anabolic pathway (Section 12.3.1). The pyruvate kinase of the hyperthermophilic archaeon *Pyrobaculum aerophilum* is not activated by FDP but by 3-phosphoglycerate.

4.1.5.3 Regulation of modified EMP pathways in archaea

There is increasing evidence that shows that phosphorylation of fructose-6-phosphate and PEP conversion to pyruvate are not regulated in archaea. There are a few reports on the regulation of archaeal glycolysis. Glyceraldehyde-3-phosphate:ferredoxin oxidoreductase of the modified EMP pathways in archaea is inhibited by ATP and pyruvate kinase is activated by 3-phosphoglycerate, not by fructose-1,6-diphosphate, in a hyperthermophilic archaeon *Pyrobaculum aerophilum*.

4.1.5.4 Global regulation

In addition to regulation of individual enzymes, glycolysis is regulated as a part of a global regulation system. CcpA (catabolite control protein A) activates transcription of the genes for glycolytic enzymes in Gram-positive bacteria such as *Bacillus subtilis* (Section 12.1.3.3) and CsrA (carbon storage regulator A) activates expression of the glycolytic genes, and represses genes for gluconeogenesis and glycogen synthesis in *Escherichia coli* (Section 12.1.9.3). In *Bacillus subtilis* and related low G + C Gram-positive bacteria, genes for enzymes catalysing glyceraldehyde-3-phosphate to phosphoenolpyruvate form an operon. This operon is repressed by CggR (central glycolytic gene regulator) under gluconeogenic conditions, and CggR activity is repressed by fructose-1,6-diphosphate (Section 12.1.3.3). In *Corynebacterium glutamicum*, the global regulator proteins RamA and RamB play an important role in expression and control of genes involved in sugar uptake, glycolysis, gluconeogenesis, and acetate, lactate and ethanol metabolism coordinating the global metabolism.

4.2 Glucose-6-phosphate synthesis: gluconeogenesis

Hexose-6-phosphates are the precursors of polysaccharide synthesis. Microbes growing on carbon sources other than sugars (Chapter 7) need to synthesize glucose-6-phosphate from their substrate. Since phosphofructokinase and pyruvate kinase do not catalyse their reverse reactions, hexose-6-phosphate cannot be synthesized by a reversal of the EMP pathway:

$$\text{fructose-6-phosphate} + \text{ATP} \longrightarrow \text{fructose-1, 6-diphosphate} + \text{ADP}$$
$$(\Delta G^{0'} = -14.2\ \text{kJ/mol fructose-6-phosphate})$$

$$\text{phosphoenolpyruvate} + \text{ADP} \longrightarrow \text{pyruvate} + \text{ATP}$$
$$(\Delta G^{0'} = -23.8\ \text{kJ/mol phosphoenolpyruvate})$$

The above reactions are exergonic, and the reverse reactions are thermodynamically unfavourable. Separate enzymes are therefore used in gluconeogenesis to overcome this problem.

4.2.1 PEP synthesis

Microbes growing on non-carbohydrate compounds produce PEP through pyruvate, oxaloacetate or malate (Chapter 7).

Phosphoenolpyruvate (PEP) synthetase is widespread in microbes:

$$\text{pyruvate} + \text{ATP} \xrightarrow{\text{PEP synthetase}} \text{PEP} + \text{AMP} + P_i$$
$$(\Delta G^{0'} = -8.4\ \text{kJ/mol pyruvate})$$

Pyruvate phosphate dikinase is another enzyme that synthesizes PEP from pyruvate. This enzyme is found in *Acetobacter xylinum* growing on ethanol and in *Propionibacterium shermanii* growing on lactate:

$$\text{pyruvate} + \text{ATP} + P_i \xrightleftharpoons{\text{pyruvate phosphate dikinase}} \text{PEP} + \text{AMP} + PP_i$$

Phosphoenolpyruvate carboxykinase produces PEP from a TCA cycle intermediate, oxaloacetate:

$$\text{oxaloacetate} + \text{ATP} \xrightleftharpoons{\text{PEP carboxykinase}} \text{PEP} + CO_2 + \text{ADP}$$

Malate can be converted to PEP either through oxaloacetate by the action of malate dehydrogenase or through pyruvate by the action of malate enzyme:

$$\text{malate} + \text{NAD}^+ + \text{H}^+ \xrightleftharpoons{\text{malate enzyme}} \text{pyruvate} + \text{NADH} + CO_2$$

When *Bacillus subtilis* is cultivated on malate, malate dehydrogenase and the phosphoenolpyruvate carboxykinase convert the substrate to PEP. NADPH is generated by $NADP^+$-dependent malate enzyme, and NAD^+-dependent malate enzyme is used to produce ATP.

Phosphoenolpyruvate carboxykinase is essential for infection in *Mycobacterium tuberculosis*, which uses TCA cycle intermediates as carbon and energy sources in host environments.

4.2.2 Fructose diphosphatase

PEP can be converted to fructose-1,6-diphosphate by EMP pathway enzymes, since they catalyse the reverse reactions. However, phosphofructokinase does not catalyse the reverse reaction and fructose-1,6-diphosphate is dephosphorylated to fructose-6-phosphate and P_i by fructose diphosphatase:

$$\text{fructose-1,6-diphosphate} \xrightarrow{\text{fructose diphosphatase}} \text{fructose-6-phosphate} + P_i$$

Three classes of fructose diphosphatase are known in prokaryotes, with class II being the most common. Many bacteria possess more than one class of the enzyme.

4.2.3 Gluconeogenesis in archaea

Methanogenic archaea do not use sugars, and most of the halophilic and hyperthermophilic archaea metabolize sugars through modified ED pathways (Section 4.4.3.2). Nevertheless, they have a similar gluconeogenic system to bacteria. Glyceraldehyde-3-phosphate:ferredoxin oxidoreductase of the modified archeal EMP pathway found in some hyperthermophiles cannot catalyse the reverse reaction (Section 4.1.4), in addition to the two enzymes that cannot catalyse the reverse reactions in the classical EMP pathway, pyruvate kinase and phosphofructokinase. Pyruvate is phosphorylated to phosphoenolpyruvate through the reactions catalysed by phosphoenolpyruvate synthase, pyruvate: phosphate dikinase or PEP carboxykinase as in bacteria.

Most archaea, including hyperthermophiles, that convert glyceraldehyde-3-phosphate to 3-phosphoglycerate in a one-step reaction catalysed by glyceraldehyde-3-phosphate:ferredoxin oxidoreductase (Figure 4.3), convert 3-phosphoglycerate to glyceraldehyde-3-phosphate in two steps via 1,3-diphosphoglycerate catalysed by the classical EMP pathway enzymes, 3-phosphoglycerate kinase and glyceraldehyde-3-phosphate dehydrogenase.

Almost all archaeal groups, as well as deeply branching bacterial lineages, do not have fructose diphosphatase. Instead they have a bifunctional enzyme, fructose diphosphate aldolase/phosphatase that has fructose diphosphatase activity as well as fructose diphosphate aldolase activity:

$$\text{glyceraldehyde-3-phosphate} + \text{dihydroxyacetone phosphate} \xrightarrow{\text{fructose diphosphate aldolase/phosphatase}} \text{glucose-6-phosphate} + P_i$$

Pyrophosphate (PP_i)-dependent phosphofructokinase is found in the hyperthermophilic archaea such as *Pyrobaculum calidifontis*, *Thermoproteus neutrophilus* and *Thermoproteus tenax*, and this enzyme catalyses the reverse reaction. Similar reactions are known in plants and bacteria. This enzyme may be a component of gluconeogenesis:

$$\text{fructose-1,6-diphosphate} + P_i \xrightleftharpoons{\text{PPi-dependent phosphofructokinase}} \text{fructose-6-phosphate} + PP_i$$

4.2.4 Regulation of gluconeogenesis

When organisms are cultivated with carbon sources other than carbohydrates, genes for gluconeogenesis are activated and glycolytic genes are repressed. The Cra (catabolite repressor/activator) protein is responsible for this control in enteric bacteria, including *Escherichia coli* (Section 12.1.3.2). When enteric bacteria are cultivated using non-carbohydrate substrates such as acetate and ethanol, the Cra protein activates the expression of genes for the enzymes of gluconeogenesis (fructose-1,6-diphosphatase) and represses the transcription of genes for glycolytic enzymes (phosphofructokinase). Similar proteins are known in other bacteria, such as Crc protein (catabolite repression control) in many strains of *Pseudomonas* spp. (Section 12.1.3.2) and CcpA (catabolite control protein A) in Gram-positive bacteria, including strains of *Bacillus, Staphylococcus, Streptococcus, Lactococcus* and *Listeria* among others (Section 12.1.3.3). Two similar regulatory proteins, CceR (central carbon and energy metabolism regulator) and AkgR (alpha-ketoglutarate regulator) have been identified in *Rhodobacter sphaeroides*.

In addition to transcriptional control, gluconeogenesis is regulated through control of fructose diphosphatase activity. When the energy status is good, this enzyme is activated by ATP to supply carbon skeletons for growth. AMP represses enzyme activity when the energy status is too low to support growth. A futile cycle is avoided by the opposite controls of glycolysis and gluconeogenesis.

4.3 Hexose monophosphate (HMP) pathway

When *Escherichia coli* grows on glucose as the sole carbon and energy source, about 72 per cent of the substrate is metabolized through the EMP pathway, and the HMP pathway consumes the remaining 28 per cent. This is because the EMP pathway cannot meet all the requirements for biosynthesis. The HMP pathway provides biosynthetic metabolism with pentose-5-phosphate, erythrose-4-phosphate and NADPH. This pathway is also called the pentose phosphate pathway. NADPH is used to supply reducing power in biosynthetic processes. $NADP^+$ is reduced only by isocitrate dehydrogenase (Section 5.2) when glucose is metabolized through the EMP pathway and TCA cycle. In a few cases, $NADP^+$ is reduced by EMP enzymes (Section 4.1.3.1). NADH is seldom used in biosynthetic reactions. Most of the NADPH needed for biosynthesis arises from the HMP pathway.

A functional HMP pathway is not known in archaea, and they obtain pentose-5-phosphate through a reverse flux of the ribulose monophosphate pathway (Section 7.10.3.1), and NADPH is provided by the non-phosphorylating $NADP^+$-dependent GAP dehydrogenase (Section 4.1.4). Aromatic amino acids are produced from 6-deoxy-5-ketofructose-1-phosphate and aspartate semialdehyde, and not from erythrose-4-phosphate (Section 6.4.4).

4.3.1 HMP pathway in three steps

For convenience, the HMP pathway can be discussed in three steps (Figure 4.4). During the initial step of the HMP pathway, glucose-6-phosphate is oxidized to ribulose-5-phosphate and CO_2, reducing $NADP^+$. Glucose-6-phosphate dehydrogenase, lactonase and 6-phosphogluconate dehydrogenase catalyse these reactions. In the following reactions, ribulose-5-phosphate is converted to ribose-5-phosphate and xylulose-5-phosphate by the action of isomerase and epimerase. Finally, the pentose-5-phosphates are transformed to glucose-6-phosphate and glyceraldehyde-3-phosphate, through carbon rearrangement by transaldolase and transketolase. A transaldolase transfers a three-carbon fragment, and a two-carbon fragment transfer is catalysed by a transketolase. HMP can be summarized as:

$$\text{glucose-6-phosphate} + 2NADP^+ \longrightarrow \text{glyceraldehyde-3-phosphate} + 3CO_2 + 2NADPH + 2H^+$$

Ribose-5-phosphate is the precursor for nucleotide synthesis, and aromatic amino acids are produced from erythrose-4-phosphate. NADPH supplies reducing power during biosynthesis (Section 6.4.4).

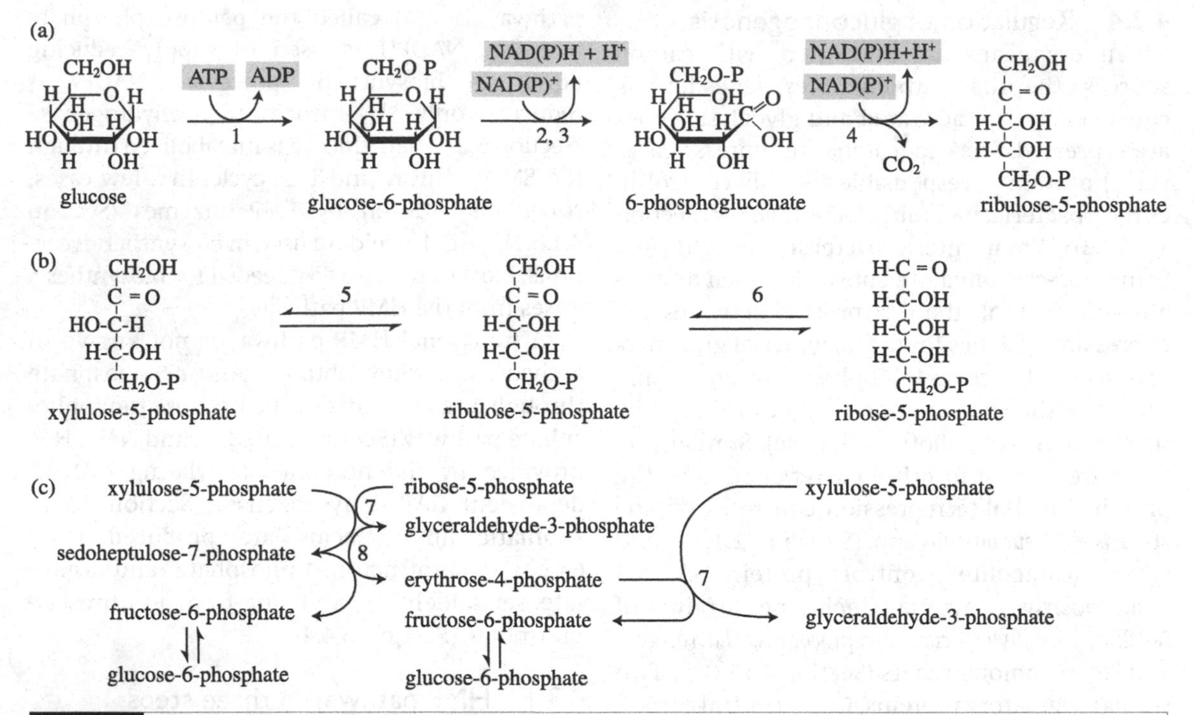

Figure 4.4 Hexose monophosphate (pentose phosphate) pathway.

Glucose is oxidized to ribulose-5-phosphate coupled to the reduction of NADPH (a). Isomerase and epimerase convert ribulose-5-phosphate to ribose-5-phosphate and xylulose-5-phosphate (b). Transaldolase and transketolase rearrange pentose-5-phosphates into glucose-6-phosphate and glyceraldehyde-3-phosphate involving erythrose-4-phosphate (c). Nucleotide synthesis starts with ribose-5-phosphate, and aromatic amino acids are produced from erythrose-4-phosphate and phosphoenolpyruvate. NADPH supplies reducing power during the biosynthetic processes.

1, hexokinase; 2, glucose-6-phosphate dehydrogenase; 3, lactonase; 4, 6-phosphogluconate dehydrogenase; 5, ribulose-5-phosphate-3-epimerase; 6, ribose-5-phosphate isomerase; 7, transketolase; 8, transaldolase.

Some eukaryotic microorganisms metabolize more glucose through the HMP pathway when they use nitrate as their nitrogen source. They use NADPH in assimilatory nitrate reduction (Section 6.2.2).

Reactions of the HMP pathway can be catalysed by alternative enzymes. An *Escherichia coli* double mutant (*talA* and *talB*) of transaldolase isoenzymes, which convert sedoheptulose-7-phosphate and glyceraldehydes-3-phosphate to erythrose-4-phosphate and fructose-6-phosphate, can grow on glucose. This reaction is bypassed by phosphofructokinase and fructose diphosphate aldolase. The former enzyme phosphorylates sedoheptulose-7-phosphate to sedoheptulose-1,7-diphosphate, that is cleaved to erythrose-4-phosphate and glyceraldehyde-3-phosphate by the fructose diphosphate aldolase.

4.3.2 Additional functions of the HMP pathway

In addition to supplying precursors and reducing power for biosynthesis from glucose, the HMP and related pathways have some other functions. The HMP pathway is the major glycolytic metabolism in microbes that (1) utilize pentoses, and (2) do not possess other glycolytic activities. The oxidative HMP cycle (Figure 4.5) is also employed for the complete oxidation of sugars in bacteria lacking a functional TCA cycle.

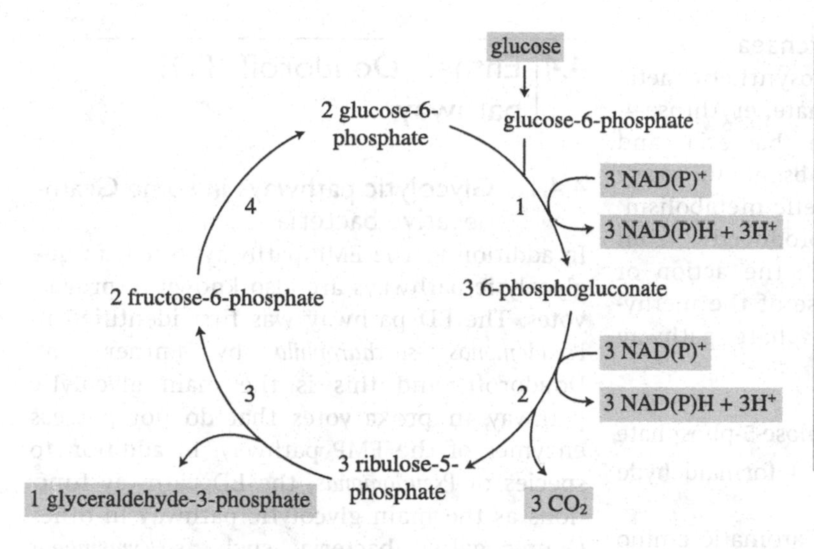

Figure 4.5 Oxidative HMP cycle.

Organisms lacking functional EMP or ED pathways, or a functional TCA cycle, oxidize glucose through the oxidative HMP cycle. Glyceraldehyde-3-phosphate is oxidized to pyruvate as in the EMP pathway.

1, glucose-6-phosphate dehydrogenase; 2, 6-phosphogluconate dehydrogenase; 3, carbon rearrangement as in the HMP pathway; 4, glucose-6-phosphate isomerase.

4.3.2.1 Utilization of pentoses

Pentoses are phosphorylated and metabolized to fructose-6-phosphate and glyceraldehyde-3-phosphate through steps 2 and 3 of the HMP pathway. This will be discussed in Chapter 7 (Section 7.2.2).

4.3.2.2 Oxidative HMP cycle

Thiobacillus novellus and *Brucella abortus* oxidize glucose completely, although they lack a functional EMP or ED pathway. Glucose is oxidized through the oxidative HMP cycle (Figure 4.5). Glyceraldehyde-3-phosphate is oxidized to pyruvate as in the EMP pathway. The HMP cycle is found in species of *Gluconobacter* which do not have a functional TCA cycle. These bacteria possess the incomplete TCA fork to meet the supply of biosynthetic precursors (Section 5.4.1).

4.3.3 Regulation of the HMP pathway

Lactonase is the only enzyme unable to catalyse the reverse reaction among the enzymes of the HMP pathway (Figure 4.4), but regulation of the pathway is exerted through control of glucose-6-phosphate dehydrogenase and 6-phosphogluconate dehydrogenase. These enzymes are inhibited, with the accumulation of NADPH and NADH.

Dehydrogenation of glucose-6-phosphate is a common reaction in the HMP and ED pathways, but is catalysed by separate enzymes in each. The enzyme involved in the ED pathway uses NAD^+ as the electron acceptor and is inhibited by ATP and PEP. On the other hand, the $NADP^+$-dependent HMP pathway enzyme is inhibited by NAD(P)H but not by ATP.

4.3.4 F_{420}-dependent glucose-6-phosphate dehydrogenase

F_{420} was first identified in methanogens, and was regarded as one of the methanogen-specific coenzymes (Section 9.4.2). This electron carrier is found not only in hyperthermophilic archaea such as species of *Sulfolobus* and *Archaeoglobus*, but also in bacteria. F_{420} was found to be involved in tetracycline biosynthesis in a *Streptomyces* sp., and a F_{420}-dependent glucose-6-phosphate dehydrogenase has been identified in *Mycobacterium smegmatis*. This enzyme is believed to be a part of the HMP pathway in this bacterium. Since F_{420} is not known in animals, this enzyme is a potential target for antibacterial drugs against medically important species of *Mycobacterium*.

In addition to F_{420}, other cofactors, originally found in methanogens, have been identified in bacteria. Coenzyme M is involved in the oxidation of propylene by *Rhodococcus rhodochrous* and a species of *Xanthobacter* (Section 7.7). Tetrahydromethanopterin occurs in a methylotroph, *Methylobacterium extorquens* (Section 7.10.2.2).

4.3.5 HMP pathway and archaea

The HMP pathway provides biosynthetic metabolism with pentose-5-phosphate, erythrose-4-phosphate and NADPH in bacteria and eukarya. This pathway is absent in many archaea. To support biosynthetic metabolism, pentose-5-phosphate is produced from hexulose-6-phosphate through the action of hexulose-6-phosphate synthase of the methylotrophic ribulose monophosphate pathway (Section 7.10.3.1):

hexulose-6-phosphate ⟶ ribulose-5-phosphate + formaldehyde

The first two reactions of aromatic amino acids producing dehydroquinate (DHQ, Section 6.4.4) are not found in *Methanocaldococcus jannaschii*. DHQ is produced not from erythrose-4-phosphate but from 6-deoxy-5-ketofructose-1-phosphate and aspartate semialdehyde in this archaeon, as below:

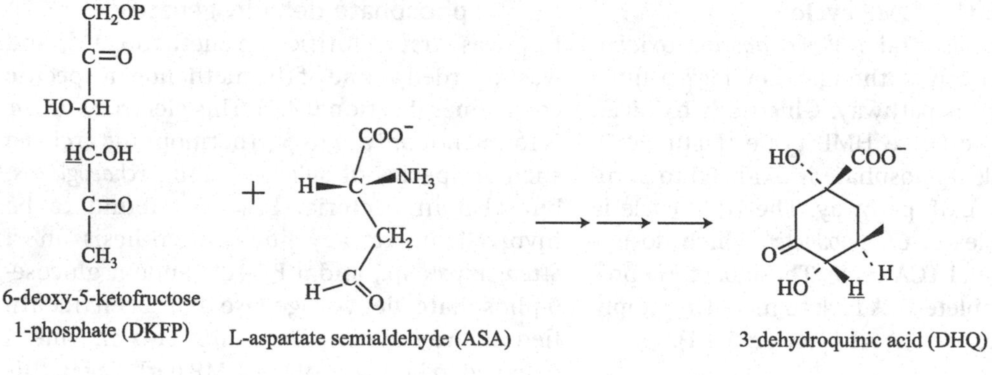

6-deoxy-5-ketofructose 1-phosphate (DKFP) + L-aspartate semialdehyde (ASA) ⟶ ⟶ ⟶ 3-dehydroquinic acid (DHQ)

NADPH is used to supply reducing power in biosynthetic processes. The main reaction to reduce $NADP^+$ to NADPH in archaea is catalysed by non-phosphorylating $NADP^+$-dependent glyceraldehyde-3-phosphate dehydrogenase in the archaeal modified EMP pathway (Section 4.1.4).

Pentoses are metabolized in archaea in a series of reactions catalysed by dehydrogenases and dehydratases producing 2-ketoglutarate (Section 7.2.3).

4.4 Entner–Doudoroff (ED) pathway

4.4.1 Glycolytic pathways in some Gram-negative bacteria

In addition to the EMP pathway, other unique glycolytic pathways are also known in prokaryotes. The ED pathway was first identified in *Pseudomonas saccharophila* by Entner and Doudoroff, and this is the main glycolytic pathway in prokaryotes that do not possess enzymes of the EMP pathway. In addition to species of *Pseudomonas*, the ED pathway functions as the main glycolytic pathway in other Gram-negative bacteria, such as *Zymomonas* and *Azotobacter* species. Enzymes for EMP and ED pathways are expressed simultaneously in *Escherichia coli*, but glucose metabolized through the ED pathway is less than 10 per cent. Gluconate is metabolized through this pathway in some Gram-negative bacteria, including *Escherichia coli* and *Vibrio cholerae*, and in some coryneform bacteria such as species of *Arthrobacter* and *Cellulomonas* (Table 4.2). The ED pathway is functional in *Vibrio cholerae* and activates prime virulence genes (*ctxA* and *tcpA*) and their regulator (*toxT*), suggesting the importance of the ED pathway in *Vibrio cholerae* pathogenesis. When a haloalkaliphilic bacterium *Alkaliflexus imshenetskii* is cultivated on cellobiose, enzymes for the EMP and ED pathways are expressed simultaneously.

Table 4.2 | Major glycolytic pathways in prokaryotes.

Organism	EMP	ED
Arthrobacter species	+	+/–[a]
Azotobacter chroococcum	+	+
Ralstonia eutropha (*Alcaligenes eutrophus*)	–	+
Bacillus subtilis	+	–
Cellulomonas flavigena	+	+/–[a]
Escherichia coli and enteric bacteria	+	+/–[a]
Pseudomonas saccharophila	–	+
Rhizobium japonicum	–	+
Thiobacillus ferrooxidans	–	+
Xanthomonas phaseoli	–	+
Thermotoga maritima	+	+[b]
Thermoproteus tenax	+[c]	+[b,d]
Halococcus saccharolyticus	+[c,e]	+[b,d,f]
Halobacterium saccharovorum	–	+[d]
Clostridium aceticum	–	+[d]
Sulfolobus acidocaldarius	–	+[d]
Alkaliflexus imshenetskii	+	+
Vibrio cholera	+	+

+, present; –, absent.
[a] When gluconate is used as energy and carbon source.
[b] Enzymes for EMP and ED pathways are expressed simultaneously.
[c] Modified EMP pathway.
[d] Modified ED pathway.
[e] Fructose.
[f] Glucose.

The majority of marine bacteria use the ED pathway for glucose catabolism, whereas small numbers rely on the EMP pathway. Bacteria with the ED pathway where $NADP^+$ is reduced are known to be more resistant against oxidative stress, which could be an important feature contributing to preferential use of the ED pathway in the oceans.

The sulfonated sulfoquinovose (SQ; 6-deoxy-6-sulfoglucose) is a structural analogue of glucose-6-phosphate (GP) and its degradation proceeds in an analogous manner to the ED pathway in *Escherichia coli* and *Pseudomonas putida* (Section 7.2).

4.4.2 Key enzymes of the ED pathway

In the first two reactions of the ED pathway, glucose-6-phosphate is converted to 6-phosphogluconate via phosphogluconolactone, as in the HMP pathway. 6-phosphogluconate is dehydrated to 2-keto-3-deoxy-6-phosphogluconate (KDPG) by 6-phosphogluconate dehydratase. KDPG aldolase splits its substrate into pyruvate and glyceraldehyde-3-phosphate (Figure 4.6). The latter is oxidized to pyruvate as in the EMP pathway. The key enzymes of this pathway are 6-phosphogluconate dehydratase and KDPG aldolase.

Two ATP are generated and one high energy phosphate bond is consumed with a net gain of one ATP per glucose oxidized in this pathway.

Genes for the ED pathway in *Pseudomonas putida* are repressed by RexR in the absence of glucose. The ED pathway intermediate, 2-keto-3-deoxy-6-phosphogluconate, binds RexR to activate gene expression. In *Rhodobacter sphaeroides*,

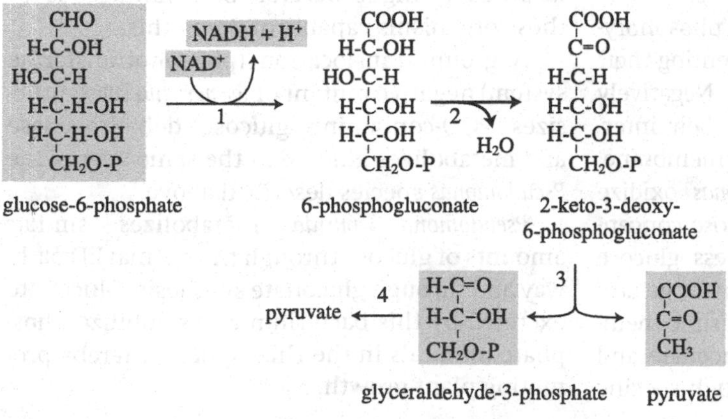

Figure 4.6 The Entner–Doudoroff (ED) pathway.

This metabolism is known only in prokaryotes, mainly Gram-negative bacteria, that do not possess the EMP pathway.

1, glucose-6-phosphate dehydrogenase; 2, 6-phosphogluconate dehydratase; 3, 2-keto-3-deoxy-6-phosphogluconate aldolase; 4, as in the EMP pathway.

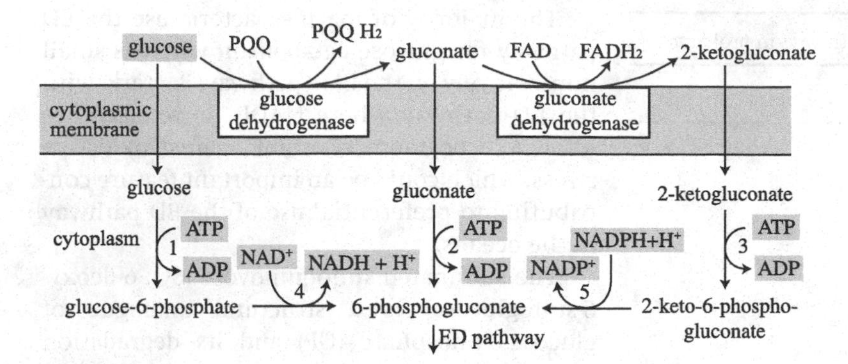

Figure 4.7 **Glucose utilization by some species of *Pseudomonas* in a high glucose environment.**

(Dawes, E. A. 1986, *Microbial Energetics*, Figure 3.5. Blackie & Son, Glasgow)

When the glucose concentration is high, some strains of *Pseudomonas* oxidize glucose in the periplasmic region. Glucose dehydrogenase is a quinoprotein containing pyrroloquinoline quinone (PQQ) and gluconate dehydrogenase containing FAD. The reduced PQQ and FAD transfer electrons to cytochrome c, and gluconate and 2-ketogluconate are transported through specific transporters to be metabolized in a similar way as in the ED pathway, when glucose is depleted.

1, hexokinase; 2, gluconate kinase; 3, 2-ketogluconate kinase; 4, glucose-6-phosphate dehydrogenase; 5, 2-keto-6-phosphogluconate reductase.

CceR (carbon and energy metabolism regulator) represses genes encoding enzymes in the ED pathway and activates those encoding the TCA cycle and gluconeogenesis. This repressor is inactivated by an ED pathway intermediate, 6-phosphogluconate.

4.4.3 Modified ED pathways

Unusually, some prokaryotes oxidize glucose and the intermediates are phosphorylated before being metabolized in a similar manner as in the ED pathway.

4.4.3.1 Extracellular oxidation of glucose by Gram-negative bacteria

Sugars are metabolized after they are phosphorylated, the latter process probably preventing their loss through membrane diffusion. Negatively charged phosphorylated sugars and their intermediates are less permeable to the membrane. However, some strains of *Pseudomonas* oxidize glucose extracellularly when the glucose concentration is high. These bacteria possess glucose dehydrogenase and gluconate dehydrogenase on the periplasmic face of the cytoplasmic membrane. When glucose is depleted, gluconate and 2-ketogluconate are transported through specific transporters and phosphorylated, consuming ATP (Figure 4.7). 2-keto-6-phosphogluconate is reduced to 6-phosphogluconate by a NADPH-dependent reductase.

The glucose dehydrogenase in these bacteria is a quinoprotein containing pyrroloquinoline quinone (PQQ, methoxatin) as a prosthetic group (Section 7.10.2) and gluconate dehydrogenase is a FAD-dependent enzyme. The reduced forms of PQQ and FAD transfer electrons to cytochrome *c* of the electron transport chain.

Gluconate and 2-ketogluconate are uncommon in nature, and few microbes use these compounds. The ability to oxidize glucose and to use its products might therefore be advantageous for those organisms capable of doing this.

A group translocation (phosphotransferase system) negative mutant of *Escherichia coli* synthesizes PQQ-containing glucose dehydrogenase and metabolizes glucose in the same way as the *Pseudomonas* species described above.

Pseudomonas putida metabolizes similar amounts of glucose through the normal ED pathway and through gluconate synthesis. Gluconate excreted by this bacterium can solubilize phosphate minerals in the rhizosphere, thereby promoting plant growth.

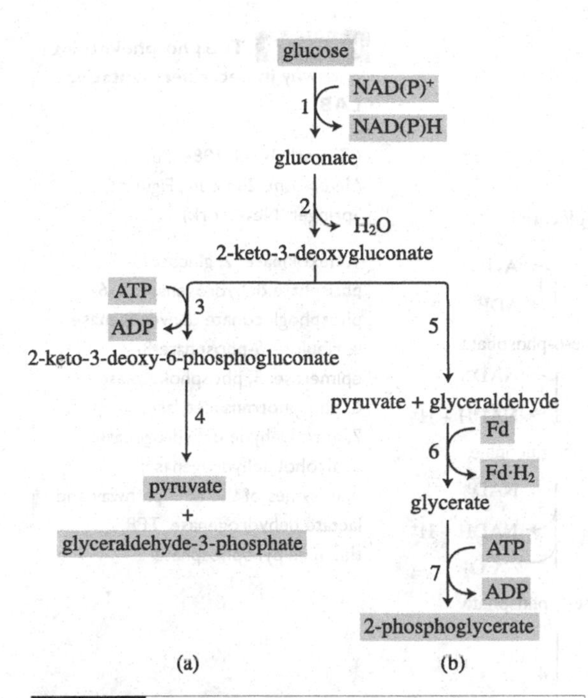

Figure 4.8 Modified ED pathways in archaea.

(a) *Halobacterium saccharovorum* and *Haloferax volcanii*.
(b) *Sulfolobus acidocaldarius* and *Thermoproteus tenax*.

1, glucose dehydrogenase or glucose:ferredoxin oxidoreductase; 2, gluconate dehydratase; 3, 2-keto-3-deoxygluconate kinase; 4, 2-keto-3-deoxy-6-phosphogluconate aldolase; 5, 2-keto-3-deoxygluconate aldolase; 6, glyceraldehyde:ferredoxin oxidoreductase; 7, glycerate kinase.

4.4.3.2 Modified ED pathways in archaea

Hyperthermophilic archaea belonging to the genera *Sulfolobus, Thermoplasma, Picrophilus* and *Thermoproteus* metabolize glucose to pyruvate and glyceraldehyde without phosphorylation, in a similar way as in the ED pathway, and halophilic archaea such as *Halobacterium saccharovorum* and *Haloferax volcanii* oxidize glucose to 2-keto-3-deoxygluconate, which is phosphorylated before metabolism through the ED pathway (Figure 4.8). Xylonate dehydratase can replace gluconate dehydratase in *Haloferax volcanii*. The archaeal glucose dehydrogenase is a NAD(P)$^+$-dependent enzyme. Some eubacteria, including *Clostridium aceticum* and *Rhodopseudomonas sphaeroides*, metabolize glucose in a similar mechanism to this halophilic archaeon.

As in the modified EMP pathway in archaea (Section 4.1.4), glyceraldehyde-3-phosphate is reduced to 3-phosphoglycerate in a one-step reaction catalysed by glyceraldehyde-3-phosphate:ferredoxin oxidoreductase (GAPOR) without generating ATP. *Thermoproteus tenax* uses the modified EMP pathway and ED pathway for carbohydrate catabolism in parallel (Section 4.1.4). The benefit of simultaneously expressing genes for two different glycolytic pathways is not yet known.

4.5 Phosphoketolase pathways

Lactate is the sole glucose fermentation product in homofermentative lactic acid bacteria (LAB), while heterofermentative LAB produce acetate and ethanol in addition to lactate from glucose (Section 8.4). The former ferment glucose through the EMP pathway, while the phosphoketolase (PK) pathway is employed in the latter and in bifidus bacteria. A heterofermentative bacterium, *Leuconostoc mesenteroides*, produces lactate and ethanol from glucose through the PK pathway, involving one PK active on xylulose-5-phosphate (Figure 4.9). Lactate and acetate are produced from glucose by *Bifidobacterium bifidum* with two PKs active on fructose-6-phosphate and xylulose-5-phosphate (Figure 4.10). Facultatively homofermentative LAB ferment pentoses and low concentrations of glucose through the PK pathway to produce lactate, acetate and ethanol (Section 8.4).

4.5.1 Glucose fermentation by *Leuconostoc mesenteroides*

Heterofermentative LAB, including *Leuconostoc mesenteroides*, ferment glucose to lactate, ethanol and carbon dioxide:

$$C_6H_{12}O_6 \rightarrow CH_3CHOHCOOH + CH_3CH_2OH + CO_2$$

These bacteria oxidize glucose-6-phosphate to ribulose-5-phosphate as in the HMP pathway, before converting it to xylulose-5-phosphate via an epimerase. Phosphoketolase splits the pentose-5-phosphate to glyceraldehyde-3-phosphate and acetyl-phosphate. Acetyl-phosphate

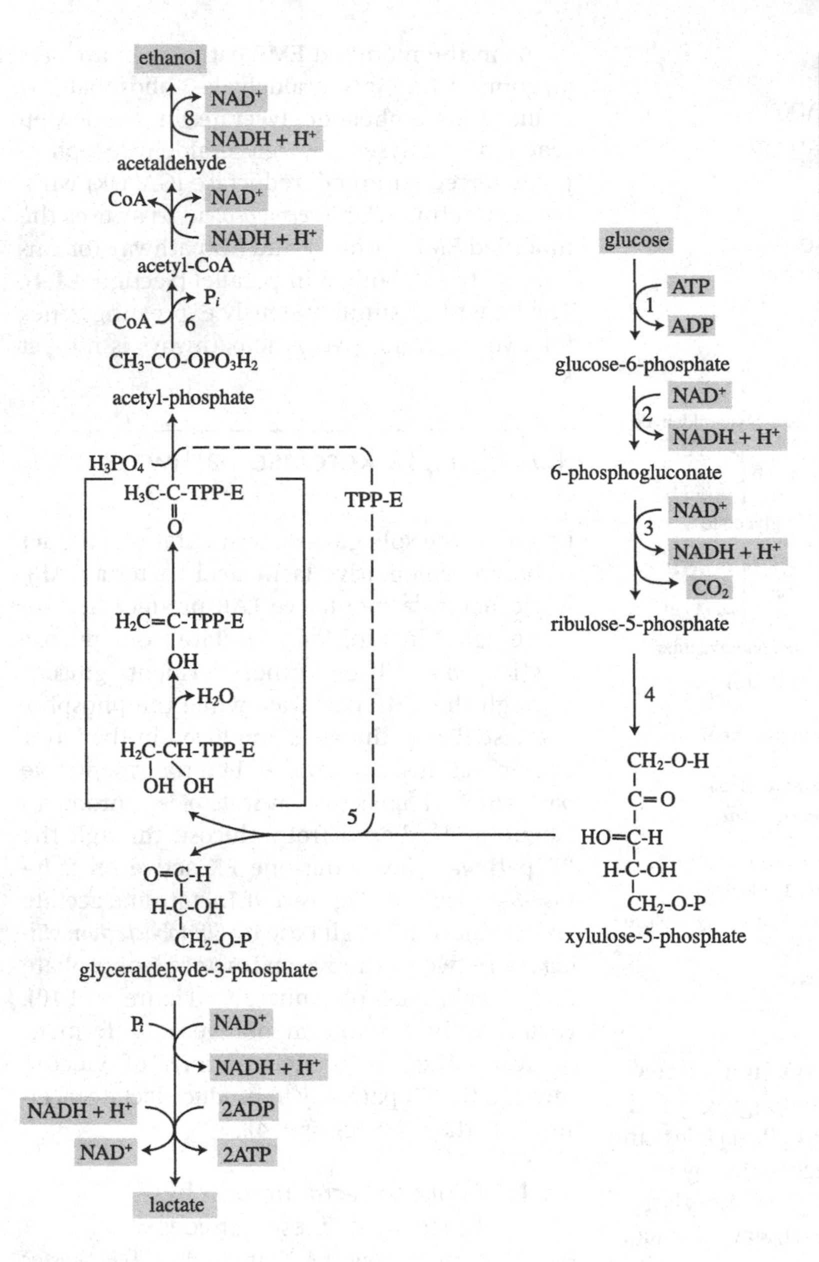

Figure 4.9 The phosphoketolase pathway in heterofermentative LAB.

(Gottschalk, G. 1986, *Bacterial Metabolism*, 2nd edn., Figure 8.3. Springer, New York)

1, hexokinase; 2, glucose-6-phosphate dehydrogenase; 3, 6-phosphogluconate dehydrogenase; 4, ribulose-5-phosphate-3-epimerase; 5, phosphoketolase; 6, phosphotransacetylase; 7, acetaldehyde dehydrogenase; 8, alcohol dehydrogenase; 9, enzymes of the EMP pathway and lactate dehydrogenase. TPP, thiamine pyrophosphate.

is reduced to ethanol to regenerate NAD^+ and pyruvate is produced from the triose-phosphate as in the latter part of the EMP pathway.

Since LAB have a restricted electron transport chain they use pyruvate and acetyl-phosphate as electron acceptors in the reactions catalysed by lactate dehydrogenase, acetaldehyde dehydrogenase and alcohol dehydrogenase, to regenerate NAD^+ from NADH (Section 8.4). For this reason *Leuconostoc mesenteroides* synthesizes one more ATP from pentoses than hexoses. In the hexose fermentation, acetyl-phosphate is reduced to ethanol to oxidize NADH (reactions 7 and 8 in Figure 4.9), which is reduced by glucose-6-phosphate dehydrogenase (reaction 2 in Figure 4.9) and 6-phosphogluconate dehydrogenase (reaction 3 in Figure 4.9). Acetyl-phosphate is used to synthesize ATP in the reaction catalysed by acetate kinase in pentose fermentation.

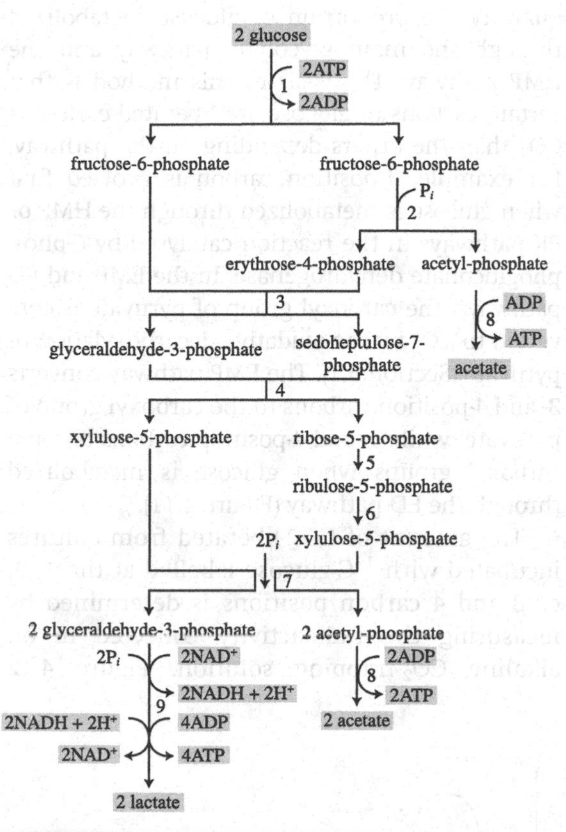

Figure 4.10 The bifidum pathway in *Bifidobacterium bifidum*.

(Gottschalk, G. 1986, *Bacterial Metabolism*, 2nd edn., Figure 8.4. Springer, New York)

Bifidobacterium bifidum ferments glucose to lactate and acetate with two phosphoketolases, each active on fructose-6-phosphate and xylulose-5-phosphate. This fermentation results in a net gain of 2.5 ATP per glucose fermented.

1, hexokinase and glucose-6-phosphate isomerase; 2, fructose-6-phosphate phosphoketolase; 3, transaldolase; 4, transketolase; 5, ribose-5-phosphate isomerase; 6, ribulose-5-phosphate-3-epimerase; 7, xylulose-5-phosphate phosphoketolase; 8, acetate kinase; 9, enzymes of the EMP pathway and lactate dehydrogenase.

Thiamine pyrophosphate (TPP) is a prosthetic group in phosphoketolase. TPP binds glycoaldehyde, and the complex is dehydrated before being phosphorylated to acetyl-phosphate. PK in *Leuconostoc mesenteroides* is active on fructose-6-phosphate as well as xylulose-5-phosphate, but has a much higher affinity for the latter. Hexose fermentation through the PK pathway results in the net gain of 1 ATP.

4.5.2 Bifidum pathway

As shown in Figure 4.10, *Bifidobacterium bifidum* ferments two molecules of glucose to two molecules of lactate and three molecules of acetate. Phosphoketolase splits fructose-6-phosphate into erythrose-4-phosphate and acetyl-phosphate. Transketolase and transaldolase rearrange erythrose-4-phosphate and a second molecule of fructose-6-phosphate to two molecules of xylulose-5-phosphate, as in the HMP pathway. Xylulose-5-phosphate is metabolized to lactate and acetate as in *Leuconostoc mesenteroides*. From the fermentation of two molecules of glucose, 7 ATP are synthesized and 2 ATP are consumed with a net gain of 2.5 ATP per glucose fermented.

In addition to heterofermentative LAB and species of *Bifidobacterium*, phosphoketolase is known in a few organisms, including an opportunistic pathogen, *Gardnerella vaginalis*, *Clostridium acetobutylicum* and the fungi *Aspergillus nidulans* and *Cryptococcus neoformans*, as a pentose utilization pathway (Section 7.2.2). Pentoses are metabolized mainly through the HMP pathway in *Cl. acetobutylicum*, but more substrate is metabolized through the phosphoketolase pathway at high substrate concentrations.

4.6 Glycolysis in archaea

Modified EMP (Section 4.1.4) or ED (Section 4.4.3.2) pathways are the main glycolytic metabolism in carbohydrate-utilizing archaea. A few reactions in archaeal gluconeogenesis are catalysed by different enzymes from the classical bacterial pathway (Section 4.2.3).

The HMP pathway is not known in archaea (Section 4.3.5) that synthesize pentose-5-phosphate through the methylotrophic ribulose monophosphate pathway (Section 7.10.3.1). Erythrose-4-phosphate is replaced with 6-phospho-5-ketofructose-1-phosphate and aspartate semialdehyde to produce aromatic amino acids in archaea (Section 6.4.4).

The main reaction to reduce $NADP^+$ to NADPH in archaea is catalysed by non-phosphorylating $NADP^+$-dependent glyceraldehyde-3-

phosphate dehydrogenase in the archaeal modified EMP pathway (Section 4.1.4).

Pentoses are metabolized in archaea in a series of reactions catalysed by dehydrogenases and dehydratases, producing 2-ketoglutarate (Section 7.2.2).

4.7 Use of radiorespirometry to determine glycolytic pathways

In addition to the EMP pathway, prokaryotes metabolize glucose in the ED and PK pathways. The glycolytic pathway that occurs in a given organism can be determined through analysis of key enzyme activities and radioactive tracer experiments. For example, if the cell-free extract of an organism shows activities of 6-phosphogluconate dehydrogenase and 2-keto-3-deoxy-6-phosphogluconate aldolase, but not that of phosphofructokinase, the given organism possesses the ED pathway.

A radiorespirometric approach is another method for determining the glycolytic pathway using ^{14}C- glucose labelled in various positions. The results of these experiments show not only what the main glycolytic pathway is, but also quantify the proportion of glucose metabolized through the main glycolytic pathway and the HMP pathway. The basis of this method is that certain carbons of glucose are liberated earlier as CO_2 than the others depending on the pathway. For example, 1-position carbon is evolved first when glucose is metabolized through the HMP or PK pathways in the reaction catalysed by 6-phosphogluconate dehydrogenase. In the EMP and ED pathways, the carboxyl group of pyruvate is converted to CO_2 by the oxidative decarboxylation of pyruvate (Section 5.1). The EMP pathway converts 3- and 4-position carbons to the carboxyl group of pyruvate, while 1- and 4-position carbons become carboxyl groups when glucose is metabolized through the ED pathway (Figure 4.11).

The amount of CO_2 liberated from cultures incubated with ^{14}C-glucose labelled at the 1, 2, or 3 and 4 carbon positions is determined by measuring the radioactivity collected in an alkaline CO_2-trapping solution. Figure 4.12

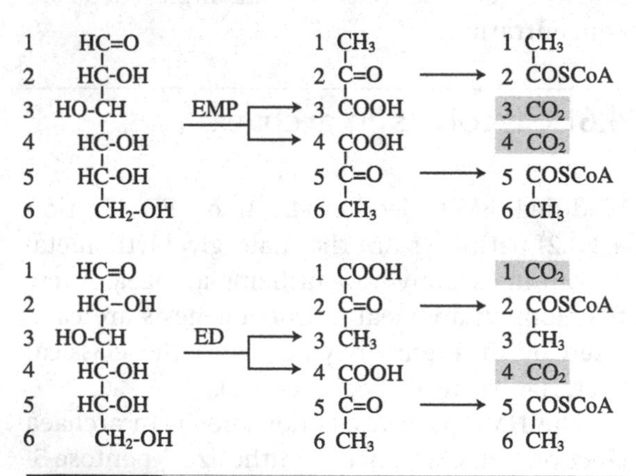

Figure 4.11 The origin of the carboxyl carbon of pyruvate from glucose when metabolized through the EMP and ED pathways.

(Gottschalk, G. 1986, *Bacterial Metabolism*, 2nd edn., Figure 5.10. Springer, New York)

Pyruvate dehydrogenase decarboxylates pyruvate to acetylCoA, generating CO_2. The carboxyl group of pyruvate originates from the 3- and 4-position carbons of glucose when metabolized through the EMP pathway, and from the 1- and 4-position carbons in the ED pathway.

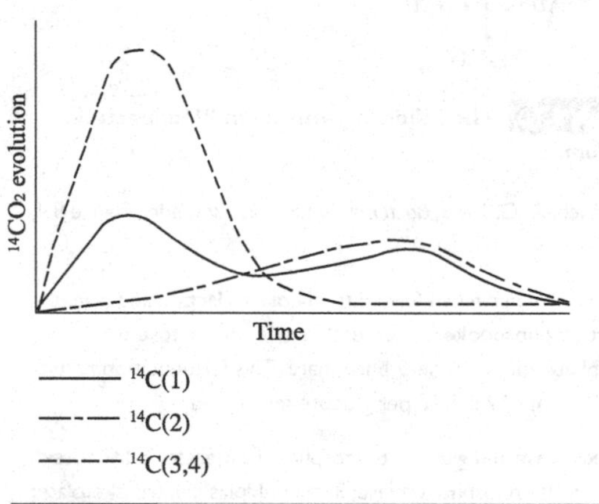

Figure 4.12 Analysis of glucose degradation in *Escherichia coli* through radiorespirometry.

(Gottschalk, G. 1986, *Bacterial Metabolism*, 2nd edn., Figure 3.15. Springer, New York)

Through determination of the radioactivity of CO_2 evolved from cultures grown on $^{14}C(1)$, $^{14}C(2)$ or $^{14}C(3)$ and $^{14}C(4)$-labelled glucose, the glycolytic pathways can be characterized. CO_2 released at an early stage from $^{14}C(1)$ is due to the reaction catalysed by 6-phosphogluconate dehydrogenase of the HMP pathway. CO_2 originating from pyruvate by the action of pyruvate dehydrogenase of the EMP pathway is from $^{14}C(3)$ and $^{14}C(4)$-labelled glucose. CO_2 from the oxidation of acetyl-CoA through the TCA cycle appears at the later stage from $^{14}C(1)$ and $^{14}C(2)$-labelled glucose.

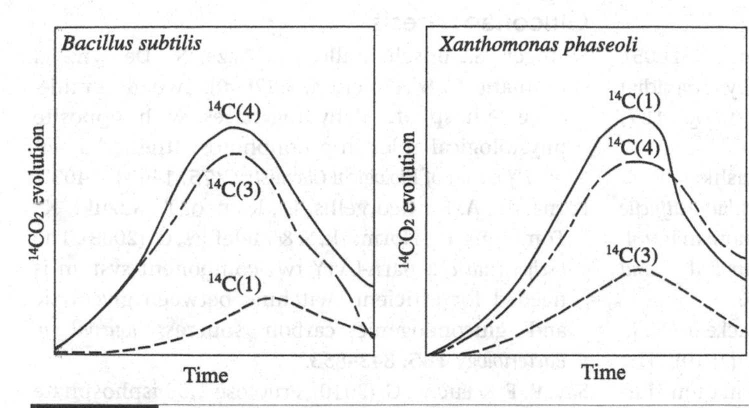

Figure 4.13 Determination of glycolysis in *Bacillus subtilis* and *Xanthomonas phaseoli* through radiorespirometry.

(Gottschalk, G. 1986, *Bacterial Metabolism*, 2nd edn., Figure 5.11. Springer, New York)

Bacillus subtilis degrades glucose through the EMP pathway, producing pyruvate with carboxyl groups originating from C3 and C4, which are released as CO_2 by the reaction catalysed by pyruvate dehydrogenase. In contrast, the pyruvates produced by *Xanthomonas phaseoli* have carboxyl groups originating from C1 and C4.

shows CO_2 evolution from cultures of *Escherichia coli* grown on glucose labelled at various positions. The culture incubated with glucose labelled at C3 and C4 releases CO_2 at an early stage, showing that this bacterium degrades glucose through the EMP pathway. On the other hand, CO_2 is released twice from C1-labelled glucose. The first release indicates the degradation of glucose through the HMP pathway, and the second release is from the TCA cycle. Through comparison of CO_2 released from the culture incubated with glucose labelled at C3 and C4 with that of C1-labelled glucose, the amount of glucose metabolized through the EMP and HMP pathways can be calculated as 72 per cent and 28 per cent, respectively (Section 4.3).

Results shown in Figure 4.13 are obtained from radioactivity measurements of CO_2 evolved from cultures of *Bacillus subtilis* and *Xanthomonas phaseoli* grown on ^{14}C(1), ^{14}C(3) and ^{14}C(4)-labelled glucose. The former organism possesses the EMP pathway as the main glycolytic pathway: the ED pathway is the main glycolytic mechanism in the latter organism. *Bacillus subtilis* evolves CO_2 faster from ^{14}C(3) and ^{14}C(4)-labelled glucose than from ^{14}C(1)-glucose, while more CO_2 is evolved from ^{14}C(1) and ^{14}C(4)-labelled glucose than ^{14}C(3)-glucose by *Xanthomonas phaseoli*.

Further Reading

Note this section contains key references only. Additional recommended references are available at www.cambridge.org/ProkaryoticMetabolism.

EMP and modified pathways

Albi, T. & Serrano, A. (2015). Two strictly polyphosphate-dependent gluco(manno)kinases from diazotrophic cyanobacteria with potential to phosphorylate hexoses from polyphosphates. *Applied Microbiology and Biotechnology* **99**, 3887–3900.

Bielen, A. A. M., Willquist, K., Engman, J., Oost, J. v. d., Niel, E. W. J. v. & Kengen, S. W. M. (2010). Pyrophosphate as a central energy carrier in the hydrogen-producing extremely thermophilic *Caldicellulosiruptor saccharolyticus*. *FEMS Microbiology Letters* **307**, 48–54.

Zhou, J., Olson, D. G., Argyros, D. A., Deng, Y., van Gulik, W. M., van Dijken, J. P. & Lynd, L. R. (2013). Atypical glycolysis in *Clostridium thermocellum*. *Applied and Environmental Microbiology* **79**, 3000–3008.

Methylglyoxal bypass

Ko, J., Kim, I., Yoo, S., Min, B., Kim, K. & Park, C. (2005). Conversion of methylglyoxal to acetol by *Escherichia coli* aldo-keto reductases. *Journal of Bacteriology* **187**, 5782–5789.

Liyanage, H., Kashket, S., Young, M. & Kashket, E. R. (2001). *Clostridium beijerinckii* and *Clostridium difficile* detoxify methylglyoxal by a novel mechanism involving glycerol dehydrogenase. *Applied and Environmental Microbiology* **67**, 2004–2010.

Ozyamak, E., Black, S. S., Walker, C. A., MacLean, M. J., Bartlett, W., Miller, S. & Booth, I. R. (2010). The critical role of *S*-lactoylglutathione formation during methylglyoxal detoxification in *Escherichia coli*. *Molecular Microbiology* **78**, 1577–1590.

Subedi, K. P., Choi, D., Kim, I., Min, B. & Park, C. (2011). Hsp31 of *Escherichia coli* K-12 is glyoxalase III. *Molecular Microbiology* **81**, 926–936.

Modified EMP pathways in Archaea

Bräsen, C., Esser, D., Rauch, B. & Siebers, B. (2014). Carbohydrate metabolism in archaea: current insights into unusual enzymes and pathways and their regulation. *Microbiology and Molecular Biology Reviews* **78**, 89–175.

Brunner, N. A., Siebers, B. & Hensel, R. (2001). Role of two different glyceraldehyde-3-phosphate dehydrogenases in controlling the reversible Embden-Meyerhof-Parnas pathway in *Thermoproteus tenax*: regulation on protein and transcript level. *Extremophiles* **5**, 101–109.

Labes, A. & Schonheit, P. (2003). ADP-dependent glucokinase from the hyperthermophilic sulfate-reducing archaeon *Archaeoglobus fulgidus* strain 7324. *Archives of Microbiology* **180**, 69–75.

Matsubara, K., Yokooji, Y., Atomi, H. & Imanaka, T. (2011). Biochemical and genetic characterization of the three metabolic routes in *Thermococcus kodakarensis* linking glyceraldehyde 3-phosphate and 3-phosphoglycerate. *Molecular Microbiology* **81**, 1300–1312.

Tjaden, B., Plagens, A., Dorr, C., Siebers, B. & Hensel, R. (2006). Phosphoenolpyruvate synthetase and pyruvate, phosphate dikinase of *Thermoproteus tenax*: key pieces in the puzzle of archaeal carbohydrate metabolism. *Molecular Microbiology* **60**, 287–298.

Verhees, C. H., Tuininga, J. E., Kengen, S. W. M., Stams, A. J. M., vanderOost, J. & de Vos, W. M. (2001). ADP-dependent phosphofructokinases in mesophilic and thermophilic methanogenic archaea. *Journal of Bacteriology* **183**, 7145–7153.

Gluconeogenesis

Fillinger, S., Boschi-Muller, S., Azza, S., Dervyn, E., Branlant, G. & Aymerich, S. (2000). Two glyceraldehyde-3-phosphate dehydrogenases with opposite physiological roles in a nonphotosynthetic bacterium. *Journal of Biological Chemistry* **275**, 14031–14037.

Pernestig, A. K., Georgellis, D., Romeo, T., Suzuki, K., Tomenius, H., Normark, S. & Melefors, O. (2003). The *Escherichia coli* BarA-UvrY two-component system is needed for efficient switching between glycolytic and gluconeogenic carbon sources. *Journal of Bacteriology* **185**, 843–853.

Say, R. F. & Fuchs, G. (2010). Fructose 1,6-bisphosphate aldolase/phosphatase may be an ancestral gluconeogenic enzyme. *Nature* **464**, 1077–1081.

Tjaden, B., Plagens, A., Dorr, C., Siebers, B. & Hensel, R. (2006). Phosphoenolpyruvate synthetase and pyruvate, phosphate dikinase of *Thermoproteus tenax*: key pieces in the puzzle of archaeal carbohydrate metabolism. *Molecular Microbiology* **60**, 287–298.

Verhees, C. H., Akerboom, J., Schiltz, E., de Vos, W. M. & van der Oost, J. (2002). Molecular and biochemical characterization of a distinct type of fructose-1,6-bisphosphatase from *Pyrococcus furiosus*. *Journal of Bacteriology* **184**, 3401–3405.

Wolfe, A. J. (2005). The acetate switch. *Microbiology and Molecular Biology Reviews* **69**, 12–50.

HMP pathway

Brouns, S. J. J., Walther, J., Snijders, A. P. L., van de Werken, H. J. G., Willemen, H. L. D. M., Worm, P., de Vos, M. G. J., Andersson, A., Lundgren, M., Mazon, H. F. M., van den Heuvel, R. H. H., Nilsson, P., Salmon, L., de Vos, W. M., Wright, P. C., Bernander, R. & van der Oost, J. (2006). Identification of the missing links in prokaryotic pentose oxidation pathways: evidence for enzyme recruitment. *Journal of Biological Chemistry* **281**, 27378–27388.

Grochowski, L. L., Xu, H. & White, R. H. (2005). Ribose-5-phosphate biosynthesis in *Methanocaldococcus jannaschii* occurs in the absence of a pentose-phosphate pathway. *Journal of Bacteriology* **187**, 7382–7389.

Orita, I., Sato, T., Yurimoto, H., Kato, N., Atomi, H., Imanaka, T. & Sakai, Y. (2006). The ribulose monophosphate pathway substitutes for the missing pentose phosphate pathway in the archaeon *Thermococcus kodakaraensis*. *Journal of Bacteriology* **188**, 4698–4704.

White, R. H. (2004). l-Aspartate semialdehyde and a 6-deoxy-5-ketohexose 1-phosphate are the precursors to the aromatic amino acids in *Methanocaldococcus jannaschii*. *Biochemistry* **43**, 7618–7627.

ED and modified ED pathways

Chavarría, M., Nikel, P. I., Pérez-Pantoja, D. & de Lorenzo, V. (2013). The Entner-Doudoroff pathway empowers *Pseudomonas putida* KT2440 with a high tolerance to oxidative stress. *Environmental Microbiology* **15**, 1772–1785.

Conway, T. (1992). The Entner-Doudoroff pathway: history, physiology and molecular biology. *FEMS Microbiology Reviews* **103**, 1–28.

Felux, A.-K., Spiteller, D., Klebensberger, J. & Schleheck, D. (2015). Entner-Doudoroff pathway for sulfoquinovose degradation in *Pseudomonas putida* SQ1. *Proceedings of the National Academy of Sciences of the USA* **112**, 4298–4305.

Gunnarsson, N., Mortensen, U. H., Sosio, M. & Nielsen, J. (2004). Identification of the Entner-Doudoroff pathway in an antibiotic-producing actinomycete species. *Molecular Microbiology* **52**, 895–902.

Patra, T., Koley, H., Ramamurthy, T., Ghose, A. C. & Nandy, R. K. (2012). The Entner-Doudoroff pathway is obligatory for gluconate utilization and contributes to the pathogenicity of *Vibrio cholerae*. *Journal of Bacteriology* **194**, 3377–3385.

Reher, M., Fuhrer, T., Bott, M. & Schonheit, P. (2010). The nonphosphorylative Entner-Doudoroff pathway in the thermoacidophilic euryarchaeon *Picrophilus torridus* involves a novel 2-keto-3-deoxygluconate-specific aldolase. *Journal of Bacteriology* **192**, 964–974.

Zaparty, M., Tjaden, B., Hensel, R. & Siebers, B. (2008). The central carbohydrate metabolism of the hyperthermophilic crenarchaeote *Thermoproteus tenax*: pathways and insights into their regulation. *Archives of Microbiology* **190**, 231–245.

PK pathways

Glenn, K. & Smith, K. S. (2015). Allosteric regulation of *Lactobacillus plantarum* xylulose 5-phosphate/fructose 6-phosphate phosphoketolase (Xfp). *Journal of Bacteriology* **197**, 1157–1163.

Sund, C. J., Liu, S., Germane, K. L., Servinsky, M. D., Gerlach, E. S. & Hurley, M. M. (2015). Phosphoketolase flux in *Clostridium acetobutylicum* during growth on l-arabinose. *Microbiology* **161**, 430–440.

Yevenes, A. & Frey, P. A. (2008). Cloning, expression, purification, cofactor requirements, and steady state kinetics of phosphoketolase-2 from *Lactobacillus plantarum*. *Bioorganic Chemistry* **36**, 121–127.

Metabolic analysis

Antoniewicz, M. (2015). Methods and advances in metabolic flux analysis: a mini-review. *Journal of Industrial Microbiology and Biotechnology* **42**, 317–325.

Klamt, S. & Stelling, J. (2003). Two approaches for metabolic pathway analysis? *Trends in Biotechnology* **21**, 64–69.

Siebers, B. & Schonheit, P. (2005). Unusual pathways and enzymes of central carbohydrate metabolism in archaea. *Current Opinion in Microbiology* **8**, 695–705.

Chapter 5

Tricarboxylic acid (TCA) cycle, electron transport and oxidative phosphorylation

Pyruvate produced from glycolysis and other metabolic pathways is metabolized in various ways, depending on the organism and growth conditions. Pyruvate is either used as a precursor for biosynthesis, or is oxidized completely to CO_2 under aerobic conditions. This chapter is devoted to the mechanisms of pyruvate oxidation, electron transport and oxidative phosphorylation.

5.1 Oxidative decarboxylation of pyruvate

Pyruvate is oxidized by the pyruvate dehydrogenase complex to acetyl-CoA and CO_2 reducing NAD^+ under aerobic conditions. The pyruvate dehydrogenase complex consists of 24 molecules of pyruvate dehydrogenase (E1) containing thiamine pyrophosphate (TPP), 24 molecules of dihydrolipoate acetyltransferase (E2) containing dihydrolipoate and 12 molecules of dihydrolipoate dehydrogenase (E3) containing flavin adenine dinucleotide (FAD). In addition, NAD^+ and coenzyme A participate in the reaction (Figure 5.1). This reaction is irreversible, and takes place in the mitochondrion in eukaryotic cells. The reaction can be summarized as

$$\text{pyruvate} + \text{CoA-SH} + NAD^+ \longrightarrow \text{acetyl-CoA} + NADH + H^+ \quad (\Delta G^{0'} = -33.5\ \text{kJ/mol pyruvate})$$

The pyruvate dehydrogenase complex shares its properties with other 2-keto acid dehydrogenase complexes that produce acyl-CoA, such as 2-ketoglutarate dehydrogenase (reaction 4 in Figure 5.2) and 2-ketobutyrate dehydrogenase (Section 7.5.6), but their activities are controlled differently.

The expression of the pyruvate dehydrogenase operon is repressed by the pyruvate-sensing PdhR, and pyruvate derepresses its expression by forming a pyruvate–PdhR complex. In addition, PdhR controls the respiratory electron transport genes, NADH dehydrogenase II (*ndh*) and one of the terminal oxidases (*cyoABCDE*). The enzyme activity is controlled by its substrate, products and the adenylate energy charge. Pyruvate and

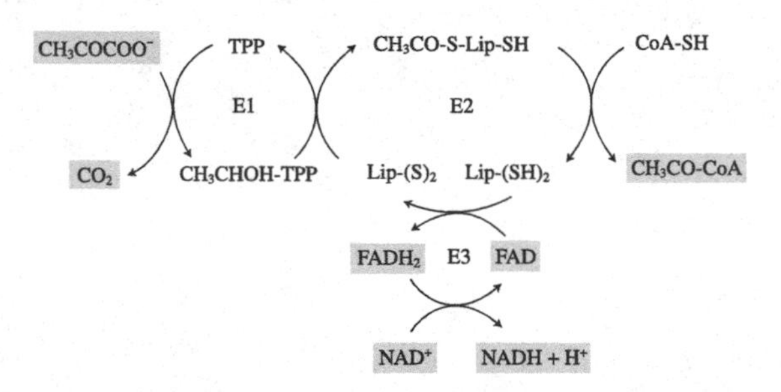

Figure 5.1 Oxidative decarboxylation of pyruvate by the pyruvate dehydrogenase complex.

Pyruvate dehydrogenase (E1) decarboxylates pyruvate, and its prosthetic group thiamine pyrophosphate (TPP) binds the resulting hydroxyethyl group, which binds to the lipoate (Lip) of dihydrolipoate acetyltransferase (E2), reducing its disulfide. E2 transfers an acetyl group to coenzyme A to form acetyl-CoA. Dihydrolipoate dehydrogenase (E3) transfers electrons from reduced lipoate to NAD^+.

AMP activate enzyme activity, and acetyl-CoA, NADH and ATP repress it. Pyruvate and acetyl-CoA are precursors for amino acid (Section 6.4.1) and fatty acid (Section 6.6.1) biosynthesis, respectively.

5.2 Tricarboxylic acid (TCA) cycle

This cyclic metabolic pathway was discovered by Krebs and his colleagues in animal tissue. It is referred to as the tricarboxylic acid (TCA) cycle, Krebs cycle or citric acid cycle. Acetyl-CoA produced by the pyruvate dehydrogenase complex is completely oxidized to CO_2, reducing NAD^+, $NADP^+$ and FAD. These reduced electron carriers are oxidized through the processes of electron transport and oxidative phosphorylation to form the proton motive force and to synthesize ATP.

The TCA cycle provides not only reducing equivalents for ATP synthesis but also precursors for biosynthesis. The TCA cycle or related metabolic processes are indispensable, providing biosynthetic precursors for all forms of cells except mycoplasmas, which take required materials from their host animal cells.

5.2.1 Citrate synthesis and the TCA cycle

Citrate synthase synthesizes citrate from acetyl-CoA and oxaloacetate (Figure 5.2). This is an exergonic reaction ($\Delta G^{0'} = -32.2$ kJ/mol acetyl-CoA) and irreversible. The reverse reaction is catalysed by a separate enzyme, ATP:citrate lyase, in the reductive TCA cycle (Section 5.4.2). Citrate is converted to isocitrate, catalysed by aconitase. Isocitrate is oxidized to 2-ketoglutarate by isocitrate dehydrogenase. In most bacteria this enzyme is $NADP^+$ dependent, but two separate enzymes are found in eukaryotes, using $NADP^+$ or NAD^+.

The 2-ketoglutarate dehydrogenase complex oxidizes its substrate to succinyl-CoA. As with the pyruvate dehydrogenase complex, this enzyme complex consists of many peptides and cofactors, and catalyses oxidative decarboxylation, producing acyl-CoA. This is another irreversible reaction in the TCA cycle. The reverse reaction is catalysed by 2-ketoglutarate synthase (2-ketoglutarate:ferredoxin oxidoreductase) in the reductive TCA cycle to fix CO_2 (Section 5.4.2). Some anaerobic fermentative bacteria do not have this enzyme. They supply the precursors for biosynthesis through the incomplete TCA fork (Section 5.4.1). Glutamate and related

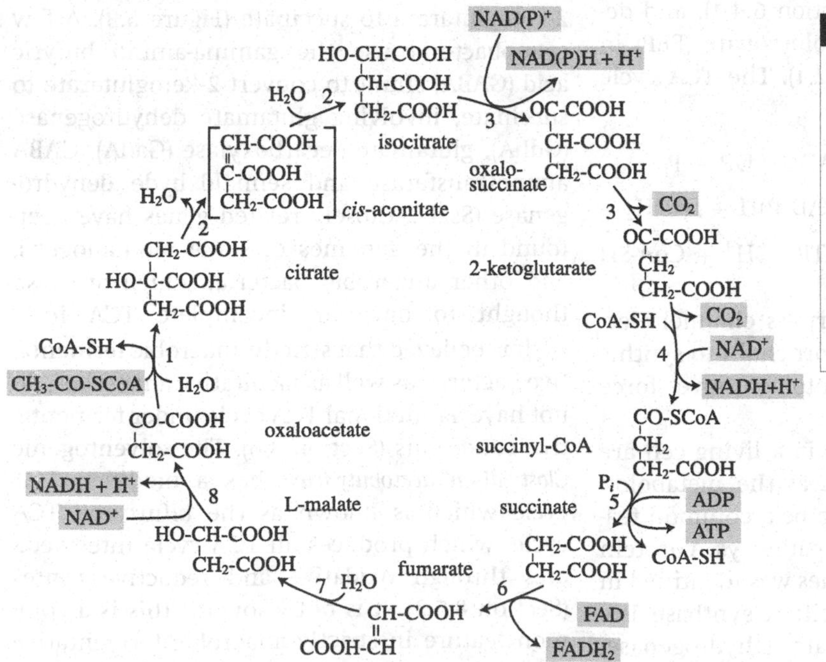

Figure 5.2 Acetyl-CoA oxidation through the tricarboxylic acid cycle.

1, citrate synthase; 2, aconitase; 3, isocitrate dehydrogenase; 4, 2-ketoglutarate dehydrogenase complex; 5, succinate thiokinase (succinyl-CoA synthetase); 6, succinate dehydrogenase; 7, fumarase (fumarate hydratase); 8, malate dehydrogenase.

amino acids are synthesized from 2-ketoglutarate (Section 6.4.3). Succinyl-CoA is the precursor for porphyrin synthesis which provides the chemical nucleus of cytochromes and chlorophyll (Section 6.7). Both carbons of acetyl-CoA are liberated as CO_2 in the two reduction reactions. Like all acyl-CoA derivatives, succinyl-CoA has a high energy bond. This energy is conserved as ATP through the succinate thiokinase (succinyl-CoA synthetase) reaction producing succinate. This is an example of substrate-level phosphorylation. Guanosine triphosphate (GTP) is synthesized in the mitochondrion by an analogous reaction in eukaryotic cells.

Succinate is oxidized to fumarate by succinate dehydrogenase. Since the redox potential of fumarate/succinate (−0.03 V) is considerably higher than NAD^+/NADH (−0.32 V), $NAD(P)^+$ cannot be reduced in this reaction. The prosthetic group of succinate dehydrogenase, FAD, is reduced. Electrons of the reduced succinate dehydrogenase are transferred to coenzyme Q of the electron transport chain (Section 5.8.2). Fumarate is hydrated to malate by fumarase before being oxidized to oxaloacetate by malate dehydrogenase, reducing NAD^+. Oxaloacetate is then ready to accept acetyl-CoA for the next round of the cycle. Oxaloacetate is used to synthesize amino acids (Section 6.4.1), and decarboxylated to phosphoenolpyruvate (PEP) in gluconeogenesis (Section 4.2.1). The TCA cycle can be summarized as:

$$CH_3\text{-CO-CoA} + 3NAD(P)^+ + FAD + ADP + P_i + 3H_2O \longrightarrow 2CO_2 + 3NAD(P)H + FADH_2 + ATP + 3H^+ + CoA\text{-}SH$$

The reduced electron carriers channel electrons to the electron transport chain to synthesize ATP through the proton motive force (Section 5.8).

The majority of proteins in a living cell are active in complexes known as the metabolon (Section 2.3.5), that seems to be a common feature especially for metabolic pathways. A protein complex of TCA cycle enzymes was identified in *Bacillus subtilis*, consisting of citrate synthase, isocitrate dehydrogenase, malate dehydrogenase, fumarase and aconitase. In addition, another protein complex is formed between malate dehydrogenase and phosphoenolpyruvate carboxykinase, linking the TCA cycle and gluconeogenesis under gluconeogenic growth conditions.

Aconitase is a bifunctional protein in bacteria as well as eukaryotes. In addition to enzyme activity, this protein has RNA binding properties when it is oxidized, losing the [4Fe–4S] cluster. This enzyme is oxidized by reactive oxygen species (Section 12.2.5) and under iron-limited conditions. The apo-protein of this enzyme is a pleiotropic post-transcriptional regulator controlling translation of various mRNAs, inhibiting their translation through binding to the 5′ untranslated region (UTR) and stabilizing them by binding to 3′ UTR.

5.2.2 Modified TCA cycle

Cyanobacteria were believed to operate an incomplete TCA fork since the proton motive force is generated through photosynthesis, and they do not possess 2-ketoglutarate dehydrogenase activity. More recently, genes encoding a novel 2-ketoglutarate decarboxylase (OgdA) and succinic semialdehyde dehydrogenase (SsaD) have been identified in many cyanobacteria. These two enzymes catalyse the oxidation of 2-ketoglutarate to succinate (Figure 5.3). A few cyanobacteria use the gamma-amino butyric acid (GABA) shunt to convert 2-ketoglutarate to succinate, involving glutamate dehydrogenase (GdhA), glutamate decarboxylase (GadA), GABA aminotransferase and semialdehyde dehydrogenase (SsaD). Closely related genes have been found in the genomes of some methanogens, and other anaerobic bacteria, which are also thought to have an incomplete TCA fork.

It is believed that strictly anaerobic fermentative bacteria, as well as facultative anaerobes, do not have a functional TCA cycle under fermentative conditions (Section 8.6). The solventogenic *Clostridium acetobutylicum* has a modified TCA cycle which is known as the bifurcated TCA cycle, which produces all TCA cycle intermediates through oxidative and reductive routes (Section 8.5.2). It is not known if this is a common feature in strictly anaerobic fermentative bacteria.

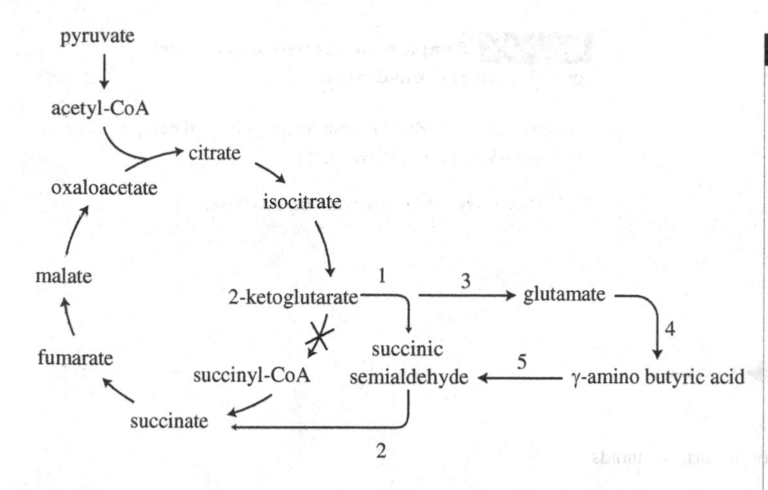

Figure 5.3 Modified TCA cycle in cyanobacteria.

(Modified from *Science* **334**: 1551–1553, 2011)

Cyanobacteria without 2-ketoglutarate dehydrogenase activity convert 2-ketoglutarate to succinate, catalysed by 2-ketoglutarate decarboxylase (1) and succinic semialdehyde dehydrogenase (2). A few cyanobacteria can convert 2-ketoglutarate to succinate through the gamma-amino butyric acid (GABA) shunt. Glutamate dehydrogenase (GdhA, 3), glutamate decarboxylase (GadA, 4), GABA aminotransferase (5) and semialdehyde dehydrogenase (SsaD, 2) are involved in this shunt.

5.2.3 Regulation of the TCA cycle

The TCA cycle is an amphibolic pathway serving anabolic needs by producing ATP as well as catabolic needs by providing precursors for biosynthesis. Consequently, this metabolism is regulated by the energy status of the cell and the availability of biosynthetic precursors. In addition, oxygen regulates the TCA cycle since the reduced electron carriers are recycled, consuming oxygen as the electron acceptor.

Oxygen controls the expression of genes for TCA cycle enzymes. Facultative anaerobes do not synthesize 2-ketoglutarate dehydrogenase under anaerobic conditions without alternative electron acceptors, such as nitrate. The activity is lower with nitrate than with oxygen as the electron acceptor. In Gram-negative bacteria, including *Escherichia coli*, the regulatory proteins FNR and Arc regulate the transcription of many genes for aerobic and anaerobic metabolism (Section 12.2.4). A FNR protein with a similar function is also known in Gram-positive *Bacillus subtilis*.

Because of their pivotal role in metabolism, most TCA cycle enzymes are expected to be present under almost any growth conditions. However, the activity of the TCA cycle can vary considerably. During growth on acetate, the flux through citrate synthase was reported to be fourfold greater than during growth on glucose. When *Escherichia coli* is cultivated using a non-carbohydrate substrate such as acetate, Cra (catabolite repressor/activator) protein (Section 12.1.3.2) activates expression of genes for TCA and glyoxylate (Section 5.3.2) cycle enzymes including citrate synthase, isocitrate dehydrogenase and isocitrate lyase. In alphaproteobacteria, including *Rhodobacter sphaeroides*, central carbon and energy metabolism regulator CceR, and 2-ketoglutarate regulator AkgR, cooperatively repress genes of the ED pathway and activate genes for gluconeogenesis (Section 4.2.4) and the TCA cycle in a similar way to the Cra protein in enteric bacteria. A similar regulator protein, HexR, was identified in the Fe(III) reducing bacterium *Shewanella oneidensis* (Section 9.2.1).

Citrate synthase is regulated to control the TCA cycle. Generally this enzyme is repressed, with the accumulation of NADH and ATP or 2-ketoglutarate. This accumulation means that the cell has enough energy and precursors for biosynthesis. Gram-negative bacteria have two different citrate synthase enzymes, one repressed by NADH and the other unaffected. Gram-positive bacteria have only one enzyme and this is not repressed by NADH. Instead, ATP inhibits this enzyme. AMP activates the citrate synthase inhibited by NADH in some bacteria.

5.3 Replenishment of TCA cycle intermediates

Some intermediates of the TCA cycle serve as precursors for biosynthesis. For efficient

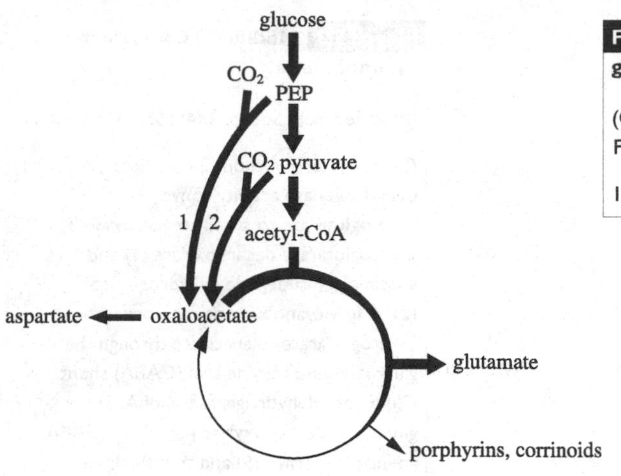

Figure 5.4 Anaplerotic sequence in bacteria growing on carbohydrates.

(Gottschalk, G. 1986, *Bacterial Metabolism*, 2nd edn., Figure 3.45. Springer, New York)

1, PEP carboxylase; 2, pyruvate carboxylase.

operation of this cyclic metabolism, the intermediates used for biosynthesis should be replenished, otherwise the concentration of oxaloacetate would be too low to start the TCA cycle. Oxaloacetate is replenished through a process called the anaplerotic sequence. When substrates not metabolized through pyruvate or PEP, such as acetate, are used, TCA cycle intermediates are replenished through other pathways, the glyoxylate cycle, the ethylmalonyl-CoA pathway and the methylaspartate cycle, depending on the organisms.

5.3.1 Anaplerotic sequence

Bacteria growing on carbohydrates synthesize oxaloacetate from pyruvate or phosphoenolpyruvate (PEP), as shown in Figure 5.4.

Many organisms, from bacteria to mammals, carboxylate pyruvate to oxaloacetate and this is catalysed by pyruvate carboxylase consuming ATP:

$$\text{pyruvate} + HCO_3^- + \text{ATP} \xrightleftharpoons{\text{pyruvate carboxylase}} \text{oxaloacetate} + \text{ADP} + P_i$$

This enzyme requires biotin. Acetyl-CoA activates this enzyme in many bacteria, as in animals, but some bacteria, such as *Pseudomonas aeruginosa*, have a pyruvate carboxylase that is not activated by acetyl-CoA.

A PEP carboxylase mutant of *Escherichia coli* is unable to grow in a glucose-mineral salts medium, but can grow when supplemented with TCA cycle intermediates. This bacterium has PEP carboxylase as the anaplerotic sequence, not pyruvate carboxylase. This property is shared by many other bacteria, including *Bacillus anthracis, Thiobacillus novellus, Acetobacter xylinum* and *Azotobacter vinelandii*:

$$\text{PEP} + HCO_3^- \xrightleftharpoons{\text{PEP carboxylase}} \text{oxaloacetate} + P_i$$

Pyruvate (PEP) carboxylase participates in CO_2 fixation in the rhodopsin-containing *Dokdonia* sp. (Section 11.6).

5.3.2 Glyoxylate cycle

Bacteria growing on carbon sources that are not metabolized through pyruvate or PEP cannot replenish TCA cycle intermediates through the anaplerotic sequence, and need even more oxaloacetate to produce PEP for gluconeogenesis (Section 4.2.1). They therefore need another mechanism, the glyoxylate cycle, for this purpose.

Escherichia coli growing on acetate synthesizes isocitrate lyase and malate synthase for a functional glyoxylate cycle (Figure 5.5). These enzymes convert two molecules of acetyl-CoA to a molecule of malate in conjunction with TCA cycle enzymes. Acetyl-CoA is converted to isocitrate through the TCA cycle, and isocitrate lyase cleaves isocitrate to succinate and glyoxylate:

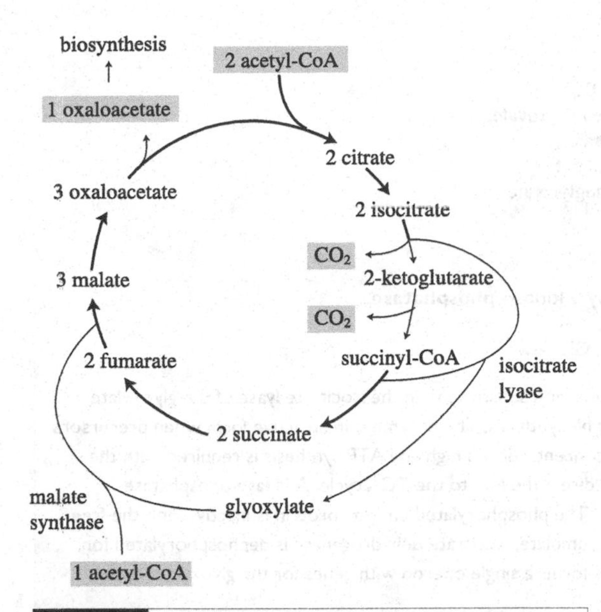

Figure 5.5 Glyoxylate cycle for replenishment of TCA cycle intermediates.

(Gottschalk, G. 1986, *Bacterial Metabolism*, 2nd edn., Figure 4.5. Springer, New York)

Bacteria growing on carbon sources that are metabolized to acetyl-CoA, but not through pyruvate or PEP, cannot replenish TCA cycle intermediates through the anaplerotic sequence. They synthesize malate from two molecules of acetyl-CoA through the glyoxylate cycle, utilizing isocitrate lyase and malate synthase, together with TCA cycle enzymes.

$$\text{isocitrate} \xrightarrow{\text{isocitrate lyase}} \text{succinate} + \text{glyoxylate}$$

Succinate is oxidized to oxaloacetate through the TCA cycle, and glyoxylate is used in the synthesis of malate by malate synthase with a second molecule of acetyl-CoA:

$$\text{glyoxylate} + \text{acetyl-CoA} \xrightarrow{\text{malate synthase}} \text{malate} + \text{CoA-SH}$$

The glyoxylate cycle can be summarized as

$$2CH_3\text{-CO-CoA} + 3H_2O + FAD \longrightarrow C_4H_6O_5 + FADH_2 + 2\text{CoA-SH}$$

5.3.2.1 Regulation of the glyoxylate cycle

When an organism is growing on carbon sources that are not metabolized through pyruvate or PEP, the TCA cycle supplies energy while the glyoxylate cycle supplies precursors for biosynthesis. Genes for isocitrate lyase and malate synthase are transcribed with the accumulation of acetyl-CoA. The Cra (catabolite repressor/activator) protein is involved in this regulation in *Escherichia coli* (Section 12.1.3.2). The activity of isocitrate lyase is inhibited by PEP, succinate and pyruvate.

Since isocitrate is a branch point of the TCA cycle and the glyoxylate cycle, the activities of isocitrate lyase and isocitrate dehydrogenase which act on this common substrate should be regulated to control the flux. The bacteria solve the problem through differences in affinity for the substrate and by controlling the activity of the enzyme with the higher affinity. The dehydrogenase has a much higher affinity (K_m = 1–2 μM) for the substrate than that of the lyase (K_m = 3 mM). When the TCA cycle is needed to generate energy, isocitrate dehydrogenase is activated, but this enzyme is inactivated when precursors for biosynthesis should be synthesized through the glyoxylate cycle. An enzyme with kinase-phosphatase activity controls the activity of isocitrate dehydrogenase (Figure 5.6). The kinase-phosphatase removes phosphate from the inactive phosphorylated isocitrate dehydrogenase, to induce flux through the TCA cycle when metabolic intermediates such as isocitrate, PEP, OAA, 2-KG and 3-phosphoglycerate are in high concentration, and when more ATP is needed with the accumulation of AMP and ADP. Under the opposite conditions, the kinase-phosphatase phosphorylates the enzyme protein to inactivate it. When NADPH accumulates, isocitrate is directed towards the glyoxylate cycle. In *Escherichia coli*, genes for the kinase-phosphatase are located in the same operon as the genes for isocitrate lyase and malate synthase.

The TCA cycle is controlled by citrate synthase activity, and isocitrate dehydrogenase activity is regulated to control the glyoxylate cycle.

5.3.3 Ethylmalonyl-CoA pathway

Acetate supports the growth of a photosynthetic bacterium, *Rhodobacter sphaeroides*, that does not possess isocitrate lyase, the key enzyme of the glyoxylate cycle. This bacterium synthesizes malate and succinyl-CoA from 3

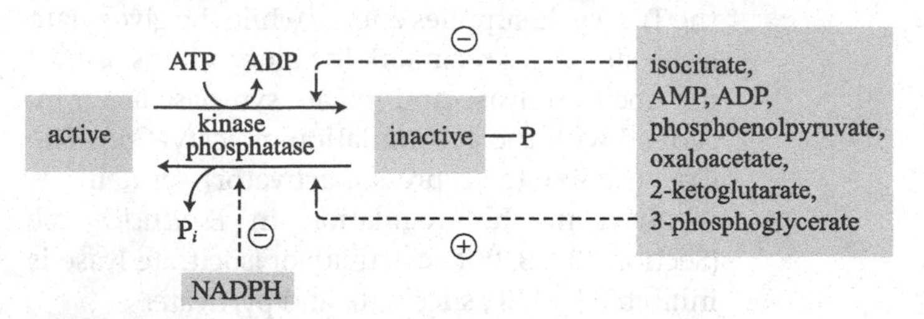

Figure 5.6 Control of isocitrate dehydrogenase activity by a kinase-phosphatase.

(Dawes, E. A. 1986, *Microbial Energetics*, Figure 3.10. Blackie & Son, Glasgow)

Since isocitrate dehydrogenase has a much higher affinity for the common substrate than the isocitrate lyase of the glyoxylate cycle, the former is inactivated when the cell needs precursors for biosynthesis. The enzyme is in an active form when precursors accumulate. Under the opposite conditions where the precursor concentration is high and ATP synthesis is required with the accumulation of AMP and ADP, the enzyme is in an active form to direct the flux to the TCA cycle. A kinase-phosphatase interchanges active and inactive forms of isocitrate dehydrogenase. The phosphorylated enzyme protein is inactive, and the free enzyme is active. When metabolic intermediates and AMP/ADP accumulate, isocitrate dehydrogenase is dephosphorylated for proper functioning of the TCA cycle. The kinase-phosphatase gene forms a single operon with genes for the glyoxylate cycle enzymes, isocitrate lyase and malate synthase.

acetyl-CoA and 2 CO_2 through the ethylmalonyl-CoA pathway (Figure 5.7). This pathway has been identified in many alphaproteobacteria, methylotrophic *Methylobacterium extorquens* and *Streptomyces coelicolor*.

This pathway has industrial potential as a new source of chemicals and enzymes that catalyse reactions to produce added-value products.

5.3.4 Methylaspartate cycle

A halophilic archaeon, *Haloarcula marismortui*, can grow on acetate as a sole carbon and energy source, but does not possess the glyoxylate cycle or ethylmalonyl-CoA pathway. This archaeon utilizes yet another cyclic metabolism to assimilate acetate, which is known as the methylaspartate cycle (Figure 5.8). Malate is produced from 2 acetyl-CoA as in the glyoxylate cycle. This cyclic metabolism combines reactions from different metabolic processes, including the glyoxylate cycle and the ethylmalonyl-CoA pathway. Acetyl-CoA is initially transformed into glutamate through reactions of the TCA cycle and glutamate dehydrogenase. Through a series of reactions, glutamate is converted to 3-methylmalyl-CoA that is metabolized to propionyl-CoA and glyoxylate as in the ethylmalonyl-CoA pathway. Propionyl-CoA carboxylation leads to methylmalonyl-CoA and subsequently to the citric acid cycle intermediate succinyl-CoA. Glyoxylate condensation with another acetyl-CoA molecule yields the final assimilation product malate.

The known haloarchaeal genomes have genes for either the glyoxylate cycle, the methylaspartate cycle or both, but not those for the ethylmalonyl-CoA pathway.

5.4 Incomplete TCA fork and reductive TCA cycle

The TCA cycle is an amphibolic metabolism serving both catabolic and anabolic needs. Parts of the TCA cycle, known as the incomplete TCA fork, are maintained in fermentative cells which cannot use NADH to produce ATP through oxidative phosphorylation in order to obtain precursors for biosynthesis. Though cyanobacteria synthesize ATP through photosynthetic electron transport (Section 11.4.3), they have a modified TCA cycle replacing 2-ketoglutarate dehydrogenase with 2-ketoglutarate decarboxylase and succinic semialdehyde dehydrogenase (Section 5.2.2). In some chemolithotrophs, CO_2 is fixed through the reductive TCA cycle (Section 10.8.2).

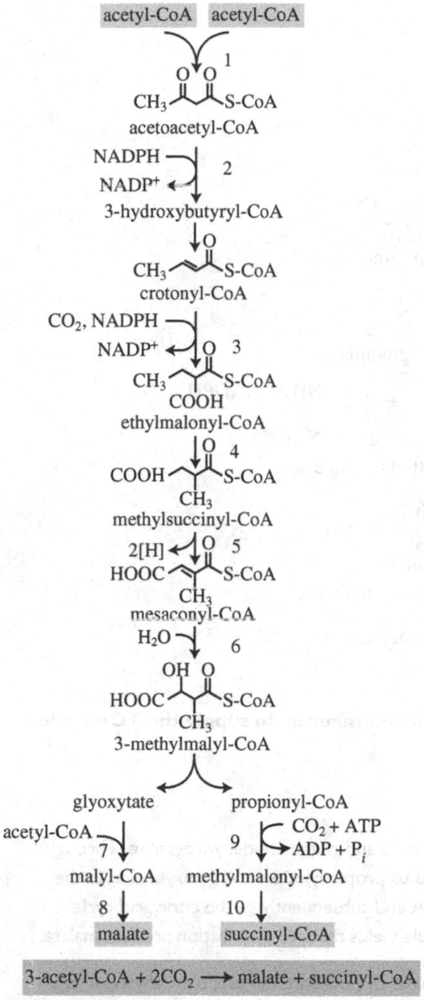

acetyl-CoA acetyl-CoA
1
acetoacetyl-CoA
NADPH → $NADP^+$ 2
3-hydroxybutyryl-CoA
crotonyl-CoA
CO_2, NADPH → $NADP^+$ 3
ethylmalonyl-CoA
4
methylsuccinyl-CoA
2[H] 5
mesaconyl-CoA
H_2O 6
3-methylmalyl-CoA
glyoxytate propionyl-CoA
acetyl-CoA 7 9 CO_2 + ATP → ADP + P_i
malyl-CoA methylmalonyl-CoA
8 10
malate succinyl-CoA

3-acetyl-CoA + $2CO_2$ ⟶ malate + succinyl-CoA

Figure 5.7 Ethylmalonyl-CoA pathway for assimilating acetyl-CoA in a photosynthetic bacterium, *Rhodobacter sphaeroides*.

(Modified from *Mol. Microbiol.* **73**: 992–1008, 2009)

This pathway has been identified in many alphaproteobacteria, methylotrophic *Methylobacterium extorquens* and *Streptomyces coelicolor*.

1, β-ketothiolase; 2, acetoacetyl-CoA reductase; 3, crotonyl-CoA carboxylase/reductase; 4, ethylmalonyl-CoA/methylmalonyl-CoA epimerase and ethylmalonyl-CoA mutase; 5, methylsuccinyl-CoA dehydrogenase; 6, mesaconyl-CoA hydratase; 7, β-methylmalonyl-CoA/malyl-CoA lyase (this enzyme catalyses the glyoxylate producing reaction as well as the subsequent consuming reaction); 8, malyl-CoA thioesterase; 9, propionyl-AoA carboxylase; 10, ethylmalonyl-CoA/methylmalonyl-CoA epimerase, ethylmalonyl-CoA mutase and methylmalonyl-CoA mutase.

5.4.1 Incomplete TCA fork

Some bacteria do not have a functional TCA cycle under certain growth conditions. Enteric bacteria, including *Escherichia coli*, do not synthesize 2-ketoglutarate dehydrogenase under fermentative conditions (Section 12.2.4) since NADH cannot be recycled through oxidative phosphorylation under these conditions. Precursors supplied by the TCA cycle are obtained through the incomplete TCA fork, consisting of an oxidative fork to produce 2-ketoglutarate and a reductive fork to synthesize succinyl-CoA (Figure 5.9).

Since succinate dehydrogenase cannot reduce fumarate to succinate, fumarate reductase replaces it in the incomplete TCA fork. Fumarate reductase is one of the anaerobic enzymes expressed under anaerobic conditions in *Escherichia coli* under control by the regulatory protein FNR. The activities of the other TCA cycle enzymes are lower under anaerobic conditions than in aerobic conditions (Section 12.2.4).

5.4.2 Reductive TCA cycle

The Calvin cycle is employed in most chemolithotrophs to fix CO_2 (Section 10.8.1), but enzymes of the Calvin cycle are not found in some chemolithotrophs, including photosynthetic green sulfur bacteria, a chemolithotrophic bacterium (*Hydrogenobacter thermophilus*) and an archaeon (*Sulfolobus acidocaldarius*). These fix CO_2 through a reversal of the TCA cycle which is referred to as the reductive TCA cycle (Section 10.8.2). As mentioned above, three TCA cycle enzymes cannot catalyse the reverse reactions and separate enzymes substitute for them:

- ATP-citrate lyase: substitutes for citrate synthase
- 2-ketoglutarate:ferredoxin oxidoreductase: substitutes for 2-ketoglutarate dehydrogenase
- fumarate reductase: substitutes for succinate dehydrogenase.

This mechanism of CO_2 fixation is known only in prokaryotes. Other CO_2 fixation pathways known only in prokaryotes are the acetyl-CoA pathway (carbon monoxide dehydrogenase pathway), the 3-hydroxypropionate pathway and the

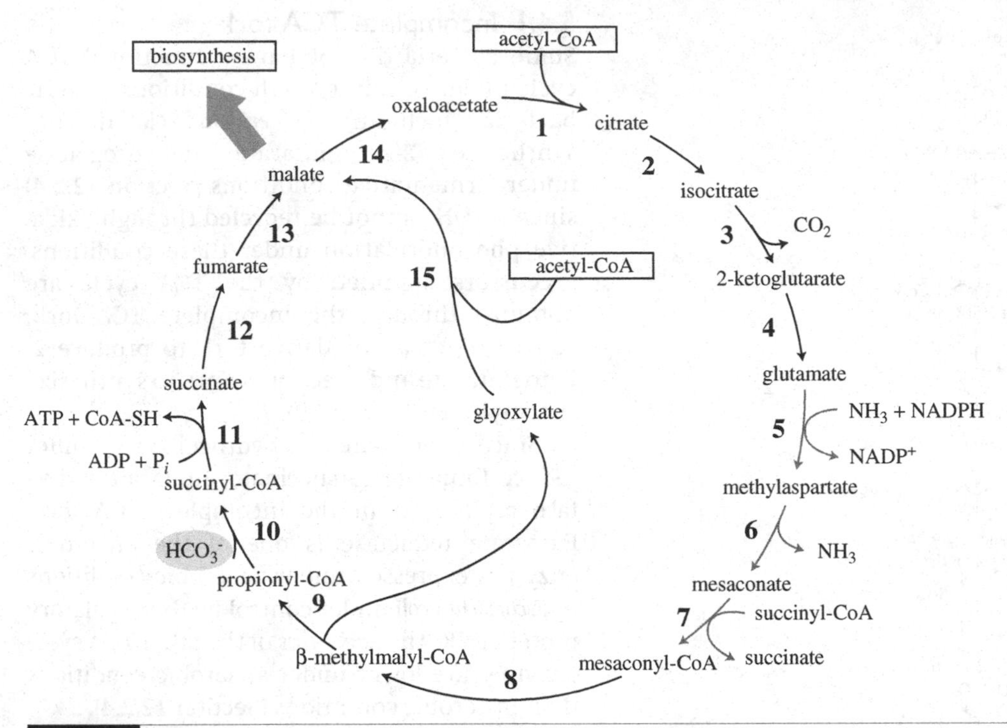

Figure 5.8 The methylaspartate cycle in a halophilic archaeon, *Haloarcula marismortui*, to supply the TCA cycle intermediate, malate, when growing on acetate.

(Modified from *Science* **331**: 334–337, 2011)

Acetyl-CoA is initially transformed into glutamate through reactions of the citric acid cycle and glutamate dehydrogenase. Through a series of reactions, glutamate is converted to β-methylmalyl-CoA that is metabolized to propionyl-CoA and glyoxylate, as in the ethylmalonyl-CoA pathway. Propionyl-CoA carboxylation leads to methylmalonyl-CoA and subsequently to the citric acid cycle intermediate succinyl-CoA. Glyoxylate condensation with another acetyl-CoA molecule yields the final assimilation product malate.

1, citrate synthase; 2, aconitase; 3, isocitrate dehydrogenase; 4, glutamate dehydrogenase; 5, glutamate mutase; 6, methylaspartate ammonia lyase; 7, CoA transferase; 8, mesaconyl-CoA dehydratase; 9, β-methylmalylnyl-CoA lyase; 10, propionyl-CoA carboxylase, ethylmalonyl-CoA/methylmalonyl-CoA epimerase, ethylmalonyl-CoA mutase and methylmalonyl-CoA mutase; 11, succinate thiokinase; 12, succinate dehydrogenase; 13, fumarase; 14, malate dehydrogenase; 15, malate synthase.

4-hydroxybutyrate cycles. These will be discussed in detail later (Section 10.8).

5.5 Energy transduction in prokaryotes

Life can be thought of as a process that transforms materials available from the environment into cellular components according to genetic information. Material transformation is also coupled to energy transduction. Energy is needed not only for growth and reproduction but also for the maintenance of viability in processes that include biosynthesis, transport, motility and many others.

Organisms use energy sources available in their environment. Light and chemical energies are converted into biological energy for growth and maintenance of viability. Photosynthesis (Chapter 11) is the process where light energy is utilized, and chemical energy is used through fermentation (Chapter 8) and respiration (Section 5.8 and Chapter 9). Organic compounds produced by photosynthesis are used by other organisms in fermentation and respiration. For

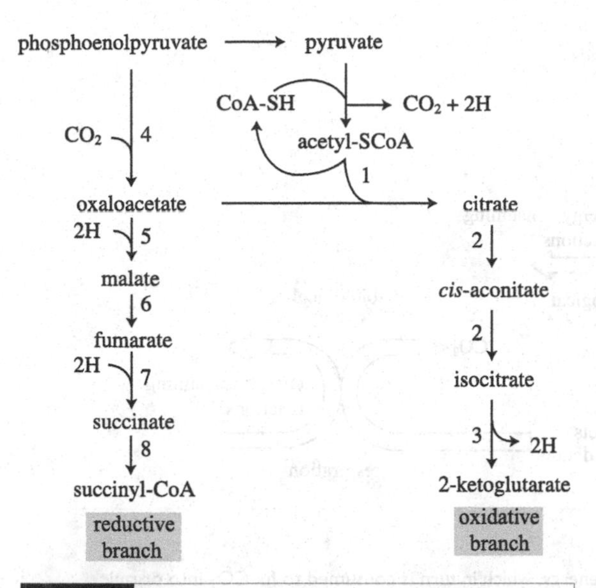

Figure 5.9 Incomplete TCA fork to supply precursors for biosynthesis in organisms that do not synthesize 2-ketoglutarate dehydrogenase.

(Dawes, E. A. 1986, *Microbial Energetics*, Figure 3.8. Blackie & Son, Glasgow)

The TCA cycle is an amphibolic metabolism generating reducing equivalents to synthesize ATP through oxidative phosphorylation as the catabolic function, and biosynthetic precursors as the anabolic function. Fermentative anaerobic bacteria are unable to recycle NADH through oxidative phosphorylation under electron acceptor-limited conditions. They do not have 2-ketoglutarate dehydrogenase activity and operate an incomplete TCA fork instead of a functional TCA cycle to supply biosynthetic precursors.

1, pyruvate dehydrogenase and citrate synthase; 2, aconitase; 3, isocitrate dehydrogenase; 4, PEP carboxylase; 5, malate dehydrogenase; 6, fumarase; 7, fumarate reductase; 8, succinyl-CoA synthetase.

this reason, photosynthesis is referred to as primary production. Reduced inorganic compounds are also used as energy sources in chemolithotrophs (Chapter 10).

Figure 5.10 shows energy transduction processes in biological systems. In these processes, free energy is conserved in the exergonic (free energy producing) reactions and consumed in the endergonic (free energy consuming) reactions. To understand these biological reactions in terms of thermodynamics, the relationship between the biological reactions and the free energy change should be understood.

5.5.1 Free energy

The free energy change in an exergonic reaction is expressed as a negative figure, and a positive figure is used to describe an endergonic reaction, since free energy leaves the system in an exergonic reaction while the system gains free energy in an endergonic reaction. The free energy change depends on the conditions of a given reaction. The standard condition is defined as the concentration of the reactants and products in one activity unit (when all are in active forms, a 1 M concentration is the same as one activity unit) and at 25°C for convenience. The free energy change at standard conditions is expressed as ΔG^0, and $\Delta G^{0'}$ and ΔG are used to describe the free energy changes under standard conditions at pH = 7, and at the given conditions, respectively. Since physiological pH is neutral, $\Delta G^{0'}$ is a frequently used term in biology. $\Delta G^{0'}$ can be calculated in various ways.

5.5.1.1 $\Delta G^{0'}$ from the free energy of formation

The free energy of formation ($\Delta G_f^{0'}$) of common compounds can be found in most chemical data handbooks. $\Delta G^{0'}$ is calculated from $\Delta G_f^{0'}$ using the following equation:

$$\Delta G^{0'} = \Sigma\Delta G_f^{0'} \text{ of products} - \Sigma\Delta G_f^{0'} \text{ of reactants}$$

For example, $\Delta G^{0'}$ is calculated in the reaction of glucose oxidation as:

$$C_6H_{12}O_6 + 6O_2 \longrightarrow 6CO_2 + 6H_2O$$

where the free energy of formation of each component is:

$$\Delta G_f^{0'} \text{glucose} = -917.22 \text{ kJ}$$

$$\Delta G_f^{0'} O_2 = 0 \text{ kJ}$$

$$\Delta G_f^{0'} H_2O = -237.18 \text{ kJ}$$

$$\Delta G_f^{0'} CO_2 = -386.02 \text{ kJ}$$

$$\Delta G^{0'} = [(-237.18 \times 6) + (-386.02 \times 6)] - (-917.22) = -2821.98 \text{ kJ/mol glucose}$$

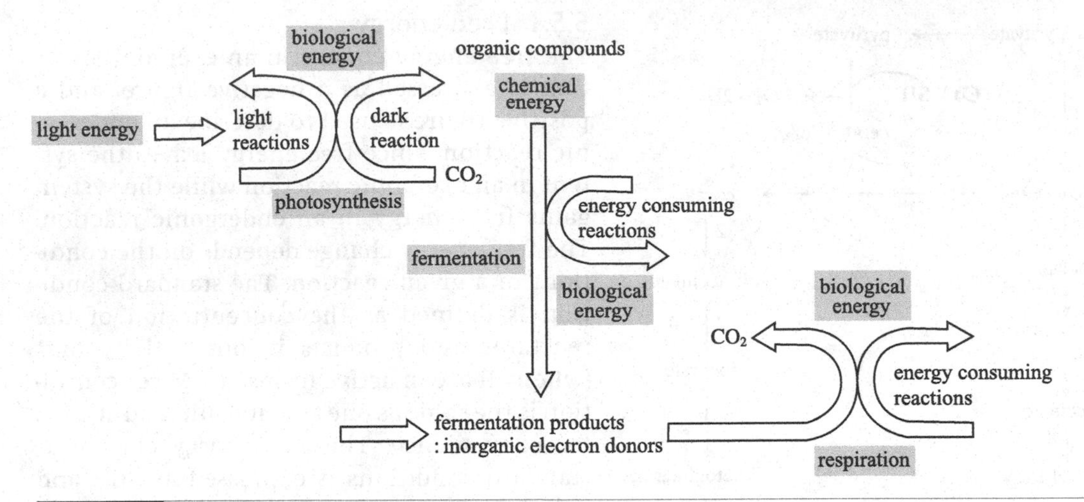

Figure 5.10 Biological energy transduction processes.

Photosynthesis: a process converting light energy into biological energy, which in turn is consumed to fix CO_2 into organic compounds. Fermentation: a process converting chemical energy into biological energy without the external supply of electron acceptors. Respiration: a process converting chemical energy into biological energy by oxidizing organic and inorganic electron donors, coupled with the reduction of externally supplied electron acceptors.

5.5.1.2 $\Delta G^{0'}$ from the equilibrium constant

The equilibrium constant ($K^{0'}_{eq}$) of a reaction, $A + B \rightleftharpoons C + D$, is expressed as

$$K^{0'}_{eq} = ([C][D]/[A][B])$$

$K^{0'}_{eq}$ shows in which direction the reaction takes place at pH = 7.0 under standard conditions.

When $K^{0'}_{eq} > 1.0$, $\Delta G^{0'} < 0$ and the reaction proceeds in the right direction.

When $K^{0'}_{eq} = 1.0$, $\Delta G^{0'} = 0$ and the velocity is the same in both directions.

When $K^{0'}_{eq} < 1.0$, $\Delta G^{0'} > 0$ and the reaction proceeds in the reverse direction.

The equilibrium constant can be used to calculate $\Delta G^{0'}$ using the following equation:

$$\Delta G^{0'} = -2.303\ RT \log K^{0'}_{eq}$$

R = gas constant(8.314 J/mol · K)
T = temperature in K (25°C = 298 K)

5.5.1.3 ΔG from $\Delta G^{0'}$

ΔG^0 and $\Delta G^{0'}$ express free energy at reactant and product concentrations of one activity unit (= 1 M). However, the concentrations inside a cell are much lower than that. ΔG at a given concentration of reactants and products can be calculated from $\Delta G^{0'}$ in a reaction of $A + B \rightleftharpoons C + D$ using

$$\Delta G = \Delta G^{0'} + 2.303RT \log([C][D]/[A][B])$$

Assuming that the concentrations of ATP, ADP and P_i are 2.25, 0.25 and 1.65 mM, respectively, ΔG of ATP hydrolysis to ADP and P_i can be calculated from $\Delta G^{0'} = -30.5$ kJ/mol ATP:

$$\Delta G = -30.5 + (8.314 \times 298 \times 2.303) \times \log([2.5 \times 10^{-3}]/[2.25 \times 10^{-3}]) = -51.8 \text{ kJ/mol ATP}$$

The figures used in the calculation are close to those values found in an actively growing bacterial culture. The free energy change of ATP hydrolysis under physiological conditions is referred to as the phosphorylation potential and is expressed as ΔGp (Section 5.6.3).

5.5.1.4 $\Delta G^{0'}$ from ΔG^0

ΔG^0 of a reaction where H^+ is involved is the free energy change at a H^+ concentration of 1 M

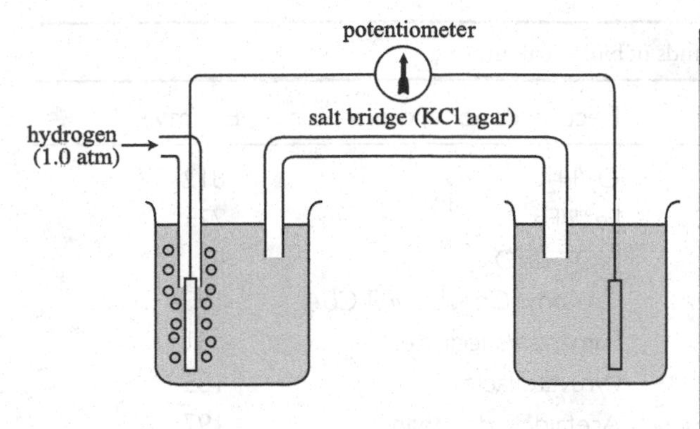

Figure 5.11 Determination of redox potential.

A vessel is filled with a solution containing 1M of each of the reduced and oxidized forms of a compound of known redox potential, and the other vessel with 1M of each of reduced and oxidized forms of the test compound. A platinum electrode is placed in each vessel. These two vessels are connected with a KCl salt bridge, and the electrodes with a potentiometer. Because of the differences in the tendency to transfer electrons to the platinum electrode in each vessel, a potential is developed. The potentiometer gives the difference in redox potential of the compounds. A solution of 1M H^+ (oxidized form) gassed with 1 atm H_2 (reduced form) is arbitrarily defined as a reaction that has a redox potential of 0 V.

(pH = 0). Biologists are more interested in $\Delta G^{0'}$ than ΔG^0. $\Delta G^{0'}$ can be calculated from ΔG^0 using

$$\Delta G^{0'} = \Delta G^0 - 2.303RT \times 7$$

5.5.2 Free energy of an oxidation/reduction reaction

Energy is generated from an oxidation/reduction reaction. Respiration is a series of oxidation/reduction reactions. The energy from respiratory oxidation/reduction reactions is conserved in biological systems. The amount of energy generated from a reaction is proportional to the difference in the oxidation/reduction potential of the reductant and oxidant.

5.5.2.1 Oxidation/reduction potential

Oxidation is defined as a reaction which loses electron(s), and reduction as a reaction that gains electron(s). Since an electron cannot 'float' in solution, oxidation and reduction reactions are coupled. A given compound can be an oxidant in one reaction and a reductant in another reaction. This property depends on the affinity of the compound for electrons, which is relative to that of other compounds. The affinity for electrons is referred to as the oxidation/reduction (redox) potential. The higher the affinity for electrons, the higher the redox potential.

Arbitrarily, the redox potential of the half reaction, $\frac{1}{2}H_2 \rightleftharpoons H^+ + e^-$, is defined as 0 V, and the relative values to this reaction are expressed as the redox potential of a given half reaction (Figure 5.11). The redox potential at standard conditions is expressed as E^0 and that at standard conditions, pH 7.0, as $E^{0'}$.

Biologists are often interested in $E^{0'}$, since physiological pH is neutral. $E^{0'}$ can be calculated from E^0 using the following equation:

$$E^{0'} = E^0 - 2.303(RT/nF) \times 7$$

R: gas constant (8.314 J/mol · K)
T= temperature in K (25°C = 298 K)
n: number of electrons involved in the reaction
F: Faraday constant (96487 J/V · mol)

$E^{0'}$ of the H^+/H_2 half reaction is calculated as

$$E^{0'} = E^0 - 2.303 \times (8.314 \times 298/1 \times 96487) \times 7$$
$$= -0.41 \text{ V}$$

When a temperature of 30°C is used, the $E^{0'}$ is calculated as 0.42 V. $E^{0'}$ values of some half reactions of biological interest are listed in Table 5.1.

5.5.2.2 Free energy from $\Delta E^{0'}$

Energy is generated from oxidation/reduction reactions, and the amount of energy is directly proportional to the redox potential difference between the reductant and the oxidant ($\Delta E^{0'}$). Free energy from an oxidation/reduction reaction can be calculated using

Table 5.1 Oxidation/reduction potential of compounds of biological interest.

Electron carrier	$E^{0\prime}$ (mV)	Electron donor and acceptor	$E^{0\prime}$ (mV)
Cytochrome f^a	365	O_2/H_2O	812
Cytochrome a^a	290	Fe^{3+}/Fe^{2+}	771
Cytochrome c^a	254	NO_3^-/NO_2	421
Ubiquinone/ubiquinol	113	Crotonyl-CoA/butyryl-CoA	190
Cytochrome b^a	77	Fumarate/succinate	31
Rubredoxin[a]	–57	Pyruvate/lactate	–185
$FMN/FMNH_2$	–190	Acetaldehyde/ethanol	–197
Cytochrome $c_3{}^a$	–205	Acetoin/2,3-butanediol	–244
$FAD/FADH_2$	–219	Acetone/isopropanol	–286
Glutathione[a]	–230	CO_2/formate	–413
NAD(P)/NAD(P)H	–320	H^+/H_2	–414
Ferredoxin[a]	–413	Gluconate/glucose	–440
		CO_2/CO	–540
		Acetate/acetaldehyde	–581
Artificial electron carrier	**$E^{0\prime}$ (mV)**		
Toluidine[a]	224	Janus green[a]	–225
DCPIP[a]	217	Neutral red[a]	–325
Phenazine methosulfate[a]	80	Benzyl viologen[a]	–359
Methylene blue[a]	11	Methyl viologen[a]	–446

[a] oxidized form/reduced form.
DCPIP, 2, 6-dichlorophenolindophenol.

$$\Delta G^{0\prime} = -nF\Delta E^{0\prime}$$

n = number of electrons involved in the reaction
F = Faraday constant (96487 J/V · mol)
$\Delta E^{0\prime}$ = oxidation/reduction potential difference between the reductant and oxidant

The $\Delta G^{0\prime}$ of succinate oxidation to fumarate with molecular oxygen can be calculated using this equation, as follows:

$$\text{succinate} + \tfrac{1}{2}O_2 \longrightarrow \text{fumarate} + H_2O$$

$E^{0\prime}$ of fumarate/succinate = +0.03 V
$E^{0\prime}$ of $\tfrac{1}{2}O_2/H_2O$ = +0.82 V
$\Delta G^{0\prime} = -2 \times 96487 \times (0.82 - 0.03)$
$= -152.45$ kJ/mol succinate

5.5.3 Free energy of osmotic pressure

Active transport can concentrate a nutrient in the cell by as much as over 100 times the external concentration (Section 3.3). Such a concentration difference produces an osmotic pressure across the semi-permeable cytoplasmic membrane. This pressure can be expressed as free energy as in the following equation. The free energy developed by the osmotic pressure is the energy needed to transport one mol of the solute under the given conditions:

$$\Delta G^{0\prime} = -2.303RT \log[S]_i/[S]_o$$

R = gas constant (8.314 J/mol · K)
T = temperature in K (25°C = 298 K)
$[S]_i$ and $[S]_o$ = solute concentration inside and outside the cell

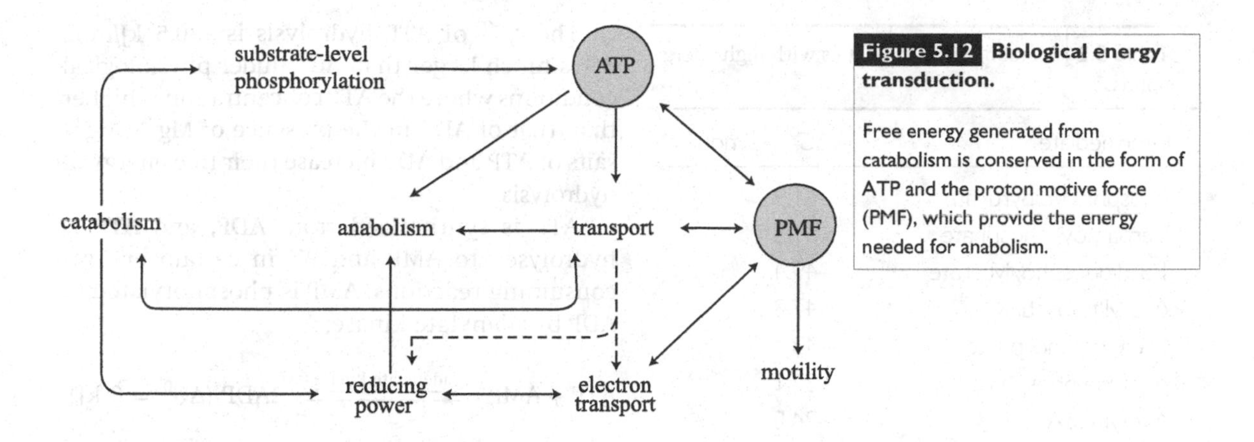

Figure 5.12 Biological energy transduction.

Free energy generated from catabolism is conserved in the form of ATP and the proton motive force (PMF), which provide the energy needed for anabolism.

5.5.4 Sum of free energy change in a series of reactions

Excessive heat would be fatal to the cell. Therefore, most biological reactions are catalysed in multiple steps. The sum of the free energy changes in each step is the same as that obtained in a one-step reaction. If a compound is metabolized through a different series of reactions, the free energy change is constant if the final product(s) are the same. For example, when glucose is metabolized to two pyruvates either through the EMP pathway or the ED pathway, the free energy change is the same.

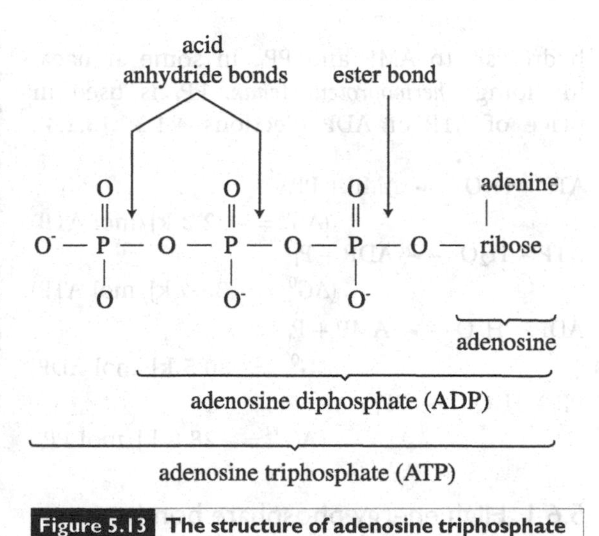

Figure 5.13 The structure of adenosine triphosphate (ATP).

5.6 Role of ATP in the biological energy transduction process

Biological reactions are divided into energy-generating catabolism and energy-consuming anabolism. Free energy generated from catabolism is conserved in the form of adenosine triphosphate (ATP) and the proton motive force (Section 5.7) which are consumed in catabolism (Figure 5.12). In this sense, it can be said that ATP and the proton motive force play a central role in biological metabolism, linking catabolism and anabolism. ATP is used to supply energy for biosynthesis and transport by the ABC pathway (Section 3.4), while active transport, motility and reverse electron transport processes (Section 10.1) consume the proton motive force. ATP is well suited for this role since the energy needed for its synthesis and released from its hydrolysis is smaller than the energy available from most of the energy-generating catabolic reactions and bigger than most of the energy-consuming anabolic reactions (Section 5.6.1). In addition, ATP is a general intermediate for nucleic acid biosynthesis.

As shown in Figure 5.13, ATP comprises adenosine with three phosphates bound to the 5′-carbon of the ribose residue. The phosphate groups are termed α, β and γ from the group nearest to ribose. ATP is hydrolysed as shown below, with the release of free energy. In most cases, ATP hydrolysis is coupled to energy-consuming reactions. However, pyrophosphate (PP_i) is hydrolysed without energy conservation in most cases by pyrophosphatase to pull the energy-consuming reactions coupled to ATP

Table 5.2 Metabolic intermediates with high energy bonds.

Intermediate	$\Delta G^{0'}$ (kJ/mol)
Phosphoenolpyruvate	−61.9
Carbamoyl phosphate	−51.5
1,3-diphosphoglycerate	−49.4
Acetyl phosphate	−47.3
Creatine phosphate	−43.1
Arginine phosphate	−38.1
Acetyl-CoA	−34.5

hydrolysis to AMP and PP_i. In some archaea, including *Thermoproteus tenax*, PP_i is used in place of ATP or ADP (Sections 4.1.3, 13.2.4).

$$ATP + H_2O \rightarrow AMP + PP_i \quad (\Delta G^{0'} = -32.2 \text{ kJ/mol ATP})$$

$$ATP + H_2O \rightarrow ADP + P_i \quad (\Delta G^{0'} = -30.5 \text{ kJ/mol ATP})$$

$$ADP + H_2O \rightarrow AMP + P_i \quad (\Delta G^{0'} = -30.5 \text{ kJ/mol ADP})$$

$$PP_i + H_2O \rightarrow 2P_i \quad (\Delta G^{0'} = -28.8 \text{ kJ/mol } PP_i)$$

5.6.1 High energy phosphate bonds

It was mentioned previously that free energy is released on the hydrolytic removal of γ-phosphate from ATP and β-phosphate from ADP. For this reason, γ–β and β–α phosphate linkages in ATP are called high energy bonds. There are several other metabolic intermediates with high energy bonds (Table 5.2).

Phosphoenolpyruvate and 1,3-diphosphoglycerate are EMP pathway intermediates and are used to synthesize ATP from ADP in substrate-level phosphorylation (SLP). SLP processes are exergonic reactions because the free energy change ($\Delta G^{0'}$) from the reaction is bigger than that for ATP synthesis. Many ATP-consuming anabolic reactions are exergonic. Since both catabolism and anabolism coupled to ATP synthesis and hydrolysis are exergonic reactions, ATP-mediated metabolic reactions are thermodynamically favourable.

The $\Delta G^{0'}$ of ATP hydrolysis is −30.5 kJ/mol. ΔG is much larger than $\Delta G^{0'}$ under physiological conditions where the ATP concentration is higher than that of ADP in the presence of Mg^{2+}. Mg^{2+} salts of ATP and ADP increase their free energy of hydrolysis.

ATP is synthesized from ADP, and ATP is hydrolysed to AMP and PP_i in certain energy-consuming reactions. AMP is phosphorylated to ADP by adenylate kinase:

$$ATP + AMP \xrightleftharpoons{\text{adenylate kinase}} 2ADP \ (\Delta G^{0'} = 0 \text{ kJ})$$

Since the $\Delta G^{0'}$ of this reaction is 0 kJ, the direction is determined by the concentration of cellular ATP, ADP and AMP, which reflects the energy status of the cell.

5.6.2 Adenylate energy charge

A high ATP concentration means that the energy status is good, and when the energy supply cannot meet demand, the ADP and AMP concentration is high. Arbitrarily, the adenylate energy charge (EC) is a term used to describe the energy status of a cell. EC can be calculated using the following equation:

$$EC = ([ATP] + ½[ADP])/([ATP]+[ADP]+[AMP])$$

The numerator of this equation is the sum of high energy phosphate bonds in the form of ATP, and the denominator the concentration of the total adenylate pool. An EC number of 1 means that the total adenylate is in the ATP form, and 0, in the AMP form.

Since the reaction catalysed by adenylate kinase is reversible with a $\Delta G^{0'}$ of 0 kJ/mol, and enzyme activity is high in a cell, the relative concentrations of ATP, ADP and AMP are determined by the EC (Figure 5.14). Many catabolic reactions are repressed by ATP and activated by ADP and/or AMP: anabolic reactions are controlled in a reverse manner (Section 12.4). The reactions are regulated not by the absolute concentration of each adenosine nucleotide, but by their ratio. Thus, overall metabolism is controlled by the EC value (Figure 5.15).

When a bacterium uses acetate as the sole carbon and energy source, acetate should be

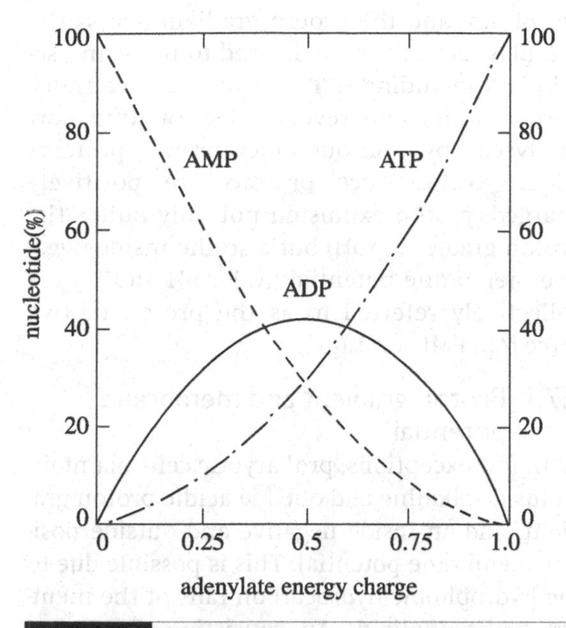

Figure 5.14 Relationship of adenylate energy charge (EC) and the relative concentrations of ATP, ADP and AMP.

(Dawes, E. A. 1986, *Microbial Energetics*, Figure 2.2. Blackie & Son, Glasgow)

Since $\Delta G^{0'}$ of the reaction catalysed by adenylate kinase is 0 kJ/mol, the direction of the reaction will be determined by the relative concentrations of ATP, ADP and AMP, which can be expressed by the EC value.

metabolized through the TCA cycle in catabolism and through the glyoxylate cycle in anabolism, to supply carbon compounds for biosynthesis (Section 5.3.2). It has been mentioned that AMP and ADP activate isocitrate dehydrogenase to metabolize the substrate through the TCA cycle (Figure 5.6, Section 5.3.2.1). The activity is not controlled by AMP and ADP per se, but by the EC value, activated at EC < 0.8 and inactivated at EC > 0.8. An anabolic enzyme, aspartate kinase (Section 6.4.1), is activated at a high EC value and repressed at a low EC value. When a bacterial culture is transferred from a rich medium to a poor medium, AMP is excreted or hydrolysed to adenosine or adenine to maintain a high EC value at a low rate of ATP synthesis.

A growing bacterial culture maintains an EC value of 0.8–0.95, and the value gradually decreases to around 0.5 and rapidly thereafter when the culture starves. Bacterial cultures

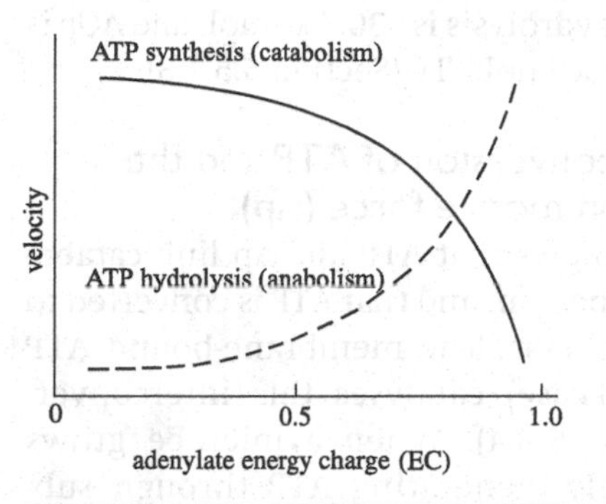

Figure 5.15 Regulation of catabolism and anabolism by the adenylate energy charge (EC).

(Dawes, E. A. 1986, *Microbial Energetics*, Figure 2.3. Blackie & Son, Glasgow)

Adenylate energy charge controls the overall growth of microbes, regulating catabolism which synthesizes ATP, and anabolism which consumes it.

with EC values less than 0.5 cannot form colonies. This is not surprising because ATP is essential for viability. Since the EC value is a unitless figure, it does not give any information on the size of the adenylate pool, the concentration of each adenosine nucleotide or the turnover velocity. Since the EC value controls the overall metabolism that is observed during growth, similar EC values are expected in fast-growing and slow-growing cultures, though a fast-growing culture has a bigger adenylate pool and a higher ATP turnover rate than a slow-growing culture.

5.6.3 Phosphorylation potential (ΔGp)

Free energy needed for ATP synthesis or released by its hydrolysis in the cell is referred to as the phosphorylation potential (ΔGp), which is determined by the concentration of ATP, ADP and P_i as in the following equation (Section 5.5.1.3):

$$\Delta Gp = \Delta G^{0'} + 2.303 \log([ADP][P_i]/[ATP])$$

In addition to the concentration of each adenosine nucleotide, ΔGp depends on the concentration of metal ions such as Mg^{2+} which bind the nucleotide. Binding of metal ions increases the ΔGp value. The concentrations of metal ions and P_i vary according to the growth conditions.

$\Delta G^{0'}$ for ATP hydrolysis is –30.5 kJ/mol, and ΔGp is around –51.8 kJ/mol ATP (Section 5.5.1.3).

5.6.4 Interconversion of ATP and the proton motive force (Δp)

Figure 5.12 shows that ATP and Δp link catabolism and anabolism, and that ATP is converted to Δp, and vice versa. The membrane-bound ATP synthase (ATPase) catalyses this interconversion (Section 5.8.4). When a microbe grows fermentatively, generating ATP through substrate-level phosphorylation, ATP is hydrolysed to increase the Δp. In contrast, Δp is consumed to synthesize ATP when respiration is the main energy conservation process. ATP and Δp are therefore consumed for different purposes (Figure 5.12).

5.6.5 Substrate-level phosphorylation (SLP)

ATP is synthesized in the cytoplasm as a result of the transfer of phosphate from metabolic intermediates with high energy phosphate bonds to ADP. 1,3-diphosphoglycerate, phosphoenolpyruvate and acyl-phosphate are the metabolic intermediates used to synthesize ATP through SLP. Succinyl-CoA conversion to succinate in the TCA cycle is another example of SLP and is catalysed by succinyl-CoA synthetase:

$$\text{succinyl-CoA} + P_i + \text{ADP(GDP)} \xrightarrow{\text{succinyl-CoA synthetase}} \text{succinate} + \text{ATP(GTP)} + \text{CoA-SH}$$

5.7 Proton motive force (Δp)

Chemical and light energies are converted to biological forms of energy through photosynthesis, fermentation and respiration (Figure 5.10). Free energy generated from oxidation–reduction reactions is conserved in the form of ATP through photosynthesis, as in respiration. This is referred to as electron transport phosphorylation (ETP) or oxidative phosphorylation. The free energy from ETP is coupled to the expulsion of protons from the cytoplasm to the periplasm, and the proton gradient across the cytoplasmic membrane is used to perform useful work including ATP synthesis, active transport, motility and reverse electron transport catalysed by various membrane proteins (Figure 5.12). Since protons are positively charged, proton expulsion not only builds the proton gradient (ΔpH) but also the inside-negative membrane potential (Δψ). ΔpH and Δψ are collectively referred to as the proton motive force (Δp, PMF, or $\Delta\tilde{\mu}_{H^+}$).

5.7.1 Proton gradient and membrane potential

With few exceptions, prokaryotic cells maintain an inside alkaline and outside acidic proton gradient, and an inside negative and outside positive membrane potential. This is possible due to the hydrophobic hydrocarbon tails of the membrane phospholipids. Δp, consisting of ΔpH and Δψ, is expressed in volts as:

$$\Delta p = \Delta + Z\Delta pH$$

Δ = membrane potential (mV)
ΔpH = pH gradient
$Z = 2.303\ RT/F$
R = gas constant
T = temperature in K (25°C = 298 K)
F = Faraday constant.

Bacterial cells maintain a Δp of 0.15–0.20 V. ETP generates a Δp that is used to synthesize ATP. In a fermentative cell, ATP synthesized through SLP is hydrolysed to develop the Δp (Figure 5.16). Most of the enzymes in the cytoplasm have an optimum pH of around neutrality while polymers, including DNA, RNA and proteins, are unstable at extremes of pH. For these reasons, the pH of the cytoplasm is maintained at around neutrality regardless of the external pH.

5.7.2 Acidophiles and alkaliphiles

Many microbes grow best at neutral pH, but some have their optimum pH on the extreme acidic or alkaline side. These are referred to as neutrophiles, acidophiles and alkaliphiles, respectively. All of them maintain an internal pH near to neutrality. Neutrophiles possess mechanisms to tolerate extreme pH values. When a neutrophilic bacterium, *Salmonella enterica* var. Typhimurium,

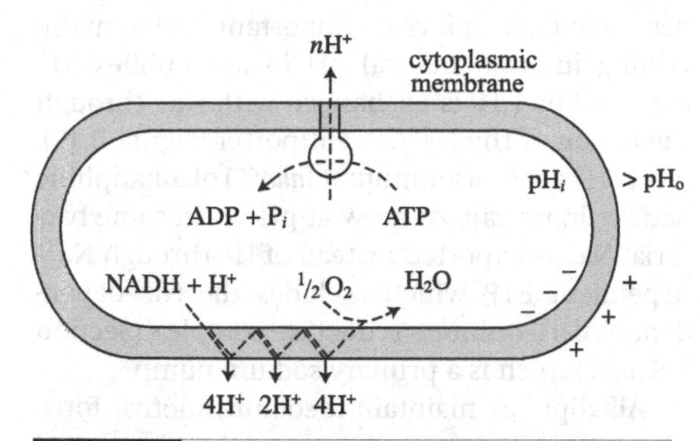

Figure 5.16 Proton motive force.

The proton motive force (Δp) is a biological form of energy facilitating ATP synthesis, transport and motility. Δp consists of an inside alkaline and outside acidic proton gradient, and an inside negative and outside positive membrane potential. Electron transport phosphorylation (ETP) during respiration and photosynthesis builds Δp (lower part). Δp is consumed to synthesize ATP. In a fermentative cell, the membrane-bound ATPase hydrolyses ATP to export H^+ which develops the Δp (upper part).

is acclimatized to a mild acidic pH (pH 5.8), over 50 acid shock proteins are synthesized to maintain the internal pH at neutrality. These acid shock proteins render the bacterium able to tolerate a more acidic pH of 3.0. Among the acid shock proteins, amino acid decarboxylases are the best known. At acidic pH in the presence of lysine, the bacterium synthesizes lysine decarboxylase (CadA) and the lysine/cadaverine antiporter (CadB). CadA decarboxylates lysine to cadaverine, consuming H^+ to raise the internal pH. CadB exports cadaverine in exchange for lysine. The CadA/CadB system is also known in other bacteria including *Escherichia coli, Vibrio cholerae* and *Bacillus cereus*. Glutamate is used in a similar way in some Gram-negative bacteria including *Escherichia coli, Proteus mirabilis* and *Yersinia enterocolitica*, and in Gram-positive bacteria such as *Listeria monocytogenes, Lactococcus lactis, Streptococcus* spp. and *Clostridium* spp. Arginine and ornithine are also used for the same purpose in *Escherichia coli, Salmonella enterica* var. Typhimurium and *Bacillus cereus*.

Urea is imported and hydrolysed by urease in the cytoplasm, yielding NH_3 and CO_2 when it is available in urease-positive bacteria. NH_3 production results in the consumption of protons. *Helicobacter pylori* is best known for its ability to grow at acidic pH values with urease. *Proteus mirabilis, Yersinia enterocolitica* and *Streptococcus salivaris* utilize urea for the same purpose.

Under acidic conditions, the expression of respiratory chain complexes increases to pump protons out of the cell more efficiently in respiratory bacteria. In non-respiratory bacteria F_1F_O-ATPase activity (Section 5.8.4) is upregulated to promote ATP-dependent H^+ efflux under acidic conditions. These responses against acidic pH are controlled by two-component systems (Section 12.1.7) such as ArsRS in *H. pylori*.

Many bacteria synthesize a Na^+/H^+ antiporter at alkaline pH values to convert the proton motive force to a sodium motive force, to maintain a neutral internal pH as in alkaliphiles (Section 5.7.4). A $K^+(Na^+)/H^+$ antiporter is responsible for adaptation of *Vibrio cholerae* to low pH. This antiporter functions as an electroneutral alkali cation/proton antiporter, importing K^+ or Na^+ coupled to the export of H^+.

5.7.3 Proton motive force in acidophiles

Acidithiobacillus ferrooxidans, Helicobacter pylori, Thermoplasma acidophilum and *Sulfolobus acidocaldarius* are acidophiles with an optimum pH of 1–4. These microbes maintain their internal pH around neutrality with a H^+ gradient of 10^3–10^6. They maintain a low or inside positive membrane potential to compensate for the large potential rendered by the H^+ gradient (Figure 5.17). The inside positive membrane potential gives an additional benefit to the organisms by preventing H^+ leakage through the membrane due to the large concentration gradient.

Fatty acids such as acetate inhibit the growth of acidophiles at low concentrations of 5–10 mM. Their optimum pH is lower than the pK_a of the fatty acids. The pK_a of acetate is 4.8. Fatty acids at this pH are mainly in undissociated forms, which are hydrophobic and thus permeable to the membrane phospholipids. The fatty acids diffuse into the cell and dissociate, lowering the internal pH. They function as protonophores (uncouplers, Section 5.8.5). The accumulation of fatty acids in the cell increases turgor pressure, which may provide another reason for growth inhibition.

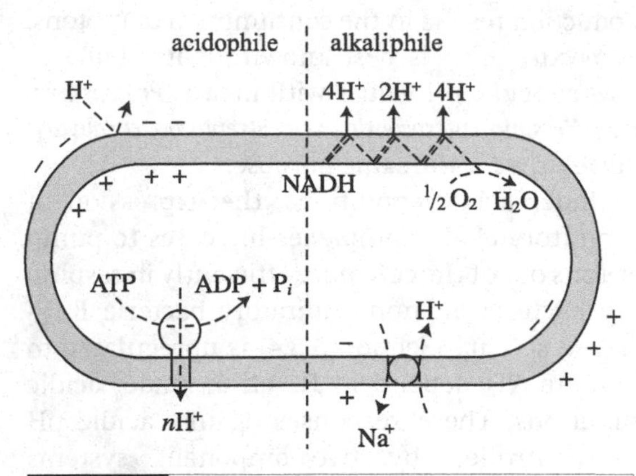

Figure 5.17 Proton motive force in acidophiles and alkaliphiles.

Acidophiles maintain a neutral internal pH at an external acidic pH with a large ΔpH. They have a low or inside positive membrane potential to provide an adequate Δp, compensating for the large ΔpH. The internal positive membrane potential renders an additional benefit to the bacteria in preventing H^+ leakage due to the large concentration gradient. Alkaliphiles grow optimally at alkaline pH with a neutral internal pH. They have a high Na^+/H^+ antiporter activity. H^+ exported by ETP is exchanged with Na^+ to prevent alkalinization of the cytoplasm. They maintain a sodium motive force instead of a proton motive force. Alkaliphiles and halophiles have Na^+-dependent ETP in addition to the Na^+/H^+ antiporter. This is referred to as a primary sodium pump.

During vinegar fermentation, acetic acid bacteria lose their viability when aeration is disrupted. The bacteria pump H^+ out of the cell through ETP with aeration, but the cytoplasm is acidified without H^+ expulsion under oxygen-limited conditions.

Saccharolytic clostridia ferment carbohydrates to fatty acids (Section 8.5), lowering the growth pH to pH 4.5. At these conditions, the cells become very resistant to physical stresses. Such resistance to low pH in the presence of fatty acids and to stress is related to changes in cell surface structure. Some saccharolytic clostridia reduce fatty acids to alcohols (Section 8.5.2).

5.7.4 Proton motive force and sodium motive force in alkaliphiles

Alkaliphiles maintain a neutral internal pH at external pH values over 10. Their growth is Na^+ dependent. Na^+ plays an important role in maintaining internal neutral pH in alkaliphiles. H^+ exported by ETP is exchanged with Na^+ through the action of the Na^+/H^+ antiporter (Figure 5.17). A Na^+/H^+ antiporter mutant (*nhaC*$^-$) of alkaliphilic *Bacillus firmus* cannot grow at pH 10. In some bacteria, Na^+ is exported instead of H^+ through Na^+-dependent ETP which includes the Na^+-dependent NADH–quinone reductase complex (Section 5.8.3.3), which is a primary sodium pump.

Alkaliphiles maintain a sodium motive force ($\Delta\tilde{\mu}_{Na^+}$) instead of a proton motive force, and they have Na^+-dependent ATPase and transporters (Section 5.8.4.1). Their motility is also Na^+ dependent. Since $1H^+$ is exchanged with nNa^+ ($n > 1.0$), the Na^+/H^+ antiporter increases the membrane potential.

A Na^+/H^+ antiporter is also known in neutrophiles including *Escherichia coli*. Mutants in the antiporter are less tolerant to alkaline pH than wild-type strains. The Na^+/H^+ antiporter renders neutrophiles with some alkaline-resistant characteristics. A sodium motive force is not only found in alkaliphiles but also in halophilic archaea. Some decarboxylases are Na^+ dependent (Section 5.8.6.2), and homoacetogens (Section 9.5.3) conserve energy through a Na^+ pump.

5.8 Electron transport (oxidative) phosphorylation

Electron carriers such as $NAD(P)^+$, FAD and PQQ are reduced during glycolysis and the TCA cycle. Electrons from these carriers enter the electron transport chain at different levels. Electron carriers are oxidized, reducing molecular oxygen to water through ETP to conserve free energy as the proton motive force (Δp) (Figure 5.24a).

5.8.1 Chemiosmotic theory

It took many years to elucidate how the free energy generated from ETP is conserved as ATP. Compounds with high energy bonds are not involved in ETP as in substrate-level phosphorylation. ATP is synthesized only with an intact membrane or membrane vesicles, and ATP synthesis is inhibited in the presence of

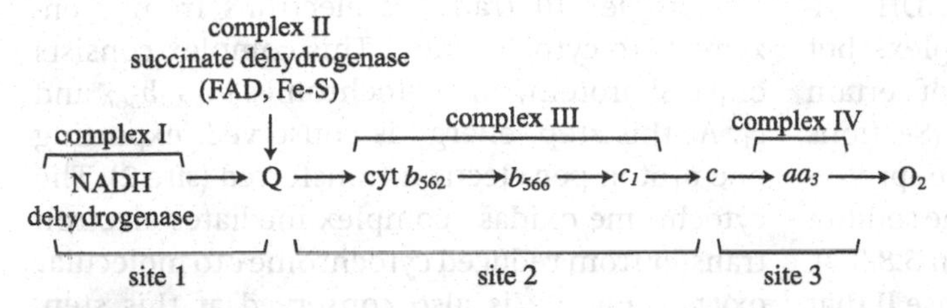

Figure 5.18 The eukaryotic electron transport chain localized in the mitochondrial inner membrane.

(Dawes, E. A. 1986, *Microbial Energetics*, Figure 7.1. Blackie & Son, Glasgow)

The electron transport chain forms four complexes in the mitochondrial inner membrane. These are referred to as complex I, II, III and IV. Complex I has the lowest redox potential and consists of NADH dehydrogenase with FMN as the prosthetic group and [Fe-S] proteins. This complex is called NADH-ubiquinone reductase. Succinate dehydrogenase of the TCA cycle contains FAD as a prosthetic group, and forms complex II (succinate-ubiquinone reductase) with [Fe-S] proteins. Complexes I and II transfer a pair of electrons to coenzyme Q from NADH and succinate, respectively. Complex III is ubiquinol-cytochrome *c* reductase, transferring electrons from reduced coenzyme Q (ubiquinol) to cytochrome *c*. This complex contains cytochrome b_{562}, cytochrome b_{566}, cytochrome c_1 and [Fe-S] protein. Complex IV is referred to as cytochrome *c* oxidase. Among the electron transport processes, three steps generate enough free energy to translocate protons out of the mitochondrial inner membrane. These are shown in the figure as sites 1, 2 and 3. They are the NADH dehydrogenase (NADH:ubiquinone oxidoreductase) which is a proton pump (site 1), the ubiquinone:cytochrome *c* reductase that moves positive charge across the membrane via the so-called Q-cycle (site 2) and cytochrome *c* oxidase, which combines proton pumping with opposite electron and proton movements (site 3).

uncouplers or ionophores. From these observations, a chemiosmotic mechanism was proposed. According to this, export of charged particles is coupled to oxidation–reduction reactions to form an electrochemical gradient which is used for ATP synthesis. H^+ are the charged particles exported, and the electrochemical gradient is the proton motive force, consisting of the H^+ gradient (ΔpH) across the membrane and the membrane potential ($\Delta\psi$). The phospholipid membrane is impermeable to H^+ and OH^- and is suitable to maintain the proton gradient. Most of the electron carriers involved in ETP are arranged in the membrane, and the membrane-bound ATP synthase synthesizes ATP from ADP and P_i, consuming the proton gradient.

5.8.2 Electron carriers and the electron transport chain

Electron carriers involved in electron transport from NADH to molecular oxygen are localized in the mitochondrial inner membrane in eukaryotic cells and in the cytoplasmic membrane in prokaryotic cells. The mitochondrial electron transport chain is shown in Figure 5.18, and bacterial electron transport systems are shown in Figure 5.22. Bacterial systems are diverse, depending on the species and strain as well as on the availability of electron acceptors.

5.8.2.1 Mitochondrial electron transport chain

Eukaryotic electron transport is discussed here as a model to compare with the process in prokaryotes. The mitochondrial electron transport chain consists of complexes I, II, III and IV. The overall reaction can be summarized as dehydrogenases (complexes I and II) and an oxidase (complex IV) connected by quinone (including complex III).

NADH dehydrogenase oxidizes NADH, reduced in various catabolic pathways, to NAD^+. This enzyme contains FMN as a prosthetic group and forms with [Fe-S] proteins a complex known as complex I or NADH-ubiquinone reductase. FMN is reduced with the oxidation of NADH and the [Fe-S] proteins mediate electron and proton transfer from $FMNH_2$ to coenzyme Q. This reaction generates enough free energy to translocate protons, and the electron transfer from NADH to ubiquinone (coenzyme Q) is referred to as site 1 of ETP. In a mitochondrion and in most bacteria, four protons per NADH oxidized are translocated by this complex, but sodium ions are exported by complex I of certain bacteria, including *Vibrio alginolyticus* (Sections 5.7.4 and 5.8.3.3). In addition to the proton pumping NADH dehydrogenase I and the sodium pumping NADH dehydrogenase (Section 5.8.3.3), some bacteria have NADH dehydrogenase II that does not translocate protons across the membrane. This membrane-bound NADH dehydrogenase II plays a crucial role in catabolic metabolism, maintaining the NADH/NAD^+ balance. In several pathogenic bacteria, this is the only enzyme performing respiratory NADH:quinone oxidoreductase activity.

As a step in the TCA cycle, succinate dehydrogenase oxidizes succinate to fumarate, reducing its prosthetic group, FAD, before electrons are transferred to coenzyme Q. This enzyme forms complex II (or the succinate-ubiquinone reductase complex) of ETP with [Fe-S] proteins, cytochrome b_{558} and low molecular weight peptides. Other dehydrogenases containing FAD as a prosthetic group reduce coenzyme Q in a similar way. These include glycerol-3-phosphate dehydrogenase and acyl-CoA dehydrogenase.

Electrons from coenzyme Q are transferred to a series of reddish-brown coloured proteins known as cytochromes. Cytochromes involved in mitochondrial electron transport are b_{562}, b_{566}, c_1, c and aa_3, as shown in Figure 5.18. Two separate protein complexes mediate electron transfer from coenzyme Q to molecular oxygen through the cytochromes. These are ubiquinol-cytochrome c reductase (complex III) and cytochrome oxidase (complex IV).

Complex III transfers electrons from coenzyme Q to cytochrome c. This complex consists of [Fe-S] protein, and cytochromes b_{562}, b_{566} and c_1. At this step, energy is conserved, exporting one proton per electron transferred (site 2). The cytochrome oxidase complex mediates electron transfer from reduced cytochrome c to molecular oxygen. Energy is also conserved at this step, translocating two protons per electron transferred (site 3). This terminal oxidase complex contains cytochrome a and cytochrome a_3.

5.8.2.2 Electron carriers

Electron transport involves various electron carriers, including flavoproteins, quinones, [Fe-S] proteins and cytochromes. Flavoproteins are proteins containing riboflavin (vitamin B_2) derivatives as their prosthetic group. They are FMN (flavin mononucleotide) and FAD (flavin adenine dinucleotide). The redox potential of the flavoproteins varies, not due to the flavin structure, but due to the differences in the protein component. Flavoproteins play the key role in the electron bifurcation reaction (Section 5.8.6.4).

Two structurally different quinones are involved in the electron transport process in bacteria, ubiquinone (UQ) and menaquinone (MQ), which serve as coenzyme Q. Quinones are lipid electron carriers, highly hydrophobic and mobile in the semi-solid lipid phase of the membrane. As shown in Figure 5.19, quinones have a side chain of 6, 8 or 10 isoprenoid units. These are named Q_6, Q_8 and Q_{10}, according to the number of isoprenoid units. UQ is found in mitochondria (Figure 5.19). Both forms of quinones are found in Gram-negative facultative anaerobes. Derivatives of UQ and MQ are known in bacteria.

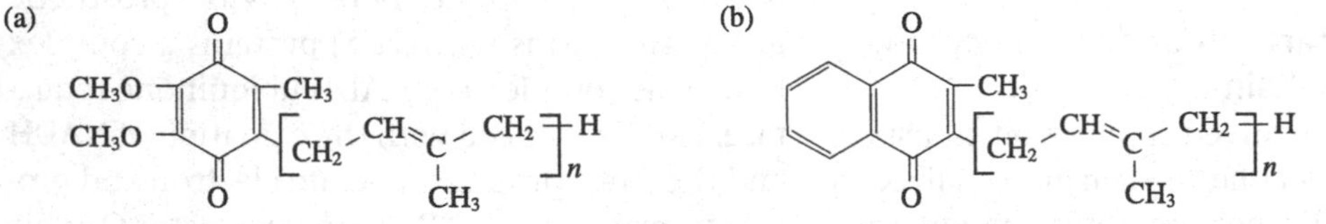

(a) (b)

Figure 5.19 Structure of (a) ubiquinone and (b) menaquinone.

(Gottschalk, G. 1986, *Bacterial Metabolism*, 2nd edn., Figure 2.9. Springer, New York)

n = 4, 6, 8 or 10.

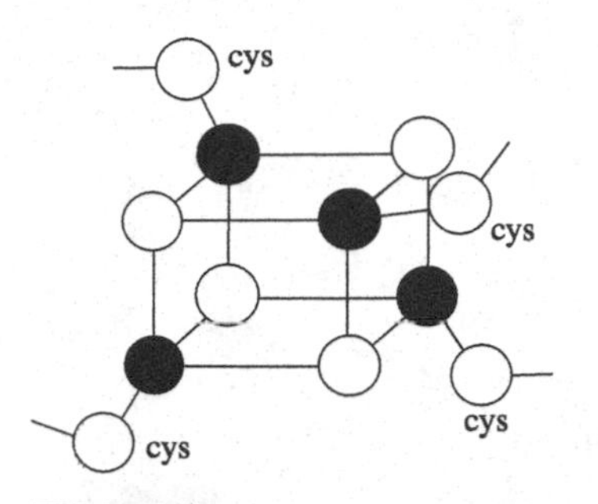

Figure 5.20 The structure of a [4Fe-4S] cluster.

(Gottschalk, G. 1986, *Bacterial Metabolism*, 2nd edn., Figure 8.8. Springer, New York)

o, sulfur atom; •, iron atom.

They include plastoquinone (demethylated UQ), demethylmenaquinone and methylmenaquinone (methylated benzene ring of MQ). Archaea are not known to have UQ and its derivatives, but have MQ and its derivatives. In addition, sulfur-containing quinones are known in halophiles and hyperthermophiles. Methanophenazine substitutes for quinones in methanogens (Section 9.4.2). The structure of coenzyme Q can be used as one characteristic in bacterial classification. Quinones can carry protons as well as electrons.

[Fe-S] proteins contain [Fe-S] cluster(s), usually [2Fe-2S] or [4Fe-4S]. The non-haem irons are attached to sulfide residues of the cysteines of the protein and acid-labile sulfur (Figure 5.20). The acid-labile sulfur is released as H_2S at an acidic pH. [Fe-S] proteins participating in electron transport can carry protons as well as electrons. There are many different [Fe-S] proteins mediating not only the electron transport process in the membrane, but also various oxidation–reduction reactions in the cytoplasm. The redox potential of different [Fe-S] proteins spans between as low as −410 mV (clostridial ferredoxin, Section 8.5) and +350 mV. Many enzymes catalysing oxidation–reduction reactions are [Fe-S] proteins, including hydrogenase, formate dehydrogenase, pyruvate:ferredoxin oxidoreductase and nitrogenase.

Cytochromes are haemoproteins. They are classified according to their prosthetic haem structures (Figure 5.21) and absorb light at 550–650 nm. Cytochrome b_{562} refers to a cytochrome *b* with the maximum wavelength absorption at 562 nm. Haem is covalently bound to the proteins in cytochrome *c*, and haems are non-covalently associated with the protein in other cytochromes. Since cytochromes carry only one electron, electron transfer from reduced coenzyme Q to cytochrome requires two steps (Section 5.8.3).

5.8.2.3 Diversity of electron transport chains in prokaryotes

As in the mitochondrial electron transport chain, the prokaryotic electron transport chain is organized with dehydrogenase and oxidase connected by quinone. However, electron transport chains in prokaryotes are much more diverse since they use diverse electron donors and electron acceptors, including those not used by any eukaryotes. Separate discussion will be made on the diversity of dehydrogenases in Chapter 7 (organic electron donors) and in Chapter 10 (inorganic electron donors, chemolithotrophs). Chapter 9 contains exclusive discussion of the electron transport chains to electron acceptors other than O_2 (anaerobic respiration). In this section, prokaryotic electron transport chains analogous to the mitochondrial system will be described.

The electron carriers involved in prokaryotic electron transport are diverse. As described earlier, menaquinone and ubiquinone and their derivatives are used as coenzyme Q in bacteria, and archaea have menaquinone and its derivatives. In addition diverse cytochromes participate in prokaryotic electron transport mainly in relation to the terminal oxidase, which leads to branched electron transport chains depending on the availability of O_2. These terminal oxidases have varying properties, such as their affinity for oxygen, transcriptional regulation and proton pumping ability.

Some bacteria such as *Paracoccus denitrificans* and *Alcaligenes eutrophus* have very similar electron transport systems to mitochondria. They have cytochrome aa_3 as the terminal oxidase while others use cytochrome *d* or *o* in its place. Cytochrome *o* has a *b*-type haem, and cytochrome *d* has a different haem structure

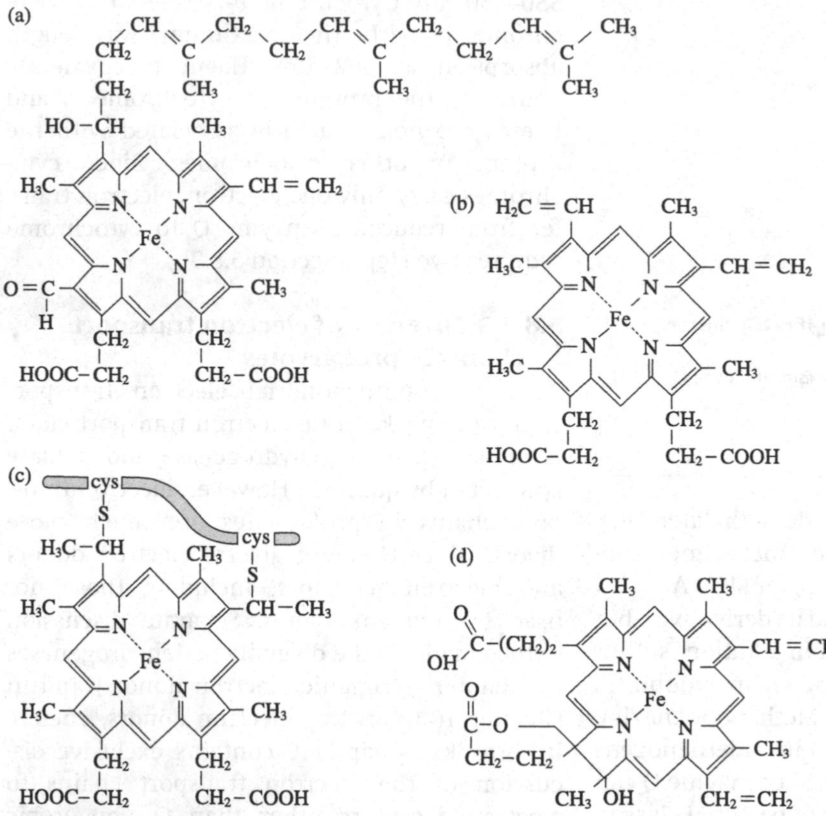

Figure 5.21 Structure of haems of cytochromes.

(Gottschalk, G. 1986, *Bacterial Metabolism*, 2nd edn., Figure 2.10. Springer, New York)

(a)–(d) show the prosthetic haems of cytochromes *a*, *b*, *c* and *d*, respectively.

(Figure 5.21). They are not only structurally different but also show different responses towards respiratory inhibitors, and form branched electron transport pathways (Figure 5.22). This diversity of electron transport systems is closely related to bacterial growth under a variety of conditions.

The terminal oxidases in bacteria have different affinities for O_2. Under O_2-limited conditions, cytochrome *d* replaces the normal terminal oxidase, cytochrome aa_3, in *Klebsiella pneumoniae* and *Haemophilus parainfluenzae*. Similarly, *Paracoccus denitrificans* and *Cupriavidus metallidurans* (formerly *Alcaligenes eutrophus*) use cytochrome *o* as their terminal oxidase. The high affinity cytochrome *d* and *o* enables the bacteria to use the electron acceptor (O_2) efficiently at low concentrations. In nitrogen-fixing *Azotobacter vinelandii*, cytochrome *d* functions as the terminal oxidase under nitrogen-fixing conditions with less energy conservation than the normal oxidase (Figure 5.22). Cytochrome *d* keeps the intracellular O_2 concentration low to protect the O_2-labile nitrogenase (Section 6.2.1).

The opportunistic pathogen *Pseudomonas aeruginosa* inhabits various environments, which is attributed to its very versatile energy metabolism with a highly branched respiratory chain terminated by multiple terminal oxidases. This bacterium has five terminal oxidases. Two of them, bo_3 oxidase and the cyanide-insensitive oxidase, use quinol as electron donor to reduce

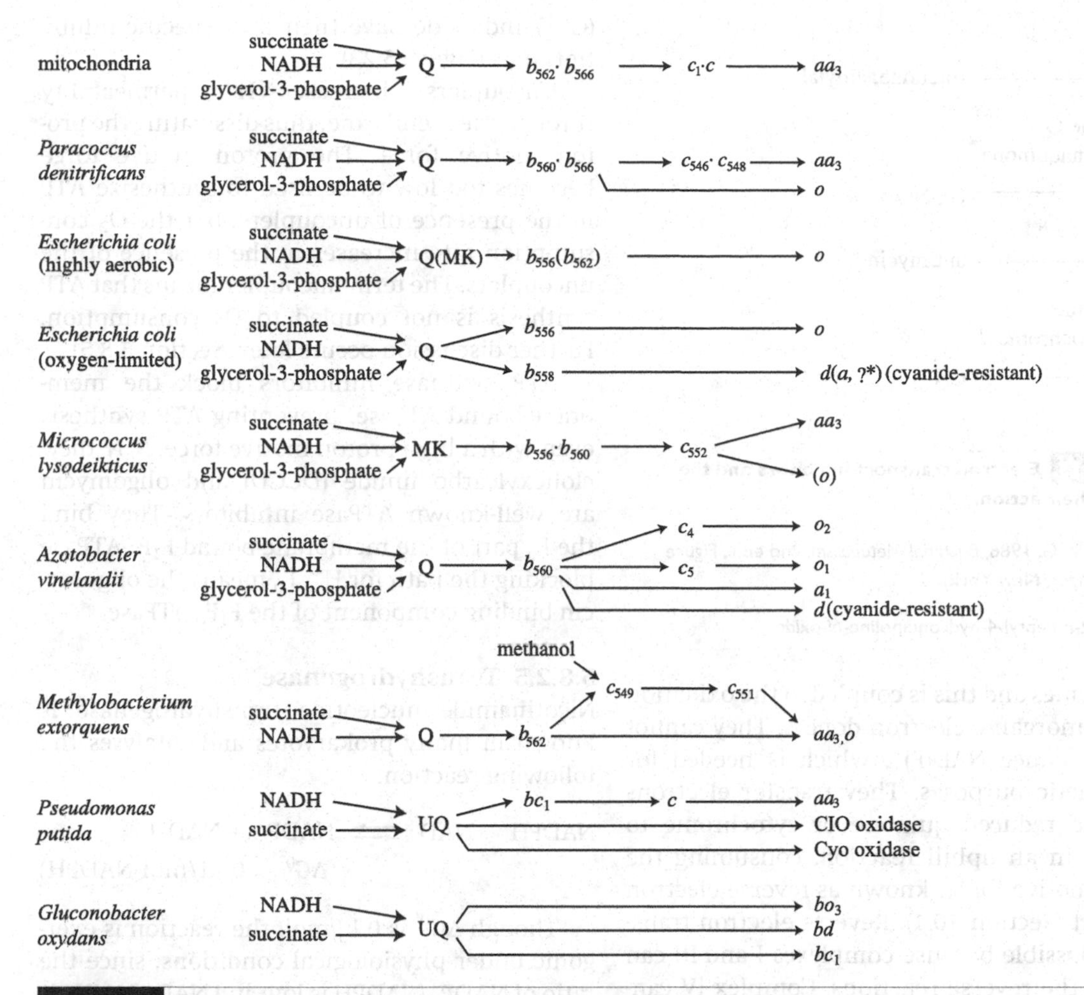

Figure 5.22 Comparison of mitochondrial and bacterial electron transport systems.

(Modified from Dawes, E. A. 1986, *Microbial Energetics*, Blackie & Son, Glasgow)

Under nitrogen-fixing conditions with high oxygen tension, *Azotobacter vinelandii* consumes oxygen rapidly without conserving energy at site I to protect nitrogenase (Section 6.2.1.4).

oxygen, and the others (*cbb*$_3$-1, *cbb*$_3$-2, and *aa*$_3$) are cytochrome *c* oxidases. Each oxidase has a specific affinity for oxygen, efficiency of energy coupling and tolerance to various stresses such as cyanide and reactive nitrogen species. The high oxygen affinity oxidase, *cbb*$_3$-1, is a constitutive enzyme and the cyanide-insensitive oxidase is induced under various stress conditions. The genes for others are expressed under iron (*bo*$_3$) and general nutrient (*aa*$_3$) starvation and low oxygen (*cbb*$_3$-2) conditions.

Though the presence of electron carriers such as quinones and cytochromes in various aerobic and facultative anaerobic archaea is confirmed, the electron transport chains reducing oxygen are not yet fully elucidated.

In branched bacterial electron transport systems, the number of sites for energy conservation is lower than that of the mitochondrial system (Section 5.8.4). This might enable a survival strategy under certain conditions, but at the expense of reduced energy conservation. Acetic acid bacteria (Section 7.11.1) have only quinol oxidases with a low H^+/O ratio (Section 5.8.4.2).

With a few exceptions, the dehydrogenases of chemolithotrophs reduce quinones or

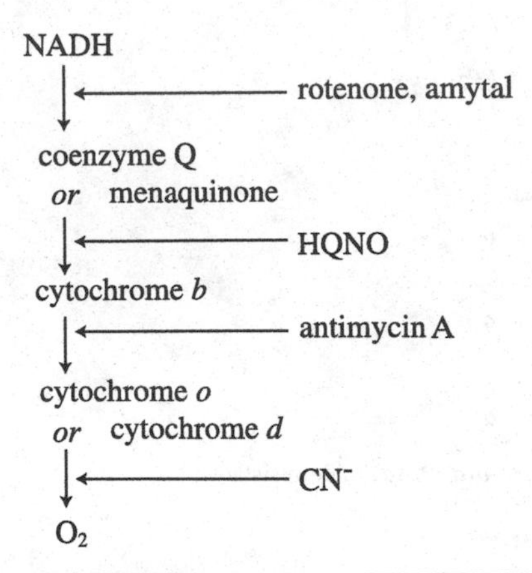

Figure 5.23 Electron transport inhibitors and the sites of their action.

(Gottschalk, G. 1986, *Bacterial Metabolism*, 2nd edn., Figure 2.17. Springer, New York)

HQNO, 2-*n*-heptyl-4-hydroquinoline-N-oxide.

cytochromes and this is coupled to the oxidation of their inorganic electron donors. They cannot directly reduce $NAD(P)^+$, which is needed for biosynthetic purposes. They transfer electrons from the reduced quinone or cytochrome to $NAD(P)^+$ in an uphill reaction, consuming the proton motive force, known as reverse electron transport (Section 10.1). Reverse electron transport is possible because complexes I and III can catalyse the reverse reactions. Complex IV cannot catalyse the reverse reaction to use water as the source of electrons. Water is used as the electron source in oxygenic photosynthesis through a different reaction (Section 11.4.3).

5.8.2.4 Inhibitors of electron transport phosphorylation (ETP)

Inhibitors of ETP are grouped into three kinds according to their mechanism of action: electron transport inhibitors, uncouplers and ATPase inhibitors.

Electron transport inhibitors interfere with the enzymes and electron carriers involved in the electron transport system. They inhibit not only ATP synthesis but also oxygen consumption. Rotenone, amytal and piericidin A inhibit NADH dehydrogenase, and 2-*n*-heptyl-4-hydroquinoline-*N*-oxide (HQNO), antimycin A, cyanide (CN^-) and azide have their own specific inhibition sites (Figure 5.23).

Uncouplers increase H^+ permeability through the membrane, thus dissipating the proton motive force. The proton motive force becomes too low to be used to synthesize ATP in the presence of uncouplers, but the O_2 consumption rate increases in the presence of the uncouplers. The term uncoupler means that ATP synthesis is not coupled to O_2 consumption. Further discussion occurs later (Section 5.8.5).

ATP synthase inhibitors block the membrane-bound ATPase, preventing ATP synthesis even with a high proton motive force. *N, N′* dicyclohexylcarbodiimide (DCCD) and oligomycin are well-known ATPase inhibitors. They bind the F_o part of the membrane-bound F_1F_o-ATPase blocking the path for H^+. F_o means the oligomycin-binding component of the F_1F_o-ATPase.

5.8.2.5 Transhydrogenase

Nicotinamide nucleotide transhydrogenase is known in many prokaryotes and catalyses the following reaction:

$$NADPH + NAD^+ \rightleftharpoons NADP^+ + NADH \quad (\Delta G^{0'} = 0 \text{ kJ/mol NADPH})$$

Though $\Delta G^{0'}$ is 0 kJ/mol, the reaction is exergonic under physiological conditions, since the ratio of $NADP^+$/NADPH is low and NAD^+/NADH is high in the cell. This reaction is coupled to H^+ extrusion, increasing the proton motive force. This reaction is referred to as site 0 of ETP.

In addition to the membrane-bound energy-dependent pyridine nucleotide transhydrogenase, a soluble energy-independent enzyme is found in certain bacteria, although its physiological function remains obscure.

5.8.3 Arrangement of electron carriers in the H^+-translocating membrane

It is not well established how protons are translocated when coupled to electron transport, but there is a consensus that the protons are translocated by exploiting the different properties of the electron carriers (Q-loop and Q-cycle) and through the proton pump. Among the electron

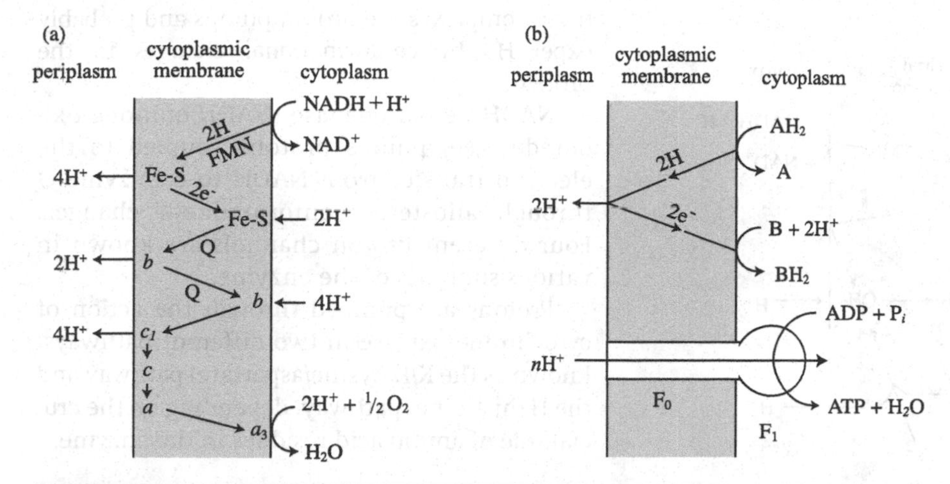

Figure 5.24 **Formation of the proton motive force and ATP synthesis through ETP.**

(Modified from Dawes, E. A. 1986, *Microbial Energetics*, Figure 7.3. Blackie & Son, Glasgow)

Since the electron carriers are arranged in the mitochondrial inner membrane and prokaryotic cytoplasmic membrane in such a way that electrons move from the inner face to the outer face of the membrane with H^+, and in the reverse direction without H^+, H^+ is exported during the electron transport process (a). The membrane-bound ATPase synthesizes ATP, consuming the proton motive force with H^+ flow to the low H^+ concentration side (b).

carriers, [Fe-S] proteins and cytochromes only carry electrons, while coenzyme Q can carry electrons as well as protons. They are arranged in such a way that H^+ are exported during ETP (Figure 5.24a). The location of these membrane proteins has been studied using various techniques, including electron microscopy, immunology and using inhibitors and proteolytic enzymes. These studies have located NADH dehydrogenase to the cytoplasmic side (matrix side in a mitochondrion) of the membrane and cytochrome *c* to the periplasmic side.

5.8.3.1 Q-cycle and Q-loop

To describe H^+ translocation at site 2 catalysed by ubiquinol–cytochrome *c* reductase (complex III), a Q-cycle and Q-loop have been proposed. According to the Q-loop mechanism, FMN of NADH dehydrogenase is reduced to $FMNH_2$ by oxidizing NADH on the inside face of the membrane. Electrons from $FMNH_2$ are transferred to a [Fe-S] protein on the outer face of the membrane, leaving $2H^+$ outside. This reaction is catalysed by complex I. The reduced [Fe-S] protein reduces coenzyme Q at the inner face of the membrane, consuming $2H^+$. The reduced coenzyme Q moves to the outer face of the membrane to reduce cytochrome *c*, leaving $2H^+$ outside again through cytochrome *b* (Figure 5.25).

The Q-loop mechanism is consistent with experimental results obtained with various bacteria, including *Escherichia coli*, where $2H^+$ are translocated by the ubiquinol–cytochrome *c* reductase (complex III). However, in a mitochondrion and in some bacteria, $4\,H^+$ are translocated by complex III. To explain how the extra $2H^+$ are exported, the Q-cycle has been proposed. In this mechanism, reduced coenzyme Q (QH_2) is oxidized to semiquinol (QH), reducing [Fe-S] protein before being fully oxidized to Q, transferring the electron to cytochrome *b*. The reduced [Fe-S] protein transfers the electron to cytochrome *c*, and the reduced cytochrome *b* returns the electron to Q which takes up $1\,H^+$ to be reduced to QH. QH is fully reduced to QH_2, taking one electron from [Fe-S] protein and consuming $1\,H^+$ at the inner face of the membrane (Figure 5.25). According to this hypothesis, $2H^+$ are translocated during a $QH_2 \rightarrow QH \rightarrow Q \rightarrow QH \rightarrow QH_2$ cycle, transferring one electron from Q to cytochrome *c*.

5.8.3.2 Proton pump

Complexes I (NADH-ubiquinone reductase) and IV (cytochrome oxidase) translocate H^+, but

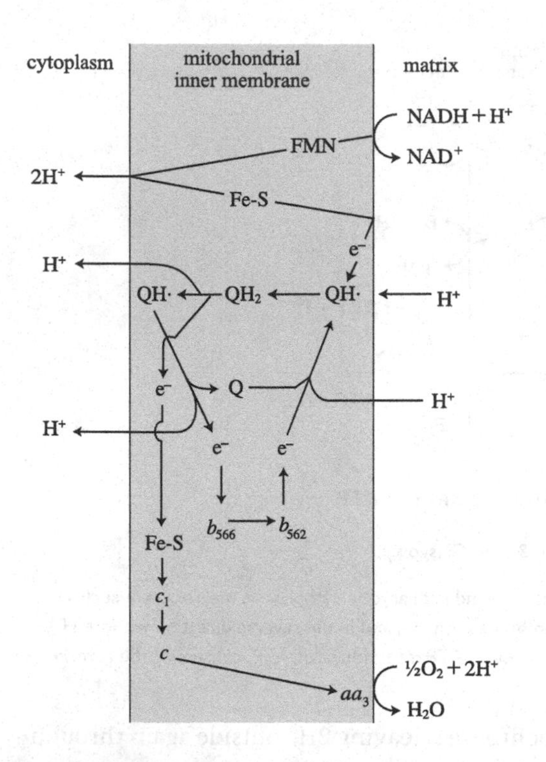

Figure 5.25 H+ translocation through the Q-cycle catalysed by the ubiquinol–cytochrome reductase complex during ETP. (Modified from Dawes, E. A. 1986, *Microbial Energetics*, Figure 7.4. Blackie & Son, Glasgow)

In some bacteria, including *Escherichia coli*, $2H^+$ are translocated by the ubiquinol–cytochrome *c* reductase (complex III). In this reaction, known as the Q-loop, coenzyme Q is reduced at the inner face of the membrane and moves to the other side of the membrane, where electrons are transferred to cytochrome *c* leaving protons outside. The Q-cycle has been proposed to explain the translocation of $4H^+$ in a mitochondrion and some bacteria. In this mechanism the reduced coenzyme Q (QH_2) is oxidized to semiquinol (QH) reducing [Fe-S] protein before being fully oxidized to Q, transferring the electron to cytochrome *b*, as illustrated in the diagram. The reduced [Fe-S] protein transfers the electron to cytochrome *c*, and the reduced cytochrome *b* returns the electron to Q which takes up $1H^+$ to be reduced to QH. QH is fully reduced to QH_2, taking one electron from [Fe-S] protein and consuming $1H^+$. During a $QH_2 \rightarrow QH \rightarrow Q \rightarrow QH \rightarrow QH_2$ cycle transferring one electron from Q to cytochrome *c*, $2H^+$ are translocated.

mechanisms similar to the Q-loop or Q-cycle are not known for these complexes. It is believed that these complexes are proton pumps and probably expel H^+ by conformational changes in the protein.

NADH dehydrogenase (NADH:quinone oxidoreductase) pumps protons coupled to the electron transfer from NADH to coenzyme Q through allosteric conformational changes. Four different proton channels are known in various subunits of the enzyme.

Protons are pumped through the action of cytochrome oxidase in two different pathways, known as the K/D (lysine/aspartate) pathway and the H (histidine) pathway, depending on the crucial role of amino acid residues in the enzyme.

5.8.3.3 Sodium pump

A NADH:quinone oxidoreductase (Na^+-NQR) converting redox reaction energy into a transmembrane sodium electrochemical potential ($\Delta\tilde{\mu}_{Na^+}$) is known in the respiratory chain of the marine bacteria *Vibrio alginolyticus* and *V. costicola* as a primary sodium pump. This enzyme with six subunits is encoded in the *nqr*-operon of the bacterial genome. This operon is widely distributed among many bacteria, including pathogenic and conditionally pathogenic bacteria such as *V. cholerae, Haemophilus influenzae, Klebsiella pneumoniae, Neisseria gonorrhoeae, Neisseria meningitidis, Yersinia pestis, Pseudomonas aeruginosa* and *Porphyromonas gingivalis*. Na^+-NQR contains the following set of prosthetic groups: one [2Fe–2S] cluster, one non-covalently bound FAD, two covalently bound FMN residues, one non-covalently bound riboflavin and also possibly one ubiquinone-8.

Na^+-NQR oxidizes NADH, reducing coenzyme Q. Two sodium ions are exported across the membrane. The energy conservation efficiency by this enzyme is approximately two-fold lower than that by H^+-translocating NADH:quinine oxidoreductase that translocates four protons per NADH oxidized. The structure, subunits and prosthetic groups of Na^+-NQR are completely different from the proton-pumping complex I, and most likely the mechanism of sodium ion pumping is quite different from that of proton pumping. A similar enzyme was found in the mitochondria of the yeast *Yarrowia lipolytica*.

5.8.4 ATP synthesis

The membrane-bound ATP synthase (ATPase) synthesizes ATP, consuming the proton motive force (Figure 5.24b).

5.8.4.1 ATP synthase

This enzyme is located in the mitochondrial inner membrane of eukaryotes or the prokaryotic cytoplasmic membrane. The enzyme consists of a membrane-embedded F_o part and a F_1 part protruding into the cytoplasmic side. F_o is a pore for H^+ to pass, and consists of three different peptides, a, b and c, and ATP synthesis and hydrolysis are catalysed by the F_1 part, consisting of five different peptides ($\alpha_3\beta_3\gamma\delta\varepsilon$) (Figure 5.26).

Oligomycin and dicyclohexylcarbodiimide (DCCD) inhibit ATP synthase by blocking H^+ movement through the pore formed by the F_o part. F_o is the oligomycin-binding site. The ATPase of the mitochondrial inner membrane and bacterial cytoplasmic membrane is referred to as F_1F_o-ATPase, F-ATPase or H^+-ATPase. The main function of this enzyme is ATP synthesis, while consuming the proton motive force. In many bacteria a similar ATPase synthesizes ATP but consumes the sodium motive force. This is referred to as a Na^+-ATPase.

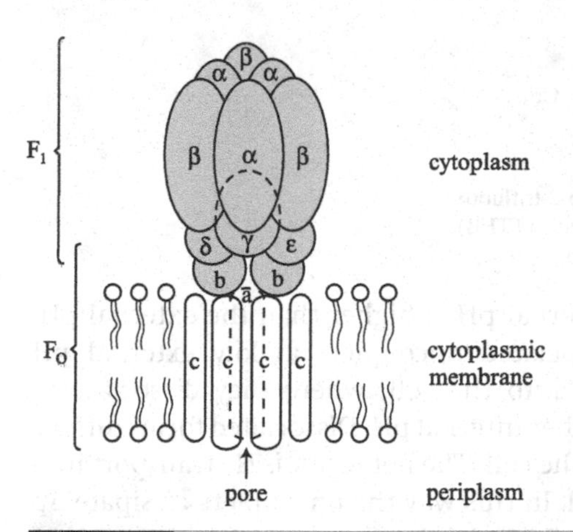

Figure 5.26 **Model of the bacterial F_1F_0-ATP synthase.**

(Dawes, E. A. 1986, *Microbial Energetics*, Figure 5.3. Blackie & Son, Glasgow)

The F_o part of ATP synthase consists of highly hydrophobic polypeptides a, b, c, forming a hydrophilic pore for H^+ movement. The catalytic F_1 part consists of five different polypeptides in the ratio of $\alpha_3\beta_3\gamma\delta\varepsilon$.

An ATPase structurally different from the F-ATPase is found at the eukaryotic cytoplasmic membrane and the membrane of organelles other than chloroplasts and mitochondria. The function of this enzyme is development of the proton motive force, consuming the ATP synthesized in mitochondria and chloroplasts. This enzyme was first identified in the vacuole of a eukaryotic cell, and named a V-ATPase. V-ATPase has a catalytic V_1 part and membrane-embedded V_o part. V_1 contains eight different peptides (A, B, C, D, E, F, G, H) and V_o contains five different peptides (a, d, c, c′, c′′). A V-type ATPase has been identified in bacteria and archaea, including *Clostridium fervidum*, *Enterococcus hirae*, *Thermus thermophilus* and others. This prokaryotic V-ATPase synthesizes ATP while consuming the sodium motive force.

During methanogenesis, energy is conserved as sodium as well as proton motive forces (Section 9.4.4). Methanogenic archaea have a unique A_1A_O-ATPase that shares structural features with the V-ATPase but functional features with the F-ATPase. This ATPase synthesizes ATP, consuming not only the proton motive force but also the sodium motive force. In addition to the A-ATPase, some methanogens have a F-ATPase that is dispensable.

The ATPase is localized on the outer membrane in the sulfur-oxidizing anaerobic thermophilic archaean *Ignicoccus hospitalis*. This archaeon has unusual cellular architecture with an inner and outer membrane (Section 2.3.3.2).

5.8.4.2 H^+/O ratio

Not only is the mechanism of coupling proton translocation to electron transport not well established, but neither is the number of protons translocated. It is generally accepted that the NADH–ubiquinone reductase complex exports four protons, ubiquinol-cytochrome *c* reductase exports two protons, and cytochrome oxidase exports two protons, with the consumption of four electrons (Figure 5.24a). Since two electrons reduce one oxygen atom, the number is referred to as the H^+/O ratio. A ratio of 10 is accepted for mitochondria, and the ratio is similar or less than 10 in prokaryotes, depending on the electron carriers involved in ETP (Figure 5.22).

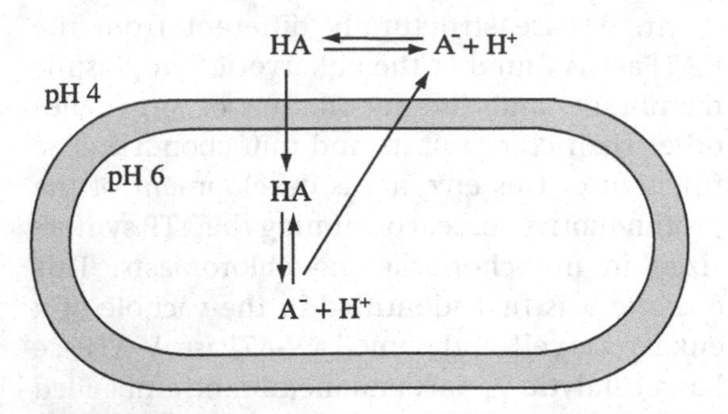

Figure 5.27 Dissipation of Δp by an uncoupler.

Uncouplers are weak acids or weak bases, hydrophobic not only in undissociated form but also in dissociated form and permeable to the membrane. When an uncoupler is equilibrated at higher internal pH and at low external pH, the concentration of the undissociated form is higher in the external medium and the concentration of the dissociated form is higher in the cell. The undissociated form diffuses into the cell and the dissociated form out of the cell, thus dissipating Δp. ATP cannot be synthesized due to the low Δp, and the cell consumes more O_2 to increase the Δp.

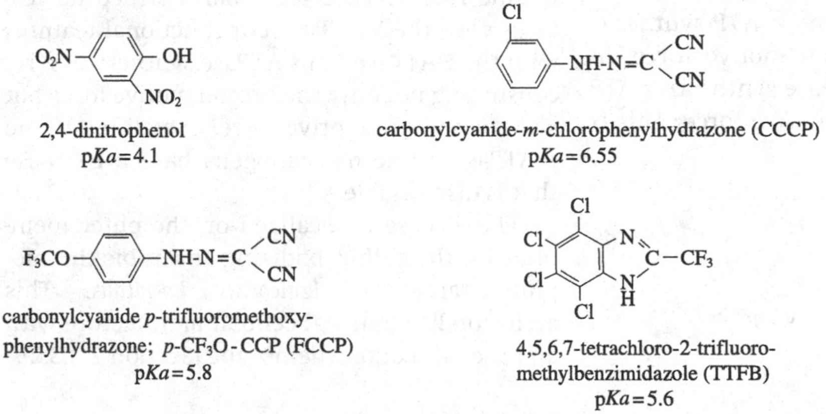

5.8.4.3 H^+/ATP stoichiometry

The membrane-bound ATP synthase catalyses ATP synthesis and hydrolysis, consuming and developing the proton motive force (Figure 5.24b). It is not known for certain how many protons move through the enzyme for one ATP to be generated (H^+/ATP stoichiometry). H^+/ATP stoichiometry depends on the phosphorylation potential (Δ*Gp*, Section 5.6.3) and the size of the proton motive force (Δp, Section 5.7.1). Various experimental results have shown a H^+/ATP stoichiometry of 2–5 with a stoichiometry of 3 generally being accepted.

This number has an important implication in terms of the minimum free energy conserved in a biological system. Energy is conserved through fermentation (substrate-level phosphorylation, SLP), respiration and photosynthesis (ETP) in biological systems. In SLP a free energy change bigger than Δ*Gp* can be conserved for thermodynamic reasons. Similarly, free energy changes can be conserved during respiration and photosynthesis when the change is bigger than Δ*Gp*/[H^+/ATP stoichiometry]. In the case of organisms using Na^+ in place of H^+, the minimum amount of energy conserved would be Δ*Gp*/[Na^+/ATP stoichiometry].

5.8.5 Uncouplers

Many compounds are known to inhibit ATP synthesis with an increase in O_2 consumption during respiration. These are known as uncouplers. They are weak acids or weak bases, and are permeable to the membrane since they are hydrophobic, not only in undissociated forms, but also in dissociated forms. In a respiring cell, the internal pH is higher than the external pH. Undissociated uncouplers at low external pH diffuse into the cell, where they dissociate at the higher internal pH. Dissociated forms diffuse out of the cell. The net result is H^+ transport into the cell. In this way the uncouplers dissipate Δp (Figure 5.27). In the presence of an uncoupler, ATP cannot be synthesized due to the low Δp, and the cell consumes more O_2 to increase Δp.

Chemicals that increase the permeability of ions are referred to as ionophores. Uncouplers are therefore ionophores for H^+. Alternatively, they can be called protonophores. Figure 5.28 shows some examples of uncouplers.

2,4-dinitrophenol
p*Ka*=4.1

carbonylcyanide-*m*-chlorophenylhydrazone (CCCP)
p*Ka*=6.55

carbonylcyanide *p*-trifluoromethoxyphenylhydrazone; *p*-CF₃O-CCP (FCCP)
p*Ka*=5.8

4,5,6,7-tetrachloro-2-trifluoromethylbenzimidazole (TTFB)
p*Ka*=5.6

Figure 5.28 Structures of some uncouplers.

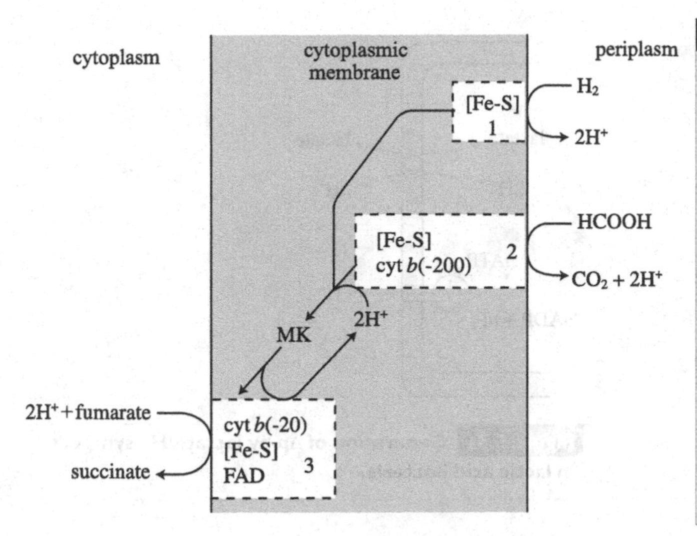

Figure 5.29 ATP synthesis through fumarate reductase.

(Gottschalk, G. 1986, *Bacterial Metabolism*, 2nd edn., Figure 8.22. Springer, New York)

A membrane-bound electron carrier is reduced, oxidizing hydrogen, leaving $2H^+$ outside. The electrons are consumed to reduce fumarate consuming $2H^+$ in the cell. Fumarate reductase generates Δp by this mechanism. Species of *Propionibacterium* have a similar energy conservation process during the fermentation of lactate to propionate (Section 8.7.1).

Numbers in boxes are the redox potentials of the cytochromes: 1, hydrogenase; 2, formate dehydrogenase; 3, fumarate reductase.

5.8.6 Primary H^+ (Na^+) pumps in fermentative metabolism

The free energy changes available from the oxidation–reduction reactions in respiration and photosynthesis are conserved in the form of Δp. In addition to O_2, prokaryotes use various other electron acceptors for their respiration. The term anaerobic respiration is used to describe the energy conservation process using electron acceptors other than O_2. Some primary H^+(Na^+) active transport processes are known in fermentative bacteria where internally supplied electron acceptors are used (Section 8.1).

5.8.6.1 Fumarate reductase

Species of *Propionibacterium* ferment lactate to propionate through the succinate–propionate pathway (Section 8.7.1). Fumarate reductase of this pathway reduces fumarate to succinate, extruding H^+. A similar energy conservation process is known in *Vibrio succinogenes*, *Desulfovibrio gigas* and *Clostridium formicoaceticum*. This membrane-bound enzyme uses H_2 and formate as the electron donors and needs menaquinone and cytochrome *b* as coenzymes (Figure 5.29). Strictly speaking, the latter example is anaerobic respiration, since the electron acceptor, fumarate, is externally supplied.

5.8.6.2 Na^+-dependent decarboxylase

Klebsiella pneumoniae (*Klebsiella aerogenes*) has a Na^+-dependent methylmalonyl-CoA decarboxylase. This enzyme is membrane bound, and exports Na^+ coupled to the decarboxylation reaction. The free energy change of decarboxylation (−30 kJ/mol methyl-malonyl-CoA) is conserved as a sodium motive force, which is converted to Δp by a Na^+/H^+ antiporter to synthesize ATP. Figure 5.30 shows a decarboxylation reaction, conserving energy as a sodium motive force.

Energy is similarly conserved by glutaconyl-CoA decarboxylase in *Acidamicoccus fermentans* and *Clostridium symbiosum*, and by succinate decarboxylase in *Veillonella parvula*. Energy conservation is not known in the decarboxylation of oxaloacetate in the citrate fermentation by *Pediococcus halophilus* and *Lactococcus lactis* (Section 8.4.6).

A hyperthermophilic archaeon, *Thermococcus onnurineus*, develops a sodium motive force coupled to the oxidation of formate to carbon dioxide and hydrogen through the action of a membrane-bound enzyme, formate dehydrogenase, hydrogenase and a Na^+/H^+ antiporter.

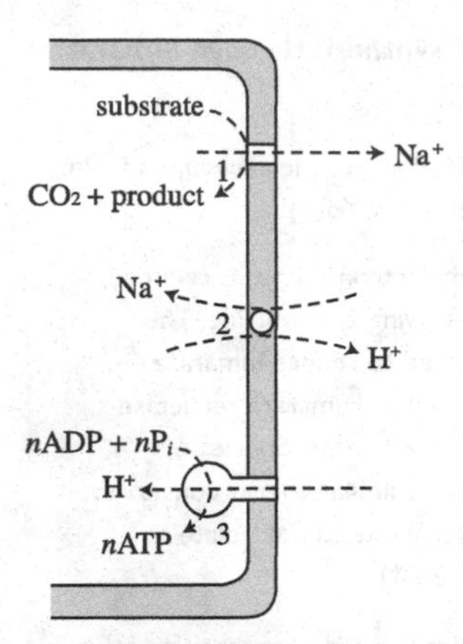

Figure 5.30 **Δp generation by Na+-dependent decarboxylase.**

1, oxaloacetate decarboxylase; 2, Na^+/H^+ antiporter; 3, F_1F_o- ATPase.

(Gottschalk, G. 1986, *Bacterial Metabolism*, 2nd edn., Figure 8.17. Springer, New York)

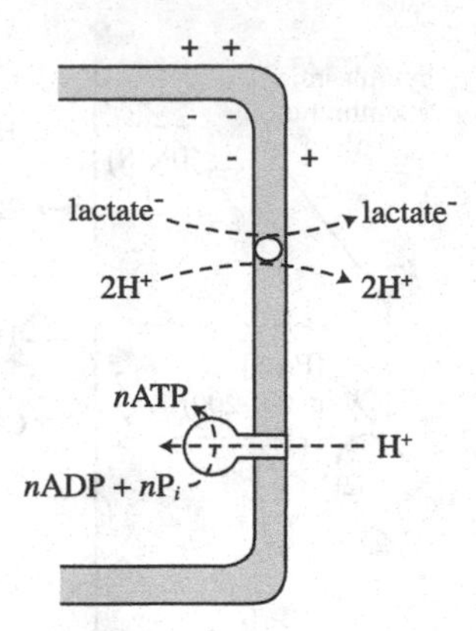

Figure 5.31 **Generation of Δp by lactate/H+ symport in lactic acid bacteria.**

5.8.6.3 Δp formation through fermentation product/H^+ symport

A lactic acid bacterium, *Lactococcus cremoris*, extrudes H^+ by a lactate/H^+ symporter, utilizing the potential energy developed by the lactate gradient due to the high internal lactate concentration in the cell (Figure 5.31). When lactate has accumulated in the environment, H^+ cannot be symported with lactate. In the lactic acid bacterial malolactic fermentation (Section 8.4.6) a malate^{2-}/lactate$^-$ antiporter imports malate in exchange for lactate. This reaction generates $\Delta\psi$, since malate^{2-} is imported exporting lactate$^-$. In addition, H^+ is consumed in the malate decarboxylation reaction (Section 3.6).

5.8.6.4 Energy conservation through electron bifurcation

Many anaerobic prokaryotes conserve energy in oxidation/reduction reactions between compounds with different redox potential through the electron bifurcation reaction. Flavin plays a pivotal role in this reaction. Flavin has two electron-carrying sites, one with a high redox potential and the other a low potential. In the clostridial fermentation butyryl-CoA dehydrogenase oxidizes two NADH to reduce crotonyl-CoA to butyryl-CoA and ferredoxin (Section 8.5.1.2). Since hydrogenase oxidizes the reduced ferredoxin, less acetyl-CoA is used as the electron acceptor. ATP is produced from acetyl-CoA. Similar electron bifurcation reactions are known in methanogens (Section 9.4.3) and homoacetogens (Section 9.5.3). The reduced ferredoxin is used to develop a H^+ (Na^+) gradient by the membrane-bound energy-converting hydrogenase or the H^+ (Na^+)-translocating ferredoxin: NAD^+ oxidoreductase, described below.

5.8.6.5 Energy-converting hydrogenase

A [FeNi] hydrogenase (Ech) complex (Section 10.5.2) is embedded in the cytoplasmic membrane of various anaerobic microorganisms, including incomplete oxidizing sulfate-reducing *Desulfovibrio* spp. (Section 9.3.2), methanogens (Section 9.4.4) and homoacetogens (Section 9.5.3). Ech complexes have a similar structure to the energy-conserving NADH:quinone oxidoreductase (complex I) of the aerobic electron transport system (Section 5.8). The export of protons or sodium ions is coupled to oxidation of reduced ferredoxin, producing H_2. These Ech complexes are referred to as energy-converting hydrogenases.

5.8.6.6 H^+(Na^+)-translocating ferredoxin: NAD^+oxidoreductase

Another membrane-bound ferredoxin oxidizing enzyme complex, the H^+(Na^+)-translocating

Table 5.3 Luminescent bacteria

Habitat	Strain	Symbiosis
Marine	*Photobacterium fischerii*	fish
Marine	*Vibrio harveyi*	none
Freshwater	*Vibrio cholerae* biotype *albensis*	none
Soil	*Xenorhabdus nematophilus*	nematode

ferredoxin:NAD^+ oxidoreductase (Rnf) complex, translocates $H^+(Na^+)$ across the membrane coupled to ferredoxin oxidation with NAD^+ reduction. In methanogens, ferredoxin:methanophenazine oxidoreductase has a similar function (Section 9.4.3.3)

5.9 Other biological energy transduction processes

Figure 5.10 illustrates energy transduction processes in microbes. Chemotrophs convert chemical energy into biological energy, and light energy is used by phototrophs. The biological energy in the form of ATP and Δp is invested in energy-requiring processes, including transport (Chapter 3), biosynthesis (Chapter 6) and motility. Not included in the figure is bioluminescence, which is found in certain bacteria.

5.9.1 Bacterial bioluminescence

Luminous bacteria convert biological energy into light. Some of these are symbionts and others are free living (Table 5.3). *Photobacterium fischerii* lives symbiotically in the light organs of certain fish, while a soil nematode is the habitat of *Xenorhabdus nematophilus*. Some members of the *Vibrio* genus are free-living luminous bacteria in marine and freshwater ecosystems.

Light is emitted when the cell density reaches certain levels. Luminescent bacteria excrete N-acyl homoserine lactone as an autoinducer. The components of the bioluminescence reaction are induced only above the threshold concentration of this autoinducer in the environment. This induction is referred to as quorum sensing (Section 12.2.8) and the reaction is strictly O_2 dependent. $FMNH_2$ oxidation provides energy for the luminescence. The bacterial luciferase forms a complex binding $FMNH_2$, O_2 and a long-chain aliphatic aldehyde, to emit light (Figure 5.32). The bacterial luciferase is a kind of monooxygenase (Section 7.7) and requires aldehyde for light emission. Aldehyde is oxidized to fatty acid by the luciferase, and reduction to aldehyde is coupled to the oxidation of NADH.

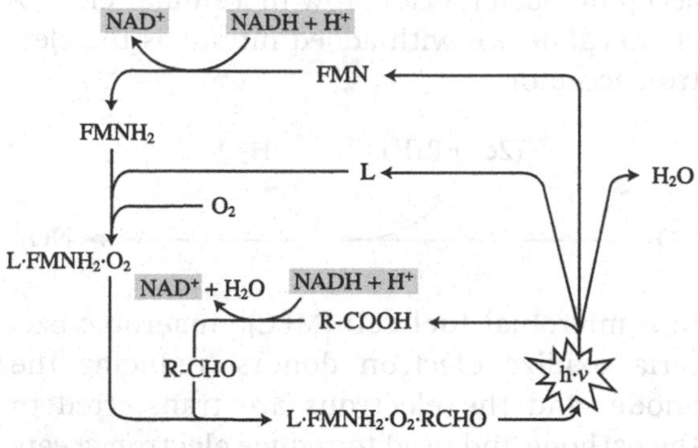

Figure 5.32 **Light emission by *Photobacterium fischerii*.**

(Gottschalk, G. 1986, *Bacterial Metabolism*, 2nd edn., Figure 5.19. Springer, New York)

L, bacterial luciferase; R-CHO, aliphatic aldehyde.

5.9.2 Electricity as an energy source

Chemotrophs use chemical energy as their energy source, and light energy is used by phototrophs. Chemical and light energy are converted into biological energy that is used for various biological functions. These include biosynthesis (chemical energy), transport (potential energy), motility (mechanical energy) and luminescence (light energy). In electric fishes, biological energy is converted into electricity.

There is some evidence that shows that electricity is used as an energy source in denitrification. Strains of *Geobacter* reduce nitrate with an electrode as the electron donor. An enrichment culture can be made using an electrode as the electron donor and nitrate as the electron acceptor. Bacteria can grow in a similar electrochemical device with added nitrate as the electron acceptor:

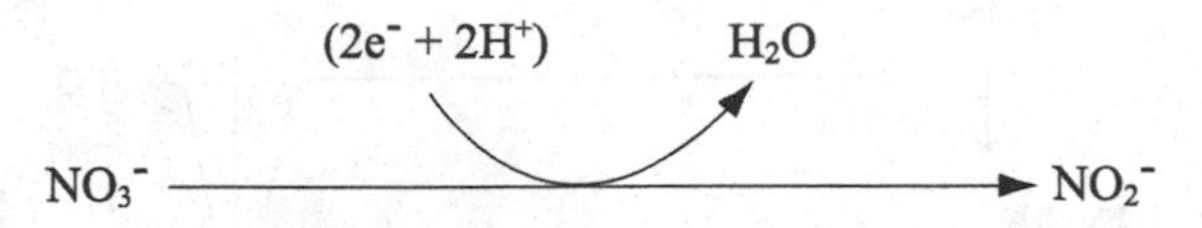

In a microbial fuel cell (MFC), anaerobic bacteria oxidize electron donors, reducing the anode, and the electrons are transferred to the cathode and used to reduce electron acceptors such as oxygen and nitrate. The electron acceptor reduction reaction at the cathode is catalysed either by inorganic catalysts or by bacteria. Electron transfer from bacterial cells to the anode and from the cathode to bacterial cells is facilitated by extracellular electron transfer mediated by cell surface electron carriers.

Direct current (DC) is applied to metal structures to protect them from microbially influenced corrosion in a process known as *galvanic corrosion protection*. In this case, microorganisms use low redox potential electrons of the DC instead of extracting electrons from metals. It is debatable if the energy in the form of electrons supplied to the cathode of MFC and in the corrosion protection process is electrical energy or (electro)chemical energy.

Further Reading

Note this section contains key references only. Additional recommended references are available at www.cambridge.org/ProkaryoticMetabolism.

TCA cycle

Austin, C. M., Wang, G. & Maier. R. J. (2015). Aconitase functions as a pleiotropic posttranscriptional regulator in *Helicobacter pylori*. *Journal of Bacteriology* **197**, 3076–3086.

Bott, M. (2007). Offering surprises: TCA cycle regulation in *Corynebacterium glutamicum*. *Trends in Microbiology* **15**, 417–425.

Hu, Y. & Holden, J. F. (2006). Citric acid cycle in the hyperthermophilic archaeon *Pyrobaculum islandicum* grown autotrophically, heterotrophically, and mixotrophically with acetate. *Journal of Bacteriology* **188**, 4350–4355.

Meyer, F. M., Gerwig, J., Hammer, E., Herzberg, C., Commichau, F. M., Völker, U. & Stülke, J. (2011). Physical interactions between tricarboxylic acid cycle enzymes in *Bacillus subtilis*: evidence for a metabolon. *Metabolic Engineering* **13**, 18–27.

Pechter, K. B., Meyer, F. M., Serio, A. W., Stülke, J. & Sonenshein, A. L. (2013). Two roles for aconitase in the regulation of tricarboxylic acid branch gene expression in *Bacillus subtilis*. *Journal of Bacteriology* **195**, 1525–1537.

van der Rest, M. E., Frank, C. & Molenaar, D. (2000). Functions of the membrane-associated and cytoplasmic malate dehydrogenases in the citric acid cycle of *Escherichia coli*. *Journal of Bacteriology* **182**, 6892–6899.

Zhang, S. & Bryant, D. A. (2011). The tricarboxylic acid cycle in cyanobacteria. *Science* **334**, 1551–1553.

Anaplerotic sequence

Alber, B. E., Spanheimer, R., Ebenau-Jehle, C. & Fuchs, G. (2006). Study of an alternate glyoxylate cycle for acetate assimilation by *Rhodobacter sphaeroides*. *Molecular Microbiology* **61**, 297–309.

Borjian, F., Han, J., Hou, J., Xiang, H., Zarzycki, J. & Berg, I. A. (2017). Malate synthase and β-methylmalyl coenzyme A lyase reactions in the methylaspartate cycle in *Haloarcula hispanica*. *Journal of Bacteriology* **199**(4).

El-Mansi, M., Cozzone, A. J., Shiloach, J. & Eikmanns, B. J. (2006). Control of carbon flux through enzymes of central and intermediary metabolism during growth of *Escherichia coli* on acetate. *Current Opinion in Microbiology* **9**, 173–179.

Ensign, S. A. (2006). Revisiting the glyoxylate cycle: alternate pathways for microbial acetate assimilation. *Molecular Microbiology* **61**, 274–276.

Erb, T. J., Brecht, V., Fuchs, G., Müller, M. & Alber, B. E. (2009). Carboxylation mechanism and stereochemistry of crotonyl-CoA carboxylase/reductase, a carboxylating enoyl-thioester reductase. *Proceedings of the National Academy of Sciences of the USA* **106**, 8871–8876.

Khomyakova, M., Bükmez, Ö., Thomas, L. K., Erb, T. J. & Berg, I. A. (2011). A methylaspartate cycle in haloarchaea. *Science* **331**, 334–337

Leroy, B., De Meur, Q., Moulin, C., Wegria, G. & Wattiez, R. (2015). New insight into the photoheterotrophic growth of the isocitrate lyase-lacking purple bacterium *Rhodospirillum rubrum* on acetate. *Microbiology* **161**, 1061–1072.

Sauer, U. & Eikmanns, B. J. (2005). The PEP-pyruvate-oxaloacetate node as the switch point for carbon flux distribution in bacteria. *FEMS Microbiology Reviews* **29**, 765–794.

Incomplete TCA fork and reductive TCA cycle

Brutinel, E. D. & Gralnick, J. A. (2012). Anomalies of the anaerobic tricarboxylic acid cycle in *Shewanella oneidensis* revealed by Tn-seq. *Molecular Microbiology* **86**, 273–283.

Juhnke, H. D., Hiltscher, H., Nasiri, H. R., Schwalbe, H. & Lancaster, C. R. D. (2009). Production, characterization and determination of the real catalytic properties of the putative 'succinate dehydrogenase' from *Wolinella succinogenes*. *Molecular Microbiology* **71**, 1088–1101.

Miura, A., Kameya, M., Arai, H., Ishii, M. & Igarashi, Y. (2008). A soluble NADH-dependent fumarate reductase in the reductive tricarboxylic acid cycle of *Hydrogenobacter thermophilus* TK-6. *Journal of Bacteriology* **190**, 7170–7177.

Energy transduction in prokaryotes

Amend, J. P. & Shock, E. L. (2001). Energetics of overall metabolic reactions of thermophilic and hyperthermophilic Archaea and Bacteria. *FEMS Microbiology Reviews* **25**, 175–243.

Schoepp-Cothenet, B., van Lis, R., Atteia, A., Baymann, F., Capowiez, L., Ducluzeau, A.-L., Duval, S., ten Brink, F., Russell, M. J. & Nitschke, W. (2013). On the universal core of bioenergetics. *Biochimica et Biophysica Acta* **1827**, 79–93.

von Stockar, U., Maskow, T., Liu, J., Marison, I. W. & Patino, R. (2006). Thermodynamics of microbial growth and metabolism: an analysis of the current situation. *Journal of Biotechnology* **121**, 517–533.

Adenosine triphosphate (ATP) and ATPase

Au, K. M., Barabote, R. D., Hu, K. Y. & Saier, M. H. J. (2006). Evolutionary appearance of H^+-translocating pyrophosphatases. *Microbiology* **152**, 1243–1247.

Capaldi, R. & Aggeler, R. (2002). Mechanism of the F_1F_o-type ATP synthase, a biological rotary motor. *Trends in Biochemical Sciences* **27**, 154–160.

Ferguson, S. A., Keis, S. & Cook, G. M. (2006). Biochemical and molecular characterization of a Na^+-translocating F_1F_o-ATPase from the thermoalkaliphilic bacterium *Clostridium paradoxum*. *Journal of Bacteriology* **188**, 5045–5054.

Hicks, D. B., Liu, J., Fujisawa, M. & Krulwich, T. A. (2010). F_1F_O-ATP synthases of alkaliphilic bacteria: lessons from their adaptations. *Biochimica et Biophysica Acta* **1797**, 1362–1377.

Junge, W. & Nelson, N. (2015). ATP synthase. *Annual Review of Biochemistry* **84**, 631–657.

Lapierre, P., Shial, R. & Gogarten, J. P. (2006). Distribution of F- and A/V-type ATPases in *Thermus scotoductus* and other closely related species. *Systematic and Applied Microbiology* **29**, 15–23.

Mulkidjanian, A. Y., Makarova, K. S., Galperin, M. Y. & Koonin, E. V. (2007). Inventing the dynamo machine: the evolution of the F-type and V-type ATPases. *Nature Reviews Microbiology* **5**, 892–899.

Schlegel, K., Leone, V., Faraldo-Gómez, J. D. & Müller, V. (2012). Promiscuous archaeal ATP synthase concurrently coupled to Na^+ and H^+ translocation. *Proceedings of the National Academy of Sciences of the USA* **109**, 947–952.

Proton (sodium) motive force, and acid and alkali tolerance

Baker-Austin, C. & Dopson, M. (2007). Life in acid: pH homeostasis in acidophiles. *Trends in Microbiology* **15**, 165–171.

Cotter, P. D. & Hill, C. (2003). Surviving the acid test: responses of Gram-positive bacteria to low pH. *Microbiology and Molecular Biology Reviews* **67**, 429–453.

Hunte, C., Screpanti, E., Venturi, M., Rimon, A., Padan, E. & Michel, H. (2005). Structure of a Na^+/H^+ antiporter and insights into mechanism of action and regulation by pH. *Nature* **435**, 1197–1202.

Kanjee, U. & Houry, W. A. (2013). Mechanisms of acid resistance in *Escherichia coli*. *Annual Review of Microbiology* **67**, 65–81.

Krulwich, T. A., Sachs, G. & Padan, E. (2011). Molecular aspects of bacterial pH sensing and homeostasis. *Nature Reviews Microbiology* **9**, 330–343.

Lund, P., Tramonti, A. & De Biase, D. (2014). Coping with low pH: molecular strategies in neutralophilic bacteria. *FEMS Microbiology Reviews* **38**, 1091–1125.

Quinn, M. J., Resch, C. T., Sun, J., Lind, E. J., Dibrov, P. & Häse, C. C. (2012). NhaP1 is a $K^+(Na^+)/H^+$ antiporter required for growth and internal pH homeostasis of *Vibrio cholerae* at low extracellular pH. *Microbiology* **158**, 1094–1105.

Rhee, J. E., Jeong, H. G., Lee, J. H. & Choi, S. H. (2006). AphB influences acid tolerance of *Vibrio vulnificus* by activating expression of the positive regulator CadC. *Journal of Bacteriology* **188**, 6490–6497.

Electron transport phosphorylation

Antonyuk, S. V., Han, C., Eady, R. R. & Hasnain, S. S. (2013). Structures of protein–protein complexes involved in electron transfer. *Nature* **496**, 123–126.

Arai, H., Kawakami, T., Osamura, T., Hirai, T., Sakai, Y. & Ishii, M. (2014). Enzymatic characterization and in vivo function of five terminal oxidases in *Pseudomonas aeruginosa*. *Journal of Bacteriology* **19**, 4206–4215.

Elling, F. J., Becker, K. W., Könneke, M., Schröder, J. M., Kellermann, M. Y., Thomm, M. & Hinrichs, K.-U. (2016). Respiratory quinones in archaea: phylogenetic distribution and application as biomarkers in the marine environment. *Environmental Microbiology* **18**, 692–707.

Fadeeva, M. S., Nunez, C., Bertsova, Y. V., Espin, G. & Bogachev, A. V. (2008). Catalytic properties of Na^+-translocating NADH:quinone oxidoreductases from *Vibrio harveyi*, *Klebsiella pneumoniae*, and *Azotobacter vinelandii*. *FEMS Microbiology Letters* **279**, 116–123.

Lunak, Z. R. & Noel, K. D. (2015). A quinol oxidase, encoded by *cyoABCD*, is utilized to adapt to lower O_2 concentrations in *Rhizobium etli* CFN42. *Microbiology* **161**, 203–212.

Magalon, A., Arias-Cartin, R. & Walburger, A. (2012). Supramolecular organization in prokaryotic respiratory systems. *Advances in Microbial Physiology* **61**, 217–266.

Marreiros, B. C., Sena, F. V., Sousa, F. M., Batista, A. P. & Pereira, M. M. (2016). Type II NADH:quinone oxidoreductase family: phylogenetic distribution, structural diversity and evolutionary divergences. *Environmental Microbiology* **18**, 4697–4709.

Richardson, D. J. (2000). Bacterial respiration: a flexible process for a changing environment. *Microbiology* **146**, 551-571.

Richhardt, J., Luchterhand, B., Bringer, S., Büchs, J. & Bott, M. (2013). Evidence for a key role of cytochrome bo_3 oxidase in respiratory energy metabolism of *Gluconobacter oxydans*. *Journal of Bacteriology* **195**, 4210–4220.

Simon, J. G., van Spanning, R. J. M. & Richardson, D. J. (2008). The organisation of proton motive and non-proton motive redox loops in prokaryotic respiratory systems. *Biochimica et Biophysica Acta* **1777**, 1480–1490.

Steuber, J., Vohl, G. Casutt, M. S., Vorburger, T., Diederichs, K. & Fritz, G. (2014). Structure of the *V. cholerae* Na^+-pumping NADH:quinone oxidoreductase. *Nature* **516**, 62–67.

Other prokaryotic energy transduction mechanisms

Biegel, E. & Müller, V. (2010). Bacterial Na^+-translocating ferredoxin:NAD^+ oxidoreductase. *Proceedings of the National Academy of Sciences of the USA* **107**, 18138–18142.

Buckel, W. & Thauer, R. K. (2013). Energy conservation via electron bifurcating ferredoxin reduction and proton/Na^+ translocating ferredoxin oxidation. *Biochimica et Biophysica Acta* **1827**, 94–113.

Hedderich, R. (2004). Energy-converting [NiFe] hydrogenases from archaea and extremophiles: ancestors of complex I. *Journal of Bioenergetics and Biomembranes* **36**: 65–75.

Kim, B. H., Lim, S. S., Daud, W. R. W., Gadd, G. M. & Chang, I. S. (2015). The biocathode of microbial electrochemical systems and microbially-influenced corrosion. *Bioresource Technology* **190**, 395–401.

Lyell, N. L., Colton, D. M., Bose, J. L., Tumen-Velasquez, M. P., Kimbrough, J. H. & Stabb, E. V. (2013). Cyclic AMP receptor protein regulates pheromone-mediated bioluminescence at multiple levels in *Vibrio fischeri* ES114. *Journal of Bacteriology* **195**, 5051–5063.

Chapter 6

Biosynthesis and growth

Chapters 4 and 5 describe and explain the anabolic reactions that supply carbon skeletons, reducing equivalents (NADPH) and adenosine 5′-triphosphate (ATP) needed for biosynthesis. This chapter summarizes how the products of such anabolic reactions are used in biosynthesis and growth, ranging from monomer synthesis to the assembly of macromolecules within cells. Chemoheterotrophs, such as *Escherichia coli*, use approximately half of the glucose consumed to synthesize cell materials, while the other half is oxidized to carbon dioxide under aerobic conditions.

6.1 Molecular composition of bacterial cells

The elemental composition of microbial cells was discussed in Chapter 2 in order to help to understand what materials the bacteria use as their nutrients. These elements make up a range of molecules with various functions. Cellular molecular composition varies depending on the strain and growth conditions. As an example, Table 6.1 lists the molecular composition of *Escherichia coli* during the logarithmic phase when grown in a glucose–mineral salts medium. The moisture content is over 70 per cent, and protein is most abundant, occupying 55 per cent of the dry cell weight, followed by RNA at about 20 per cent. It is understandable that proteins are abundant since they catalyse cellular reactions. The DNA content is least variable, while the RNA content is higher at a higher growth rate. Not shown in Table 6.1 are storage materials, such as poly-β-hydroxybutyrate and glycogen, which vary profoundly within cells depending on growth conditions, and can comprise up to 70 per cent of the cell dry weight (Section 13.2).

The biosynthetic process, known as anabolism, can be discussed in three steps:

monomer biosynthesis
polymerization of monomers
assembly of polymers into cellular structure.

Figure 6.1 summarizes the catabolism that supplies carbon skeletons for monomer synthesis, followed by their polymerization and assembly into cell structure. For anabolism, nitrogen, sulfur, phosphorus and certain other elements are needed in addition to the carbon skeletons.

6.2 Assimilation of inorganic nitrogen

Many cell constituents are nitrogenous compounds and include amino acids and nucleic acid bases. Nitrogen exists in various redox states, ranging between −5 and +3. Organic nitrogen is used preferentially over inorganic nitrogen by almost all microbes. When organic nitrogen, ammonia or nitrate is not available, some prokaryotes (within the bacteria and archaea) can reduce gaseous nitrogen to

Table 6.1 Molecular composition of an *Escherichia coli* cell[a].

Component	Content (%)	Average molecular weight	Molecules/cell	Variety
Protein	55.0	40 000	2 360 000	1050
RNA	20.5			
23S rRNA		1 000 000	18 700	1
16S rRNA		500 000	18 700	1
5S rRNA		39 000	18 700	1
tRNA		25 000	205 000	60
mRNA		1 000 000	1380	400
DNA	3.1	2.5×10^9	2.13	1
Lipid	9.1	705	22 000 000	4[b]
LPS	3.4	4346	1 200 000	1[b]
Peptidoglycan	2.5	?	1	1
Glycogen	2.5	1 000 000	4360	1
Low molecular weight organics	2.9	?	?	?
Inorganics	1.0	?	?	?

Dry weight of a single cell, 0.28 picogram (pg).
Moisture content of the cell, 0.67 pg. Total weight of a single cell, 0.95 pg.
[a] Composition of the cells in the log phase. Cells were grown in glucose–mineral salts medium at 37°C (mass doubling time 40 min).
[b] Phospholipids in four groups regardless of the fatty acid composition. LPS, lipopolysaccharide.

ammonia to meet their nitrogen requirements (Table 6.2). This process is known as nitrogen fixation and is not found in eukaryotes.

6.2.1 Nitrogen fixation

Nitrogen is incorporated into cell constituents through transamination reactions using glutamate or glutamine as the amino group donor (Section 6.2.3). Glutamate and glutamine are synthesized from ammonia. Gaseous nitrogen (N_2) is very stable as it possesses a triple bond. When reduced or oxidized forms of nitrogen (fixed nitrogen) are not available, some prokaryotes reduce the structurally stable N_2 to ammonia to use as the nitrogen source, investing a large amount of energy in the form of ATP and reduced electron carriers. Ammonia fixed in this way serves as the nitrogen source for all forms of organisms, just as photosynthesis supplies organic materials as the energy source for many organisms. Biological N_2 fixation is estimated to be as high as 1.3×10^{14} g a year, which is more than twice the amount fixed industrially and naturally by lightning (5×10^{13} g). The fixed nitrogen returns to gaseous nitrogen through nitrification (Section 10.2) and denitrification (Section 9.1), which, together with biological nitrogen fixation, constitute the nitrogen cycle (Figure 6.2).

6.2.1.1 N_2-fixing organisms

A wide variety of prokaryotes have the ability to fix N_2 (Table 6.2). These include certain photosynthetic prokaryotes, anaerobic and aerobic bacteria, and archaea. Some fix N_2 in a symbiotic relationship with plants, while others can do so in the free-living state. Bacteria belonging to the genus *Rhizobium* are well known as symbiotic

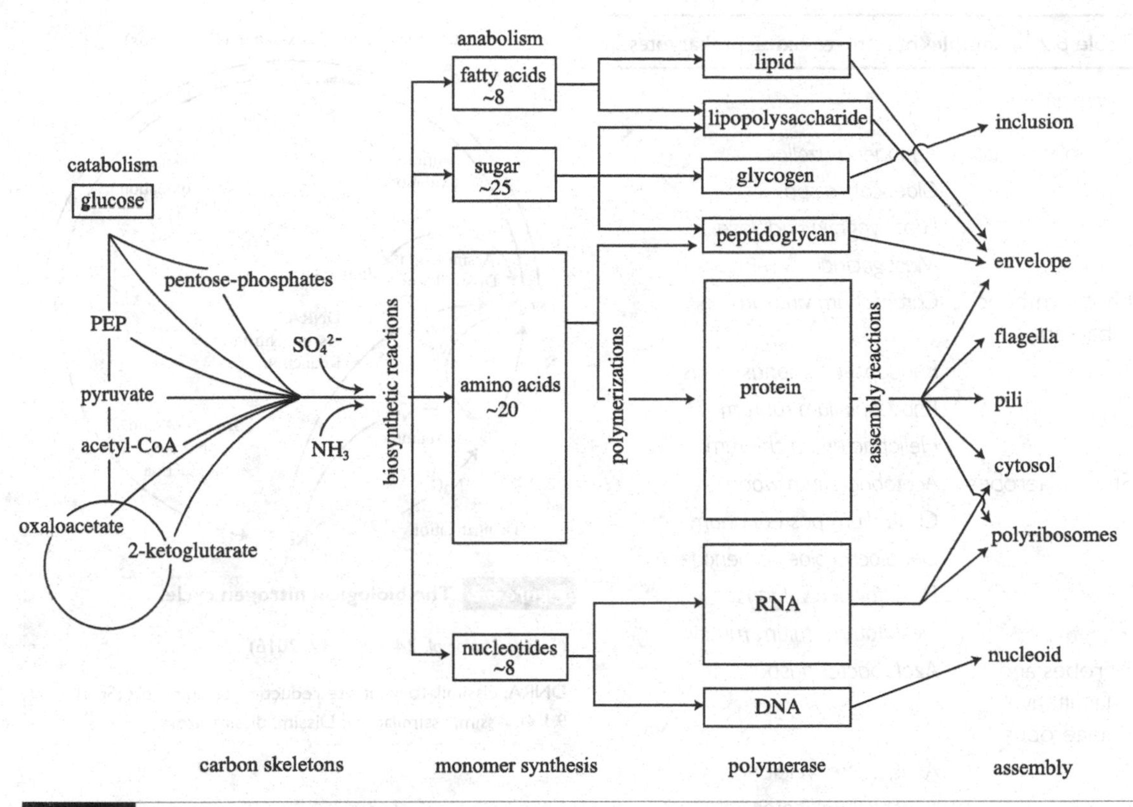

Figure 6.1 Catabolism and anabolism.

Monomers are synthesized from carbon skeletons and inorganic substances, such as ammonia, sulfate and phosphate, that are obtained from the growth medium or cellular environment. The monomers are polymerized into proteins, nucleic acids, polysaccharides, phospholipids and other macromolecules. Catabolism supplies not only carbon skeletons, but also ATP and NADPH required for anabolic processes. The size of the squares in the diagram represents the relative content in the *Escherichia coli* cell, summarized in Table 6.1. The numbers of individual monomers are shown in the squares.

nitrogen fixers with legumes. Alder trees host nitrogen-fixing *Frankia alni*, an actinomycete-like organism. *Anabaena azollae* is an example of a symbiotic nitrogen-fixing cyanobacterium associated with a fern, *Azolla*. *nifH*, a gene of the *nif* regulon, has been found in many uncultured phylogenetically diverse bacteria and archaea, suggesting that the diversity of N_2-fixing microbes is greatly underestimated.

6.2.1.2 Biochemistry of N_2 fixation

Nitrogen reduction to ammonia can be expressed as:

$$N_2 + 3H_2 + 2H^+ \longrightarrow 2NH_4^+ \quad (\Delta G^{0'} = -39.3\text{kJ/mol } NH_4^+)$$

This is an exergonic reaction, but requires a high activation energy due to the stable triple bond in N_2. For this reason, the industrial N_2 fixation process employs a high temperature (300–600°C) and high pressure (200–800 atm). Nitrogen-fixing microbes produce the enzyme nitrogenase, which reduces nitrogen under normal physiological conditions.

NITROGENASE

N_2 fixation is catalysed by nitrogenase. Nitrogenase is a complex protein consisting of azoferredoxin and molybdoferredoxin in a 2:1 ratio. Both of the enzymes are [Fe-S] proteins, and molybdenum is contained in molybdoferredoxin. Azoferredoxin is a homodimer of a protein

Table 6.2 Examples of nitrogen-fixing prokaryotes.

Bacteria	
Cyanobacteria	*Anabaena azollae*
	Gloeocapsa spp.
	Leptolyngbya nodulosa
	Mastigocladus laminosus
Photosynthetic bacteria	*Chromatium vinosum*
	Rhodopseudomonas viridis
	Rhodospirillum rubrum
	Heliobacterium chlorum
Strict anaerobes	*Acetobacterium woodii*
	Clostridium pasteurianum
	Dehalococcoides ethenogenes
	Desulfovibrio vulgaris
	Desulfotomaculum ruminis
Aerobes and facultative anaerobes	*Azotobacter paspali*
	Azotobacter vinelandii
	Azospirillum lipoferum
	Bacillus polymyxa
	Beijerinkia indica
	Burkholderia kururiensis
	Derxia gummosa
	Frankia alni
	Gluconacetobacter diazotrophicus
	Halobacterium halobium
	Klebsiella pneumoniae
	Methylococcus capsulatus
	Methylosinus trichosporium
	Mycobacterium flavum
	Pseudomonas azotogensis
	Rhizobium japonicum
	Thiobacillus ferrooxidans
Archaea	
Methanogens	*Methanosarcina barkeri*
	Methanococcus maripaludis
	Methanobacterium thermoautotrophicum

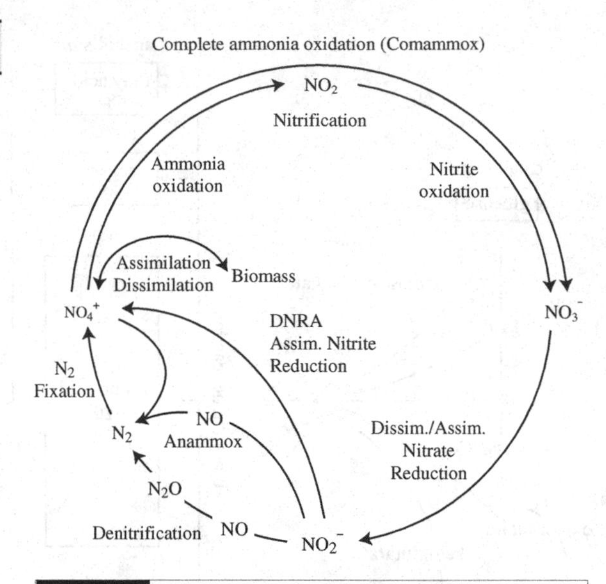

Figure 6.2 **The biological nitrogen cycle.**

(*Trends Microbiol.* **24**: 699–712, 2016)

DNRA, dissimilatory nitrate reduction to ammonia (Section 9.1.4); Assim, assimilatory; Dissim, dissimilatory.

containing a [4Fe–4S] cluster, and molybdoferredoxin consists of two molecules of two proteins containing two molybdenum and 28 iron and sulfur atoms (Table 6.3). Molybdoferredoxin reduces nitrogen with the reducing equivalents provided by azoferredoxin. Based on their functions, molybdoferredoxin is termed dinitrogenase, and azoferredoxin is termed dinitrogenase reductase. Since the redox potential of azoferredoxin is low (−0.43 V), ferredoxin or flavodoxin is believed to be the electron donor for its reduction. In this process, dissociated azoferredoxin from molybdoferredoxin is reduced, accepting the electrons from low redox potential ferredoxin or flavodoxin, followed by ATP binding. Molybdoferredoxin binds N_2 before forming a nitrogenase complex with the ATP-reduced azoferredoxin complex. At this point, electrons are transferred from azoferredoxin to molybdoferredoxin with ATP hydrolysis. Since azoferredoxin is a one-electron carrier, the reduction of a nitrogen molecule requires six oxidation–reduction cycles with the hydrolysis of at least 16 ATP molecules (Figure 6.3).

Table 6.3 | The nitrogenase complex of *Rhizobium* spp.

Characteristics	Azoferredoxin (dinitrogenase reductase)	Molybdoferredoxin (dinitrogenase)
Molecular weight	70 000	230 000
Subunit	2	4[a]
Iron	8[b]	28
Molybdenum	0	2
Acid-labile sulfide	8	28

[a] Two subunits each of two peptides with molecular weights of 55 000 and 60 000.
[b] One [4Fe–4S] for each subunit.

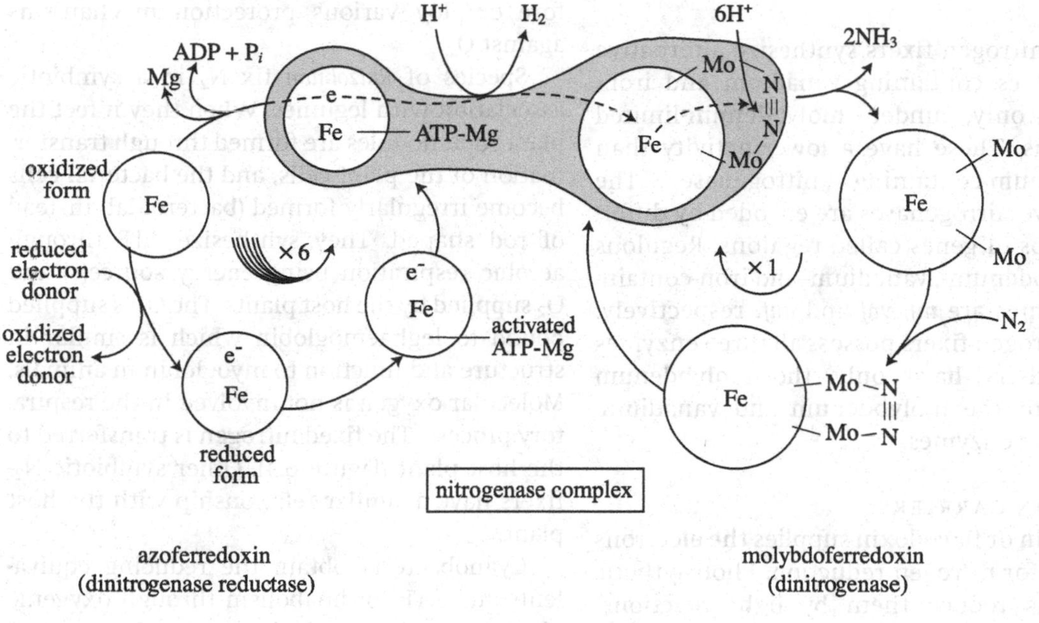

Figure 6.3 N_2 **reduction by the nitrogenase complex.**

(Sprent, J. I. 1979, *The Biology of Nitrogen-fixing Organisms*, Figure 2.1. McGraw-Hill, Maidenhead)

Azoferredoxin (dinitrogenase reductase) is reduced, coupled with the oxidation of ferredoxin or flavodoxin, and binds ATP. Molybdoferredoxin (dinitrogenase) binds N_2 and forms a nitrogenase complex with the reduced azoferredoxin–ATP complex. Electrons required to reduce N_2 are transferred from azoferredoxin to molybdoferredoxin, and this reaction repeats six times to reduce one molecule of N_2.

Nitrogenase can reduce various other substances in addition to dinitrogen (Table 6.4). The ability to reduce acetylene to ethylene is exploited in a simple nitrogenase assay method. Protons are reduced by the nitrogenase complex to H_2 during normal N_2 fixation.

Table 6.4 Substances reduced by the nitrogenase complex.

Substrate	Product(s)
N_2	$2NH_4^+$
N_3^-	N_2, NH_4^+
N_2O	N_2
HCN	CH_4, NH_4^+, CH_3NH_2
CH_3CN	C_2H_6, NH_4^+
CH_2CHCN	C_3H_6, NH_4^+, C_3H_8
C_2H_2	C_2H_4
$2H^+$	H_2

Most nitrogen fixers synthesize alternative nitrogenases containing vanadium and iron, or iron only, under molybdenum-limited conditions. These have a lower activity than molybdenum-containing nitrogenase. The alternative nitrogenases are encoded by different groups of genes called regulons. Regulons for molybdenum-, vanadium- and iron-containing enzymes are *nif*, *vnf* and *anf*, respectively. Some nitrogen fixers possess all three enzymes while others have only the molybdenum enzyme or the molybdenum and vanadium-containing enzymes.

ELECTRON CARRIERS

Ferredoxin or flavodoxin supplies the electrons required for nitrogen reduction. Photosythetic organisms reduce them by light reactions, and obligate anaerobes reduce ferredoxin by pyruvate:ferredoxin oxidoreductase or hydrogenase. Aerobes reduce them through a reverse electron transport mechanism (Section 10.1) using reduced pyridine nucleotides. Hydrogen produced from the nitrogenase reaction is used to reduce ferredoxin by the action of the hydrogenase (Section 10.5.2).

6.2.1.3 Bioenergetics of N_2 fixation

The Y_{ATP} of *Klebsiella pneumoniae* was measured to be 4.2 ± 0.2 g/mol ATP under nitrogen-fixing conditions and 10.9 ± 1.5 g/mol ATP with ammonia. These figures show that nitrogenase consumes 29 ATP to reduce one dinitrogen. This figure is much higher than the predicted ATP consumption by the nitrogenase complex. This might be because of the energy consumed in the reverse electron transport process to reduce the low redox potential electron carriers.

6.2.1.4 Molecular oxygen and N_2 fixation

All nitrogenases known to date are inactivated irreversibly by molecular oxygen. O_2 is required to synthesize the ATP needed for N_2 fixation through aerobic respiration, and photosystem II (Section 11.4) produces O_2 in N_2-fixing cyanobacteria. To avoid irreversible inactivation of nitrogenase by O_2, N_2-fixing organisms therefore employ various protection mechanisms against O_2.

Species of *Rhizobium* fix N_2 in a symbiotic association with legumes. When they infect the plant root, nodules are formed through transformation of the plant cells, and the bacterial cells become irregularly formed (bacteroidal) instead of rod shaped. They synthesize ATP through aerobic respiration using energy source(s) and O_2 supplied by the host plants. The O_2 is supplied bound to leghaemoglobin which is similar in structure and function to myoglobin in animals. Molecular oxygen is not involved in the respiratory process. The fixed nitrogen is transferred to the host plant (Figure 6.4). Other symbiotic N_2-fixers have a similar relationship with the host plant.

Cyanobacteria obtain the reducing equivalents and ATP for anabolism through oxygenic photosynthesis, generating molecular oxygen. Heterocystous cyanobacteria such as *Anabaena* and *Nostoc* spp. transform 5–10 per cent of normal vegetative cells within the filaments into heterocysts which lack oxygenic photosystem II to protect nitrogenase from molecular oxygen produced by photosystem II. Heterocysts fix nitrogen using electrons transported from neighbouring vegetative cells, and in return the heterocysts supply fixed nitrogen (Figure 6.5). These metabolite exchanges are mediated through communication channels that penetrate the septum between adjacent cells.

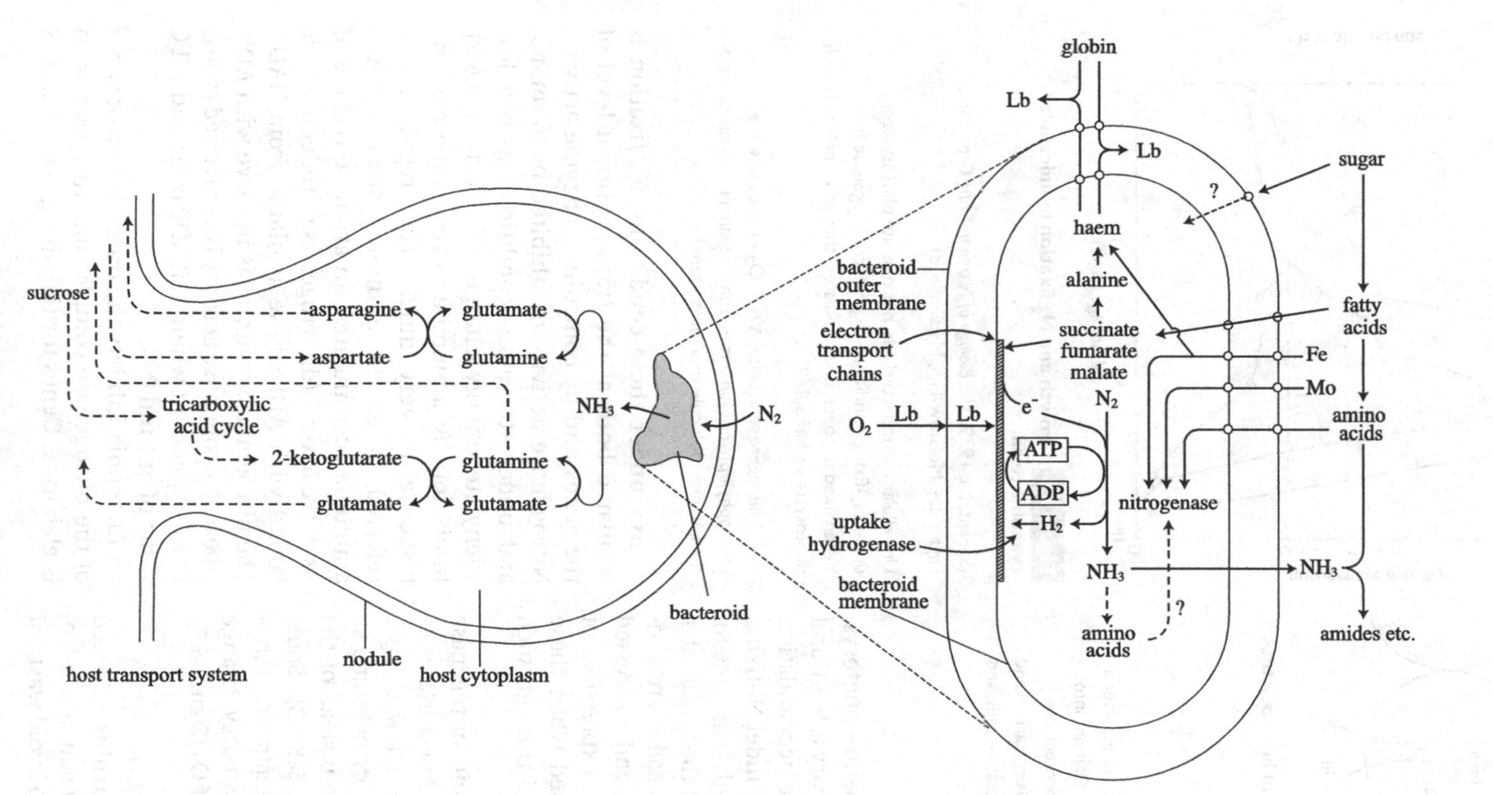

Figure 6.4 **Roles of the host plant and *Rhizobium* in symbiotic N_2 fixation.**

When *Rhizobium* infects the legume plant the host root cells form nodules, and the bacterial cells become irregular bacteroids. The host supplies the carbon and energy source, and O_2 in the leghaemoglobin (Lb) bound form. The bacterium fixes and supplies nitrogen to the host plant.

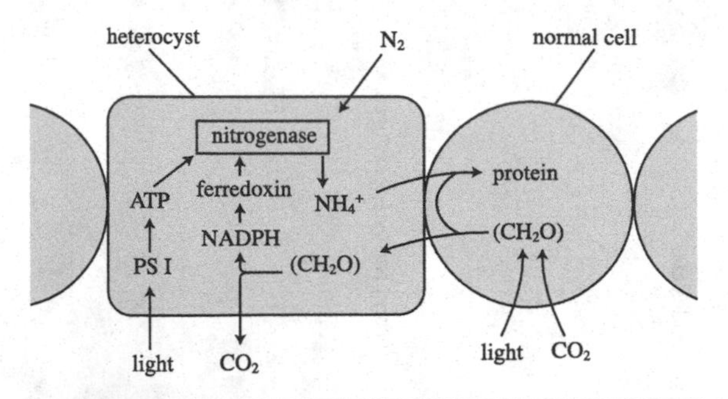

Figure 6.5 **N_2 fixation in heterocysts of heterocystous cyanobacteria.**

(Gottschalk, G. 1986, *Bacterial Metabolism*, 2nd edn., Figure 10.3. Springer, New York)

Under N_2-fixing conditions, heterocystous cyanobacteria transform 5–10 per cent of the cells in the filament into heterocysts which lack photosystem II to protect the nitrogenase from O_2. Heterocysts fix N_2 using carbon and energy sources obtained from normal vegetative cells, and supply fixed nitrogen to the vegetative cells.

PS I, photosystem I.

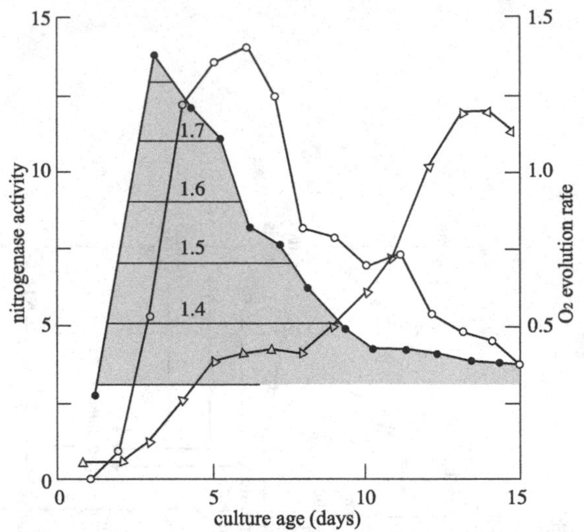

Figure 6.6 **Growth and N_2 fixation in unicellular cyanobacteria.**

(Sprent, J. I. 1979, *The Biology of Nitrogen-fixing Organisms*, Figure 2.4. McGraw-Hill, Maidenhead)

Unicellular cyanobacteria do not operate photosystem II under N_2-fixing conditions to protect nitrogenase from O_2. When fixed nitrogen is accumulated, they grow normally with photosystems I and II.

○ – ○, nitrogenase activity; Δ – Δ, O_2 evolution; ● – ●, chlorophyll/phycocyanin ratio (phycocyanin is a constituent of the antenna molecule of photosystem II).

Unicellular cyanobacteria operate photosystem I only when nitrogenase activity is high, and photosystem II appears with the accumulation of fixed nitrogen within the cell. Under N_2-fixing conditions, O_2 is not generated (Figure 6.6). Nearly all genes in the genome of the unicellular cyanobacterium *Synechococcus elongatus* are controlled by the circadian clock and expressed rhythmically (Section 11.4.3.1). Expression of genes for N_2-fixation is repressed when those for photosynthesis are expressed through such control by the circadian clock.

Aerobic bacteria protect their nitrogenase through different mechanisms. *Azotobacter vinelandii* keeps the O_2 concentration low using a high affinity terminal oxidase, cytochrome *d*, under nitrogen-fixing conditions instead of the normal cytochrome *o* (Section 5.8.2.3). Some nitrogenases of aerobes such as *Azotobacter chroococcum* and *Derxia gummosa* reversibly change their structure in the presence of O_2 (Figure 6.7).

6.2.1.5 Regulation of N_2 fixation

When an organism faces a surplus of fixed nitrogen or is starved of energy sources, nitrogenase activity is not needed. To avoid waste of energy under these conditions, N_2 fixation is regulated both at the transcriptional level of the genes and by controlling enzyme activity. Nitrogenase activity is inhibited by ammonia and under starvation conditions with a low adenylate energy charge (EC, Section 5.6.2). Inhibition by ammonia is reversible and the response is very quick. This regulation is referred to as the ammonia switch. When ammonia accumulates, an arginine residue of azoferredoxin (dinitrogenase reductase) is bound with ADP-ribose available from NAD^+. The nitrogenase complex is inactive with ADP-ribose. Nitrogenase activity is controlled to less than 10 per cent when ATP/ADP is around 1 (EC value of around 0.6).

Ammonia inhibits the transcription of genes of the nitrogenase complex and this regulation is elaborate. Genes for the nitrogenase complex

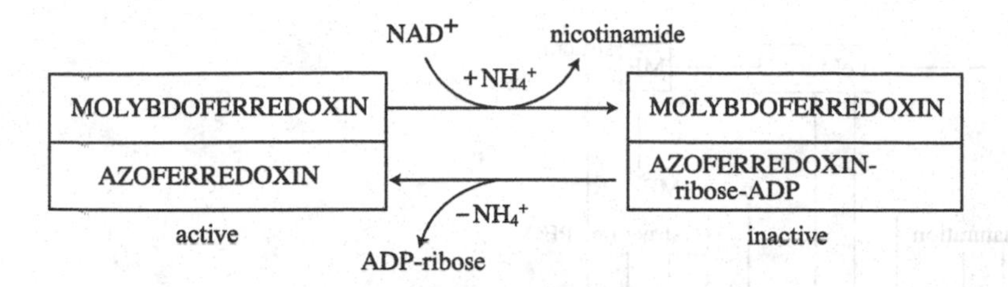

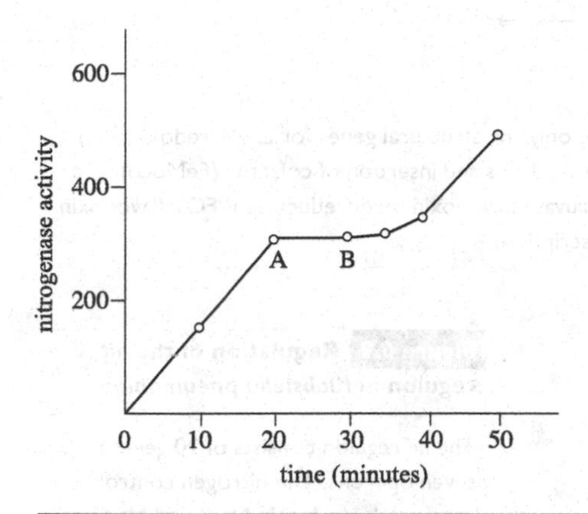

Figure 6.7 Relationship between dissolved O_2 concentration and nitrogenase activity in a free-living N_2-fixing bacterium.

(Sprent, J. I. 1979, *The Biology of Nitrogen-fixing Organisms*, Figure 2.5. McGraw-Hill, Maidenhead)

A free-living bacterium, *Azotobacter chroococcum*, fixes N_2 under low dissolved O_2 (DO) conditions, and nitrogenase activity disappears immediately when the DO concentration increases (A). The enzyme becomes active when the DO concentration decreases (B). At high DO concentrations, the nitrogenase is protected.

(*nif* regulon) consist of seven operons with 20 genes in *Klebsiella pneumoniae* (Figure 6.8). In addition to these genes, nitrogen control genes, *ntr* (Section 12.2.2), are also involved in their regulation.

Nitrogen control genes consist of *ntrA, B* and *C*. NtrA is a sigma factor of the RNA polymerase (σ^N, σ^{54}, Sections 6.11.1, 12.1.1), NtrC and NtrB are NR_I and NR_{II}, respectively (Section 12.2.2). They regulate the transcription and activity of enzymes related to ammonia metabolism. When the ammonia concentration is low, four molecules of UMP bind with NR_{II} which in turn phosphorylates NR_I. The phosphorylated NR_I activates NtrA(σ^N) to transcribe *nifA* and *nifL*. NifA is an activator for the transcription of other *nif* genes, NifL is a repressor protein (Figure 6.9).

NtrA, NtrB and NtrC regulate not only the *nif* regulon, but also other enzymes related to ammonia metabolism such as glutamine synthetase, and the transport and utilization of arginine, proline and histidine. The regulation of many operons with different functions by a single regulator is referred to as a global control system or multigene system. This is discussed later (Section 12.2).

6.2.2 Nitrate reduction

Many microbes use ammonia and nitrate as their nitrogen source when organic nitrogen is not available. Nitrate is reduced to ammonia before being incorporated into cell materials. These reactions are catalysed by nitrate reductase and nitrite reductase. They are different from a similar reaction that can occur under anaerobic conditions, where nitrate is used as an electron acceptor in an anaerobic respiratory process known as dissimilatory nitrate reduction (Section 9.1). The use of nitrate as a nitrogen source is referred to as assimilatory nitrate reduction:

$$NO_3^- + XH_2 \xrightarrow{\text{nitrate reductase}} NO_2^- + H_2O + X$$

$$NO_2^- + 3NADH + H^+ \xrightarrow{\text{nitrate reductase}} NH_3 + 3NAD^+ + 2H_2O$$

NAD(P)H is used as the reducing equivalent by the assimilatory nitrate reductase in many microbes. Reduced cytochrome provides electrons for nitrate reductase in some strains of *Pseudomonas*. Nitrite is further reduced to

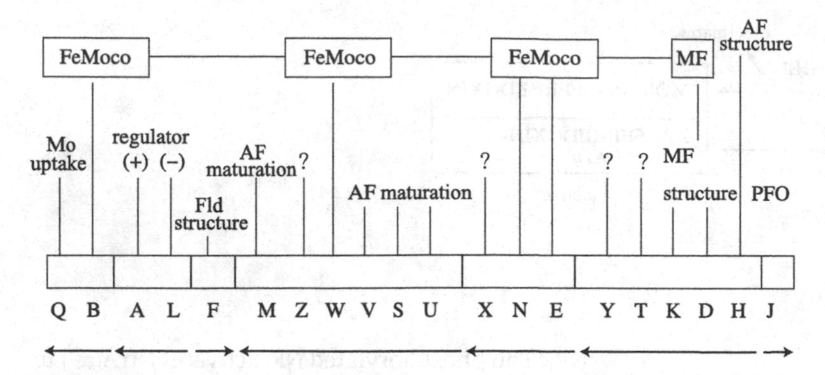

Figure 6.8 Structure of the *nif* regulon in *Klebsiella pneumoniae*.

The *nif* regulon consists of seven operons with 20 genes. The genes include not only the structural genes for azoferredoxin (AF) and molybdoferredoxin (MF), but also genes for related functions including the synthesis and insertion of cofactor (FeMoco), and proteins for electron metabolism in N_2 fixation and for regulation. They are pyruvate:flavodoxin oxidoreductase (PFO), flavodoxin (Fld) and *nifA* and *nifL*. The arrows at the bottom indicate the direction of transcription.

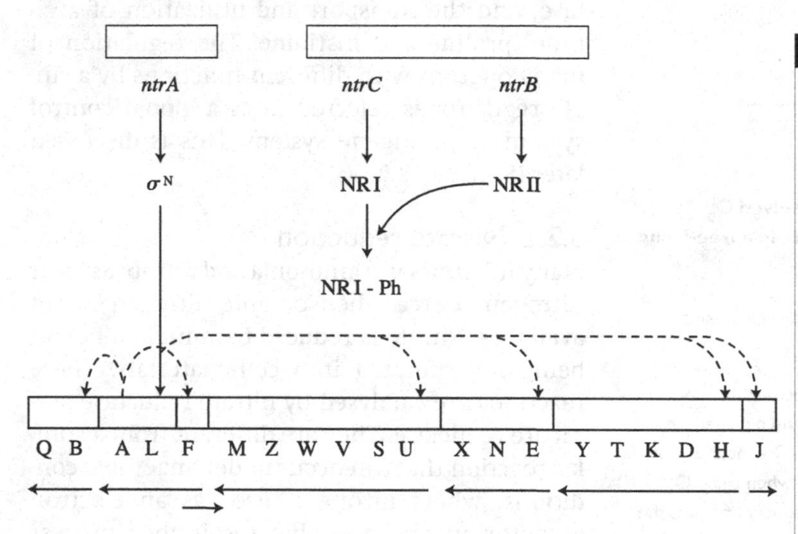

Figure 6.9 Regulation of the *nif* regulon in *Klebsiella pneumoniae*.

The *nif* regulon consists of 20 genes in seven operons. The nitrogen control gene products, NtrA, NtrB and NtrC, regulate the expression of the *nif* gene: *nifA* and *nifL* participate in the regulation. When the ammonia concentration is low, NtrB (NR_{II}) is uridylylated, and NR_{II} $(UMP)_4$ phosphorylates NR_I. The phosphorylated NR_I binds the enhancer region of *nifA* and *nifL* to induce transcription by NtrA(σ^N). NifA and NifL regulate the expression of the remaining *nif* genes as activator and repressor, respectively. When ammonia accumulates, NR_I–P_i is dephosphorylated, losing the ability to bind the enhancer region.

ammonia by nitrite reductase in a one-step reaction. NADH is the co-substrate of the enzyme.

Dissimilatory nitrate reduction is strongly inhibited by O_2, with a few exceptions in aerobic denitrifiers (Section 9.1.4), but O_2 does not inhibit assimilatory nitrate reduction, which is inhibited by ammonia (Section 12.2.2). Nitrate is reduced to N_2 by most denitrifying bacteria, but some bacteria, including *Desulfovibrio gigas*, reduce NO_3^- to NH_4^+. Since this reaction is coupled to the formation of a proton motive force and not inhibited by NH_4^+, this reaction is called dissimilatory nitrate reduction (Section 9.1.4).

6.2.3 Ammonia assimilation

Ammonia is assimilated as glutamate by means of two different reactions:

$$\text{2-ketoglutarate} + NH_3 + NADPH + H^+ \xrightarrow{\text{glutamate dehydrogenase}} \text{glutamate} + NADP^+ + H_2O$$

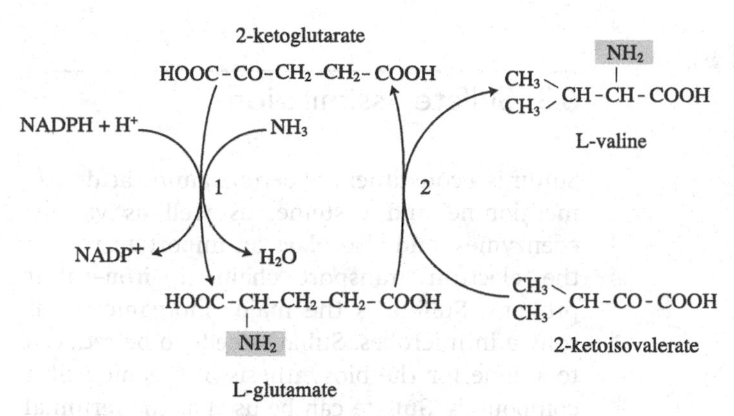

Figure 6.10 Ammonia assimilation by (1) glutamate dehydrogenase and (2) transaminase.

(Gottschalk, G. 1986, *Bacterial Metabolism*, 2nd edn., Figure 3.2. Springer, New York)

and

$$\text{glutamate} + NH_3 + ATP \xrightarrow{\text{glutamine synthetase}} \text{glutamine} + ADP + P_i$$

$$\text{glutamine} + NADPH + H^+ + \text{2-ketoglutarate} \xrightarrow{\text{glutamate synthase}} \text{2glutamate} + NADP^+$$

At a high ammonia concentration, glutamate dehydrogenase can assimilate ammonia without consuming ATP, since this enzyme has a low affinity for the substrate (K_m = 0.1 M). ATP is consumed in the assimilation of ammonia at low concentrations by the action of glutamine synthetase, which has a high substrate affinity (K_m = 0.1 mM). The amino group of glutamine is transferred to 2-ketoglutarate by glutamate synthase. This enzyme is generally called glutamine: 2-oxoglutarate aminotranferase (GOGAT).

Glutamate and glutamine donate amino groups in various synthetic reactions catalysed by transaminases (Figure 6.10). In *Escherichia coli*, low specificity transaminase A, B and C synthesize more than ten amino acids. In addition to their role in amino acid synthesis, glutamate and glutamine are also used as $-NH_2$ donors in the biosynthetic reactions of various other cell constituents, including nucleic acid bases, *N*-acetyl-glucosamine and the *N*-acetyl-muramic acid of peptidoglycan. In *Escherichia coli* about 85 per cent of organic nitrogen originates from glutamate and the remaining 15 per cent arises from glutamine.

Since glutamine synthetase consumes ATP to assimilate ammonia at low concentrations, its expression as well as its activity is tightly controlled according to ammonia availability, to avoid ATP consumption under ammonia-rich conditions. An enzyme with activity to uridylylate/deuridylyate PII protein (uridylyltransferase/uridylyl-removing enzyme, UTase/UR, *glnD* product) senses ammonia availability, and this signal is transduced to control the gene expression (Section 12.2.2) and enzyme activity through adenylylation of the enzyme molecule (Section 12.3.2.3). The enzyme is adenylylated by adenylyltransferase when glutamine is accumulated and the adenylylated enzyme is less active than the native form. Enzyme activity is not completely inhibited under ammonia-rich conditions, because glutamine should be synthesized as an amino group donor for the biosynthesis of nucleotides and some amino acids. Glutamine synthetase activity is also controlled through cumulative feedback inhibition (Section 12.3.1) by various metabolites synthesized from glutamine (Figure 6.11). The regulation involved is very complex and discussed in Chapter 12 (Section 12.2.2).

Many microbes can use toxic cyanide as the sole nitrogen source. This toxic compound is converted to ammonia in various ways including oxidation, reduction and hydrolysis. An alkaliphilic bacterium, *Pseudomonas pseudoalcaligenes*, is able to grow with cyanide as

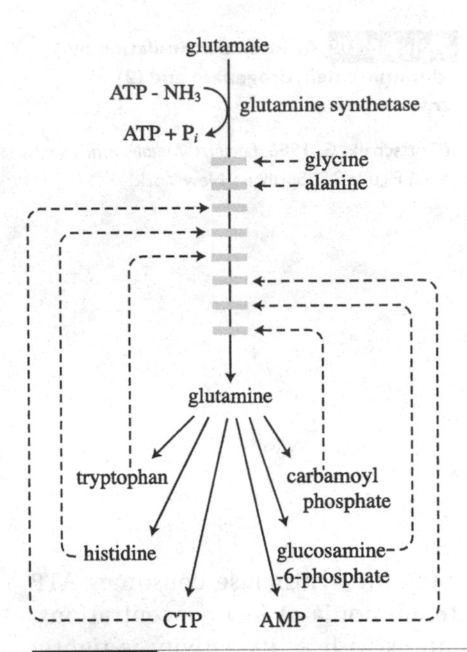

Figure 6.11 Regulation of glutamine synthetase activity by amino compounds synthesized through transamination from glutamine.

Amino compounds synthesized through transamination from glutamine modify glutamine synthetase to a less active form through adenylylation (Section 12.3.2.3).

the sole nitrogen source. Cyanide reacts with oxaloacetate spontaneously to produce a cyanohydrin (2-hydroxynitrile), which is further converted to ammonia. To provide oxaloacetate for these reactions under cyanotrophic conditions, cyanide-sensitive malate dehydrogenase is replaced with a malate:quinone oxidoreductase, and enzymes of TCA and glyoxylate cycles are higher than those observed with other nitrogen sources. In addition to cyanide, this bacterium is able to grow with several cyano derivatives, such 2- and 3-hydroxynitriles.

6.3 Sulfate assimilation

Sulfur is a constituent of certain amino acids, e.g. methionine and cysteine, as well as various coenzymes, and also plays an important role in the electron transport chain in iron-sulfur proteins. Sulfate is the major inorganic sulfur source in microbes. Sulfate needs to be reduced to sulfide for the biosynthesis of organic sulfur compounds. Sulfate can be used as the terminal electron acceptor in a group of anaerobic bacteria known as sulfate-reducing bacteria (Section 9.3). This process is referred to as dissimilatory sulfate reduction, whereas assimilatory sulfate reduction is used to describe the process that uses sulfate as the sulfur source. Sulfate is activated by adenosine-5′-phosphosulfate (APS) with ATP (Figure 6.12) as in dissimilatory sulfate reduction (Section 9.3). The redox potential ($E^{0\prime}$) of HSO_4^-/HSO_3^- is −454 mV, which is lower than any known natural electron carriers. Unlike in dissimilatory reduction, APS is further activated to adenosine-3′-phosphate-5′-phosphosulfate (PAPS):

adenosine-5′-phosphosulfate (APS) —(APS kinase; ATP → ADP)→ adenosine-3′-phosphate-5′-phosphosulfate (PAPS)

Sulfate transported into the cell is converted to adenine-5′-phosphosulfate (APS) in a reaction catalysed by ATP sulfurylase (sulfate adenyltransferase). PP_i produced in the reaction is hydrolysed to $2P_i$ by phosphatase to pull the reaction towards the synthesis of APS. APS is phosphorylated to adenosine-3′-phosphate-5′-phosphosulfate (PAPS) by APS kinase, consuming ATP. PAPS reductase reduces PAPS to sulfite, liberating AMP-3′-phosphate, using electrons from reduced thioredoxin. Sulfite reductase further reduces sulfite to sulfide using NADPH as a coenzyme.

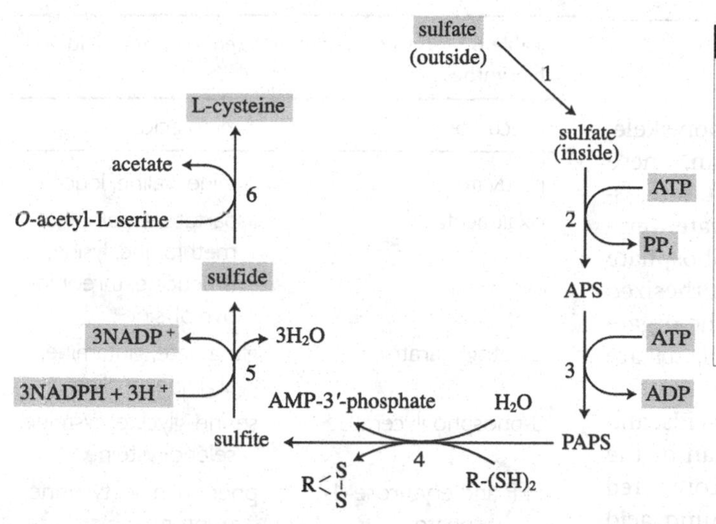

Figure 6.12 Assimilatory sulfate reduction and cysteine synthesis.

(Gottschalk, G. 1986, *Bacterial Metabolism*, 2nd edn., Figure 3.4. Springer, New York)

1, active transport; 2, ATP sulfurylase; 3, APS phosphokinase; 4, PAPS reductase; 5, sulfite reductase; 6, *O*-acetyl-L-serine sulfhydrylase. APS, adenosine-5′-phosphosulfate; PAPS, adenosine-3′-phosphate-5′-phosphosulfate; $R(SH)_2$, reduced thioredoxin.

O-acetyl-L-serine sulfhydrylase incorporates sulfide into cysteine (Figure 6.12). Just as glutamate and glutamine function as $-NH_2$ donors, cysteine is used as a –SH donor in biosynthesis.

Soil bacteria belonging to the genus *Pseudomonas* reduce adenine-5′-phosphosulfate (APS) by assimilatory APS reductases, before APS is phosphorylated to PAPS as in dissimilatory sulfate reduction (Section 9.3.1). The sulfite reductase of this group of bacteria uses reduced ferredoxin instead of NADPH as the electron donor.

A hyperthermophilic archaeon, *Aeropyrum pernix*, utilizes *O*-phospho-L-serine in place of *O*-acetyl-L-serine to assimilate sulfide to cysteine, catalysed by *O*-phosphoserine sulfhydrylase. In methanogens, *O*-phospho-L-serine is transferred to $tRNA^{cys}$ before being converted to Cys-$tRNA^{cys}$:

Many bacteria oxidize organic sulfonate, thiols, sulfide or thiophene to sulfate to use as their sulfur source in the absence of sulfate or sulfur-containing amino acids. Many marine bacteria lack genes for assimilatory sulfate reduction. They use organic sulfur compounds such as methionine and 3-dimethylsulphoniopropionate released by algae. Recalcitrant organic sulfur compounds such as dibenzothiophene are also used as a sulfur source by some bacteria. These properties have been studied as a means of removing sulfur from petroleum and coal. Genes for enyzmes utilizing organic sulfonate are controlled by a regulatory protein, SsuR, in *Burkholderia cenocepacia* and in *Rhodobacter capsulatus*.

In *Escherichia coli* a number of proteins have been identified that are produced under sulfate-limited conditions. These are called the sulfate starvation induced (SSI) stimulon.

serine ($NH_2-CH(COOH)-CH_2OH$) —[ATP → ADP; serine kinase]→ phosphoserine ($NH_2-CH(COOH)-CH_2O-P$)

hyperthermophilic archaea: phosphoserine —[H_2S → P_i; phosphoserine sulfhydrylase]→ cysteine ($NH_2-CH(COOH)-CH_2-SH$)

methanogenic archaea: phosphoserine —[phosphoseryl-tRNA synthetase]→ phosphoseryl-$tRNA^{Cys}$ ($NH_2-CH(C(=O)-O-tRNA^{Cys})-CH_2O-P$) —[$H_2S$ → P_i; cysteinyl-tRNA synthase]→ cysteinyl-$tRNA^{Cys}$ ($NH_2-CH(C(=O)-O-tRNA^{Cys})-CH_2-SH$)

6.4 Amino acid biosynthesis

Amino acids are synthesized using carbon skeletons available from central metabolism. These are pyruvate, oxaloacetate, 2-ketoglutarate, 3-phosphoglycerate, phosphoenolpyruvate, erythrose-4-phosphate and ribose-5-phosphate (Table 6.5). Some amino acids are synthesized by different pathways depending on the organism. For convenience, those of *Escherichia coli* are discussed below.

Proteins consist mainly of L-amino acids, and D forms are found in the peptidoglycan of the bacterial cell wall. L-amino acids are converted to the corresponding D forms by amino acid racemase or amino acid epimerase in bacteria. In some cases, D-amino acids are synthesized by stereospecific amination of 2-ketoacids. They are indispensible and disruption of their synthesis leads to cell death.

Table 6.5 Carbon skeletons used for amino acid biosynthesis.

Precursor	Amino acid
pyruvate	alanine, valine, leucine
oxaloacetate	aspartate, asparagine, methionine, lysine, isoleucine, threonine, pyrrolysine
2-ketoglutarate	glutamate, glutamine, arginine, proline
3-phosphoglycerate	serine, glycine, cysteine, selenocysteine
PEP and erythrose-4-phosphate	phenyalanine, tyrosine, tryptophan
ribose-5-phosphate	histidine

6.4.1 The pyruvate and oxaloacetate families

Pyruvate and oxaloacetate are converted to alanine and aspartate through reactions catalysed by transaminase. In these reactions, glutamate is used as the $-NH_2$ donor. Asparagine synthetase synthesizes asparagine from aspartate and ammonia, consuming energy in the form of ATP in a similar reaction to that catalysed by glutamine synthetase:

$$\underset{\text{pyruvate}}{CH_3-CO-COOH} \xrightarrow[\text{transaminase}]{\text{L-glutamate} \searrow\nearrow \text{2-ketoglutarate}} \underset{\text{alanine}}{CH_3-CH(NH_2)-COOH}$$

$$\underset{\text{oxaloacetate}}{HOOC-CH_2-CO-COOH} \xrightarrow[\text{transaminase}]{\text{L-glutamate} \searrow\nearrow \text{2-ketoglutarate}} \underset{\text{aspartate}}{HOOC-CH_2-CH(NH_2)-COOH}$$

$$\underset{\text{aspartate}}{HOOC-CH_2-CH(NH_2)-COOH} + NH_3 + ATP \xrightarrow{\text{asparagine synthetase}} \underset{\text{asparagine}}{H_2N-CO-CH_2-CH(NH_2)-COOH} + ADP + P_i$$

Threonine, methionine and lysine are produced from aspartate in addition to asparagine (Figure 6.13). An intermediate of lysine biosynthesis, diaminopimelate, is not found in protein, but is a precursor of peptidoglycan, and so is ornithine, an intermediate in arginine biosynthesis (Figure 6.19). Most prokaryotes employ the diaminopimelate pathway to synthesize lysine (Figure 6.13), but yeasts and fungi synthesize this amino acid from 2-ketoglutarate through the 2-aminoadipate pathway (Figure 6.14).

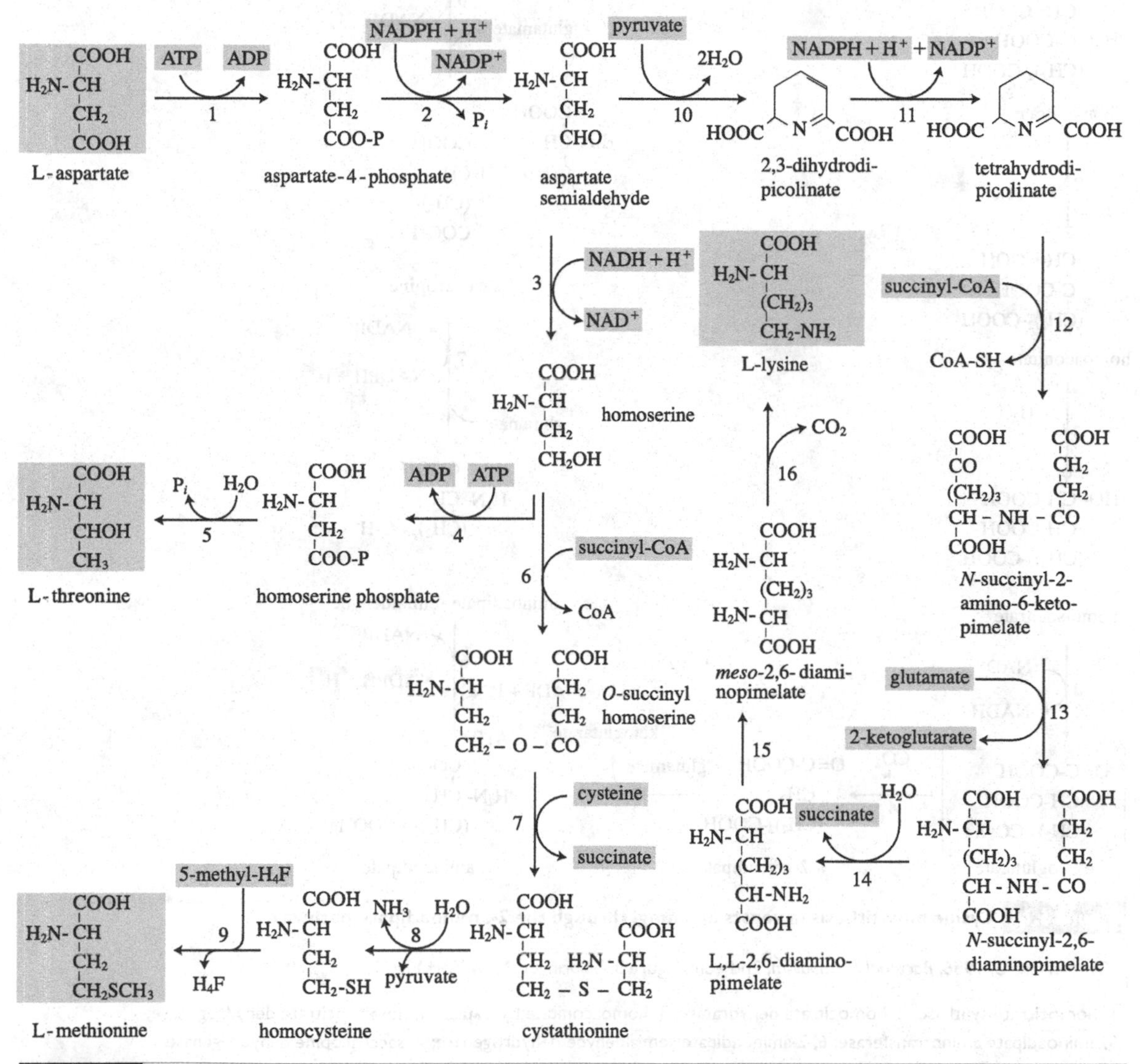

Figure 6.13 **Biosynthesis of threonine, methionine and lysine from the common precursor, aspartate.**

(Gottschalk, G. 1986, *Bacterial Metabolism*, 2nd edn., Figure 3.6. Springer, New York)

1, aspartate kinase; 2, aspartate semialdehyde dehydrogenase; 3, homoserine dehydrogenase; 4, homoserine kinase; 5, threonine synthase; 6, homoserine acyltransferase; 7, cystathionine synthase; 8, cystathionine lyase; 9, homocysteine:5-methyl-tetrahydrofolate methyltransferase; 10, dihydrodipicolinate synthase; 11, dihydrodipicolinate reductase; 12, tetrahydrodipicolinate succinylase; 13, glutamate:succinyl-diaminopimelate aminotransferase; 14, succinyl-diaminopimelate desuccinylase; 15, diaminopimelate epimerase; 16, diaminopimelate decarboxylase. H4F, tetrahydrofolate.

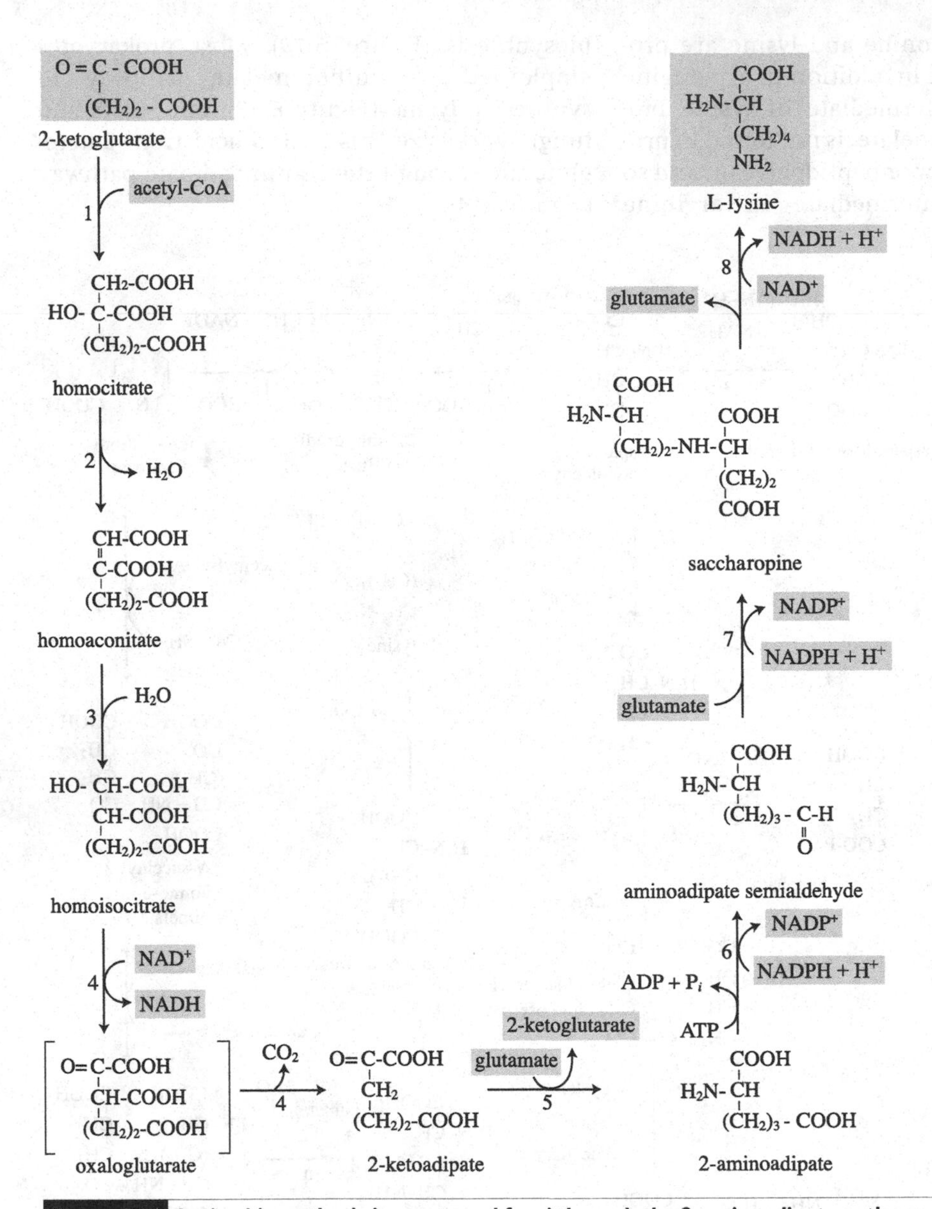

Figure 6.14 **Lysine biosynthesis in yeasts and fungi through the 2-aminoadipate pathway.**

(Gottschalk, G. 1986, *Bacterial Metabolism*, 2nd edn., Figure 3.7. Springer, New York)

1, homocitrate synthase; 2, homocitrate dehydratase; 3, homoaconitate hydratase; 4, homoisocitrate dehydrogenase; 5, aminoadipate aminotransferase; 6, 2-aminoadipate semialdehyde dehydrogenase; 7, saccharopine dehydrogenase (glutamate-forming); 8, saccharopine dehydrogenase (lysine-forming).

Analyses of genome sequences and biochemical studies have uncovered significant diversity in the enzymes and metabolic intermediates that are used for methionine biosynthesis in three reactions: acylation (reaction 6, Figure 6.13), sulfurylation (reaction 7, Figure 6.13) and methylation (reaction 9, Figure 6.13). In *Escherichia coli*, succinyl-CoA is used by homoserine acyltransferase, but acetyl-CoA is used by many bacteria including *Thermotoga maritima, Bacillus cereus* and *Agrobacterium tumefaciens*, to produce *O*-acetylhomoserine instead

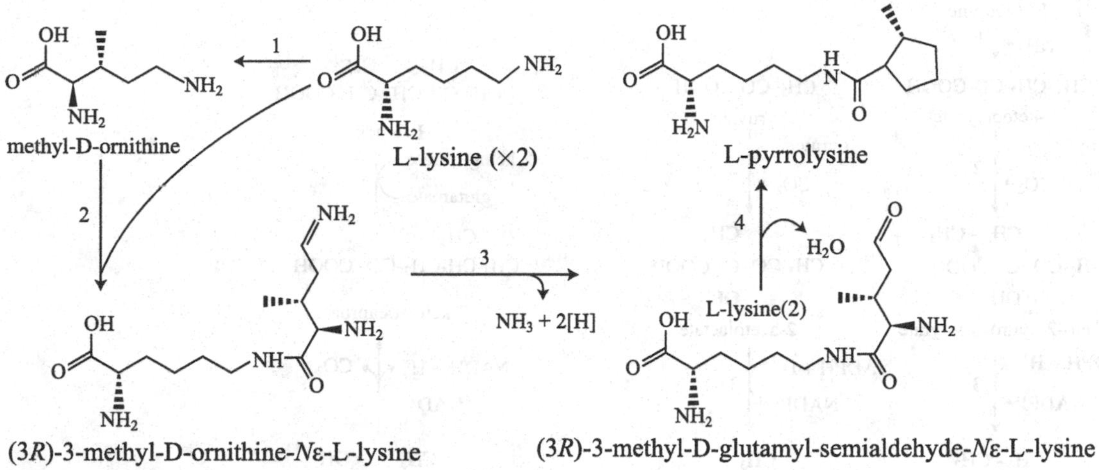

methyl-D-ornithine
L-lysine (×2)
L-pyrrolysine
$NH_3 + 2[H]$
L-lysine(2)
H_2O
(3*R*)-3-methyl-D-ornithine-*N*ε-L-lysine
(3*R*)-3-methyl-D-glutamyl-semialdehyde-*N*ε-L-lysine

Figure 6.15 Pyrrolysine biosynthesis in methanogens.

(Modified from *Nature* **471**: 647–650, 2011)

1, 3-methylornithine synthase; 2, 3-methylornithine-L-lysine ligase; 3, 3-methylornithine-*N*ε-L-lysine dehydrogenase; 4, spontaneous dehydration.

of *O*-succinylhomoserine. *O*-acetylhomoserine is sulfurylated to homocysteine in a one-step reaction catalysed by *O*-acetylhomoserine thiolase using sulfide:

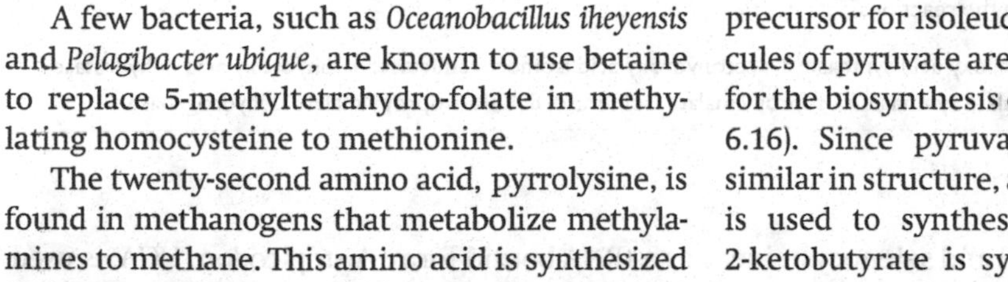

$$\underset{\text{acetyl homoserine}}{COOH-CH(NH_2)-CH_2-CH_2-O-C-CH_3} \xrightarrow[\text{acylhomoserine thiolase}]{H_2S \;\; \rightarrow \;\; CH_3COOH} \underset{\text{homocysteine}}{COOH-CH(NH_2)-CH_2-CH_2-SH}$$

A few bacteria, such as *Oceanobacillus iheyensis* and *Pelagibacter ubique*, are known to use betaine to replace 5-methyltetrahydro-folate in methylating homocysteine to methionine.

The twenty-second amino acid, pyrrolysine, is found in methanogens that metabolize methylamines to methane. This amino acid is synthesized from two molecules of lysine (Figure 6.15). The pyrrolysine operon, *pylTSBCD*, encodes $tRNA^{Pyl}$ (*pylT*), pyrrolysin-tRNA synthetase (*pylS*) and three enzymes (*pylBCD*) synthesizing the amino acid. It is interesting to note that this amino acid uses a stop codon, UAG, on mRNA for translation (Section 6.12). $tRNA^{Pyl}$ is called $tRNA_{CUA}$. How the stop codon is recognized by pyl-$tRNA^{Pyl}$ will be described later in Section 6.12.

Threonine produced from aspartate is deaminated to 2-ketobutyrate, which is used as the precursor for isoleucine biosynthesis. Two molecules of pyruvate are condensed to 2-acetolactate for the biosynthesis of valine and leucine (Figure 6.16). Since pyruvate and 2-ketobutyrate are similar in structure, a series of the same enzymes is used to synthesize isoleucine and valine. 2-ketobutyrate is synthesized by two different routes in *Geobacter sulfurreducens*: either from threonine or from acetyl-CoA and pyruvate through citramalate.

6.4.2 The phosphoglycerate family

The EMP pathway intermediate 3-phosphoglycerate is converted to serine and then further to

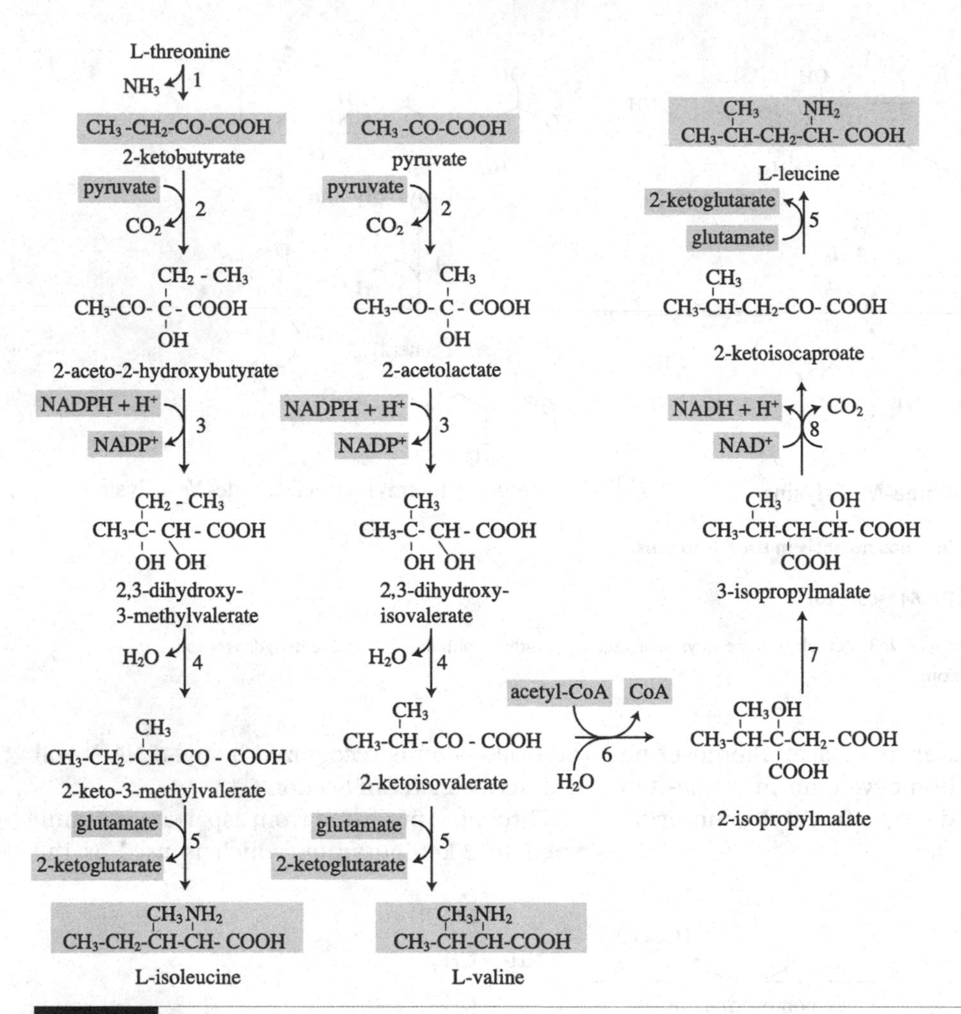

Figure 6.16 Biosynthesis of isoleucine, valine and leucine.

(Gottschalk, G. 1986, *Bacterial Metabolism*, 2nd edn., Figure 3.8. Springer, New York)

Isoleucine and valine are synthesized from 2-ketobutyrate and pyruvate, respectively. Since the precursors are similar in structure, the same enzymes catalyse the reactions.

1, threonine dehydratase; 2, acetohydroxy acid synthase; 3, acetohydroxy acid isomeroreductase; 4, dihydroxy acid dehydratase; 5, transaminase; 6, alpha-isopropylmalate synthase; 7, isopropylmalate isomerase; 8, beta-isopropylmalate dehydrogenase.

glycine, cysteine (Figure 6.17) and selenocysteine (Figure 6.18).

The twenty-first amino acid, selenocysteine (Sec), is found in all domains of life as a constituent of enzymes, including glutathione peroxidase, thioredoxin reductase, formate dehydrogenase and glycine reductase. This amino acid is very reactive and not found in a free form, but synthesized as a tRNA-bound form. A stop codon, UGA, is used for selenocysteine. The synthesis of this amino acid begins with formation of seryl-$tRNA^{sec}$ (seryl-$tRNA_{UGA}$) by the action of serine-tRNA ligase (Figure 6.18). Separately, selenide is activated to selenomonophosphate (Se-P). In bacteria, selenocysteinyl-$tRNA^{sec}$ synthase combines seryl-$tRNA^{sec}$ and Se-P to selenocysteinyl-$tRNA^{sec}$. On the other hand, seryl-tRNA^{sec} is activated to phosphoseryl-tRNA^{sec} before the synthase reacts in archaea.

6.4.3 The 2-ketoglutarate family

Glutamate synthesized from 2-ketoglutarate through the reactions catalysed by glutamate dehydrogenase or GOGAT is the precursor for the synthesis of proline, arginine and glutamine (Figure 6.19). *N*-acetylornithine deacetylase (reaction 9, Figure 6.19) has not been detected in coryneform bacteria, *Pseudomonas aeruginosa* and the yeast *Saccharomyces cerevisiae*. Instead, the reaction is catalysed by *N*-acetylglutamate-acetylornithine acetyltransferase in these organisms, coupling reactions 6 and 9 in Figure 6.19.

6.4.4 Aromatic amino acids

The benzene ring of aromatic amino acids is formed from shikimate, which is produced from the condensation of erythrose-4-phosphate and phosphoenolpyruvate. Shikimate is further metabolized to phenylpyruvate and *p*-hydroxyphenylpyruvate, before being transaminated to phenylalanine and tyrosine, respectively. Transaminase catalyses these reactions using glutamate as the $-NH_2$ donor. Tryptophan is synthesized from indole-3-glycerol phosphate, catalysed by tryptophan synthase using serine as the $-NH_2$ donor (Figure 6.20).

Indole-3-glycerol phosphate is formed through the condensation of anthranilate and 5-phospho-D-ribosyl-1-pyrophosphate (PRPP). PRPP is synthesized from ribose-5-phosphate

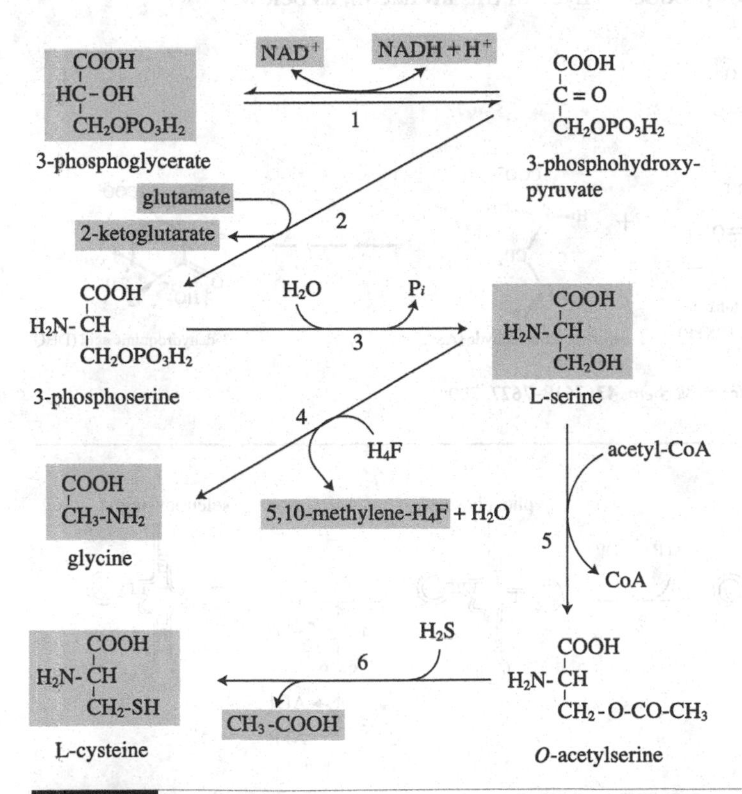

Figure 6.17 **Serine, glycine and cysteine biosynthesis from 3-phosphoglycerate.**

(Gottschalk, G. 1986, *Bacterial Metabolism*, 2nd edn., Figure 3.9. Springer, New York)

1, phosphoglycerate dehydrogenase; 2, phosphoserine aminotransferase; 3, phosphoserine phosphatase; 4, serine hydroxymethyltransferase; 5, serine transacetylase; 6, *O*-acetylserinesulfhydrylase.

H_4F, tetrahydrofolate.

taking pyrophosphate from ATP catalysed by PRPP synthetase as below:

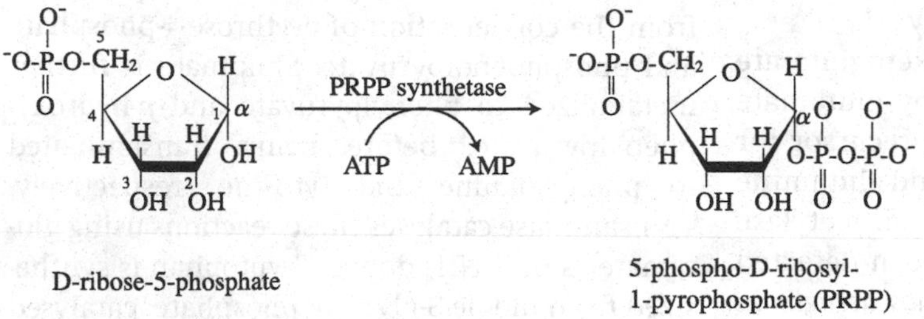

D-ribose-5-phosphate → (PRPP synthetase; ATP → AMP) → 5-phospho-D-ribosyl-1-pyrophosphate (PRPP)

PRPP is the precursor for the synthesis of histidine and nucleotides. PRPP synthetase is regulated by a feedback inhibition mechanism by its biosynthetic products.

The hexose monophosphate pathway is not known in archaea (Section 4.3.5), and they produce aromatic amino acids in a different way from bacteria. Enzymes catalysing the first two reactions of aromatic amino acid biosynthesis producing dehydroquinate (DHQ, Figure 6.20) are not found in *Methanocaldococcus jannaschii*. DHQ is produced not from erythrose-4-phosphate but from 6-deoxy-5-ketofructose-1-phosphate and aspartate semialdehyde in this archaeon, as below:

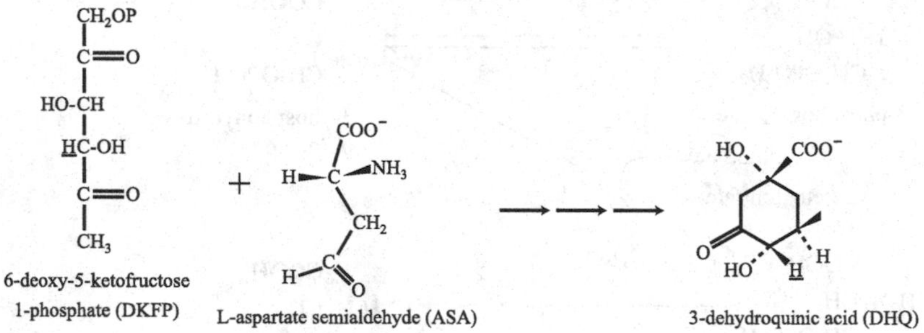

6-deoxy-5-ketofructose 1-phosphate (DKFP) + L-aspartate semialdehyde (ASA) → → → 3-dehydroquinic acid (DHQ)

(Modified from *Biochem.* **43**: 7618–7627, 2004)

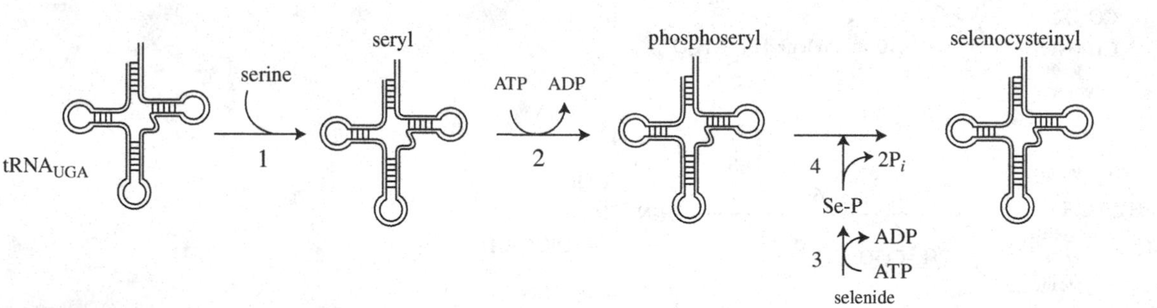

Figure 6.18 Selenocysteinyl-tRNA synthesis in archaea.

(Modified from *Extremophiles* **16**: 793–803, 2012)

The synthesis begins with formation of seryl-$tRNA^{sec}$ (seryl-$tRNA_{UCA}$) by serine-tRNA ligase (1). Separately, selenide is activated to selenomonophosphate (Se-P) by selenide kinase (3). In bacteria, selenocysteinyl-$tRNA^{sec}$ synthase (4) combines seryl-$tRNA^{sec}$ and Se-P to selenocysteinyl-$tRNA^{sec}$. On the other hand, seryl-$tRNA^{sec}$ is activated to phosphoseryl-$tRNA^{sec}$ by seryl-$tRNA^{sec}$ kinase in archaea before the synthase reacts (2).

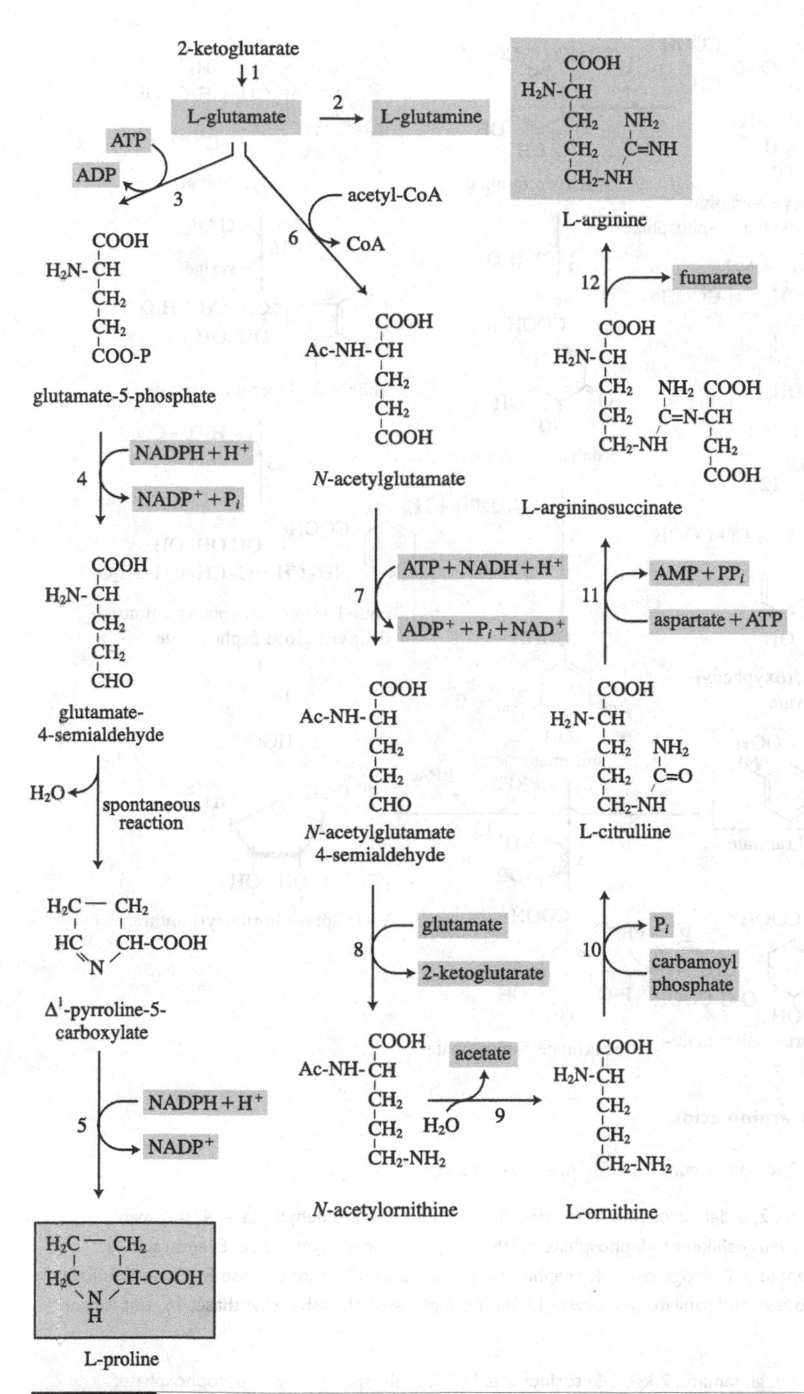

Figure 6.19 **Biosynthesis of glutamate, glutamine, proline and arginine from 2-ketoglutarate.**

(Gottschalk, G. 1986, *Bacterial Metabolism*, 2nd edn., Figure 3.10. Springer, New York)

1, glutamate dehydrogenase or glutamate synthase; 2, glutamate synthetase; 3, glutamate kinase; 4, glutamate semialdehyde dehydrogenase; 5, Δ^1-pyrroline-5-carboxylate reductase; 6, amino acid acetyltransferase; 7, *N*-acetylglutamate kinase and *N*-acetylglutamate semialdehyde dehydrogenase; 8, *N*-acetylornithine transaminase; 9, *N*-acetylornithine deacetylase; 10, ornithine transcarbamoylase; 11, argininosuccinate synthetase; 12, argininosuccinate lyase.

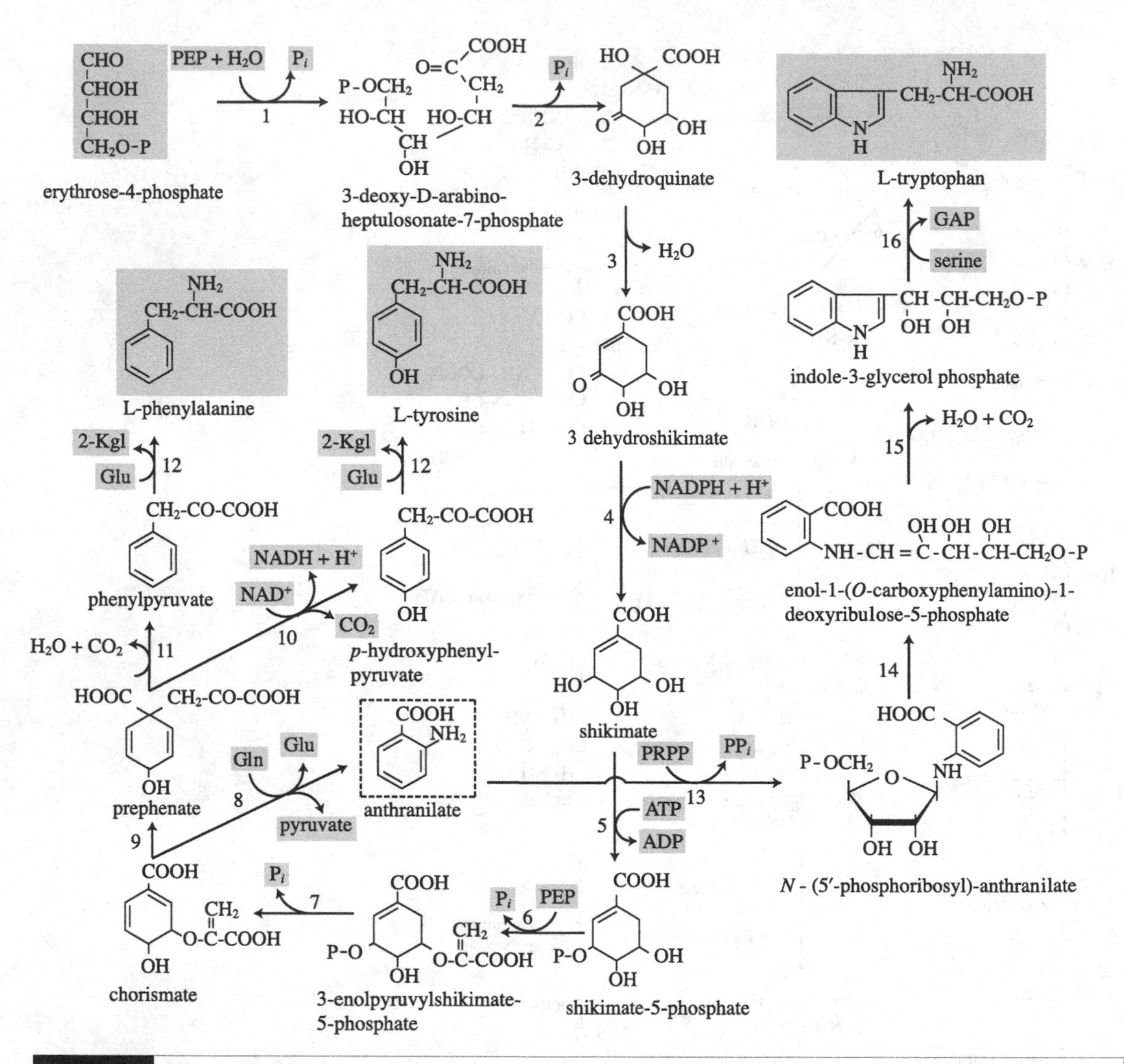

Figure 6.20 **Biosynthesis of aromatic amino acids.**

(Gottschalk, G. 1986, *Bacterial Metabolism*, 2nd edn., Figure 3.11. Springer, New York)

1, 3-deoxy-D-arabinoheptulosonate synthase; 2, 5-dehydroquinate synthase; 3, 3-dehydroquinate dehydratase; 4, shikimate dehydrogenase; 5, shikimate kinase; 6, enolpyruvylshikimate-5-phosphate synthase; 7, chorismate synthetase; 8, anthranilate synthase; 9, chorismate mutase; 10, prephenate dehydrogenase; 11, prephenate dehydratase; 12, transaminase B; 13, anthranilate phosphoribosyl transferase; 14, phosphoribosyl–anthranilate isomerase; 15, indole-3-glycerol phosphate synthase; 16, tryptophan synthase.

PEP, phosphoenolpyruvate; gln, glutamine; glu, glutamate; 2-kgl, 2-ketoglutarate; PRPP, 5-phosphoribosyl-1-pyrophosphate; GAP, glyceraldehyde-3-phosphate.

The hyperthermophilic archaea *Archaeoglobus fulgidus* and *Nanoarchaeum equitans* have a trifunctional enzyme, AroQ, to replace three enzymes of aromatic amino acid biosynthesis. AroQ has activities of chorismate mutase (reaction 9, Figure 6.20), prephenate dehydrogenase (reaction 10, Figure 6.20) and prephenate dehydratase (reaction 11, Figure 6.20).

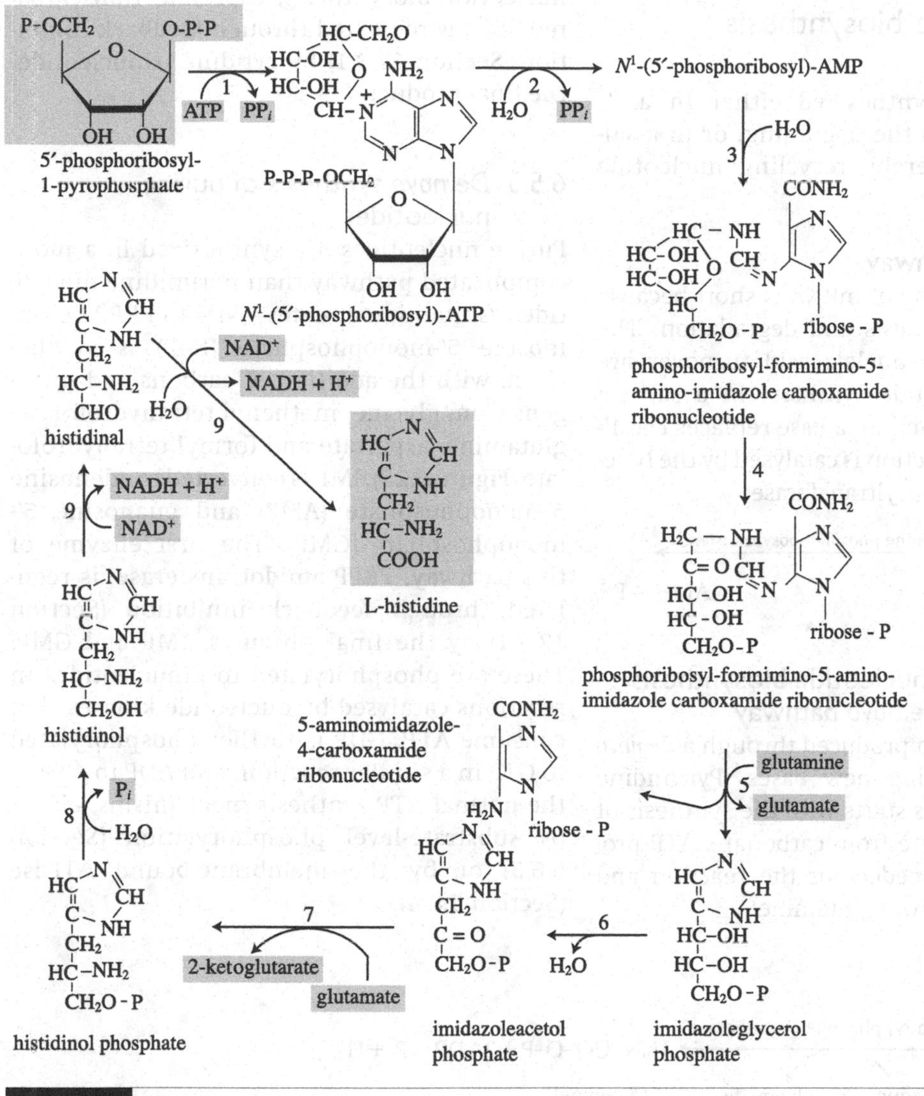

Figure 6.21 **Histidine biosynthesis.**

(Gottschalk, G. 1986, *Bacterial Metabolism*, 2nd edn., Figure 3.13. Springer, New York)

1, phosphoribosyl; ATP pyrophosphorylase; 2, phosphoribosyl-ATP pyrophosphohydrolase; 3, phosphoribosyl-AMP cyclohydrolase; 4, phosphoribosyl formimino-5-aminoimidazole carboxamide ribonucleotide isomerase; 5, phosphoribosylformimino-5-aminoimidazole carboxamide ribonucleotide:glutamine aminotransferase and cyclase; 6, imidazoleglycerol phosphate dehydratase; 7, histidinol phosphate transaminase; 8, histidinol phosphatase; 9, histidinol dehydrogenase (catalyses both reactions).

6.4.5 Histidine biosynthesis

Histidine is produced from PRPP (Figure 6.21).

6.4.6 Regulation of amino acid biosynthesis

Biosynthesis of amino acids is regulated according to their concentration. Enzyme activity is regulated by feedback inhibition while the transcription of genes is regulated by various mechanisms. This subject is discussed later (Section 12.1).

6.5 Nucleotide biosynthesis

Nucleotides are synthesized either in a *de novo* pathway from the beginning, or in a salvage pathway, thereby recycling nucleotide bases.

6.5.1 Salvage pathway

Normally the half-life of mRNA is short because of continuous synthesis and degradation. The bases arising from nucleic acid turnover are recycled to nucleotide synthesis in a salvage pathway. In this reaction, a base replaces the 1′-PP_i of PRPP. This reaction is catalysed by the base-specific phosphoribosyltransferase:

$$\text{adenine} + \text{PRPP} \xrightarrow{\text{adenine phosphoribosyltransferase}} \text{AMP} + PP_i$$

6.5.2 Pyrimidine nucleotide biosynthesis through a *de novo* pathway

Nucleotides are also produced through a *de novo* pathway synthesizing new bases. Pyrimidine nucleotide synthesis starts with the synthesis of carbamoyl phosphate from carbonate. ATP provides the energy needed for the reaction and the $-NH_2$ group is from glutamine:

$$HCO_3^- + 2ATP \xrightarrow[\text{glutamine} \rightarrow \text{glutamate}]{\text{carbamoyl phosphate synthetase}} \underset{\text{carbamoyl phosphate}}{H_2N\text{-}CO\text{-}O\text{-}P} + 2ADP + P_i + H^+$$

As shown in Figure 6.22, orotate is synthesized as a precursor of pyridine nucleotides before binding to PRPP by the action of phosphoribosyltransferase, as in the salvage pathway described previously.

Mononucleotides are phosphorylated in two steps to trinucleotides by nucleotide kinases, consuming ATP, before being used in RNA synthesis. The first enzyme of pyridine nucleotide biosynthesis, aspartate transcarbamoylase, is regulated through feedback inhibition (Section 12.3.1) by cytidine trinucleotide, the final product.

6.5.3 *De novo* synthesis of purine nucleotides

Purine nucleotides are synthesized in a more complicated pathway than pyrimidine nucleotides. Glutamine donates $-NH_2$ to PRPP before inosine 5′-monophosphate (IMP) is synthesized, with the addition of carbons and nitrogens from glycine, methenyl tetrahydrofolate, glutamine, aspartate and formyl tetrahydrofolate (Figure 6.23). IMP is converted to adenosine 5′-monophosphate (AMP) and guanosine 5′-monophosphate (GMP). The first enzyme of this pathway, PRPP amidotransferase, is regulated through feedback inhibition (Section 12.3.1) by the final products AMP and GMP. These are phosphorylated to dinucleotides in reactions catalysed by nucleotide kinases that consume ATP. GDP is further phosphorylated to GTP in a similar reaction, and ADP to ATP in the normal ATP synthesis mechanisms, either by substrate-level phosphorylation (Section 5.6.5) or by the membrane-bound ATPase (Section 5.8.4).

6.5.4 Synthesis of deoxynucleotides

RNA synthesis requires ATP, guanosine 5′-triphosphate (GTP), cytidine 5′-triphosphate (CTP) and uridine 5′-triphosphate (UTP) as building blocks, and deoxynucleotide triphosphates are the substrates for DNA polymerase. Ribonucleotides are reduced to deoxynucleotide by ribonucleotide reductase. Ribonucleotide diphosphates are the substrate for the reductase:

P-P-OCH2 base OH OH — ribonucleotide reductase → P-P-OCH2 base OH H $+ H_2O$

thioredoxin (reduced) thioredoxin (oxidized)

$NADP^+$ ← thioredoxin reductase — $NADPH + H^+$

(Gottschalk, G. 1986, *Bacterial Metabolism*, 2nd edn., Figure 3.20. Springer, New York)

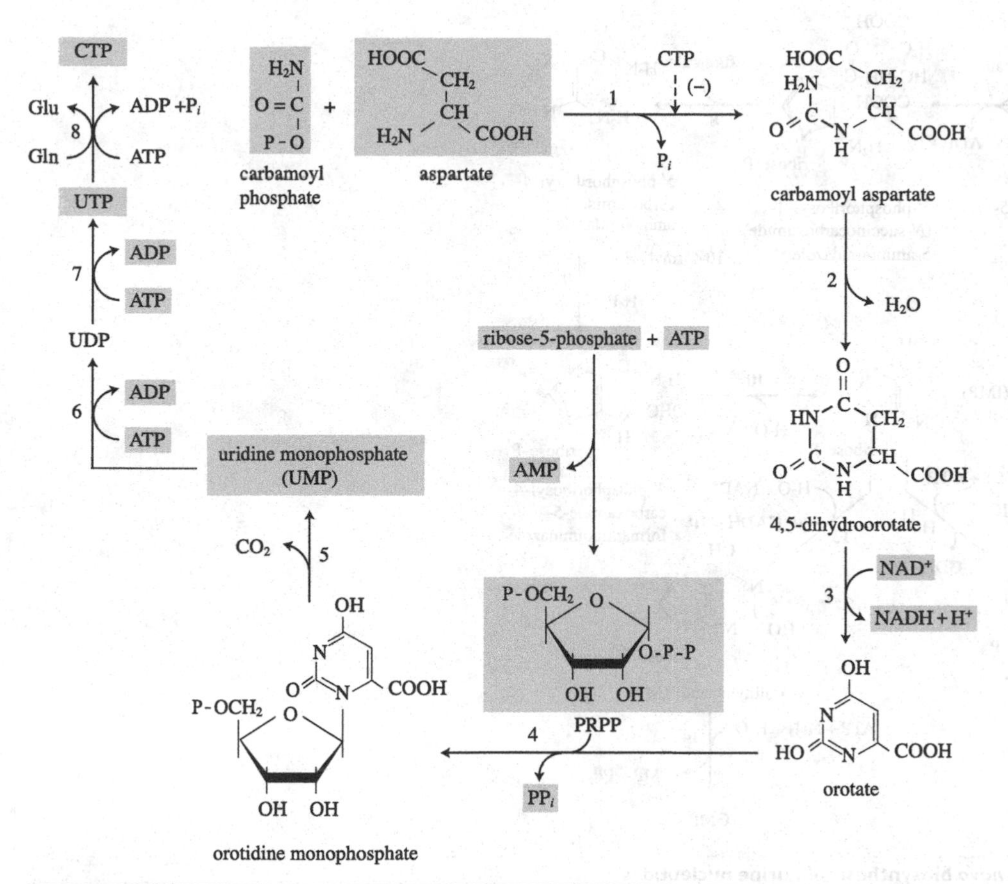

Figure 6.22 ***De novo* biosynthesis of pyridine nucleotides.**

(Gottschalk, G. 1986, *Bacterial Metabolism*, 2nd edn., Figure 3.16. Springer, New York)

1, aspartate transcarbamoylase; 2, dihydroorotase; 3, dihydroorotate dehydrogenase; 4, orotate phosphoribosyl transferase; 5, orotidine-5-phosphate decarboxylase; 6, nucleoside monophosphate kinase; 7, nucleoside diphosphate kinase; 8, CTP synthetase.

Gln, glutamine: Glu, glutamate.

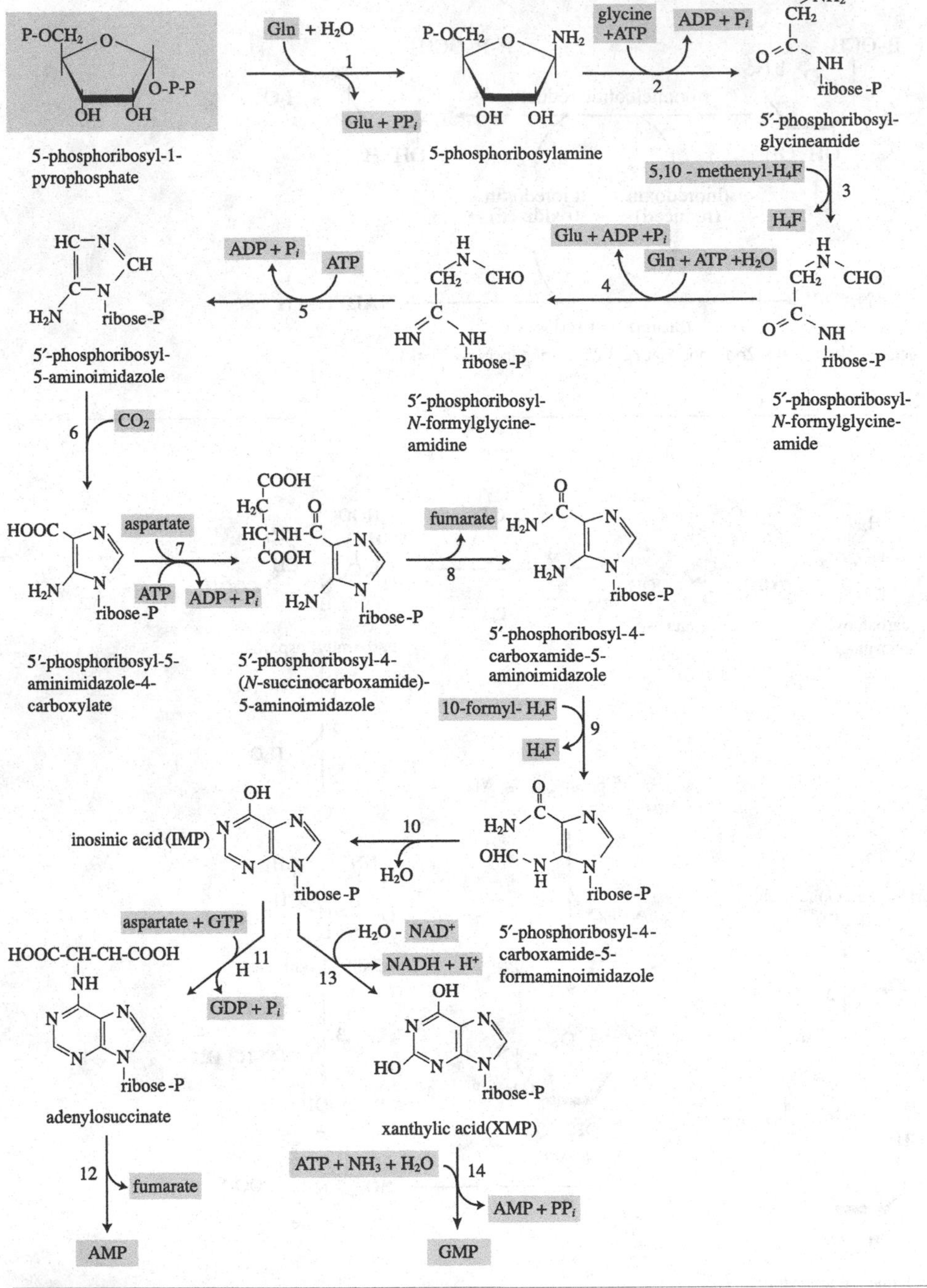

Figure 6.23 ***De novo* biosynthesis of purine nucleotides.**

(Gottschalk, G. 1986, *Bacterial Metabolism*, 2nd edn., Figure 3.19. Springer, New York)

1, amidophosphoribosyltransferase; 2, phosphoribosylglycineamide synthetase; 3, phosphoribosylglycineamide formyltransferase; 4, phosphoribosyl-formylglycineamidine synthetase; 5, phosphoribosyl-aminoimidazole synthetase; 6, phosphoribosyl-aminoimidazole carboxylase; 7, phosphoribosylaminoimidazole succinocarboxamide synthetase; 8, adenylosuccinate lyase; 9, phosphoribosylaminoimidazole-carboxamide formyltransferase; 10, IMP cyclohydrolase; 11, adenylosuccinate synthetase; 12, adenylosuccinate lyase; 13, IMP dehydrogenase; 14, GMP synthetase.

H_4F, tetrahydrofolate; Gln, glutamine: Glu, glutamate.

Table 6.6 Ribonucleotide reductases and their characteristics.

Enzyme	Ia	Ib	II	III
Organism	eukaryotes, fac G(-) anaerobes	fac G(-) anaerobes	other bacteria, halophilic archaea, thermophilic archaea	methanogens, fac G(-) anaerobes
Condition	aerobic	aerobic	aerobic/anaerobic	anaerobic
Radical	tyrosine	tyrosine	coenzyme B_{12}	glycine
Electron donor	thioredoxin I glutaredoxin	thioredoxin I glutaredoxin	thioredoxin I glutaredoxin	formate

fac G(-) anaerobes, facultatively anaerobic Gram-negative bacteria; coenzyme B_{12}, 5′-deoxyadenosylcobalamin.
Source: Trends Biochem. Sci. **22**: 81–85, 1997.

Since ribose is very stable, this reaction requires free radicals. There are three distinct classes of enzyme with different stable free radical amino acids. Type I (Ia and Ib) enzymes require O_2 to generate a tyrosine radical for the reaction. A cofactor, 5′-deoxyadenosylcobalamin (Section 8.7), is the radical in class II enzymes while the class III enzymes are anaerobic and employ a glycyl radical for the reaction (Table 6.6). Eukaryotic cells have class Ia enzymes, and class Ia, Ib and III enzymes are found in *Escherichia coli*. Many aerobic bacteria have class II enzymes, and the anaerobic class III enzymes are functional in anaerobic bacteria and anaerobic methanogenic archaea. Halophilic and thermophilic archaea employ class II enzymes. The anaerobic ribonucleotide reductase (class III) utilizes formate as the electron donor for the reaction, while the other enzymes take electrons from reduced thioredoxin or glutaredoxin coupling, with the oxidation of NADPH. Deoxyuridine diphosphate (dUDP) is dephosphorylated before being methylated to deoxythymidine phosphate (dTMP) in a reaction catalysed by thymidylate synthase. Deoxyribonucleotides are used in DNA synthesis.

dUDP → (dUDP phosphatase; P_i) → dUMP (deoxyribose-P) → (thymidylate synthase; 5,10-methylene-H_4F → dihydrofolate) → dTMP (deoxyribose-P)

(Gottschalk, G. 1986, *Bacterial Metabolism*, 2nd edn., Figure 3.20. Springer, New York)

The transcription of genes for ribonucleotide reductase is controlled by a regulator protein, DnaA, that regulates DNA replication for efficient supply of substrates for DNA polymerase.

6.6 Lipid biosynthesis

Phospholipids are essential cellular components as a major part of the cell membrane. Bacterial phospholipids are based on acylglyceride with an ester link between glycerol and fatty acids, as in eukaryotic cells. The archaeal membrane contains phospholipids with an ether linkage between polyalcohol and polyisoprenoid alcohols (Section 2.3.4). Fatty acids and polyisoprenoid alcohols are synthesized from acetyl-CoA.

6.6.1 Fatty acid biosynthesis

Acetyl-CoA is converted to acyl-acyl carrier protein (acyl-ACP) through the action of seven enzymes. In eukaryotes, these enzymes form a complex called type I fatty acid synthase, but such a complex is not found in prokaryotes (type II fatty acid synthase). Enzymes directly involved are 3-ketoacyl-ACP synthase,

3-ketoacyl ACP reductase, 3-hydroxyacyl-ACP dehydratase and enoyl-ACP reductase. Isoenzymes are identified in all of them except 3-ketoacyl-ACP reductase (FabG). These isoenzymes have specific functions. Three isoenzymes of 3-ketoacyl-ACP synthase are I (FabB), II (FabF) and III (FabH). FabA (3-hydroxydecanoyl-ACP dehydratase) and FabZ (3-hydroxyacyl-ACP dehydratase) have a similar function. Enoyl-ACP reductase also has three isozymes: I (FabI), II (FabK) and III (FabL). Fatty acid synthesis is initiated by 3-ketoacyl-ACP synthase III (FabH) catalysing the formation of acetoacetyl-ACP from malonyl-ACP and acetyl-ACP. The other two 3-ketoacyl-ACP synthases do not react with acetyl-CoA, but catalyse the elongation reaction. Mutants (*fabA*$^-$ and *fabB*$^-$) synthesize saturated fatty acids normally, but the synthesis of unsaturated fatty acids is impaired. They are involved in unsaturated fatty acid synthesis.

6.6.1.1 Saturated acyl-ACP

Acetyl-CoA is carboxylated to malonyl-CoA in a reaction catalysed by acetyl-CoA carboxylase, and malonyl transacylase transfers the malonyl group to ACP. Acetyl-CoA carboxylase is a complex enzyme consisting of one molecule each of biotin carboxylase and biotin carboxyl carrier protein (BCCP), and two molecules of carboxyltransferase:

$$\text{BCCP} + \text{HCO}_3^- + \text{ATP} \xrightarrow{\text{biotin carboxylase}} \text{BCCP-COO}^- + \text{ADP} + \text{P}_i$$

$$\text{acetyl-CoA} + \text{BCCP-COO}^- \xrightarrow{\text{carboxyltransferase}} \text{malonyl-CoA} + \text{BCCP}$$

$$\text{malonyl-CoA} + \text{ACP} \xrightarrow{\text{malonyl transferase}} \text{malonyl-ACP} + \text{CoA-SH}$$

Malonyl-ACP and acetyl-CoA condense to acetoacetyl-ACP, replacing the carboxyl group of malonyl-ACP with an acetyl group, catalysed by 3-ketoacyl-ACP synthase III (FabH). Acetoacetyl-ACP is reduced to 3-hydroxybutyryl-ACP by the action of 3-ketoacyl-ACP reductase (FabG) that uses NADPH as a coenzyme. Crotonyl-ACP is formed from 3-hydroxybutyryl-ACP through a dehydration reaction catalysed by 3-hydroxyacyl-ACP dehydratase (FabZ). Enoyl-ACP reductase (FabI) reduces crotonyl-ACP to butyryl-ACP using NAD(P)H as a coenzyme. Butyryl-ACP condenses with malonyl-ACP to start the next cycle, catalysed probably by 3-ketoacyl-ACP reductase II (FabF) (Figure 6.24). The initiating 3-ketoacyl-ACP synthase III (FabH) does not catalyse the reverse reaction, and its regulation determines the rate of fatty acid synthesis.

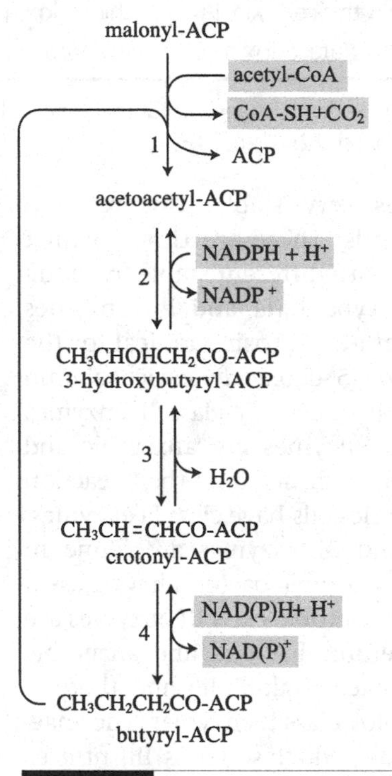

Figure 6.24 Synthesis of acyl-ACP.

(Modified from *Annu. Rev. Biochem.* **74**: 791–831, 2005)

1, 3-ketoacyl-ACP synthase (FabH for acetyl-CoA and malonyl-ACP, and FabF for further elongation); 2, 3-ketoacyl-ACP reductase (FabG); 3, 3-hydroxyacyl-ACP dehydratase (FabZ); 4, enoyl-ACP reductase (FabI).

6.6.1.2 Branched acyl-ACP

Branched fatty acids are synthesized in two different ways. Branched building blocks such as isobutyryl-ACP or methylmalonyl-ACP result in branched fatty acids from similar reactions to those in the straight-chain fatty acid biosynthetic pathway. The other pathway that synthesizes branched fatty

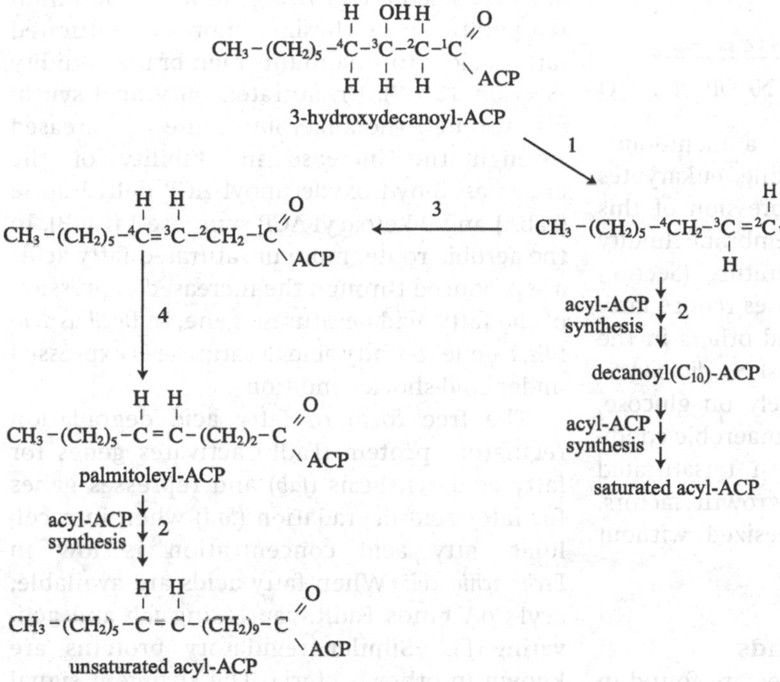

$CH_3-(CH_2)_5-{}^4CH(H)-{}^3C(H)(OH)-{}^2CH_2-{}^1C(=O)-ACP$

3-hydroxydecanoyl-ACP

1 →

$CH_3-(CH_2)_5-{}^4CH_2-{}^3CH={}^2CH-{}^1C(=O)-ACP$

3 ←

$CH_3-(CH_2)_5-{}^4CH={}^3CH-{}^2CH_2-{}^1C(=O)-ACP$

acyl-ACP synthesis ↓ 2

decanoyl(C_{10})-ACP

acyl-ACP synthesis ↓

saturated acyl-ACP

↓ 4

$CH_3-(CH_2)_5-CH=CH-(CH_2)_7-C(=O)-ACP$

palmitoleyl-ACP

acyl-ACP synthesis ↓ 2

$CH_3-(CH_2)_5-CH=CH-(CH_2)_9-C(=O)-ACP$

unsaturated acyl-ACP

Figure 6.25 Unsaturated acyl-ACP synthesis by the anaerobic route.

(Modified from *Annu. Rev. Biochem.* **74**: 791–831, 2005)

The dual-function 3-hydroxydecanoyl-ACP dehydratase (FabA) removes a water molecule from 3-hydroxydecanoyl-ACP, producing *trans*-2,3-enoyl-ACP (1). For the synthesis of saturated fatty acid residues the double bond is reduced by 3-ketoacyl-ACP reductase (FabG, 2). On the other hand, the dual-function 3-hydroxydecanoyl-ACP dehydratase (FabA) isomerizes *trans*-2,3-enoyl-ACP to *cis*-3,4-enoyl-ACP (3) that cannot be reduced by enoyl-ACP reductase (FabI), but is condensed with malonyl-ACP (4) catalysed by hydroxyacyl-ACP synthase I (FabB). FabA and FabB are the key enzymes of the synthesis of unsaturated fatty acids, and their gene expression is controlled by the FabR protein. The enzymes are unstable under physiological temperatures, and become stable, producing more unsaturated fatty acids, at lower temperatures.

acids involves methylation of unsaturated fatty acids, as in the formation of cyclopropane fatty acids (Section 6.6.1.4).

6.6.1.3 Unsaturated acyl-ACP

Many unsaturated fatty acid residues are found in biological membranes. They are synthesized by two different mechanisms. These are the aerobic route, found both in eukaryotes and prokaryotes, and the anaerobic route which occurs in some bacteria. In the anaerobic route the double bond is formed during fatty acid biosynthesis (Figure 6.25). In saturated fatty acid synthesis, 3-hydroxyacyl-ACP dehydratase (FabA) dehydrates 3-hydroxyacyl-ACP to *trans*-2,3-enoyl-ACP that can be reduced by enoyl-ACP reductase (FabI). For unsaturated fatty acid synthesis, *trans*-2,3-enoyl-ACP is isomerized to *cis*-3,4-enoyl-ACP by the bifunctional 3-hydroxyacyl-ACP dehydratase (FabA). Enoyl-ACP reductase (FabI) cannot reduce *cis*-3,4-enoyl-ACP. This *cis*-3,4-enoylACP becomes a substrate for the elongation, catalysed by a separate 3-ketoacyl-ACP synthase (FabF). FabF is expressed constitutively and is unstable under physiological temperatures. More unsaturated fatty acids are produced at a reduced growth temperature when this enzyme becomes stable (Section 12.2.7).

Saturated fatty acids are oxidized in an aerobic route to produce unsaturated fatty acids. Acyl-ACP oxidase catalyses this reaction, consuming O_2, to oxidize NADPH to water:

$$\text{palmitoyl-ACP} + \text{NADPH} + H^{+} + O_2 \xrightarrow{\text{acyl-ACP oxidase (fatty acid desaturase)}} \text{palmitoyleyl-ACP} + \text{NADP}^{+} + 2H_2O$$

The fatty acid desaturase is a membrane-bound enzyme in bacteria while eukaryotes have soluble enzymes. The expression of this enzyme is activated when the membrane fluidity becomes low at a low temperature (Section 12.2.7). Some fatty acid desaturases reduce fatty acid residues bound to ACP, and others in the form of phospholipids. The yeast *Saccharomyces cerevisiae* can grow fermentatively on glucose, but cannot grow under strictly anaerobic conditions unless supplemented with unsaturated fatty acids and ergosterol as growth factors. These lipids cannot be synthesized without molecular oxygen in yeast.

6.6.1.4 Cyclopropane fatty acids

Cyclopropane fatty acid residues are found in the bacterial membrane (Section 2.3.4). These are synthesized through the methylation of unsaturated fatty acids catalysed by cyclopropane fatty acid synthase, using *S*-adenosylmethionine as the methyl group donor. These fatty acid residues are called lactobacillic acid, since they were first identified in *Lactobacillus arabinosus*:

```
                     H H          cyclopropane                              H H
                     | |       fatty acid synthase                          | |
  CH2-O-CO-(CH2)9-C=C-(CH2)5-CH3  ------------------->  CH2-O-(CH2)9-C-C-(CH2)5-CH3
   |                                                     |              \ /
 H-C-O-                                                 H-C-O-           CH2
   |                                                     |
 -O-CH2                                                 -O-CH2

      cis-vaccenic acid residue                         lactobacillic acid residue
```

6.6.1.5 Regulation of fatty acid biosynthesis

The fatty acid composition of a membrane varies depending on growth conditions and culture age in a given bacterium. The rate of fatty acid synthesis is determined by the activity of the initiating enzyme, 3-ketoacyl-ACP synthase III (FabH). The length of the fatty acid depends on regulation of 3-ketoacyl-ACP synthase II (FabF) activity. Bacteria growing at a sub-optimum temperature synthesize more unsaturated fatty acids to maintain membrane fluidity (Section 12.2.7). Unsaturated fatty acid synthesis through the anaerobic route is increased through the increase in stability of the enzymes, 3-hydroxydecanoyl ACP dehydratase (FabA) and 3-ketoacyl-ACP synthase I (FabB). In the aerobic route, more unsaturated fatty acids are produced through the increased expression of the fatty acid desaturase gene. In *Bacillus subtilis*, a gene for fatty acid desaturase is expressed under cold-shock conditions.

The free form of fatty acid degradation regulatory protein (FadR) activates genes for fatty acid synthesis (*fab*) and represses genes for fatty acid degradation (*fad*) when intracellular fatty acid concentration is low in *Escherichia coli*. When fatty acids are available, acyl-CoA binds FadR, repressing *fab* and activating *fad*. Similar regulatory proteins are known in other bacteria. The stringent signal molecule ppGpp (Section 12.2.1) represses the expression of *fab* genes. Another regulatory protein, FabR, represses genes for unsaturated fatty acid (UFA) synthesis in the presence of UFA-CoA or UFA-ACP.

Escherichia coli synthesizes cyclopropane fatty acid (CFA) in the stationary phase. CFA-negative mutants are less resistant to freezing, which suggests that CFA is involved in survival of the bacterium. The CFA synthase gene is recognized by the stationary phase sigma factor of RNA polymerase (σ^S) (Section 12.1.1).

6.6.2 Phospholipid biosynthesis

The EMP pathway intermediate, dihydroxyacetone phosphate, is reduced to glycerol-3-phosphate oxidizing NADPH:

$$\underset{\text{dihydroxyacetone phosphate}}{CH_2OH{-}C(=O){-}CH_2{-}O{-}P(=O)(OH){-}OH} \xrightarrow[\text{NADPH + H}^+ \;\; \text{NADP}^+]{\text{glycerol-3-phosphate dehydrogenase}} \underset{\text{glycerol-3-phosphate}}{CH_2OH{-}CHOH{-}CH_2{-}O{-}P(=O)(OH){-}OH}$$

Glycerol-3-phosphate acyltransferase then synthesizes phosphatidic acid, consuming two acyl-ACPs. Phosphatidic acid serves as a precursor for the synthesis of phospholipids and triglycerides:

$$\underset{\text{glycerol-3-phosphate}}{CH_2{-}OH \,|\, CHOH \,|\, CH_2{-}O{-}P(=O)(OH){-}OH} + 2R{-}CO{-}ACP \xrightarrow[2ACP]{\text{glycerol phosphate acyltransferase}} \underset{\text{phosphatidic acid}}{CH_2{-}O{-}CO{-}R \,|\, CH{-}O{-}CO{-}R \,|\, CH_2{-}O{-}P(=O)(OH){-}OH}$$

The bacterial cytoplasmic membrane contains phosphatidylethanolamine, phosphatidylserine, phosphatidylcholine, phosphatidylinositol, phosphatidylglycerol and others. These phospholipids are phosphatidic acid derivatives containing alcohol esters linked to the phosphate residue. Phosphatidic acid is activated to CDP-diacylglycerol, consuming cytidine 5′-triphosphate (CTP) to receive alcohols:

$$\underset{\text{phosphatidic acid}}{CH_2{-}CO{-}R \,|\, CH{-}O{-}CO{-}R \,|\, CH_2O{-}P(=O)(OH){-}OH} + CTP \xrightarrow{\text{phosphatidate cytidyltransferase}} \underset{\text{CDP-diacylglycerol}}{CH_2{-}O{-}CO{-}R \,|\, CH{-}O{-}CO{-}R \,|\, CH_2O{-}P(=O)(OH){-}O{-}P(=O)(OH){-}O{-}\text{cytidine}} + PP_i$$

Cytidine 5′-monophosphate (CMP) is replaced with alcohols such as serine and inositol in a reaction catalysed by alcohol-specific phosphatidyl transferase. Phosphatidylserine is decarboxylated to phosphatidylethanolamine, before being synthesized to phosphatidylcholine through methylation using S-adenosylmethionine (SAM) as the source of the methyl group. Similar reactions are employed to produce phosphatidylinositol, phosphatidylglycerol and cardiolipin (Figure 6.26).

As in phosphatidylcholine synthesis, C1 units such as methyl groups are transferred in various reactions involving the C1 carriers tetrahydrofolate (H_4F) and SAM. H_4F participates in the reactions which add or remove all forms of C1 units except carbonate, including methyl ($-CH_3$), methylene ($-CH_2-$), methenyl ($-CH=$), formyl ($-CHO$) and formimino ($-CH=NH$), as shown in Figure 6.27. SAM functions as a $-CH_3$ donor. Methanogens have their own C1 carriers such as coenzyme M, tetrahydromethanopterin (H_4MTP) and methanofuran (MF). Some of these are found in other archaea and in some eubacteria (Section 9.4.2).

H_4F has a structure consisting of glutamate, *p*-aminobenzoate and folate with reduced pteridine (Figure 6.27). Methionine adenosyltransferase condenses methionine and ATP to produce SAM:

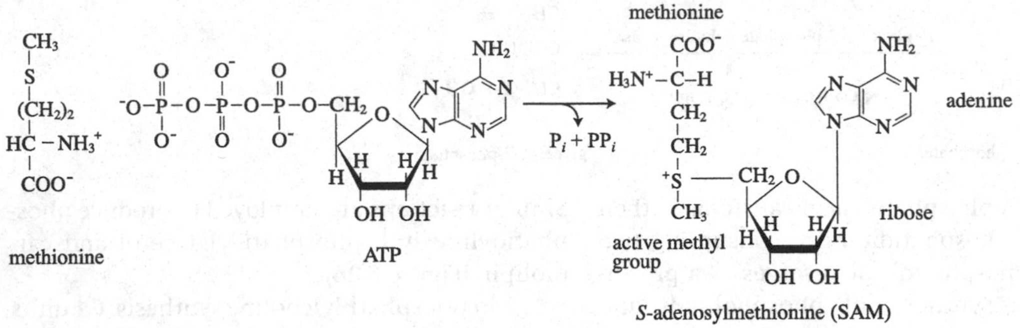

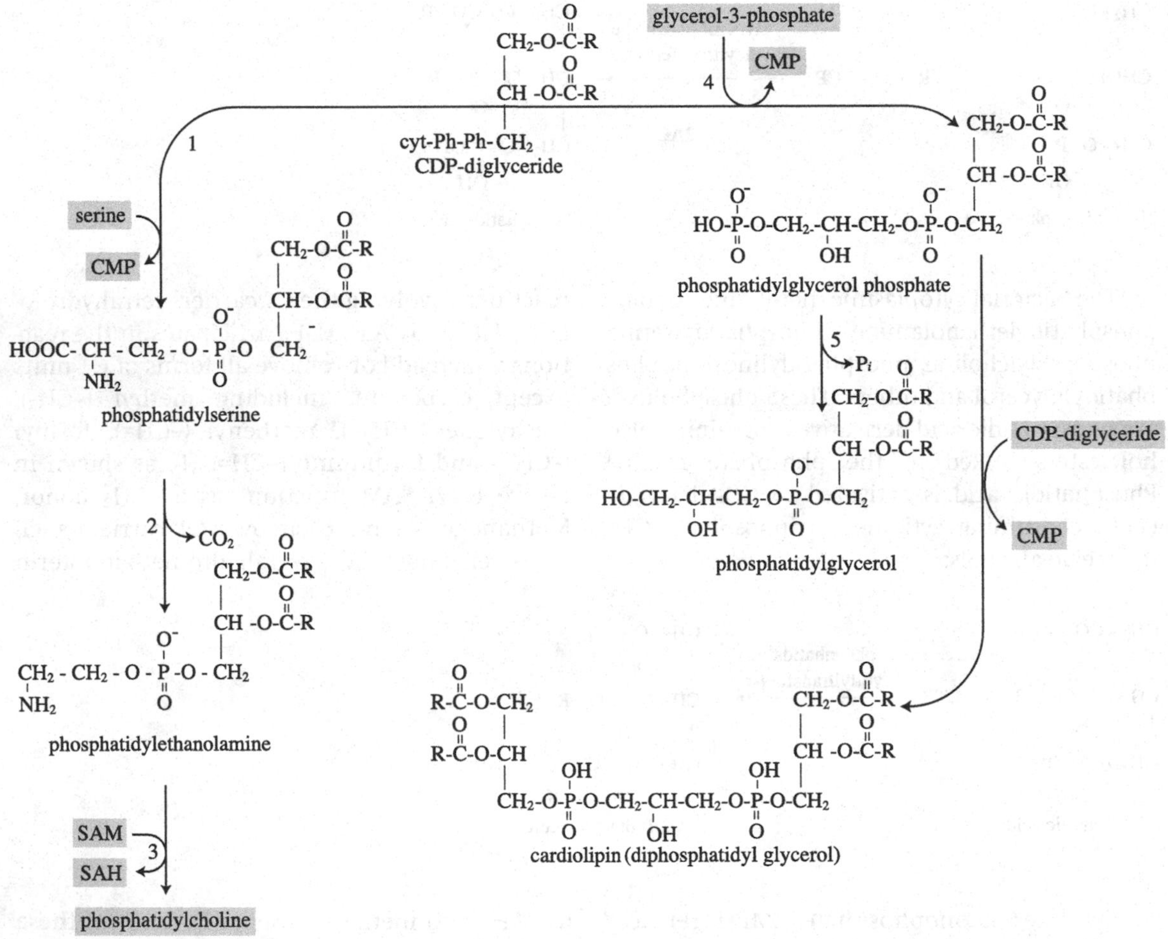

Figure 6.26 Phospholipid biosynthesis.

Phosphatidic acid is activated to CDP-diglyceride in a reaction catalysed by phosphatidate cytidyltransferase, before an alcohol replaces CMP.

1, CDP-diacylglyceride:serine *O*-phosphatidyltransferase; 2, phosphatidylserine decarboxylase; 3, phosphatidylethanolamine methyltransferase; 4, CDP-diacylglyceride:glycerol-3-phosphate 3-phosphatidyltransferase; 5, phosphatidylglycerol phosphatase.

SAM, *S*-adenosylmethionine; SAH, *S*-adenosylhomocysteine; CMP, cytidine 5′-monophosphate.

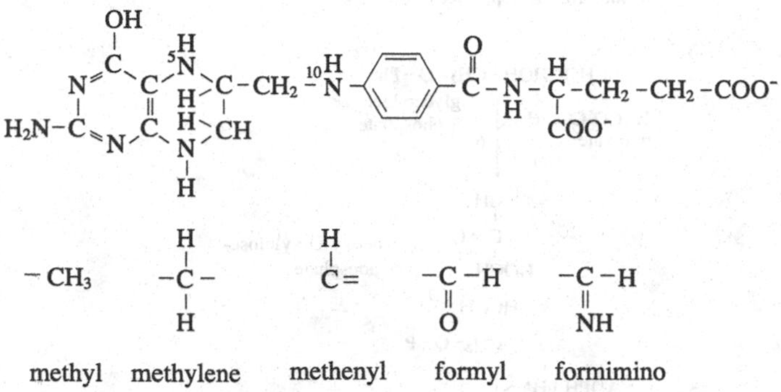

$-CH_3$ (methyl), $-CH_2-$ (methylene), $-CH=$ (methenyl), $-CH=O$ (formyl), $-CH=NH$ (formimino)

Figure 6.27 Coenzyme tetrahydrofolate and C1 units carried by tetrahydrofolate.

6.6.3 Isoprenoid biosynthesis

Various isoprenoid compounds are found in microbial cells, including alcohol residues of archaeal cytoplasmic membrane phospholipids, quinones, and carotenoids. These are synthesized through isopentenyl pyrophosphate (IPP). Two pathways are known to produce IPP. Acetyl-CoA is the starting material in the mevalonate pathway, which operates in archaea and eukaryotes. Most bacteria produce IPP from glyceraldehyde-3-phosphate through the mevalonate-independent pathway (Figure 6.28).

6.7 Haem biosynthesis

Many proteins contain haem, including cytochromes and chlorophylls. 5-aminolevulinate is the common precursor for the synthesis of tetrapyrroles. This precursor is synthesized either from glutamyl-tRNA or from glycine and succinyl-CoA. Aminolevulinate synthase is found only in *Alphaproteobacteria*:

$$\text{succinyl-CoA} + \text{glycine} \xrightarrow[CO_2]{\text{aminolevulinate synthase}} \text{5-aminolevulinate}$$

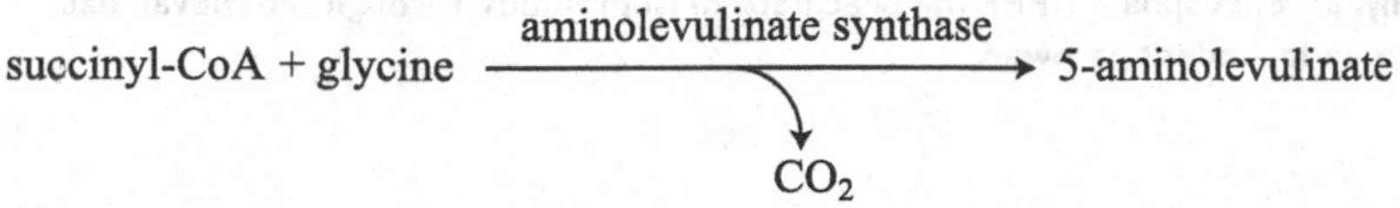

In prokaryotes other than *Alphaproteobacteria*, the haem precursor is synthesized from glutamyl-tRNA:

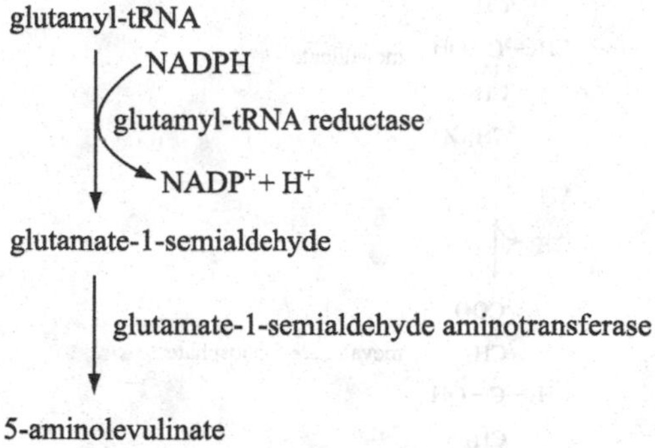

glutamyl-tRNA
↓ glutamyl-tRNA reductase (NADPH → $NADP^+ + H^+$)
glutamate-1-semialdehyde
↓ glutamate-1-semialdehyde aminotransferase
5-aminolevulinate

Two 5-aminolevulinate molecules condense to porphobilinogen, which is deaminated to uroporphyrinogen III with the basic structure of a tetrapyrrole. Corrinoids and coenzyme F_{430} are synthesized from uroporphyrinogen III. This intermediate is further metabolized to protoporphyrin IX to synthesize chlorophylls and bacteriochlorophylls. Cytochromes and phycobilins are derived from protohaem (Figure 6.29).

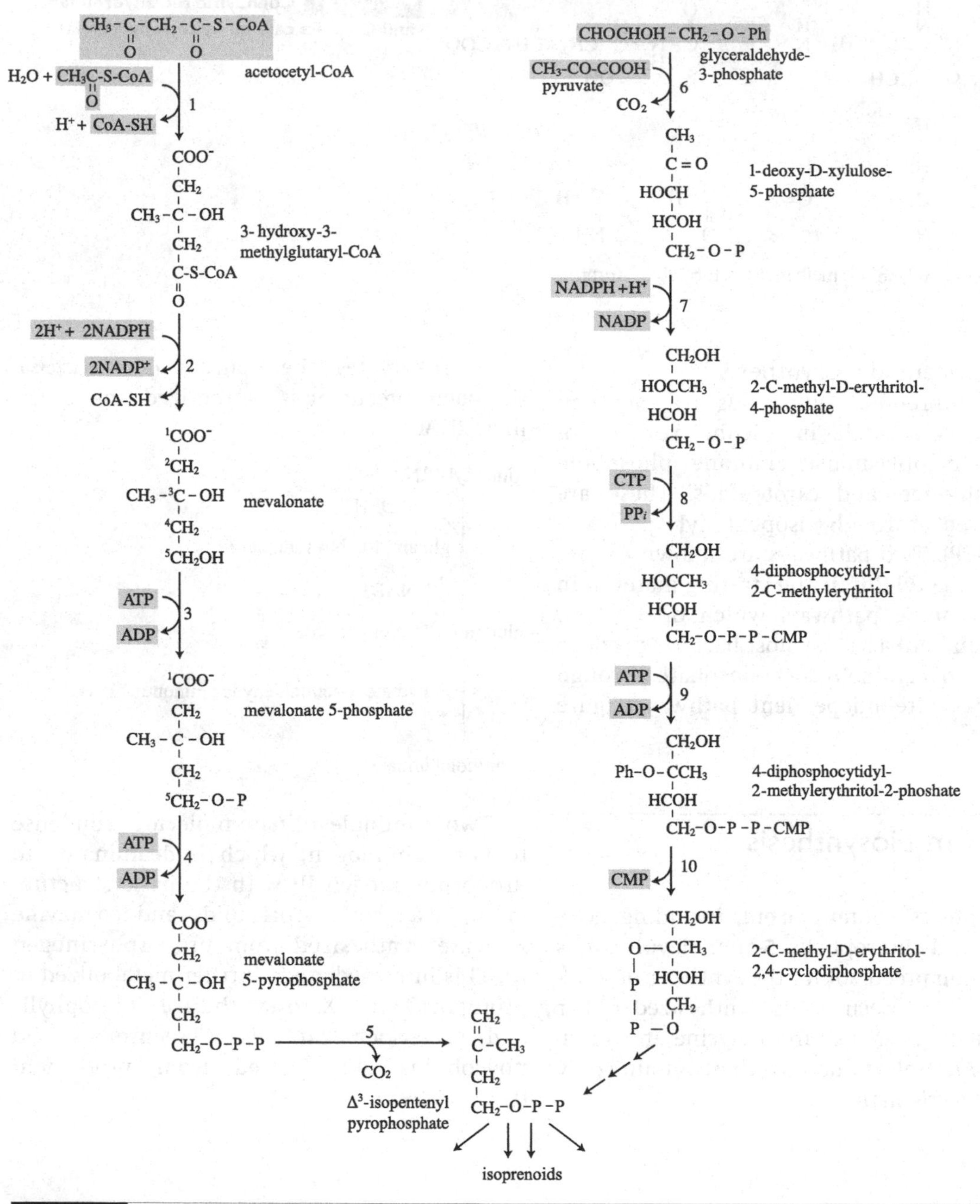

Figure 6.28 Synthesis of isopentenyl pyrophosphate (IPP), the precursor of isoprenoids, through the mevalonate pathway or through the mevalonate-independent pathway.

(*Mol. Microbiol.* **37**: 703–716, 2000)

1, hydroxylmethylglutaryl-CoA synthase; 2, hydroxymethylglutarate reductase; 3, mevalonate kinase; 4, phosphomevalonate kinase; 5, diphosphomevalonate decarboxylase; 6, 1-deoxy-D-xylulose-5-phosphate synthase; 7, 1-deoxy-D-xylulose-5-phosphate reductoisomerase; 8, 4-diphosphocytidyl-2C-methyl-D-erythritol synthase; 9, 4-diphosphocytidyl-2C-methyl-D-erythritol kinase; 10, 2C-methyl-D-erythritol-2,4-cyclodiphosphate synthase.

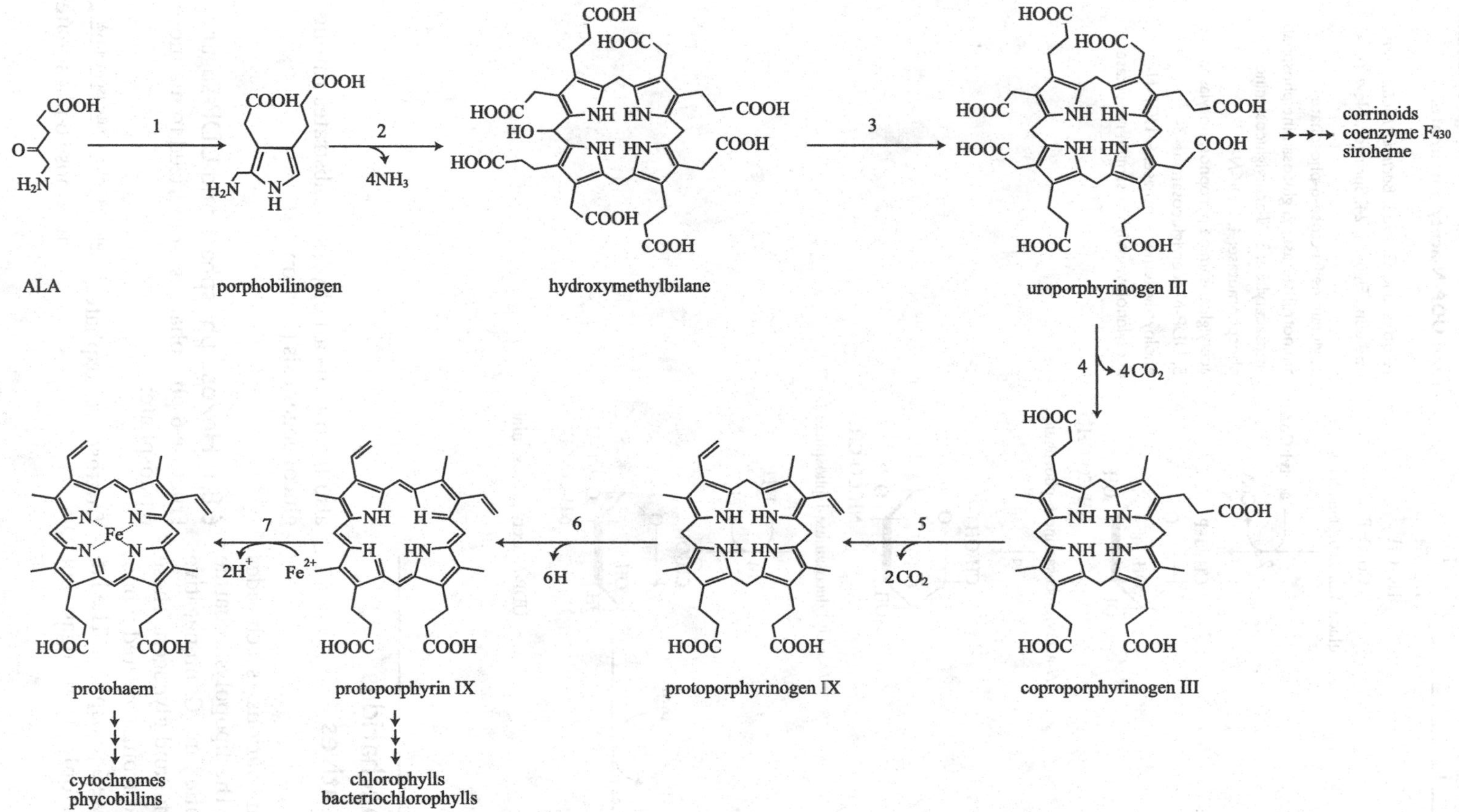

ALA → (1) → porphobilinogen → (2, $4NH_3$) → hydroxymethylbilane → (3) → uroporphyrinogen III → corrinoids, coenzyme F_{430}, siroheme

uroporphyrinogen III → (4, $4CO_2$) → coproporphyrinogen III → (5, $2CO_2$) → protoporphyrinogen IX → (6, 6H) → protoporphyrin IX → (7, Fe^{2+}, $2H^+$) → protohaem

protoporphyrin IX → chlorophylls, bacteriochlorophylls

protohaem → cytochromes, phycobillins

Figure 6.29 Haem biosynthetic pathway from 5-aminolevulinic acid (ALA) to protohaem.

(Modified from *Microbiol.* **148**: 2273–2282, 2002)

1, 5-aminolevulinate dehydratase; 2, porphobilinogen deaminase; 3, uroporphyrinogen III synthase; 4, uroporphyrinogen III decarboxylase; 5, coproporphyrinogen III oxidase; 6, protoporphyrinogen IX oxidase; 7, ferrochelatase. ALA, 5-aminolevulinate.

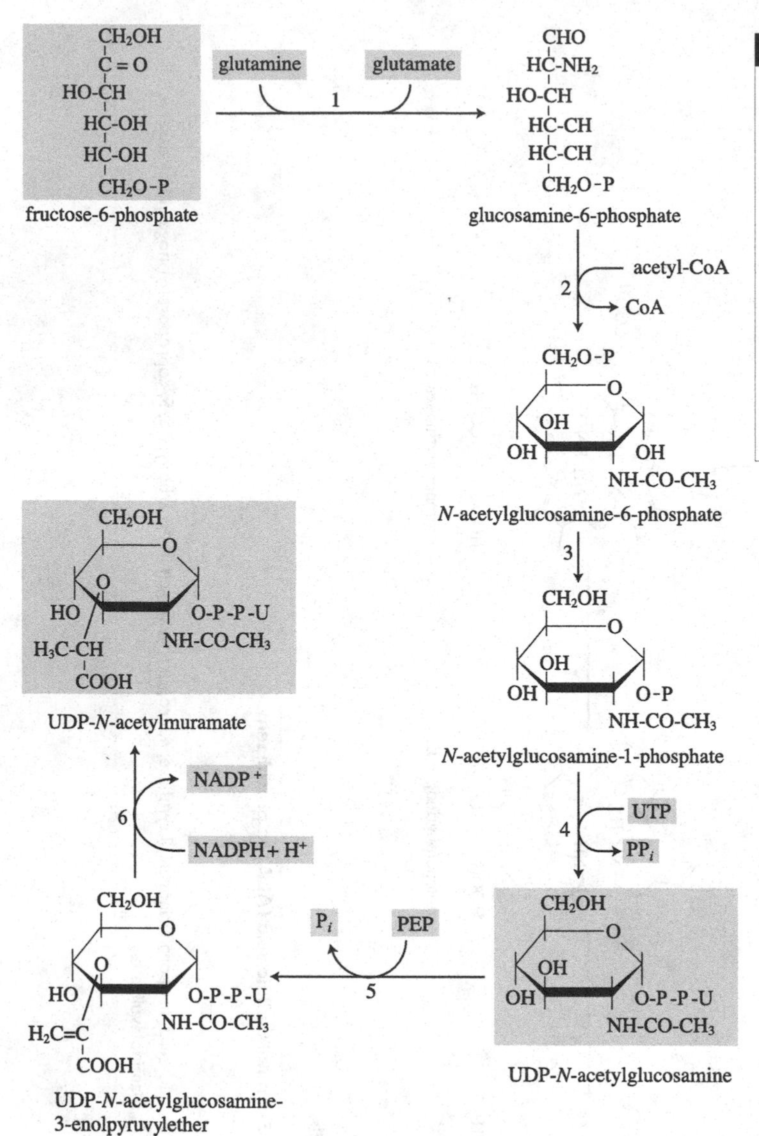

Figure 6.30 Synthesis of peptidoglycan monomers, UDP-*N*-acetylglucosamine and UDP-*N*-acetylmuramate.

(Gottschalk, G. 1986, *Bacterial Metabolism*, 2nd edn., Figure 3.24. Springer, New York)

1, glutamine:fructose-6-phosphate aminotransferase; 2, glucosamine phosphate transacetylase; 3, *N*-acetylglucosamine phosphomutase; 4, UDP-*N*-acetylglucosamine pyrophosphorylase; 5, UDP-*N*-acetylglucosamine-3-enolpyruvylether synthase; 6, UDP-*N*-acetylenolpyruvylglucosamine reductase.

6.8 Synthesis of saccharides and their derivatives

Microbial cells contain various saccharides located in the cell wall, the lipopolysaccharide of the outer membrane in Gram-negative bacteria, capsular material and glycogen. These polymers are synthesized from activated monomers derived from glucose-6-phosphate. The latter can be produced not only from sugars but also from non-carbohydrate substrates through gluconeogenesis (Section 4.2).

6.8.1 Hexose phosphate and UDP-sugar

Fructose-6-phosphate is isomerized to mannose-6-phosphate:

$$\text{fructose-6-phosphate} \xrightarrow{\text{mannose-6-phosphate isomerase}} \text{mannose-6-phosphate}$$

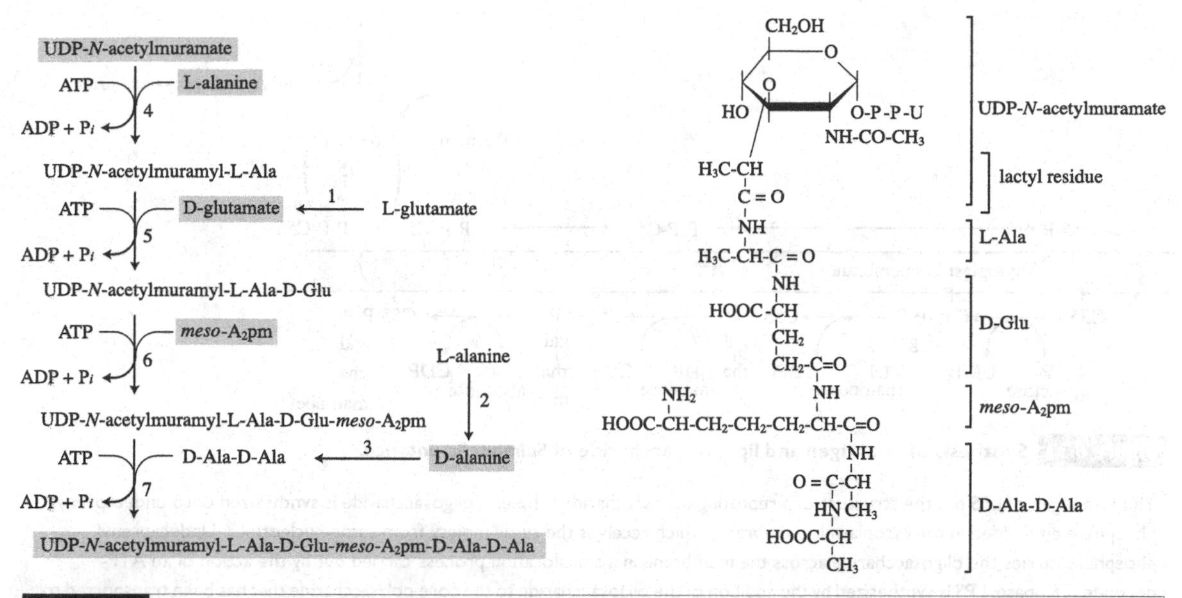

Figure 6.31 Synthesis of UDP-*N*-acetylmuramylpentapeptide through a non-ribosomal peptide synthesis process, adding amino acids to the lactyl group of UDP-*N*-acetylmuramate.

(Modified from *Microbiol. Mol. Biol. Rev.* **63**: 174–229, 1999)

The precursor for the bacterial cell wall peptidoglycan synthesis, UDP-*N*-acetylmuramylpentapeptide, is made through a non-ribosomal peptide synthesis process. Amino acids are not activated and mRNA is not required. The amino acid sequence is determined by the enzyme specificity, and ATP is consumed to provide energy needed for the formation of peptide bonds. Different amino acids are found in the second and third positions depending on the bacterial species.

1, glutamate racemase; 2, alanine racemase; 3, D-alanine-D-alanine ligase; 4, UDP-*N*-acetylmuramate-alanine ligase; 5, UDP-*N*-acetylmuramyl-alanine-D-glutamate ligase; 6, UDP-*N*-acetylmuramyl-alanyl-D-glutamate-2,6-diaminopimelate ligase; 7, D-alanyl-D-alanine adding enzyme.

meso-A_2pm, 2,6-diaminopimelate.

Polysaccharides containing galactose are synthesized from UDP-galactose, which is converted from glucose-6-phosphate in three steps:

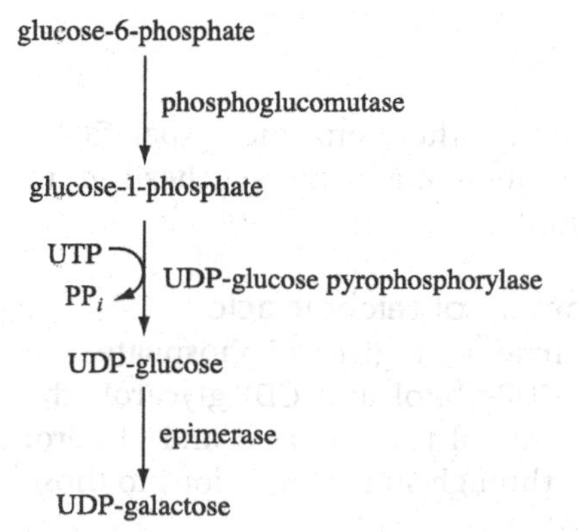

6.8.2 Monomers of peptidoglycan

Fructose-6-phosphate is used to synthesize the peptidoglycan monomers, uridine diphosphate (UDP)-*N*-acetylglucosamine and UDP-*N*-acetylmuramate. The precursor is aminated to glucosamine-6-phosphate using glutamate as the amine group donor, before being acetylated to *N*-acetylglucosamine-6-phosphate. The latter is activated to UDP-*N*-acetylglucosamine, condensing with UTP. PEP is used to add enolpyruvate to this intermediate before being reduced to UDP-*N*-acetylmuramate, consuming NADPH (Figure 6.30).

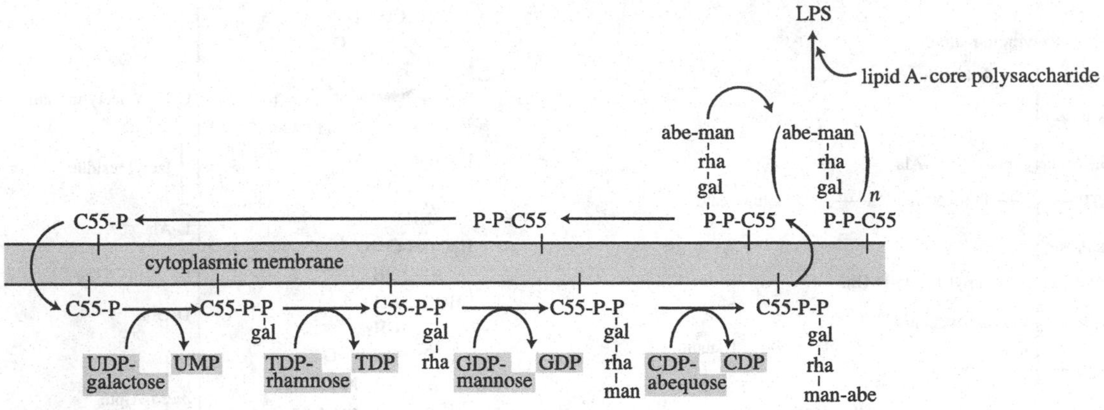

Figure 6.32 Synthesis of O-antigen and lipopolysaccharide in *Salmonella enterica*.

The O-antigen of LPS has the structure of a repeating oligosaccharide. The unit oligosaccharide is synthesized onto undecaprenyl phosphate embedded in the cytoplasmic membrane, which receives the sugar moiety from sugar-nucleotides. Undecaprenyl phosphate carries the oligosaccharide across the membrane in a translocation process carried out by the action of an ATP-dependent flippase. LPS is synthesized by the addition of the oligosaccharide to the core polysaccharide that has been transported to the periplasm bound to lipid A. LPS is transported to the outer membrane by the LPS exporter, a member of the ATP-binding cassette (ABC) family. C55-P, undecaprenyl phosphate.

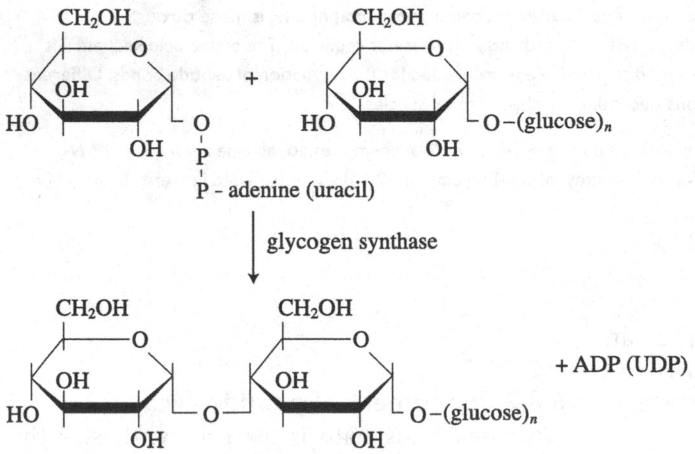

Figure 6.33 Formation of the α-1,4 linkage in glycogen by glycogen synthase.

(Gottschalk, G. 1986, *Bacterial Metabolism*, 2nd edn., Figure 3.28. Springer, New York)

Amino acids are added to UDP-*N*-acetylmuramate to synthesize UDP-*N*-acetylmuramylpentapeptide. L-alanine, D-glutamate, *meso*-diaminopimelate and D-alanyl-D-alanine form a peptide on the lactyl group of UDP-*N*-acetylmuramate, consuming ATP (Figure 6.31). This reaction is a non-ribosomal peptide synthesis process independent from mRNA and the ribosomes. The amino acid sequence is determined by the enzyme specificity. Peptidoglycan monomers are synthesized in the cytoplasm.

6.8.3 Monomers of teichoic acid

Ribose-phosphate and glycerol-phosphate are activated to CDP-ribitol and CDP-glycerol, the precursors of ribitol teichoic acid and glycerol teichoic acid, through similar reactions to those

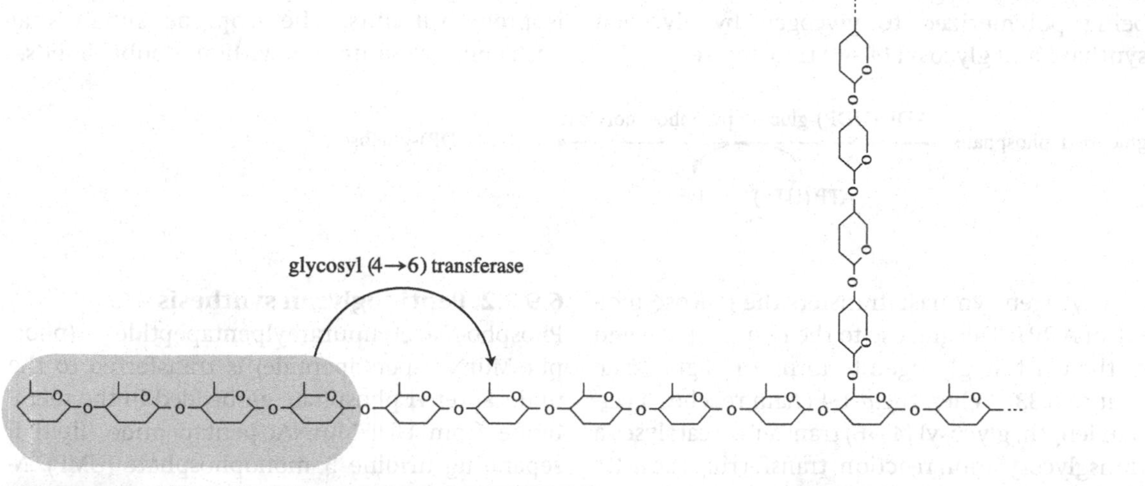

glycosyl (4→6) transferase

Figure 6.34 Formation of the α-1,6 side chain in glycogen through the transglycosylation reaction catalysed by glycosyl (4→6) transferase.
(Gottschalk, G. 1986, *Bacterial Metabolism*, 2nd edn., Figure 3.29. Springer, New York)

in the formation of UDP-sugars, consuming CTP instead of UTP.

6.8.4 Precursor of lipopolysaccharide, O-antigen

Lipopolysaccharide consists of lipid A, core polysaccharide and O-antigen (Section 2.3.3.2). O-antigen has a structure based on repeating oligosaccharide. Sugar-nucleotides are added to undecaprenyl (bactoprenol) phosphate embedded in the cytoplasmic membrane (Figure 6.32). Undecaprenyl phosphate is a carrier to transport the oligosaccharide across the membrane (Section 6.9.2.1) through the action of an ATP-dependent flippase. O-antigen ligase transfers the oligosaccharide to the core polysaccharide - lipid A in the periplasm to synthesize LPS. Core polysaccharide-lipid A is synthesized in the cytoplasm, and crosses the cytoplasmic membrane. The hydrophobic lipid A is the carrier of the core polysaccharide.

6.9 Polysaccharide biosynthesis and the assembly of cell surface structures

Polysaccharides in bacterial cells include glycogen, a storage material, and structural polymers such as peptidoglycan and teichoic acid in the cell wall, and LPS in the outer membrane. The precursors are synthesized in the cytoplasm and peptidoglycan and LPS are synthesized after the precursors are transported across the cytoplasmic membrane. For this reason, the synthesis and assembly of the cell wall and the outer membrane are closely related to the transport of their precursors.

6.9.1 Glycogen synthesis

UDP-glucose is the precursor for glycogen synthesis in eukaryotes. Glucose-1-phosphate is activated to ADP-glucose in prokaryotes before being polymerized to glycogen by glycogen synthase and glycosyl (4→6) transferase:

$$\text{glucose-1-phosphate} \xrightarrow[\text{ATP (UTP)} \quad \text{PP}_i]{\text{ADP (UDP)-glucose pyrophosphorylase}} \text{ADP (UDP)-glucose}$$

Glycogen synthase transfers the glucose moiety of ADP (UDP)-glucose to the non-reducing end of the existing glycogen to form α-1, 4-glucoside (Figure 6.33). When the α-1,4 chain reaches a certain length, glycosyl (4→6) transferase catalyses a transglycosylation reaction, transferring the α-1,4 chain to form an α-1,6 linkage (Figure 6.34).

6.9.2 Peptidoglycan synthesis and cell wall assembly

6.9.2.1 Transport of cell wall precursor components through the membrane

The cell wall consists of peptidoglycan and teichoic acid in Gram-positive bacteria, and of peptidoglycan in Gram-negative bacteria. Various proteins are also associated with the cell wall, especially in Gram-positive bacteria. Monomers of the polymeric compounds are synthesized in the cytoplasm before being transported to the periplasm to be polymerized. They are hydrophilic in nature, and cannot diffuse through the cytoplasmic membrane. To overcome this, the hydrophobic membrane compound, undecaprenyl phosphate (Figure 6.35), carries them through the membrane in a translocation process by the action of an ATP-dependent flippase.

$$\begin{matrix} CH_3 \\ \\ CH_3 \end{matrix}\!\!>\! C=CH\text{-}CH_2\text{-}(CH_2\text{-}\overset{CH_3}{\overset{|}{C}}=CH\text{-}CH_2)_9\text{-}CH_2\text{-}\overset{CH_3}{\overset{|}{C}}=CH\text{-}CH_2\text{-}O\text{-}P$$

Figure 6.35 The structure of undecaprenyl (bactoprenyl) phosphate.

Structurally similar dolichol phosphates function as glycan carriers across the cytoplasmic membrane in archaea, like the bacterial undecaprenyl phosphate. Dolichol phosphate is a polyisoprenoid phosphate containing 11–12 isoprene subunits. The isoprene subunits at both ends are saturated, without double bonds.

6.9.2.2 Peptidoglycan synthesis

Phospho-*N*-acetylmuramylpentapeptide (phospho-MurNAc-pentapeptide) is transferred to the undecaprenyl phosphate embedded in the membrane from UDP-MurNAc-pentapeptide (lipid I) separating uridine 5′-monophosphate (UMP). *N*-acetylglucosamine (GlcNAc) transferase forms undecaprenyl-GlcNAc-MurNAc-pentapeptide pyrophosphate (lipid II), separating UDP from UDP-GlcNAc. In Gram-positive bacteria, lipid II is further modified by the addition of amino acids to the third amino acid position, which is lysine in the pentapeptide. Different amino acids are added depending on the species. Glycyl-tRNA is consumed to add five glycyl units in *Staphylococcus aureus*. Undecaprenyl-GlcNAc-MurNAc-pentapeptide-(gly)$_5$ pyrophosphate is transported through the membrane by the action of the lipid II flippase, MurJ, using the undecaprenyl residue as a carrier. GlcNAc-*N*-acetylmuramylpentapeptide-(gly)$_5$ is transferred to the existing peptidoglycan, liberating undecaprenyl pyrophosphate through the action of transglycosylase, and a transpeptidase cross-links the neighbouring chains. One of the cross-linking enzymes, penicillin-binding protein 4, localizes at the division septum bound to teichoic acid in the Gram-positive bacterium *Staphylococcus aureus*. A phosphatase converts undecaprenyl pyrophosphate to undecaprenyl phosphate, which starts another round of the same series of reactions (Figure 6.36).

The intermediate, lipid II, has another essential function in cell division in the multiprotein machinery responsible for cell division, the divisome complex. Cell division and cell wall biosynthesis in prokaryotes are driven by partially overlapping multiprotein machineries whose activities are tightly controlled and coordinated (Section 6.14.1).

Inhibitors of peptidoglycan synthesis, ristocetin and vancomycin, inhibit transglycosylase, and bacitracin interferes with the dephosphorylation of undecaprenyl pyrophosphate. β-lactam antibiotics inhibit transpeptidation and carboxypeptidation reactions.

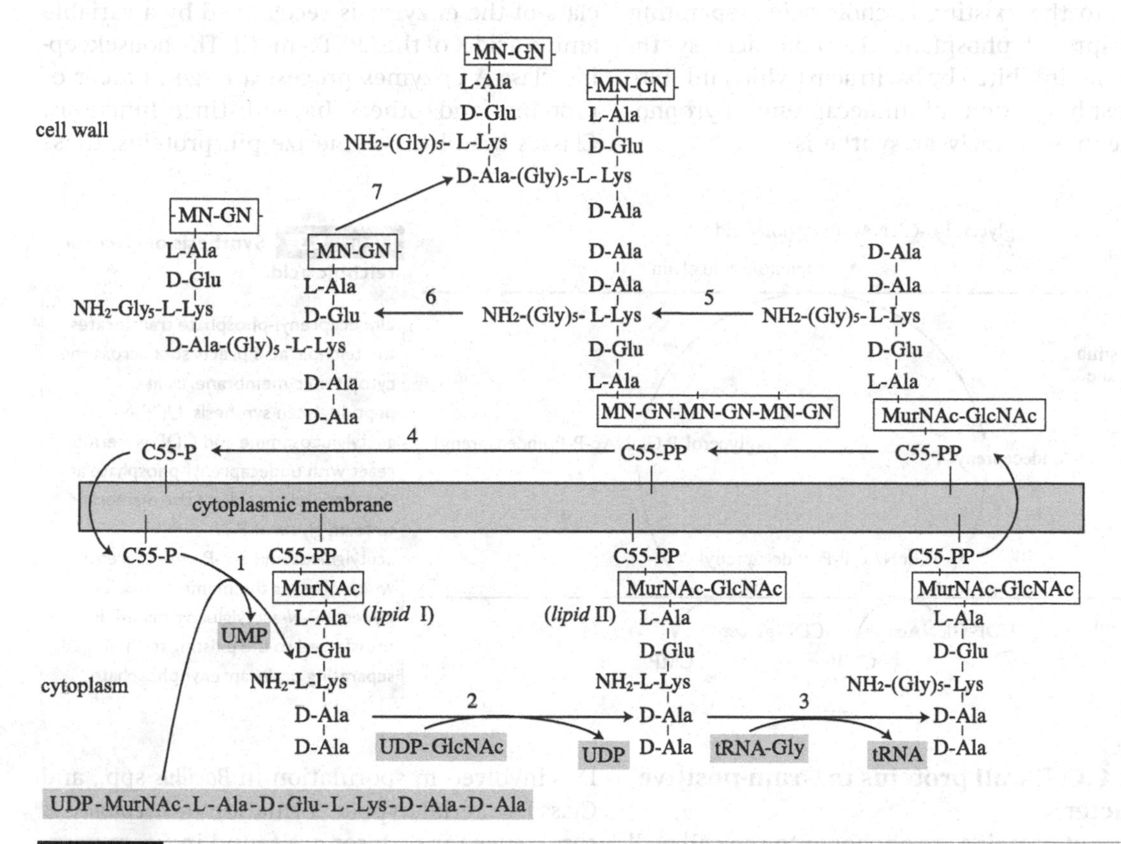

Figure 6.36 Peptidoglycan synthesis in the Gram-positive bacterium *Staphylococcus aureus*.

(*Microbiol. Mol. Biol. Revs.* **63**: 174–229, 1999)

Peptidoglycan precursors, *N*-acetylglucosamine-UDP (GlcNAc) and UDP-*N*-acetylmuramate (MurNAc)-pentapeptide, are synthesized in the cytoplasm (Figures 6.30 and 6.31). UMP is separated from UDP-MurNAc-pentapeptide, transferring phospho-MurNAc-pentapeptide to undecaprenyl phosphate to form MurNAc-pentapeptide undecaprenyl pyrophosphate, which is known as lipid I (1). Subsequently, GlcNAc is transferred to MurNAc-pentapeptide undecaprenyl pyrophosphate from UDP-GlcNAc, separating UDP to form undecaprenyl-GlcNAc-MurNAc-pentapeptide pyrophosphate, also known as lipid II (2). Five glycyl groups bind the third amino acid, lysine, in the pentapeptide, consuming glycyl-tRNAs (3) before being translocated to the outer leaflet of the membrane to be transported through the membrane by the action of the lipid II flippase, MurJ. GlcNAc-MurNAc-pentapeptide-(gly)$_5$ forms a β-1,4-glucoside with the existing peptidoglycan by a transglycosylation reaction (5), and a transpeptidase cross-link (6). Inhibitors of peptidoglycan synthesis, ristocetin and vancomycin, inhibit transglycosylase, and bacitracin interferes with the dephosphorylation of undecaprenyl pyrophosphate (4). β-lactam antibiotics inhibit (6) transpeptidation and (7) carboxypeptidation reactions.

1, phospho-*N*-acetylmuramoyl pentapeptide transferase; 2, *N*-acetylglucosamine transferase; 3, glycyl transferase; 4, undecaprenyl diphosphatase; 5, transglycosylase; 6, transpeptidase; 7, carboxypeptidase.

C55–P, undecaprenylphosphate.

6.9.2.3 Teichoic acid synthesis

Teichoic acid is synthesized in a similar way to peptidoglycan in Gram-positive bacteria (Figure 6.37). GlcNAc is transferred from UDP-GlcNAc to undecaprenyl phosphate before taking glycerol-phosphate from CDP-glyceride to form glycerol-P-*N*-acetylglucosamine-P-P-undecaprenyl.

Glycerol-P-*N*-acetylglucosamine-P is transferred to the existing teichoic acid, separating undecaprenyl phosphate. Teichoic acid synthesis is not inhibited by bacitracin (which inhibits dephosphorylation of undecaprenyl pyrophosphate in peptidoglycan synthesis).

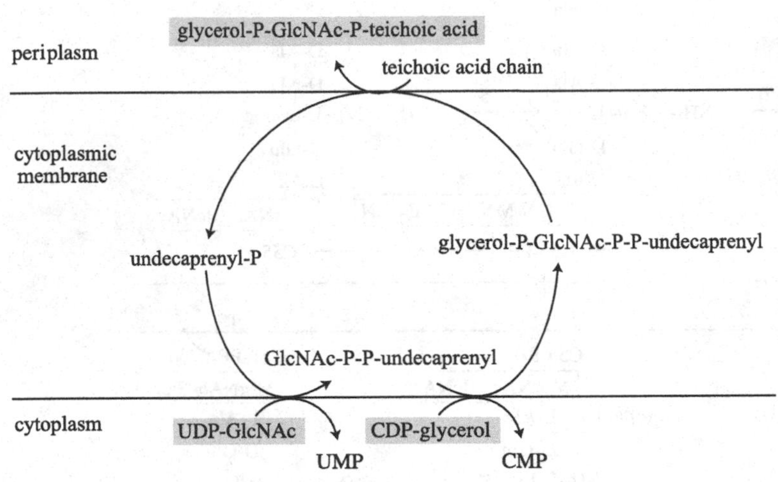

Figure 6.37 Synthesis of glycerol teichoic acid.

Undecaprenyl-phosphate translocates the teichoic acid precursors across the cytoplasmic membrane, as in peptidoglycan synthesis. UDP-*N*-acetylglucosamine and CDP-glyceride react with undecaprenyl-phosphate at the cytoplasmic side of the membrane to form glycerol-P-*N*-acetylglucosamine-P-P-undecaprenyl, which crosses the membrane before glycerol-P-*N*-acetylglucosamine-P is transferred to the existing teichoic acid, separating undecaprenyl-phosphate.

6.9.2.4 Cell wall proteins in Gram-positive bacteria

A range of proteins are anchored to the cell wall of Gram-positive bacteria enabling them to effectively interact with their environment, including enzymes and virulence-related proteins. Most of these are covalently bonded to the peptidoglycan. They are transported through the membrane before being bonded to the peptidoglycan by an enzyme called sortase. Sortase cleaves the substrates at a conserved LPXTG motif near the *C*-terminal and covalently links them to a pentaglycine crossbridge in lipid II, attached to undecaprenyl phosphate. The complex is then transferred to the existing peptidoglycan by transglycosylases and transpeptidases (Figure 6.38). Some proteins are polymerized by sortases to form pili. A multicellular bacterium, *Streptomyces coelicolor*, forms spores on the tips of aerial hyphae that are coated with proteins known as chaplins. They are substrates of sortases that act as developmental check points.

Most Gram-positive bacteria contain multiple sortases. They are grouped into six classes (A–F) according to their amino acid sequence. Each class of the enzyme is recognized by a variable amino acid X of the LPXTG motif. The housekeeping class A enzymes process a large number of proteins, and others have distinct functions. Classes B and C polymerize pili proteins. Class D is involved in sporulation in *Bacillus* spp., and Class E in aerial hyphae formation in *Streptomyces* spp. Genes for sortases are found in Gram-negative bacteria and archaea, but their role is obscure.

6.9.2.5 Cell wall assembly

The cell wall determines the bacterial cell shape (Section 2.3.3.3). Each bacterium has its own shape that may change according to the environment. Cell wall assembly is closely controlled to preserve and reproduce the cell shape in every generation. The incorporation of the precursor, GlcNAc-MurNAc-pentapeptide-(gly)$_5$, should take place at precise locations over a defined time period. It starts with the hydrolysis of peptidoglycan by transglycosylase and

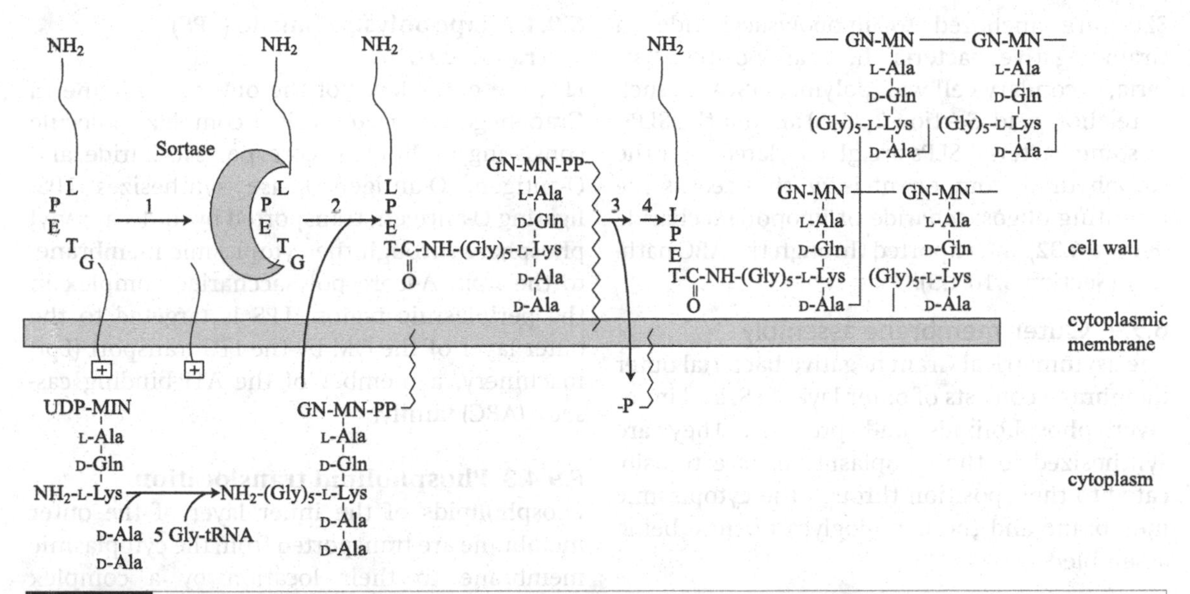

Figure 6.38 Cell wall protein anchoring in Gram-positive bacteria.

(*Trends Microbiol.* 8: 148–151, 2000)

The process can be divided into four distinct steps. (1) The full length precursor is exported from the cytoplasm via an amino-terminal leader peptide. (2) The protein is prevented from release into the extracellular milieu by the charged tail and hydrophobic domain. (3) The protein is cleaved by a sortase between the threonyl and glycyl residues of the LPXTG motif, with the formation of a thioester between the conserved cysteine of sortase and the threonine carboxyl. (4) The newly liberated carboxy terminus of threonine is transferred via an amide bond exchange to an amino group found at lipid II.

endopeptidases to accommodate newly synthesized precursor during cell growth (Figure 6.36). Various peptidoglycan hydrolases are known with functions in cell wall growth as well as others including invasion of prey cells, immune detection, intercellular communication, and competitor lysis. Specialized hydrolases enlarge the pores in the peptidoglycan for the assembly of large trans-envelope complexes such as flagella, or for sporulation or spore germination. Transglycosylase and transpeptidase incorporate the precursor into the existing peptidoglycan on the site cleaved by the hydrolytic enzymes. These enzymes form a complex known as the penicillin binding protein (PBP) complex. A new *Escherichia coli* cell synthesizes peptidoglycan in a dispersed mode along the cell cylinder, driving elongation before a zonal mode in mid-cell where the septum is formed. This switch is mediated by a cytoskeletal protein, MreB. This is followed by septation guided by another cytoskeletal protein, FstZ. On the other hand, in the coccal *Staphylococcus aureus* cell, peptidoglycan is synthesized in the middle of the cell where division takes place.

6.9.3 S-layer

The S-layer is a protein or glycoprotein layer found on the surface of almost all prokaryotic cells, with indispensable roles in natural habitats (Section 2.3.3.1). This two-dimensional layer consists of one or more proteins or glycoproteins known as S-layer protein (SLP). In spite of the conserved structure there is a wide diversity in the amino acid sequence. SLP genes are linked to genes encoding components for their modification and secretion. SLPs are translocated through the general secretion pathway (Section 3.10.1.1) in Gram-positive bacteria, while Gram-negative bacteria employ type I or II secretion pathways (Section 3.10.2). After secretion, SLPs anchor to the cell surface and spontaneously form a crystalline structure. SLPs have domains for anchoring and forming the crystalline array. Secreted

SLPs are anchored to lipopolysaccharide in Gram-negative bacteria. In Gram-positive bacteria, secondary cell wall polymers (SCWP) such as teichoic acid (Section 2.3.3.3) anchor the SLPs. In some bacteria, SLPs are glycosylated after the carbohydrate component is synthesized as the repeating oligosaccharide of lipopolysaccharide (Figure 6.32) and exported through the ABC pathway (Section 3.10.1.3).

6.9.4 Outer membrane assembly

The asymmetrical Gram-negative bacterial outer membrane consists of outer layer LPS, and inner layer phospholipids and proteins. They are synthesized in the cytoplasm and are translocated to their position through the cytoplasmic membrane and the peptidoglycan before being assembled.

6.9.4.1 Protein translocation

Proteins present in the OM are either lipoproteins or integral proteins. The former are anchored to the OM with an *N*-terminal lipid tail, stabilizing the OM structure by linking to the peptidoglycan layer. Whereas some integral OM proteins span the membrane in the form of α-helices entirely composed of hydrophobic amino acids like cytoplasmic membrane proteins, other OMPs, including porins, form β-barrels composed of antiparallel amphipathic β-strands. The hydrophobic residues in these β-strands are exposed to the lipid environment of the membrane, whereas the hydrophilic residues point towards the interior of the protein, which is the aqueous channel in the case of porins. Precursor polypeptides of the lipoproteins and integral proteins in α-helical form are translocated through the cytoplasmic membrane, either through the GSP or ABC pathways (Section 3.10.1), before being folded in the periplasmic region and targeted to their destination in a process similar to the chaperone/usher pathway (Section 3.10.2.1). Polypeptides of β-strand OMPs are translocated through the GSP to the periplasm, where chaperones guide them to an outer membrane protein complex, the β-barrel assembly machinery (BAM), that assembles and inserts the substrate into OM.

6.9.4.2 Lipopolysaccharide (LPS) translocation

LPS, the outer layer of the outer membrane in Gram-negative bacteria, is a complex molecule consisting of lipid A, core polysaccharide and O-antigen. O-antigen ligase synthesizes LPS, ligating O-antigen, transported by undecaprenyl phosphate through the cytoplasmic membrane, to the lipid A–core polysaccharide complex in the periplasmic region. LPS is targeted to the outer layer of the OM by the LPS transport (Lpt) machinery, a member of the ATP-binding cassette (ABC) family.

6.9.4.3 Phospholipid translocation

Phospholipids of the inner layer of the outer membrane are transported from the cytoplasmic membrane to their location by a complex machinery, including a cytoplasmic protein (MlaD) forming a ring associated with an ABC transporter complex, a soluble periplasmic lipid-binding protein (MlaC) and an outer membrane protein complex. MlaC ferries phospholipids from MlaD to the outer membrane protein complex.

6.9.5 Cytoplasmic membrane (CM) assembly

The CM consists of phospholipids and proteins. Protein constitutes up to 65% of the CM, with various functions including solute transport and electron transport for energy conservation. CM proteins are embedded in the membrane through the SRP (signal recognition particle) pathway, one of two general secretion pathways (GSP, Section 3.10.1.1). There is some evidence which suggests that a membrane protein, YidC in *Escherichia coli*, facilitates GSP-independent membrane integration, folding and assembly of energy-transducing membrane protein complexes. Phospholipids form the inner layer of the CM after they are synthesized, and are flipped to the outer layer by the action of flippase.

The phospholipids of the archaeal CM are a diether (archaeol) or tetraether (caldarchaeol) composed of isoprenoid hydrocarbon side chains linked to glycerol-1-phosphate via an ether bond (Section 2.3.4.3), with a similar head group

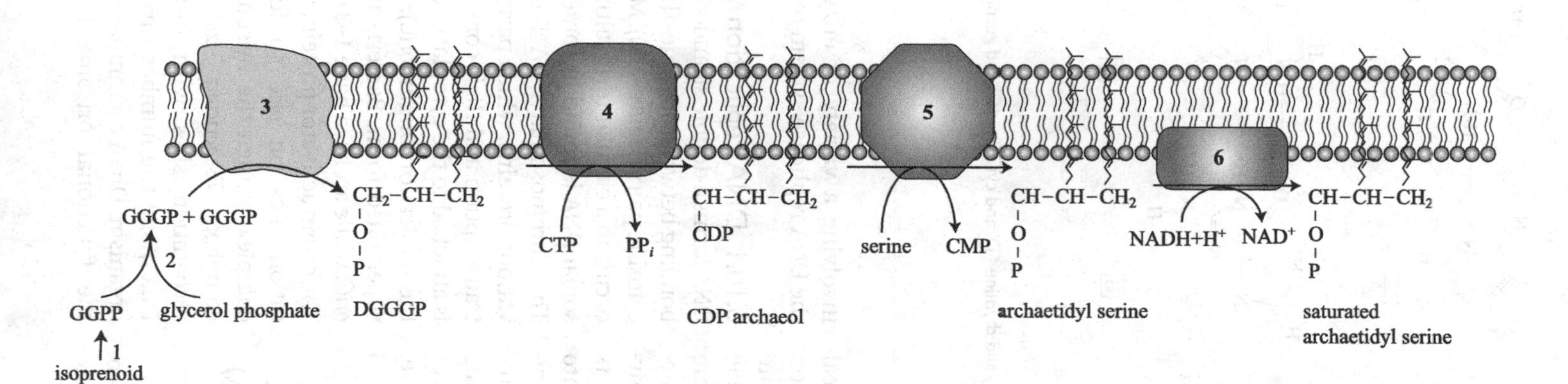

Figure 6.39 Biosynthesis of archaeal cytoplasmic membrane phospholipids

(Modified from *Front. Microbiol.* **5**: 641, 2014)

The diether (archaeol) or tetraether (caldarchaeol) phospholipids of the archaeal cytoplasmic membrane contain long-chain alcohols synthesized from isopentenyl pyrophosphate in the cytoplasm, catalysed by geranylgeranyl diphosphate (GGPP) synthase (1). The geranylgeranyl-glyceryl-phosphate (GGGP) synthase (2) combines GGPP and glycerol-1-phosphate to form geranylgeranyl-glyceryl-phosphate (GGGP). The membrane-bound archaeol synthase (3) recruits GGGP and GGPP to form archaeol and inserts it into the existing membrane. Archaeol is activated to CDP-archaeol by CDP archaeol synthase (4). CMP is replaced with an alcohol, such as serine, by the action of archaetidyl serine synthase (5). Finally, the isoprenoid alcohols are reduced to the fully saturated form by geranylgeranyl reductase (6).

IPP, isopentenylpyrophosphate; GGPP, geranylgeranyl diphosphate; GGGP, 3-*O*-geranylgeranyl-*sn*-glyceryl-1-phosphate,

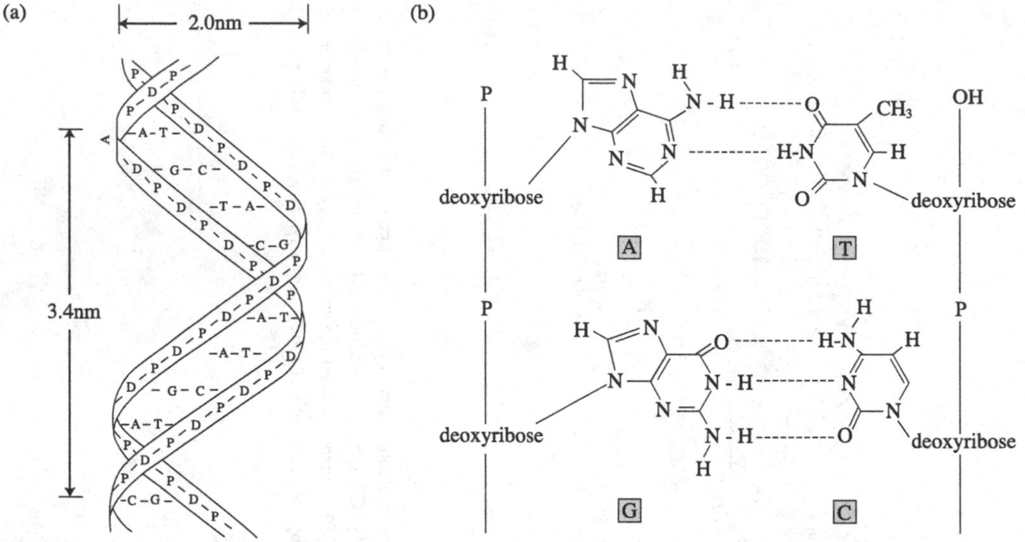

Figure 6.40 **The DNA double helix.**

(a) The double helix. (b) Specific pairing between adenine (A) and thymine (T), and cytosine (C) and guanine (G) through hydrogen bonding.

D, deoxyribose; P, phosphate.

structure to bacterial phospholipids. Archaeal phospolipid synthesis begins with condensation of four molecules of isopentenyl pyrophosphate (IPP) to geranylgeranyl diphosphate (GGPP) that forms geranylgeranyl-glyceryl-phosphate (GGGP) with glyceryl-phosphate. A membrane-bound enzyme, archaeol (digeranylgeranyl-glyceryl-phosphate, DGGGP) synthase converts GGGP and GGPP to archaeol and inserts it into the existing membrane (Figure 6.39). The membrane-bound CDP archaeol synthase activates the membrane-embedded archaeol, consuming CTP, before an alcohol such as serine replaces CDP by the action of CDP archaeol synthase. Finally the isoprenoid units are fully reduced to a saturated state.

6.10 Deoxyribonucleic acid (DNA) replication

DNA carries genetic information in its nucleotide sequence and its synthesis is complex, involving a variety of enzymes referred to as the DNA replicase system or replisome.

6.10.1 DNA replication

DNA forms a double-stranded helix with hydrogen bonding between adenine–thymine and guanine–cytosine pairs (Figure 6.40). Most prokaryotic DNA occurs in a circular form, although linear chromosomal DNA is found in some *Streptomyces* species. The chromosomal DNA forms complexes with various proteins, which participate in replication, transcription and their control. The chromosome is attached to the cytoplasmic membrane through the replication origin (*oriC*), and the DNA is in a supercoiled form. Bacteria usually have a single *oriC*, while archaea have 1–3 origins of replication. The origin activator protein, DnaA, binds its recognition sites on *oriC* to form a DNA–protein complex termed the orisome or pre-replication complex. *Escherichia coli* has nine DnaA recognition sites with different affinities for DnaA, and the number varies depending on the organism. DnaA accumulates during growth, and the functional orisome is formed to initiate

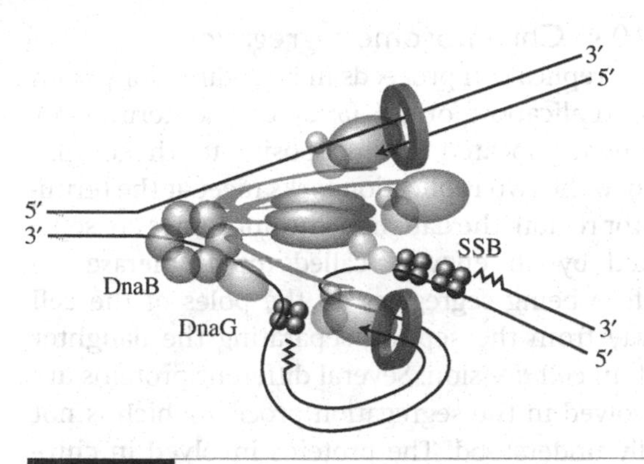

Figure 6.41 DNA replication by the replisome in *Escherichia coli*.

(Modified from *Front. Microbiol.* **6**: 562, 2015)

When DnaA concentration becomes sufficient to bind all its recognition sites, the functional orisome is formed to trigger the onset of chromosome duplication by unwinding *oriC* DNA, consuming ATP and assisting with loading of DNA helicase (DnaB) and helicase loader (DnaC) on to the single-stranded DNA. Single-strand binding proteins stabilize the newly formed replication bubble and interact with the DnaG primase. DnaG recruits the replicative DNA polymerase III, and replication begins. All these proteins form the replisome that disassembles at the end of the replication process. In *E. coli* a single replisome contains DNA polymerase III complexes.

DnaB, helicase; DnaG, primase; SSB, single strand DNA binding protein.

replication when its concentration is sufficient to bind all nine sites. Bacterial orisomes trigger the onset of chromosome duplication by unwinding *oriC* DNA, consuming ATP and assisting with loading of DNA helicase (DnaB) and helicase loader (DnaC) on to the single-stranded DNA. Single-strand DNA binding proteins stabilize the newly formed replication bubble and interact with the DnaG primase. DnaG recruits the replicative DNA polymerase III and replication begins. All these proteins form a complex during the replication process and disassemble at the end of the replication process. This complex is termed the replisome. Figure 6.41 shows DNA replication in prokaryotes.

6.10.1.1 RNA primer

DNA polymerase requires a template for base pairing and a primer to add the nucleotide. Single-strand DNA is used as the template. To provide primer, RNA polymerase (primase) synthesizes oligoribonucleotides of 10–30 bases by means of complementary base pairing with the template. This is referred to as the RNA primer. DNA polymerase adds deoxynucleotide to the primer (consuming deoxynucleoside triphosphate) based on the base pairing with the template in the direction 5′–3′.

6.10.1.2 Okazaki fragment

Helicase (DnaB) separates double-stranded DNA into single strands, the leading strand in the direction 3′–5′ and the other, the lagging strand, in the direction 5′–3′. DNA polymerase adds new nucleotide only to the 3′-OH of the existing DNA molecules. For this reason, DNA synthesis takes place continuously on the leading strand, while discontinuous DNA synthesis occurs on the lagging strand to a length of 1000–2000 nucleotides. This is referred to as the Okazaki fragment.

6.10.1.3 DNA polymerase

Escherichia coli has three separate DNA polymerases, I, II and III. All of them have DNA polymerase activity in the direction 5′–3′ and exonuclease activity removing nucleotide at the 3′ end. DNA polymerase I and III have exonuclease activity removing nucleotide at the 5′ end. DNA polymerase III synthesizes the Okazaki fragment onto the RNA primer, before the RNA primer is removed and replaced with DNA by the action of DNA polymerase I with exonuclease activity. DNA ligase links Okazaki fragments after the primer RNA is completely replaced with DNA.

6.10.1.4 Replication–transcription conflicts

DNA replication and DNA transcription use the same template and occur concurrently, possibly leading to conflict. When genes on the lagging strand are transcribed, the replisome collides head-on with the transcription complex, and co-directional conflict is expected with the replication of genes on the leading strand. Co-directional conflict is less frequent than head-on conflict but inevitable because the replisome moves 10–20 times faster than the transcription complex. Conflicts are commonly prevented by removing the RNA polymerase from the DNA through various mechanisms. When these prevention mechanisms fail, conflicts are resolved through yet more mechanisms.

6.10.2 Spontaneous mutation

Prokaryotes grow much faster than higher organisms and DNA is replicated at speeds of up to 15 000 bases per minute. Errors in DNA replication result in mutations. Mutation rates in prokaryotes are low, being around 10^{-10} errors per base inserted, even at this high replication speed. This high accuracy in the replication process is possible because of the exact base pairing and the 'proof reading' by DNA polymerase III, with exonuclease activity replacing misinserted nucleotides. Failure in these processes will result in spontaneous mutation.

6.10.3 Post-replicational modification

Bacterial DNA contains 5-methyl cytosine and 6-methyl adenine. These are the products of the methylation reactions catalysed by DNA methyltransferase after DNA is synthesized. This process is referred to as post-replicational modification. The DNA is methylated to protect it from restriction enzymes which degrade foreign DNA invading the cell.

Several hundred restriction enzymes are known in prokaryotes. They recognize specific sequences in the DNA to hydrolyse and to methylate. They are grouped into types I, II and III according to their properties. Types I and III not only hydrolyse the DNA sequence they recognize, but also methylate adenine and cytosine within the recognized sequence. Type II restriction enzymes have hydrolysing activity but not methylating activity. For this reason, type I and III enzymes are referred to as restriction–modification enzymes. *S*-adenosylmethionine serves as the methyl group donor for the DNA modification reaction. In addition to the function of restriction–modification, these enzymes may participate in other processes such as DNA replication, cell cycle events and regulation of gene expression. Topoisomerase converts the modified DNA into a super-coiled form.

6.10.4 Chromosome segregation

DNA replication proceeds in both directions from the replication origin (*oriC*) to the terminator sequence located 180° opposite to the origin. When the two replication forks meet at the terminator region, the daughter chromosomes are separated by an enzyme called topoisomerase IV, before being segregated to the poles of the cell away from the septum separating the daughter cell in cell division. Several different proteins are involved in the segregation process which is not fully understood. The proteins involved in chromosome segregation include SpoOJ in *Bacillus subtilis* and the ParB protein in *Caulobacter crescentus*. They form a complex with *oriC*.

6.11 Transcription

Transcription is the process of RNA synthesis from the DNA template. Major RNA types include mRNA, rRNA and tRNA, with specific functions in expression of the genetic information. They are all synthesized in a similar way before being modified to their own specific forms through post-transcriptional processing.

6.11.1 RNA synthesis

RNA is synthesized through transcription, a process as complex as replication. RNA polymerase (DNA-dependent RNA polymerase) catalyses the formation of phosphodiester bonds between ribonucleotides, consuming GTP, CTP, ATP or UTP, depending on the base pairing within DNA. The synthesis proceeds in the 5′–3′ direction as in replication.

Certain regions in DNA have a strong affinity for RNA polymerase and are referred to as promoters (Section 12.1.1). RNA polymerase recognizes and binds the promoter region of DNA to start RNA synthesis. RNA polymerase consists of five subunits in the ratio of $\alpha_2\beta\beta'\sigma$ in bacteria. The σ-factor of the enzyme recognizes the promoter region of DNA, and the enzyme complex binds to it. At this point, the σ-factor is separated from the complex and the core enzyme $\alpha_2\beta\beta'$ moves along the DNA synthesizing RNA according to a base-pairing mechanism

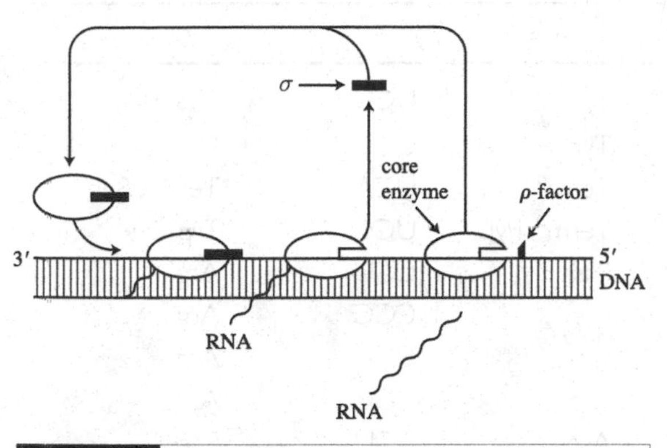

Figure 6.42 RNA synthesis.

(Gottschalk, G. 1986, *Bacterial Metabolism*, 2nd edn., Figure 3.40. Springer, New York)

The RNA polymerase complex consists of a core enzyme ($\alpha_2\beta\beta'$) and σ-factor. The enzyme complex binds DNA after the σ-factor recognizes the promoter region on the DNA. When RNA synthesis starts, σ-factor is separated from the core enzyme, which moves along the DNA synthesizing RNA according to the base-pairing mechanism. Many proteins are involved in the initiation, elongation and termination of the transcription process. An example is ρ-factor, which terminates the transcription.

(Figure 6.42). Multiple σ-factors are known in bacteria which recognize different promoter regions. Proteins known as activators and repressors control the activity of some promoters (Section 12.1.1).

Transcription is terminated either by an intrinsic input mechanism involving a specific DNA region known as a termination site, or by an extrinsic input mechanism exerted by a protein, ρ-factor.

6.11.2 Post-transcriptional processing

The transcripts of tRNA and rRNA are longer than the functional RNAs and the latter contain methylated nucleotides. The transcripts are methylated before being cut by endonuclease to their functional size (Figure 6.43). Unusual bases are found in tRNA. The nucleotides in tRNA are further modified in a mechanism called post-transcriptional processing.

It should be noted that unlike eukaryotic transcripts, most prokaryotic transcripts of mRNA do not contain an intron, and translation starts while transcription is in progress. mRNA editing is not known in prokaryotes.

6.12 Translation

Translation is the process in which the genetic information passed on to mRNA is used to make proteins. As the genetic information carried by DNA is passed to RNA during transcription, the information in mRNA is translated into protein through an amino acid

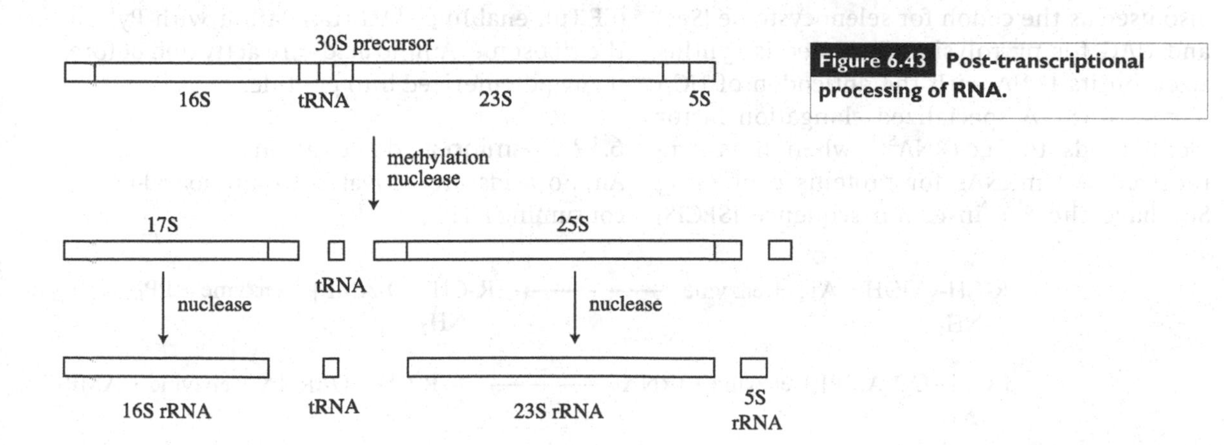

Figure 6.43 Post-transcriptional processing of RNA.

Table 6.7 | Genetic code of mRNA.

UUU	Phe	UCU	Ser	UAU	Tyr	UGU	Cys
UUC	Phe	UCC	Ser	UAC	Tyr	UGC	Cys
UUA	Leu	UCA	Ser	UAA	Term[a]	UGA	Term[a], Sec[b]
UUG	Leu	UCG	Ser	UAG	Term[a], Pyl[c]	UGG	Trp
CUU	Leu	CCU	Pro	CAU	His	CGU	Arg
CUC	Leu	CCC	Pro	CAC	His	CGC	Arg
CUA	Leu	CCA	Pro	CAA	Gln	CGA	Arg
CUG	Leu	CCG	Pro	CAG	Gln	CGG	Arg
AUU	Ile	ACU	Thr	AAU	Asn	AGU	Ser
AUC	Ile	ACC	Thr	AAC	Asn	AGC	Ser
AUA	Ile	ACA	Thr	AAA	Lys	AGA	Arg
AUG	Met[d]	ACG	Thr	AAG	Lys	AGG	Arg
GUU	Val	GCU	Ala	GAU	Asp	GGU	Gly
GUC	Val	GCC	Ala	GAC	Asp	GGC	Gly
GUA	Val	GCA	Ala	GAA	Glu	GGA	Gly
GUG	Val	GCG	Ala	GAG	Glu	GGG	Gly

[a] UAA, UAG and UGA are non-coding codons where peptide synthesis stops. They are referred to as termination or nonsense codons.
[b] Sec – selenocysteine.
[c] Pyl – pyrrolysine.
d AUG encodes *N*-formylmethionine at the beginning of mRNA in bacteria.

sequence encoded by the sequence of bases in the mRNA (Table 6.7). Methionine and tryptophan use a single codon while others use multiple codons among the 20 canonical amino acids. Three are used as termination or nonsense codons. Among them, UGA is also used as the codon for selenocysteine (Sec) and UAG for pyrrolysine (Pyl). Sec is synthesized on its tRNA with the anticodon of UCA (Figure 6.18). A specialized elongation factor (SelB) binds to Sec-tRNA^{Sec} when it is synthesized. All mRNAs for proteins containing Sec have the Sec insertion sequence (SECIS) element. The SECIS element directs the SelB-Sec-tRNA^{Sec} complex to recode on the ribosome of UGA codons in selenoprotein mRNAs. On the other hand, Pyl does not need a special elongation factor nor an insertion sequence. Pyl-tRNAPyl is bound by elongation factor Tu (EF-Tu), enabling UAG translation with Pyl on the ribosome. Amino acids are activated before being polymerized into peptide.

6.12.1 Amino acid activation

Amino acids are activated to aminoacyl-tRNA, consuming ATP:

$$\underset{\displaystyle NH_2}{\text{R-CH-COOH}} + \text{ATP} + \text{enzyme} \longrightarrow [\underset{\displaystyle NH_2}{\text{R-CH-CO-AMP}}] + \text{enzyme} + PP_i$$

$$[\underset{\displaystyle NH_2}{\text{R-CH-CO-AMP}}] + \text{enzyme} + \text{tRNA} \longrightarrow \underset{\displaystyle NH_2}{\text{R-CH-CO-tRNA}} + \text{enzyme} + \text{AMP}$$

enzyme: aminoacyl-tRNA synthetase

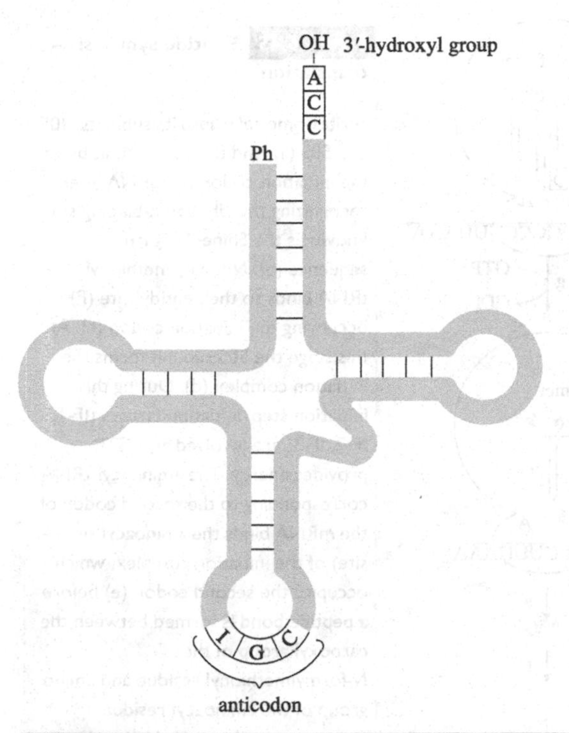

Figure 6.44 Structure of a tRNA.

(Gottschalk, G. 1986, *Bacterial Metabolism*, 2nd edn., Figure 3.43. Springer, New York)

Hairpin structures are formed in a tRNA through internal hydrogen bonding. The amino acid is attached to the ribose residue of the terminal adenosine nucleotide by the action of aminoacyl-tRNA synthetase. The anticodon is located at the bottom, recognizing the mRNA codon.

The aminoacyl-tRNA synthetase catalysing this reaction recognizes not only amino acids but also tRNA. More than one tRNA is needed for each amino acid since most of the amino acids are encoded by multiple codons, except methionine, tryptophan, selenocysteine and pyrrolysine (Table 6.7), and a tRNA is needed for each codon base pairing with its specific anticodon (Figure 6.44).

6.12.2 Synthesis of peptide: initiation, elongation and termination

The coding region in mRNA starts with AUG. Peptide synthesis starts with methionine in eukaryotes and with *N*-formylmethionine in bacteria. AUG (TAC on DNA) is referred to as the initiation codon. Though peptide synthesis is a continuous process, translation can be described for convenience as initiation, elongation and termination steps.

6.12.2.1 Ribosomes

Ribosomes are the sites of peptide synthesis. The prokaryotic ribosome is of size 70S, consisting of a 50S large subunit and 30S small subunit. The large subunit contains 5S and 23S rRNA and about 36 proteins, while the small 30S subunit consists of 16S rRNA and about 21 proteins. The number of ribosomal proteins differs depending on the species. The eukaryotic ribosome is of size 80S, consisting of 60S and 40S subunits. Mitochondria and chloroplasts have ribosomes similar to prokaryotes.

6.12.2.2 Initiation and elongation

The 30S subunit recognizes the ribosome binding site, known as the Shine–Dalgarno sequence, to bind the initiation codon on the mRNA. *N*-formylmethionyl-tRNA binds to the peptidyl site (P) occupying the initiation codon before the 50S subunit binds the 30S subunit–mRNA complex to form the initiation complex. Various proteins, including the initiation factors IF-1, IF-2 and IF-3, participate during this initiation step, and energy is provided through the hydrolysis of GTP. The efficiency of translation of individual mRNAs is defined by the structures of the mRNA translation initiation regions, which interact with small molecules, proteins or antisense RNAs to control the translation (Sections 12.1.4, 12.1.6 and 12.1.9).

The initiation complex takes up the aminoacyl-tRNA corresponding to the second codon of the mRNA to its aminoacyl site (A site), which occupies the second codon. A peptide bond is formed between the carboxyl group of the *N*-formylmethionyl residue and amino group of the aminoacyl residue occupying the A site. When the peptide bond is formed, the ribosome moves along the mRNA to position the dipeptide at the P site. These steps are repeated during the elongation process where proteins, including the elongation factors Tu, Ts and G, play important roles and energy is supplied as GTP (Figure 6.45).

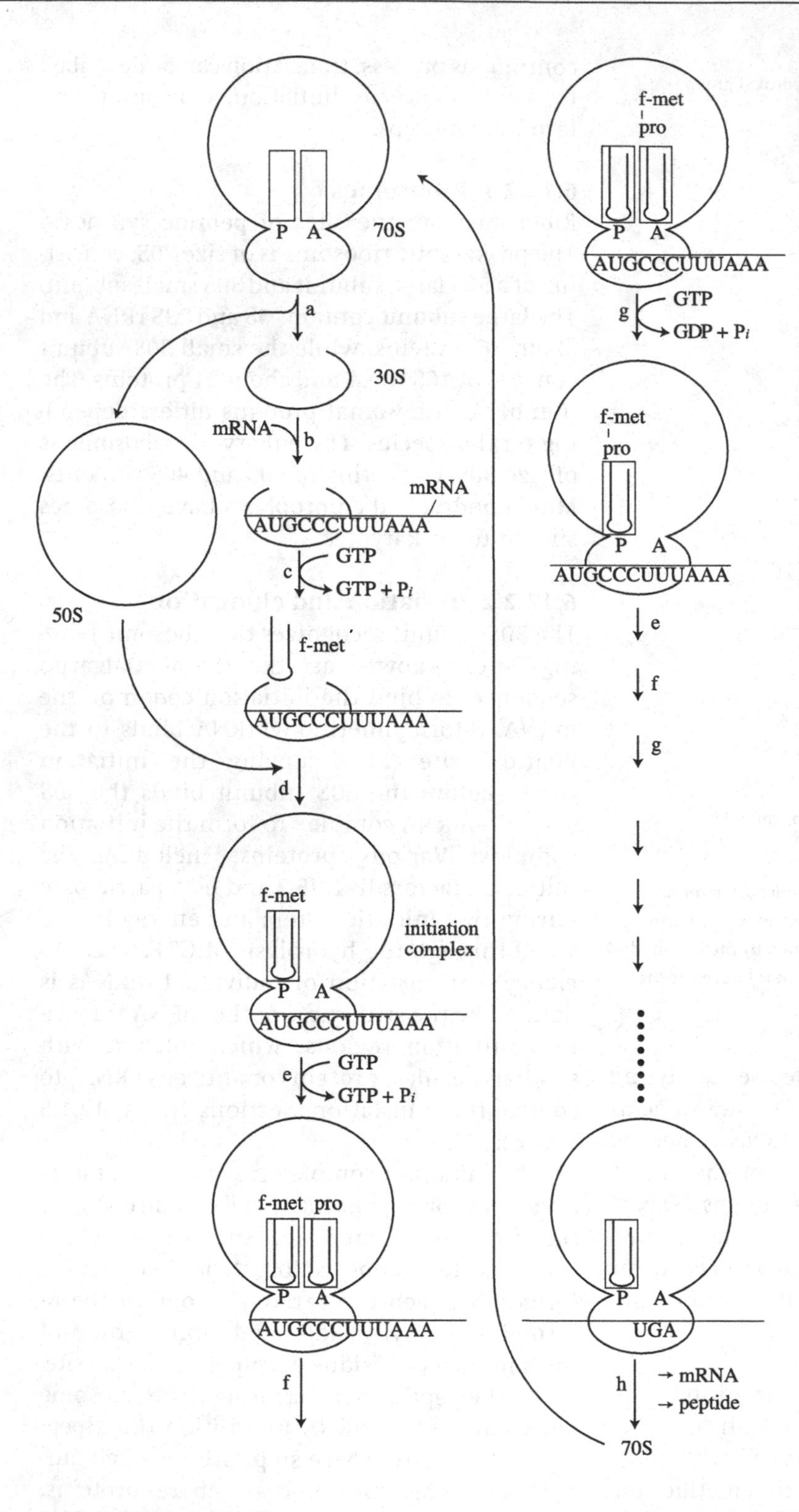

Figure 6.45 Peptide synthesis – translation.

A ribosome splits into its subunits, 30S and 50S (a), and the 30S subunit binds the initiation codon on mRNA after recognizing the ribosome-binding site, known as the Shine–Dalgarno sequence (b). *N*-formylmethionyl-tRNA binds to the peptidyl site (P) occupying the initiation codon (c). At this stage the 50S subunit forms the initiation complex (d). During this initiation step, initiation factors (IF-1, IF-2, IF-3) are involved and GTP provides energy. The aminoacyl-tRNA corresponding to the second codon of the mRNA binds the aminoacyl site (A site) of the initiation complex, which occupies the second codon (e) before a peptide bond is formed between the carboxyl group of the *N*-formylmethionyl residue and amino group of the aminoacyl residue, occupying the A site (f). When the peptide bond is formed, the ribosome moves along the mRNA to position the dipeptide to the P site (g). During the repeat of this elongation process, elongation factors (Tu, Ts, G) are needed and energy is supplied in the form of GTP. When the ribosome reaches the termination codon, releasing factors (R1, R2, S) separate the peptide from the ribosome (h).

6.12.2.3 Termination

When the ribosome reaches the termination codon, releasing factors, R1, R2 and S, separate the peptide from the ribosome. Many ribosomes bind a single mRNA during the translation process, and the term polysome is used to describe such an mRNA complex with multiple ribosomes.

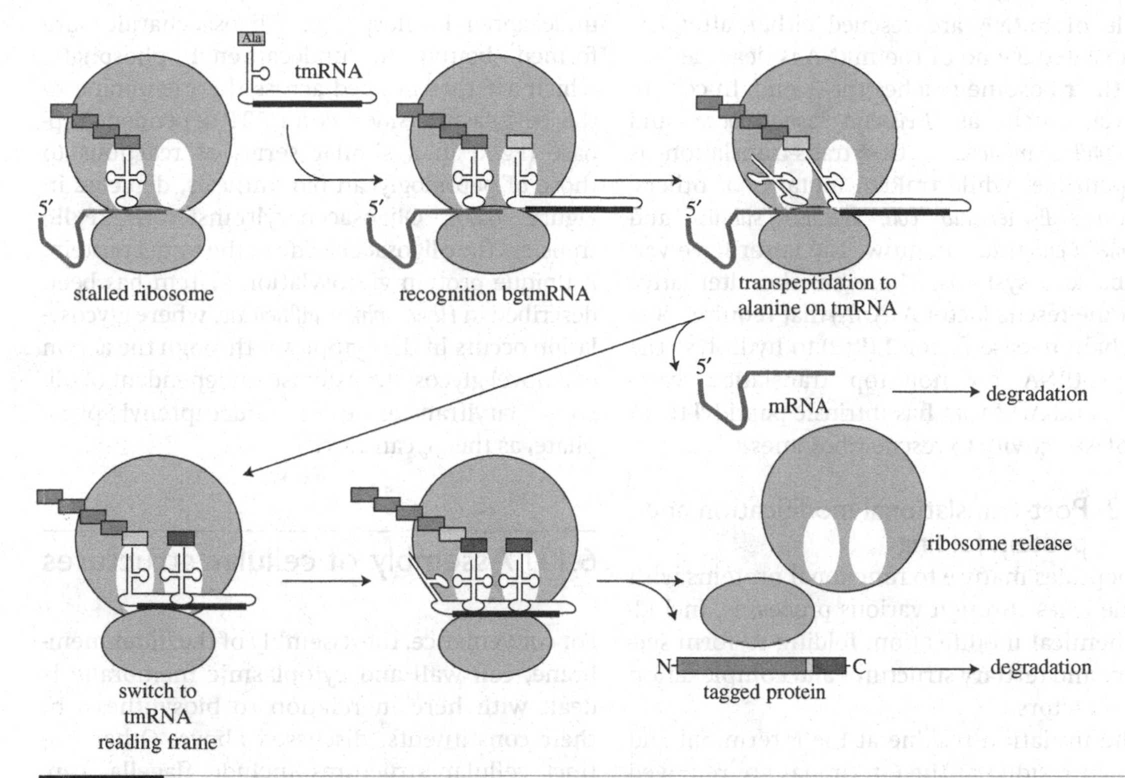

Figure 6.46 Ribosome rescue by trans-translation with tmRNA.

(Modified from *Curr. Opin. Microbiol.* **10**: 169–175, 2007)

Ribosomes are stalled on mRNAs without a stop codon and the ribosome cannot dissociate from a non-stop translation complex that holds the peptidyl-tRNA, the ribosome and the mRNA tightly together. The tmRNA aminoacylated with alanine recognizes the stalled ribosome and enters the A-site. Translation continues along the reading frame of tmRNA to reach a stop codon and the non-stop translation complex is disassembled, releasing the ribosome. The tagged protein and truncated mRNA are degraded.

6.12.2.4 Ribosome rescue by transfer-messenger RNA (tmRNA)

Ribosomes are stalled on mRNAs for various reasons, including premature termination of transcription and a missing stop codon in damaged mRNA, and bacterial ribosomes frequently translate to the 3′ end of an mRNA without terminating at a stop codon. Without a stop codon the ribosome cannot dissociate from a non-stop translation complex that holds the peptidyl-tRNA, the ribosome and the mRNA tightly together. This interaction is necessary for accurate translation. The accumulation of non-stop complexes is toxic, and without mechanisms to rescue these ribosomes the ability of the cell to synthesize proteins is rapidly diminished. This stalled ribosome is rescued by a complex of a specialized small RNA molecule, called SsrA (small stable RNA A) or transfer-messenger RNA (tmRNA) and small protein B (SmpB) structurally mimicking a tRNA molecule. The tmRNA aminoacylated with alanine interacts with the stalled ribosome. A tmRNA molecule has the ability to function similarly to both a tRNA and an mRNA, with a specialized reading frame that encodes a short peptide 8–35 amino acids long, depending on the organism. Translation continues along the reading frame of tmRNA to reach a stop codon. This process is known as trans-translation. The translation complex is disassembled, rescuing the ribosome. The tagged peptide and the damaged mRNA are degraded (Figure 6.46). Ribosomes stalled in the middle of mRNA are rescued either after the untranslated 3′-end of the mRNA is degraded, or after the ribosome reaches the 3′-end. In certain bacteria, such as *Neisseria gonorrhoeae* and *Haemophilus influenzae*, the trans-translation is indispensible, while tmRNA mutants of others, including *Escherichia coli, Bacillus subtilis* and *Salmonella enterica*, can grow. The latter have various backup systems. These are the alternative ribosome-rescue factor A (ArfA) that requires peptide chain release factor 2 (RF2) to hydrolyse the peptidyl-tRNA on non-stop translation complexes, and ArfB that has intrinsic peptidyl-tRNA hydrolase activity to rescue ribosomes.

6.12.3 Post-translational modification and protein folding

The peptides mature to functional proteins with unique roles through various processes, including chemical modification, folding to form secondary and tertiary structures and complexation with cofactors.

The initiation residue at the *N*-terminal and terminal residue at the *C*-terminal are removed from the peptides released from the ribosome. At this stage, unusual amino acids are formed through chemical modification. Extracellular and membrane proteins are exported, losing the signal peptide (Section 3.10.1).

During protein folding, small molecular weight chaperones, e.g. trigger factor, bind the nascent peptide to prevent misfolding during the translation process. Small proteins can be spontaneously folded into native forms. Large proteins are folded into native forms through a process aided by a chaperone or chaperonin (Figure 6.47), both of which require ATP for their function.

Some cell surface proteins in bacteria and archaea are glycosylated. These include S-layer proteins (Section 2.3.3.1), flagellin proteins of *Helicobacter pylori* and *Campylobacter jejuni* and pilin proteins in *Neisseria gonorrhoeae* and *Pseudomonas aeruginosa*. Protein glycosylation is best known in *C. jejuni*. Proteins responsible for protein glycosylation are encoded in the *pgl* (protein glycosylation) gene cluster. Monosaccharides are activated to UDP derivatives before being transferred to the membrane-bound undecaprenyl phosphate. Oligosaccharides are formed bound to undecaprenyl phosphate, which are then flipped across the membrane to the periplasmic side via an ATP-dependent flippase (PglK) in a similar series of reactions to those of peptidoglycan biosynthesis, depicted in Figure 6.36. Oligosaccharyltransferase (PglB) transfers the oligosaccharide to the target protein. A unique protein glycosylation system has been described in *Haemophilus influenzae*, where glycosylation occurs in the cytoplasm through the action of a novel glycosyltransferase, independent of oligosaccharyltransferase or undecaprenyl phosphate, as the glycan carrier

6.13 Assembly of cellular structures

For convenience, the assembly of the outer membrane, cell wall and cytoplasmic membrane is dealt with here in relation to biosynthesis of their constituents, discussed above. Other distinct cellular structures include flagella, capsules, ribosomes and the nucleoid.

6.13.1 Flagella

A flagellum consists of a basal structure, hook and main filament, all of which are composed of protein. The basal structure includes the basal body, flagellar motor, switch and the exporter apparatus (Section 2.3.1). The basal body has a rod structure traversing the periplasmic space and surrounded with three ring structures. These are the cytoplasmic membrane MS ring, the periplasmic P ring and the outer membrane L ring (Section 2.3.1). The flagellar motor can be subdivided into the stator and the rotor. The stator is attached to the peptidoglycan layer to surround the basal body, and the rotor is attached to the MS ring. A flagellum extends from the cytoplasm to the cell surface and has to be exported and embedded in position. Proteins constituting the flagellum are not found in the cytoplasm, suggesting that they are exported as soon as they are synthesized. Cytoplasmic flagellar proteins are embedded through the signal recognition particle pathway (SRP, Section 3.10.1.1). The proteins located

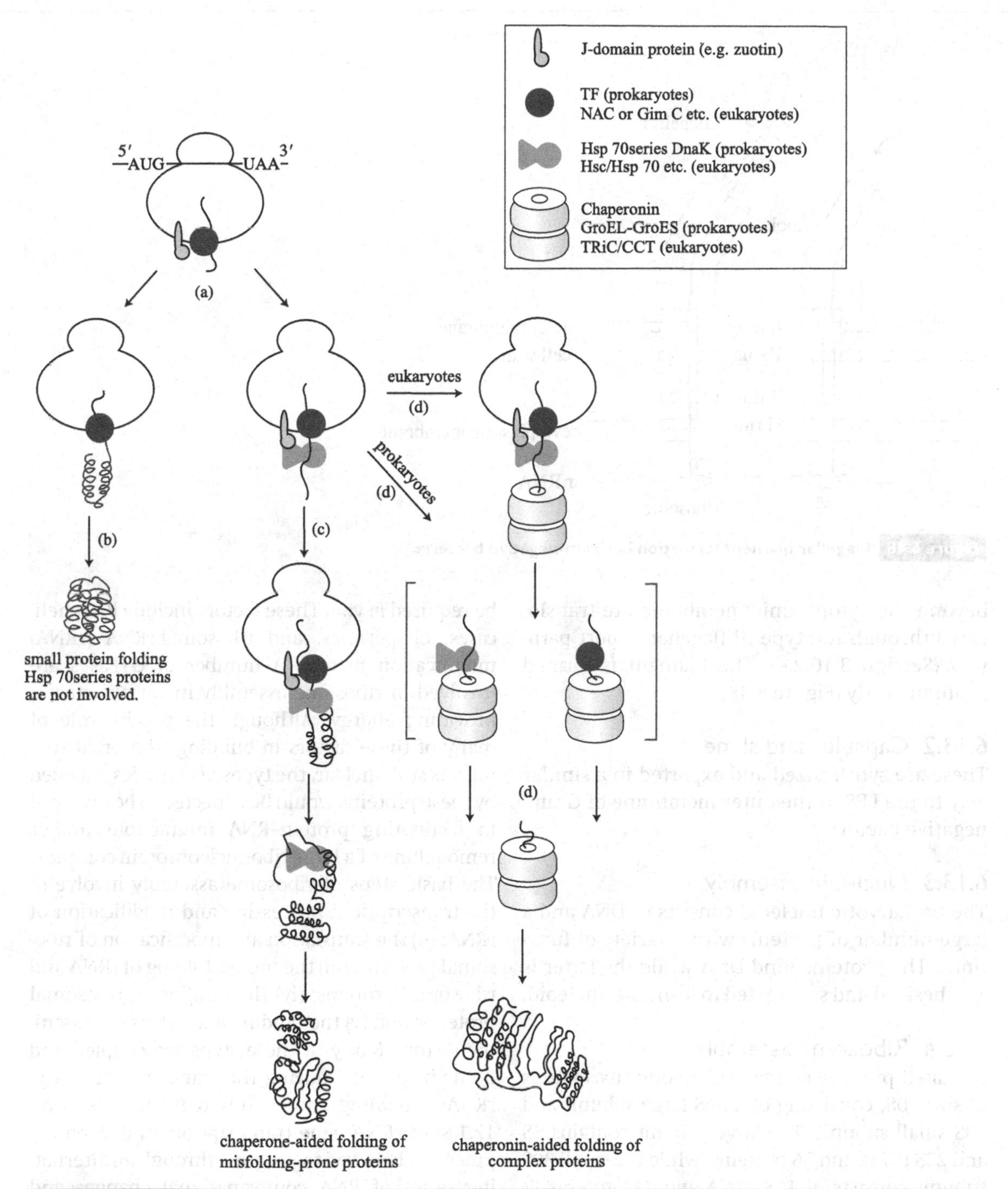

Figure 6.47 Protein folding.

(*Curr. Opin. Struct. Biol.* **10**: 26–33, 2000)

During peptide synthesis on a ribosome, misfolding is prevented by small molecular weight chaperones, such as trigger factor (TF) (a), while small proteins can fold into native forms spontaneously (b). For large proteins, folding is aided by a chaperone such as DnaK (c) or by a chaperonin such as GroEL (d). ATP is consumed to supply the energy needed for the function of chaperone and chaperonin.

Thiol:disulfide oxidoreductase forms disulfide bonds, oxidizing specific cysteine residues within the peptide. Disulfide bonds of extracellular proteins are formed by periplasmic enzymes.

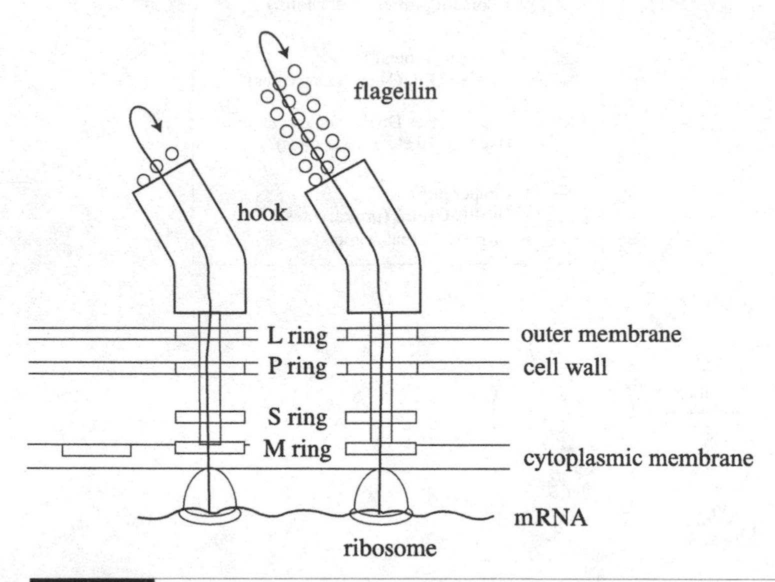

Figure 6.48 Flagellar filament formation in Gram-negative bacteria.

beyond the cytoplasmic membrane are translocated through the type III flagellar export pathway (Section 3.10.2.4). The filament is formed spontaneously (Figure 6.48).

6.13.2 Capsules and slime

These are synthesized and exported in a similar way to the LPS of the outer membrane of Gram-negative bacteria.

6.13.3 Nucleoid assembly

The prokaryotic nucleoid consists of DNA and a large number of proteins with a variety of functions. The proteins bind DNA while the latter is synthesized and segregated to form the nucleoid.

6.13.4 Ribosome assembly

As stated previously, the prokaryotic ribosome is of size 70S, consisting of a 50S large subunit and 30S small subunit. The large subunit contains 5S and 23S rRNA and 36 proteins, while the small 30S subunit consists of 16S rRNA and 21 proteins. It was widely accepted that the information for assembling these large macromolecular machines resided within the components of the ribosomes themselves. However, the non-physiological conditions required for ribosome assembly (high salt, high temperatures) and the slow kinetics of assembly *in vitro* indicated that additional factors would be required *in vivo*. These factors include RNA helicases, chaperones, and ribosomal RNA (rRNA) modification factors. A number of GTPases are involved in ribosome assembly in various stages, providing energy. Although the precise role of many of these factors in building ribosomal subunits is still unclear, the types of activities encoded by these proteins would be expected to be involved in facilitating protein–RNA interactions and/or remodelling of a large ribonucleoprotein complex. The basic steps of ribosome assembly involve (i) the transcription, processing and modification of rRNA; (ii) the translation and modification of ribosomal proteins; (iii) the proper folding of rRNA and ribosomal proteins; (iv) the binding of ribosomal proteins; and (v) the binding and release of assembly factors. Many of these steps are coupled and occur in parallel during the transcription of the rRNAs, enabling autogenous regulation (Section 12.1.8) of rRNA gene transcription. The assembly process is believed to progress through an alternating series of RNA conformational changes and protein-binding events, whereby RNA folding gains are stabilized by the binding of ribosomal proteins, incrementally driving the RNA structure to the final native state.

Since each subunit of the ribosome consists of many different proteins, proper assembly is possible with binding of the proteins in the right sequence. For example, the ribosomal proteins of the 30S subunit are divided into three classes: primary, secondary and tertiary binding proteins, depending on their binding sequence. The primary binding proteins bind directly to 16S rRNA in the absence of other ribosomal proteins. Secondary binding proteins require the prior binding of one of the primary binding proteins. Tertiary binding proteins require the prior binding of one or more secondary binding proteins.

6.14 Growth

We have already discussed the reactions that supply ATP, NADPH and carbon skeletons, and their utilization in biosynthetic processes. When a bacterium is inoculated into a liquid medium, exponential growth is observed after a lag phase. When the substrate is used up or metabolic waste accumulates, the culture enters the stationary phase. The growth rate of microbes is influenced by various environmental factors, including nutrient concentration, temperature, pH, etc. Catabolic and anabolic pathways are coordinated in such a way that each cellular component increases in the same proportion for balanced growth at any given condition. Balanced growth is possible due to elaborate metabolic regulation, that will be discussed later (Chapter 12).

In this section, discussion will be made of how microbial growth is related to energy transduction. Microbes consume energy not only for biosynthesis but also for their maintenance during growth.

Energy expenditure for biosynthesis:

Transport
Nitrogen and sulfur transformations
Monomer synthesis
Polymerization
Assembly of cell structures

Energy expenditure for maintenance:

pH and osmotic homeostasis
Turnover
Motility

6.14.1 Cell division

Prokaryotes propagate asexually. During propagation, a prokaryotic cell divides at the appropriate time and at the correct location in the cell, and each of the progeny receives a complete complement of genes and cell constituents.

6.14.1.1 Binary fission

Prokaryotes rely on binary fission with few exceptions. During growth, the cell size increases. In rod-shaped bacteria, the cell elongates. When it reaches about twice its starting length, the cell divides in the middle by binary fission. Concurrent with growth, the genetic material of the cell replicates and segregates into the progeny cells with the cell constituents, in a controlled manner. After replication, each chromosome segregates to the polar position with the aid of the chromosome partition protein (MukB in *Escherichia coli*), the structural maintenance of chromosome complex (SMC), the rod-shape determining protein (TreB) and other DNA-binding proteins.

For cell division to occur, the division apparatus, known as the divisome, assembles at the site of future cytoplasmic cleavage. The filamentous temperature-sensitive protein Z (FtsZ), a structural homologue of the eukaryotic cytoskeletal element tubulin, assembles into a ring-like structure (Z-ring) at the centre of the cell. A temperature-sensitive mutant of *ftsZ* cannot divide at high temperature to form filaments. Other components of the divisome assemble at the Z-ring. These components redirect cell wall growth, and prevent damage to the DNA while the cell envelope invaginates. These include FtsA, I, K, L, N, Q, W proteins, ZipA (Z interacting protein A), MinC (septum site-determining protein C), MinD (septum site-determining protein D), MinE (cell division topological specificity factor E) and SulA (cell division inhibitor A). FtsZ has no intrinsic affinity for the phospholipid membrane and assembly of a Z-ring in *Escherichia coli*, the cells minimally requiring the presence of either FtsA or ZipA for Z-ring formation. Both FtsA and ZipA are membrane associated and interact with a small domain (C-core) at the extreme C-terminus of FtsZ. Finally, the cell divides to form two approximately equivalent offspring. Although many of the genes involved in cell division have been identified, the mechanisms of action of these gene products are still under investigation. FtsZ is highly conserved among prokaryotes propagating through binary fission and is one of the first proteins to assemble at the future cell division site. In some prokaryotes, some genes for binary fission, including *ftsZ*, are not found. Such organisms multiply by different mechanisms, such as multiple intracellular offspring, multiple fission and by budding. In some bacteria, these alternative reproductive strategies are essential for propagation, whereas in others they are used conditionally.

To prevent untimely division of the mother cell, binary fission must be properly coordinated with cell wall growth, chromosome replication, and chromosome segregation. The concentration of FtsZ does not change either throughout the cell cycle, or under different growth conditions in *Escherichia coli* and *Bacillus subtilis*. In these bacteria, cell division must therefore be regulated at the level of Z-ring assembly. Many proteins are known to interact directly with FtsZ, affecting the assembly, organization and stability of the Z-ring, but the role of individual proteins is not yet fully elucidated.

Master cell-cycle regulators are known in differentiating bacteria such as CtrA (cell-cycle transcriptional regulator) in *Caulobacter crescentus*, Spo0A in *Bacillus subtilis* and AdpA in *Streptomyces coelicolor*. These determine the timing of replication initiation with respect to the developmental stage, to ensure the maintenance of nucleoid position and other events during cell differentiation.

6.14.1.2 Multiple intracellular offspring

Some Gram-positive bacteria (known as firmicutes) form endospores when conditions are unsuitable for growth and this has been best described in *Bacillus subtilis*. The sporulation process is initiated through a modified form of binary fission. Instead of dividing at the mid-cell point, a sporulating cell divides near one pole. Asymmetrical cell division produces a small forespore and a large parent cell. During this division, approximately one third of one of the chromosomes is trapped in the forespore. The DNA translocase SpoIIIE, which is located in the division septum, then pumps the rest of the chromosome into the forespore. The other copy of the chromosome is retained in the parent cell. Although the spore is formed at only one pole, *Bacillus subtilis* prepares for division at both poles. FtsZ rings assemble at both poles and ring-shaped invaginations can be observed near both cell poles early in sporulation. It has been suggested, however, that construction of the cell-division apparatus at both poles allows the cell a 'second chance' if the first asymmetrical division fails to capture a chromosome. Some *Bacillus subtilis* mutants have a 'disporic' phenotype. In these cells, sequential bipolar cell division occurs and the two copies of the chromosome are partitioned into the polar forespores, leaving the parental cell devoid of genetic material, which results in the arrest of sporulation. Therefore, it seems that the mechanisms for bipolar division are present in endospore formers. Once asymmetrical division and chromosome translocation are complete, the forespore is engulfed by the parent cell and the internalized forespore matures into a spore. Details of sporulation are discussed later (Section 13.3.1).

Some related firmicutes produce multiple endospores within a parental cell (also known as the mother cell), while binary fission is the normal mode of growth. These include *Anaerobacter polyendosporus*, an anaerobic bacterium isolated from rice paddy soil. Under proper laboratory conditions, this bacterium can produce up to seven endospores. On the other hand, *Metabacterium polyspora*, a yet-to-be-cultured bacterium found in the gastrointestinal tract of the guinea pig, forms multiple (up to nine) endospores in its normal life cycle. After the spores germinate, some vegetative cells propagate through binary fission, but the majority sporulate in *Metabacterium polyspora*. The process of endospore formation in *Metabacterium polyspora* differs from that in *Bacillus subtilis* and many other endospore-forming bacteria. The asymmetrical cell division of *Metabacterium polyspora* normally takes place at both cell poles. DNA is partitioned into both polar compartments, but some DNA is also retained in the parental cell. After engulfment, the forespores can undergo division to produce multiple forespores that grow and mature into multiple endospores. Unlike sporulating *Bacillus subtilis*, which contains two copies of its genome, *Metabacterium polyspora* must contain three or more copies of its genome. Morphologically similar symbionts have been found in various other rodent species, although no *Metabacterium*-like symbiont has been maintained in culture.

Another group of Gram-positive bacteria that use a modified pathway to form endospores are the segmented filamentous bacteria (SFB). SFB have been found in the intestinal tracts of various animals, but the SFB of rodents are by far the best characterized, although they cannot be maintained under laboratory conditions. SFB develop as a multicellular filament that is anchored to the gastrointestinal tract surface. Once it is firmly attached, the SFB grows and divides, eventually forming a filament that can be up to 1 mm long, although most filaments are roughly 100 μm in length. The cell-division programme is initiated at the unattached end of the filament and is sequentially triggered in cells that are closer to the holdfast. The smaller cell is then engulfed by the larger parental cell and, after engulfment, the internalized cell divides. At this stage, the intracellular offspring have one of two fates: either differentiation or maturation to form a spore. Differentiation produces holdfast protrusions. These active offspring are released into the lumen of the intestine after the mother cell lyses and they attach to the intestinal surface to establish new filaments within the host. Maturation results in two intracellular offspring cells that are encased in a common spore coat, which forms an endospore. The endospore provides an effective dispersal mechanism for the SFB. In fact, exposure to airborne endospores alone can establish an SFB population in a host. These alternative forms of offspring (either active or dormant) allow the SFB to maintain local populations and to survive inhospitable environments before colonizing amenable hosts.

Yet another mode of intracellular multiple offspring propagation is known in a yet-to-be-cultured firmicute, *Epulopiscium fishelsoni* (Section 2.3.5), that colonizes the intestinal tracts of certain species of surgeonfish. This cigar-shaped bacterium can reach more than 0.6 mm in length, the largest bacterium identified so far. These bacteria produce multiple intracellular offspring that resemble the large endospores that are formed by *Metabacterium polyspora*. However, *Epulopiscium fishelsoni* produces active (rather than dormant) offspring. Usually two offspring are produced, although up to 12 offspring per parent cell have been observed in certain strains.

6.14.1.3 Multiple offspring by multiple fission

Three lineages of prokaryotes, the cyanobacteria (Section 11.1.1), the proteobacteria, including aerobic anoxygenic phototrophic bacteria (Section 11.1.3), and the actinobacteria, propagate through multiple rounds of fission of an enlarged multinucleoid spherical cell, or synchronous division at many sites along the length of a multinucleoid filament. This leads to the formation of multiple vegetative offspring or spores. A multiple-fission reproductive phase is often induced by depletion of a nutrient, and might be used to aid offspring dispersal.

Among cyanobacteria, *Pleurocapsa* and *Stanieria* genera propagate through multiple fission to produce offspring known as baeocytes (Greek, 'small cells'), with limited binary fission. The life cycle begins with a baeocyte, which is a spherical cell 1–2 μm in diameter. The baeocyte produces an extracellular matrix that is known as the F-layer and accretion of this layer continues throughout the life cycle. This matrix, probably a polysaccharide, aids attachment of the baeocyte to a solid surface. During vegetative growth, the attached cell enlarges up to 30 μm in diameter. As the cell grows, its genomic DNA replicates and the nucleoids segregate in the cytoplasm. In subsequent reproductive stages, a rapid succession of cytoplasmic fissions leads to multiple baeocyte formation. The number of baeocytes produced (4 to more than 1000) depends on the volume of the reproductive-phase parent cell. Multiple fission is not accompanied by a notable increase in total cytoplasmic volume, and each round of division produces sequentially smaller offspring cells, which distinguishes this process from binary fission. Eventually the extracellular matrix tears open, releasing the baeocytes. It is not known what triggers the reproductive cycle in these organisms. Species of the genus *Synechocystis* use a combination of asymmetrical cell division and multiple fission for reproduction.

Multiple spores in the aerial mycelium of streptomycetes are well documented. A spore germinates to give rise to a vegetative cell whose tip growth produces branched filaments that only septate occasionally, forming long cells that contain multiple nucleoids. In response to nutrient depletion, some streptomycetes alter their pattern of growth to produce aerial mycelia and dispersible spores. A complex extracellular signalling cascade, which might provide check points to ensure the coordination of secondary metabolism in a developing colony, controls the transition from vegetative growth to reproduction. In some of the aerial filaments, synchronous cell division occurs at regular intervals to produce uninucleoid cells which develop into spores. Mechanisms that act on the Z-ring assembly in other bacteria (such as nucleoid occlusion and the Min system) do not seem to be as conserved in the streptomycetes.

Bdellovibrio spp. are tiny predatory *Deltaproteobacteria* that invade the periplasm of the prey bacterium and systematically consume it. Attack-phase *Bdellovibrio* cells are highly motile as they search for susceptible prey. Once host-contact is made, *Bdellovibrio* cells penetrate the outer membrane of a prey cell. Concealed within the host, *Bdellovibrio* lyses the cell wall, which reduces the prey to a sphaeroplast, and then the *Bdellovibrio* assimilates organic compounds from the prey cytoplasm. *Bdellovibrio* grows in the prey periplasm, but this growth is not accompanied by cell division - instead, the cell elongates. Once the host cytoplasm is consumed, the *Bdellovibrio* reproduces: the filament divides through multiple fission, and cells differentiate into motile attack-phase cells.

6.14.1.4 Budding

Budding is known in different groups of bacteria, including the firmicutes, the cyanobacteria, the prosthecate proteobacteria, the aerobic anoxygenic phototrophic bacteria (Section 11.1.3) and the *Planctomycetes*. Although mechanisms that control bud formation in these and other bacteria have been proposed, the molecular mechanisms that regulate bud formation are not fully characterized.

The budding prosthecate *Alphaproteobacteria* include *Caulobacter crescentus, Hyphomonas, Pedomicrobium* and *Ancalomicrobium* spp. They are morphologically diverse. Developmental progression in *Caulobacter crescentus*, through DNA replication, cell division and polar organelle development, is well defined. Key global response regulators, such as CtrA (cell cycle transcriptional regulator) and GcrA (cell cycle regulator), which govern progression, have been identified. It is possible that the regulatory mechanisms that have been discovered in *Caulobacter crescentus* are conserved in other prosthecate *Alphaproteobacteria* and could be involved in bud-site determination and development.

6.14.2 Growth yield

Generally, microbial cell number or cell mass (G) is proportional to the amount of substrate consumed (C), which can be expressed as

$$G = KC$$

where K is a constant depending on the substrate and expressed as 'g cell yield/mol substrate consumed'. When *Escherichia coli* is cultivated on a glucose–mineral salts medium, 0.5 g cell mass is generated from 1 g glucose consumed. In this case, K is 90 g cell yield/mol glucose. Y is used instead of K to define the molar growth yield. $Y_{glucose}$ is used to define the molar growth yield on glucose.

When the exact number of ATP molecules generated from metabolism of the substrate is known, $Y_{glucose}$ can be converted to gram cell yield/ATP generated (Y_{ATP}). Homolactic acid fermentation generates 2 mol ATP from 1 mol glucose fermented, and Y_{ATP} is half of $Y_{glucose}$. Y_{ATP} varies from 4.7 in *Zymomonas anaerobia* to 20.9 in *Lactobacillus casei* (Table 6.8). In the branched

Table 6.8 Growth yield of fermentative microorganisms.

Microorganism	Substrate	$Y_{substrate}$	ATP/mol substrate	Y_{ATP}
Streptococcus faecalis	Glucose	21.8	2.0	10.9
	Gluconate	18.7	1.8	10.4
	2-ketogluconate	19.5	2.3	8.5
	Ribose	21.0	1.67	12.6
	Arginine	10.2	1.0	10.2
	Pyruvate	10.4	1.0	10.4
Streptococcus agalactiae	Glucose	20.8	2.25	9.3
	Pyruvate	7.5	0.72	10.4
Streptococcus pyogenes	Glucose	25.5	2.6	9.8
Streptococcus lactis	Glucose	19.5	2.0	9.8
Lactobacillus plantarum	Glucose	20.4	2.0	10.2
	Galactose	32.5	2.97	10.9
Lactobacillus casei	Glucose	42.9	2.05	20.9
	Mannitol	40.5	2.22	18.2
	Citrate	18.2	0.96	19.0
Bifidobacterium bifidum	Glucose	37.4	2.85	13.1
	Lactose	52.8	5.08	10.4
	Galactose	27.8	2.80	9.9
	Mannitol	27.8	2.35	11.8
Saccharomyces cerevisiae	Glucose	20.4	2.0	10.2
Saccharomyces rosea	Glucose	23.2	2.0	11.6
Zymomonas mobilis	Glucose	8.5	1.0	8.5
Zymomonas anaerobia	Glucose	5.9	1.0	5.9
Sarcina ventriculi	Glucose	30.5	2.62	11.7
Aerobacter aerogenes	Glucose	30.6	3.0	10.2
	Fructose	35.1	3.0	11.7
	Mannitol	27.0	2.5	10.8
	Gluconate	27.5	2.5	11.0
Aerobacter cloacae	Glucose	27.1	2.27	11.9
Escherichia coli	Glucose	33.6	3.0	11.2
Ruminococcus flavefaciens	Glucose	29.1	2.75	10.6
Actinomyces israeli	Glucose	24.7	2.0	12.3
Clostridium tetanomorphum	Glutamate	6.8	0.62	10.9
Clostridium aminobutyricum	4-aminobutyrate	7.6	0.5	15.2
	4-hydroxybutyrate	8.9	0.5	17.8
Clostridium glycolicum	Ethyleneglycol	7.7	0.5	15.4
Clostridium kluyveri	Crotonate	4.8	0.5	9.6
	Ethanol + Acetate	–	–	9.2
Clostridium pasteurianum	Sucrose	73.1	6.64	11.0
Clostridium thermoaceticum	Glucose	50.0	3.0	16.6

fermentation pathway (Section 8.1), the growth conditions determine the fermentation route and therefore differences in ATP generation. *Clostridium butyricum* ferments glucose to acetate and butyrate. Under low partial pressures of hydrogen, this bacterium produces more acetate, generating one ATP per acetate produced:

pyruvate ⟶ acetyl-CoA ⟶ (P_i → CoA-SH) acetyl-phosphate ⟶ (ADP → ATP) acetate

Butyrate is the main product under high partial pressures of hydrogen, generating less ATP than from acetate production:

2 pyruvate ⟶ butyryl-CoA ⟶ butyryl-phosphate ⟶ butyrate

6.14.3 Theoretical maximum Y_{ATP}

As discussed earlier, energy in the forms of ATP (GTP) and NAD(P)H is invested to synthesize monomers and to polymerize them into cell materials. It is possible to calculate how much energy is needed to synthesize 1 g of cell material through summing the investment monomer

Table 6.9 ATP and precursors required to produce the monomers needed for the formation of one gram of *Escherichia coli* cells.

Cell constituent	Content (μmol/g dry wt)	Precursor	ATP	NADH	NADPH	C1	NH_3	S
			Requirement for ATP and precursors (μmol/μmol)					
Amino acids								
alanine	488	1 pyr	0	0	1	0	1	0
arginine	281	1 2-kg	7	−1	4	0	4	0
asparagine	229	1 oaa	3	0	1	0	2	0
aspartate	229	1 oaa	0	0	1	0	1	0
cysteine	87	1 pga	4	−1	5	0	1	1
glutamate	250	1 2-kg	0	0	1	0	1	0
glutamine	250	1 2-kg	1	0	1	0	2	0
glycine	282	1 pga	0	−1	1	−1	1	0
histidine	90	1 penP	6	−3	1	1	3	0
isoleucine	276	1 oaa, 1 pyr	2	0	5	0	1	0
leucine	428	2 pyr, 1 acCoA	0	−1	2	0	1	0
lysine	326	1 oaa, 1 pyr	2	0	4	0	2	0
methionine	146	1 oaa	7	0	8	1	1	1
phenylalanine	176	1 eryP, 2 pep	1	0	2	0	1	0
proline	210	1 2-kg	1	0	3	0	1	0
serine	205	1 pga	0	−1	1	0	1	0
threonine	241	1 oaa	2	0	3	0	1	0

Table 6.9 | (cont.)

Cell constituent	Content (μmol/g dry wt)	Precursor	ATP	NADH	NADPH	C1	NH_3	S
			Requirement for ATP and precursors (μmol/μmol)					
tryptophan	54	1 penP, 1 eryP, 1 pep	5	−2	3	0	2	0
tyrosine	131	1 eryP, 2 pep	1	−1	2	0	1	0
valine	402	1 pyr	0	0	2	0	1	0
Ribonucleotides								
ATP	165	1 penP, 1 pga	11	−3	1	1	5	0
GTP	203	1 penP, 1 pga	13	−3	0	1	5	0
CTP	126	1 penP, 1 oaa	9	0	1	0	3	0
UTP	136	1 penP, 1 oaa	7	0	1	0	2	0
Deoxyribonucleotides								
dATP	24.7	1 penP, 1 pga	11	−3	2	1	5	0
dGTP	25.4	1 penP, 1 pga	13	−3	1	1	5	0
dCTP	25.4	1 penP, 1 oaa	9	0	2	0	3	0
dTTP	24.7	1 penP, 1 oaa	10.5	0	3	1	2	0
Lipids								
glycerol phosphate	129	1 triosP	0	0	1	0	0	0
serine	129	1 pga	0	−1	1	0	0	0
average fatty acid	258	8.2 acCoA	7.2	0	14	0	0	0
Lipopolysaccharide								
UDP-glucose	15.7	1 gluP	1	0	0	0	0	0
CDP-ethanolamine	23.5	1 pga	3	−1	1	1	1	0
fatty acid	47	7 acCoA	6	0	11.5	0	0	0
CMP-KDO	23.5	1 penP, 1 pep	2	0	0	0	0	0
CDP-heptose	23. 5	1.5 gluP	1	0	−4	0	0	0
UDP-glucosamine	15.7	1 fruP	2	0	0	1	1	0
Peptidoglycan								
UDP-GlcNAc	27.6	1 fruP, 1 acCoA	3	0	0	0	1	0
UDP-MurNAc	27.6	1 fruP, 1 pep, 1acCoA	4	0	1	0	1	0
alanine	55.2	1 pyr	0	0	1	0	1	0
diaminopimelate	27.6	1 oaa, 1 pyr	2	0	3	0	2	0
glutamate	27.6	1 2-kg	0	0	1	0	1	0
Polyamine								
ornithine equivalent	59.3	1 2-kg	2	0	3	0	2	0

acCoA, acetyl-CoA; eryP, erythrose-4-phosphate; fruP, fructose-6-phosphate; gluP, glucose-6-phosphate; 2-kg, 2-ketoglutarate; oaa, oxaloacetate; penP, pentose-5-phosphate; pep, phosphoenolpyruvate; pga, 3-phosphoglycerate; pyr, pyruvate; triosP, triose phosphate; KDO, 2-keto-3-deoxyoctonate; UDP-GlcNAc, UDP-*N*-acetylgucosamine; UDP-MurNAc, UDP-*N*-acetylmuramic acid.

Table 6.10 ATP requirement for polymer synthesis by *Escherichia coli* growing in glucose-mineral salts medium.

Polymer	ATP required for the synthesis of polymer to make one gram dry wt cell material (mmol/g dry wt)
Polysaccharide	2.1
Protein	
Glucose to amino acids	1.4
Polymerization (translation)	19.1
Lipid	0.1
RNA	
Glucose to ribonucleotides	3.5
Polymerization (transcription)	0.9
DNA	
Glucose to deoxyribonucleotides	0.9
Polymerization	0.2
Transport	5.2
RNA turnover	1.4
Overall	34.8

synthesis (Table 6.9) and polymerization (Table 6.10). From this figure, the theoretical maximum Y_{ATP} (Y^{max}_{ATP}) can be calculated.

Table 6.11 summarizes (Y^{max}_{ATP}) under various nutritional conditions ranging from chemolithotrophic to heterotrophic metabolism, in a complex medium with added glucose. As expected, the lowest (Y^{max}_{ATP}) of 6.5 g/mol ATP is calculated for chemolithotrophic growth fixing carbon dioxide, while the (Y^{max}_{ATP}) is 31.9 g/mol ATP in heterotrophic metabolism in a complex medium. This difference shows that a large amount of energy is consumed for the synthesis of monomers and their polymerization. The values for heterotrophic metabolism on glucose, pyruvate, malate and acetate are 28.8 g, 13.5 g, 15.4 g and 10.0 g cell/mol ATP, respectively. This difference is due to energy consumption by gluconeogenesis.

The (Y^{max}_{ATP}) values in Table 6.11 only account for direct energy requirements for growth. The actual Y_{ATP} is much lower than these figures since there is energy expenditure for motility, pH and salt homeostasis, and other non-growth-related activities. This is known as the maintenance energy. Since less energy is consumed to synthesize storage materials such as glycogen and polyhydroxyalkanoate (Section 13.2), the growth yield is higher when the organism grows under conditions suitable for their biosynthesis.

6.14.4 Growth yield using different electron acceptors and maintenance energy

Growth yield is dependent on metabolism, as shown above. For example $Y_{glucose}$ values for *Proteus mirabilis* are 58.1, 30.1 and 14.0 g dry wt/mol glucose under aerobic, denitrifying and fermentative conditions, respectively. This difference is partly due to the number of ATP molecules generated. Since prokaryotic electron transport chains are diverse, with different H^+/O ratios and H^+/ATP stoichiometries, which vary depending on the energy status of the cell (Section 5.8.4), it is difficult to calculate the ATP yield in respiratory

Table 6.11 ATP requirement for growth and (Y_{ATP}^{max}) under different nutritional conditions.

Polymer	Cellular content (g/100 g dry wt)	ATP requirement (mmol/g cell) in medium						
		A	B	C	D	E	F	G
Polysaccharide	16.6	2.06	2.06	7.18	7.18	5.10	9.20	19.50
Protein	52.4	19.14	20.50	19.14	33.94	28.50	42.70	90.70
Lipid	9.4	0.14	0.14	2.70	2.70	2.50	5.00	17.20
RNA	15.7	2.40	4.37	4.62	7.13	7.00	10.10	17.84
DNA	3.2	0.57	1.05	0.99	1.59	1.30	1.90	3.36
mRNA turnover		1.39	1.39	1.39	1.39	1.39	1.39	1.39
Subtotal		25.70	29.51	36.02	53.93	45.79	70.29	149.99
Transport		5.74	5.20	11.55	20.00	20.00	30.60	5.20
Total		31.44	34.71	47.57	73.93	65.79	100.80	155.19
(Y_{ATP}^{max}) (g cell/mol ATP)		31.80	28.81	21.02	13.53	15.20	9.92	6.44

Cellular content is based on *Escherichia coli*.
Medium:
A, glucose + amino acids – nucleic acid bases + mineral salts.
B, glucose + mineral salts.
C, pyruvate + amino acids + nucleic acid bases + mineral salts.
D, pyruvate + mineral salts.
E, malate + mineral salts.
F, acetate + mineral salts.
G, carbon dioxide + mineral salts (chemolithotrophic metabolism).

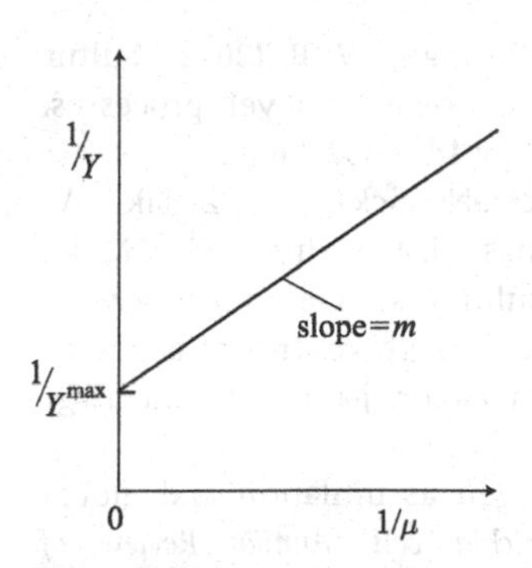

Figure 6.49 Relationship between growth rate, growth yield and maintenance energy.

Growth yield was determined in a chemostat at various dilution (growth) rates (μ). The inverse values are plotted as above to obtain the maintenance energy (m) and maximum growth yield (Y^{max}) at the given conditions.

metabolism. For this reason Y_{ATP} is calculated from fermentation.

Growth yield in respiratory metabolism can be estimated using the following equation: $1/Y = m(1/\mu) + (1/Y^{max})$ where Y is the growth yield, m is the maintenance energy, μ is the growth rate and Y^{max} is the maximum growth yield. The inverse values of the growth yield obtained from a chemostat operated at different growth rates are plotted against the inverse of the growth rate, as in Figure 6.49. From this exercise, the maintenance energy can be determined. As shown above, the theoretical maximum Y_{ATP} is more than the actual Y_{ATP}. Assuming that the total energy available from metabolism is the sum of the energy for growth and the maintenance energy, it can be seen that more energy is used for the purpose of maintenance.

6.14.5 Maintenance energy

As shown in Figure 6.49, more maintenance energy is required at a lower growth rate. The maintenance energy required is determined by the various growth conditions that influence the growth rate, e.g. pH and salt homeostasis and the half-life of cellular polymers, among others. Organisms producing fatty acids require more maintenance energy, since fatty acids behave as uncouplers. When growing under substrate excess conditions, a lower cell yield is expected with incomplete substrate oxidation and uncoupled oxidation of the substrate. This fact should be considered in the determination of maintenance energy.

Further Reading

Note this section contains key references only. Additional recommended references are available at www.cambridge.org/ProkaryoticMetabolism.

Nitrogen fixation

Bothe, H., Schmitz, O., Yates, M. G. & Newton, W. E. (2010). Nitrogen fixation and hydrogen metabolism in cyanobacteria. *Microbiology and Molecular Biology Reviews* **74**, 529–551.

Boyd, E. S, Costas, A. M. G., Hamilton, T. L., Mus, F. & Peters, J. W. (2015). Evolution of molybdenum nitrogenase during the transition from anaerobic to aerobic metabolism. *Journal of Bacteriology* **197**, 1690–1699.

Houlton, B. Z., Wang, Y. P., Vitousek, P. M. & Field, C. B. (2008). A unifying framework for dinitrogen fixation in the terrestrial biosphere. *Nature* **454**, 327–330.

Hu, Y. & Ribbe, M. W. (2016). Biosynthesis of the metalloclusters of nitrogenases. *Annual Review of Biochemistry* **85**, 455–483.

McRose, D. L., Zhang, X., Kraepiel, A. M. L. & Morel, F. M. M. (2017). Diversity and activity of alternative nitrogenases in sequenced genomes and coastal environments. *Frontiers in Microbiology* **8**, 267.

Omairi-Nasser, A., Mariscal, V., Austin, J. R. & Haselkorn, R. (2015). Requirement of Fra proteins for communication channels between cells in the filamentous nitrogen-fixing cyanobacterium *Anabaena* sp. PCC 7120. *Proceedings of the National Academy of Sciences of the USA* **112**, E4458–E4464.

Prell, J. & Poole, P. (2006). Metabolic changes of rhizobia in legume nodules. *Trends in Microbiology* **14**, 161–168.

Seefeldt, L. C., Hoffman, B. M. & Dean, D. R. (2012). Electron transfer in nitrogenase catalysis. *Current Opinion in Chemical Biology* **16**, 19–25.

Ward, B. B. & Jensen, M. M. (2014). The microbial nitrogen cycle. *Frontiers in Microbiology* **5**, 00553.

Zhang, C. C., Laurent, S., Sakr, S., Peng, L. & Bedu, S. (2006). Heterocyst differentiation and pattern formation in cyanobacteria: a chorus of signals. *Molecular Microbiology* **59**, 367–375.

Assimilation of inorganic nitrogen and sulfur

Acera, F., Carmona, M. I., Castillo, F., Quesada, A. & Blasco, R. (2017). A cyanide-induced 3-cyanoalanine nitrilase in the cyanide-assimilating bacterium *Pseudomonas pseudoalcaligenes* strain CECT 5344. *Applied and Environmental Microbiology* **83**, e00089–17.

Bender, R. A. (2010). A NAC for regulating metabolism: the nitrogen assimilation control protein (NAC) from *Klebsiella pneumoniae*. *Journal of Bacteriology* **192**, 4801–4811.

Lewis, T. A., Glassing, A., Harper, J. & Franklin, M. J. (2013). Role for ferredoxin: NAD(P)H oxidoreductase (FprA) in sulfate assimilation and siderophore biosynthesis in pseudomonads. *Journal of Bacteriology* **195**, 3876–3887.

Liu, Y., Beer, L. L. & Whitman, W. B. (2012). Sulfur metabolism in archaea reveals novel processes. *Environmental Microbiology* **14**, 2632–2644.

Lochowska, A., Iwanicka-Nowicka, R., Zielak, A., Modelewska, A., Thomas, M. S. & Hryniewicz, M. M. (2011). Regulation of sulfur assimilation pathways in *Burkholderia cenocepacia* through control of genes by the SsuR transcription factor. *Journal of Bacteriology* **193**, 1843–1853.

Reitzer, L. (2003). Nitrogen assimilation and global regulation in *Escherichia coli*. *Annual Review of Microbiology* **57**, 155–176.

Tripp, H. J., Kitner, J. B., Schwalbach, M. S., Dacey, J. W. H., Wilhelm, L. J. & Giovannoni, S. J. (2008). SAR11 marine bacteria require exogenous reduced sulphur for growth. *Nature* **452**, 741–744.

van Heeswijk, W. C., Westerhoff, H. V. & Boogerd, F. C. (2013). Nitrogen assimilation in *Escherichia coli*: putting molecular data into a systems perspective.

Microbiology and Molecular Biology Reviews **77**, 628–695.

Ye, R. W. & Thomas, S. M. (2001). Microbial nitrogen cycles: physiology, genomics and applications. *Current Opinion in Microbiology* **4**, 307–312.

Amino acid synthesis

Fazius, F., Zaehle, C. & Brock, M. (2013). Lysine biosynthesis in microbes: relevance as drug target and prospects for β-lactam antibiotics production. *Applied Microbiology and Biotechnology* **97**, 3763–3772.

Ferla, M. P. & Patrick, W. M. (2014). Bacterial methionine biosynthesis. *Microbiology* **160**, 1571–1584.

Hove-Jensen, B., Andersen, K. R., Kilstrup, M., Martinussen, J., Switzer, R. L. & Willemoës, M. (2017). Phosphoribosyl diphosphate (PRPP): biosynthesis, enzymology, utilization, and metabolic significance. *Microbiology and Molecular Biology Reviews* **81**, e00040-16.

Itoh, Y., Bröcker, M. J., Sekine, S.-i., Hammond, G., Suetsugu, S., Söll, D. & Yokoyama, S. (2013). Decameric SelA•tRNASec ring structure reveals mechanism of bacterial selenocysteine formation. *Science* **340**, 75–78.

Krzycki, J. A. (2013). The path of lysine to pyrrolysine. *Current Opinion in Chemical Biology* **17**, 619–625.

Kulis-Horn, R. K., Persicke, M. & Kalinowski, J. (2014). Histidine biosynthesis, its regulation and biotechnological application in *Corynebacterium glutamicum*. *Microbial Biotechnology* **7**, 5–25.

Mir, R., Jallu, S. & Singh, T. P. (2015). The shikimate pathway: review of amino acid sequence, function and three-dimensional structures of the enzymes. *Critical Reviews in Microbiology* **41**, 172–189.

Radkov, A. & Moe, L. (2014). Bacterial synthesis of d-amino acids. *Applied Microbiology and Biotechnology* **98**, 5363–5374.

Risso, C., Van Dien, S. J., Orloff, A., Lovley, D. R. & Coppi, M. V. (2008). Elucidation of an alternate isoleucine biosynthesis pathway in *Geobacter sulfurreducens*. *Journal of Bacteriology* **190**, 2266–2274.

White, R. H. (2004). l-Aspartate semialdehyde and a 6-deoxy-5-ketohexose 1-phosphate are the precursors to the aromatic amino acids in *Methanocaldococcus jannaschii*. *Biochemistry* **43**, 7618–7627.

Nucleotide synthesis

Buckel, W. & Golding, B. T. (2006). Radical enzymes in anaerobes. *Annual Review of Microbiology* **60**, 27–49.

Martin, J. E. & Imlay, J. A. (2011). The alternative aerobic ribonucleotide reductase of *Escherichia coli*, NrdEF, is a manganese-dependent enzyme that enables cell replication during periods of iron starvation. *Molecular Microbiology* **80**, 319–334.

West, T. P. (2014). Pyrimidine nucleotide synthesis in *Pseudomonas nitroreducens* and the regulatory role of pyrimidines. *Microbiological Research* **169**, 954–958.

Monomer synthesis – lipids

Behrouzian, B. & Buist, P. H. (2002). Fatty acid desaturation: variations on an oxidative theme. *Current Opinion in Chemical Biology* **6**, 577–582.

Broussard, T. C., Price, A. E., Laborde, S. M. & Waldrop, G. L. (2013). Complex formation and regulation of *Escherichia coli* acetyl-CoA carboxylase. *Biochemistry* **52**, 3346–3357.

Chang, W.-c., Song, H., Liu, H.-w. & Liu, P. (2013). Current developments in isoprenoid precursor biosynthesis and regulation. *Current Opinion in Chemical Biology* **17**, 571–579.

Köcher, S., Breitenbach, J., Müller, V. & Sandmann, G. (2009). Structure, function and biosynthesis of carotenoids in the moderately halophilic bacterium *Halobacillus halophilus*. *Archives of Microbiology* **191**, 95–104.

Pini, C., Godoy, P., Bernal, P., Ramos, J.-L. & Segura, A. (2011). Regulation of the cyclopropane synthase *cfaB* gene in *Pseudomonas putida* KT2440. *FEMS Microbiology Letters* **321**, 107–114.

Schujman, G. E. & de Mendoza, D. (2008). Regulation of type II fatty acid synthase in Gram-positive bacteria. *Current Opinion in Microbiology* **11**, 148–152.

Schweizer, H. & Choi, K.-H. (2011). *Pseudomonas aeruginosa* aerobic fatty acid desaturase DesB is important for virulence factor production. *Archives of Microbiology* **193**, 227–234.

Villanueva, L., Damste, J. S. S. & Schouten, S. (2014). A re-evaluation of the archaeal membrane lipid biosynthetic pathway. *Nature Reviews Microbiology* **12**, 438–448.

Zhang, Y. M. & Rock, C. O. (2008). Membrane lipid homeostasis in bacteria. *Nature Reviews Microbiology* **6**, 222–233.

Monomer synthesis – others

Dailey, H. A., Dailey, T. A., Gerdes, S., Jahn, D., Jahn, M., O'Brian, M. R. & Warren, M. J. (2017). Prokaryotic heme biosynthesis: multiple pathways to a common essential product. *Microbiology and Molecular Biology Reviews* **81**, e00048-16.

Fontecave, M., Atta, M. & Mulliez, E. (2004). *S*-adenosylmethionine: nothing goes to waste. *Trends in Biochemical Sciences* **29**, 243–249.

Kranz, R. G., Richard-Fogal, C., Taylor, J.-S. & Frawley, E. R. (2009). Cytochrome c biogenesis: mechanisms for covalent modifications and trafficking of heme and for heme-iron redox control. *Microbiology and Molecular Biology Reviews* **73**, 510–528.

Roessner, C. A. & Scott, A. I. (2006). Fine-tuning our knowledge of the anaerobic route to cobalamin (vitamin B_{12}). *Journal of Bacteriology* **188**, 7331–7334.

Sanders, C., Turkarslan, S., Lee, D.-W. & Daldal, F. (2010). Cytochrome *c* biogenesis: the Ccm system. *Trends in Microbiology* **18**, 266–274.

Cell surface polymer synthesis

D'Elia, M. A., Henderson, J. A., Beveridge, T. J., Heinrichs, D. E. & Brown, E. D. (2009). The *N*-acetylmannosamine transferase catalyzes the first committed step of teichoic acid assembly in *Bacillus subtilis* and *Staphylococcus aureus*. *Journal of Bacteriology* **191**: 4030–4034.

Garufi, G., Hendrickx, A. P., Beeri, K., Kern, J. W., Sharma, A., Richter, S. G., Schneewind, O. & Missiakas, D. (2012). Synthesis of lipoteichoic acids in *Bacillus anthracis*. *Journal of Bacteriology* **194**: 4312–4321.

Guan, Z., Naparstek, S., Kaminski, L., Konrad, Z. & Eichler, J. (2010). Distinct glycan-charged phosphodolichol carriers are required for the assembly of the pentasaccharide N-linked to the *Haloferax volcanii* S-layer glycoprotein. *Molecular Microbiology* **78**, 1294–1303.

Pasquina, L. W., Santa Maria, J. P. & Walker, S. (2013). Teichoic acid biosynthesis as an antibiotic target. *Current Opinion in Microbiology* **16**: 531–537.

Perez, C., Gerber, S., Boilevin, J., Bucher, M., Darbre, T., Aebi, M., Reymond, J.-L. & Locher, K. P. (2015). Structure and mechanism of an active lipid-linked oligosaccharide flippase. *Nature* **524**, 433–438.

Sham, L.-T., Butler, E. K., Lebar, M. D., Kahne, D., Bernhardt, T. G. & Ruiz, N. (2014). MurJ is the flippase of lipid-linked precursors for peptidoglycan biogenesis. *Science* **345**, 220–222.

Cell wall and S-layer assembly

Albers, S.-V. & Meyer, B. H. (2011). The archaeal cell envelope. *Nature Reviews Microbiology* **9**, 414–426.

Cava, F., Kuru, E., Brun, Y. V. & de Pedro, M. A. (2013). Modes of cell wall growth differentiation in rod-shaped bacteria. *Current Opinion in Microbiology* **16**, 731–737.

Duong, A., Capstick, D. S., Di Berardo, C., Findlay, K. C., Hesketh, A., Hong, H.-J. & Elliot, M. A. (2012). Aerial development in *Streptomyces coelicolor* requires sortase activity. *Molecular Microbiology* **83**, 992–1005.

Egan, A. J. F., Cleverley, R. M., Peters, K., Lewis, R. J. & Vollmer, W. (2017). Regulation of bacterial cell wall growth. *FEBS Journal* **284**, 851–867.

Fagan, R. P. & Fairweather, N. F. (2014). Biogenesis and functions of bacterial S-layers. *Nature Reviews Microbiology* **12**, 211–222.

Frirdich, E. & Gaynor, E. C. (2013). Peptidoglycan hydrolases, bacterial shape, and pathogenesis. *Current Opinion in Microbiology* **16**, 767–778.

Hanson, B. R. & Neely, M. N. (2012). Coordinate regulation of Gram-positive cell surface components. *Current Opinion in Microbiology* **15**, 204–210.

Lee, T. K. & Huang, K. C. (2013). The role of hydrolases in bacterial cell-wall growth. *Current Opinion in Microbiology* **16**, 760–766.

Sobhanifar, S., King, D. T. & Strynadka, N. C. J. (2013). Fortifying the wall: synthesis, regulation and degradation of bacterial peptidoglycan. *Current Opinion in Structural Biology* **23**, 695–703.

Spirig, T., Weiner, E. M. & Clubb, R. T. (2011). Sortase enzymes in Gram-positive bacteria. *Molecular Microbiology* **82**, 1044–1059.

Wang, Y.-T., Missiakas, D. & Schneewind, O. (2014). GneZ, a UDP-GlcNAc 2-epimerase, is required for S-layer assembly and vegetative growth of *Bacillus anthracis*. *Journal of Bacteriology* **196**, 2969–2978.

Wirth, R., Bellack, A., Bertl, M., Bilek, Y., Heimerl, T., Herzog, B., Leisner, M., Probst, A., Rachel, R., Sarbu, C., Schopf, S. & Wanner, G. (2011). The mode of cell wall growth in selected archaea is similar to the general mode of cell wall growth in bacteria as revealed by fluorescent dye analysis. *Applied and Environmental Microbiology* **77**, 1556–1562.

Wu, C., Huang, I. H., Chang, C., Reardon-Robinson, M. E., Das, A. & Ton-That, H. (2014). Lethality of sortase depletion in *Actinomyces oris* caused by excessive membrane accumulation of a surface glycoprotein. *Molecular Microbiology* **94**, 1227–1241.

Outer membrane assembly

Cuthbertson, L., Mainprize, I. L., Naismith, J. H. & Whitfield, C. (2009). Pivotal roles of the outer membrane polysaccharide export and polysaccharide copolymerase protein families in export of extracellular polysaccharides in Gram-negative bacteria. *Microbiology and Molecular Biology Reviews* **73**, 155–177.

Dong, H., Xiang, Q., Gu, Y., Wang, Z., Paterson, N. G., Stansfeld, P. J., He, C., Zhang, Y., Wang, W. & Dong, C. (2014). Structural basis for outer membrane lipopolysaccharide insertion. *Nature* **511**, 52–56.

Knowles, T. J., Scott-Tucker, A., Overduin, M. & Henderson, I. R. (2009). Membrane protein architects: the role of the BAM complex in outer membrane protein assembly. *Nature Reviews Microbiology* **7**, 206–214.

Rigel, N. W. & Silhavy, T. J. (2012). Making a beta-barrel: assembly of outer membrane proteins in Gram-negative bacteria. *Current Opinion in Microbiology* **15**, 189–193.

Ruiz, N., Kahne, D. & Silhavy, T. J. (2009). Transport of lipopolysaccharide across the cell envelope: the long road of discovery. *Nature Reviews Microbiology* **7**, 677–683.

Tommassen, J. (2010). Assembly of outer-membrane proteins in bacteria and mitochondria. *Microbiology* **156**, 2587–2596.

Replication and chromosome segregation

Beattie, T. R. & Reyes-Lamothe, R. (2015). A replisome's journey through the bacterial chromosome. *Frontiers in Microbiology* **6**, 562.

Denamur, E. & Matic, I. (2006). Evolution of mutation rates in bacteria. *Molecular Microbiology* **60**, 820–827.

Gao, F. (2015). Bacteria may have multiple replication origins. *Frontiers in Microbiology* **6**, 324.

Hayes, F. & Barilla, D. (2006). The bacterial segrosome: a dynamic nucleoprotein machine for DNA trafficking and segregation. *Nature Reviews Microbiology* **4**, 133–143.

Kelman, L. M. & Kelman, Z. (2004). Multiple origins of replication in archaea. *Trends in Microbiology* **12**, 399–401.

Kelman, L. M. & Kelman, Z. (2014). Archaeal DNA replication. *Annual Review of Genetics* **48**, 71–97.

Kuzminov, A. (2013). The chromosome cycle of prokaryotes. *Molecular Microbiology* **90**, 214–227.

McHenry, C. S. (2011). DNA replicases from a bacterial perspective. *Annual Review of Biochemistry* **80**, 403–436.

Michel, B. & Sandler, S. J. (2017). Replication restart in bacteria. *Journal of Bacteriology* **199**, e00102-17.

Reyes-Lamothe, Nicolas, R., E. & Sherratt, D. J. (2012). Chromosome replication and segregation in bacteria. *Annual Review of Genetics* **46**, 121–143.

Robinson, A. O. & van Oijen, A. M. (2013). Bacterial replication, transcription and translation: mechanistic insights from single-molecule biochemical studies. *Nature Reviews Microbiology* **11**, 303–315.

Wolański, M., Jakimowicz, D. & Zakrzewska-Czerwińska, J. (2014). Fifty years after the replicon hypothesis: cell-specific master regulators as new players in chromosome replication control. *Journal of Bacteriology* **196**, 2901–2911.

Transcription and post-transcription modification

Borukhov, S. & Severinov, K. (2002). Role of the RNA polymerase sigma subunit in transcription initiation. *Research in Microbiology* **153**, 557–562.

Grohmann, D. & Werner, F. (2011). Recent advances in the understanding of archaeal transcription. *Current Opinion in Microbiology* **14**, 328–334.

Lee, D. J., Minchin, S. D. & Busby, S. J. W. (2012). Activating transcription in bacteria. *Annual Review of Microbiology* **66**, 125–152.

Lewis, P. J., Doherty, G. P. & Clarke, J. (2008). Transcription factor dynamics. *Microbiology* **154**, 1837–1844.

Nickels, B. E. & Dove, S. L. (2011). NanoRNAs: a class of small RNAs that can prime transcription initiation in bacteria. *Journal of Molecular Biology* **412**, 772–781.

Ray-Soni, A., Bellecourt, M. J. & Landick, R. (2016). Mechanisms of bacterial transcription termination: all good things must end. *Annual Review of Biochemistry* **85**, 319–347.

Sankar, T. S., Wastuwidyaningtyas, B. D., Dong, Y., Lewis, S. A. & Wang, J. D. (2016). The nature of mutations induced by replication–transcription collisions. *Nature* **535**, 178–181.

Stuart, K. & Panigrahi, A. K. (2002). RNA editing: complexity and complications. *Molecular Microbiology* **45**, 591–596.

Translation and protein folding

Cobucci-Ponzano, B., Rossi, M. & Moracci, M. (2012). Translational recoding in archaea. *Extremophiles* **16**, 793–803.

Ivanova, N. N., Schwientek, P., Tripp, H. J., Rinke, C., Pati, A., Huntemann, M., Visel, A., Woyke, T., Kyrpides, N. C. & Rubin, E. M. (2014). Stop codon reassignments in the wild. *Science* **344**, 909–913.

Jarrell, K. F., Ding, Y., Meyer, B. H., Albers, S.-V., Kaminski, L. & Eichler, J. (2014). *N*-linked glycosylation in archaea: a structural, functional, and genetic analysis. *Microbiology and Molecular Biology Reviews* **78**, 304–341.

Keiler, K. C. (2015). Mechanisms of ribosome rescue in bacteria. *Nature Reviews Microbiology* **13**, 285–297.

Lin, Z. & Rye, H. S. (2006). GroEL-mediated protein folding: making the impossible, possible. *Critical Reviews in Biochemistry and Molecular Biology* **41**, 211–239.

Ling, J., O' Donoghue, P. & Soll, D. (2015). Genetic code flexibility in microorganisms: novel mechanisms and impact on physiology. *Nature Reviews Microbiology* **13**, 707–721.

McGary, K. & Nudler, E. (2013). RNA polymerase and the ribosome: the close relationship. *Current Opinion in Microbiology* **16**, 112–117.

Petry, S., Weixlbaumer, A. & Ramakrishnan, V. (2008). The termination of translation. *Current Opinion in Structural Biology* **18**, 70–77.

Schmeing, T. M. & Ramakrishnan, V. (2009). What recent ribosome structures have revealed about the mechanism of translation. *Nature* **461**, 1234–1242.

Shieh, Y.-W., Minguez, P., Bork, P., Auburger, J. J., Guilbride, D. L., Kramer, G. & Bukau, B. (2015). Operon structure and cotranslational subunit association direct protein assembly in bacteria. *Science* **350**, 678–680.

Assembly of cellular structures

Kuzminov, A. (2013). The chromosome cycle of prokaryotes. *Molecular Microbiology* **90**, 214–227.

Li, H. & Sourjik, V. (2011). Assembly and stability of flagellar motor in *Escherichia coli*. *Molecular Microbiology* **80**, 886–899.

Marraffini, L. A., DeDent, A. C. & Schneewind, O. (2006). Sortases and the art of anchoring proteins to the envelopes of Gram-positive bacteria. *Microbiology and Molecular Biology Reviews* **70**, 192–221.

Ruiz, N., Kahne, D. & Silhavy, T. J. (2006). Advances in understanding bacterial outer-membrane biogenesis. *Nature Reviews Microbiology* **4**, 57–66.

Shajani, Z., Sykes, M. T. & Williamson, J. R. (2011). Assembly of bacterial ribosomes. *Annual Review of Biochemistry* **80**, 501–526.

Whitfield, C. (2006). Biosynthesis and assembly of capsular polysaccharides in *Escherichia coli*. *Annual Review of Biochemistry* **75**, 39–68.

Cell division and growth

Angert, E. R. (2005). Alternatives to binary fission in bacteria. *Nature Reviews Microbiology* **3**, 214–224.

Bisson-Filho, A. W., Hsu, Y.-P., Squyres, G. R., Kuru, E., Wu, F., Jukes, C., Sun, Y., Dekker, C., Holden, S., VanNieuwenhze, M. S., Brun, Y. V. & Garner, E. C. (2017). Treadmilling by FtsZ filaments drives peptidoglycan synthesis and bacterial cell division. *Science* **355**, 739–743.

Busiek, K. K. & Margolin, W. (2015). Bacterial actin and tubulin homologs in cell growth and division. *Current Biology* **25**, R243–R254.

den Blaauwen, T. (2013). Prokaryotic cell division: flexible and diverse. *Current Opinion in Microbiology* **16**, 738–744.

Desmond-Le Quemener, E. & Bouchez, T. (2014). A thermodynamic theory of microbial growth. *ISME Journal* **8**, 1747–1751.

Duda, V. I., Suzina, N. E., Polivtseva, V. N., Gafarov, A. B., Shorokhova, A. P. & Machulin, A. V. (2014). Transversion of cell polarity from bi- to multipolarity is the mechanism determining multiple spore formation in *Anaerobacter polyendosporus* PS-1T. *Microbiology–Moscow* **83**, 608–615.

Duggin, I. G., Aylett, C. H. S., Walsh, J. C., Michie, K. A., Wang, Q., Turnbull, L., Dawson, E. M., Harry, E. J., Whitchurch, C. B., Amos, L. A. & Lowe, J. (2015). CetZ tubulin-like proteins control archaeal cell shape. *Nature* **519**, 362–365.

Erickson, H. P., Anderson, D. E. & Osawa, M. (2010). FtsZ in bacterial cytokinesis: Cytoskeleton and force generator all in one. *Microbiology and Molecular Biology Reviews* **74**, 504–528.

Härtel, T. & Schwille P. (2014). ESCRT-III mediated cell division in *Sulfolobus acidocaldarius* – a reconstitution perspective. *Frontiers in Microbiology* **5**, 257.

Lindas, A.-C. & Bernander, R. (2013). The cell cycle of archaea. *Nature Reviews Microbiology* **11**, 627–638.

Pinho, M. G., Kjos, M. & Veening, J.-W. (2013). How to get (a)round: mechanisms controlling growth and division of coccoid bacteria. *Nature Reviews Microbiology* **11**, 601–614.

Samson, R. Y. & Bell, S. D. (2011). Cell cycles and cell division in the archaea. *Current Opinion in Microbiology* **14**, 350–356.

Chapter 7

Heterotrophic metabolism on substrates other than glucose

It has been described previously how glucose and mineral salts can support the growth of certain heterotrophs. In this case, the organisms obtain ATP, NADPH and carbon skeletons for biosynthesis through central metabolism. Almost all natural organic compounds can be utilized through microbial metabolism. In this chapter, the bacterial metabolism of organic compounds other than glucose under aerobic conditions is discussed, and Chapters 8 and 9 describe anaerobic metabolism. Since central metabolism is reversible in one way or another, it can be assumed that an organism can use a compound if that compound is converted to intermediates of central metabolism. Some bacteria can use an extensive variety of organic compounds as sole carbon and energy sources, while some organisms can only use limited numbers of organic compounds; for example, *Bacillus fastidiosus* can use only urate.

7.1 Hydrolysis of polymers

Plant and animal cells consist mainly of polymers. They include polysaccharides, such as starch and cellulose, as well as proteins, nucleic acids, and many others. Such polymers cannot be easily transported into microbial cells but are first hydrolysed to monomers or oligomers by extracellular enzymes, before being transported into the cell.

7.1.1 Starch hydrolysis

Starch is a glucose polymer consisting of amylose and amylopectin. The former has a straight-chain structure with α-1,4-glucoside bonds, while the latter has side chains with α-1,6-glucoside bonds. Starch is the commonest storage material in plants, and many prokaryotes produce amylase so that they can utilize it as their energy and carbon source.

Amylases are classified according to their mode of action. α-amylase is an endoglucanase that randomly hydrolyses α-1,4-glucoside bonds to produce a mixture of dextrin, maltose and glucose, but does not hydrolyse α-1,6-bonds. Many bacteria, including species of *Bacillus*, *Pseudomonas* and *Clostridium*, produce this enzyme. In fact, *Bacillus stearothermophilus* is used to produce α-amylase industrially.

β-amylase is an exoglucanase and removes maltose units from the non-reducing end of amylose. Another exoglucanase, glucoamylase, hydrolyses glucose units from the non-reducing end of amylose. Many bacteria, including *Bacillus* and *Pseudomonas* spp., produce this enzyme. Glucoamylase is produced industrially using certain fungi.

Amylases hydrolyse the α-1,4-glucoside bond of amylose, but cannot hydrolyse the α-1,6-glucoside bond. The α-1,6-glucoside bond of amylopectin is hydrolysed by pullulanase and isoamylase. Pullulanase hydrolyses pullulan, an α-glucan produced by the fungus *Aureobasidium pullulans*, and isoamylase attacks the α-1,6-glucoside bonds of amylopectin and glycogen. These are referred to as debranching enzymes. *Aerobacter aerogenes, Bacillus cereus* and several other bacteria produce pullulanase; *Bacillus amyloliquefaciens* can produce isoamylase. Some hyperthermophilic bacteria and archaea including *Thermoanaerobacter ethanolicus, Pyrococcus furiosus, Sulfolobus acidocaldarius, Thermotoga neapolitana* and *Thermococcus litoralis*, produce an enzyme that can hydrolyse not only α-1,4 but also α-1,6-glucoside bonds. This enzyme is called an amylopullulanase or neopullulanase and is extremely thermostable. A similar enzyme is produced by a halophilic archaeon, *Halorubrum* sp.

7.1.2 Cellulose hydrolysis

Cellulose is one of the most abundant organic compounds in nature and is a glucose homopolymer like starch. Although starch is easily hydrolysed, the structural material cellulose is more resistant to hydrolysis. Cellulose is a straight chain of β-1,4-linked glucose units without any side chains, unlike starch. Because of the absence of side chains in cellulose, extensive hydrogen bonding between the cellulose molecules forms a crystalline structure. At least three different enzymes are required for the complete hydrolysis of crystalline cellulose. These are endo-β-glucanase, exo-β-glucanase (β-glucan cellobiohydrolase) and cellobiase (β-glucosidase). Some bacteria, such as *Cellulomonas flavigena* and *Clostridium thermocellum*, produce all three classes of enzyme. These form a cellulosome complex with other proteins. These bacteria produce multiple endo-β-glucanases and exo-β-glucanases, but it is not known if they are produced by separate genes or if some of them are partially degraded forms of the enzymes. Some bacteria can only use amorphous cellulose. They lack β-glucan cellobiohydrolase, an enzyme responsible for the digestion of crystalline cellulose.

Cellulases hydrolyse β-1,4-glucoside bonds, which are hidden within the crystalline structure. The enzymes are too big for direct contact with the hydrolytic sites and various hypotheses have been proposed to explain how the crystalline cellulose is hydrolysed by the enzyme. According to one hypothesis, endo-β-1,4-glucanase hydrolyses the amorphous region of the substrate, generating large numbers of non-reducing ends for β-glucan cellobiohydrolase, a member of the exo-β-1,4-glucanase enzyme class, to remove cellobiose units (Figure 7.1).

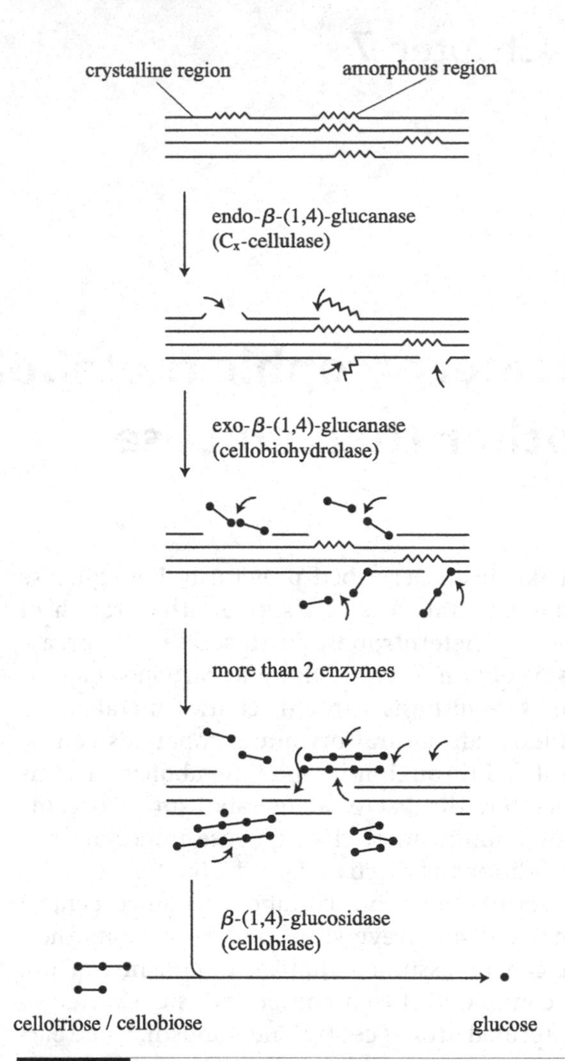

Figure 7.1 Hydrolysis of crystalline cellulose by exo-β-1,4-glucanase.

(Gottschalk, G. 1986, *Bacterial Metabolism*, 2nd edn., Figure 6.2. Springer, New York)

Among the three classes of cellulose-degrading enzymes, endo-β-1,4-glucanase hydrolyses amorphous regions to generate more sites for the exo-β-1,4-glucanase. Then, according to one hypothesis, β-glucan cellobiohydrolase, a member of the exo-β-1,4-glucanase class, not only removes cellobiose units from non-reducing ends but also decrystallizes the crystalline substrate. Cellobiase hydrolyses the cellobiose to glucose.

$$\text{crystalline cellulose} \xrightarrow[\text{formation of unstable glucopyranose ring}]{C_1} \text{reactive cellulose} \xrightarrow[\text{hydrolysis}]{C_x} \text{soluble sugar}$$

Figure 7.2 Initiation of cellulose degradation by a decrystallizing enzyme.

An alternative hypothesis for the decrystallization of cellulose by an enzyme possessing decrystallizing activity but little or no hydrolytic activity. A protein, swollenin, is found in many microorganisms with this property. The decrystallized cellulose is hydrolysed by hydrolytic enzymes.

Through the concerted action of these enzymes, crystalline cellulose is completely hydrolysed. Cellobiose is either hydrolysed to glucose by cellobiase or transported into the cell.

Another hypothesis has also been proposed to explain digestion of crystalline cellulose. A C_1 enzyme decrystallizes the native cellulose to the amorphous form, with little or no hydrolytic activity before the hydrolytic enzymes attack the β-glucose bonds (Figure 7.2). X-ray crystallography has shown the existence of this enzyme. An enzyme with strong cellulose-binding activity and low hydrolytic ability has been found in bacteria such as *Cellulomonas flavigena*, *Clostridium thermocellum* and *Clostridium cellulovorans*, and in fungi such as *Trichoderma reesei*. This hypothesis has been substantiated by the finding of a C_1-like protein, swollenin. This is found in a cellulolytic fungus, *Trichoderma reesei*, that reduces cellulose crystallinity. Swollenin is similar to the plant protein, expansin, that disrupts hydrogen bonds between plant cell wall polysaccharides without hydrolysing them. Swollenin is found in phylogenetically diverse bacteria, fungi, and other organisms, including non-cellulolytic organisms, most of which colonize plant surfaces. These loosen plant cell walls with limited lytic activity. This protein has strong synergistic effects with hydrolytic enzymes.

It is claimed that polysaccharide monooxygenases found in both fungi and bacteria decrystallize and depolymerize cellulose through an oxidative mechanism involving hydroxylation of cellulose at the C1 or C4 carbon. This reaction requires both molecular oxygen and an extracellular electron source (Section 7.7) which can be derived from cellobiose dehydrogenase (CDH) or small-molecule reductants present in lignocellulosic biomass.

7.1.3 Other polysaccharide hydrolases

In addition to starch and cellulose, other natural polysaccharides include hemicellulose, pectin and chitin. Hemicellulose is a heteropolysaccharide consisting of various pentoses and hexoses, and their derivatives, linked with β-glycoside bonds. Many microorganisms, including cellulolytic species, can use this heteropolysaccharide as a carbon and energy source with the extracellular enzymes collectively called hemicellulase. Xylose is the most abundant monosaccharide in most hemicelluloses, and hemicellulase is sometimes referred to as xylanase.

Pectin is a methyl ester of α-1,4-polygalacturonate. Pectin esterase hydrolyses the ester bond to produce methanol, and endo- and exo-type pectinases degrade α-1,4-polygalacturonate to galacturonate. *Bacillus polymyxa*, *Erwinia carotovora* and several other bacteria can produce these enzymes. Industrially, these enzymes can be used to clarify fruit juices such as apple.

Chitin has the structure of poly-β-1,4-*N*-acetylglucosamine and is the major constituent of fungal cell walls and the exoskeletons of insects and crustaceans. This polysaccharide is the second most abundant in the biosphere after cellulose. Chitin has a crystalline structure and degradation requires more than one enzyme. Bacterial chitinases degrade their substrate with the aid of chitin-binding proteins that are lytic polysaccharide monooxygenases. Chitinase is produced by many soil bacteria, including *Chromobacterium violaceum*, *Serratia marcescens*, *Serratia plymuthica*, *Serratia liquefaciens*, *Aeromonas hydrophila*, *Enterobacter agglomerans*, *Pseudomonas aeruginosa*, *Pseudomonas chitinovorans*, *Bacillus circulans*, *Vibrio proteolyticus*, *Streptomyces lividans* and *Streptomyces griseus*, and the archaeon, *Sulfolobus tokodaii*. The enzyme hydrolyses β-1,4 bonds to produce *N*-acetylglucosamine. Chitin deacetylase removes acetate from chitin to chitosan, which is hydrolysed to glucosamine by chitinase. Chitin deacetylase is found in fungi such as *Absidia cierulea*, *Colletotrichum lindemuthanum* and *Mucor rouxii*. Chitinase producers can inhibit the growth of plant pathogenic fungi and can be used as biocontrol agents.

Various other polysaccharides are found in nature, including those constituting the saccharide portion of proteins and lipids. These are also hydrolysed by extracellular enzymes that are produced by many microorganisms. Polysaccharide-hydrolysing enzymes are classified according to their amino acid sequence.

Polysaccharide hydrolases are recognized as virulence factors of bacterial pathogens and include chitinases, galactosidases and pullulanases among others. Substrates of some of these enzymes are not present in mammalian cells. They may degrade glycosides of glycoproteins.

7.1.4 Disaccharide phosphorylases

Some disaccharides are imported into the cell and utilized through non-hydrolytic enzymes. These phosphorylate a monosaccharide of the disaccharide using inorganic phosphate, liberating the other monosaccharide in the free form. These enzymes are referred to as phosphorylases. Since hexokinase is not involved in metabolism of the phosphorylated sugar, one less ATP is consumed in the metabolism of disaccharides by phosphorylases than by hydrolases:

$$\text{cellobiose} + P_i \xrightarrow{\text{cellobiose phosphorylase}} \text{glucose-1-phosphate} + \text{glucose}$$

$$\text{maltose} + P_i \xrightarrow{\text{maltose phosphorylase}} \text{glucose-1-phosphate} + \text{glucose}$$

$$\text{sucrose} + P_i \xrightarrow{\text{sucrose phosphorylase}} \text{glucose-1-phosphate} + \text{fructose}$$

Sucrose phosphorylase is found in *Pseudomonas saccharophila*, and cellobiose is phosphorylated in *Cellulomonas flavigena*. Many more phosphorylases are recognized through genome sequence analyses.

Some fungi synthesize trehalose as their reserve material and in some bacteria this disaccharide serves as an osmoprotectant. The disaccharide is a glucose dimer with an α-1,1-linkage. Many bacteria can use trehalose as their sole carbon and energy source.

Trehalose is metabolized in different ways depending on the organism. Trehalase hydrolyses it to two molecules of glucose, while trehalose phosphorylase cleaves it to glucose and glucose-1-phosphate. In other organisms, trehalose is phosphorylated to trehalose-6-phosphate by the action of trehalose kinase before being cleaved to glucose-6-phosphate and glucose by a hydrolase. Some plant-associated bacteria, such as *Sinorhizobium meliloti*, *Mesorhizobium loti* and *Agrobacterium tumefaciens*, metabolize trehalose through 3-ketotrehalose. These bacteria use isomers of sucrose (leucrose, palatinose and trehalulose) in a similar way.

7.1.5 Hydrolysis of proteins, nucleic acids and lipids

All cells have intracellular polymer-hydrolysing enzymes for turnover of cellular polymers. These enzymes are not involved in the use of polymers available in the environment as their energy and carbon sources. As mentioned previously, microorganisms that utilize extracellular polymers produce extracellular enzymes.

Extracellular proteases are classified into acidic, neutral and alkaline enzymes, according to their optimum pH. Subtilisin produced by *Bacillus licheniformis* is an alkaline protease with an optimum pH of 8–11. Neutral proteases are produced by *Bacillus megaterium* and *Pseudomonas aeruginosa*. Most alkaline proteases are non-specific and hydrolyse any peptide bonds. Many acidic proteases are specific and hydrolyse the peptide bonds of specific amino acids.

DNA and RNA can serve as carbon and energy sources in bacteria that produce extracellular DNase and RNase, respectively. *Staphylococcus aureus* produces extracellular DNase, and RNase is known in *Bacillus subtilis*, *Streptomyces* spp. and other bacteria. Lipids are utilized by

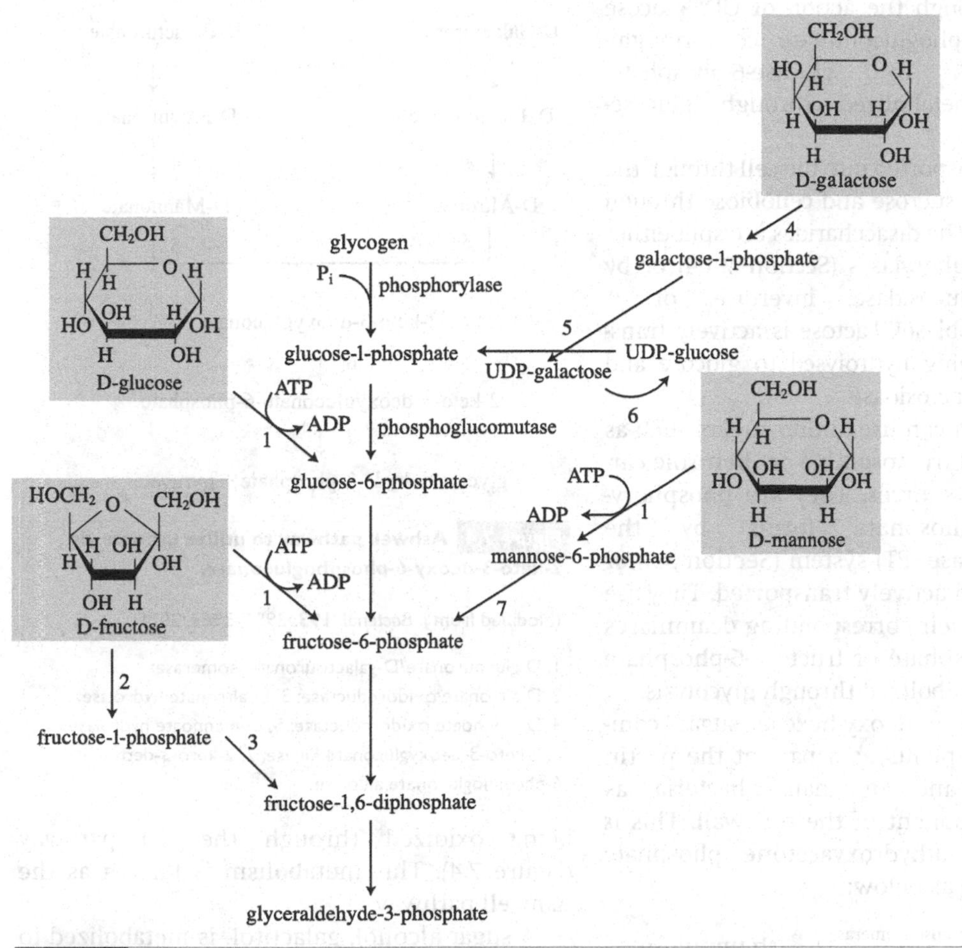

Figure 7.3 Hexose metabolism to EMP pathway intermediates.

1, hexokinase; 2, group translocation system, or ketohexokinase; 3, 1-phosphofructokinase; 4, galactokinase; 5, glucose: galactose-1-phosphate uridylyltransferase; 6, uridine diphosphate glucose epimerase; 7, mannose-6-phosphate isomerase.

microorganisms after hydrolysis by extracellular lipase to fatty acids and glycerol.

7.2 Utilization of sugars

Extracellular polysaccharide hydrolysis generates various hexoses, pentoses and their derivatives. These substances support the growth of many bacteria and archaea.

7.2.1 Hexose utilization

Figure 7.3 shows how common hexoses are converted to central metabolic intermediates. Fructose is phosphorylated either to fructose-1-phosphate during group translocation, or to fructose-6-phosphate by hexokinase after active transport into the cell. Fructose-1-phosphate is further phosphorylated to fructose-1,6-diphosphate by 1-phosphofructokinase.

Mannose is actively transported into the cell before hexokinase converts it to mannose-6-phosphate. Phosphomannoisomerase isomerizes mannose-6-phosphate to fructose-6-phosphate. Galactokinase transfers phosphate from ATP to galactose to form galactose-1-phosphate, and glucose:galactose-1-phosphate uridylyltransferase transfers UMP from UDP-glucose to galactose-1-phosphate to form UDP-galactose and glucose-1-phosphate. UDP-galactose is converted to

UDP-glucose through the action of UDP-glucose epimerase. Phosphoglucomutase converts glucose-1-phosphate to glucose-6-phosphate. Glycogen is metabolized through glucose-1-phosphate.

Maltose is transported into the cell through the ABC system, and sucrose and cellobiose through active transport. The disaccharides are split either by specific phosphorylases (Section 7.1.4) or by hydrolases, α-glucosidase, invertase, or β-glucosidase (cellobiase). Lactose is actively transported before being hydrolysed to glucose and galactose by β-galactosidase.

Many bacteria can use amino sugars such as glucosamine and fructosamine as their sole carbon and energy sources. They are phosphorylated to 6-phosphate forms by the phosphotransferase (PT) system (Section 3.5) or by a kinase when actively transported. They are deaminated by their corresponding deaminases to glucose-6-phosphate or fructose-6-phosphate before being metabolized through glycolysis.

Rhamnose is a deoxy-hexose sugar commonly found in plants as a part of the pectin polysaccharide and in many bacteria as a common component of the cell wall. This is metabolized to dihydroxyacetone phosphate and lactaldehyde as below:

$$\text{rhamnose} \xrightarrow{\text{rhamnose isomerase}} \text{rhamnulose}$$

$$\text{rhamnulose} + \text{ATP} \xrightarrow{\text{rhamnulose kinase}} \text{rhamnulose-1-phosphate} + \text{ADP}$$

$$\text{rhamnulose-1-phosphate} \xrightarrow{\text{rhamnulose-1-phosphate aldolase}} \text{dihydroxyacetone phosphate} + \text{lactaldehyde}$$

Lactaldehyde is converted either to propanediol by a reductase or to lactate by a dehydrogenase. Propanediol metabolism is decribed below (Section 7.4).

Hexouronates, such as glucouronate and galactouronate, are used as sole carbon and energy sources by enteric bacteria, including *Escherichia coli*. They are converted to an Entner-Doudoroff pathway intermediate, 2-keto-3-deoxy-6-phosphogluconate, by an isomerase, oxidoreductase and hydrolase before being oxidized through the ED pathway (Figure 7.4). This metabolism is known as the Ashwell pathway.

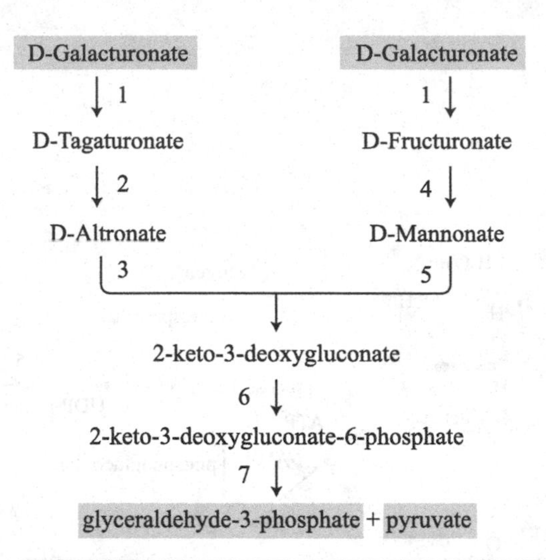

Figure 7.4 Ashwell pathway to utilize uronate via 2-keto-3-deoxy-6-phosphogluconate.

(Modified from *J. Bacteriol.* **193**: 3956–3963, 2011)

1, D-glucouronate/D-galactouronate isomerase; 2, D-altronate oxidoreductase; 3, D-altronate hydrolase; 4, D-mannoate oxidoreductase; 5, D-mannoate hydrolase; 6, 2-keto-3-deoxygluconate kinase; 7, 2-keto-3-deoxy-6-phosphogluconate aldolase.

A sugar alcohol, galactitol, is metabolized to dihydroxyacetone phosphate and glyceraldehyde-3-phosphate by the pathogenic bacterium *Salmonella enterica*. This sugar alcohol is imported into the cell through group translocation before being reduced to tagatose-6-phosphate. Tagatose-6-phosphate is phosphorylated and split to trioses, as in the EMP pathway. This metabolism is related to the virulence of this bacterium.

The sulfonated sulfoquinovose (SQ; 6-deoxy-6-sulfoglucose), a structural analogue of glucose-6-phosphate (GP), is the polar headgroup of the sulfolipid in photosynthetic membranes. SQ is metabolized in analogy to the ED pathway in *Pseudomonas putida*, producing pyruvate and 3-sulfolactaldehyde (SLA), while *Escherichia coli* and *Klebsiella oxytoca* cleave SQ to dihydroxyacetone phosphate and SLA in analogy to the EMP pathway (Figure 7.5). Pyruvate and dihydroxyacetone phosphate enter central metabolism,

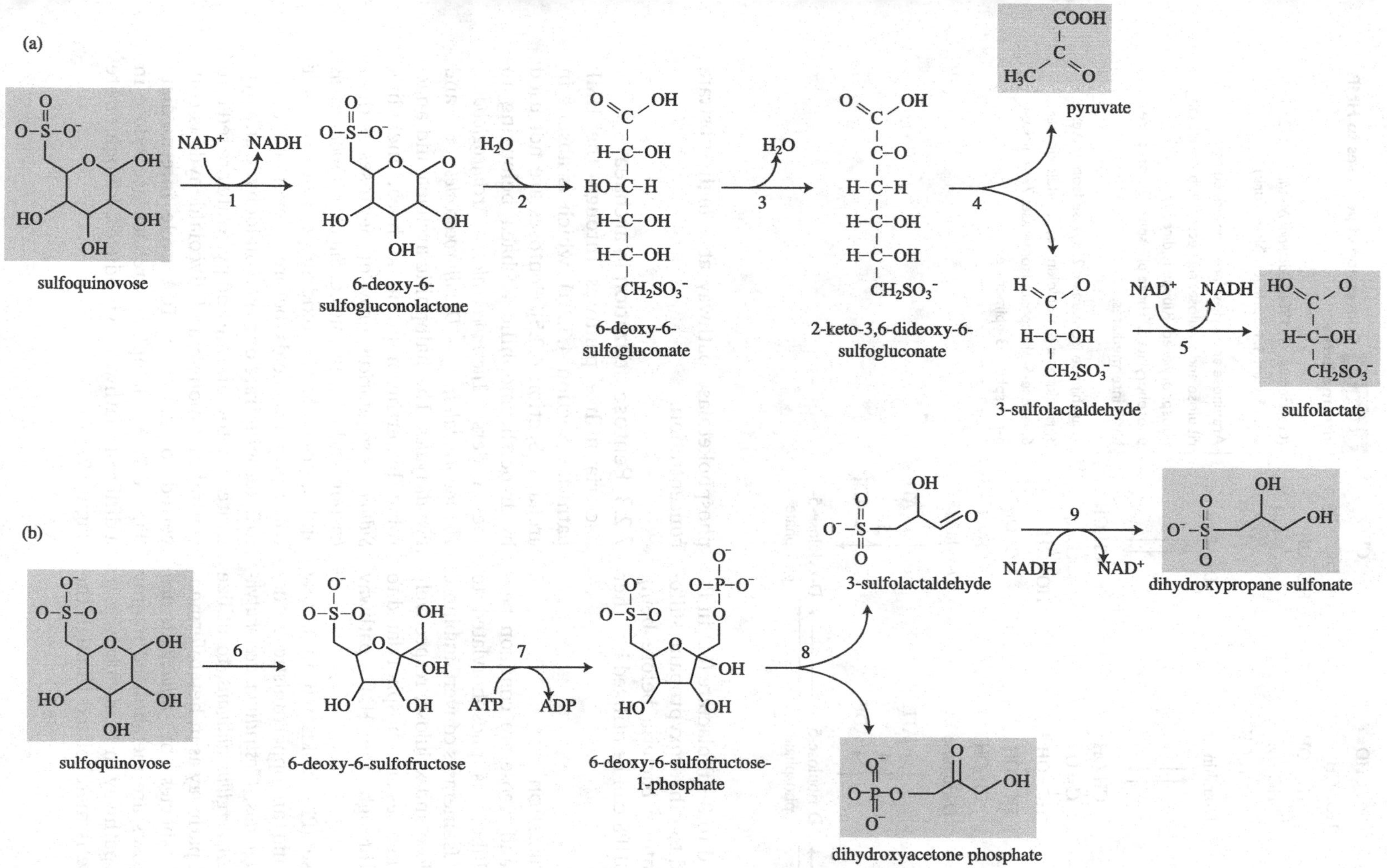

Figure 7.5 **Sulfoquinovose metabolism in (a) *Pseudomonas putida*** (*PNAS* **112**: E4298–E4305, 2015) **and in (b) enteric bacteria** (Modified from *Nature* **507**: 114–117, 2014)

1, sulfoquinovose dehydrogenase; 2, 6-deoxy-6-sulfogluconolactone (SGL) lactonase; 3, 6-deoxy-6-sulfogluconate (SG) dehydratase; 4, 2-keto-3,6-dideoxy-6-sulfogluconate (KDSG) aldolase; 5, 3-sulfolactaldehyde (SLA) dehydrogenase; 6, sulfoquinovose isomerase; 7, 6-deoxy-6-sulfofructose (SF) kinase; 8, 6-deoxy-6-sulfofructose-1-phosphate (SFP) aldolase; 9, 3-sulfolactaldehyde (SLA) reductase.

SQ, sulfoquinovose; SGL, 6-deoxy-6-sulfogluconolactone; SG, 6-deoxy-6-sulfogluconate; KDSG, 2-keto-3,6-dideoxy-6-sulfogluconate; SLA, 3-sulfolactaldehyde; SF, 6-deoxy-6-sulfofructose; SFP, 6-deoxy-6-sulfofructose-1-phosphate; DHAP, dihydroxyacetone phosphate.

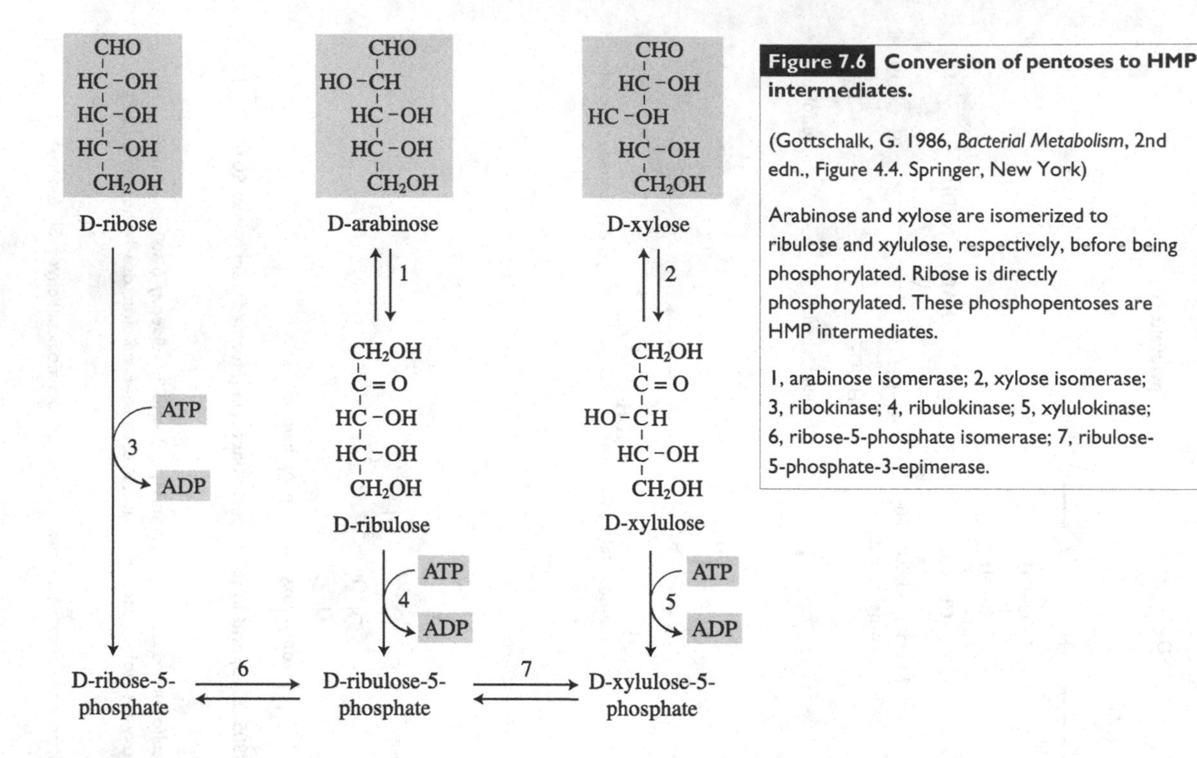

Figure 7.6 Conversion of pentoses to HMP intermediates.

(Gottschalk, G. 1986, *Bacterial Metabolism*, 2nd edn., Figure 4.4. Springer, New York)

Arabinose and xylose are isomerized to ribulose and xylulose, respectively, before being phosphorylated. Ribose is directly phosphorylated. These phosphopentoses are HMP intermediates.

1, arabinose isomerase; 2, xylose isomerase; 3, ribokinase; 4, ribulokinase; 5, xylulokinase; 6, ribose-5-phosphate isomerase; 7, ribulose-5-phosphate-3-epimerase.

and SLA is oxidized to 3-sulfolactate (SL) in *Ps. putida* and reduced to dihydroxypropane sulfonate (DHPS) in enteric bacteria before being excreted. SL and DHPS can be utilized by other bacteria.

7.2.2 Pentose utilization

Xylose, arabinose and ribose are common pentoses in nature. Ribose is phosphorylated to ribose-5-phosphate. Isomerases convert arabinose and xylose to ribulose and xylulose, respectively (Figure 7.6). These ketoses are phosphorylated to be metabolized through the HMP pathway (Section 4.3).

Phosphoketolase (PK) is known in a few organisms, including an opportunistic pathogen, *Gardnerella vaginalis*, *Clostridium acetobutylicum* and a fungus *Aspergillus nidulans*, to utilize pentoses in the PK pathway as in heterofermentative LAB and species of *Bifidobacterium* (Section 4.5). Pentoses are metabolized mainly through the HMP pathway in *Cl. acetobutylicum*, but more substrate is metabolized through the phosphoketolase pathway at a high substrate concentration.

7.2.3 Pentose utilization in archaea

Bacteria utilize pentoses through the HMP pathway (Section 4.3.2.1), which is absent in archaea (Section 4.3.5). Pentoses are not used by hyperthermophilic archaea belonging to the orders *Thermococcales*, *Archaeoglobales*, *Thermoproteales*, *Desulfurococcales* and *Pyrodictyales*. The halophilic archaea and anaerobic hyperthermophilic archaea, including *Sulfolobus solfataricus*, grow on pentoses as their carbon and energy source. *Sulfolobus solfataricus* and *S. acidocaldarius* oxidize pentoses to their corresponding acids before converting them to 2-ketoglutarate or to glycolaldehyde and pyruvate. 2-ketoglutarate and pyruvate are fed into central metabolism and glycolaldehyde is converted to malate, a TCA cycle intermediate (Figure 7.7a). Halophilic archaea use xylose in a different pathway to produce 2-ketoglutarate (Figure 7.7b).

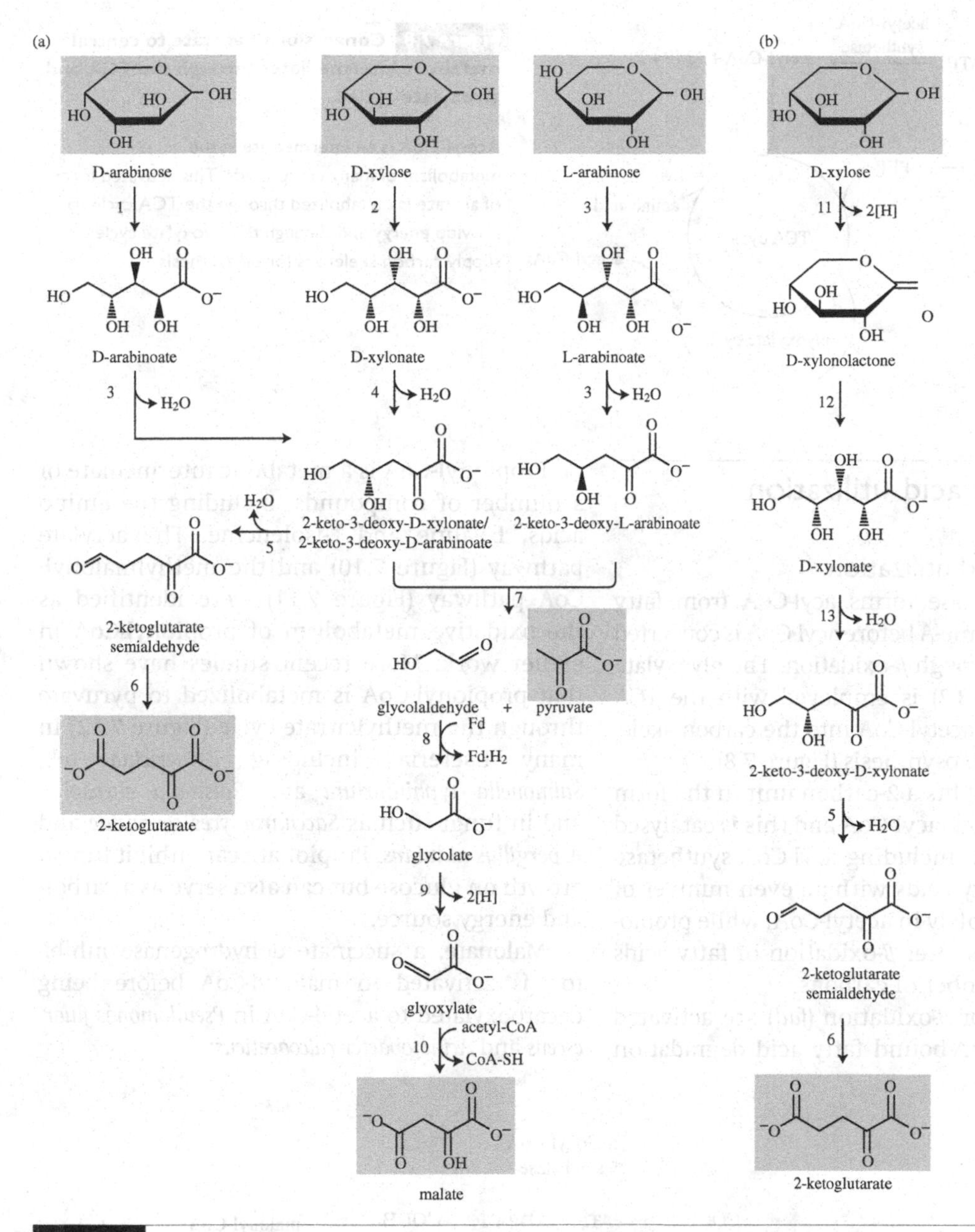

Figure 7.7 **Pentose utilization in (a) *Sulfolobus* spp. and (b) halophilic archaea.**

(Modified from *Microbiol. Mol. Biol. Rev.* **78**: 89–175, 2014)

The HMP pathway is not known in archaea and some archaea use pentoses in a series of reactions catalysed by dehydrogenases and dehydratases, producing 2-ketoglutarate or glycolaldehyde and pyruvate.

1, arabinose dehydrogenase; 2, glucose dehydrogenase; 3, arabinoate dehydratase; 4, xylonate dehydratase; 5, 2-keto-3-deoxyarabinoate (2-keto-3-deoxyxylonate) dehydratase; 6, 2-ketoglutarate semialdehyde dehydrogenase; 7, 2-keto-3-deoxy-(6-phospho) gluconate aldolase; 8, glycolaldehyde: ferredoxin oxidoreductase; 9, glycolate dehydrogenase; 10, malate synthase; 11, xylose dehydrogenase; 12, xylono-1,4-lactone lactonase; 13, xylonate dehydratase.

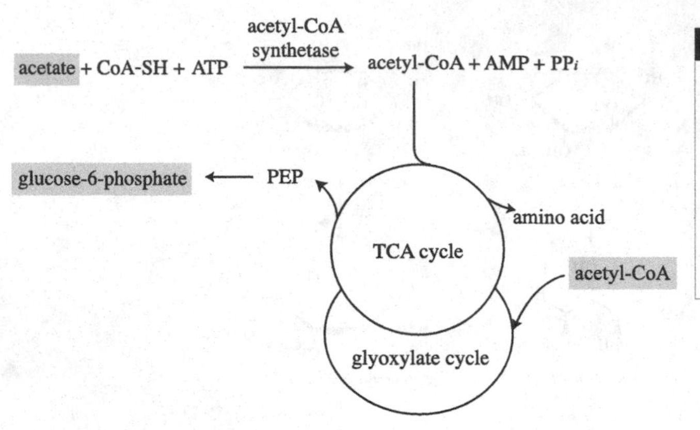

Figure 7.8 Conversion of acetate to central metabolic intermediates through the TCA and glyoxylate cycles.

Acetyl-CoA is an intermediate in the metabolism of many compounds. This activated form of acetate is metabolized through the TCA cycle to provide energy and through the glyoxylate cycle to supply carbon skeletons for biosynthesis.

7.3 Organic acid utilization

7.3.1 Fatty acid utilization

Acyl-CoA synthetase forms acyl-CoA from fatty acids and coenzyme-A before acyl-CoA is converted to acetyl-CoA through β-oxidation. The glyoxylate cycle (Section 5.3.2) is employed with the TCA cycle to convert acetyl-CoA into the carbon skeletons needed for biosynthesis (Figure 7.8).

β-oxidation splits a 2-carbon unit in the form of acetyl-CoA from acyl-CoA and this is catalysed by five enzymes, including acyl-CoA synthetase (Figure 7.9). Fatty acids with an even number of carbons result solely in acetyl-CoA, while propionyl-CoA remains after β-oxidation of fatty acids with an odd number of carbons.

The genes for β-oxidation (*fad*) are activated by the acetyl-CoA-bound fatty acid degradation regulatory protein (FadR) and repressed by the free form of FadR in *Escherichia coli*. FadR regulates genes for fatty acid synthesis (*fab*) in the opposite manner to *fad* genes (Section 6.6.1.5).

Propionyl-CoA is a metabolic intermediate of a number of compounds, including the amino acids, L-valine and L-isoleucine. The acrylate pathway (Figure 7.10) and the methylmalonyl-CoA pathway (Figure 7.11) were identified as the oxidative metabolism of propionyl-CoA in earlier work. More recent studies have shown that propionyl-CoA is metabolized to pyruvate through the methylcitrate cycle (Figure 7.12) in many bacteria, including *Escherichia coli*, *Salmonella typhimurium* and *Ralstonia eutropha*, and in fungi such as *Saccharomyces cerevisiae* and *Aspergillus nidulans*. Propionate can inhibit fungal growth on glucose but can also serve as a carbon and energy source.

Malonate, a succinate dehydrogenase inhibitor, is activated to malonyl-CoA before being decarboxylated to acetyl-CoA in *Pseudomonas fluorescens* and *Acinetobacter calcoaceticus*:

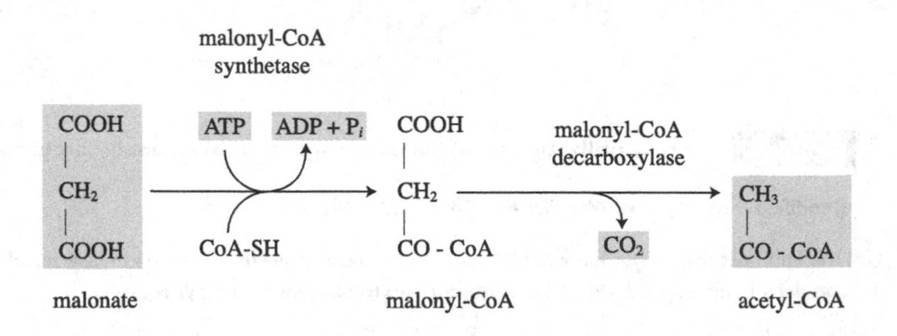

An acyl carrier protein (ACP) replaces coenzyme A in malonate metabolism in some bacteria, such as *Klebsiella pneumoniae* and the anaerobe *Malonomonas rubra*:

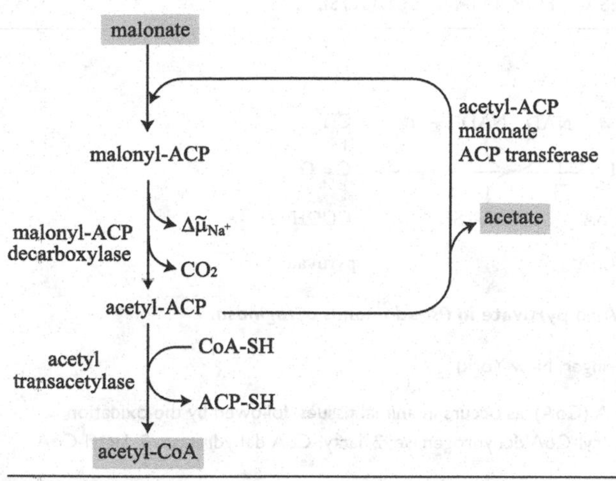

$CH_3-CH_2-CH_2-CH_2-CH_2-CH_2-CH_2-CH_2-CH_2-CH_2-CH_2-CH_2-CH_2-CH_2-CH_2-COOH$

1: CoA-SH, ATP → AMP, PP_i

$CH_3-CH_2-CH_2-CH_2-CH_2-CH_2-CH_2-CH_2-CH_2-CH_2-CH_2-CH_2-CH_2-CH_2-CH_2-CO-CoA$

2: FAD → $FADH_2$

$CH_3-CH_2-CH_2-CH_2-CH_2-CH_2-CH_2-CH_2-CH_2-CH_2-CH_2-CH_2-CH_2-CH=CH-CO-CoA$

3: H_2O

$CH_3-CH_2-CH_2-CH_2-CH_2-CH_2-CH_2-CH_2-CH_2-CH_2-CH_2-CH_2-CH_2-CH(OH)-CH_2-CO-CoA$

4: NAD^+ → $NADH + H^+$

$CH_3-CH_2-CH_2-CH_2-CH_2-CH_2-CH_2-CH_2-CH_2-CH_2-CH_2-CH_2-CH_2-C(=O)-CH_2-CO-CoA$

5: CoA-SH

$CH_3-CH_2-CH_2-CH_2-CH_2-CH_2-CH_2-CH_2-CH_2-CH_2-CH_2-CH_2-CH_2-CO-CoA + CH_3-CO-CoA$ (acetyl-CoA)

H_2O, CoA-SH; FAD, NAD^+ → $FADH_2$, $NADH + H^+$

$CH_3-CH_2-CH_2-CH_2-CH_2-CH_2-CH_2-CH_2-CH_2-CH_2-CH_2-CO-CoA + CH_3-CO-CoA$

↓

$CH_3-CH_2-CH_2-CH_2-CH_2-CH_2-CH_2-CH_2-CH_2-CO-CoA + CH_3-CO-CoA$

↓

$CH_3-CH_2-CH_2-CH_2-CH_2-CH_2-CH_2-CO-CoA + CH_3-CO-CoA$

↓

$CH_3-CH_2-CH_2-CH_2-CH_2-CO-CoA + CH_3-CO-CoA$

↓

$CH_3-CH_2-CH_2-CO-CoA + CH_3-CO-CoA$

↓

$CH_3-CO-CoA + CH_3-CO-CoA$

Figure 7.9 Palmitate degradation to acetyl-CoA through *β*-oxidation.

(Gottschalk, G. 1986, *Bacterial Metabolism*, 2nd edn., Figure 6.5. Springer, New York)

1, acyl-CoA synthetase; 2, fatty acyl-CoA dehydrogenase; 3, 3-hydroxyacyl-CoA hydrolase; 4, 3-hydroxyacyl-CoA dehydrogenase; 5, acetyl-CoA acetyltransferase.

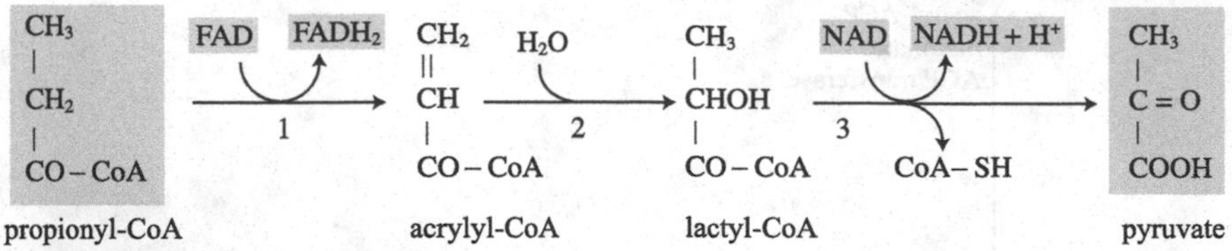

CH_3–CH_2–CO–CoA (propionyl-CoA) —1 (FAD → $FADH_2$)→ CH_2=CH–CO–CoA (acrylyl-CoA) —2 (H_2O)→ CH_3–CHOH–CO–CoA (lactyl-CoA) —3 (NAD → NADH + H^+; CoA–SH)→ CH_3–C=O–COOH (pyruvate)

Figure 7.10 Acrylate pathway oxidizing propionyl-CoA to pyruvate in *Pseudomonas aeruginosa*.

(Gottschalk, G. 1986, *Bacterial Metabolism*, 2nd edn., p. 150. Springer, New York)

Pseudomonas aeruginosa oxidizes valine to propionyl-coenzyme A (CoA), as occurs in animal tissues, followed by the oxidation of propionyl-CoA to acrylyl-CoA, lactyl-CoA, and pyruvate. 1, acyl-CoA dehydrogenase; 2, lactyl-CoA dehydratase; 3, lactyl-CoA dehydrogenase.

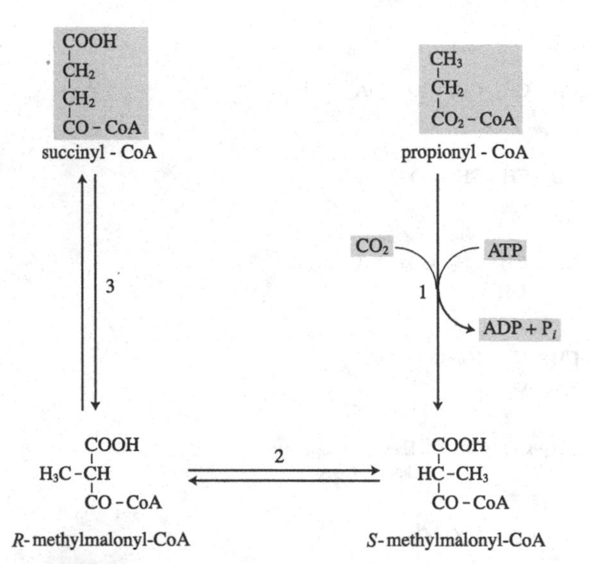

Figure 7.11 Conversion of propionyl-CoA to succinyl-CoA through the methylmalonyl-CoA pathway.

(Gottschalk, G. 1986, *Bacterial Metabolism*, 2nd edn., Figure 6.6. Springer, New York)

1, propionyl-CoA carboxylase; 2, methylmalonyl-CoA racemase; 3, methylmalonyl-CoA mutase.

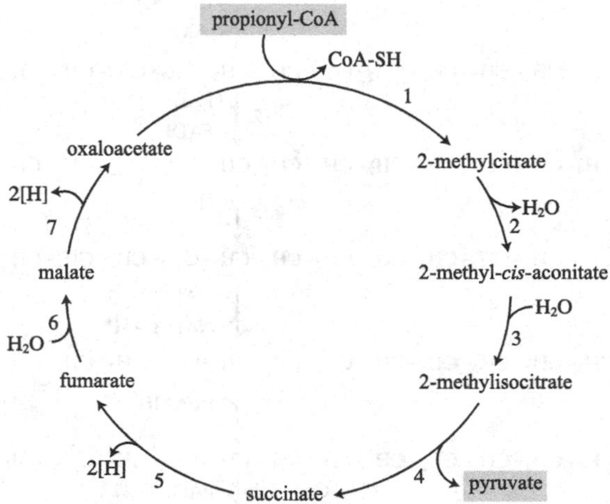

Figure 7.12 Oxidation of propionyl-CoA to pyruvate through the methylcitrate cycle.

(*Microbiology* **147**: 2203–2214, 2001)

1, methylcitrate synthase; 2, 2-methyl-*cis*-aconitate dehydratase; 3, 2-methyl-*cis*-aconitate hydratase; 4, methylisocitrate lyase; 5, succinate dehydrogenase; 6, fumarase; 7, malate dehydrogenase.

7.3.2 Organic acids more oxidized than acetate

Acetate can serve as the sole carbon source through the TCA and glyoxylate cycles. C2 compounds more oxidized than acetate cannot be metabolized in the same way. These compounds include glycolate, glyoxylate and oxalate. Glycolate is generated from the dephosphorylation of phosphoglycolate that is produced during photorespiration (Section 10.8.1.2). Purine degradation results in glyoxylate. *Escherichia coli* and *Pseudomonas* spp. use these substances through the dicarboxylic acid cycle to generate energy, and through the glycerate pathway to supply carbon skeletons for biosynthesis (Figure 7.13). The dicarboxylic acid cycle is similar to the glyoxylate cycle described in Chapter 5 (Section 5.3.2). Since phosphoenolpyruvate, an intermediate of the dicarboxylic acid cycle, is used for biosynthesis, it is replenished through the glycerate pathway.

Paracoccus denitrificans converts glyoxylate through the 3-hydroxyaspartate pathway to oxaloacetate, a TCA cycle intermediate (Figure 7.14).

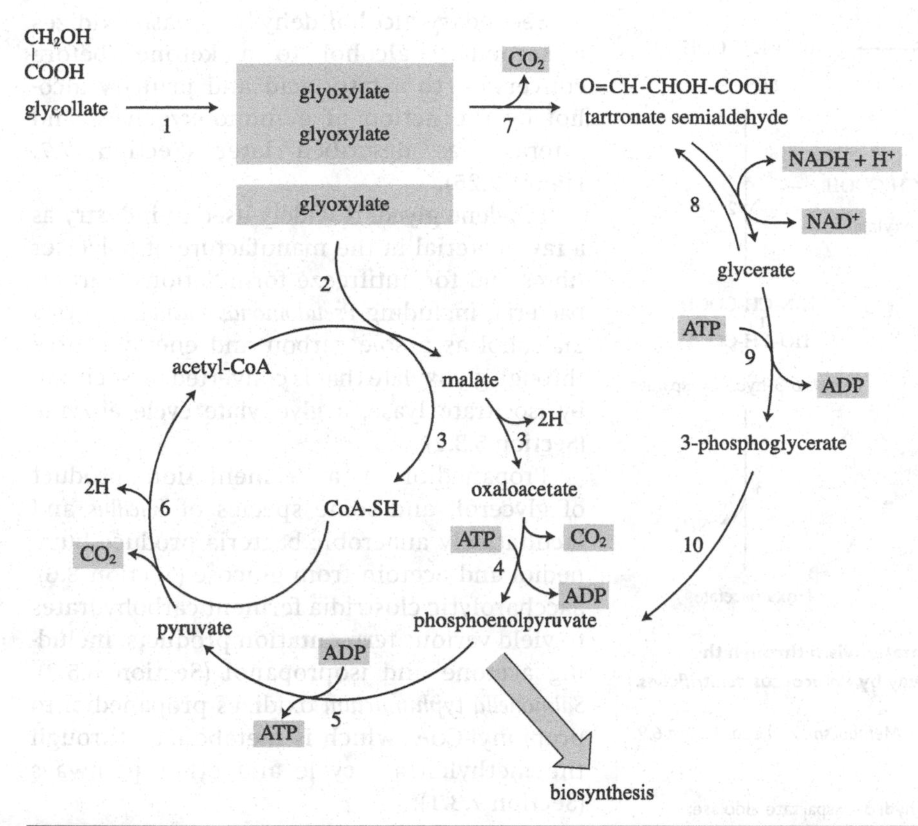

Figure 7.13 The catabolic dicarboxylic acid cycle and anabolic glycerate pathway for glyoxylate utilization.

Certain bacteria, including *Escherichia coli* and species of the genus *Pseudomonas*, oxidize glyoxylate through the dicarboxylic acid cycle and convert the substrate to phosphoenolpyruvate through the glycerate pathway.

1, glycolate dehydrogenase; 2, malate synthase; 3, malate dehydrogenase; 4, PEP carboxykinase; 5, pyruvate kinase; 6, pyruvate dehydrogenase; 7, glyoxylate carboligase; 8, tartronate semialdehyde reductase; 9, glycerate kinase; 10, phosphoglycerate mutase and enolase.

In the 3-hydroxyaspartate pathway, glyoxylate is converted to glycine before condensing with a second glyoxylate molecule to yield *erythro*-3-hydroxyaspartate. *Erythro*-3-hydroxyaspartate dehydratase deaminates this to oxaloacetate. As in the deamination of amino acids with a hydroxyl group (Section 7.5.3), this enzyme is referred to as a dehydratase, although it deaminates the substrate.

Pseudomonas oxalaticus uses oxalate as sole carbon and energy source. This most oxidized dicarboxylic acid is activated to oxalyl-CoA before a part of it is oxidized to CO_2 as an energy source, and the remaining part is reduced to glyoxylate for biosynthesis (Figure 7.15). *Bacillus oxalophilus* and *Methylobacterium extorquens* also use oxalate as their sole carbon and energy source.

A methylotroph, *Methylobacterium extorquens*, reduces oxalyl-CoA to glyoxylate to incorporate it to the serine–isocitrate pathway (Section 7.10.3.2).

7.4 Utilization of alcohols and ketones

Anaerobic fermentative microorganisms produce alcohols such as ethanol, butanol and others. Alcohol dehydrogenase and aldehyde dehydrogenase oxidize primary alcohols to fatty acids:

$$R\text{-}CH_2\text{-}CH_2OH + NAD^+ \xrightarrow{\text{alcohol dehydrogenase}} R\text{-}CH_2\text{-}CHO + NADH + H^+$$

$$R\text{-}CH_2\text{-}CHO + NAD^+ + H_2O \xrightarrow{\text{aldehyde dehydrogenase}} R\text{-}CH_2\text{-}COOH + NADH + H^+$$

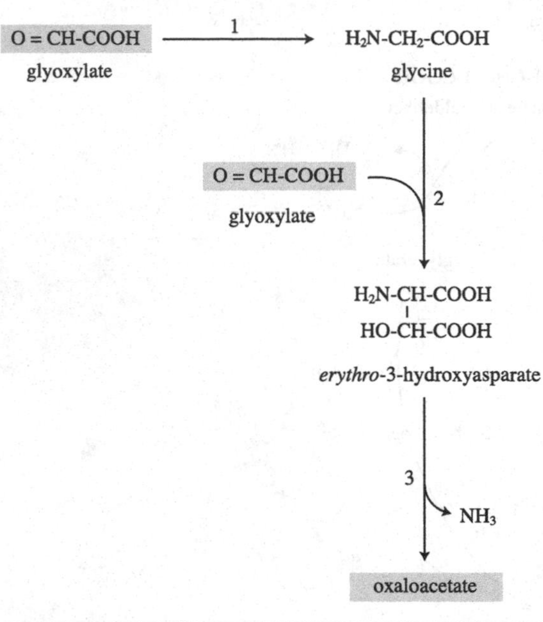

Figure 7.14 Glyoxylate metabolism through the 3-hydroxyaspartate pathway by *Paracoccus denitrificans*.

(Gottschalk, G. 1986, *Bacterial Metabolism*, 2nd edn., Figure 6.9. Springer, New York)

1, transaminase; 2, *erythro*-3-hydroxyaspartate aldolase; 3, *erythro*-3-hydroxyaspartate dehydratase.

Secondary alcohol dehydrogenase oxidizes a secondary alcohol to a ketone, before conversion to a fatty acid and primary alcohol by the action of a monooxygenase and esterase, as described later (Section 7.7, Figure 7.25).

Ethylene glycol is widely used in industry as a raw material in the manufacture of polyester fibres and for antifreeze formulations. Various bacteria, including *Pseudomonas putida*, use this dialcohol as a sole carbon and energy source through glyoxylate that is converted to isocitrate by isocitrate lyase, a glyoxylate cycle enzyme (Section 5.3.2)

Propanediol is a fermentation product of glycerol, and some species of *Bacillus* and facultatively anaerobic bacteria produce butanediol and acetoin from glucose (Section 8.6). Saccharolytic clostridia ferment carbohydrates to yield various fermentation products, including acetone and isopropanol (Section 8.5.2). *Salmonella typhimurium* oxidizes propanediol to propionyl-CoA, which is metabolized through the methylcitrate cycle and other pathways (Section 7.3.1):

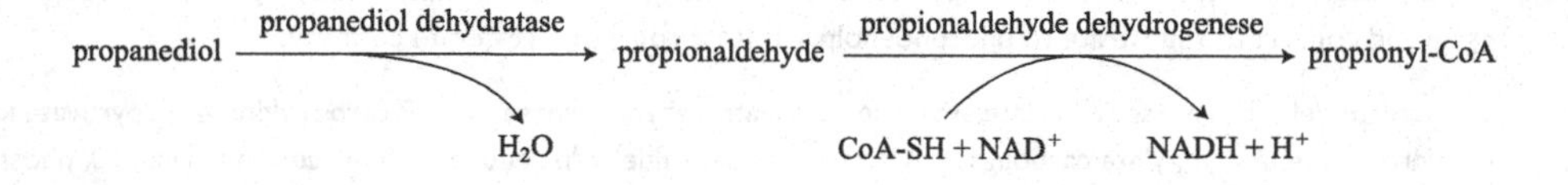

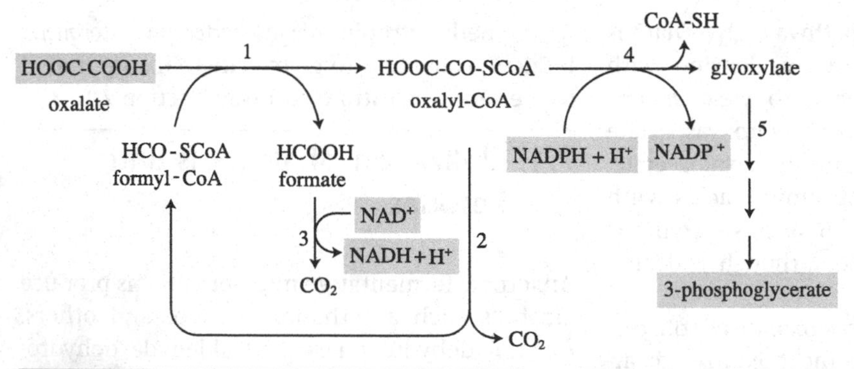

Figure 7.15 Oxalate metabolism in *Pseudomonas oxalaticus*.

(Gottschalk, G. 1986, *Bacterial Metabolism*, 2nd edn., Figure 6.10. Springer, New York)

Oxalate is activated by coenzyme A transferase (1) to oxalyl-CoA, that is either decarboxylated to formyl-CoA (2) or reduced to glyoxylate (4). Glyoxylate is used for biosynthesis through the glycerate pathway (5) and formyl-CoA is oxidized via formate (3).

These enzymes and cofactors are encapsulated within a cellular microcompartment with a multiprotein shell to localize the reactive intermediate, propionaldehyde, and minimize cellular toxicity and DNA damage.

Butanediol is oxidized to acetaldehyde and acetyl-CoA by species of *Bacillus*, *Ralstonia eutropha* (*Alcaligenes eutrophus*) and *Pelobacter carbinolicus*, through the action of diol dehydrogenase and the acetoin dehydrogenase complex. The acetoin dehydrogenase enzyme complex is a keto acid dehydrogenase like the pyruvate dehydrogenase complex (Section 5.1):

$$\text{butanediol} \xrightarrow[\text{NAD}^+ \;\to\; \text{NADH} + \text{H}^+]{\text{diol dehydrogenase}} \text{acetoin} \xrightarrow[\text{CoA-SH} + \text{NAD}^+ \;\to\; \text{NADH} + \text{H}^+]{\text{acetoin dehydrogenase complex}} \text{acetaldehyde} + \text{acetyl-CoA}$$

A *Xanthobacter* sp. can oxidize isopropanol to acetone before carboxylating it to acetoacetate. Acetone is carboxylated by *Rhodobacter capsulatus*, *Rhodomicrobium vannielii* and *Thiosphaera pantotropha*:

$$\text{isopropanol} \xrightarrow[\text{NAD}^+ \;\to\; \text{NADH} + \text{H}^+]{\text{secondary alcohol dehydrogenese}} \text{acetone} \xrightarrow[\text{ATP} + \text{CO}_2 \;\to\; \text{AMP} + 2\text{P}_i]{\text{acetone carboxylase}} \text{acetoacetate}$$

Glycerol is converted to dihydroxyacetone phosphate, either first being phosphorylated by glycerol kinase and then oxidized by glycerol-3-phosphate oxidase, or first being oxidized by glycerol dehydrogenase and then phosphorylated by dihydroxyacetone kinase. In *Azospirillum brasilense*, glycerol is oxidized by an alcohol dehydrogenase containing quinone that is induced by glycerol through the two-component system (Section 12.1.7). Glycerol is a preferred substrate over glucose in the haloarchaeon *Haloferax volcanii*.

Various bacteria use acetone through carboxylation to actoacetate:

$$\text{acetone} + \text{CO}_2 + \text{ATP} + 2\text{H}_2\text{O} \longrightarrow \text{acetoacetate} + \text{AMP} + 2\text{P}_i$$

7.5 Amino acid utilization

Nutrient broth is a common medium for cultivation of many bacteria in the laboratory and contains peptone and beef extract. These substances largely consist of amino acids and peptides. Organisms growing on such a nutrient medium transport and metabolize amino acids and peptides through the central metabolic pathways. Amino acids are used for protein synthesis and are deaminated to the corresponding 2-keto acids. The 2-keto acids are oxidized to acyl-CoA by 2-keto acid dehydrogenases, for use as carbon and energy sources. They are deaminated through different mechanisms depending on their nature.

7.5.1 Oxidative deamination

Amino acids are deaminated either by amino acid oxidase reducing its prosthetic flavin or by amino acid dehydrogenase reducing $NAD(P)^+$. Amino acid oxidases have a low specificity for the substrate and a single enzyme can oxidize up to ten different amino acids. Since bacterialcell walls contain D-amino acids, bacteria have L-amino acid as well as D-amino acid oxidase:

$$\text{R-CHNH}_2\text{-COOH} + \text{H}_2\text{O} \longrightarrow \text{R-CO-COOH} + \text{NH}_3 + 2e^- + 2\text{H}^+$$

Amino acid dehydrogenase oxidizes L-alanine or L-glutamate to pyruvate and 2-ketoglutarate, respectively. Since transaminases convert pyruvate and 2-ketoglutarate to alanine and glutamate, all amino acids can be deaminated by the combination of transaminase and amino acid dehydrogenase:

$$\text{glutamate} + NAD^+ + H_2O \xrightleftharpoons{\text{glutamate dehydrogenase}} \text{2-ketoglutarate} + NADH + NH_4^+$$

$$\text{alanine} + NAD^+ + H_2O \xrightleftharpoons{\text{alanine dehydrogenase}} \text{pyruvate} + NADH + NH_4^+$$

7.5.2 Transamination

Transamination is an enzymic reaction that transfers the $-NH_2$ group from amino acids to 2-keto acids. As shown above, alanine and glutamate dehydrogenases deaminate their substrate. When coupling transamination and dehydrogenation of alanine or glutamate, an amino acid is oxidized to the corresponding 2-keto acid, reducing $NAD(P)^+$:

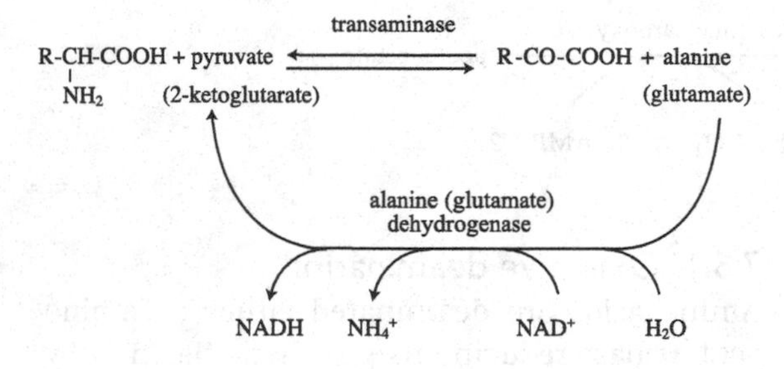

7.5.3 Amino acid dehydratase

Amino acid dehydratases deaminate serine and threonine, removing the hydroxyl group (-OH) at the same time:

$$\underset{\text{serine}}{CH_2OH\text{-}CHNH_2\text{-}COOH} \xrightarrow{\text{serine dehydratase}} \underset{\text{pyruvate}}{CH_3\text{-}CO\text{-}COOH} + NH_3$$

$$\underset{\text{threonine}}{CH_3\text{-}CHOH\text{-}CHNH_2\text{-}COOH} \xrightarrow{\text{threonine dehydratase}} \underset{\text{2-ketobutyrate}}{CH_3\text{-}CH_2\text{-}CO\text{-}COOH} + NH_3$$

As stated above (Section 7.3.2), these enzymes are referred to as dehydratases although they deaminate their substrates. The reactions take place in two steps, the first step being a dehydration reaction:

$$R\text{-}CHOH\text{-}CHNH_2\text{-}COOH \xrightarrow{-H_2O} R\text{-}CH{=}CNH_2\text{-}COOH \xrightarrow{+H_2O,\ -NH_3} R\text{-}CH_2\text{-}CO\text{-}COOH$$

Isoleucine biosynthesis starts with the deamination of threonine to 2-ketobutyrate, which is catalysed by threonine dehydratase (Section 6.4.1). *Escherichia coli* has separate threonine dehydratases for isoleucine synthesis and the use of threonine as a carbon and energy source. As expected, they are regulated differently.

Aspartate and histidine are deaminated in similar reactions to those catalysed by dehydratase. Unlike dehydrogenases and oxidases, aspartase and histidase form double bonds between two and three carbons. Water does not take part in these reactions, nor do electron carriers:

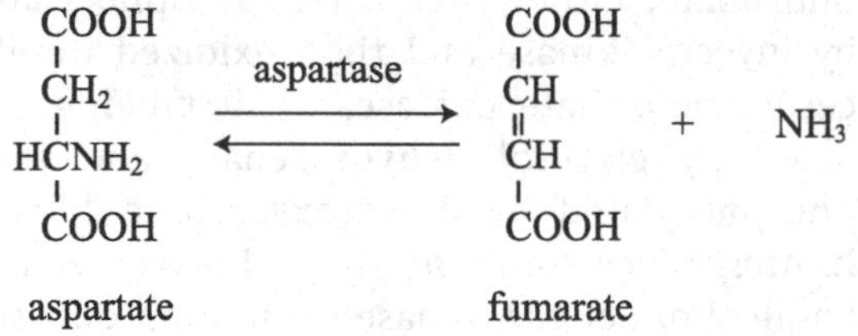

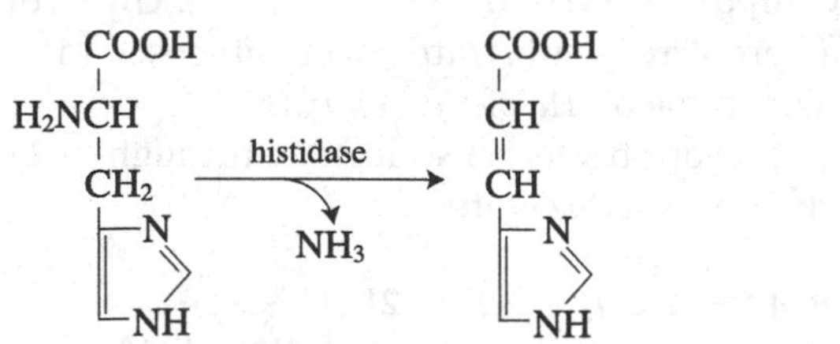

$$\underset{\text{aspartate}}{HOOC\text{-}CH_2\text{-}HCNH_2\text{-}COOH} \underset{}{\xrightleftharpoons{\text{aspartase}}} \underset{\text{fumarate}}{HOOC\text{-}CH{=}CH\text{-}COOH} + NH_3$$

$$HOOC\text{-}H_2NCH\text{-}CH_2\text{-}(\text{imidazole}) \xrightarrow[-NH_3]{\text{histidase}} HOOC\text{-}CH{=}CH\text{-}(\text{imidazole})$$

7.5.4 Deamination of cysteine and methionine

Transmethylase removes the methyl group from methionine to yield homocysteine. Desulfhydrase removes amino and sulfide groups simultaneously from cysteine and homocysteine to produce pyruvate and 2-ketobutyrate, respectively. Desulfhydrase is known in many aerobic and facultative anaerobic bacteria, including *Escherichia coli*, *Proteus vulgaris* and *Bacillus subtilis*:

$$\underset{\text{cysteine}}{CH_2\,SH\text{-}CHNH_2\text{-}COOH} + H_2O \xrightarrow{\text{cysteine desulfhydrase}} \underset{\text{pyruvate}}{CH_3\text{-}CO\text{-}COOH} + NH_3 + H_2S$$

$$\underset{\text{homocysteine}}{CH_2\,SH\text{-}CH_2\text{-}CHNH_2\text{-}COOH} + H_2O \xrightarrow{\text{homocysteine desulfhydrase}} \underset{\text{2-ketobutyrate}}{CH_3\text{-}CH_2\text{-}CO\text{-}COOH} + NH_3 + H_2S$$

7.5.5 Deamination products of amino acids

Deamination of amino acids yields various organic acids:

glycine → glyoxylate
alanine → pyruvate
cysteine → pyruvate
aspartate → oxaloacetate, fumarate
asparagine → oxaloacetate, fumarate
glutamate → 2-ketoglutarate
glutamine → 2-ketoglutarate
threonine → 2-ketobutyrate
methionine → 2-ketobutyrate
serine → pyruvate
histidine → urocanate
valine → 2-ketoisovalerate
leucine → 2-ketoisocaproate
isoleucine → 2-keto-3-methylvalerate.

Pyruvate, oxaloacetate, fumarate and 2-ketoglutarate are intermediates of central metabolism and can be used for both anabolic and catabolic purposes. Glyoxylate is metabolized through the dicarboxylic acid cycle–glycerate pathway (Section 7.3.2, Figure 7.13) or through the 3-hydroxyaspartate pathway (Section 7.3.2, Figure 7.14), depending on the organism. Urocanate from histidine deamination is metabolized through glutamate, as shown in Figure 7.16. A dehydrogenase complex oxidizes 2-ketobutyrate to propionyl-CoA, which is metabolized through the methylcitrate cycle and other routes (Section 7.3.1).

Amino acids with a side chain are oxidized to 2-ketoisovalerate, 2-ketoisocaproate and 2-keto-3-methylvalerate. They are further oxidized to the corresponding acyl-CoA by the 2-keto acid dehydrogenase complex, before acyl-CoA dehydrogenase forms a double bond between two and three carbons. The same enzyme catalyses each of these reactions. Unsaturated fatty acids with

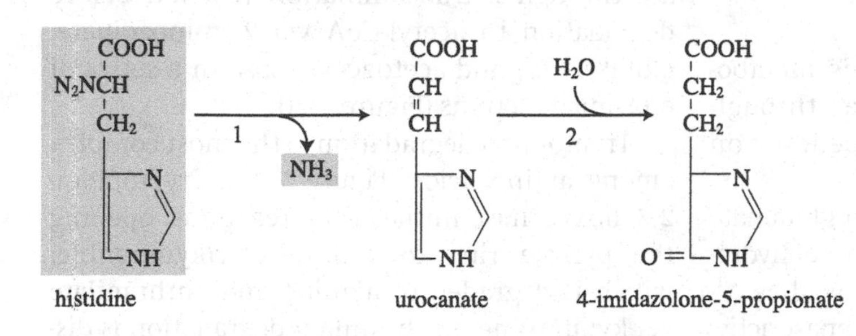

Figure 7.16 Histidine degradation.

(Gottschalk, G. 1986, *Bacterial Metabolism*, 2nd edn., Figure 6.3. Springer, New York)

1, histidase; 2, urocanase; 3, imidazolone propionase; 4, formiminoglutamate hydrolase.

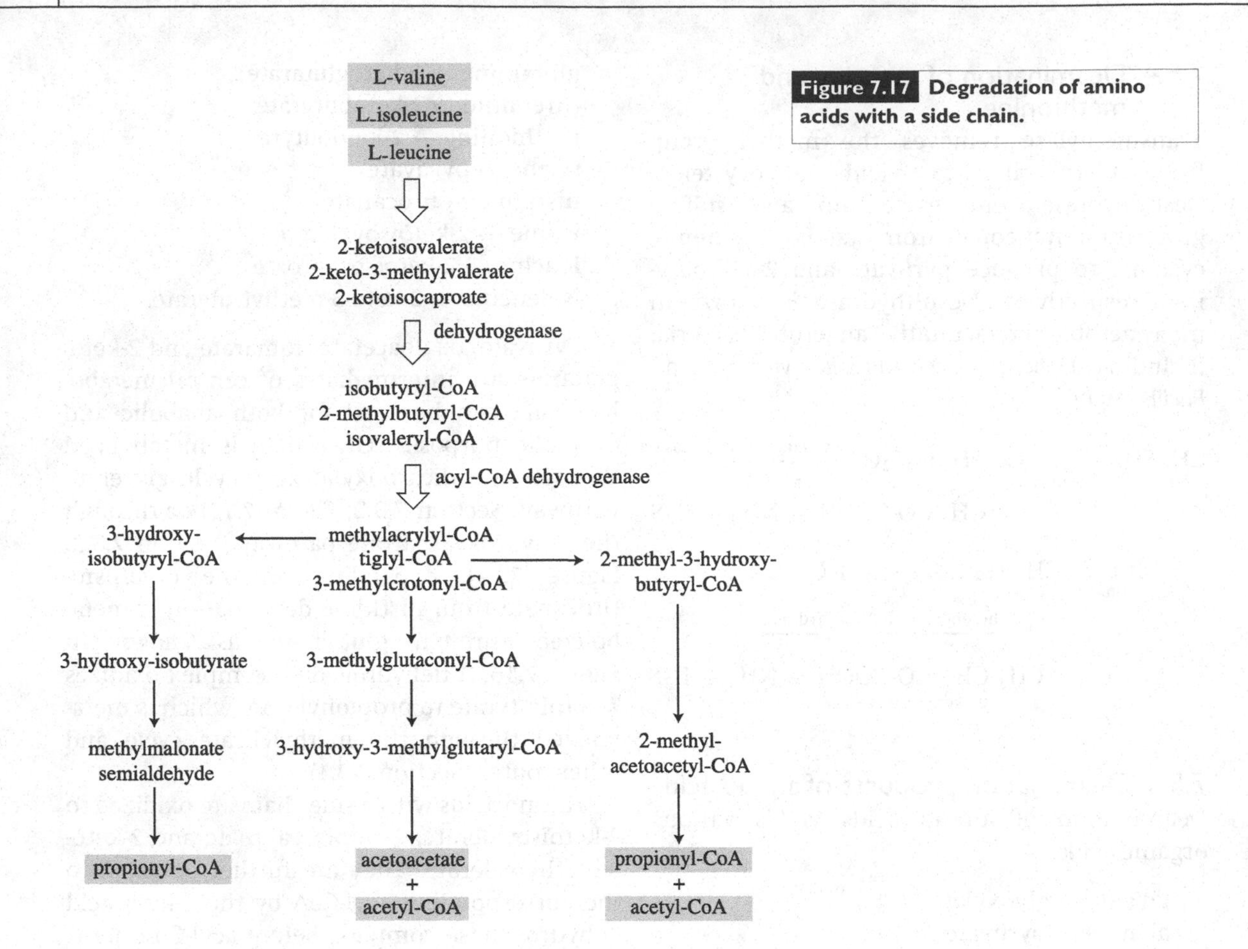

Figure 7.17 Degradation of amino acids with a side chain.

side chains are metabolized by a separate enzyme to propionyl-CoA, acetyl-CoA and acetoacetate (Figure 7.17). Acetoacetate yields two acetyl-CoA through acetoacetyl-CoA.

7.5.6 Other amino acids

As shown in Figure 7.18, arginine is metabolized to glutamate or succinate through a number of different pathways, depending on the organism.

Enteric bacteria convert proline to glutamate through reactions catalysed by proline dehydrogenase, which has proline oxidase as well as Δ^1-pyrroline-5-carboxylate (P5C) dehydrogenase activity. FAD is reduced from the first reaction of proline oxidation, and the same enzyme oxidizes P5C to glutamate, reducing NAD^+ (Figure 7.19). This enzyme binds to the cytoplasmic membrane when proline is available as the electron donor, but in the absence of proline, the *put* (proline dehydrogenase and proline permease) operon is the binding site of the enzyme on the chromosome which inhibits transcription of the operon (Section 12.1.8).

Lysine is converted to 2-keto-6-aminocaproate through a transamination reaction before degradation to acetyl-CoA via 2-aminoadipate, glutaryl-CoA and acetoacetyl-CoA, in a series of enzymic reactions (Figure 7.20).

Tryptophan degradation is the most complex among amino acids (Figure 7.21). Tryptophan-2,3-dioxygenase initiates the reactions, opening the pyrrole ring to yield formylkynurenine, which is degraded to alanine and anthranilate via kynurenine. Anthranilate degradation is discussed in Section 7.8, together with phenylalanine, tyrosine and aromatic hydrocarbons.

Amino acids are the building blocks of proteins and their synthesis consumes energy. Their degradation is tightly regulated through catabolite repression (Section 12.1.3), nitrogen regulation (Section 12.2.2) and other mechanisms.

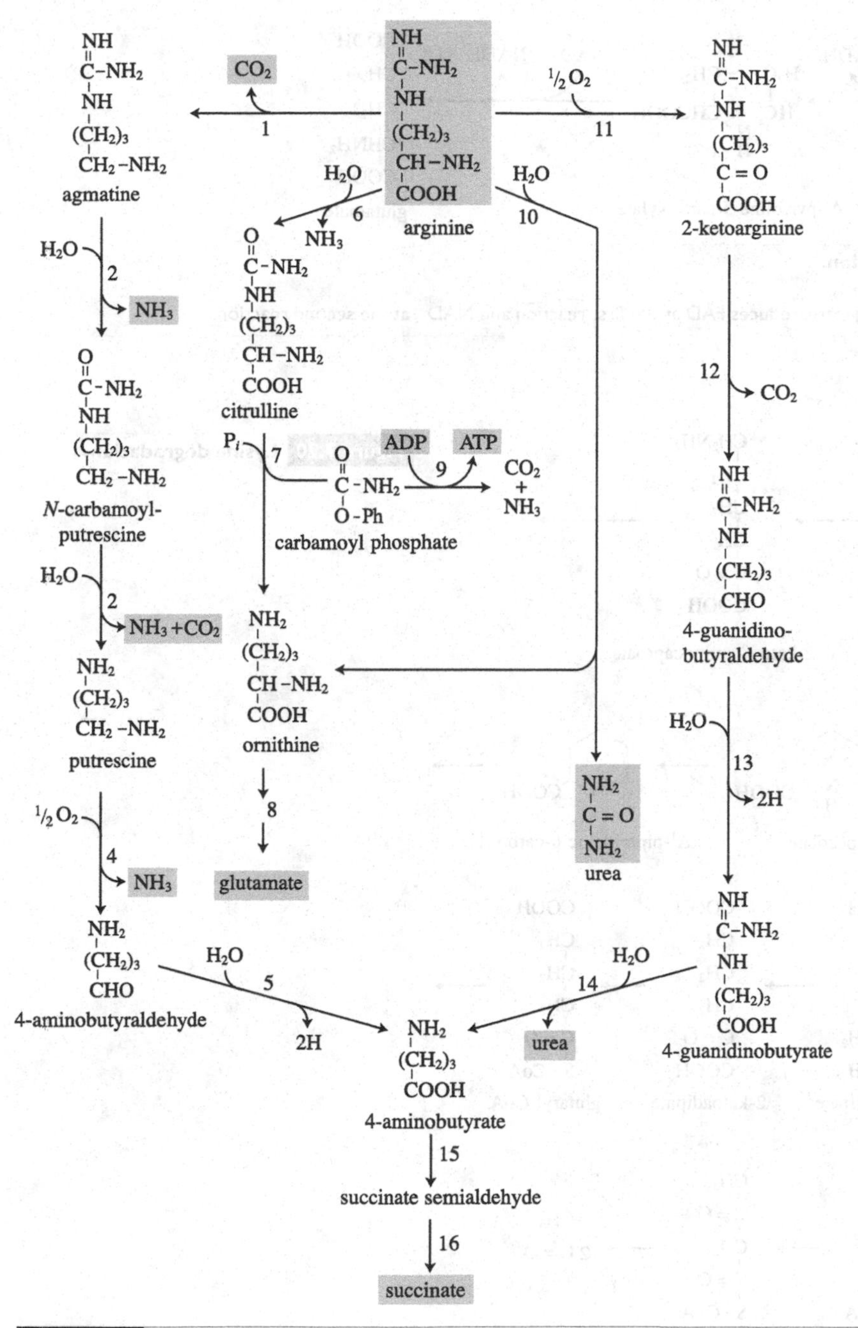

Figure 7.18 Arginine degradation.

(Gottschalk, G. 1986, *Bacterial Metabolism*, 2nd edn., Figure 6.4. Springer, New York)

Arginine is metabolized through various pathways, depending upon the organism. Enteric bacteria, including *Escherichia coli*, start the reaction with arginine decarboxylase (1), and arginine oxidase (11) catalyses the first reaction in *Pseudomonas putida*. Anaerobic bacteria such as *Clostridium perfringens* and lactic-acid bacteria deiminate arginine (6), and *Bacillus subtilis* removes urea from arginine by the action of arginase (10).

1, arginine decarboxylase; 2, agmatine deiminase; 3, *N*-carbamoylputrescine hydrolase; 4, putrescine oxidase; 5, aminobutyraldehyde dehydrogenase; 6, arginine deiminase; 7, ornithine carbamoyltransferase; 8, reverse reaction of arginine biosynthesis (enzymes 6, 7, 8, 9 in Figure 6.17); 9, carbamate kinase; 10, arginase; 11, arginine oxidase; 12, 2-ketoarginine decarboxylase; 13, 4-guanidinobutyraldehyde oxidoreductase; 14, guanidinobutyrase; 15, transaminase; 16, succinate semialdehyde dehydrogenase.

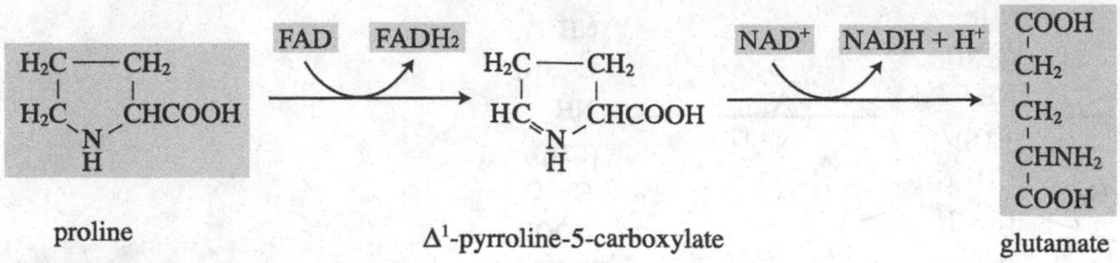

FAD → $FADH_2$; NAD^+ → NADH + H^+

proline → Δ^1-pyrroline-5-carboxylate → glutamate

Figure 7.19 Proline degradation.

A single enzyme, proline dehydrogenase, reduces FAD at the first reaction and NAD^+ at the second reaction.

Figure 7.20 Lysine degradation.

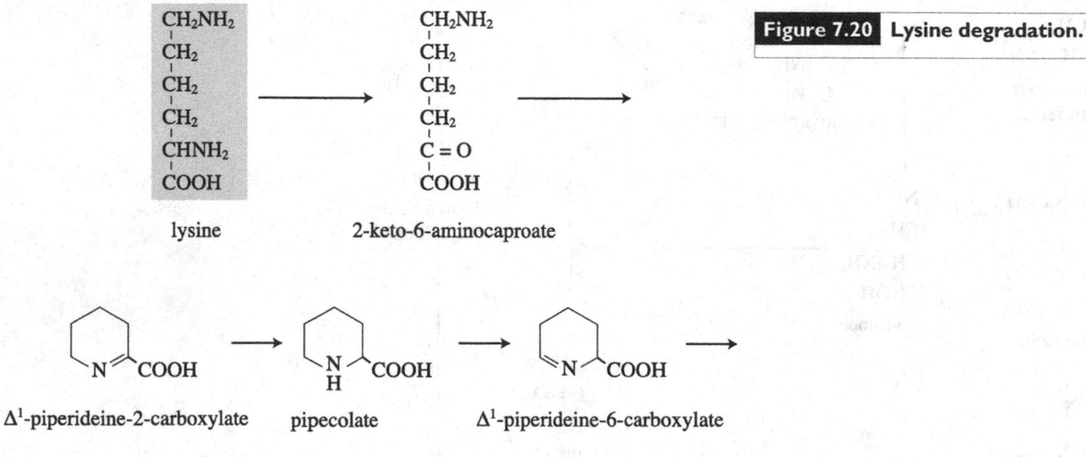

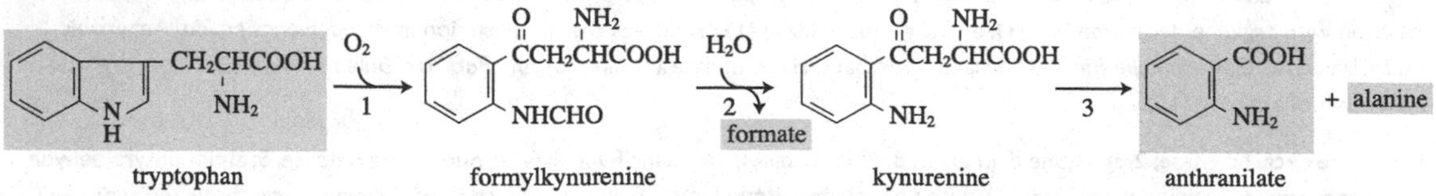

2-aminoadipate-6-semialdehyde → 2-aminoadipate → 2-ketoadipate → glutaryl-CoA →

glutaconyl-CoA → crotonyl-CoA → acetoacetyl-CoA → 2 acetyl-CoA

tryptophan —(O_2, 1)→ formylkynurenine —(H_2O, 2, formate)→ kynurenine —(3)→ anthranilate + alanine

Figure 7.21 Tryptophan degradation.

1, tryptophan-2,3-dioxygenase; 2, formylkynureninase; 3, kynureninase.

7.6 Degradation of nucleic acid bases

Nucleases hydrolyse RNA and DNA to ribonucleoside monophosphate and deoxyribonucleoside monophosphate. They are further hydrolysed to bases and ribose (deoxyribose) by nucleotidase and nucleosidase:

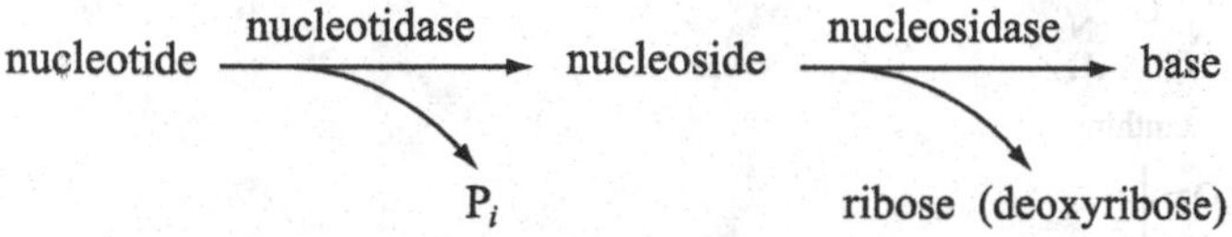

nucleotide —(nucleotidase)→ nucleoside (+ P_i) —(nucleosidase)→ base (+ ribose (deoxyribose))

Guanine is deaminated to xanthine in a one-step reaction by guanine deaminase, while adenine requires two reactions to be oxidized to xanthine via hypoxanthine, catalysed by adenine deaminase and xanthine dehydrogenase (or xanthine oxidase). Hypoxanthine and xanthine are oxidized either by xanthine dehydrogenase or by xanthine oxidase, depending on the organism. Xanthine dehydrogenase or oxidase not only oxidizes hypoxanthine to xanthine but also catalyses the next reaction to urate. Xanthine dehydrogenase uses $NAD(P)^+$ or ferredoxin as its electron acceptor, while xanthine oxidase reduces molecular oxygen, generating superoxide.

Urate is degraded to glyoxylate and urea through a series of reactions (Figure 7.22). Glyoxylate is metabolized through the dicarboxylic acid cycle–glycerate pathway (Figure 7.13) or the hydroxyaspartate pathway (Figure 7.14), as described earlier (Section 7.3.2). Some organisms use part of the pathway shown in Figure 7.22 to use purine bases as their carbon or nitrogen sources. Pyrimidine bases are degraded either through barbiturate or through reactive peroxy ureidoacrylate, depending on the organism, as shown in Figure 7.23.

7.7 Oxidation of aliphatic hydrocarbons

Many prokaryotic and eukaryotic microbes can use aliphatic hydrocarbons, especially those that are liquid at ambient temperature (Table 7.1). Methane is metabolized through a specialized pathway and is described in a separate section (Section 7.10).

Hydrocarbons are water insoluble and differ from water-soluble substrates in terms of transport. Since microbes cannot thrive in a pure oil phase, microbes use hydrocarbons as their carbon and energy source at the water–oil interface. Hydrophobic glycolipids are found on the cell surfaces of bacteria and fungi that use hydrocarbons. This glycolipid solubilizes the hydrocarbon before it is transported into the cell. Many bacteria, including *Acinetobacter calcoaceticus*, produce a surfactant to improve hydrocarbon transport as an emulsion.

Hydrocarbon monooxygenase oxidizes a hydrocarbon to a primary alcohol at the cytoplasmic membrane (Figure 7.24). This enzyme can oxidize two different substrates using one molecule of oxygen. This kind of enzyme is

Table 7.1 Some examples of hydrocarbon-utilizing microbes.

Bacteria
Acinetobacter calcoaceticus
Arthrobacter paraffineus
Arthrobacter simplex
Corynebacterium glutamicum
Mycobacterium smegmatis
Nocardia petroleophila
Pseudomonas aeruginosa
Pseudomonas fluorescens
Fungi
Candida lipolytica
Torulopsis colliculosa
Cephalosporium roseum
Hormoconis (Cladosporium) resinae

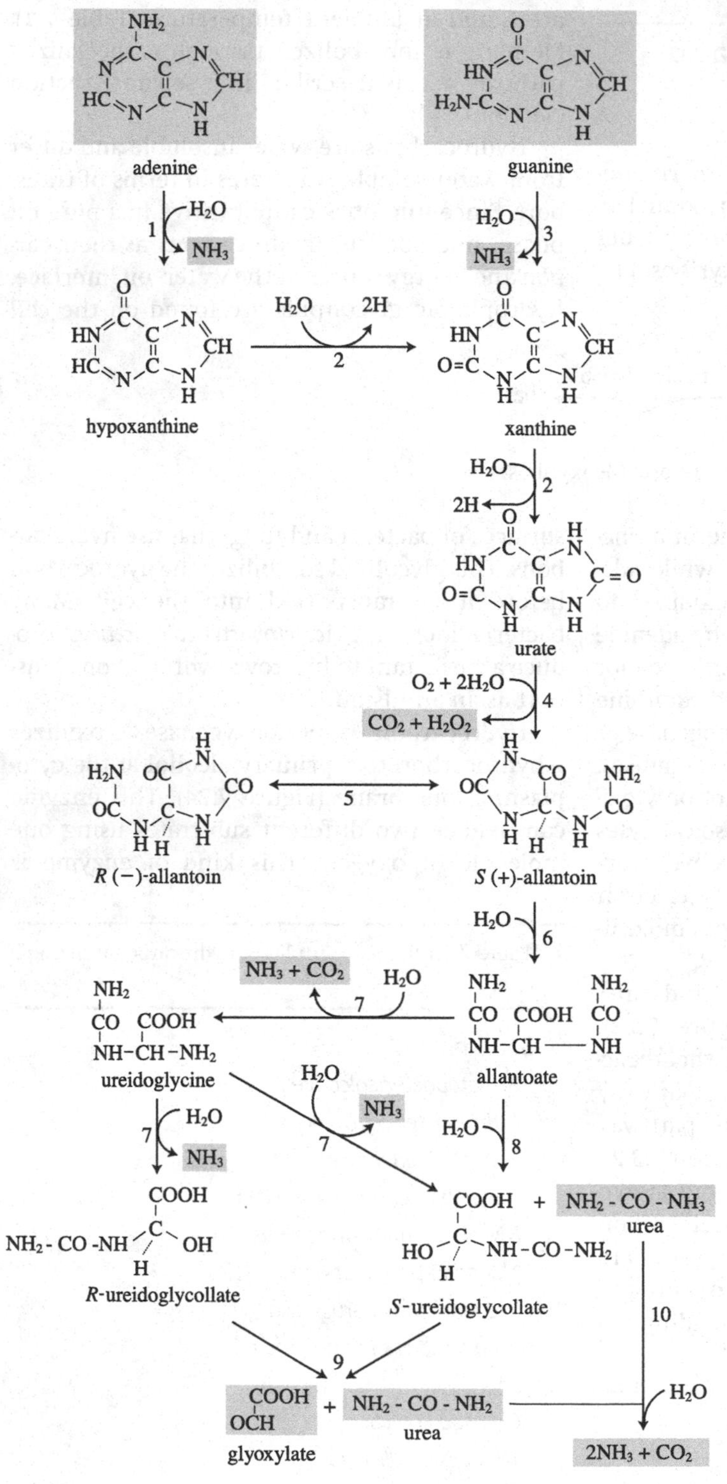

Figure 7.22 Degradation of purine bases.

1, adenine deaminase; 2, xanthine dehydrogenase or xanthine oxidase; 3, guanine deaminase; 4, uricase; 5, allantoin racemase; 6, S(+)-allantoinase; 7, allantoate amidohydrolase; 8, allantoicase; 9, ureidoglycolase; 10, urease.

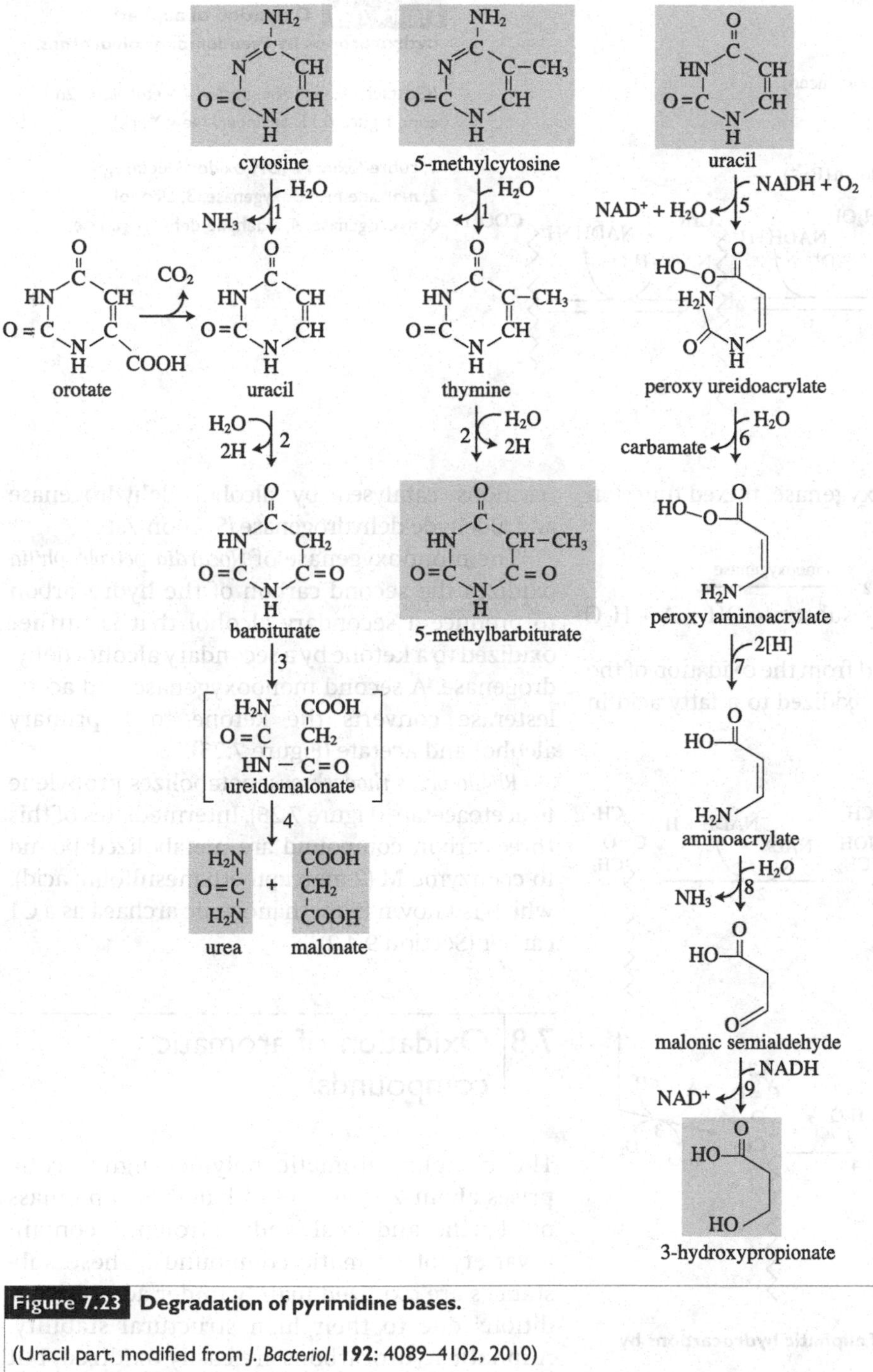

Figure 7.23 Degradation of pyrimidine bases.

(Uracil part, modified from *J. Bacteriol.* **192**: 4089–4102, 2010)

Uracil is metabolized either through barbiturate or through reactive peroxy ureidoacrylate, depending on the organism.

1, cytosine deaminase; 2, uracil dehydrogenase; 3, barbiturase; 4, ureidomalonase; 5, pyrimidine oxygenase; 6, ureidoacrylate peracid hydrolase; 7, aminoacrylate peracid reductase; 8, aminoacrylate hydrolase; 9, 3-hydroxypropionate dehydrogenase.

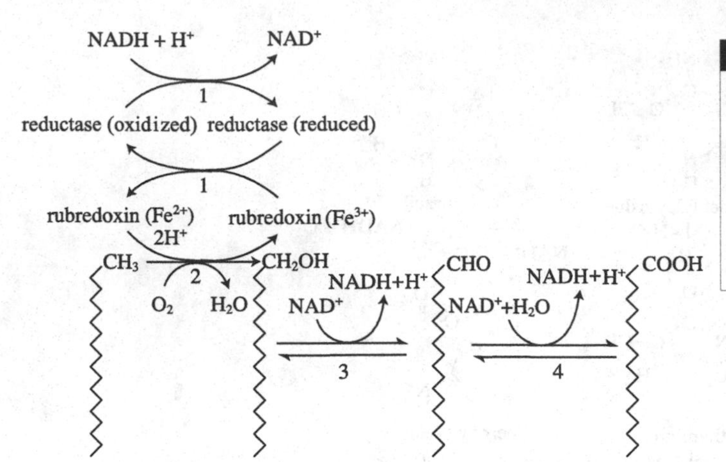

Figure 7.24 Oxidation of aliphatic hydrocarbons by *Pseudomonas oleovorans.*

(Gottschalk, G. 1986, *Bacterial Metabolism*, 2nd edn., Figure 6.11. Springer, New York)

1, rubredoxin: NADH oxidoreductase; 2, *n*-alkane monooxygenase; 3, alcohol dehydrogenase; 4, aldehyde dehydrogenase.

referred to as a monooxygenase, mixed function oxidase or hydroxylase:

$$\text{substrate-H} + AH_2 + O_2 \xrightarrow{\text{monooxygenase}} \text{substrate-OH} + A + H_2O$$

The alcohol produced from the oxidation of the hydrocarbon is further oxidized to a fatty acid in reactions catalysed by alcohol dehydrogenase and aldehyde dehydrogenase (Section 7.4).

The monooxygenase of *Nocardia petroleophilia* oxidizes the second carbon of the hydrocarbon to produce a secondary alcohol that is further oxidized to a ketone by a secondary alcohol dehydrogenase. A second monooxygenase and acetylesterase converts the ketone to a primary alcohol and acetate (Figure 7.25).

Rhodococcus rhodochrous metabolizes propylene to acetoacetate (Figure 7.26). Intermediates of this three-carbon compound are metabolized bound to coenzyme M (2-mercaptoethanesulfonic acid), which is known in methanogenic archaea as a C1 carrier (Section 9.4.2).

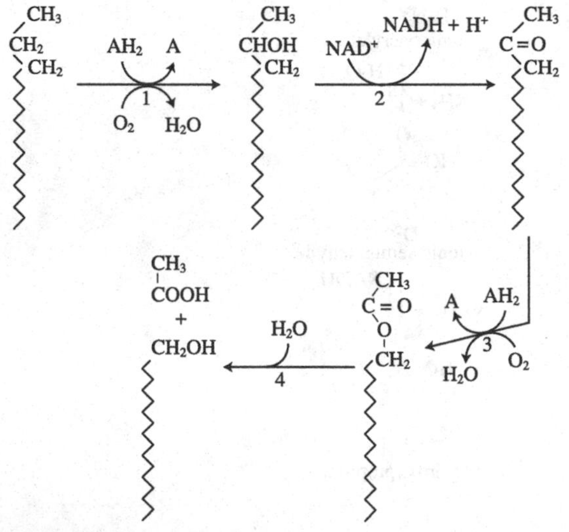

Figure 7.25 Oxidation of aliphatic hydrocarbons by *Nocardia petroleophila.*

(Gottschalk, G. 1986, *Bacterial Metabolism*, 2nd edn., Figure 6.12. Springer, New York)

1, monooxygenase; 2, secondary alcohol dehydrogenase; 3, monooxygenase; 4, acetylesterase.

7.8 Oxidation of aromatic compounds

The complex aromatic polymer lignin comprises about 25 per cent of land-based biomass on Earth, and coal and petroleum contain a variety of aromatic compounds. These substances are oxidized mainly under aerobic conditions due to their high structural stability. Aliphatic hydrocarbons are easily oxidized, but the aromatic portion of petroleum is persistent in natural ecosystems. Aromatic hydrocarbon degradation is best known in *Pseudomonas* spp., e.g. *Pseudomonas acidovorans* and *Pseudomonas putida*.

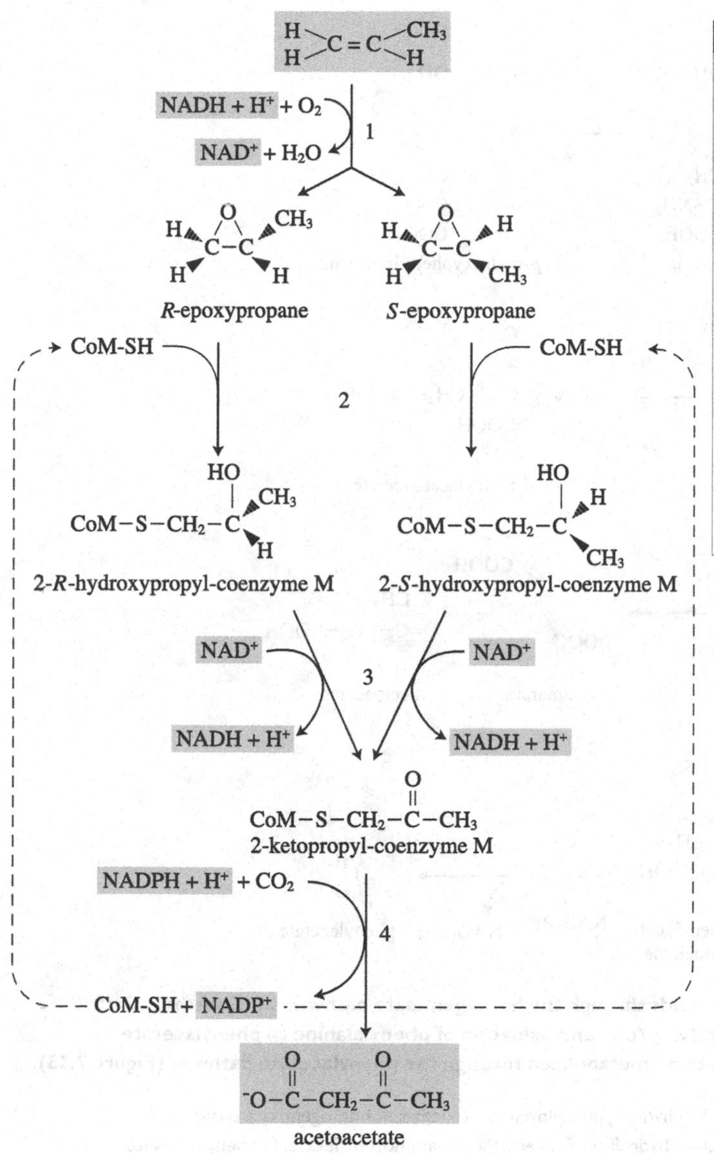

Figure 7.26 Propylene oxidation by *Rhodococcus rhodochrous*.

(Modified from *J. Bacteriol.* **182**: 2629–2634, 2000)

This is a rare example of bacterial metabolism involving coenzyme M, a common coenzyme in methanogenic archaea. Other archaeal coenzymes found in bacteria are F_{420} in tetracycline-producing *Streptomyces* species and *Mycobacterium smegmatis* (Section 4.3.4), and tetrahydromethanopterin in methylotrophs (Section 7.10.2).

1, alkane monooxygenase; 2, epoxyalkane: coenzyme M transferase; 3, hydroxypropyl-CoM dehydrogenase; 4, NADPH:2-ketopropyl-CoM dehydrogenase.

7.8.1 Oxidation of aromatic amino acids

Phenylalanine is utilized through either the homogentisate pathway (Figure 7.27a) after being oxidized to tyrosine or the phenylacetate pathway under aerobic conditions (Figure 7.33), depending on the organism.

In the homogentisate pathway, phenylalanine monooxygenase oxidizes phenylalanine to tyrosine that is deaminated to *p*-hydroxyphenylpyruvate. Dioxygenases are involved in the following reactions (Figure 7.27a). Enzymes that incorporate both atoms of molecular oxygen into one substrate are referred to as dioxygenases or oxidases. Homogentisate oxidase opens up the benzene ring, finally producing fumarate and acetoacetate. Phenylalanine is converted to phenylacetate (Figure 7.27b) that is metabolized through the phenylacetate pathway (Figure 7.33).

7.8.2 Benzene ring cleavage

The metabolism of aromatic compounds can be divided into two steps. In the first step monooxygenases incorporate hydroxyl groups into the benzene ring. Through this step, aromatic

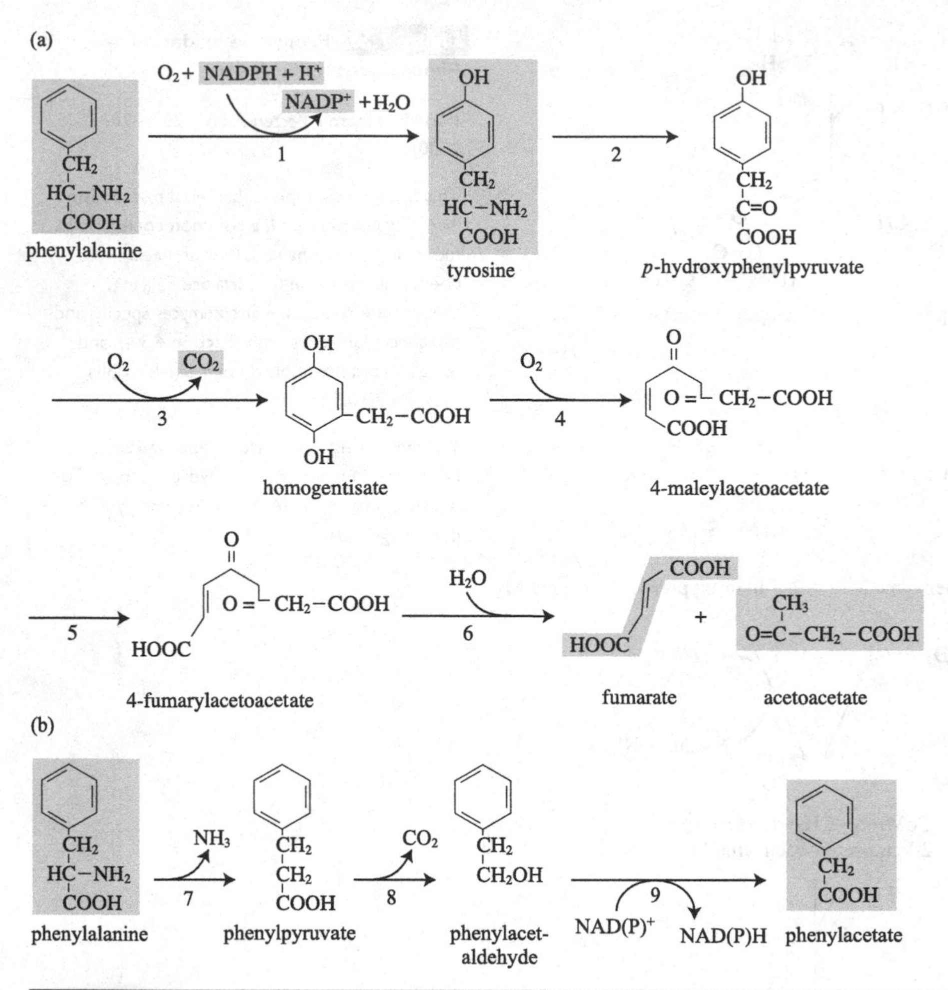

Figure 7.27 **Degradation of aromatic amino acids through the homogentisate pathway** (a) (Gottschalk, G. 1986, *Bacterial Metabolism*, 2nd edn., Figure 6.13. Springer, New York) **and oxidation of phenylalanine to phenylacetate** (b) (modified from J. *Bacteriol.* 196: 483–492, 2014) **to be metabolized through the phenylacetate pathway (Figure 7.33).**

1, phenylalanine monooxygenase; 2, transaminase; 3, *p*-hydroxyphenylpyruvate oxidase; 4, homogentisate oxidase; 5, maleylacetoactate isomerase; 6, fumarylacetoacetate hydrolase; 7, phenylalanine aminotransferase; 8, phenylpyruvate decarboxylase; 9, phenylacetaldehyde dehydrogenase.

compounds are converted to one of four intermediates: protocatechuate, catechol, gentisate and phenylacetate. Aromatics with a hydroxyl group are mainly converted to protocatechuate (Figure 7.28), and catechol is derived from aromatic hydrocarbons, aromatic compounds with amino groups and lignin monomers (Figure 7.29). Some bacteria generate gentisate from naphthalene, 3-hydroxybenzoate, phenol derivatives, 3,6-dichloro-2-methoxybenzoate and other substances. Styrene, ethylbenzene and phenylalanine are converted to phenylacetate in some bacteria. The benzene rings of these intermediates are opened up by oxygenases in four different pathways, depending on the organisms. These are *ortho* and *meta* cleavage, and the gentisate and phenylacetate pathways.

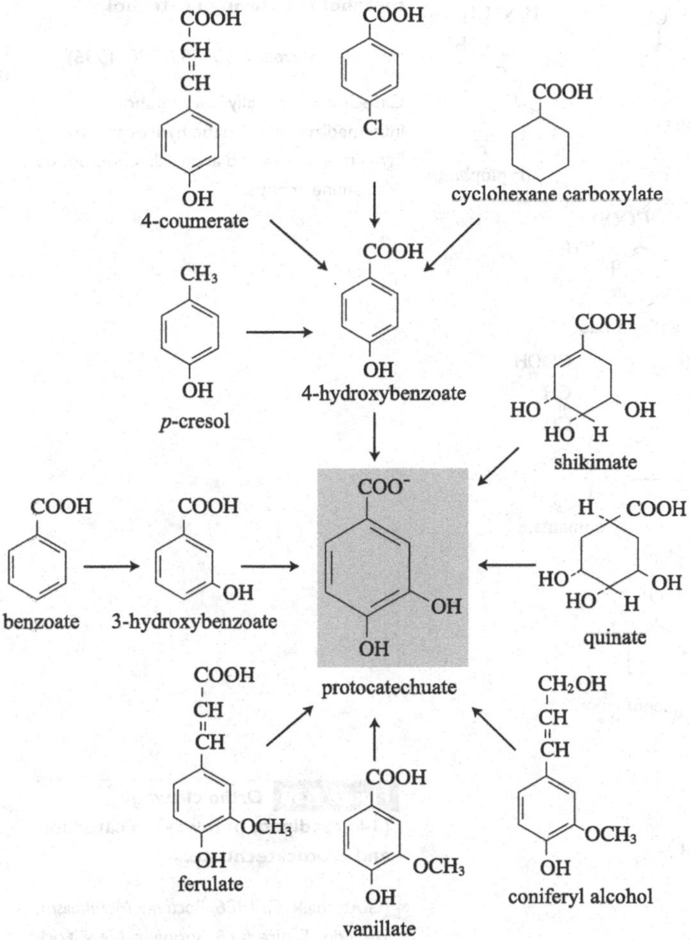

Figure 7.28 Aromatic compounds metabolized through protocatechuate.

(*Ann. Rev. Microbiol.* **50**: 553–590, 1996)

Hydroxylated aromatic hydrocarbons are generally metabolized through protocatechuate.

Ring fission is catalysed by dioxygenases and termed *ortho* cleavage (Figure 7.30) when it occurs between the hydroxyl groups and *meta* cleavage (Figure 7.31) when it occurs adjacent to one of the hydroxyls. Another ring cleavage pathway, the gentisate pathway, is responsible for the oxidation of aromatic compounds with hydroxyl groups at *para* positions (Figure 7.32). After the ring fission reactions, the products are metabolized to succinate, acetyl-CoA, pyruvate and acetaldehyde. Bacteria with the *ortho* cleavage pathway have the genes in their chromosome while the genes of *meta* cleavage are found on plasmids. These plasmids are referred to as degradative plasmids, to differentiate them from those with other functions such as antibiotic resistance.

Gentisate and its derivatives are degraded to pyruvate and fumarate through the gentisate pathway initiated by gentisate dioxygenase (Figure 7.32). Depending on the organism, maleylpyruvate is either isomerized to fumarylpyruvate or degraded directly.

Various bacteria metabolize phenylalanine, ethylbenzene, styrene and tropate through yet another pathway, the phenylacetate pathway (Figure 7.33). Phenylacetate is activated to phenylacetyl-CoA before the benzene ring is cleaved by a monooxygenase to produce 2 acetyl-CoA and 1 succinyl-CoA. This pathway is involved in polychlorinated biphenyl degradation in *Burkholderia xenovorans*.

7.8.3 Oxygenase and aromatic compound oxidation

Since hydrocarbon degradation is initiated by an oxygenase in aerobic organisms, it was believed that molecular oxygen is essential for the degradation of aromatic compounds. However, many

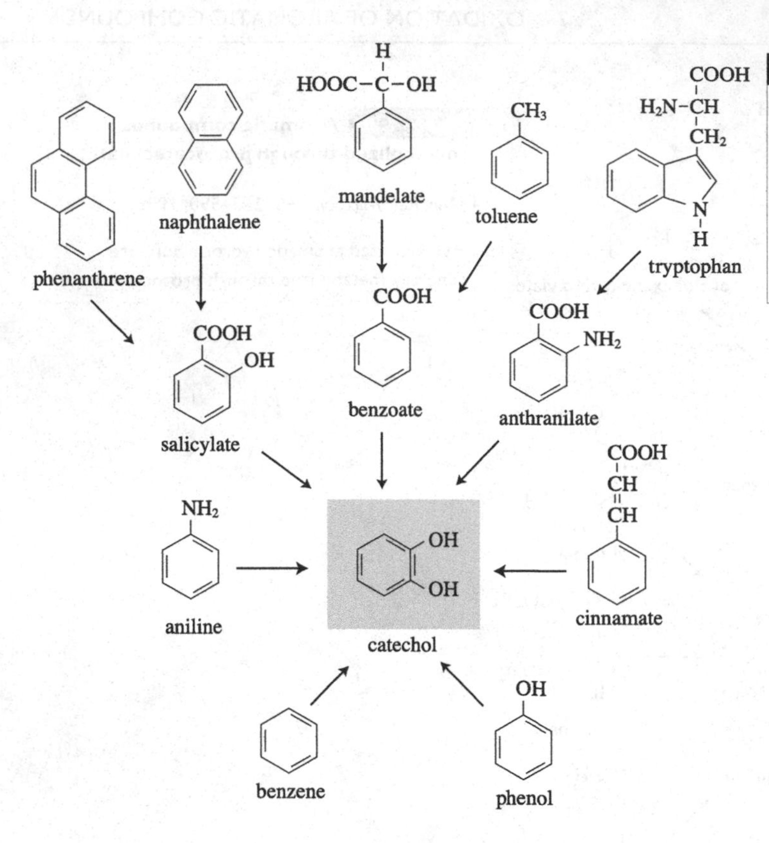

Figure 7.29 Aromatic compounds metabolized through catechol.

(*Ann. Rev. Microbiol.* **50**: 553–590, 1996)

Catechol is generally the metabolic intermediate of aromatic hydrocarbons, lignin monomers and aromatic compounds with amine groups.

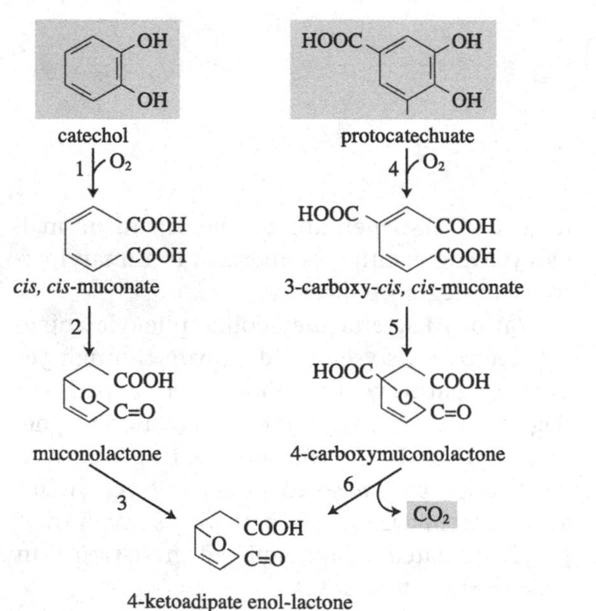

Figure 7.30 ***Ortho*** **cleavage (3-ketoadipate pathway) of catechol and protocatechuate.**

(Gottschalk, G. 1986, *Bacterial Metabolism*, 2nd edn., Figure 6.16. Springer, New York)

1, catechol 1,2-dioxygenase; 2, muconate-lactonizing enzyme; 3, muconolactone isomerase; 4, protocatechuate 3,4-dioxygenase; 5, 3-carboxymuconate-lactonizing enzyme;
6, 4-carboxymuconolactone decarboxylase;
7, 4-ketoadipate enol-lactone hydrolase;
8, 3-ketoadipate succinyl-CoA transferase;
9, 3-ketoadipate-CoA thiolase.

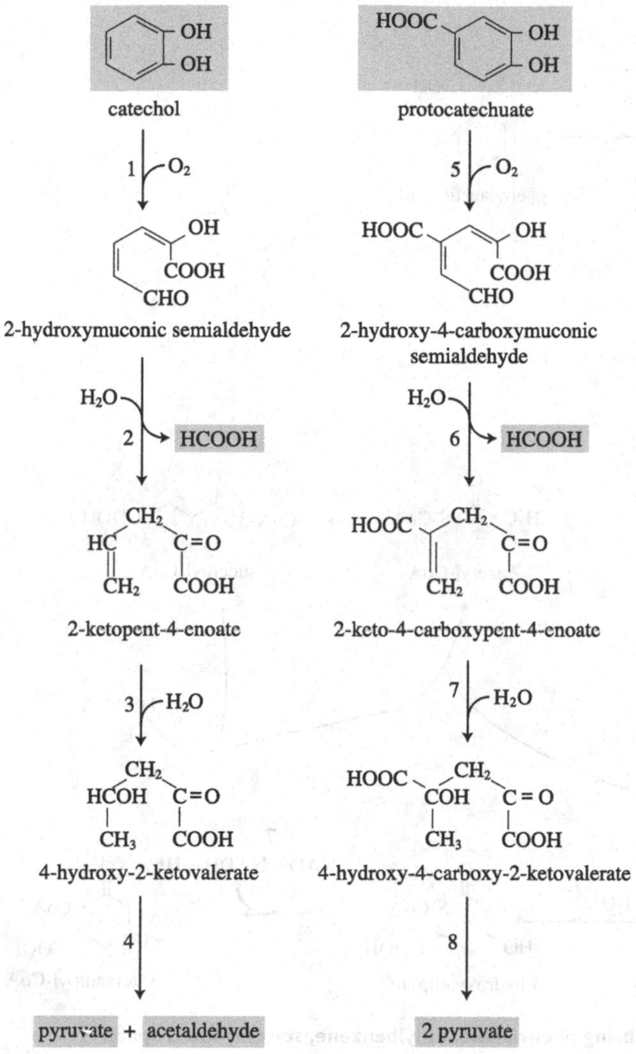

catechol → (1, O_2) → 2-hydroxymuconic semialdehyde → (H_2O; 2; HCOOH) → 2-ketopent-4-enoate → (3, H_2O) → 4-hydroxy-2-ketovalerate → (4) → pyruvate + acetaldehyde

protocatechuate → (5, O_2) → 2-hydroxy-4-carboxymuconic semialdehyde → (H_2O; 6; HCOOH) → 2-keto-4-carboxypent-4-enoate → (7, H_2O) → 4-hydroxy-4-carboxy-2-ketovalerate → (8) → 2 pyruvate

Figure 7.31 ***Meta*** **cleavage of catechol and protocatechuate.**

(Gottschalk, G. 1986, *Bacterial Metabolism*, 2nd edn., Figure 6.17. Springer, New York)

1, catechol 2,3-dioxygenase; 2, 2-hydroxymuconic semialdehyde hydrolase; 3, 2-ketopent-4-enoic acid hydrolase; 4, 4-hydroxy-2-ketovalerate aldolase; 5, protocatechuate 4,5-dioxygenase; 6, 2-hydroxy-4-carboxymuconic semialdehyde hydrolase; 7, 2-keto-4-carboxy pent-4-enoic acid hydrolase; 8, 4-hydroxy-4-carboxy-2-ketovalerate aldolase.

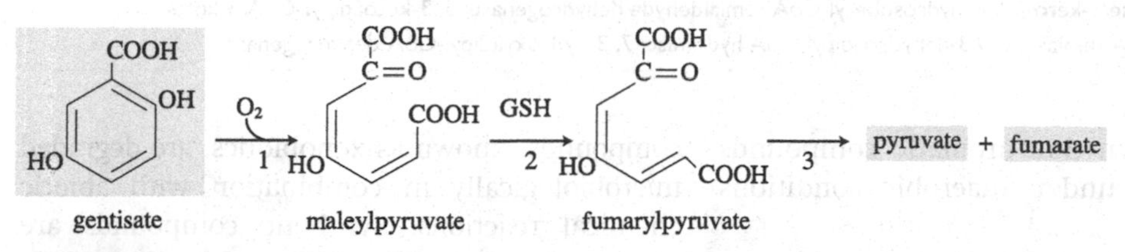

Figure 7.32 **The gentisate pathway.**

(*J. Bacteriol.* **183**: 700–708, 2001)

Naphthalene and other aromatic hydrocarbons are degraded through gentisate. Maleylpyruvate is converted to pyruvate and fumarate either directly or through fumarylpyruvate.

1, gentisate dioxygenase; 2, maleylpyruvate isomerase; 3, fumarylpyruvate hydrolase.

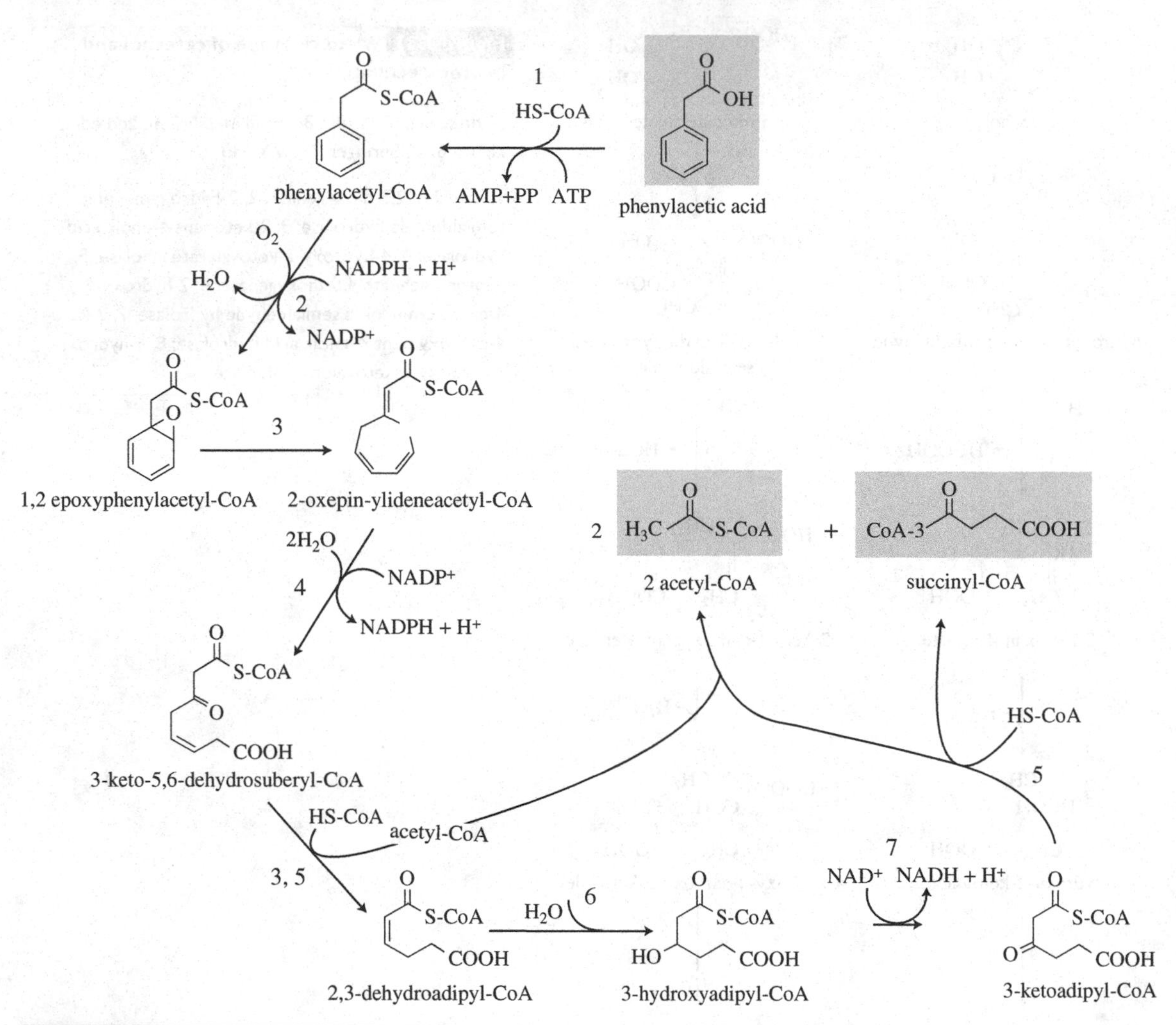

Figure 7.33 Phenylacetate pathway in bacteria utilizing phenylanine, ethylbenzene, styrene and tropate.

(Modified from *Proc. Nat. Acad. Sci. USA* **107**: 14390–14395, 2010)

1, phenylacetate-CoA ligase; 2, ring 1,2-phenylacetyl-CoA epoxidase; 3, ring 1,2-epoxyphenylacetyl-CoA isomerase; 4, oxepin-CoA hydrolase/3-keto-5,6-dehydrosuberyl-CoA semialdehyde dehydrogenase; 5, 3-ketoadipyl-CoA/3-keto-5,6-dehydrosuberyl-CoA thiolase; 6, 2,3-dehydroadipyl-CoA hydratase; 7, 3-hydroxyadipyl-CoA dehydrogenase.

studies have shown that aromatic compounds can be degraded under anaerobic conditions (Section 9.9).

7.9 Utilization of natural and anthropogenic xenobiotics

Various natural and synthetic compounds are persistent in the natural environment. These compounds, known as xenobiotics, are degraded microbiologically in combination with abiotic chemical reactions. Synthetic compounds are often resistant to microbial attack, since their exposure in nature is too short for microorganisms to be able to design new enzyme structures capable of degrading them. Synthetic compounds are metabolized by enzymes involved in the metabolism of structurally related compounds of natural origin. Examples are organophosphonates, such as the widely used herbicide glyphosate, and plastics such as polyethylene. Glyphosate is dephosphonated by the C–P lyase that cleaves the C–P bond of natural phosphonates. Plastic polymers are hydrolysed by lipases and esterases after they weather due to exposure to sunlight, oxygen and physical stress. These enzymes are either of broad specificity or mutated. Species belonging to the genus *Sphingopyxis* oxidize polyethylene glycol to glyoxylate by the actions of dehydrogenases. Organic nitro-compounds and chlorinated organic compounds are metabolized similarly under aerobic conditions, or after they are reduced to corresponding amines or dechlorinated as electron acceptors under anaerobic conditions (Section 9.7).

7.10 Utilization of methane and methanol

Methane, methanol and methylamines are naturally available in large quantities. Some bacteria and yeasts are known to use them as their sole carbon and energy sources. These organisms are referred to as methylotrophs and metabolize the C1 compounds through pathways not known in multicarbon compound metabolism. The term 'methylotroph' is used to refer to all C1-utilizing organisms in a broad sense, and also used to describe C1-utilizing organisms that cannot use methane in a narrow sense. Most of them are strict aerobes, but some can respire on nitrate under anaerobic conditions. Anaerobic methane oxidation is discussed later (Section 9.9.2)

7.10.1 Methanotrophy and methylotrophy

Methylotrophs are divided into methanotrophs and methylotrophs according to their ability to use methane. Methanotrophs use C1 compounds but do not use multicarbon compounds, with a few exceptions including bacteria of the genus *Methylocella*. One strain of *Methylocella silvestris* utilizes propane as well as methane. Methylotrophs do not use methane. Based on their carbon assimilation metabolism, methylotrophs are divided into heterotrophic methylotrophs and autotrophic methylotrophs. Autotrophic methylotrophs assimilate carbon dioxide through the Calvin cycle, while heterotrophic methylotrophs assimilate formaldehyde through the ribulose monophosphate pathway or the serine–isocitrate lyase pathway (Table 7.2). Methanotrophs are not known in eukaryotes, and methylotrophic yeasts assimilate formaldehyde through the xylulose monophosphate pathway.

Table 7.2 Classification of methylotrophs according to their characteristics.

Physiological characteristics	Carbon assimilation pathway	Organism	C1 compounds assimilated
Heterotrophic methylotrophy	serine–isocitrate lyase pathway	*Methylobacterium extorquens* AM1	methanol, MMA, formate
		Pseudomonas MA	MMA
		Methylbacterium organophilum	methane, methanol
	ribulose monophosphate pathway	*Arthrobacter* P1	MMA, DMA, TMA
		Bacillus PM6	MMA, DMA, TMA, TMO, tetramethylammonium
Autotrophic methylotrophy	Calvin cycle	Group 1. Phototrophic *Rhodopseudomonas* spp.	methanol, CO, formate

Table 7.2 (cont.)

Physiological characteristics	Carbon assimilation pathway	Organism	C1 compounds assimilated
		Group 2. Chemoautotrophic	methanol, MMA, formate
		Thiobacillus A2, *Paracoccus denitrificans, Pseudomonas carboxydovorans*	methanol, MMA, formate, CO
		Group 3. *Pseudomonas oxalaticus*	formate
Obligate methylotrophy	serine–isocitrate lyase pathway	*Methylobacterium* spp.	methane
		Methylocystis spp.	methane
		Methylomonas methanooxidans	methane, methanol
		Methylosinus trichosporium	methane, methanol
	ribulose monophosphate pathway	*Methylomonas methanica*	methane, methanol
		Methylophilus methylotrophus	methanol, MMA, DMA
		Methanococcus capsulatus	TMA

MMA, monomethylamine; DMA, dimethylamine; TMA, trimethylamine; TMO, trimethyl-*N*-oxide; CO, carbon monoxide.

Table 7.3 Characteristics of methanotrophs.

Methanotroph	Morphology	Flagella	Resting cell	Intracellular membrane structure	Carbon assimilation pathway	G + C content (%)
Methylomonas	rod	polar	cyst-like body	I	RMP pathway	50–54
Methylobacter	rod	polar	thick-walled cyst	II	RMP pathway	50–54
Methylocella	rod	none	none	none	SIL pathway	60
Methyloacida	rod	none	none	none	?	?
Methylococcus	coccus	none	cyst-like body	I	RMP pathway	62
Methylosinus	rod, vibrioid	polar tuft	exospore	II	SIL pathway	62–66
Methylocystis	vibrioid	none	PHB-rich cyst	II	SIL pathway	?
Methylobacterium	rod	none	none	II	SIL pathway	58–66

I, multilayer membrane structure throughout the cell; II, double-layer membrane structure under the cell surface; RMP pathway, ribulose monophosphate pathway; SIL pathway, serine–isocitrate lyase pathway; PHB, polyhydroxybutyrate.

7.10.2 Methanotrophy

7.10.2.1 Characteristics of methanotrophs

Methanotrophs use C1 compounds as their carbon and energy source, and are unable to use multicarbon compounds, except bacteria of the genus *Methylocella*. For this reason, they are referred to as obligate methylotrophs. In addition to their use of C1 compounds, they have some other characteristics. All are Gram-negative and have extensive intracellular membrane structures similar to nitrifiers. This membrane structure can be used in the classification of methanotrophs (Table 7.3). Spore and cyst forms of resting cells are known in all obligate methylotrophs except species of *Methylobacterium*.

7.10.2.2 Dissimilation of methane by methanotrophs

Methane monooxygenase oxidizes methane to methanol using NADH as the cosubstrate (Figure 7.34). Electrons from methanol oxidation are channelled to the electron transport system for ATP synthesis. Formaldehyde or carbon dioxide is assimilated for biosynthesis of cell materials.

Methanol dehydrogenase reduces pyrroloquinoline quinone (PQQ, Figure 7.35) coupled to the oxidation of methanol to formaldehyde. Two different methanol dehydrogenases are known, one containing calcium and the other containing rare earth elements (REEs) such as lanthanum (La), cerium (Ce), praseodymium (Pr) and neodymium (Nd) as the essential cofactor (Section 2.1). In addition to methanol dehydrogenase, PQQ with a redox potential of +0.12 V serves as the coenzyme of glucose dehydrogenase in *Pseudomonas putida* (Section 4.4.3) and alcohol dehydrogenase in acetic acid bacteria (Section 7.11.1).

Formaldehyde is oxidized to carbon dioxide either in free or in bound form. Formaldehyde dehydrogenase and formate dehydrogenase oxidize formaldehyde in the free form to CO_2 via formate in Gram-positive methylotrophs such as *Amycolatopsis methanolica*, and in the autotrophic methylotroph *Paracoccus denitrificans*:

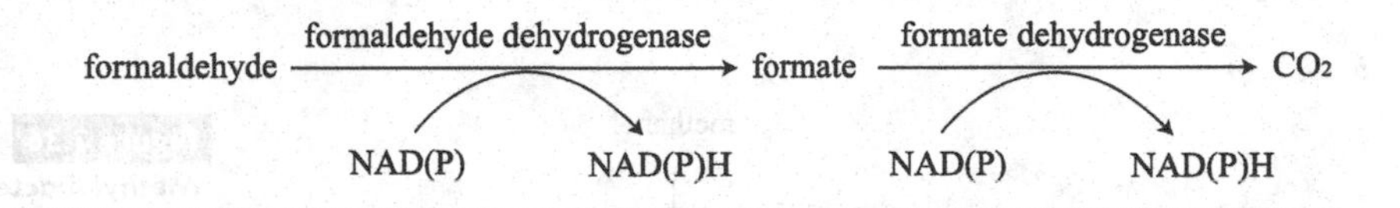

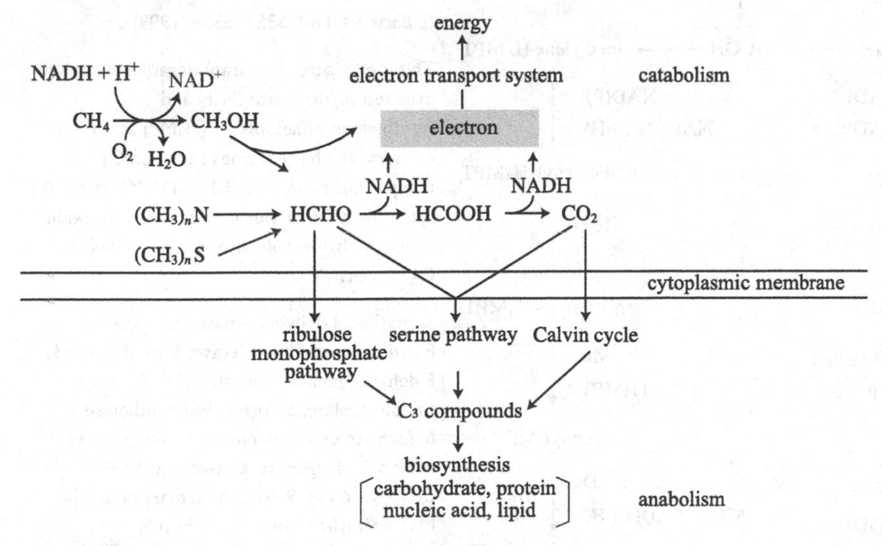

Figure 7.34 **Metabolism of one-carbon compounds.**

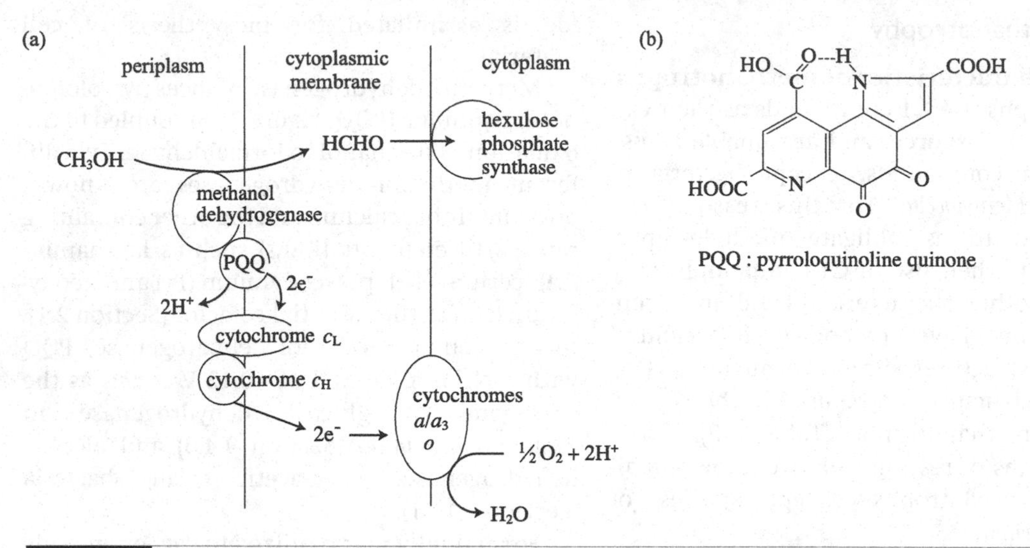

Figure 7.35 (a) Methanol dehydrogenase and the electron transport chain of a methylotroph, and (b) the structure of PQQ.

Cytochrome c_L, cytochrome *c* of low potential; cytochrome c_H, cytochrome *c* of high potential.

On the other hand, formaldehyde is bound to tetrahydrofolate (H_4F) or to tetrahydromethanopterin (H_4MTP) before being oxidized to CO_2 in many Gram-negative bacteria that convert formaldehyde to cell materials through the ribulose monophosphate or serine–isocitrate lyase pathway (Figure 7.36). H_4MTP is regarded as a methanogen-specific coenzyme (Section 9.4.2). *Methylobacterium extorquens* has both sets of enzymes that catalyse oxidation of C1 compounds

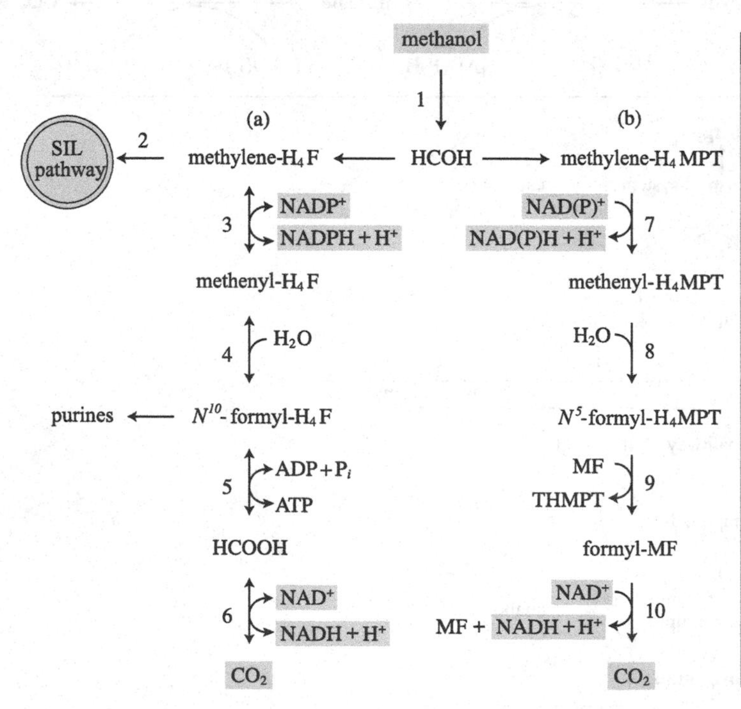

Figure 7.36 Methanol oxidation by *Methylobacterium extorquens*.

(*J. Bacteriol.* **180**: 5351–5356, 1998)

This methylotrophic Gram-negative bacterium uses tetrahydrofolate (H_4F) and tetrahydromethanopterin (H_4MP) as C1 carriers, and has enzymes active on C1 compounds carried by H_4F and H_4MP. H_4MP (b) is the main C1 carrier in this bacterium, while other methylotrophs use H_4F (a) as their main C1 carrier.

1, methanol dehydrogenase; 2, serine H_4 F hydroxymethyltransferase; 3, methylene H_4 F dehydrogenase; 4, methenyl H_4F cyclohydrolase; 5, formyl H_4F synthetase; 6, formate dehydrogenase; 7, methylene H_4 MTP dehydrogenase; 8, methenyl H_4MTP cyclohydrolase; 9, formyl methanofuran H_4 MTP formyltransferase; 10, formyl methanofuran (formyl-MF) dehydrogenase.

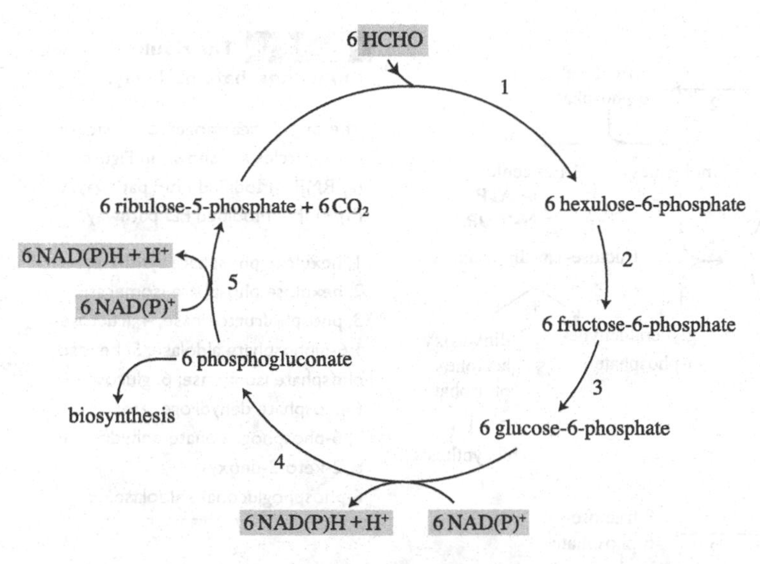

Figure 7.37 Oxidation of formaldehyde through the ribulose monophosphate cycle in *Methylophilus methylotrophus*.

1, hexulose phosphate synthase;
2, hexulose-6-phosphate isomerase;
3, fructose-6-phosphate isomerase;
4, glucose-6-phosphate dehydrogenase;
5, 6-phosphogluconate dehydrogenase.

bound either to H_4F or H_4MTP. Since the enzyme activities are higher for C1 compounds bound to H_4MTP, it is believed that formaldehyde is oxidized after being bound to H_4MTP in this bacterium (Figure 7.36b). H_4F is used in methylotrophs that employ the serine–isocitrate lyase pathway to convert formaldehyde to cell materials (Figure 7.36a)

Methylophilus methylotrophus does not possess enzymes that oxidize formaldehyde in the free form or bound to C1 carriers. This bacterium oxidizes formaldehyde to CO_2 through the ribulose monophosphate cycle (Figure 7.37).

7.10.3 Carbon assimilation by methylotrophs

Obligate methylotrophs and heterotrophic methylotrophs employ either the ribulose monophosphate (RMP) pathway to assimilate formaldehyde or the serine–isocitrate lyase (SIL) pathway to assimilate formaldehyde and CO_2. Autotrophic methylotrophs fix CO_2 through the Calvin cycle. Methylotrophic yeasts growing on methanol employ yet another novel pathway, the xylulose monophosphate (XMP) pathway.

7.10.3.1 Ribulose monophosphate (RMP) pathway

The RMP pathway that assimilates formaldehyde as triose phosphate is a collection of four different pathways all sharing the first two reactions. RMP accepts formaldehyde to form hexulose-6-phosphate that is isomerized to fructose-6-phosphate (Figure 7.38).

Fructose-6-phosphate is cleaved through two alternative routes, one involving fructose-1,6-diphosphate aldolase (RMP–EMP variant, Figure 7.38a) and the other 2-keto-3-deoxy-6-phosphogluconate aldolase (RMP–ED variant, Figure 7.38b). The resulting triose phosphate and pyruvate are used in assimilatory metabolism. The remaining triose phosphate is used to regenerate ribulose-5-phosphate through carbon rearrangement with two molecules of fructose-6-phosphate. As shown in Figure 7.39, the carbon rearrangement takes place in two different ways, depending on the organism: one involves transaldolase and transketolase (TA variant, Figure 7.39a), while the other involves fructose-1,6-diphosphate aldolase (FDA variant, Figure 7.39b). Parts of this pathway are used to convert fructose-6-phosphate to pentose-5-phosphate in those archaea that lack a functional HMP pathway (Section 4.3.5).

Each variant of the RMP pathway can be summarized as:

EMP–TA variant:	3HCHO + ATP → glyceraldehyde-3-phosphate
EMP–FDA variant:	3HCHO + 2ATP → dihydroxyacetone-phosphate
ED–TA variant:	3HCHO → pyruvate + NAD(P)H
ED–FDA variant:	3HCHO + ATP → pyruvate + NAD(P)H

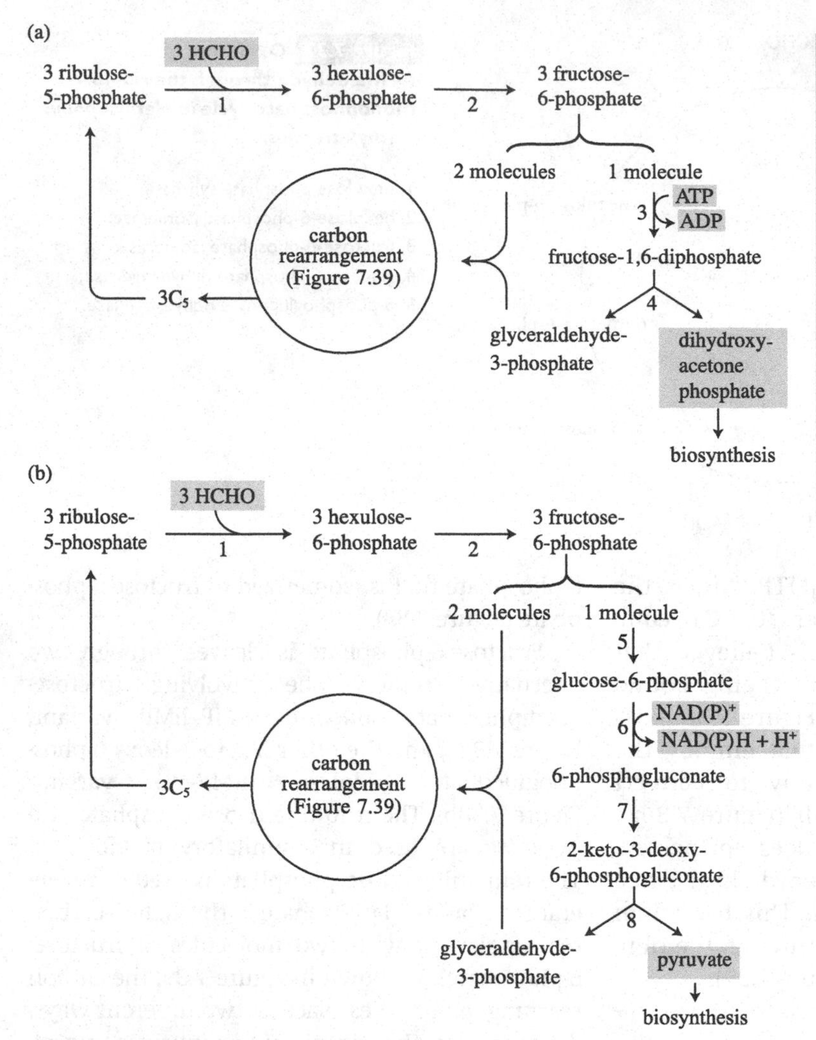

Figure 7.38 The ribulose monophosphate pathway.

The carbon rearrangement reactions in the circles are shown in Figure 7.39.
(a) RMP – Modified EMP pathway.
(b) RMP – Modified ED pathway.

1, hexulose phosphate synthase; 2, hexulose phosphate isomerase; 3, phosphofructokinase; 4, fructose-1,6-diphosphate aldolase; 5, hexose phosphate isomerase; 6, glucose-6-phosphate dehydrogenase; 7, 6-phosphogluconate dehydratase; 8, 2-keto-3-deoxy-6-phosphogluconate aldolase.

7.10.3.2 Serine–isocitrate lyase (SIL) pathway

Formaldehyde is the source of all the carbons of triose phosphate synthesized through the RMP pathway, but *Methylosinus trichosporium* uses formaldehyde and CO_2 to synthesize 2-phosphoglycerate via acetyl-CoA. This metabolic pathway is referred to as the serine–isocitrate lyase (SIL) pathway (Figure 7.40). SIL can be considered in two parts: (1) acetyl-CoA synthesis from formaldehyde and CO_2 through 2-phosphoglycerate, and (2) acetyl-CoA conversion to serine via glyoxylate (Figure 7.41).

Formaldehyde forms methylene-H_4F with tetrahydrofolate (H_4F) before condensing with glycine to serine. An aminotransferase converts serine to hydroxypyruvate, coupling the amination of glyoxylate to glycine. Hydroxypyruvate is reduced to PEP to be carboxylated to oxaloacetate, catalysed by PEP carboxylase. Oxaloacetate is reduced and activated to malyl-CoA to be cleaved to glyoxylate and acetyl-CoA. Acetyl-CoA is oxidized to glyoxylate in the next round of reactions (Figure 7.41).

On the other hand, *Methylobacterium extorquens* condenses two molecules of acetyl-CoA to acetoacetyl-CoA which is converted to succinyl-CoA through the ethylmalonyl-CoA pathway (Section 5.3.3). Succinyl-CoA is converted to malate as in the TCA cycle.

Each step of the SIL pathway can be summarized as:

$$HCHO + CO_2 + 2ATP + 2NAD(P)H + 2H^+ + CoA\text{-}SH \longrightarrow CH_3CO\text{-}CoA + 2ADP + 2P_i + 2NAD(P)^+$$

$$CH_3CO\text{-}CoA + NAD^+ + FAD + 2H_2O \longrightarrow CHO\text{-}COOH + NADH + H^+ + FADH_2$$

$$CHO\text{-}COOH + HCHO + NAD(P)H + H^+ + ATP \longrightarrow 3\text{-phosphoglycerate} + ADP + NAD(P)^+$$

$$+)\ 2HCHO + CO_2 + 3ATP + 2NAD(P)H + 2H^+ + FAD \longrightarrow 3\text{-phosphoglycerate} + 2NAD(P)^+ + FADH_2 + 3ADP + 2P_i + H_2O$$

7.10.3.3 Xylulose monophosphate (XMP) pathway

Methylotrophic yeasts assimilate methanol through the xylulose monophosphate (XMP) pathway (Figure 7.42) which is different from the bacterial C1 metabolic pathways. Dihydroxyacetone phosphate is synthesized from formaldehyde in this pathway for biosynthetic purposes.

Dihydroxyacetone synthase condenses formaldehyde with xylulose-5-phosphate to produce glyceraldehyde-3-phosphate and dihydroxyacetone. Dihydroxyacetone synthase is a kind of transketolase that uses formaldehyde as its substrate. Since dihydroxyacetone is an important intermediate, this metabolism is alternatively referred to as the dihydroxyacetone pathway. Dihydroxyacetone is phosphorylated to dihydroxyacetone phosphate by triokinase. One third of the triose phosphate is used for biosynthesis, while the remaining molecules condense with equivalent molecules of glyceraldehyde-3-phosphate to produce fructose-6-phosphate. Carbon rearrangement is similar to that of the RMP pathway (Figure 7.39), and

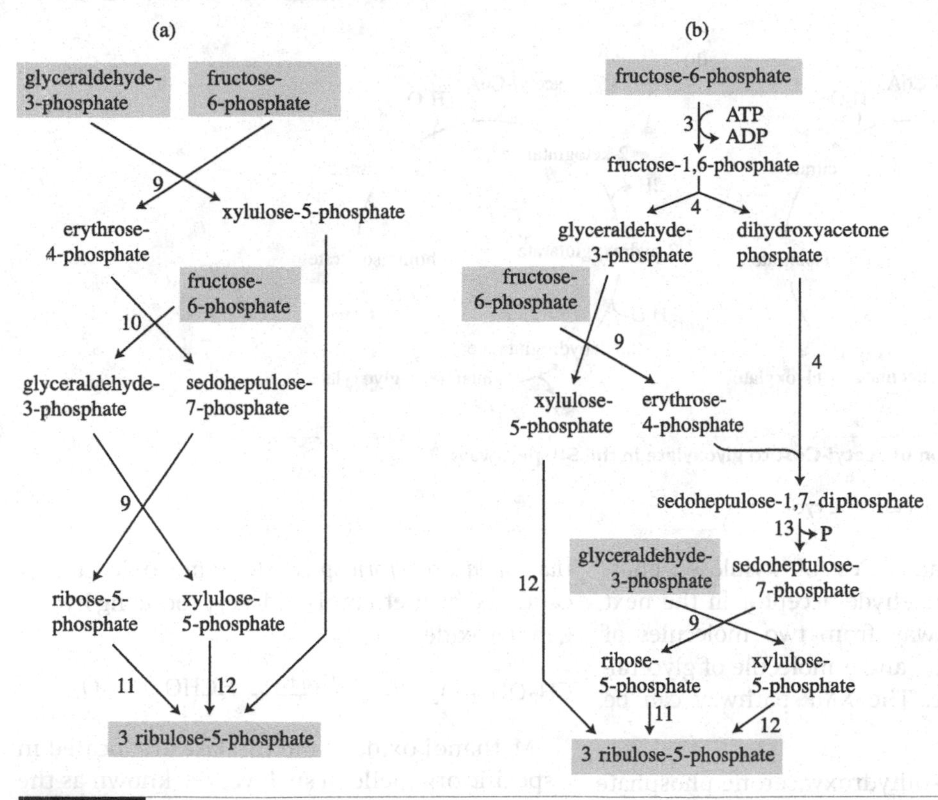

Figure 7.39 Two alternative routes of carbon rearrangement in the RMP pathway (the circled part of Figure 7.38).

(a) TA variant involving transketolase and transaldolase. (b) FDA variant involving transketolase and fructose-1,6-diphosphatase aldolase.

3, 4, as in Figure 7.34; 9, transketolase; 10, transaldolase; 11, ribose-5-phosphate isomerase; 12, ribulose-5-phosphate-3-epimerase.

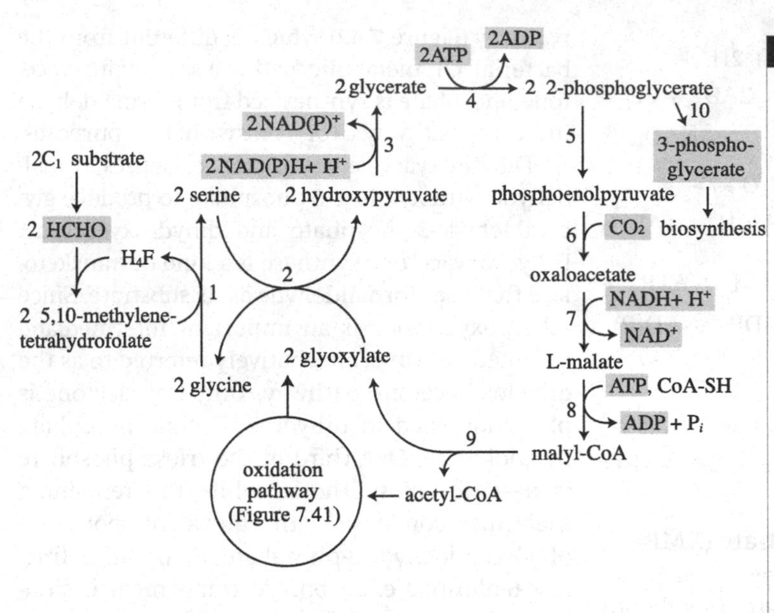

Figure 7.40 The serine–isocitrate lyase (SIL) pathway.

The SIL pathway is divided into two steps for convenience. These are (1) acetyl-CoA synthesis from formaldehyde and CO_2 through 2-phosphoglycerate, and (2) acetyl-CoA conversion to serine via glyoxylate. The circle shows acetyl-CoA oxidation to glyoxylate, as detailed in Figure 7.41.

1, serine hydroxymethyltransferase; 2, serine: glyoxylate aminotransferase; 3, hydroxypyruvate reductase; 4, glycerate kinase; 5, PEP hydratase; 6, PEP carboxylase; 7, malate dehydrogenase; 8, malyl-CoA synthetase; 9, malyl-CoA lyase; 10, phosphoglycerate mutase.

H_4F, tetrahydrofolate.

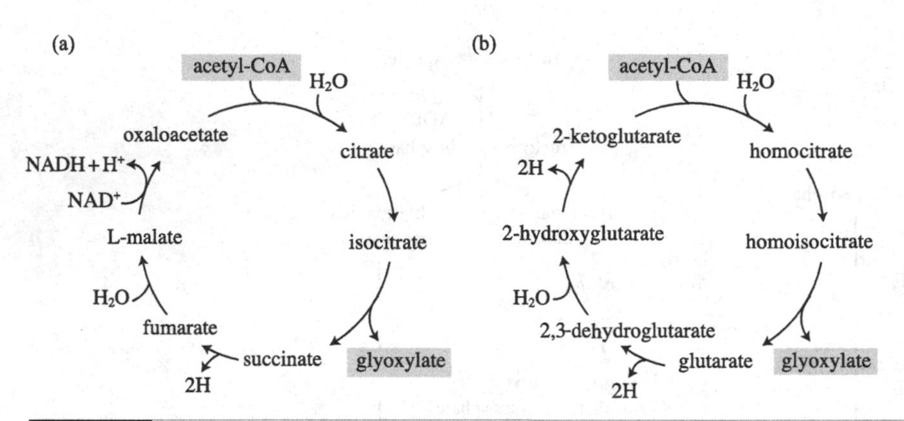

Figure 7.41 Oxidation of acetyl-CoA to glyoxylate in the SIL pathway.

(a) Glyoxylate cycle; (b) homocitrate cycle.

generates three molecules of xylulose-5-phosphate as the formaldehyde acceptor in the next round of the pathway from two molecules of fructose-6-phosphate, and a molecule of glyceraldehyde-3-phosphate. The XMP pathway can be summarized as

$$3HCHO + 3ATP \longrightarrow \text{dihydroxyacetone phosphate} + 3ADP + 2P_i$$

Methane utilizers are not known in eukaryotes and methanol utilization is restricted to a few yeast strains, including species of *Candida*, *Hansenula* and *Torulopsis*. Methanol oxidation is catalysed by methanol oxidase, producing hydrogen peroxide:

$$CH_3OH + O_2 \xrightarrow{\text{methanol oxidase}} HCHO + H_2O_2$$

Methanol oxidase and catalase are located in a specific organelle in such yeasts, known as the peroxisome.

7.10.4 Energy efficiency in C1 metabolism

Microorganisms assimilate C1 compounds through the RMP, SIL and XMP pathways, and the Calvin cycle. However, energy requirements as ATP and reduced electron carriers differ between the pathways. Since the starting materials and the products are different, calculations can be made normalizing pyruvate as the final product (Table 7.4).

It can be seen that the EMP–TA variant of the RMP pathway is the most efficient pathway while the least efficient is the Calvin cycle.

7.11 Incomplete oxidation

During heterotrophic metabolism under aerobic conditions, part of the substrate is converted into cell materials and the remainder is oxidized to carbon dioxide, providing energy for growth. Certain microbes excrete metabolic intermediates in large quantities due to a lack of enzymes for complete oxidation, or because enzymes are repressed under the given conditions. Examples are acetate production by acetic acid bacteria and acetoin and butanediol production by members of the *Bacillus* genus.

7.11.1 Acetic acid bacteria

Acetic acid bacteria are aerobic bacteria that produce acetate from ethanol. When ethanol is completely consumed, species of *Acetobacter* utilize acetate while acetate is not consumed by species belonging to the genus *Gluconobacter*.

As discussed earlier (Section 4.3.2.2), *Gluconobacter* spp. metabolize sugars through the oxidative HMP cycle, and the resulting glyceraldehyde-3-phosphate is oxidized to acetate

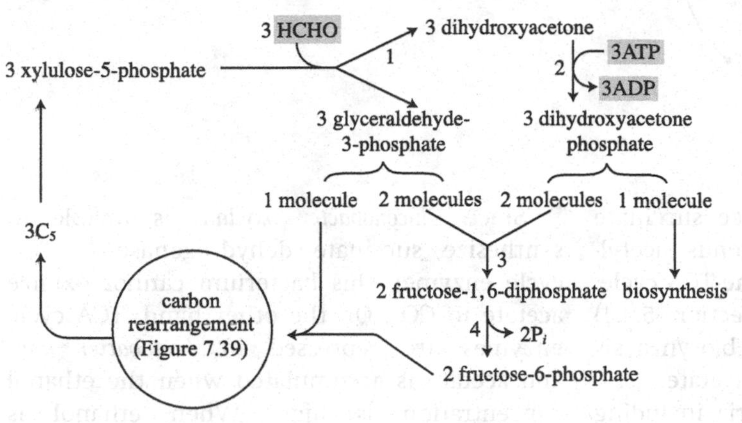

Figure 7.42 Xylulose monophosphate pathway of methylotrophic yeasts.

The carbon rearrangement marked by a circle is similar to that of the ribulose monophosphate pathway shown in Figure 7.35.

1, dihydroxyacetone synthase; 2, dihydroxyacetone kinase (triokinase); 3, fructose-1,6-diphosphate aldolase; 4, fructose diphosphatase.

Table 7.4 Energy efficiency in C1 assimilation.

Pathway	Variant	Substrate	Electron carriers and ATP		
			NAD(P)H	$FADH_2$	ATP
Calvin cycle		$3CO_2$	−5	0	−7
RMP pathway	EMP–TA	3HCHO	+1	0	+1
	EMP–FDA	3HCHO	+1	0	0
	ED–TA	3HCHO	+1	0	0
	ED–FDA	3HCHO	+1	0	−3
SIL pathway		2HCHO + CO_2	−2	+1	−2
XMP pathway		3HCHO	+1	0	1

+, generated; −, consumed.

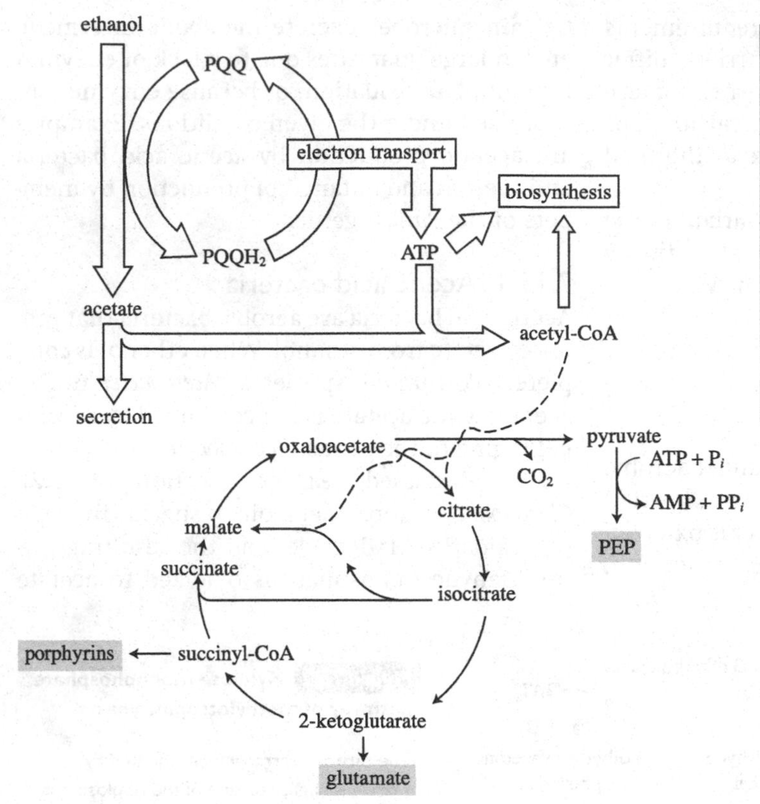

Figure 7.43 Assimilation of acetate as the carbon source by *Acetobacter aceti* growing on ethanol.

(Gottschalk, G. 1986, *Bacterial Metabolism*, 2nd edn., Figure 6.24. Springer, New York)

PQQ reduced from acetate oxidation is reoxidized through the electron transport phosphorylation process, conserving the energy. Acetate is activated to acetyl-CoA, which is metabolized through the glyoxylate cycle to supply carbon skeletons for biosynthesis.

via pyruvate and acetyl-CoA. Since succinate dehydrogenase is absent in this genus, acetyl-CoA cannot be oxidized through the TCA cycle. The incomplete TCA pathway (Section 5.4.1) operates to supply precursors for biosynthesis where PEP is carboxylated to oxaloacetate.

All members of acetic acid bacteria, including *Gluconobacter oxydans* and *Acetobacter aceti*, oxidize ethanol to acetate:

$$CH_3CH_2OH + PQQ \xrightarrow{\text{alcohol dehydrogenase}} CH_3CHO + PQQH_2$$

$$CH_3CHO + PQQ + H_2O \xrightarrow{\text{aldehyde dehydrogenase}} CH_3COOH + PQQH_2$$

As shown in the reactions, these enzymes are quinoproteins containing pyrroloquinoline quinone (PQQ). Electrons from reduced PQQ are transferred to coenzyme Q of the bacterial electron transport chain. The mid-point redox potential of PQQ is + 0.12 V, which is much higher than that of $NAD^+/NADH$, which is −0.34 V (Section 7.10.2.2).

Since *Gluconobacter oxydans* is unable to synthesize succinate dehydrogenase (a TCA cycle enzyme), this bacterium cannot oxidize acetate to CO_2. On the other hand, TCA cycle enzymes are repressed in *Acetobacter aceti* and acetate is accumulated when the ethanol concentration is high. When ethanol is exhausted, genes for the TCA cycle enzymes are expressed in *Acetobacter aceti* in order to utilize the acetate. This bacterium requires amino acids to grow on ethanol and acetate, probably due to a limited ability to generate carbon compounds for biosynthesis (Figure 7.43).

Acetic acid bacteria oxidize various alcohols and ketones. This property is exploited for oxidation of D-sorbitol to L-sorbose in the ascorbic acid production process.

7.11.2 Acetoin and butanediol

The expression of TCA cycle enzymes is repressed in *Bacillus* species, including *Bacillus polymyxa*, *Bacillus subtilis* and others, growing on carbohydrates. They do not oxidize the substrate

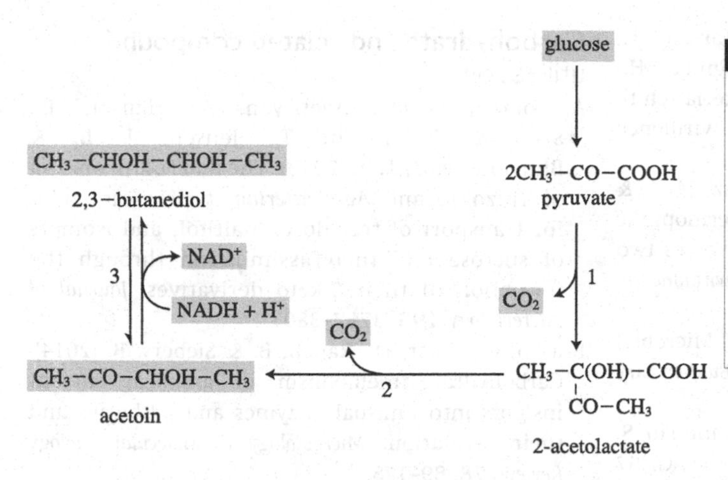

Figure 7.44 Formation of acetoin and 2,3-butanediol by *Bacillus subtilis*.

(Gottschalk, G. 1986, *Bacterial Metabolism*, 2nd edn., Figure 6.26. Springer, New York.)

Acetoin and 2,3-butanediol are produced during vegetative cell growth, and used to supply energy in the sporulation process.

1, 2-acetolactate synthase; 2, 2-acetolactate decarboxylase; 3, 2,3-butanediol dehydrogenase.

completely during vegetative cell growth, accumulating acetoin and 2,3-butanediol (Figure 7.44). This metabolism is similar to fermentation in facultative anaerobic enteric bacteria (Section 8.6.2). The enteric bacteria ferment carbohydrates to these compounds under anaerobic conditions, while the *Bacillus* spp. do this under aerobic conditions.

When carbohydrate is exhausted, *Bacillus* species sporulate using acetoin and 2,3-butanediol as energy sources through the TCA cycle (Section 7.4). 2,3-butanediol is a compound of industrial interest as a raw material, with two functional groups for polymer synthesis.

7.11.3 Other products of aerobic metabolism

Coryneform bacteria such as *Corynebacterium glutamicum* and *Brevibacterium flavum* excrete glutamate in large quantities under aerobic nutrient-rich conditions. The enzymes of the TCA cycle have low activities in this case. This is exploited industrially to produce monosodium glutamate.

Further Reading

Note this section contains key references only. Additional recommended references are available at www.cambridge.org/ProkaryoticMetabolism.

Depolymerization of polymers

Artzi, L., Bayer, E. A. & Morais, S. (2017). Cellulosomes: bacterial nanomachines for dismantling plant polysaccharides. *Nature Reviews Microbiology* **15**, 83–95.

Ballschmiter, M., Armbrecht, M., Ivanova, K., Antranikian, G. & Liebl, W. (2005). AmyA, an α-amylase with β-cyclodextrin-forming activity, and AmyB from the thermoalkaliphilic organism *Anaerobranca gottschalkii*: two α-amylases adapted to their different cellular localizations. *Applied and Environmental Microbiology* **71**, 3709–3715.

Beeson, W. T., Vu, V. V., Span, E. A., Phillips, C. M. & Marletta, M. A. (2015). Cellulose degradation by polysaccharide monooxygenases. *Annual Review of Biochemistry* **84**, 923–946.

Bertoldo, C. & Antranikian, G. (2002). Starch-hydrolyzing enzymes from thermophilic archaea and bacteria. *Current Opinion in Chemical Biology* **6**, 151–160.

Chimileski, S., Dolas, K., Naor, A., Gophna, U. & Papke, R. T. (2014). Extracellular DNA metabolism in *Haloferax volcanii*. *Frontiers in Microbiology* **5**, 57.

Choi, K.-H. & Cha, J. (2015). Membrane-bound amylopullulanase is essential for starch metabolism of *Sulfolobus acidocaldarius* DSM639. *Extremophiles* **19**, 909–920.

Collins, T., Gerday, C. & Feller, G. (2005). Xylanases, xylanase families and extremophilic xylanases. *FEMS Microbiology Reviews* **29**, 3–23.

Frederiksen, R. F., Paspaliari, D. K., Larsen, T., Storgaard, B. G., Larsen, M. H., Ingmer, H., Palcic, M. M. & Leisner, J. J. (2013). Bacterial chitinases and chitin-binding proteins as virulence factors. *Microbiology* **159**, 833–847.

Gao, J., Bauer, M. W., Shockley, K. R., Pysz, M. A. & Kelly, R. M. (2003). Growth of hyperthermophilic archaeon *Pyrococcus furiosus* on chitin involves two family 18 chitinases. *Applied and Environmental Microbiology* **69**, 3119–3128.

Jayani, R. S., Saxena, S. & Gupta, R. (2005). Microbial pectinolytic enzymes: a review. *Process Biochemistry* **40**, 2931–2944.

Jiao, Y.-L., Wang, S.-J., Lv, M.-S., Fang, Y.-W. and Liu, S. (2013). An evolutionary analysis of the GH57 amylopullulanases based on the DOMON glucodextranase-like domains. *Journal of Basic Microbiology* **53**, 231–239.

Khalikova, E., Susi, P. & Korpela, T. (2005). Microbial dextran-hydrolyzing enzymes: fundamentals and applications. *Microbiology and Molecular Biology Reviews* **69**, 306–325.

LaRowe, D. E. & Van Cappellen, P. (2011). Degradation of natural organic matter: a thermodynamic analysis. *Geochimica et Cosmochimica Acta* **75**, 2030–2042.

Pinchuk, G. E., Ammons, C., Culley, D. E., Li, S. M., McLean, J. S., Romine, M. F., Nealson, K. H., Fredrickson, J. K. & Beliaev, A. S. (2008). Utilization of DNA as a sole source of phosphorus, carbon, and energy by *Shewanella* spp.: ecological and physiological implications for dissimilatory metal reduction. *Applied and Environmental Microbiology* **74**, 1198–1208.

Saloheimo, M., Paloheimo, M., Hakola, S., Pere, J., Swanson, B., Nyyssönen, E., Bhatia, A., Ward, M. & Penttilä, M. (2002). Swollenin, a *Trichoderma reesei* protein with sequence similarity to the plant expansins, exhibits disruption activity on cellulosic materials. *European Journal of Biochemistry* **269**, 4202–4211.

Siroosi, M., Amoozegar, M., Khajeh, K., Fazeli, M. & Habibi Rezaei, M. (2014). Purification and characterization of a novel extracellular halophilic and organic solvent-tolerant amylopullulanase from the haloarchaeon, *Halorubrum* sp. strain Ha25. *Extremophiles* **18**, 25–33.

Staufenberger, T., Imhoff, J. F. & Labes, A. (2012). First crenarchaeal chitinase found in *Sulfolobus tokodaii*. *Microbiological Research* **167**, 262–269.

Carbohydrate and related compound utilization

Ampomah, O. Y., Avetisyan, A., Hansen, E., Svenson, J., Huser, T., Jensen, J. B. & Bhuvaneswari, T. V. (2013). The *thuEFGKAB* operon of rhizobia and *Agrobacterium tumefaciens* codes for transport of trehalose, maltitol, and isomers of sucrose and their assimilation through the formation of their 3-keto derivatives. *Journal of Bacteriology* **195**, 3797–3807.

Bräsen, C., Esser, D., Rauch, B. & Siebers, B. (2014). Carbohydrate metabolism in archaea: current insights into unusual enzymes and pathways and their regulation. *Microbiology & Molecular Biology Reviews* **78**, 89–175.

Csiszovszki, Z., Krishna, S., Orosz, L., Adhya, S. & Semsey, S. (2011). Structure and function of the D-galactose network in enterobacteria. *mBio* **2**, e00053-11.

Felux, A.-K., Spiteller, D., Klebensberger, J. & Schleheck, D. (2015). Entner–Doudoroff pathway for sulfoquinovose degradation in *Pseudomonas putida* SQ1. *Proceedings of the National Academy of Sciences of the USA* **112**, E4298–E4305.

Johnsen, U., Sutter, J.-M., Zaiß, H. & Schönheit, P. (2013). L-Arabinose degradation pathway in the haloarchaeon *Haloferax volcanii* involves a novel type of l-arabinose dehydrogenase. *Extremophiles* **17**, 897–909.

Nolle, N., Felsl, A., Heermann, R. & Fuchs, T. M. (2017). Genetic characterization of the galactitol utilization pathway of *Salmonella enterica* serovar *Typhimurium*. *Journal of Bacteriology* **199**, e00595-16.

Orita, I., Sato, T., Yurimoto, H., Kato, N., Atomi, H., Imanaka, T. & Sakai, Y. (2006). The ribulose monophosphate pathway substitutes for the missing pentose phosphate pathway in the archaeon *Thermococcus kodakaraensis*. *Journal of Bacteriology* **188**, 4698–4704.

Qian, Z., Wang, Q., Tong, W., Zhou, C., Wang, Q. & Liu, S. (2010). Regulation of galactose metabolism through the HisK:GalR two-component system in *Thermoanaerobacter tengcongensis*. *Journal of Bacteriology* **192**, 4311–4316.

Sund, C. J., Liu, S., Germane, K. L., Servinsky, M. D., Gerlach, E. S. & Hurley, M. M. (2015). Phosphoketolase flux in *Clostridium acetobutylicum* during growth on l-arabinose. *Microbiology* **161**, 430–440.

Suvorova, I. A., Tutukina, M. N., Ravcheev, D. A., Rodionov, D. A., Ozoline, O. N. & Gelfand, M. S. (2011). Comparative genomic analysis of the hexuronate metabolism genes and their regulation in *Gammaproteobacteria*. *Journal of Bacteriology* **193**, 3956–3963.

Zeng, L. & Burne, R. A. (2015). NagR differentially regulates the expression of the glms and *nagAB* genes required for amino sugar metabolism by *Streptococcus mutans*. *Journal of Bacteriology* **197**, 3533–3544.

Organic acid utilization

Bramer, C. O. & Steinbuchel, A. (2001). The methylcitric acid pathway in *Ralstonia eutropha*: new genes identified involved in propionate metabolism. *Microbiology* **147**, 2203–2214.

Claes, W. A., Puhler, A. & Kalinowski, J. (2002). Identification of two *prpDBC* gene clusters in *Corynebacterium glutamicum* and their involvement in propionate degradation via the 2-methylcitrate cycle. *Journal of Bacteriology* **184**, 2728–2739.

Ensign, S. A. (2006). Revisiting the glyoxylate cycle: alternate pathways for microbial acetate assimilation. *Molecular Microbiology* **61**, 274–276.

Kretzschmar, U., Ruckert, A., Jeoung, J. H. & Gorisch, H. (2002). Malate:quinone oxidoreductase is essential for growth on ethanol or acetate in *Pseudomonas aeruginosa*. *Microbiology* **148**, 3839–3847.

Meyer, F. M. & Stülke, J. (2013). Malate metabolism in *Bacillus subtilis*: distinct roles for three classes of malate-oxidizing enzymes. *FEMS Microbiology Letters* **339**, 17–22.

Palacios, S. & Escalante-Semerena, J. C. (2004). 2-methylcitrate-dependent activation of the propionate catabolic operon (prpBCDE) of *Salmonella enterica* by the PrpR protein. *Microbiology* **150**, 3877–3887.

Sahin, N. (2003). Oxalotrophic bacteria. *Research in Microbiology* **154**, 399–407.

Savvi, S., Warner, D. F., Kana, B. D., McKinney, J. D., Mizrahi, V. & Dawes, S. S. (2008). Functional characterization of a vitamin B_{12}-dependent methylmalonyl pathway in *Mycobacterium tuberculosis*: implications for propionate metabolism during growth on fatty acids. *Journal of Bacteriology* **190**, 3886–3895.

Schneider, K., Skovran, E. & Vorholt, J. A. (2012). Oxalyl-coenzyme A reduction to glyoxylate is the preferred route of oxalate assimilation in *Methylobacterium extorquens* AM1. *Journal of Bacteriology* **194**, 3144–3155.

Suvorova, I. A., Ravcheev, D. A. & Gelfand, M. S. (2012). Regulation and evolution of malonate and propionate catabolism in Proteobacteria. *Journal of Bacteriology* **194**, 3234–3240.

Van Bogaert, I. N. A., Groeneboer, S., Saerens, K. & Soetaert, W. (2011). The role of cytochrome P450 monooxygenases in microbial fatty acid metabolism. *FEBS Journal* **278**, 206–221.

Wolfe, A. J. (2005). The acetate switch. *Microbiology and Molecular Biology Reviews* **69**, 12–50.

Alcohol utilization

Bizzini, A., Zhao, C., Budin-Verneuil, A., Sauvageot, N., Giard, J.-C., Auffray, Y. & Hartke, A. (2010). Glycerol is metabolized in a complex and strain-dependent manner in *Enterococcus faecalis*. *Journal of Bacteriology* **192**, 779–785.

de Faveri, D., Torre, P., Molinari, F. & Converti, A. (2003). Carbon material balances and bioenergetics of 2,3-butanediol bio-oxidation by *Acetobacter hansenii*. *Enzyme and Microbial Technology* **33**, 708–719.

Havemann, G. D. & Bobik, T. A. (2003). Protein content of polyhedral organelles involved in coenzyme B12-dependent degradation of 1,2-propanediol in *Salmonella enterica* serovar *typhimurium* LT2. *Journal of Bacteriology* **185**, 5086–5095.

Hirota-Mamoto, R., Nagai, R., Tachibana, S., Yasuda, M., Tani, A., Kimbara, K. & Kawai, F. (2006). Cloning and expression of the gene for periplasmic poly(vinyl alcohol) dehydrogenase from *Sphingomonas* sp. strain 113P3, a novel-type quinohaemoprotein alcohol dehydrogenase. *Microbiology* **152**, 1941–1949.

Lehman, B. P., Chowdhury, C. & Bobik, T. A. (2017). The N terminus of the PduB protein binds the protein shell of the Pdu microcompartment to its enzymatic core. *Journal of Bacteriology* **199**(8), e00785-16.

Mern, D. S., Ha, S.-W., Khodaverdi, V., Gliese, N. & Gorisch. H. (2010). A complex regulatory network controls aerobic ethanol oxidation in *Pseudomonas aeruginosa*: indication of four levels of sensor kinases and response regulators. *Microbiology* **156**, 1505–1516.

Mückschel, B., Simon, O., Klebensberger, J., Graf, N., Rosche, B., Altenbuchner, J., Pfannstiel, J., Huber, A. & Hauer, B. (2012). Ethylene glycol metabolism by *Pseudomonas putida*. *Applied and Environmental Microbiology* **78**, 8531–8539.

Rosier, C., Leys, N., Henoumont, C., Mergeay, M. & Wattiez, R. (2012). Purification and characterization of the acetone carboxylase of *Cupriavidus metallidurans* strain CH34. *Applied and Environmental Microbiology* **78**, 4516–4518.

Sherwood, K. E., Cano, D. J. & Maupin-Furlow, J. A. (2009). Glycerol-mediated repression of glucose metabolism and glycerol kinase as the sole route of glycerol catabolism in the haloarchaeon *Haloferax volcanii*. *Journal of Bacteriology* **191**, 4307–4315.

Amino acid and nucleic acid base utilization

Bender, R. A. (2012). Regulation of the histidine utilization (Hut) system in bacteria. *Microbiology and Molecular Biology Reviews* **76**, 565–584.

Colabroy, K. L. & Begley, T. P. (2005). Tryptophan catabolism: identification and characterization of a new degradative pathway. *Journal of Bacteriology* **187**, 7866–7869.

Gerth, M. L., Ferla, M. P. & Rainey, P. B. (2012). The origin and ecological significance of multiple branches for histidine utilization in *Pseudomonas aeruginosa* PAO1. *Environmental Microbiology* **14**, 1929–1940.

Hoschle, B., Gnau, V. & Jendrossek, D. (2005). Methylcrotonyl-CoA and geranyl-CoA carboxylases are involved in leucine/isovalerate utilization (Liu) and acyclic terpene utilization (Atu), and are encoded by liuB/liuD and atuC/atuF, in *Pseudomonas aeruginosa*. *Microbiology* **151**, 3649–3656.

Kim, K.-S., Pelton, J. G., Inwood, W. B., Andersen, U., Kustu, S. & Wemmer, D. E. (2010). The Rut pathway for pyrimidine degradation: novel chemistry and toxicity problems. *Journal of Bacteriology* **192**, 4089–4102.

Millett, E. S., Efimov, I., Basran, J., Handa, S., Mowat, C. G. & Raven, E. L. (2012). Heme-containing dioxygenases involved in tryptophan oxidation. *Current Opinion in Chemical Biology* **16**, 60–66.

Moses, S., Sinner, T., Zaprasis, A., Stöveken, N., Hoffmann, T., Belitsky, B. R., Sonenshein, A. L. & Bremer, E. (2012). Proline utilization by *Bacillus subtilis*: uptake and catabolism. *Journal of Bacteriology* **194**, 745–758.

Hydrocarbon utilization

Basu, A., Apte, S. K. & Phale, P. S. (2006). Preferential utilization of aromatic compounds over glucose by *Pseudomonas putida* CSV86. *Applied and Environmental Microbiology* **72**, 2226–2230.

Doyle, E., Muckian, L., Hickey, A. M. & Clipson, N. (2008). Microbial PAH degradation. *Advances in Applied Microbiology* **65**, 27–66.

Funhoff, E. G., Bauer, U., Garcia-Rubio, I., Witholt, B. & van Beilen, J. B. (2006). CYP153A6, a soluble P450 oxygenase catalyzing terminal-alkane hydroxylation. *Journal of Bacteriology* **188**, 5220–5227.

George, K. W. & Hay, A. G. (2011). Bacterial strategies for growth on aromatic compounds. *Advances in Applied Microbiology* **74**, 1–33.

Khajamohiddin, S., Repalle, E. R., Pinjari, A. B., Merrick, M. & Siddavattam, D. (2008). Biodegradation of aromatic compounds: an overview of meta-fission product hydrolases. *Critical Reviews in Microbiology* 34, 13–31.

Krishnakumar, A. M., Sliwa, D., Endrizzi, J. A., Boyd, E. S., Ensign, S. A. & Peters, J. W. (2008). Getting a handle on the role of coenzyme M in alkene metabolism. *Microbiology and Molecular Biology Reviews* **72**, 445–456.

Mallick, S., Chakraborty, J. & Dutta, T. K. (2010). Role of oxygenases in guiding diverse metabolic pathways in the bacterial degradation of low-molecular-weight polycyclic aromatic hydrocarbons: a review. *Critical Reviews in Microbiology* **37**, 64–90.

Rojo, F. (2009). Degradation of alkanes by bacteria. *Environmental Microbiology* **11**, 2477–2490.

Teufel, R., Mascaraque, V., Ismail, W., Voss, M., Perera, J., Eisenreich, W., Haehnel, W. & Fuchs, G. (2010). Bacterial phenylalanine and phenylacetate catabolic pathway revealed. *Proceedings of the National Academy of Sciences of the USA* **107**, 14390–14395.

van Hamme, J. D., Singh, A. & Ward, O. P. (2003). Recent advances in petroleum microbiology. *Microbiology and Molecular Biology Reviews* **67**, 503–549.

Vangnai, A. S., Sayavedra-Soto, L. A. & Arp, D. J. (2002). Roles for the two 1-butanol dehydrogenases of *Pseudomonas butanovora* in butane and 1-butanol metabolism. *Journal of Bacteriology* **184**, 4343–4350.

Utilization of natural and anthropogenic xenobiotics

Fenner, K., Canonica, S., Wackett, L. P. & Elsner, M. (2013). Evaluating pesticide degradation in the environment: blind spots and emerging opportunities. *Science* **341**: 752–758.

Gewert, B., Plassmann, M. M. & MacLeod, M. (2015). Pathways for degradation of plastic polymers floating in the marine environment. *Environmental Science: Processes and Impacts* **17**, 1513–1521.

Kivisaar, M. (2009). Degradation of nitroaromatic compounds: a model to study evolution of metabolic pathways. *Molecular Microbiology* **74**, 777–781.

Mattes, T. E., Alexander, A. K. & Coleman, N. V. (2010). Aerobic biodegradation of the chloroethenes: pathways, enzymes, ecology, and evolution. *FEMS Microbiology Reviews* **34**, 445–475.

McGrath, J. W., Chin, J. P. & Quinn, J. P. (2013). Organophosphonates revealed: new insights into the microbial metabolism of ancient molecules. *Nature Reviews Microbiology* **11**, 412–419.

Reisch, C. R., Stoudemayer, M. J., Varaljay, V. A., Amster, I. J., Moran, M. A. & Whitman, W. B. (2011). Novel pathway for assimilation of dimethylsulphoniopropionate widespread in marine bacteria. *Nature* **473**, 208–211.

Shah, A. A., Hasan, F., Hameed, A. & Ahmed, S. (2008). Biological degradation of plastics: a comprehensive review. *Biotechnology Advances* **26**, 246–265.

Singh, B., Kaur, J. & Singh, K. (2012). Microbial remediation of explosive waste. *Critical Reviews in Microbiology* **38**, 152–167.

Sivan, A. (2011). New perspectives in plastic biodegradation. *Current Opinion in Biotechnology* **22**, 422–426.

Wei, R., Oeser, T. & Zimmermann, W. (2014). Synthetic polyester-hydrolyzing enzymes from thermophilic actinomycetes. *Advances in Applied Microbiology* **89**, 267–305.

Methylotrophy

Balasubramanian, R., Smith, S. M., Rawat, S., Yatsunyk, L. A., Stemmler, T. L. & Rosenzweig, A. C. (2010). Oxidation of methane by a biological dicopper centre. *Nature* **465**, 115–119.

Chistoserdova, L., Kalyuzhnaya, M. G. & Lidstrom, M. E. (2009). The expanding world of methylotrophic metabolism. *Annual Review of Microbiology* **63**, 477–499.

Crombie, A. T. & Murrell, J. C. (2014). Trace-gas metabolic versatility of the facultative methanotroph *Methylocella silvestris*. *Nature* **509**, 148–151.

Dedysh, S. N., Smirnova, K. V., Khmelenina, V. N., Suzina, N. E., Liesack, W. & Trotsenko, Y. A. (2005). Methylotrophic autotrophy in *Beijerinckia mobilis*. *Journal of Bacteriology* **187**, 3884–3888.

Firsova, J., Doronina, N., Lang, E., Spröer, C., Vuilleumier, S. H. & Trotsenko, Y. (2009). *Ancylobacter dichloromethanicus* sp. nov. – a new aerobic facultatively methylotrophic bacterium utilizing dichloromethane. *Systematic and Applied Microbiology* **32**, 227–232.

Hu, B. & Lidstrom, M. (2012). CcrR, a TetR family transcriptional regulator, activates the transcription of a gene of the ethylmalonyl coenzyme A pathway in *Methylobacterium extorquens* AM1. *Journal of Bacteriology* **194**, 2802–2808.

Martinez-Gomez, N. C., Nguyen, S. & Lidstrom, M. E. (2013). Elucidation of the role of the methylene-tetrahydromethanopterin dehydrogenase MtdA in the tetrahydromethanopterin-dependent oxidation pathway in *Methylobacterium extorquens* AM1. *Journal of Bacteriology* **195**, 2359–2367.

Pol, A., Barends, T. R. M., Dietl, A., Khadem, A. F., Eygensteyn, J., Jetten, M. S. M. & Op den Camp, H. J. M. (2014). Rare earth metals are essential for methanotrophic life in volcanic mudpots. *Environmental Microbiology* **16**, 255–264.

Semrau, J. D., DiSpirito, A. A. & Murrell, J. C. (2008). Life in the extreme: thermoacidophilic methanotrophy. *Trends in Microbiology* **16**, 190–193.

Theisen, A. R., Ali, M. H., Radajewski, S., Dumont, M. G., Dunfield, P. F., McDonald, I. R., Dedysh, S. N., Miguez, C. B. & Murrell, J. C. (2005). Regulation of methane oxidation in the facultative methanotroph *Methylocella silvestris* BL2. *Molecular Microbiology* **58**, 682–692.

Wood, A. P., Aurikko, J. P. & Kelly, D. P. (2004). A challenge for 21st century molecular biology and biochemistry: what are the causes of obligate autotrophy and methanotrophy? *FEMS Microbiology Reviews* **28**, 335–352.

Incomplete oxidation

Adachi, O., Moonmangmee, D., Toyama, H., Yamada, M., Shinagawa, E. & Matsushita, K. (2003). New developments in oxidative fermentation. *Applied Microbiology and Biotechnology* **60**, 643–653.

Holscher, T. & Gorisch, H. (2006). Knockout and overexpression of pyrroloquinoline quinone biosynthetic genes in *Gluconobacter oxydans* 621H. *Journal of Bacteriology* **188**, 7668–7676.

Richhardt, J., Luchterhand, B., Bringer, S., Büchs, J. & Bott, M. (2013). Evidence for a key role of cytochrome bo_3 oxidase in respiratory energy metabolism of *Gluconobacter oxydans*. *Journal of Bacteriology* **195**, 4210–4220.

Chapter 8

Anaerobic fermentation

Anaerobic conditions are maintained in some ecosystems where the rate of oxygen supply is lower than that of oxygen consumption. Organic compounds are removed from anaerobic ecosystems through the concerted action of fermentative and anaerobic respiratory microorganisms. In microbiology, the term 'fermentation' can be used to describe either microbial processes that produce useful products or a form of anaerobic microbial growth using internally supplied electron acceptors and generating ATP mainly through substrate-level phosphorylation (SLP).

8.1 Electron acceptors used in anaerobic metabolism

8.1.1 Fermentation and anaerobic respiration

Respiration refers to the reduction of oxygen by electrons from the electron transport chains coupled to the generation of a proton motive force through electron transport phosphorylation (ETP; Section 5.8). Under anaerobic conditions, some microorganisms grow using an ETP process, with externally supplied oxidized compounds other than oxygen as the terminal electron acceptor. This type of growth is referred to as anaerobic respiration. In a fermentative process, ATP is generated through SLP, with the oxidation of electron donors coupled to the reduction of electron carriers such as $NAD(P)^+$ or flavin adenine dinucleotide (FAD). The reduced electron carriers are reoxidized, reducing the metabolic intermediate.

This chapter describes the fermentation processes carried out by various anaerobic prokaryotes. In fermentation, ATP is generated not only through SLP, but also by other mechanisms such as the reactions catalysed by fumarate reductase and Na^+-dependent decarboxylase and lactate/H^+ symport, as described earlier (Section 5.8.6).

8.1.2 Hydrogen in fermentation

The product formed/substrate consumed ratio is constant in some fermentations, such as the ethanol (Section 8.3) and homolactate (Section 8.4) fermentations, while it is variable in others such as the clostridial fermentation (Section 8.5). Fermentation processes with a constant ratio are referred to as linear pathways, while the others are referred to as branched fermentative pathways (Figure 8.1). A branched pathway yields more ATP and more oxidized products than a linear pathway. To produce more oxidized products, a proportion of the reduced electron carriers, such as NAD(P)H, should be oxidized coupled with the reduction of H^+ to H_2. The formation of products in a branched pathway is dictated by the growth conditions, especially the hydrogen partial pressure.

Anaerobic metabolism involves various electron donors and acceptors with different redox potentials. Examples include crotonyl-CoA reduction to butyryl-CoA ($E^{0'} = +0.19$ V) coupled to NADH oxidation ($E^{0'} = -0.32$ V) and lactate oxidation ($E^{0'} = -0.19$ V) coupled to NAD^+ reduction. Butyryl-CoA dehydrogenase oxidizes two NADH to reduce crotonyl-CoA and ferredoxin ($E^{0'} = -0.50$ V) in an electron bifurcation reaction

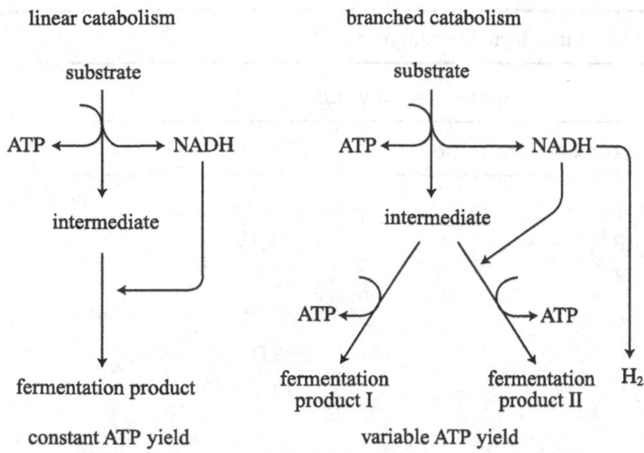

Figure 8.1 Linear and branched pathways in anaerobic fermentation.

(*Bacteriol. Rev.* **41**: 100–180, 1977)

Ethanol fermentation is an example of a linear fermentative pathway, while more than one product is formed in the clostridial acetate–butyrate fermentation. In a branched fermentative pathway, more ATP can be generated when protons are used as the electron acceptor.

(Section 8.5.1.2), while the *Acetobacterium woodii* lactate dehydrogenase oxidizes lactate and ferredoxin to reduce two NAD^+ in an electron confurcation reaction (Sections 8.5.1.3 and 9.5.2.2). The enzymes catalysing electron bifurcation and confurcation reactions contain a flavin that has two electron-carrying sites, one with a high redox potential and the other with a low redox potential. Electron bifurcation and confurcation reactions play important roles in anaerobic metabolism, conserving energy and overcoming thermodynamic barriers (Section 5.8.6.4).

8.2 Molecular oxygen and anaerobes

As discussed in Chapter 2 (Section 2.2.2), microbes are classified according to their response to molecular oxygen (O_2), into aerobes, facultative anaerobes and obligate anaerobes. Aerotolerant obligate anaerobes are distinguished from strict anaerobes among the obligate anaerobes. O_2 comprises about 20 per cent of air, and air-saturated liquid media contains about 7–8 mg/l O_2 at ambient temperature. Microaerophiles use O_2 as their electron acceptor at a low dissolved O_2 concentration of 0.1–3 mg/l O_2, but they cannot grow above this concentration. Strict anaerobes and microaerophiles are inhibited by molecular oxygen or its metabolites. Several hypotheses have been proposed to explain the inhibitory mechanism.

Molecular oxygen reacts with reduced flavoproteins, Fe–S proteins and cytochromes to be reduced to hydrogen peroxide (H_2O_2) or superoxide ($O_2^{\cdot-}$). These are very strong oxidants with high redox potentials [$E^{0'}(O_2^{\cdot-}/H_2O_2) = +0.98$ V, $E^{0'}(H_2O_2/H_2O) = +1.35$ V] and destroy cellular polymers such as DNA, RNA, proteins and other essential components. Aerobes and facultative anaerobes therefore possess enzymes which detoxify them. These enzymes are superoxide dismutase (SOD) and catalase (Table 8.1). Peroxidase is another enzyme that removes hydrogen peroxide. The reactions are:

$$2O_2^{\cdot-} + 2H^+ \xrightarrow{\text{SOD}} H_2O_2 + O_2$$

$$2H_2O_2 \xrightarrow{\text{catalase}} 2H_2O_2 + O_2$$

$$H_2O_2 + RH_2 \xrightarrow{\text{peroxidase}} 2H_2O + R$$

Strict anaerobes have been thought to be sensitive to O_2 because they do not possess SOD and catalase. This is true in some cases, but these enzyme activities have been identified in some strict anaerobes, and genes for these enzymes and related proteins have been found in their genomes. For example, SOD was found in *Clostridium butyricum*. Methanogenic archaea are one of the most O_2-sensitive groups of organisms. *Methanosarcina barkeri* possesses SOD and catalase activities, and the SOD gene has been identified in *Methanobacterium bryantii* and *Methanobacterium thermoautotrophicum*. Sulfate-reducing bacteria are less sensitive to O_2 than methanogens, and *Desulfovibrio gigas*, *Desulfovibrio vulgaris* and several other *Desulfovibrio* species have been shown to possess SOD and catalase activities.

Table 8.1 | Superoxide dismutase and catalase activities in selected prokaryotes.

Organisms	Specific activity (U/mg)	
	Superoxide dismutase	Catalase
Aerobes and facultative anaerobes		
Escherichia coli	1.8	6.1
Salmonella typhimurium	1.4	2.4
Rhizobium japonicum	2.6	0.7
Micrococcus radiodurans	7.0	289.0
Pseudomonas species	2.0	22.5
Aerotolerant anaerobes		
Eubacterium limosum	11.6	0
Enterococcus (Streptococcus) faecalis	0.8	0
Lactococcus (Streptococcus) lactis	1.4	0
Clostridium oroticum	0.6	0
Lactobacillus plantarum	0	0
Strict anaerobes		
Veillonella alcalescens	0	0
Clostridium pasteurianum	0	0
Clostridium butyricum[a]	1.4	0
Clostridium sticklandii	0	0
Butyrivibrio fibrisolvens	0	0.1
Methanosarcina barkeri	+[b]	40.0
Desulfovibrio gigas	3.4	52.6
Pyrococcus furiosus	0	0

[a] A strong peroxidase activity (5.8 U/mg) is found in this bacterium.
[b] Superoxide dismutase has been purified in this archaeon.

In addition to SOD and catalase, species of *Desulfovibrio* have proteins that detoxify reactive oxygen species including superoxide. These are Fe-containing electron carriers such as rubredoxin, desulfoferredoxin, neelaredoxin and rubrerythrin. Similar proteins are present in the hyperthermophilic anaerobic fermentative bacterium *Thermotoga maritime*. Desulfoferrodoxin is also responsible for detoxification of reactive oxygen species in *Clostridium acetobutyricum*. A similar

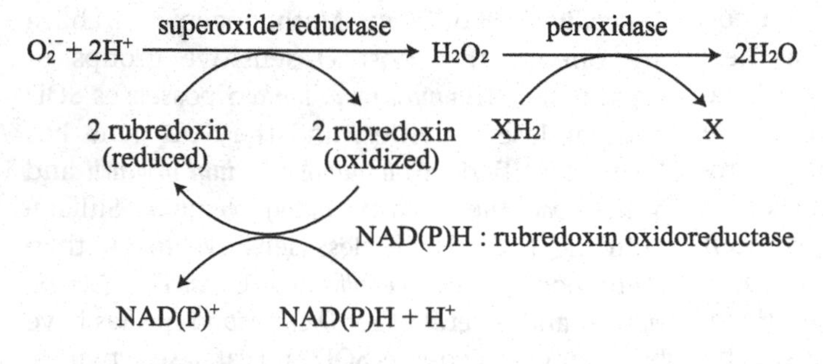

(modified from *Science*, **286**: 306–309, 1999)

protein to neelaredoxin in an archaeon, *Archaeoglobus fulgidus*, functions as a SOD, and another similar protein, superoxide reductase, destroys superoxide directly to water, and not through O_2, in *Pyrococcus furiosus*. Methanoferredoxin of *Methanosarcina mezei* is another superoxide reductase.

The microaerobic and nitrate-ammonifying *Wolinella succinogenes* does not have catalase but has different defence mechanisms against oxidative and nitrosative stresses. This bacterium has SOD and employs two periplasmic multihaem *c*-type cytochromes, cytochrome *c* peroxidase and cytochrome *c* nitrite reductase, to detoxify hydrogen peroxide and hydroxylamine, respectively. Two peroxiredoxin isoenzymes protect the cells against different organic hydroperoxides.

Though some strict anaerobes have SOD and catalase activities (Table 8.1), they do not grow under aerobic conditions in the laboratory. This may be due to the nature of the molecular oxygen. Dissolved oxygen increases the redox potential of the solution and a high redox potential inhibits the growth of some strict anaerobes. Methanogens grow at a redox potential lower than −0.3 V. Sulfide is an essential component of some enzymes, and molecular oxygen oxidizes this to form a disulfide. The organisms might not be able to grow with such inactivated enzymes. One hypothesis proposes that growth is impossible due to a lack of reducing equivalents for biosynthesis, since electrons are exhausted to reduce oxygen. The strictly anaerobic sulfate-reducing bacterium *Desulfovibrio vulgaris* and the metal-reducing bacterium *Geobacter sulfurreducens* have cytoplasmic membrane-bound oxygen reductases that reduce oxygen with electrons supplied by a periplasmic hydrogenase. *Clostridium acetobutyricum* has NADH peroxidase and NADH oxidase to reduce hydrogen peroxide and molecular oxygen. In an anaerobic hyperthermophilic archaeon, *Pyrococcus furiosus*, reduced ferredoxin in the modified EMP pathway (Section 4.1.4) is the electron donor for similar reactions. It is therefore likely that multiple mechanisms are responsible for growth inhibition by oxygen.

8.3 Ethanol fermentation

Saccharomyces cerevisiae ferments carbohydrates through the EMP pathway to ethanol, and the ED pathway is used by *Zymomonas mobilis*. Pyruvate is decarboxylated to acetaldehyde, which is used as the electron acceptor. Acetaldehyde is reduced to ethanol, which consumes the electrons generated during the glycolytic process, where ATP is generated through SLP (Figure 8.2). *Saccharomyces cerevisiae* generates

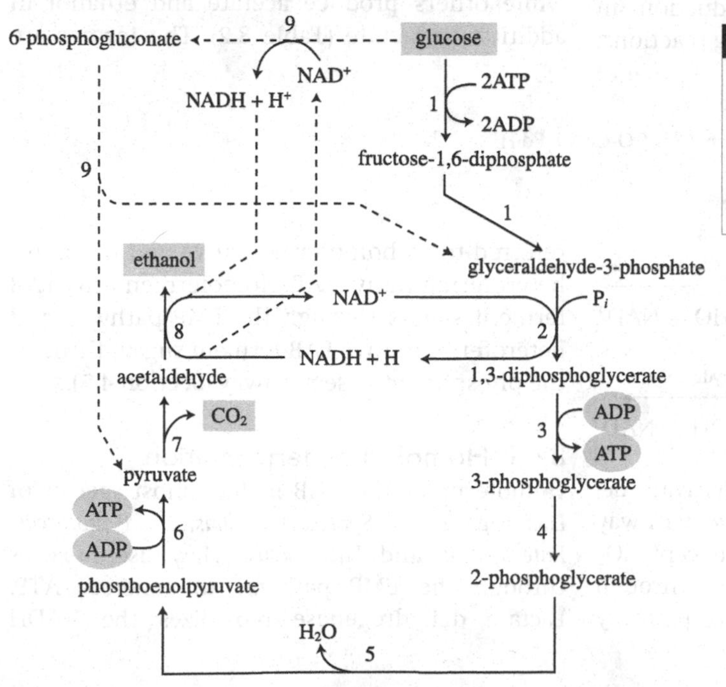

Figure 8.2 Ethanol fermentation by *Saccharomyces cerevisiae* and *Zymomonas mobilis*.

1–6, EMP pathway (in solid lines); 7, pyruvate decarboxylase; 8, alcohol dehydrogenase; 9, ED pathway (in dotted lines).

2 ATP from 1 hexose molecule but a single ATP results from 1 hexose molecule in *Zymomonas mobilis*:

$$CH_3\text{-}CO\text{-}COOH \xrightarrow{\text{pyruvate decarboxylase}} CH_3\text{-}CHO + CO_2$$

$$CH_3\text{-}CHO + NADH + H^+ \xrightarrow{\text{alcohol dehydrogenase}} 7CH_3\text{-}CH_2OH + NAD^+$$

Zymomonas mobilis consumes molecular oxygen with glucose but not with ethanol, but the electron transport chain has not been characterized. A NADH:CoQ oxidoreductase mutant strain grew better than the wild-type strain under aerobic conditions, minimizing accumulation of the toxic intermediate, acetaldehyde.

Pyruvate decarboxylase has thiamine pyrophosphate as a prosthetic group, as in pyruvate dehydrogenase. Pyruvate decarboxylase is known mainly in eukaryotes. In addition to *Zymomonas mobilis*, this enzyme is found in a facultative anaerobe, *Erwinia amylovora*, and in a strictly anaerobic acidophile, *Sarcina ventriculi*. Pyruvate decarboxylase is a key enzyme of ethanol fermentation.

It should be noted that ethanol is produced through different reactions in saccharolytic clostridia, heterofermentative lactic acid bacteria and enteric bacteria. These bacteria oxidize pyruvate to acetyl-CoA before reducing it to ethanol. They do not possess pyruvate decarboxylase. Ethanol production in clostridia is catalysed by the following reactions:

$$CH_3\text{-}CO\text{-}COOH + Fd \xrightarrow[\text{CoA-SH} \rightarrow CO_2]{\text{pyruvate:ferredoxin oxidoreductase}} CH_3\text{-}CO\text{-}CoA + Fd\cdot H_2$$

$$CH_3\text{-}CO\text{-}CoA + NADH + H^+ \xrightarrow{\text{aldehyde dehydrogenase}} CH_3\text{-}CHO + NAD^+$$

$$CH_3\text{-}CHO + NADH + H^+ \xrightarrow{\text{aldehyde dehydrogenase}} CH_3\text{-}CH_2OH + NAD^+$$

Ethanol fermentation through pyruvate decarboxylase in a linear fermentative pathway does not produce any by-products except CO_2 and water, while ethanol fermentation through acetyl-CoA is a branched fermentative pathway and produces various fermentation products, such as lactate, acetate and H_2.

Thermophilic anaerobes ferment various carbohydrates, including cellulose and pentoses, through acetyl-CoA to ethanol. Among them are *Thermoanaerobacter brockii* (formally *Thermoanaerobium brockii*), *Thermoanaerobacterium saccharolyticum*. *Thermoanaerobacter ethanolicus* and *Clostridium thermocellum* (Section 8.5.1.6). It is interesting to note that in these thermophilic bacteria a bifunctional enzyme, AdhE, reduces acetyl-CoA to ethanol coupled to the oxidation of NADPH. AdhE has aldehyde dehydrogenase and alcohol dehydrogenase domains.

8.4 Lactate fermentation

Lactate is a common fermentation product in many facultative and obligate anaerobes. Some bacteria produce lactate as a major fermentation product and these are referred to as lactic acid bacteria (LAB). Most LAB have a limited ability to synthesize monomers for biosynthesis and vitamins which are needed as growth factors. LAB are regarded as obligate anaerobes, but they can use oxygen, synthesizing cytochromes when hemin (haemin) is provided in the medium. Some LAB produce only lactate from sugars while others produce acetate and ethanol in addition to lactate (Table 8.2). The former are referred to as homofermentative and the latter heterofermentative LAB. Homofermentative LAB ferment sugars through the EMP pathway and heterofermentative LAB ferment sugars through the phosphoketolase pathway (Section 4.5).

8.4.1 Homolactate fermentation

Homofermentative LAB include most species of *Lactobacillus*, *Sporolactobacillus*, *Pediococcus*, *Enterococcus* and *Lactococcus*. They use hexoses through the EMP pathway to generate ATP. Lactate dehydrogenase reoxidizes the NADH

Table 8.2 Representative LAB and their fermentation mode.

Genus and species	Fermentation mode	
	Homofermentative	Heterofermentative
Lactobacillus		
L. delbrueckii	+	−
L. lactis	+	−
L. bulgaricus	+	−
L. casei	+	−
L. curvantus	+	−
L. plantarum	+	−
L. brevis	−	+
L. fermentum	−	+
Sporolactobacillus		
S. inulinus	+	−
Enterococcus		
E. faecalis	+	−
Lactococcus		
L. cremoris	−	+
L. lactis	+	−
Leuconostoc		
L. mesenteroides	−	+
L. dextranicum	−	+
Pediococcus		
P. damnosus	+	−
Bifidobacterium[a]		
B. bifidum	−	+

[a] These bacteria are phylogenetically classified apart from the lactic acid bacteria.

reduced during the EMP pathway, using pyruvate as the electron acceptor (Figure 8.3). As fermentation proceeds lactate is accumulated, lowering the intracellular pH. Lactate dehydrogenase is active in acidic conditions, producing lactate as the major product. Under alkaline conditions, homofermentative LAB produce large quantities of acetate and ethanol. Lactate dehydrogenase deficient mutants of three homofermentative LAB, *Lactococcus lactis*, *Enterococcus faecalis* and *Streptococcus pyogenes*, grow on glucose under microaerobic conditions, producing acetate, ethanol, acetoin and butanediol. Similarly, homofermentative LAB use hydroxycinnamic acids as the electron acceptors, diverting carbon flux from lactate to acetate.

8.4.2 Heterolactate fermentation

Species of *Leuconostoc* and *Bifidobacterium* produce ethanol and acetate in addition to lactate. They employ a unique glycolytic pathway known as the phosphoketolase pathway (Section 4.5). As shown in Figure 4.9, heterofermentative LAB like *Leuconostoc mesenteroides* oxidize glucose-6-phosphate to ribulose-5-phosphate. Epimerase converts ribulose-5-phosphate to xylulose-5-phosphate before cleavage to glyceraldehyde-3-phosphate and acetyl-phosphate by the action of

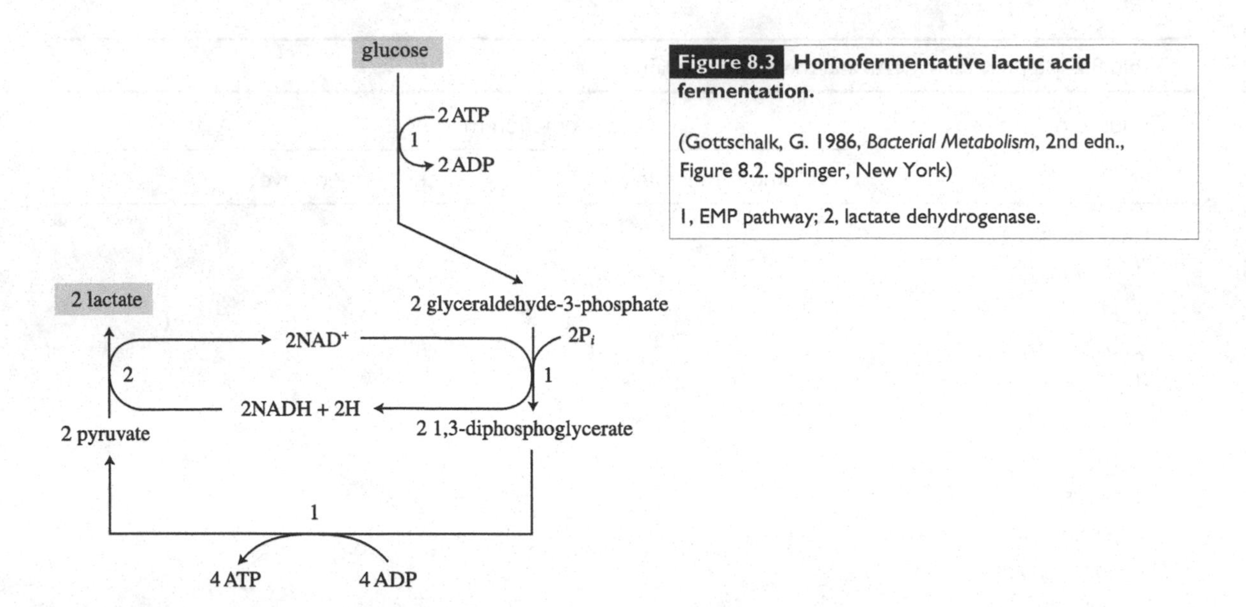

Figure 8.3 Homofermentative lactic acid fermentation.

(Gottschalk, G. 1986, *Bacterial Metabolism*, 2nd edn., Figure 8.2. Springer, New York)

1, EMP pathway; 2, lactate dehydrogenase.

phosphoketolase. Glyceraldehyde-3-phosphate is metabolized to lactate as in the homolactate fermentation, generating ATP. Acetyl-phosphate is reduced to ethanol, acting as the electron acceptor to oxidize the NADH reduced in the glucose-6-phosphate oxidation process. One ATP per hexose is available from this fermentation.

Pentoses are converted to xylulose-5-phosphate without reducing NAD^+. In this case, acetyl-phosphate is not used as the electron acceptor but is used to synthesize ATP. *Leuconostoc mesenteroides* synthesizes 1 ATP from a molecule of hexose and 2 ATP from a molecule of pentose.

Bifidobacterium bifidum ferments two molecules of hexose to two molecules of lactate and three molecules of acetate, employing two separate phosphoketolases, one active on fructose-6-phosphate and the other on xylulose-5-phosphate (Section 4.5.2). This bacterium synthesizes 5 ATP from two molecules of glucose. Since hexose-6-phosphate is not metabolized through a reductive process, acetyl-phosphate is used to synthesize ATP, as in pentose metabolism by *Leuconostoc mesenteroides*.

8.4.3 Biosynthesis in lactic acid bacteria (LAB)

LAB require many amino acids, vitamins, nucleic acid bases and other substances as growth factors because they cannot synthesize them. However, they can synthesize a few monomers and polymers for biosynthesis from acetyl-CoA. Acetyl-CoA and acetyl-phosphate are not intermediates in homolactate fermentation. Pyruvate is oxidized to acetyl-CoA through different routes, depending on the strain. The pyruvate dehydrogenase multienzyme complex oxidizes pyruvate to acetyl-CoA in most species of *Lactococcus* and *Enterococcus*, as in aerobic bacteria. *Enterococcus faecalis*, *Bifidobacterium bifidum* and *Lactobacillus casei* rely on pyruvate:formate lyase, an enzyme found in anaerobic metabolism in facultative anaerobic enteric bacteria (Section 8.6):

$$\underset{\text{pyruvate}}{CH_3\text{-}CO\text{-}COOH} \xrightarrow[\text{CoA-SH}]{\text{pyruvate:formate lyase}} \underset{\text{acetyl CoA}}{CH_3\text{-}CO\text{-}CoA} + \underset{\text{formate}}{HCOOH}$$

Pyruvate oxidase and phosphotransacetylase convert pyruvate to acetyl-CoA in other LAB, including *Lactobacillus delbrueckii* and *Lactobacillus plantarum*:

$$\underset{\text{pyruvate}}{CH_3\text{-}CO\text{-}COOH} \xrightarrow[\text{FAD} + P_i \quad \text{FADH}_2]{\text{pyruvate oxidase}} \underset{\text{acetyl phosphate}}{CH_3\text{-}CO\text{-}P + CO_2}$$

$$CH_3\text{-CO-O-Ph} + \text{CoA-SH} \xrightarrow{\text{phosphotransacetylase}} 6CH_3\text{-CO-CoA} + P_i$$

$FADH_2$ is reoxidized by lactate dehydrogenase:

$$\text{pyruvate} + FADH_2 \xrightarrow{\text{lactate dehydrogenase}} \text{lactate} + FAD$$

8.4.4 Oxygen metabolism in LAB

LAB are aerotolerant obligate anaerobes and possess superoxide dismutase and peroxidase to detoxify superoxide ($O_2^{\cdot-}$) and hydrogen peroxide (H_2O_2), respectively. Peroxidase generates sulfite radicals from sulfite, which are lethal to cells. Sulfite can therefore be used for the selective elimination of LAB from peroxidase-negative aerobic organisms:

$$H_2O_2 + XH_2\ (NADH + H^+) \xrightarrow{\text{peroxidase}} 2H_2O + X\ (NAD^+)$$

Catalase activity is negative but becomes positive in some LAB, such as *Enterococcus faecalis*, *Lactobacillus brevis* and *Lactobacillus plantarum*, when hemin is provided. These organisms have the genetic information to synthesize the catalase apoprotein but not hemin. The $Y_{glucose}$ of *Enterococcus faecalis* is 22 g under anaerobic conditions but increases up to 52 g under aerobic conditions. This is because *Enterococcus faecalis* can synthesize ATP through an electron transport phosphorylation (ETP) process using O_2 as an electron acceptor.

Lactobacillus plantarum can use oxygen as well as nitrate as terminal electron acceptors when both hemin and quinine are provided. The nitrate is reduced to ammonia.

In addition to ETP, peroxidase (or oxidase) increases the cell yield, oxidizing NADH with O_2 as electron acceptor. When (per)oxidase oxidizes NADH in heterofermentative LAB, extra ATP can be synthesized through substrate-level phosphorylation from acetyl-phosphate that is not needed as an electron acceptor. This reaction is catalysed by acetate kinase.

Similarly, homofermentative *Lactococcus lactis* oxidizes pyruvate to acetate, or condenses it to acetoin, instead of using it as an electron acceptor under aerobic conditions when H_2O-forming NADH oxidase oxidizes NADH to NAD^+, reducing O_2.

8.4.5 Lactate/H$^+$ symport

When the intracellular lactate concentration is higher than that of the medium, lactate is exported with $2H^+$, generating a proton motive force, in *Lactococcus cremoris* (Section 5.8.6.3).

8.4.6 LAB in fermented food

Dairy products and fermented vegetables, such as *kimchi*, in Far Eastern countries, are typical LAB fermented foods. In addition to lactate production, LAB produce flavours that are specific for each food type.

Citrate is produced in the initial phase of soybean sauce fermentation and converted to acetate in the later stages by LAB. *Pediococcus halophilus*, isolated from maturing soybean sauce, ferments citrate to acetate and formate (Figure 8.4). In this fermentation, citrate lyase cleaves citrate to acetate and oxaloacetate (OAA). OAA is decarboxylated to pyruvate. It is not known if energy is conserved during the decarboxylation reaction of OAA in this bacterium, and this reaction is not coupled to proton (sodium) motive force generation as in other LAB. Since pyruvate is not needed to dispose of electrons, it is converted to acetyl-CoA through a reaction catalysed by pyruvate:formate lyase. The acetyl-CoA is used to synthesize ATP. Some enteric bacteria, including *Salmonella typhimurium* and *Klebsiella pneumoniae*, ferment citrate through a similar metabolism (Figure 8.16, Section 8.6.3). Citrate fermentation by *Pediococcus halophilus* can be summarized as:

$$\text{citrate} + ADP + P_i \longrightarrow 2\text{acetate} + \text{formate} + ATP$$

Milk contains about 1.5 g/l citrate, which is converted to diacetyl, acetoin, acetate and lactate during the butter fermentation by LAB such as *Lactococcus lactis* subsp. *diacetylactis*, *Leuconostoc cremoris*, *Leuconostoc oenos* and *Leuconostoc mesenteroides* (Figure 8.5). Citrate is metabolized to acetate and pyruvate as in *Pediococcus halophilus*. When the NADH/NAD$^+$ ratio is high, pyruvate is reduced to lactate, and acetoin is produced through 2-acetolactate when the ratio is low. A similar reaction is found in some *Bacillus* species (Section 7.11.2) and in enteric bacteria (Figure 8.14, Section 8.6.2). ATP

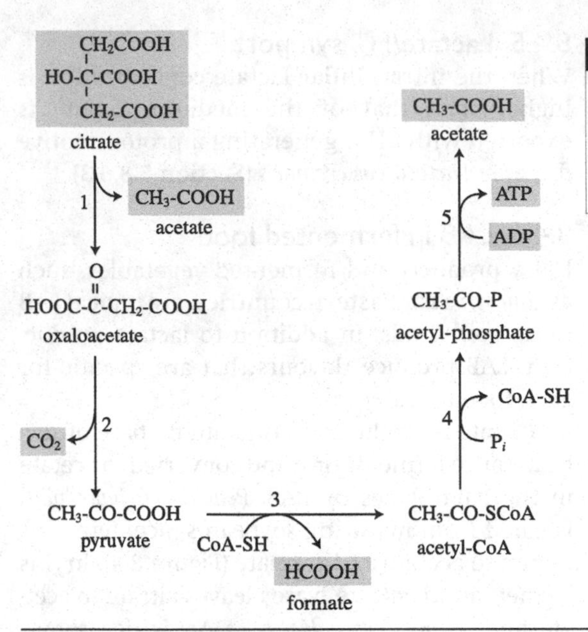

Figure 8.4 Citrate conversion to acetate by *Pediococcus halophilus* in the soybean sauce fermentation.

1, citrate lyase; 2, oxaloacetate decarboxylase; 3, pyruvate:formate lyase; 4, phosphotransacetylase; 5, acetate kinase.

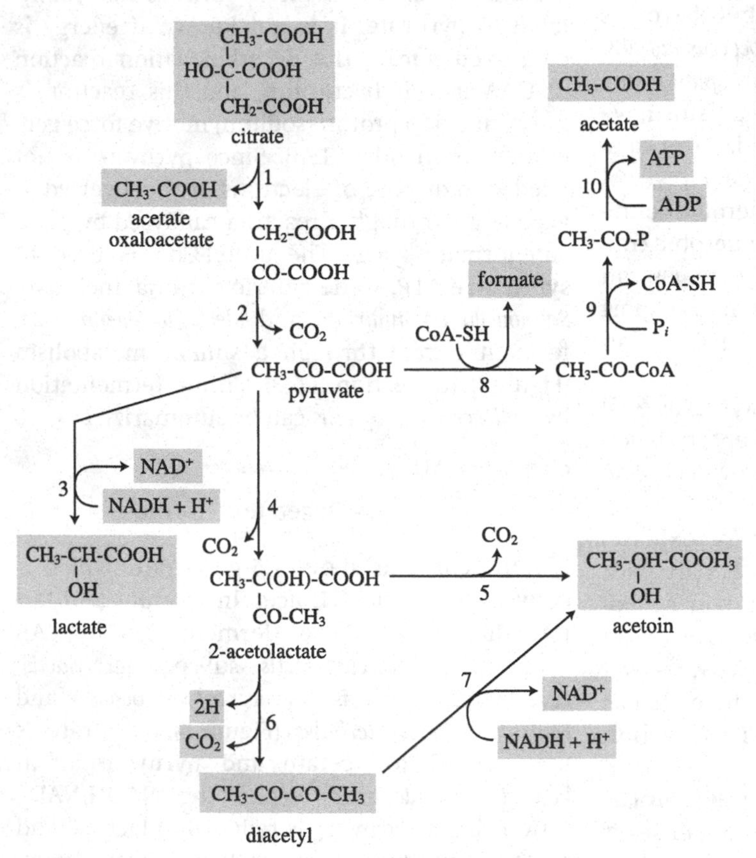

Figure 8.5 Conversion of citrate to acetoin by lactic acid bacteria during butter and wine maturation.

1, citrate lyase; 2, oxaloacetate decarboxylase; 3, lactate dehydrogenase; 4, 2-acetolactate synthase; 5, 2-acetolactate decarboxylase; 6, spontaneous chemical reaction; 7, acetoin dehydrogenase; 8, pyruvate dehydrogenase complex; 9, phosphotransacetylase; 10, acetate kinase.

is synthesized by acetate kinase during this fermentation, but energy is not conserved in the decarboxylation of OAA. A proton motive force is generated through citrate/lactate exchange, which is a similar transport system to the malate/lactate exchange (Section 5.8.6).

Lactobacillus plantarum, *Lactobacillus casei*, *Leuconostoc mesenteroides*, *Leuconostoc oenos* and *Lactococcus lactis* convert malate to lactate in a process known as the malolactic fermentation:

$$\text{malate}^{2-} \longrightarrow \text{lactate}^{-} + CO_2$$

Since these bacteria cannot use malate as their carbon and energy source, the malolactic fermentation is possible in the presence of fermentable substrates such as glucose. A proton motive force is generated in this fermentation through H^+ consumption in the decarboxylation of malate to lactate and through malate^{2-}/lactate$^-$ exchange (Section 5.8.6).

Glycerol is dehydrated to 3-hydroxypropionaldehyde that serves as an electron acceptor in heterofermentative *Lactobacillus* species. Since these bacteria preferentially use this electron acceptor, pyruvate is oxidized to acetate, synthesizing ATP, through the reaction catalysed by acetate kinase:

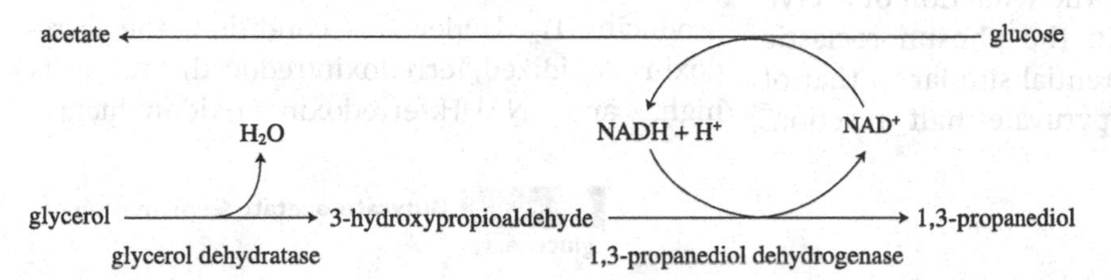

Consequently, the growth yield of heterofermentative LAB is higher on glucose with glycerol than on glucose alone. A similar fermentation is found in the metabolism of glycerol by enteric bacteria (Figure 8.15, Section 8.6.2).

8.4.7 Lactic acid bacteria as a probiotic

LAB are commonly described as probiotics, which can aid maintainance of a healthy intestinal microbiome in animals, including humans, and many commercial probiotic food products and supplements are available. In addition, LAB isolated from the vaginas of healthy women have been used to reduce vaginal ailments such as vulvovaginal candidiasis caused by *Candida glabrata*. These include *Lactobacillus rhamnosus*, *Lactobacillus reuteri*, *Lactobacillus fermentum* and *Lactobacillus jensenii*, among other species.

8.5 Butyrate and acetone–butanol–ethanol fermentations

Spore-forming anaerobic Gram-positive bacteria without sulfate-reducing ability are classified in the genus *Clostridium*. They are divided into saccharolytic and proteolytic clostridia according to their preferred electron donor. Saccharolytic clostridia ferment carbohydrates to butyrate and acetate, and proteinaceous compounds are fermented by proteolytic clostridia. The latter organisms are mostly pathogenic. Clostridial fermentation is a typical branched fermentative pathway. In addition to clostridia, species belonging to the genera *Butyrivibrio*, *Eubacterium* and *Fusobacterium* also produce butyrate (Table 8.3). All of these are obligate anaerobes.

Table 8.3 Anaerobes producing butyrate as their main fermentation product.

Butyribacterium methylotrophicum
Butyrivibrio fibrisolvens
Clostridium butyricum
C. kluyveri
C. pasteurianum
Eubacterium limosum
Fusobacterium nucleatum

8.5.1 Butyrate fermentation

Clostridium butyricum transports glucose by group translocation before metabolizing it to pyruvate. Pyruvate is oxidized to acetyl-CoA through a reaction known as a phosphoroclastic reaction, catalysed by pyruvate:ferredoxin oxidoreductase. This is different from the reaction catalysed by pyruvate dehydrogenase (Section 5.1). Hydrogenase can oxidize the reduced ferredoxin in this reaction to produce H_2:

$$\text{ferredoxin}\cdot H_2 \xrightarrow{\text{hydrogenase}} \text{ferredoxin} + H_2$$

All three groups of hydrogenase (Section 10.5.2), [FeFe], [NiFe] and [Fe], are found in clostridia.

8.5.1.1 Phosphoroclastic reaction

Pyruvate:ferredoxin oxidoreductase contains thiamine pyrophosphate as a prosthetic group, like pyruvate dehydrogenase (Section 5.1), and catalyses the following reactions:

$$CH_3\text{-CO-COOH} + \text{TPP-enzyme} \rightleftharpoons CH_3\text{-CHOH-TPP-enzyme} + CO_2$$

$$CH_3\text{-CHOH-TPP-enzyme} + \text{ferredoxin} + \text{CoA-SH} \rightleftharpoons \text{TPP-enzyme} + CH_3\text{-CO-CoA} + \text{ferredoxin}\cdot H_2$$

As shown above, the phosphoroclastic reaction is reversible, and this enzyme can reduce acetyl-CoA to pyruvate, while pyruvate dehydrogenase does not catalyse the reduction of acetyl-CoA. Ferredoxin used in the phosphoroclastic reaction has a redox potential similar to that of the acetyl-CoA + CO_2/pyruvate half reaction, while the pyridine nucleotide (NAD^+/NADH + H^+) has a higher redox potential:

$2H^+/H_2$	$E^{0'} = -0.41$ V
Fd(oxidized)/Fd(reduced)	$E^{0'} = -0.41$ V
NAD^+/NADH + H^+	$E^{0'} = -0.32$ V
Acetyl-CoA + CO_2/pyruvate	$E^{0'} = -0.42$ V

Since ferredoxin has a redox potential similar to that of the $2H^+/H_2$ half reaction, hydrogenase activity to oxidize the reduced ferredoxin depends on the hydrogen partial pressure, and so does the phosphoroclastic reaction. Ferredoxin is a [Fe–S] protein (Section 5.8.2.2).

8.5.1.2 Butyrate formation

Acetyl-CoA produced from the phosphoroclastic reaction is metabolized either to acetate through acetyl-phosphate or to butyrate through aceto-acetyl-CoA (Figure 8.6). The NADH reduced during glycolysis is oxidized, reducing acetoacetyl-CoA to butyrate. When the hydrogen partial pressure is low, hydrogenase oxidizes reduced ferredoxin producing H_2. Under this condition, the ferredoxin (oxidized)/ferredoxin(reduced) ratio is high, and NADH:ferredoxin oxidoreductase

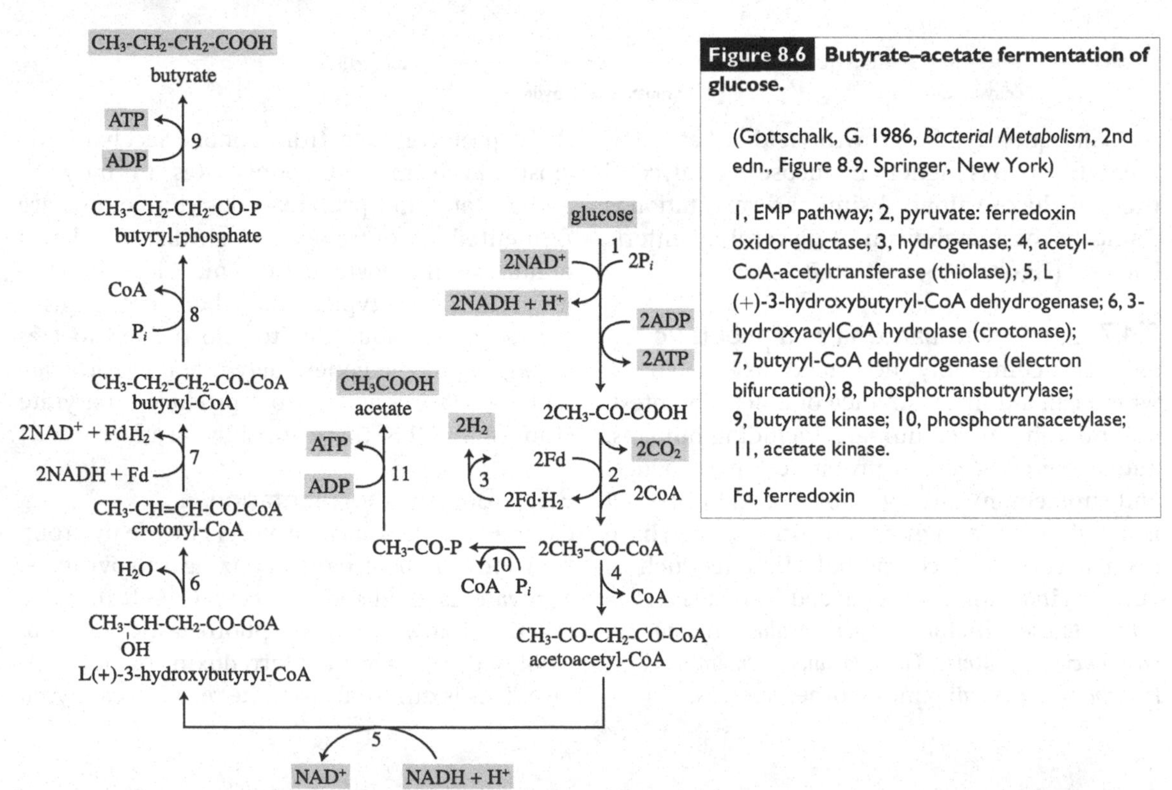

Figure 8.6 Butyrate–acetate fermentation of glucose.

(Gottschalk, G. 1986, *Bacterial Metabolism*, 2nd edn., Figure 8.9. Springer, New York)

1, EMP pathway; 2, pyruvate: ferredoxin oxidoreductase; 3, hydrogenase; 4, acetyl-CoA-acetyltransferase (thiolase); 5, L (+)-3-hydroxybutyryl-CoA dehydrogenase; 6, 3-hydroxyacylCoA hydrolase (crotonase); 7, butyryl-CoA dehydrogenase (electron bifurcation); 8, phosphotransbutyrylase; 9, butyrate kinase; 10, phosphotransacetylase; 11, acetate kinase.

Fd, ferredoxin

couples NADH oxidation to ferredoxin reduction. Since NADH is oxidized through these reactions, acetyl-CoA is not needed as the electron acceptor, and it is converted to acetyl-phosphate on which acetate kinase reacts to synthesize ATP:

$$NADH + H^+ \xrightleftharpoons{\text{NADH:ferredoxin oxidoreductase}} Fd{\cdot}H_2 \xrightleftharpoons{\text{hydrogenase}} H_2$$

In an undisturbed culture of *Clostridium butyricum*, 100 mol glucose is fermented to 76 mol butyrate and 42 mol acetate. However, when H_2 is continuously removed by shaking, the butyrate/acetate ratio becomes 1. This is due to the fact that the equilibrium shifts to the right in the above reactions. Since kinases synthesize ATP from acetyl-phosphate and butyryl-phosphate, 4 ATP are synthesized from glucose fermentation to acetate and 3 ATP from the butyrate fermentation.

The reduction of crotonyl-CoA to butyryl-CoA catalysed by butyryl-CoA dehydrogenase requires not only NADH but also ferredoxin. In this reaction an electron transferring flavoprotein (Etf) is involved. Flavins have two electron-carrying sites: one with a high redox potential and the other with a low redox potential. Out of four electrons from the oxidation of 2NADH, two electrons are transferred to the low redox potential site and the other two to the high redox potential site. Electrons from the former are used to reduce ferredoxin while the latter reduce crotonyl-CoA to butyryl-CoA. This is known as electron bifurcation (Sections 5.8.6.4 and 8.1.2) and is also involved in other butyrate-forming anaerobes, methanogens (Section 9.4.3) and homoacetogens (Section 9.5):

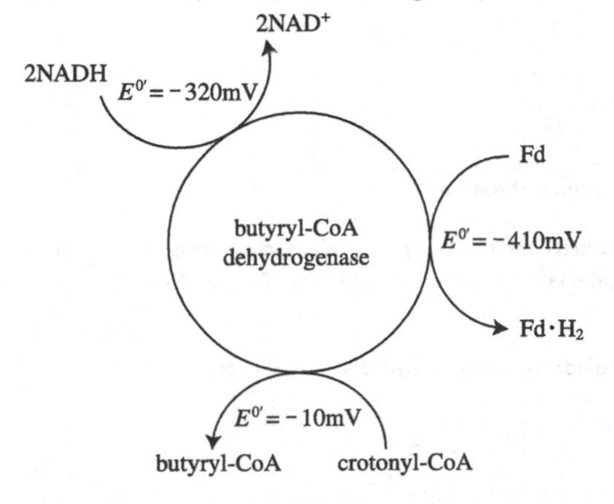

8.5.1.3 Lactate fermentation by *Clostridium butyricum*

Silage is pasturage (grass or other vegetation suitable for grazing animals) fermented by LAB. Stale silage contains butyrate, which can be formed from lactate and acetate by *Clostridium butyricum* and *Clostridium tyrobutyricum*. They do not ferment lactate alone, but produce butyrate from lactate and acetate (Figure 8.7). As shown in Table 8.4, 65 mmol butyrate and 59 mmol H_2 are produced from 100 mmol lactate and 32 mmol acetate.

As shown in the reactions below, when lactate and acetate are consumed in a 1:1 ratio, a net ATP gain is not possible:

$$\begin{array}{rl} & 2\text{lactate} + \text{ADP} \longrightarrow \text{butyrate} + \text{ATP} + 4\text{H} \\ +) & 2\text{lactate} + \text{ATP} + 4\text{H} \longrightarrow \text{butyrate} + \text{ADP} \\ \hline & 2\text{lactate} + 2\text{acetate} \longrightarrow 2\text{butyrate} \end{array}$$

A net gain of ATP is only possible when more lactate is consumed than acetate, and this becomes feasible when some of the electrons from lactate oxidation are consumed to produce H_2. Lactate dehydrogenase and pyruvate: ferredoxin oxidoreductase are involved in the lactate oxidation process. The redox potential of the pyruvate/lactate half reaction ($E^{0\prime}$ = –0.19 V) is too high for lactate oxidation to be coupled to NAD^+ reduction ($E^{0\prime}$ of $NAD^+/NADH$ = –0.32 V). Lactate oxidation is coupled to ferredoxin oxidation to reduce two NAD^+ in *Acetobacterium woodii* (Section 9.5.2.2). This reaction is referred to as electron confurcation (Section 8.1.2). It is believed lactate is oxidized

Table 8.4 Fermentation balance in *Clostridium butyricum* (mmol/100 mmol substrate consumed).

	Lactate + acetate	Pyruvate	Glucose
Acetate	–32[a]	33	42
Butyrate	65	33	76
CO_2	100	93	188
H_2	59	30	235

[a] – means consumption.

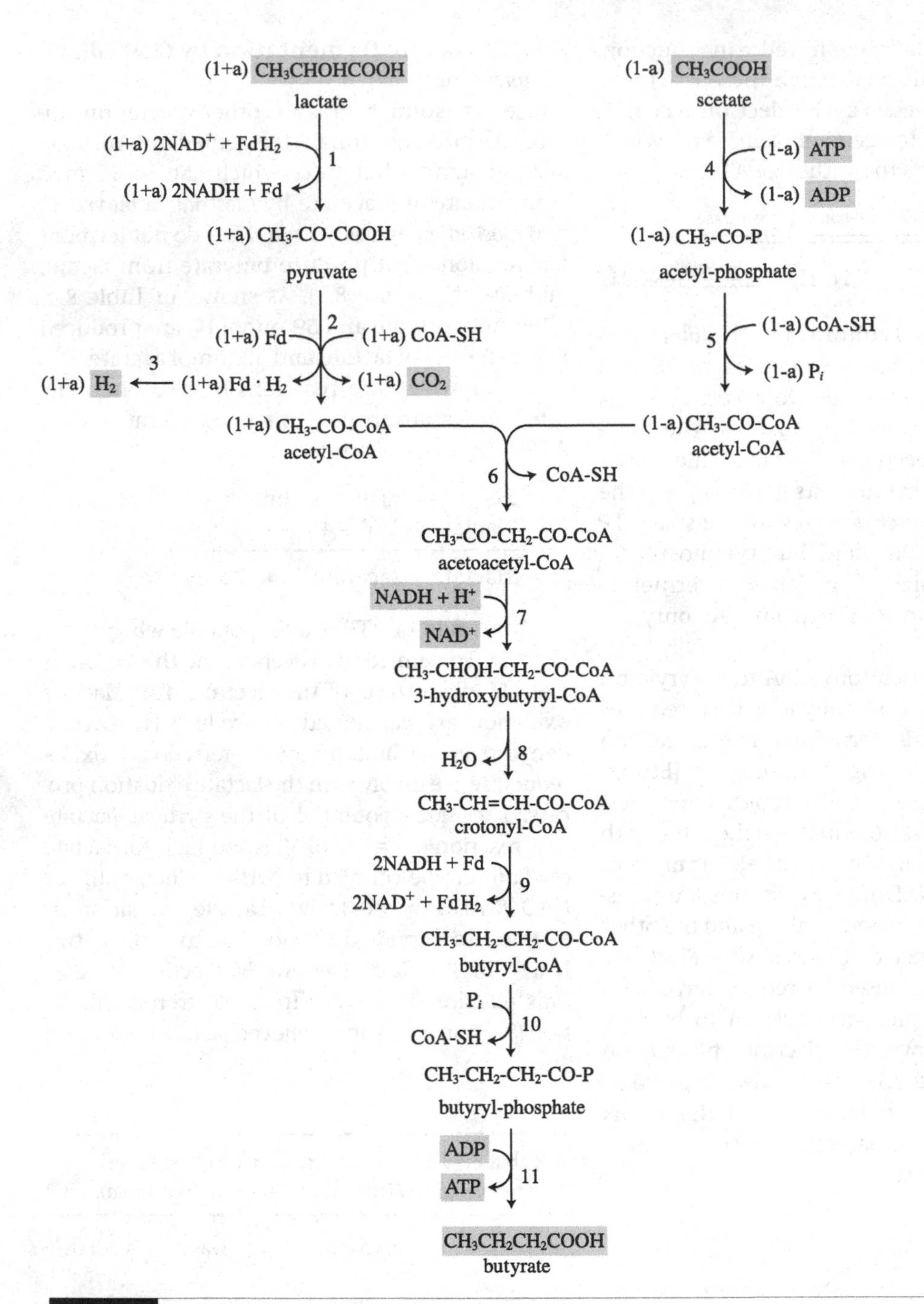

Figure 8.7 Lactate–acetate fermentation to butyrate by *Clostridium butyricum*.

When (1 + a) lactate is oxidized to (1 + a) acetyl-CoA, (1 – a) acetate are activated to (1 – a) acetyl-CoA, consuming (1 – a) ATP. Butyrate (1) is produced consuming [(1 + a) + (1 – a) = 2] acetyl-CoA and 4 H, with generation of 1 ATP. The net ATP generated is {[1 – (1 – a)] = a}, with 2a H_2 evolution.

1, lactate dehydrogenase (electron confurcation); 2, pyruvate:ferredoxin oxidoreductase; 3, hydrogenase; 4, acetate kinase; 5, phosphotransacetylase; 6–11, butyrate fermentation pathway.

in clostridia by electron confurcation. Ferredoxin is reduced by pyruvate:ferredoxin oxidoreductase and electron bifurcation by butyryl-CoA dehydrogenase. NADH is oxidized in the reduction of acetoacetyl-CoA to butyrate. Some of the electrons of the reduced ferredoxin from pyruvate oxidation are used for H_2 evolution. From the fermentation balance it can be seen that 59/2 mmol butyrate are produced solely from lactate and 59/2 mmol ATP are synthesized from 100 mmol lactate consumed, since four electrons can be generated in the conceptual conversion of lactate to butyrate. The lactate–acetate fermentation by *Clostridium butyricum* is summarized as

$$\text{acetate} + 3\text{lactate} + \text{ADP} + P_i \longrightarrow 2\text{butyrate} + 2H_2 + \text{ATP}$$

Similarly, *Clostridium kluyveri* ferments ethanol and acetate to butyrate and caproate, as in the following reaction:

$$6\text{ethanol} + 3\text{lactate} \longrightarrow 3\text{butyrate} + \text{caproate} + 2H_2 + 4H_2O + H^+$$

Just as *Clostridium butyricum* ferments lactate with acetate, *Clostridium kluyveri* ferments ethanol in the presence of acetate. Those anaerobes fermenting lactate and ethanol with acetate should obtain all the carbon skeletons needed for biosynthesis from these substrates. Pyruvate is provided by lactate dehydrogenase in *Clostridium butyricum*, and through the phosphoroclastic reaction catalysed by pyruvate:ferredoxin oxidoreductase from acetyl-CoA in *Clostridium kluyveri*:

$$CH_3\text{-CO-CoA} + \text{Fd}{\cdot}H_2 + CO_2 \xrightleftharpoons{\text{pyruvate: ferredoxin oxidoreductase}} CH_3\text{-CO-COOH} + \text{CoA-SH} + \text{Fd}$$

Glucose-6-phosphate is synthesized from pyruvate through gluconeogenesis, and pyruvate carboxylase converts pyruvate to oxaloacetate to obtain TCA cycle intermediates through the incomplete TCA fork (Figure 8.8; Section 5.4.1):

$$\text{pyruvate} + HCO_3^- + \text{ATP} \xrightleftharpoons{\text{pyruvate carboxylase}} \text{oxaloacetate} + \text{ADP} + P_i$$

8.5.1.4 Glycerol fermentation by *Clostridium butyricum*

Clostridium butyricum metabolizes glycerol disproportionately, oxidizing it to acetate and butyrate and the resulting electrons are consumed, reducing the remaining glycerol to 1,3-propaneiol (Figure 8.9).

8.5.1.5 *Clostridium butyricum* as a probiotic

Microbiome analysis indicates that propionate- and butyrate-producing bacteria play important roles in the human colon in promoting host health. Their suggested health-promoting effects include anti-lipogenic and anti-inflammatory properties, and the lowering of cholesterol levels. In addition, the short-chain fatty acids may exhibit anti-bacterial effects against undesirable bacteria. Bacteria that produce such fatty acids possess resistance mechanisms against them in order to survive.

Saccharolytic clostridia and lactic acid bacteria produce acetate, butyrate and lactate in high concentrations. Their pK_a values are 4.82, 4.75 and 3.86, respectively. As the fermentation proceeds, the acidic fermentation products accumulate, with a decrease in pH to values near or lower than the pK_a. At these conditions, undissociated forms of the acids accumulate. These are toxic to cells since the undissociated acids are hydrophobic and permeable to the cytoplasmic membrane, dissipating the proton motive force. However, the producing organisms do have some mechanisms of resistance to such acids in their undissociated forms. Acid resistance in *Clostridium acetobutylicum* is determined by a non-coding RNA and a GTPase. *Clostridium butyricum* spores have been used as a probiotic for over half a century in the Far East. It is believed that the fatty acids produced by this bacterium can control undesirable bacteria in the intestine, in a similar manner to the lactate produced by probiotic lactic acid bacteria.

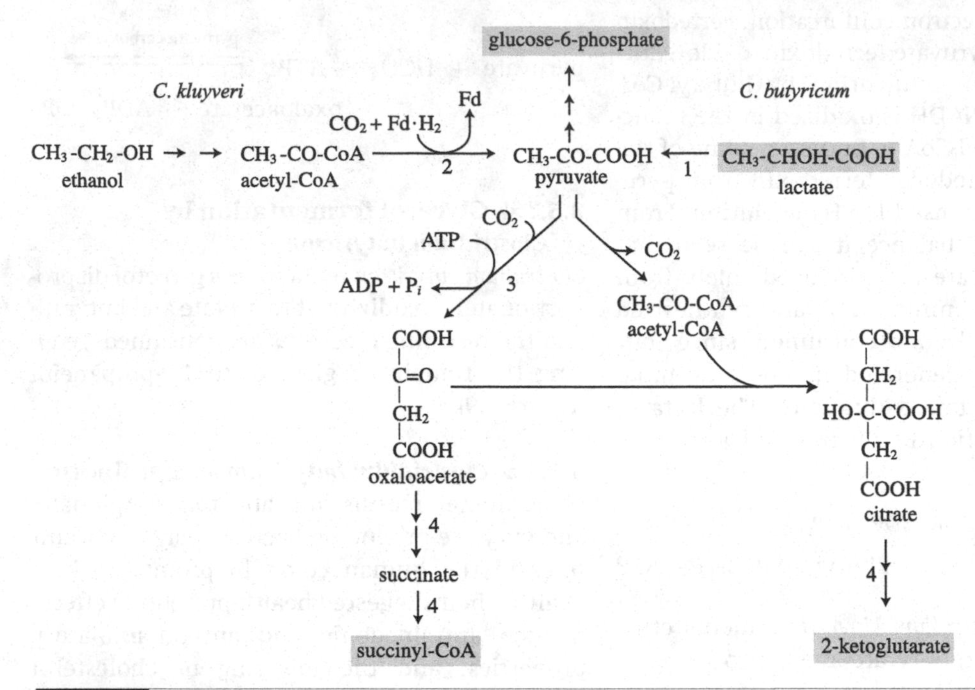

Figure 8.8 Supply of carbon skeletons through the incomplete TCA fork from lactate and ethanol in *Clostridium butyricum* and *Clostridium kluyveri*.

1, lactate dehydrogenase (electron confurcation); 2, pyruvate:ferredoxin oxidoreductase; 3, pyruvate carboxylase; 4, incomplete TCA fork.

8.5.1.6 Non-butyrate clostridial fermentation

Not all saccharolytic clostridia produce butyrate as their fermentation product. As described earlier (Section 8.3), some anaerobic fermentative bacteria ferment carbohydrates to acetate, ethanol, lactate, CO_2 and H_2. These include saccharolytic clostridia such as *Clostridium sphenoides*, and cellulolytic clostridia such as *Clostridium thermocellum*, *Clostridium cellulolyticum*, *Clostridium josui* and *Clostridium cellulovorans*.

8.5.2 Acetone–butanol–ethanol fermentation

Most butyrate-producing saccharolytic clostridia form small amounts of butanol, and a few strains such as *Clostridium acetobutylicum*, *Clostridium beijerinckii*, *Clostridium saccharobutylicum* and *Clostridium saccharoperbutylacetonicum* produce butanol in high concentration with acetone (or isopropanol) and ethanol. At the beginning of the fermentation they produce butyrate and acetate, disposing of the excess electrons to reduce H^+ to H_2, just like *Clostridium butyricum*. The solvent production starts when the acidic products accumulate (Figure 8.10).

During the solventogenic phase, sugars are fermented directly to solvents, and the acidic products are also converted to solvents. Acetate and butyrate are activated to acetyl-CoA and butyryl-CoA through the reactions catalysed by acetoacetyl-CoA:acetate coenzyme A transferase or kinase and phosphotransacetylase. The acyl-CoAs are reduced to ethanol and butanol by aldehyde dehydrogenase and alcohol dehydrogenase. Acetoacetate is decarboxylated to acetone. Acetone is further reduced to isopropanol in *Clostridium beijerinckii*, since its alcohol dehydrogenase is active not only on aldehydes but also on ketones (Figure 8.11).

For the onset of solventogenesis, electron flux as well as carbon flux should be diverted. More electrons are needed to produce butanol and ethanol than the acidic products. The electrons used

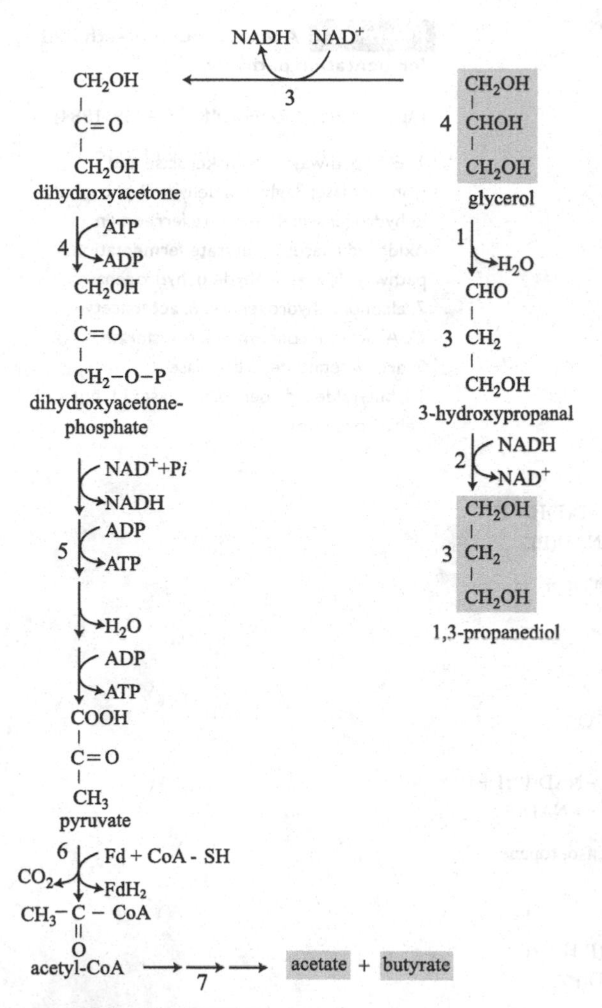

Figure 8.9 Glycerol fermentation by *Clostridium butyricum*.

1. glycerol dehydratase; 2, 1,3-propanediol dehydrogenase; 3, glycerol dehydrogenase; 4. dihydroxyacetone kinase; 5, parts of the EMP pathway; 6, pyruvate: ferredoxin oxidoreductase; 7, parts of the butyrate fermentation pathway.

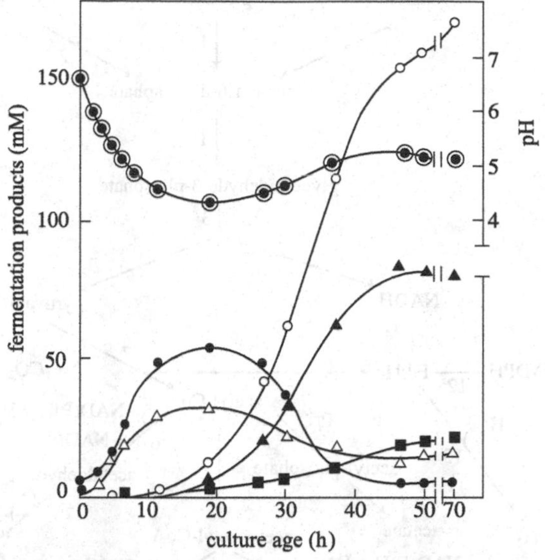

Figure 8.10 Butanol fermentation by *Clostridium acetobutylicum*.

(Gottschalk, G. 1986, *Bacterial Metabolism*, 2nd edn., Figure 8.10. Springer, New York)

This bacterium ferments sugar to butyrate and acetate in the initial stage of growth. When the acids accumulate with a decrease in pH, the bacterium switches its metabolism to produce solvents such as butanol, acetone and ethanol, thus consuming the acids.

○, butanol; ▲, acetone; ■, ethanol; ●, butyrate; Δ, acetate; ◉ pH.

to reduce H^+ to H_2 during the acidogenic phase are used by aldehyde dehydrogenase and alcohol dehydrogenase during the solventogenic phase. In butanol-producing clostridia, $NAD(P)^+$:ferredoxin oxidoreductase is active, exchanging electrons between $NAD(P)^+$ and ferredoxin. During the acidogenic phase this enzyme is active in reducing ferredoxin, oxidizing NAD(P)H, and catalyses the reverse reaction during solventogenesis. H_2 produced during acidogenesis is taken up by the bacteria for solvent production. Solventogenic clostridia have a H_2-producing hydrogenase as well as an uptake hydrogenase:

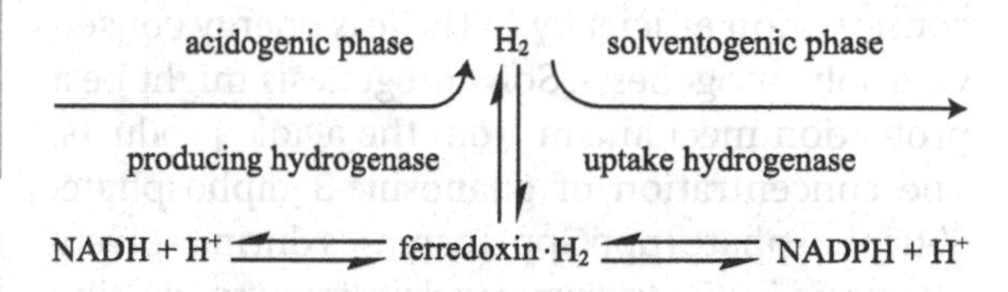

This is a typical branched fermentative pathway. Electrons from glycolysis are disposed of, reducing H^+ to H_2, and acyl-CoAs are not used as electron acceptors but used to synthesize ATP during acidogenesis through acyl-phosphate. Their reduction to solvent does not involve ATP synthesis. It is not known why and how the bacteria switch from acidogenesis with a high energy

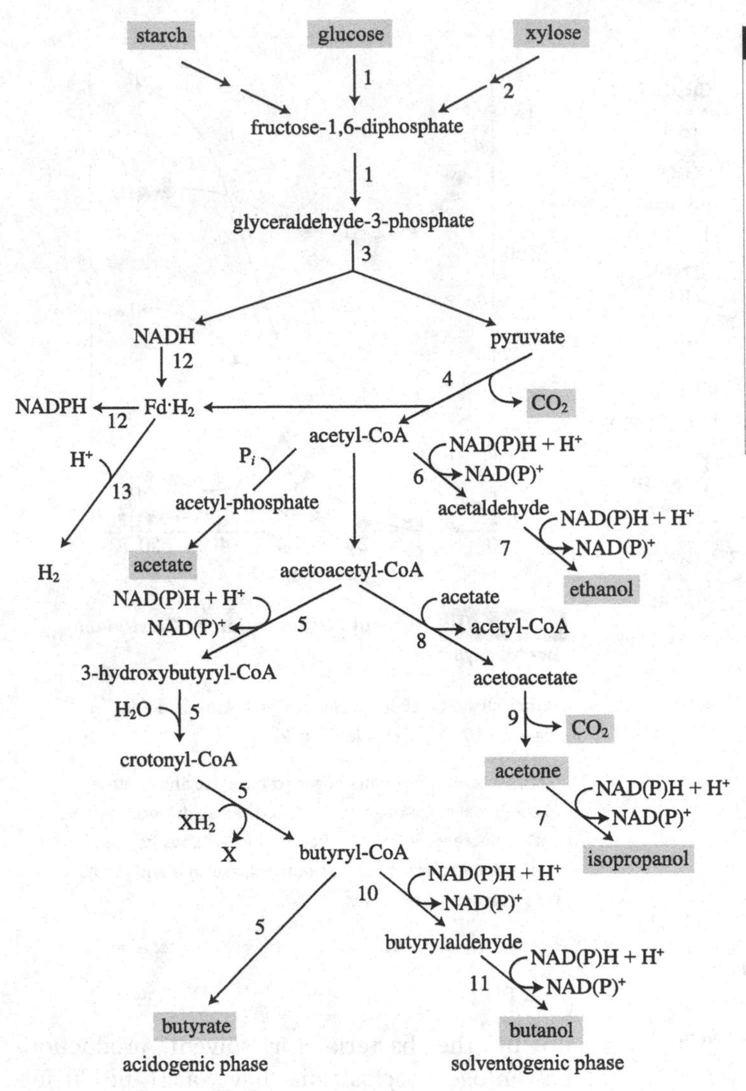

Figure 8.11 Acetone–butanol–ethanol fermentation pathway.

(*Appl. Environ. Microbiol.* **48**: 764–769, 1984)

1, EMP pathway; 2, transketolase and transaldolase; 3, glyceraldehyde-3-phosphate dehydrogenase; 4, pyruvate:ferredoxin oxidoreductase; 5, butyrate fermentation pathway; 6, acetaldehyde dehydrogenase; 7, alcohol dehydrogenase; 8, acetoacetyl-CoA: acetate coenzyme A transferase; 9, acetoacetate decarboxylase; 10, butyraldehyde dehydrogenase; 11, butanol dehydrogenase.

conservation efficiency to the less energy conserving solventogenesis. Solventogenesis might be a protection mechanism from the acidic products. The concentration of guanosine-3′-diphosphate-5′-triphosphate (pppGpp) increases during solventogenesis in *Clostridium acetobutylicum*, suggesting that genes for solventogenic enzymes are under stringent control (Section 12.2.1). A plasmid carries some genes for solventogenic metabolism in *Clostridium acetobutylicum*.

Acidogenic reactions from acyl-CoAs to acids are reversible. When the acidic products accumulate, the acyl-CoA concentration increases. Under these conditions the free coenzyme A concentration becomes too low for efficient metabolism. Solventogenesis can be explained as a coenzyme A recovery mechanism. Solventogenesis is also accompanied by sporulation. For this reason solventogenesis can be considered to be a secondary metabolism. Disruption of the sporulation protein blocks sporulation but not solventogenesis, and mother cell-specific prespore sigma factor σ^E mutants do not produce either solvent or spores, while disruption of the prespore-specific σ^G gene abolishes sporulation with solvent production. Solventogenesis starts with the onset of the sporulation process. Alcohol dehydrogenase gene expression is under the control of a redox-sensing transcriptional repressor Rex. Rex mutants produce more alcohols with reduced acetone formation.

Clostridium acetobutylicum does not ferment lactate only, but uses it with carbohydrate, increasing butyrate and butanol production with a decrease in acetone production. Lactate oxidation to pyruvate ($E^{0'} = -0.19$ V) is coupled with the reduction of crotonyl-CoA to butyryl-CoA ($E^{0'} = +0.19$ V). It is not known if lactate is oxidized through the electron confurcation reaction in this bacterium, as occurs in *Clostridium butyricum* (Section 8.5.1.3) and in *Acetobacterium woodii* (Section 9.5.2.2).

The *Clostridium acetobutylicum* genome sequence lacks obvious homologues for many of the enzymes of the TCA cycle, including citrate synthase, 2-ketoglutarate dehydrogenase, succinyl-coenzyme A (CoA) synthetase and fumarate reductase/succinate dehydrogenase. However, tracer experiments have shown that all TCA cycle intermediates are formed from glucose 5 via 2-ketoglutarate or through a reductive direction via fumarate (Figure 8.12). It is not yet known if other strictly anaerobic fermentative bacteria have a similar TCA cycle in order to supply carbon skeletons for biosynthesis.

8.5.3 Fermentation balance

Unlike linear fermentative pathways such as the ethanol and homo-fermentative lactic acid fermentations, saccharolytic clostridial fermentation produces a variety of products in a branched fermentative pathway (Table 8.5).

In a fermentation, electrons generated during ATP-synthesizing metabolism should be properly disposed of using metabolic intermediates as electron acceptors. In complex fermentations, such as that carried out by *Clostridium acetobutylicum*, it is difficult to judge if the electron balance is even. To make it simple, the oxidation/reduction (O/R)

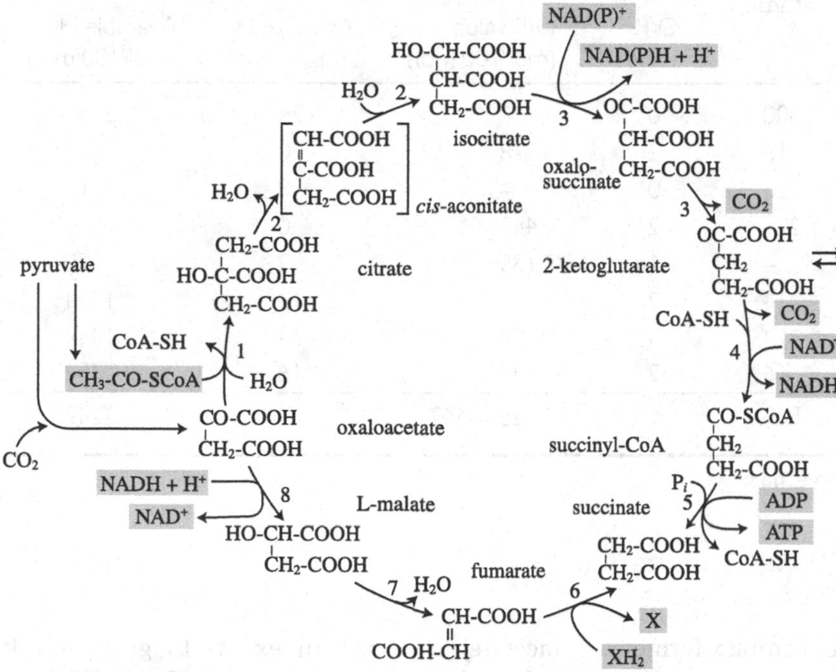

Figure 8.12 Complete bifurcated TCA cycle in *C. acetobutylicum*.

(Modified from *J. Bacteriol.* **192**: 4452–4461, 2010)

In this bifurcated TCA cycle, 2-ketoglutarate is produced from oxaloacetate, and succinate is produced reductively from fumarate or oxidatively from 2-ketoglutarate.

1–5, enzymes of the TCA cycle; 6–8, enzymes of the reductive TCA cycle.

Table 8.5 | Sugar fermentation by selected *Clostridium* species (mmol product/100 mmol sugar consumed).

Product	*C. butyricum*	*C. perfringens*	*C. acetobutylicum*
Butyrate	76	34	4
Acetate	42	60	14
Lactate	–	33	–
CO_2	188	176	221
H_2	235	214	135
Ethanol	–	26	7
Butanol	–	–	56
Acetone	–	–	22

–, not produced.

Table 8.6 | Carbon and electron recoveries and oxidation–reduction balance in the acetone–butanol–ethanol fermentation.

Substrate and product	Mol/ 100 mol substrate	Mol carbon	O/R balance		Balance of available hydrogen	
			O/R value	O/R value (mol/100 mol)	Available H	Available H (mol/100 mol)
Glucose	100	600	0	–	24	2400
Butyrate	4	16	−2	−8	20	80
Acetate	14	28	0	–	8	112
CO_2	221	221	+2	+442	0	–
H_2	135	–	−1	−135	2	270
Ethanol	7	14	−2	−14	12	1344
Butanol	56	224	−4	−224	24	1344
Acetone	22	66	−2	−44	16	352
Total		569		−425, +442		2242

Carbon recovery: 569/600 × 100 = 94.8%.

O/R balance = 442/425 = 1.04.

H recovery: 2242/2400 × 100 = 93%.

–, not applicable.

balance can be calculated in complex fermentations, as shown in Table 8.6. Arbitrarily, the O/R values of formaldehyde (CH_2O) and its multiples, for example hexoses and pentoses, are taken as zero. Each 2H in excess is expressed as −1, and +1 for a lack of 2H. For example, ethanol with a formula of C_2H_6O added to H_2O gives $C_2H_8O_2$. In comparison with $C_2H_4O_2$ (acetate), 4 H are in excess to give an O/R value of −2. Similarly, acetate ($C_2H_4O_2$) has an O/R value of 0, and carbon dioxide, $C(-H_2O)O$ gives a value of +2 (Table 8.6).

Since the cell yield in a fermentation is low and the cell mass has an O/R value of about zero, the O/R value of the fermentation products should be similar to that of the substrate.

8.6 Mixed acid and butanediol fermentation

8.6.1 Mixed acid fermentation

Some Gram-negative facultative anaerobic bacteria ferment glucose, producing various products including lactate, acetate, succinate, formate, CO_2 and H_2. These include species of *Escherichia*, *Salmonella*, *Shigella* and *Enterobacter* (Figure 8.13).

Glucose is metabolized through the EMP pathway. Phosphoenolpyruvate (PEP) carboxylase synthesizes oxaloacetate from PEP before being reduced to succinate. Pyruvate is either reduced to lactate by lactate dehydrogenase, or cleaved to acetyl-CoA and formate by pyruvate: formate lyase. According to the availability of electrons, acetyl-CoA is either reduced to ethanol or used to synthesize ATP.

Strictly anaerobic bacteria such as *Anaerobiospirillum succiniciproducens* and *Actinobacillus succinogenes* ferment carbohydrate mainly to succinate through a similar metabolism. In this case the succinate yield is as high as the amount of carbohydrate fermented. PEP carboxylase (reaction 9, Figure 8.13) fixes a large amount of CO_2 in this fermentation.

8.6.2 Butanediol fermentation

Some *Erwinia*, *Klebsiella* and *Serratia* species produce 2,3-butanediol in addition to lactate and ethanol from pyruvate, the EMP pathway product. Pyruvate is the substrate for one of three enzymes in these bacteria. These are lactate dehydrogenase, pyruvate:formate lyase and 2-acetolactate synthase (Figure 8.14). The reactions catalysed by these enzymes are similar to those of the mixed acid fermentation, except for 2-acetolactate synthase. This enzyme condenses two molecules of pyruvate to 2-acetolactate that is further decarboxylated and reduced to 2,3-butanediol. A similar metabolism is found in *Bacillus polymyxa* during vegetative growth (Section 7.11.2) and in lactic acid bacteria fermenting citrate (Section 8.4.6).

The first enzyme of this metabolism, 2-acetolactate synthase, is best characterized in Gram-negative facultative bacteria. This enzyme has thiamine pyrophosphate as a cofactor to catalyse the following reactions:

$$CH_3\text{-CO-COOH} + \text{E-TPP} \longrightarrow CH_3\text{-CHOH-TPP-E} + CO_2$$

$$CH_3\text{-CHOH-TPP-E} + CH_3\text{-CO-COOH} \longrightarrow \begin{matrix} CH_3\text{-C=O} \\ CH_3\text{-COH-COOH} \end{matrix} + \text{E-TPP}$$

2-acetolactate

Under anaerobic conditions, 2,3-butanediol-producing facultative anaerobes produce acidic products, lowering the external and intracellular pH. 2-acetolactate synthase, which catalyses the first reaction to produce 2,3-butanediol, has an optimum at pH 6.0. When the intracellular pH drops, this enzyme becomes active to divert carbon flux from acid production to the neutral solvent. An enzyme catalysing the same reaction catalyses the first reaction of valine synthesis from pyruvate (Section 6.4.1). The enzyme involved in valine synthesis has an optimum at pH 8.0, and also catalyses the condensation reaction of 2-ketobutyrate and pyruvate to synthesize isoleucine. This enzyme is referred to as the pH 8.0 enzyme, while the enzyme involved in 2,3-butanediol synthesis is referred to as the pH 6.0 enzyme. Enzymes catalysing 2,3-butanediol form an operon with a regulator gene. The regulator protein activates the transcription of genes for the enzymes and the regulator gene is activated by acetate and quorum-sensing signalling molecules (Section 12.2.8).

Klebsiella pneumoniae, *Klebsiella oxytoca* and *Enterobacter aerogenes* ferment glycerol to various products, including 2,3-butanediol (Figure 8.15). They oxidize a part of glycerol to pyruvate, and dispose of the resulting electrons to reduce the remaining glycerol to 1,3-propanediol. Pyruvate is metabolized as in the 2,3-butanediol fermentation. Glycerol is reduced to 1,3-propanediol by lactic acid bacteria while oxidizing carbohydrate (Section 8.4.6). These diols are important petrochemical intermediates. The facultative anaerobe *Enterococcus faecalis* ferments glycerol, as

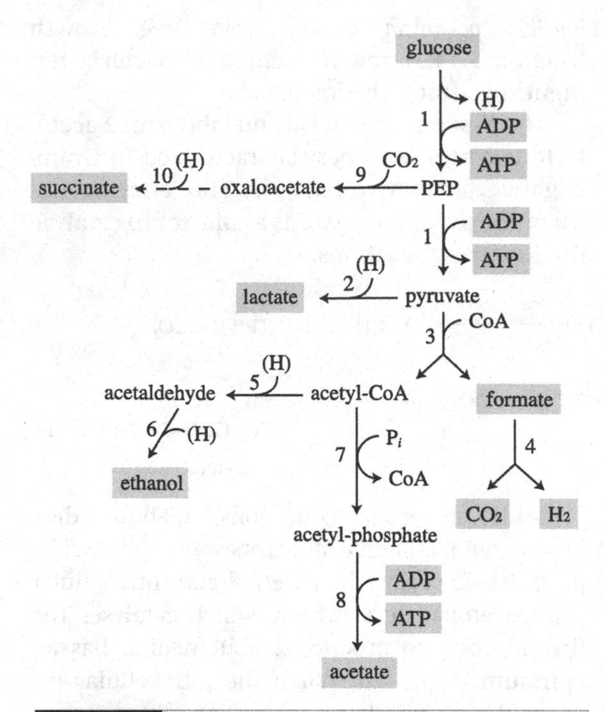

Figure 8.13 **Mixed acid fermentation by some Gram-negative facultative anaerobic bacteria.**

(Gottschalk, G. 1986, *Bacterial Metabolism*, 2nd edn., Figure 8.15. Springer, New York)

Facultative anaerobes belonging to the genera *Escherichia*, *Salmonella*, *Shigella*, *Enterobacter* and others, ferment sugars to lactate, acetate, formate, succinate and ethanol in the absence of electron acceptors.

1, EMP pathway; 2, lactate dehydrogenase; 3, pyruvate:formate lyase; 4, formate:hydrogen lyase; 5, acetaldehyde dehydrogenase; 6, alcohol dehydrogenase; 7, phosphotransacetylase; 8, acetate kinase; 9, phosphoenolpyruvate (PEP) carboxylase; 10, enzymes of the TCA cycle.

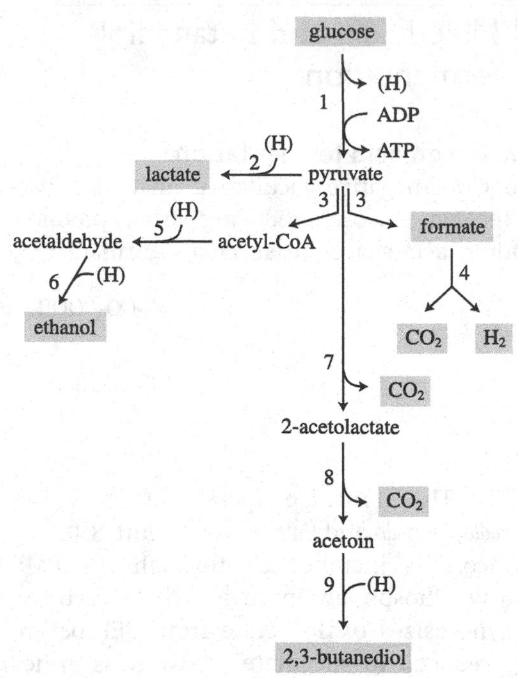

Figure 8.14 **Butanediol fermentation by some Gram-negative facultative anaerobic bacteria.**

(Gottschalk, G. 1986, *Bacterial Metabolism*, 2nd edn., Figure 8.15. Springer, New York)

Facultative anaerobes belong to the genera *Erwinia*, *Klebsiella* and *Serratia* and produce 2,3-butanediol in addition to lactate and ethanol.

1, EMP pathway; 2, lactate dehydrogenase; 3, pyruvate:formate lyase; 4, formate:hydrogen lyase; 5, acetaldehyde dehydrogenase; 6, alcohol dehydrogenase; 7, 2-acetolactate synthase; 8, 2-acetolactate decarboxylase; 9, 2,3-butanediol dehydrogenase.

shown in Figure 8.15, under anaerobic conditions, but metabolizes glycerol in two pathways under aerobic conditions. Glycerol is metabolized as in Figure 8.15 or phosphorylated to glycerol-3-phosphate before being oxidized to glyceraldehyde-3-phosphate.

8.6.3 Citrate fermentation by facultative anaerobes

Since a TCA cycle enzyme, 2-ketoglutarate dehydrogenase, is not expressed in facultative anaerobes under fermentative conditions, they cannot oxidize citrate through the TCA cycle. However, citrate is metabolized in several different pathways depending on the organism (Figure 8.16). *Salmonella typhimurium* and *Klebsiella pneumoniae* cleave citrate to acetate and oxaloacetate through the action of citrate lyase. Oxaloacetate is decarboxylated to pyruvate (Figure 8.16a). Pyruvate is oxidized as in glucose metabolism, synthesizing ATP through the action of acetate kinase. This reaction is similar to citrate oxidation by *Pediococcus halophilus* (Figure 8.4). Oxaloacetate decarboxylase in these bacteria is a Na^+-dependent enzyme and generates a sodium gradient, while the same

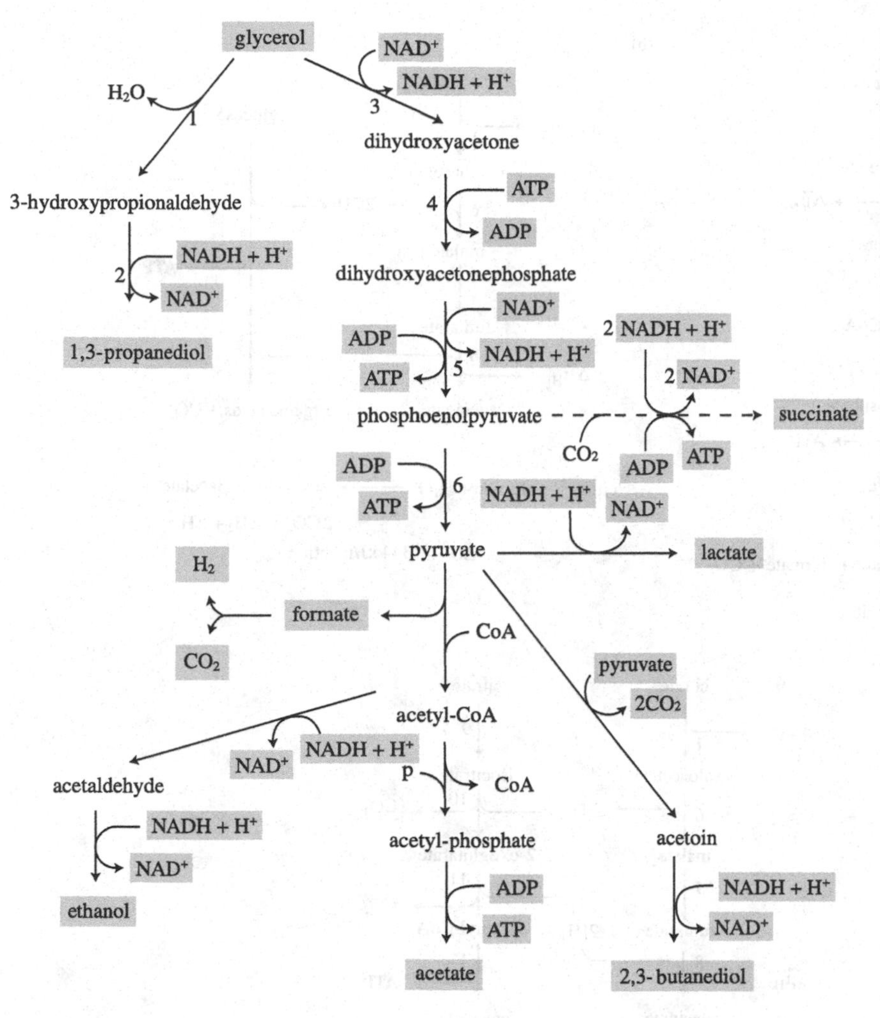

Figure 8.15 Glycerol fermentation by *Klebsiella pneumoniae*.

(*Appl. Microbiol. Biotechnol.* **50**: 24–29, 1998)

Glycerol is metabolized though a similar pathway to the butanediol fermentation, as well as being used as an electron acceptor to be reduced to 1,3-propanediol.

1, glycerol dehydratase; 2, 1,3-propanediol dehydrogenase; 3, glycerol dehydrogenase; 4, dihydroxyacetone kinase; 5, 6, enzymes in Figure 8.14.

reaction in lactic acid bacteria is not coupled to energy conservation.

Citrate is used as an electron acceptor to oxidize glucose in certain strains of *Escherichia coli* that possess genes for the citrate/succinate antiporter CitT and citrate lyase. These are activated by the two-component system CitA/CitB, which comprises the membrane-bound sensor kinase CitA and the response regulator CitB (Figure 8.16b).

Providencia rettgeri metabolizes citrate to succinate. Genes for all the TCA cycle enzymes are expressed in this bacterium under fermentative conditions, but citrate is not oxidized through the functional TCA cycle, since electrons from the TCA cycle in the form of NAD(P)H cannot be disposed of under the given conditions (Figure 8.16 c). Instead, this bacterium metabolizes citrate to succinate partly through the forward reaction of the TCA cycle and also through the reverse reaction to balance the oxidation and reduction. Energy is conserved in this metabolism through the action of fumarate reductase in the form of a proton motive force.

8.6.4 Anaerobic enzymes

Enteric bacteria oxidize pyruvate in a reaction catalysed by the pyruvate dehydrogenase complex under aerobic conditions. This enzyme is not expressed under anaerobic conditions, and NADH inhibits its activity. On the other hand, pyruvate:formate lyase functions only under anaerobic conditions, since it is expressed under fermentative conditions and irreversibly

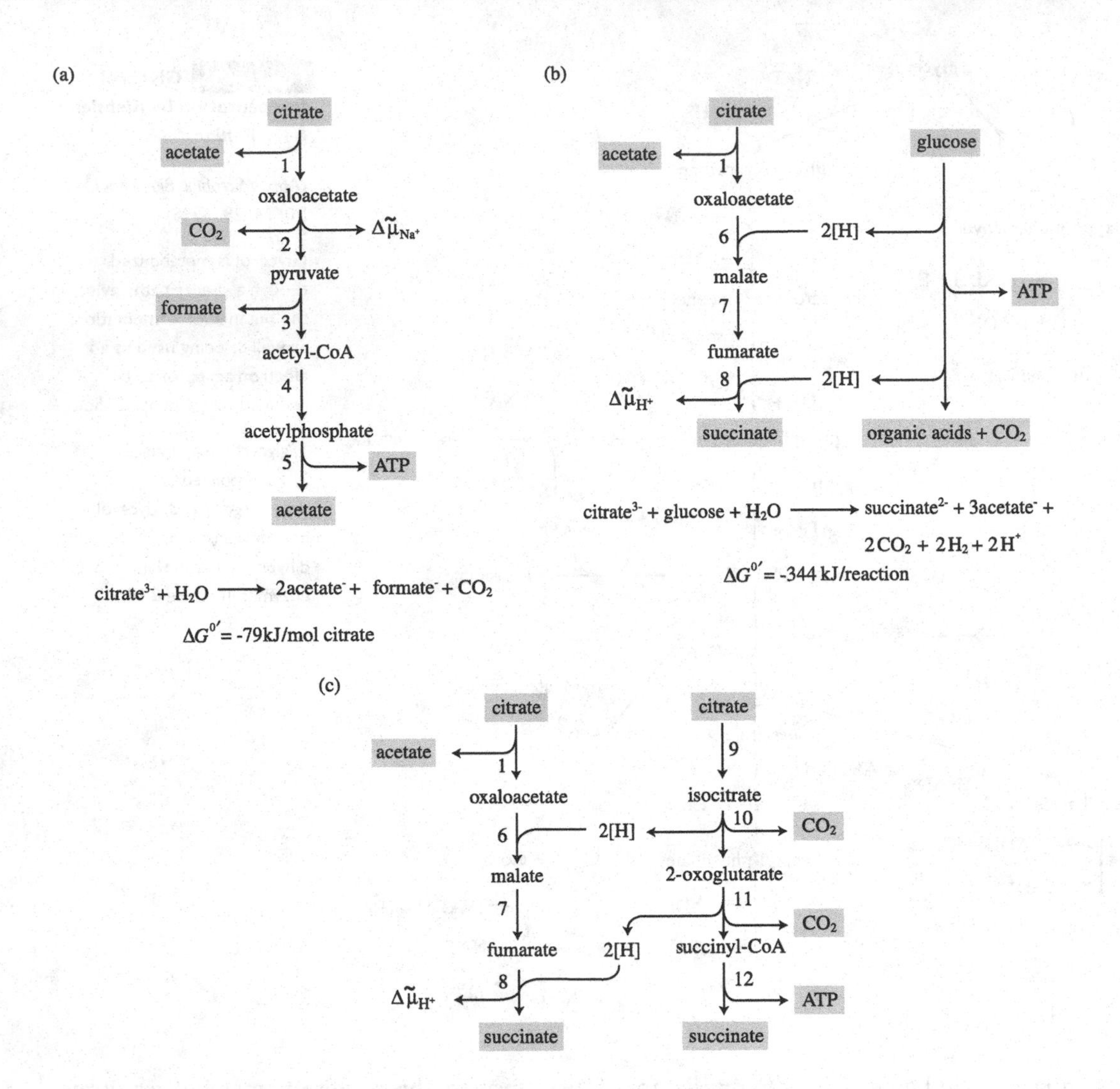

Figure 8.16 Citrate fermentation by enteric bacteria.

(*Arch. Microbiol.* **167**: 78–88, 1997)

(a) In the absence of electron acceptors, *Klebsiella pneumoniae* and *Salmonella typhimurium* ferment citrate to acetate, as in *Pediococcus halophilus* (Figure 8.4). (b) *Escherichia coli* reduces citrate as an electron acceptor with a fermentable substrate.
(c) *Providencia rettgeri* metabolizes citrate to succinate in a similar metabolism to that of the incomplete TCA fork.

1, citrate lyase; 2, oxaloacetate decarboxylase; 3, pyruvate:formate lyase; 4, phosphotransacetylase; 5, acetate kinase; 6, malate dehydrogenase; 7, fumarase; 8, fumarate reductase; 9, aconitase; 10, isocitrate dehydrogenase; 11, 2-ketoglutarate dehydrogenase complex; 12, succinate thiokinase.

inactivated by molecular oxygen. This enzyme can oxidize pyruvate to acetyl-CoA, required in anabolism, without reducing NAD^+ under fermentative conditions.

Formate:hydrogen lyase cleaves the formate produced by pyruvate:formate lyase to CO_2 and H_2 under fermentative conditions. This enzyme is a complex of formate dehydrogenase II (FDH_{II}) and

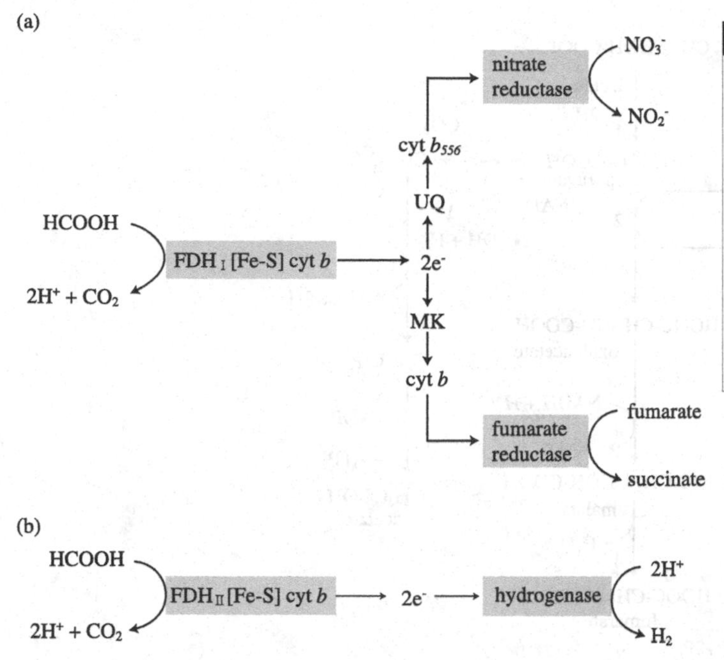

Figure 8.17 Formate metabolism in facultative anaerobes.

(Gottschalk, G. 1986, *Bacterial Metabolism*, 2nd edn., Figure 8.16. Springer, New York)

Facultative anaerobic enteric bacteria oxidize pyruvate in a reaction catalysed by pyruvate: formate lyase without reducing pyridine nucleotides. Formate dehydrogenase oxidizes formate to reduce nitrate or fumarate through anaerobic respiration (a) or to reduce protons (b).

hydrogenase. When electron acceptors such as nitrate or fumarate are present, FDH_I oxidizes formate to CO_2, transferring the electrons to nitrate reductase or fumarate reductase via NADH for energy conservation through anaerobic respiration. Formate metabolism under anaerobic conditions is shown in Figure 8.17. These enzymes are present in species of *Escherichia* and *Enterobacter*, but not in species of *Shigella* and *Erwinia*:

$$HOOC^- + H^+ \xrightarrow{\text{formate:hydrogen lyase}} CO_2 + H_2$$

$$HOOC^- + NAD^+ \xrightarrow{\text{formate dehydrogenase I}} CO_2 + NADH$$

Enzymes expressed only under anaerobic conditions in facultative anaerobes are referred to as anaerobic enzymes. They include pyruvate:formate lyase, hydrogenase, nitrate reductase and fumarate reductase. Their expression is controlled by the FNR protein (Section 12.2.4).

8.7 Propionate fermentation

Species of the genera *Propionibacterium*, *Clostridium propionicum* and *Megasphaera elsdenii* ferment carbohydrate or lactate to propionate, acetate and CO_2:

$$3\text{glucose} \longrightarrow 4\text{propionate} + 2\text{acetate} + 2CO_2$$

$$3\text{lactate} \longrightarrow 2\text{propionate} + \text{acetate} + CO_2$$

Lactate is the preferred substrate over carbohydrate in most propionate producers. They ferment glucose or lactate to propionate through either the acrylate pathway or the succinate-propionate pathway. Human gut bacteria, including *Roseburia inulinivorans*, produce propionate from deoxy sugars such as fucose and rhamnose. This bacterium produces butyrate from glucose, and propionate from fucose via propanediol, which is known as the propanediol pathway (Section 8.7.3). Spore-forming *Propionispora vibrioides* ferments sugar alcohols such as mannitol, sorbitol and xylitol to propionate and acetate through an unknown pathway. *Propionispora vibrioides* ferments the aliphatic polyester poly (propylene adipate) to propionate.

8.7.1 Succinate–propionate pathway

Species belonging to the genus *Propionibacterium* ferment lactate to propionate via succinate through this pathway (Figure 8.18). Lactate

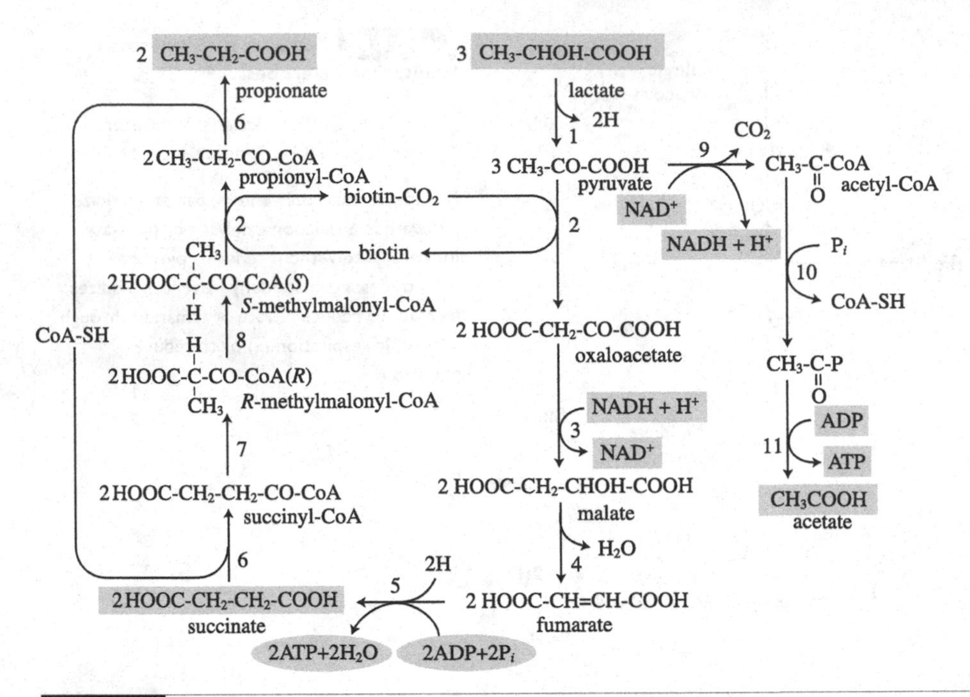

Figure 8.18 Succinate–propionate pathway in the genus *Propionibacterium*, which ferments lactate to propionate.

(Gottschalk, G. 1986, *Bacterial Metabolism*, 2nd edn., Figure 8.18. Springer, New York)

1, lactate dehydrogenase; 2, methylmalonyl-CoA:pyruvate transcarboxylase; 3, malate dehydrogenase; 4, fumarase; 5, fumarate reductase; 6, coenzyme A transferase; 7, methylmalonyl-CoA mutase; 8, methylmalonyl-CoA racemase; 9, pyruvate dehydrogenase; 10, phosphotransacetylase; 11, acetate kinase.

dehydrogenase with flavin oxidizes lactate to pyruvate. Two molecules of pyruvate are reduced to propionate as the electron acceptor, while one pyruvate is oxidized to acetate through acetyl-CoA synthesizing 1 ATP. Some enzymes of this pathway are of interest. They are transcarboxylase requiring biotin, fumarate reductase coupled to ATP synthesis and methylmalonyl-CoA mutase that uses coenzyme B_{12}. In this fermentation 3 ATP are synthesized fermenting 3 lactate, one by acetate kinase and two by fumarate reductase.

Methylmalonyl-CoA:pyruvate transcarboxylase transfers a carboxyl group from methylmalonyl-CoA to pyruvate to form propionyl-CoA and oxaloacetate, as in reaction 2 of Figure 8.16. Biotin is involved in this reaction as a cofactor.

Fumarate reductase couples the reduction of fumarate, using NADH as the electron donor, to the formation of a proton motive force, as described previously (Section 5.8.6.1).

When haem is provided, *Bacteroides fragilis* ferments carbohydrates to propionate and acetate through the succinate–propionate pathway with functional cytochromes. However, without haem, lactate and acetate are produced from the carbohydrate fermentation because fumarate reductase cannot function in the absence of functional cytochromes. As in lactic acid bacteria of the *Lactococcus* genus, this bacterium synthesizes the apoprotein of cytochromes.

Methylmalonyl-CoA mutase catalyses the carbon-rearranging reaction to convert succinyl-CoA to methylmalonyl-CoA. This reaction requires coenzyme B_{12}. Coenzyme B_{12} has the structure of 5′-deoxyadenosylcobalamin (Figure 8.19). This coenzyme is required for various enzyme reactions involving carbon

cyanocobalamin   5′-deoxyadenosylcobalamin

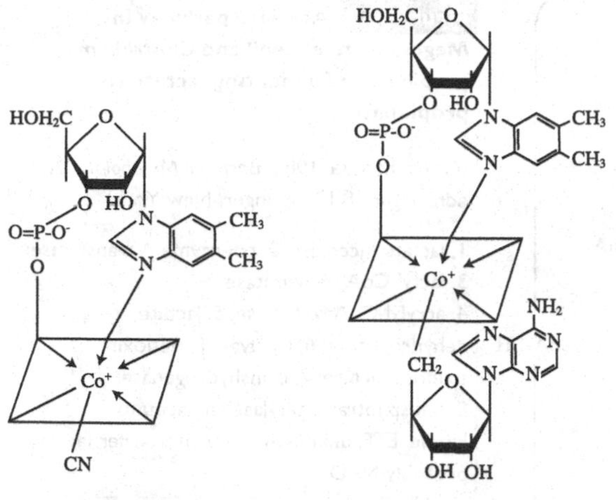

Figure 8.19 Structure of cyanocobalamin and coenzyme B_{12} (5′-deoxyadenosylcobalamin).

(Gottschalk, G. 1986, *Bacterial Metabolism*, 2nd edn., Figure 8.21. Springer, New York)

rearrangement. Vitamin B_{12} is produced commercially using *Propionibacterium shermanii* and related bacteria, since they synthesize coenzyme B_{12} in large concentrations for methylmalonyl-CoA mutase.

Propionate producers utilizing the succinate–propionate pathway excrete a small amount of succinate into the medium. In this case, methylmalonyl-CoA:pyruvate transcarboxylase cannot produce propionyl-CoA because the concentration of methylmalonyl-CoA becomes too low. To replace the excreted succinate, pyruvate or phosphoenolpyruvate is carboxylated to oxaloacetate which is reduced to succinate via malate:

$$\text{pyruvate} + \text{ATP} + P_i \xrightarrow{\text{pyruvate orthophosphate dikinase}} \text{PEP} + \text{AMP} + PP_i$$

$$\text{PEP} + CO_2 + \text{ADP} \xrightarrow{\text{PEP carboxykinase}} \text{oxaloacetate} + \text{ATP}$$

$$\text{PEP} + CO_2 + P_i \xrightarrow{\text{PEP carboxytransphosphorylase}} \text{oxaloacetate} + PP_i$$

Since pyrophosphatase activity is low in species of *Propionibacterium*, the pyrophosphate produced in the above reactions is used to phosphorylate sugars, thus conserving the energy carried by this inorganic compound (Section 13.2.4).

8.7.2 Acrylate pathway

The majority of the propionate producers ferment carbohydrate or lactate through the succinate–propionate pathway, but *Clostridium propionicum* and *Megasphaera elsdenii* ferment lactate to propionate via lactyl-CoA and acrylyl-CoA. Electrons for this reductive metabolism are supplied from the oxidation of lactate to acetate (Figure 8.20). The ATP yield of the acrylate pathway is 1 ATP/3 lactate, much lower than the 1 ATP/lactate in the succinate–propionate pathway.

8.7.3 Propanediol pathway

A human gut bacterium, *Roseburia inulinivorans*, ferments the deoxy sugar fucose, producing propionate with acetate and butyrate. This bacterium ferments glucose to acetate and butyrate. Fucose is isomerized to fuculose before being phosphorylated. Fuculophosphate is split into dihydroxyacetone phosphate and lactaldehyde. The latter is converted to propionate via propanediol, and the former to acetate and butyrate (Figure 8.21).

8.8 Fermentation of amino acids and nucleic acid bases

In addition to carbohydrates and organic acids, amino acids and nucleic acid bases are fermented under anaerobic conditions. Many strains of *Clostridium* ferment these nitrogenous compounds. These are distinguished as proteolytic clostridia, from the saccharolytic clostridia that ferment carbohydrates. Proteolytic clostridia are mainly pathogenic to animals including humans.

Many anaerobic bacteria, including proteolytic clostridia, can ferment individual amino acids

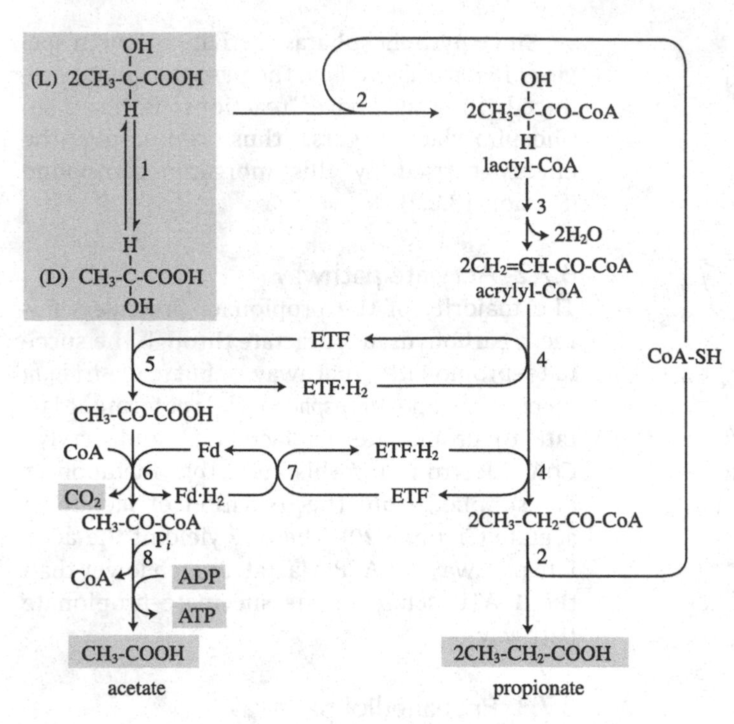

Figure 8.20 **Acrylate pathway in *Megasphaera elsdenii* and *Clostridium propionicum* fermenting lactate to propionate.**

(Gottschalk, G. 1986, *Bacterial Metabolism*, 2nd edn., Figure 8.19. Springer, New York)

1, lactate racemase; 2, coenzyme A transferase; 3, lactyl-CoA dehydratase; 4, acrylyl-CoA reductase; 5, lactate dehydrogenase; 6, pyruvate:ferredoxin oxidoreductase; 7, transhydrogenase; 8, phosphotransacetylase and acetate kinase. ETF, unknown electron transfer factor, probably NAD^+.

but a mixture is a better substrate in many cases. When amino acid mixtures are fermented, some amino acids are oxidized with ATP synthesis using other amino acids as the electron acceptors. The fermentation of amino acid mixtures is referred to as the Stickland reaction (Section 8.8.2).

8.8.1 Fermentation of individual amino acids

Amino acids are deaminated during their fermentation, as in oxidation under aerobic conditions. Reductive deamination is the commonest anaerobic amino acid metabolism (shown in the reaction below), although some amino acids are deaminated through oxidative deamination or transamination reactions, depending on the organism:

$$R\text{-}CHNH_2\text{-}COOH + 2H \longrightarrow RCH_2\text{-}COOH + NH_3$$

Clostridium propionicum deaminates alanine to pyruvate that is fermented to propionate and acetate as in the acrylate pathway (Figure 8.20).

Clostridium litoralis, *Clostridium sticklandii*, *Eubacterium acidaminophilum* and *Peptostreptococcus micros* ferment glycine according to the following stoichiometry:

$$4\text{glycine} + 2H_2O \longrightarrow 3\text{acetate} + 2CO_2 + 4NH_3$$

Glycine is oxidatively decarboxylated, forming methylenetetrahydrofolate (methylene-H_4F). Methylene-H_4F is further oxidized to CO_2 and the resulting electrons are transferred to thioredoxin before being consumed, reducing glycine in a reaction catalysed by glycine reductase (Figure 8.22). Glycine reductase is a selenium-containing protein reducing glycine to acetyl-phosphate using reduced thioredoxin as the electron donor. ATP is synthesized through substrate-level phosphorylation in the reactions catalysed by formyltetrahydrofolate synthetase and acetate kinase.

Threonine is a substrate for fermentative growth of *Clostridium propionicum* and *Peptostreptococcus prevotii*. Threonine dehydratase deaminates threonine to 2-ketobutyrate,

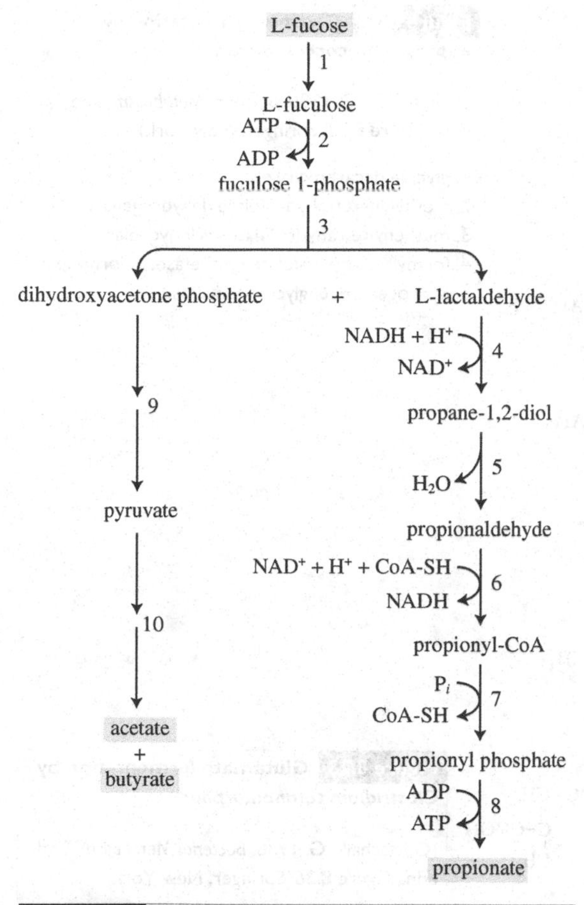

Figure 8.21 Fucose fermentation by *Roseburia inulinivorans* producing propionate.

(Modified from *J. Bacteriol.* **188**(12): 4340–4349, 2006)

1, fucose isomerase; 2, fuculokinase; 3, fuculophosphate aldolase; 4, propanediol oxidoreductase; 5, propanediol dehydratase; 6, propanealdehyde dehydrogenase; 7, phosphotransacylase; 8, propionate kinase; 9, parts of EMP pathway; 10, as in *Clostridium butyricum* fermentation.

before being oxidized to propionyl-CoA in a reaction catalysed by 2-ketobutyrate:ferredoxin oxidoreductase, which is a similar reaction to pyruvate oxidation by pyruvate: ferredoxin oxidoreductase. Hydrogenase oxidizes the reduced ferredoxin, producing hydrogen, and propionyl-CoA is converted to propionate, producing ATP:

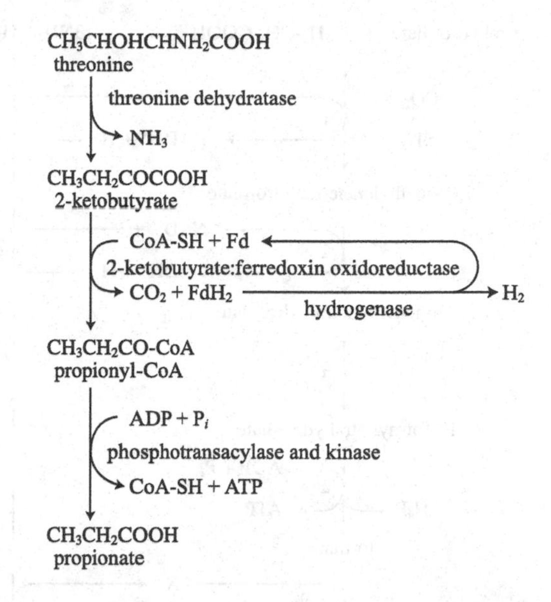

Glutamate is fermented through different pathways depending on the bacterial species. *Clostridium tetanomorphum* splits glutamate to acetate and pyruvate through a series of reactions initiated by coenzyme B_{12}-containing glutamate mutase (Figure 8.23). Pyruvate is metabolized to acetate and butyrate, as in the saccharolytic clostridial butyrate fermentation with ATP synthesis by kinases (Figure 8.6). On the other hand, *Acidaminococcus fermentans* and *Peptostreptococcus asaccharolyticus* ferment glutamate to acetate and butyrate via glutaconyl-CoA (Figure 8.24). The Na^+-dependent glutaconyl-CoA decarboxylase generates a sodium motive force (Section 5.8.6.2), and the kinases synthesize ATP.

Many obligate and facultative anaerobes, including *Clostridium novyi* and *Bacteroides melaninogenicus*, ferment aspartate to acetate and succinate. The former converts aspartate to alanine in a reaction catalysed by aspartate decarboxylase. Alanine is fermented as shown above. *Bacteroides melaninogenicus* deaminates aspartate to fumarate, one third of which is oxidized to acetate and two-thirds are reduced to succinate, consuming the electrons generated from the oxidation reactions. This bacterium possesses menaquinone and cytochrome *b* for the generation of a proton motive force by fumarate reductase, as occurs in propionate producers (Section 8.7.1).

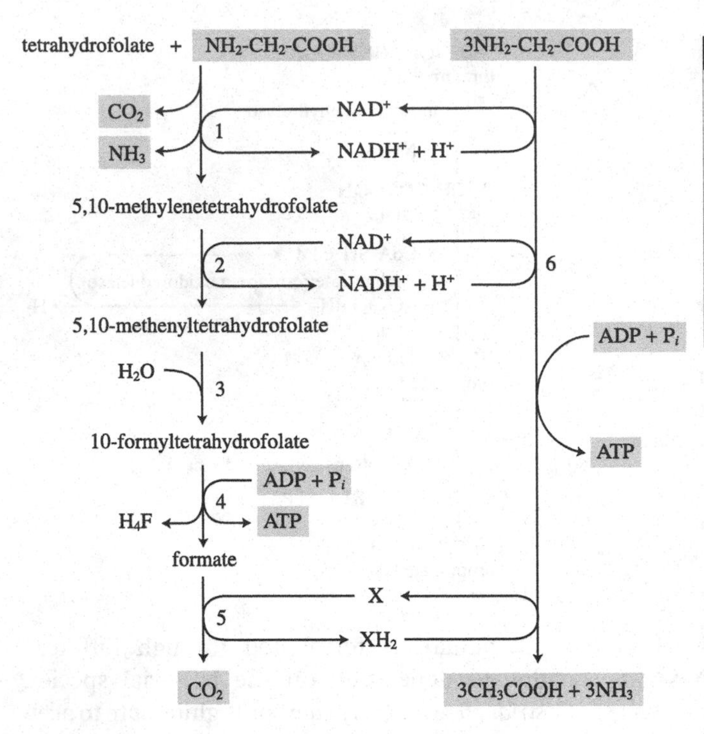

Figure 8.22 Glycine fermentation by *Peptostreptococcus micros.*

(Gottschalk, G. 1986, *Bacterial Metabolism*, 2nd edn., Figure 8.33. Springer, New York)

1, glycine decarboxylase;
2, methylenetetrahydrofolate dehydrogenase;
3, methenyltetrahydrofolate cyclohydrolase;
4, formyltetrahydrofolate synthetase; 5, formate dehydrogenase; 6, glycine reductase.

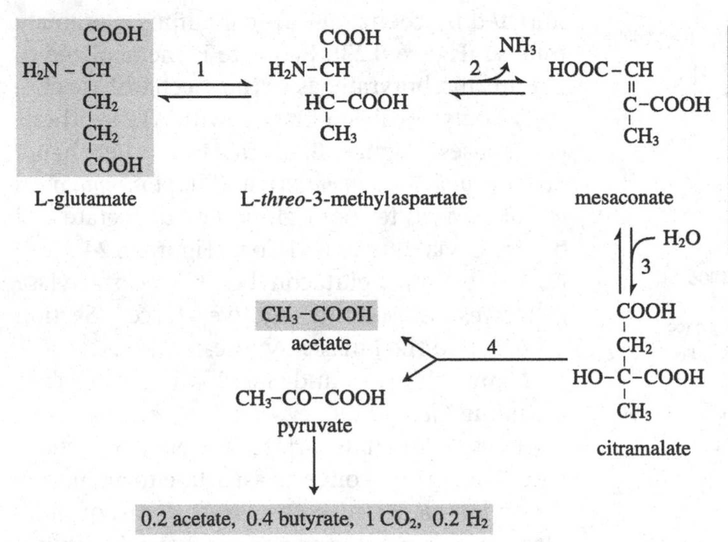

Figure 8.23 Glutamate fermentation by *Clostridium tetanomorphum.*

(Gottschalk, G. 1986, *Bacterial Metabolism*, 2nd edn., Figure 8.36. Springer, New York)

1, glutamate mutase; 2, 3-methylaspartase;
3, citramalate dehydratase; 4, citramalate lyase.

Lysine is fermented to acetate and butyrate by *Clostridium sticklandii* and *Clostridium subterminale* (Figure 8.25). Lysine is converted to 3-keto-5-aminohexanoate before being split to acetoacetate and 3-aminobutyryl-CoA, with the addition of acetyl-CoA. Acetoacetate is converted to acetate with the synthesis of ATP and 3-aminobutyryl CoA is reduced to butyrate.

Arginine is fermented as shown in Figure 8.26. This fermentation was first discovered in *Mycoplasma* spp. and later in *Clostridium sticklandii*, *Clostridium botulinum* and *Halobacterium salinarium*. In this fermentation, arginine deiminase converts arginine to citrulline, which is split to ornithine and carbamoyl phosphate. Ornithine is metabolized to acetate and butyrate, and carbamate kinase synthesizes ATP from carbamoyl phosphate. Asimilar fermentation by heterofermentative bacteria, such as *Leuconostoc oenos*,

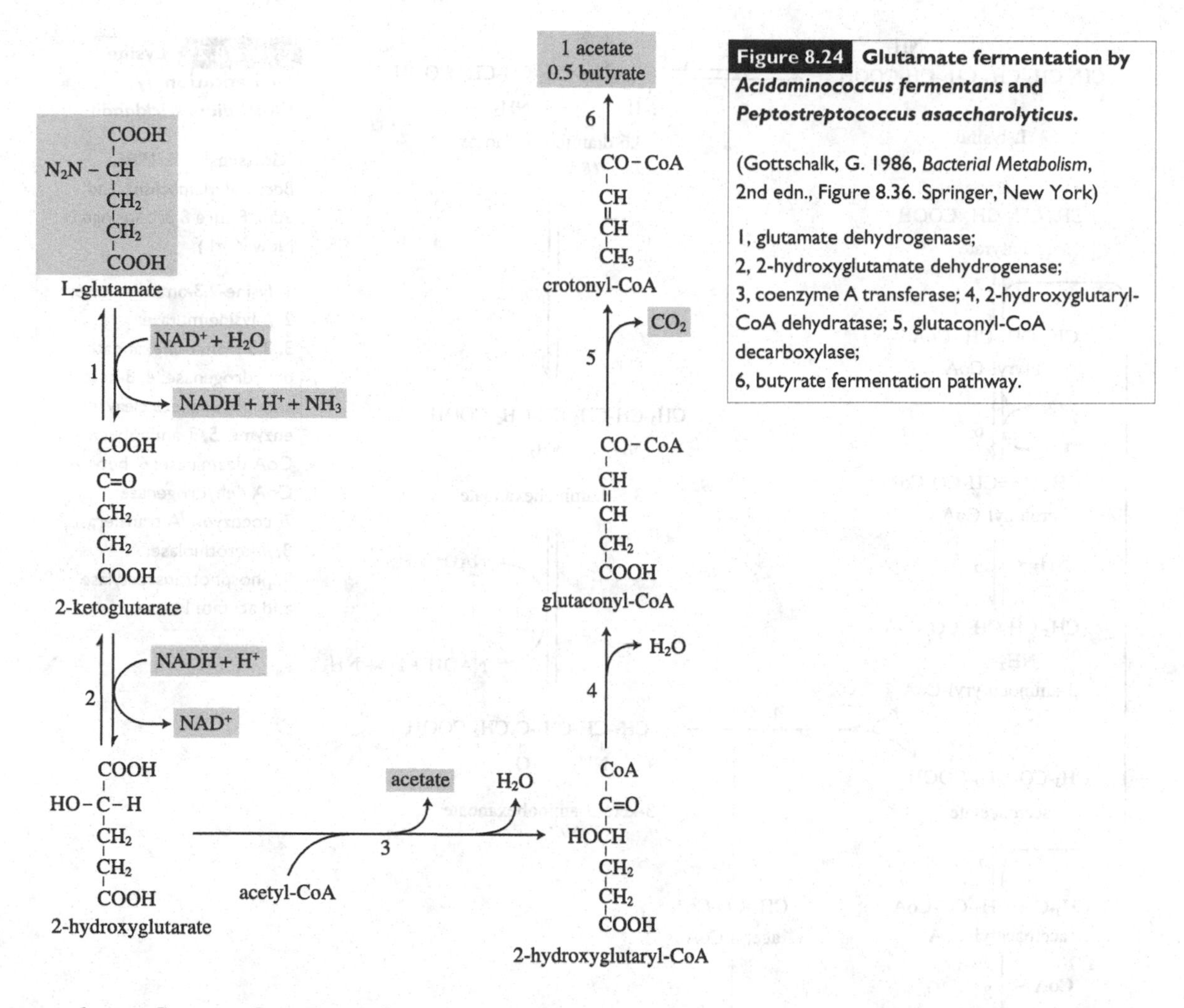

Figure 8.24 Glutamate fermentation by *Acidaminococcus fermentans* and *Peptostreptococcus asaccharolyticus*.

(Gottschalk, G. 1986, *Bacterial Metabolism*, 2nd edn., Figure 8.36. Springer, New York)

1, glutamate dehydrogenase;
2, 2-hydroxyglutamate dehydrogenase;
3, coenzyme A transferase; 4, 2-hydroxyglutaryl-CoA dehydratase; 5, glutaconyl-CoA decarboxylase;
6, butyrate fermentation pathway.

produces flavour from arginine during wine maturation.

In addition to those discussed above, all natural amino acids are fermented by many anaerobic bacteria, including proteolytic clostridia.

Enriched cultures of acetate-reducing bacteria oxidize leucine to isovalerate which is coupled with acetate reduction to propionate or butyrate.

8.8.2 Stickland reaction

Many anaerobic bacteria ferment single amino acids but grow better with amino acid mixtures. In this case, some amino acids serve as electron donors while others serve as electron acceptors. Alanine and valine are used as electron donors, and glycine and proline are used as electron acceptors. Depending on the mixture, leucine and aromatic amino acids can be used as electron donors or acceptors (Table 8.7).

Generally, amino acids used as electron donors are oxidatively deaminated and the resulting 2-ketoacids are further oxidized by 2-ketoacid:ferredoxin oxidoreductase to acyl-CoAs. ATP is synthesized from acyl-CoA through substrate-level phosphorylation (SLP):

$$\text{R-CH-NH}_2\text{-COOH} \xrightarrow[\text{2H, NH}_3]{\text{H}_2\text{O}} \text{R-CO-COOH} \xrightarrow[\text{CO}_2\text{, H}_2]{\text{CoA-SH}} \text{R-CO-CoA} \xrightarrow[\text{ADP} \rightarrow \text{ATP}]{} \text{R-COOH}$$

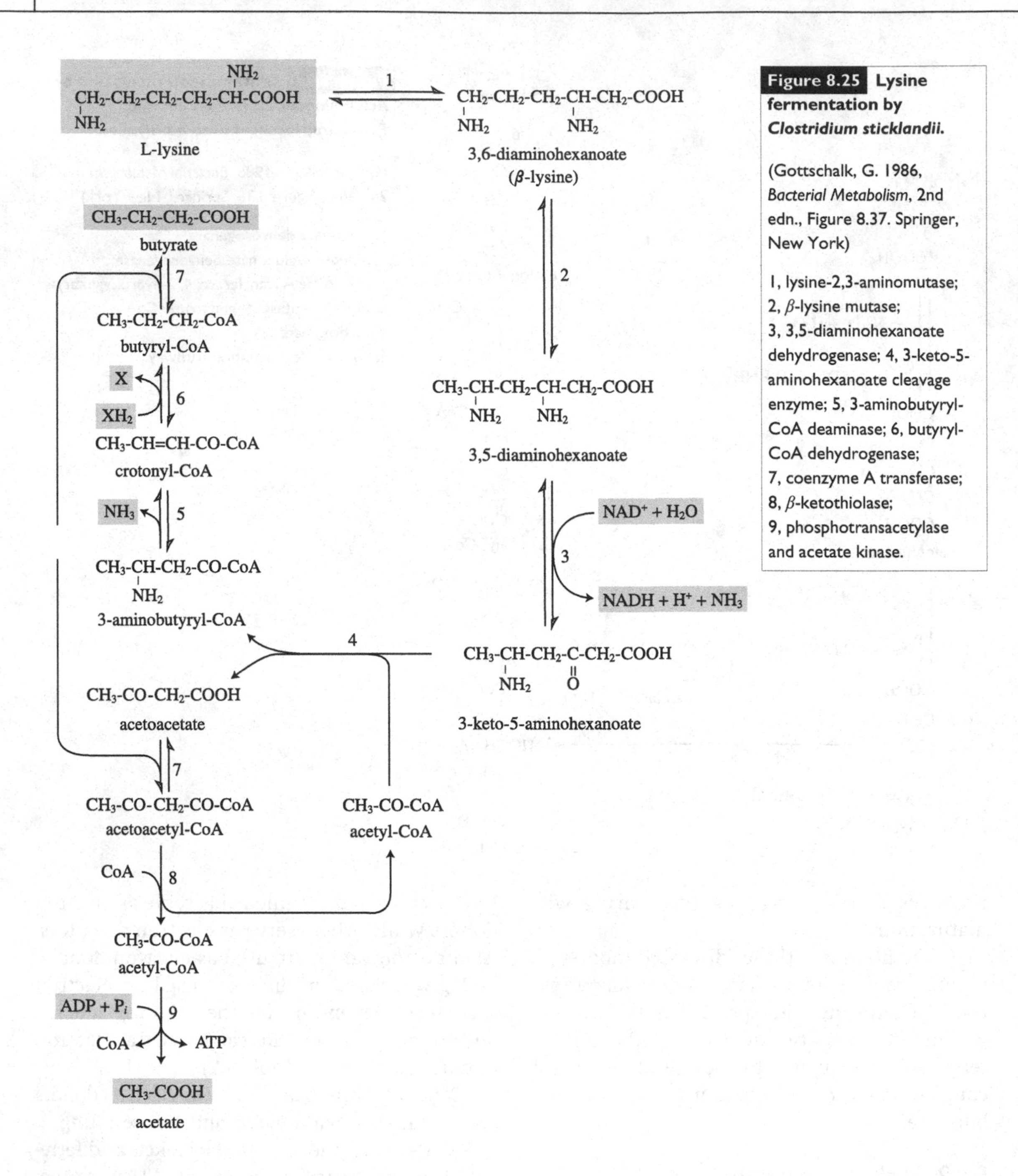

Figure 8.25 Lysine fermentation by *Clostridium sticklandii.*

(Gottschalk, G. 1986, *Bacterial Metabolism*, 2nd edn., Figure 8.37. Springer, New York)

1, lysine-2,3-aminomutase; 2, β-lysine mutase; 3, 3,5-diaminohexanoate dehydrogenase; 4, 3-keto-5-aminohexanoate cleavage enzyme; 5, 3-aminobutyryl-CoA deaminase; 6, butyryl-CoA dehydrogenase; 7, coenzyme A transferase; 8, β-ketothiolase; 9, phosphotransacetylase and acetate kinase.

Electrons from the oxidation reactions are disposed of, reducing the electron-accepting amino acids.

During the reduction of the electron acceptors in Stickland reactions, ATP may be synthesized through SLP, such as in the reaction catalysed by glycine reductase, but not in all cases. Some results have suggested electron transport phosphorylation from the reduction

Table 8.7 Amino acids used as electron donors and acceptors in the Stickland reaction.

Electron donor		Electron acceptor	
Amino acid	Products	Amino acid	Products
Alanine	acetate + NH_3 + CO_2	Glycine	acetate + NH_3
Leucine	3-methylbutyrate + NH_3 + CO_2	Proline	5-aminovalerate
Isoleucine	2-methylbutyrate + NH_3 + CO_2	Phenylalanine	phenylpropionate + NH_3
Valine	2-methylpropionate + NH_3 + CO_2	Tryptophan	indolepropionate + NH_3
Phenylalanine	phenylacetate + NH_3 + CO_2	Ornithine	5-aminovalerate + NH_3
Tryptophan	indoleacetate + NH_3 + CO_2	Leucine	4-methylvalerate + NH_3
Histidine	glutamate + NH_3 + CO_2	Betaine	acetate + trimethylamine
		Sarcosine	acetate + monomethylamine

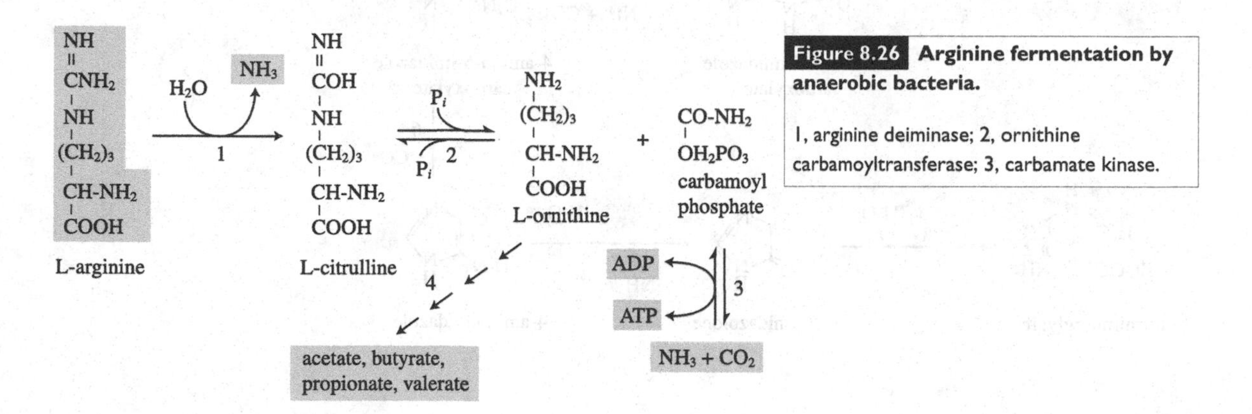

Figure 8.26 **Arginine fermentation by anaerobic bacteria.**

1, arginine deiminase; 2, ornithine carbamoyltransferase; 3, carbamate kinase.

of proline and phenylalanine as electron acceptors.

8.8.3 Fermentation of purine and pyrimidine bases

Clostridium acidiurici, *Clostridium cylindrosporum*, *Clostridium purinolyticum* and other species ferment purine bases to acetate, CO_2 and NH_3, and pyrimidine bases are fermented by *Clostridium glycolicum* and *Clostridium oroticum*.

The purine bases adenine, guanine and urate, are converted to formiminoglycine (Figure 8.27) and to acetate with ATP synthesis (Figure 8.28)

Many strains of *Clostridium* ferment pyrimidine bases, although they are less common than purine fermenters. *Clostridium glycolicum* ferments uracil to β-alanine, CO_2 and NH_3, and orotate is fermented by *Clostridium oroticum* to acetate, CO_2 and NH_3.

8.9 Fermentation of dicarboxylic acids

Some anaerobes ferment dicarboxylic acids, conserving free energy from the decarboxylation

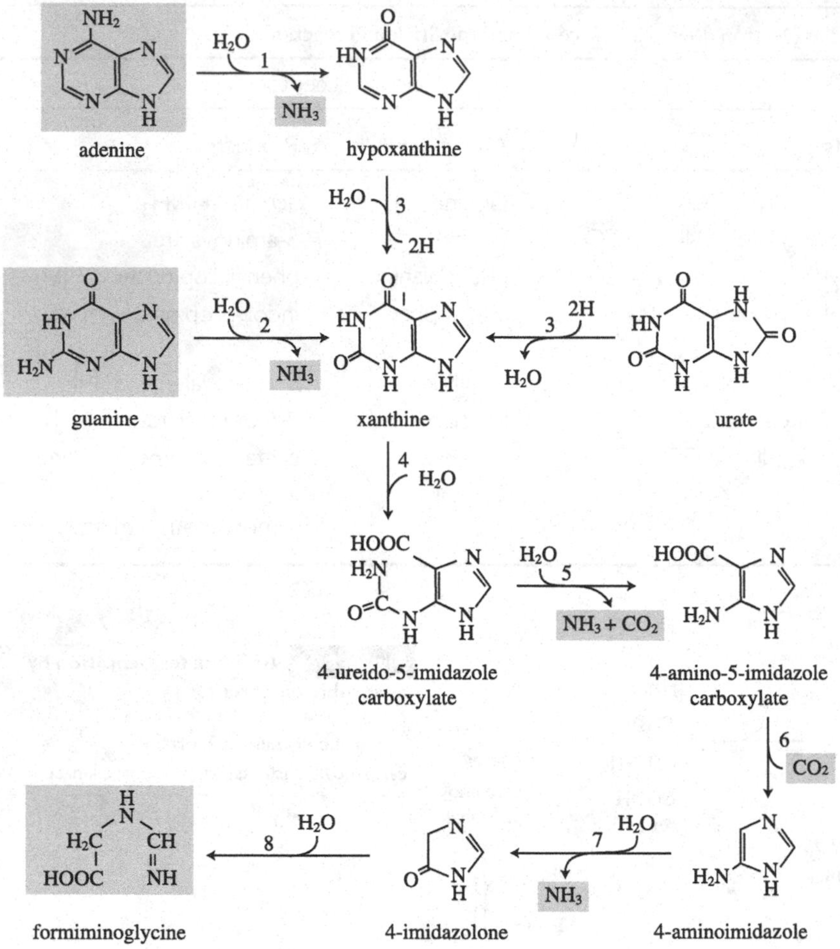

Figure 8.27 Fermentation of purine bases by *Clostridium purinolyticum*.

(Gottschalk, G. 1986, *Bacterial Metabolism*, 2nd edn., Figure 8.38. Springer, New York)

1, adenine deaminase;
2, guanine deaminase;
3, xanthine dehydrogenase;
4, xanthine amidohydrolase;
5, 4-ureido-5-imidazole carboxylate amidohydrolase;
6, 4-amino-5-imidazole carboxylate decarboxylase;
7, 4-aminoimidazole deaminase;
8, 4-imidazolonase.

reactions in the form of a sodium motive force. *Succinispira mobilis* oxidizes succinate to acetate. Malonate and oxalate are fermented by *Malonomonas rubra* and *Oxalobacter formigenes*, respectively. *Oxalobacter formigenes* can remove oxalate in the human gut, which may prevent the development of kidney stones.

8.10 Hyperthermophilic archaeal fermentation

Many prokaryotes have been isolated from hot springs and underwater hydrothermal vents. Some of them can grow at a temperature over 100°C, and most of these are archaea. Some of them are aerobic, but the majority are anaerobes. Many of these organisms use reduced sulfur compounds for their energy metabolism. These are referred to as sulfur-dependent archaea. Among them, strains of the genus *Archeaoglobus* conserve energy through anaerobic respiration using sulfate as the electron acceptor. Others reduce sulfur not as an electron acceptor, but as an electron sink (Section 9.3). They use sulfur to dispose of electrons without conserving energy. This process is referred to as fermentative sulfur reduction.

Some hyperthermophilic archaea, including *Pyrococcus furiosus*, *Pyrococcus woesei* and *Thermococcus litoralis*, ferment protein and carbohydrates in the absence of sulfur as the electron

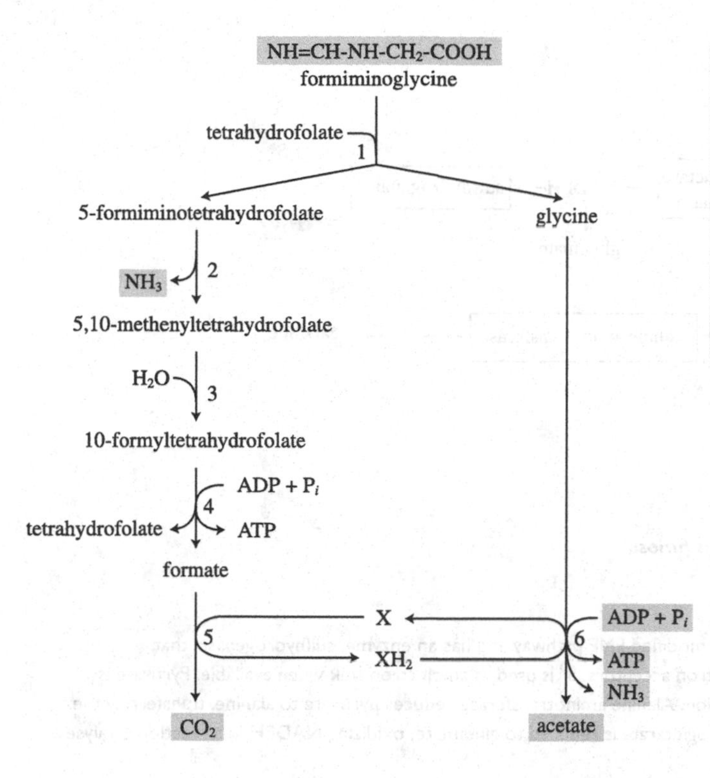

Figure 8.28 Fermentative degradation of formiminoglycine.

(Gottschalk, G. 1986, *Bacterial Metabolism*, 2nd edn., Figure 8.39. Springer, New York)

1, glycine formimino transferase; 2, formiminotetrahydrofolate cyclodeaminase; 3, methenyltetrahydrofolate cyclohydrolase; 4, formyltetrahydrofolate synthase; 5, formate dehydrogenase; 6, glycine reductase.

sink. *Pyrococcus furiosus* ferments carbohydrates in a modified EMP pathway (Section 4.1.4). In this pathway, ferredoxin oxidoreductase oxidizes glyceraldehyde-3-phosphate and pyruvate, reducing ferredoxin. Electrons from the reduced ferredoxin are transferred to NADPH before being consumed to produce H_2S, H_2 and alanine (Figure 8.29). This archaeon possesses a unique enzyme, sulfhydrogenase, that can reduce either H^+ or S^0. This organism uses pyruvate as an electron acceptor in a two-step reaction different from that carried out in bacteria. Alanine aminotransferase catalyses the amino group transfer reaction from glutamate to pyruvate, producing 2-ketoglutarate and alanine. 2-ketoglutarate is reduced by glutamate dehydrogenase consuming electrons in the form of NADPH.

Pyruvate:ferredoxin oxidoreductase oxidizes pyruvate to acetyl-CoA. Acetyl-CoA synthetase synthesizes ATP from acetyl-CoA. In this reaction acetyl-phosphate is not involved, unlike in most bacteria:

$$\text{acetyl-CoA} + \text{ADP} + P_i \xrightarrow{\text{acetyl-CoA synthetase}} \text{acetate} + \text{CoA-SH} + \text{ATP}$$

It is interesting to note that a hyperthermophilic archaeon, *Thermococcus onnurineus*, can grow on carbon monoxide or formate, producing hydrogen. Under standard conditions, CO (CO_2/CO, $E^{0'} = -540$ mV) conversion to H_2 (H^+/H_2, $E^{0'} = -414$ mV) is an exergonic reaction, but the redox potential of CO_2/formate ($E^{0'} = -413$ mV) is close to that of H^+/H_2. The ΔG for formate metabolism ranges between -8 and -20 kJ mol^{-1} under the physiological conditions where the organism is found.

8.11 Degradation of xenobiotics under fermentative conditions

The petrochemical industry produces many anthropogenic compounds unknown in natural ecosystems. These are resistant to microbial metabolism and referred to as xenobiotics. Generally they are better substrates for aerobic metabolism, but some of them are susceptible to anaerobic respiratory

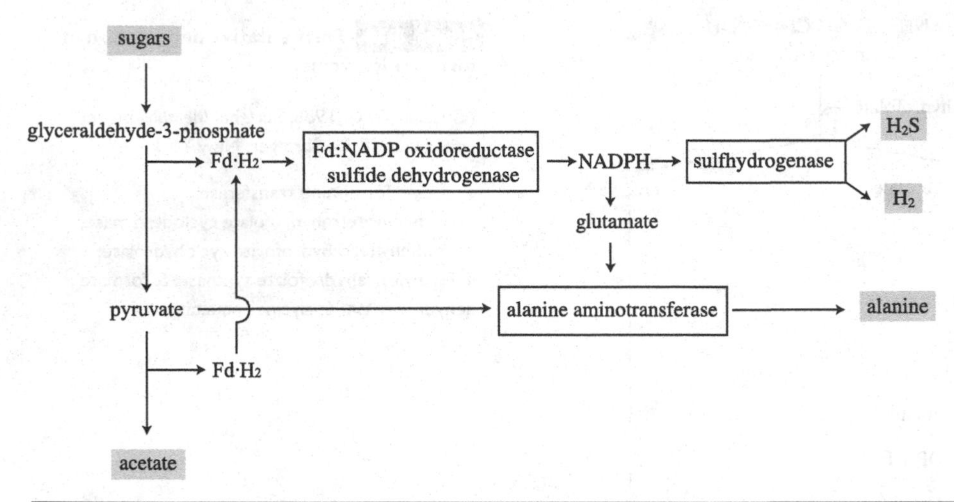

Figure 8.29 Sugar fermentation by *Pyrococcus furiosus*.

(*FEMS Microbiol. Rev.* **18**: 119–137, 1996)

This archaeon metabolizes carbohydrates through a modified EMP pathway and has an enzyme, sulfhydrogenase, that reduces H^+ and S^0. Pyruvate or H^+ are used as electron acceptors. S^0 is used as an electron sink when available. Pyruvate is reduced as an electron acceptor in a two-step reaction. Alanine aminotransferase reduces pyruvate to alanine, transferring the amino group from glutamate, and the resulting 2-ketoglutarate is reduced to glutamate, oxidizing NADPH in a reaction catalysed by glutamate dehydrogenase.

metabolism (Section 9.9). Some xenobiotics can be fermented. For example, *Pelobacter venetianus* ferments polyethylene glycol to ethanol and acetate.

Further Reading

Note this section contains key references only. Additional recommended references are available at www.cambridge.org/ProkaryoticMetabolism.

Oxygen toxicity

Atack, J. M. & Kelly, D. J. (2006). Structure, mechanism and physiological roles of bacterial cytochrome *c* peroxidases. *Advances in Microbial Physiology* **52**, 73–106.

Brioukhanov, A. L., Netrusov, A. I. & Eggen, R. I. L. (2006). The catalase and superoxide dismutase genes are transcriptionally up-regulated upon oxidative stress in the strictly anaerobic archaeon *Methanosarcina barkeri*. *Microbiology* **152**, 1671–1677.

Fournier, M., Zhang, Y., Wildschut, J. D., Dolla, A., Voordouw, J. K., Schriemer, D. C. & Voordouw, G. (2003). Function of oxygen resistance proteins in the anaerobic, sulfate-reducing bacterium *Desulfovibrio vulgaris* Hildenborough. *Journal of Bacteriology* **185**, 71–79.

Henningham, A., Döhrmann, S., Nizet, V. & Cole, J. N. (2015). Mechanisms of group A *Streptococcus* resistance to reactive oxygen species. *FEMS Microbiology Reviews* **39**, 488–508.

Imlay, J. A. (2006). Iron-sulphur clusters and the problem with oxygen. *Molecular Microbiology* **59**, 1073–1082.

Jennings, M. E., Schaff, C. W., Horne, A. J., Lessner, F. H. & Lessner, D. J. (2014). Expression of a bacterial catalase in a strictly anaerobic methanogen significantly increases tolerance to hydrogen peroxide but not oxygen. *Microbiology* **160**, 270–278.

Krätzer, C., Welte, C., Dörner, K., Friedrich, T. & Deppenmeier, U. (2011). Methanoferrodoxin represents a new class of superoxide reductase containing an iron-sulfur cluster. *FEBS Journal* **278**, 442–451.

Ramel, F., Amrani, A., Pieulle, L., Lamrabet, O., Voordouw, G., Seddiki, N., Brèthes, D., Company, M., Dolla, A. & Brasseur, G. (2013). Membrane-bound oxygen reductases of the anaerobic sulfate-reducing *Desulfovibrio vulgaris* Hildenborough: roles in oxygen defence and electron link with periplasmic hydrogen oxidation. *Microbiology* **159**, 2663–2673.

Riebe, O., Fischer, R.-J., Wampler, D. A., Kurtz, Jr., D. M. & Bahl, H. (2009). Pathway for H_2O_2 and O_2 detoxification in *Clostridium acetobutylicum*. *Microbiology* **155**, 16–24.

Zhao, X. & Drlica, K. (2014). Reactive oxygen species and the bacterial response to lethal stress. *Current Opinion in Microbiology* **21**, 1–6.

Ethanol fermentation

Balodite, E., Strazdina, I., Galinina, N., McLean, S., Rutkis, R., Poole, R. K. & Kalnenieks, U. (2014). Structure of the *Zymomonas mobilis* respiratory chain: oxygen affinity of electron transport and the role of cytochrome c peroxidase. *Microbiology* **160**, 2045–2052.

Kalnenieks, U. (2006). Physiology of *Zymomonas mobilis*: some unanswered questions. *Advances in Microbial Physiology* **51**, 73–117.

Kalnenieks, U., Galinina, N., Strazdina, I., Kravale, Z., Pickford, J. L., Rutkis, R. & Poole, R. K. (2008). NADH dehydrogenase deficiency results in low respiration rate and improved aerobic growth of *Zymomonas mobilis*. *Microbiology* **154**, 989–994.

Zheng, T., Olson, D. G., Murphy, S. J., Shao, X., Tian, L. & Lynd, L. R. (2017). Both adhE and a separate NADPH-dependent alcohol dehydrogenase gene, *adhA*, are necessary for high ethanol production in *Thermoanaerobacterium saccharolyticum*. *Journal of Bacteriology* **199**, e00542-16.

Lactate fermentation

Baureder, M. & Hederstedt, L. (2013). Heme proteins in lactic acid bacteria. *Advances in Microbial Physiology*. **62**, 1–43.

Filannino, P., Di Cagno, R., Addante, R., Pontonio, E. & Gobbetti, M. (2016). Metabolism of fructophilic lactic acid bacteria from the *Apis mellifera* L. bee gut: phenolic acids as external electron acceptors. *Applied and Environmental Microbiology* **82**, 6899–6911.

Klijn, A., Mercenier, A. & Arigoni, F. (2005). Lessons from the genomes of bifidobacteria. *FEMS Microbiology Reviews* **29**, 491–509.

Martin, M. G., Sender, P. D., Peiru, S., de Mendoza, D. & Magn, C. (2004). Acid-inducible transcription of the operon encoding the citrate lyase complex of *Lactococcus lactis* biovar diacetylactis CRL264. *Journal of Bacteriology* **186**, 5649–5660.

Papadimitriou, K., Alegría, Á., Bron, P. A., de Angelis, M., Gobbetti, M., Kleerebezem, M., Lemos, J. A., Linares, D. M., Ross, P., Stanton, C., Turroni, F., van Sinderen, D., Varmanen, P., Ventura, M., Zúñiga, M., Tsakalidou, E. & Kok, J. (2016). Stress physiology of lactic acid bacteria. *Microbiology & Molecular Biology Reviews* **80**, 837–890.

Smit, G., Smit, B. A. & Engels, W. J. M. (2005). Flavour formation by lactic acid bacteria and biochemical flavour profiling of cheese products. *FEMS Microbiology Reviews* **29**, 591–610.

Vido, K., le Bars, D., Mistou, M. Y., Anglade, P., Gruss, A. & Gaudu, P. (2004). Proteome analyses of heme-dependent respiration in *Lactococcus lactis*: involvement of the proteolytic system. *Journal of Bacteriology* **186**, 1648–1657.

Probiotics

Becattini, S., Littmann, E. R., Carter, R. A., Kim, S. G., Morjaria, S. M., Ling, L., Gyaltshen, Y., Fontana, E., Taur, Y., Leiner, I. M. & Pamer, E. G. (2017). Commensal microbes provide first line defense against *Listeria monocytogenes* infection. *Journal of Experimental Medicine*. doi 10.1084/jem.20170495

Karst, S. M. (2016). The influence of commensal bacteria on infection with enteric viruses. *Nature Reviews Microbiology* **14**, 197–204.

Kim, Y.-G., Sakamoto, K., Seo, S.-U., Pickard, J. M., Gillilland, M. G., Pudlo, N. A., Hoostal, M., Li, X., Wang, T. D., Feehley, T., Stefka, A. T., Schmidt, T. M., Martens, E. C., Fukuda, S., Inohara, N., Nagler, C. R. & Núñez, G. (2017). Neonatal acquisition of *Clostridia* species protects against colonization by bacterial pathogens. *Science* **356**, 315–319.

Petrova, M. I., van den Broek, M., Balzarini, J., Vanderleyden, J. & Lebeer, S. (2013). Vaginal microbiota and its role in HIV transmission and infection. *FEMS Microbiology Reviews* **37**, 762–792.

Rauch, M. & Lynch, S. V. (2012). The potential for probiotic manipulation of the gastrointestinal microbiome. *Current Opinion in Biotechnology* **23**, 192–201.

Santos, C. M. A., Pires, M. C. V., Leão, T. L., Hernández, Z. P., Rodriguez, M. L., Martins, A. K. S., Miranda, L. S., Martins, F. S. & Nicoli, J. R. (2016). Selection of *Lactobacillus* strains as potential probiotics for vaginitis treatment. *Microbiology* **162**, 1195–1207.

Sharon, G., Sampson, T. R., Geschwind, D. H. & Mazmanian, S. K. (2016) The central nervous system and the gut microbiome. *Cell* **167**, 915–932.

Butyrate and butanol

Amador-Noguez, D., Feng, X.-J., Fan, J., Roquet, N., Rabitz, H. & Rabinowitz, J. D. (2010). Systems-level metabolic flux profiling elucidates a complete, bifurcated tricarboxylic acid cycle in *Clostridium acetobutylicum*. *Journal of Bacteriology* **192**, 4452–4461.

Calusinska, M., Happe, T., Joris, B. & Wilmotte, A. (2010). The surprising diversity of clostridial hydrogenases: a comparative genomic perspective. *Microbiology* **156**, 1575–1588.

Louis, P. & Flint, H. J. (2017). Formation of propionate and butyrate by the human colonic microbiota. *Environmental Microbiology* **19**, 29–41.

Peters, J. W., Miller, A.-F., Jones, A. K., King, P. W. & Adams, M. W. W. (2016). Electron bifurcation. *Current Opinion in Chemical Biology* **31**, 146–152.

Saint-Amans, S., Girbal, L., Andrade, J., Ahrens, K. & Soucaille, P. (2001). Regulation of carbon and electron flow in *Clostridium butyricum* VPI 3266 grown on glucose-glycerol mixtures. *Journal of Bacteriology* **183**, 1748–1754.

Thauer, R. K. (2015). My lifelong passion for biochemistry and anaerobic microorganisms. *Annual Review of Microbiology* **69**, 1–30.

Tracy, B. P., Jones, S. W. & Papoutsakis, E. T. (2011). Inactivation of σ^E and σ^G in *Clostridium acetobutylicum* illuminates their roles in clostridial-cell-form biogenesis, granulose synthesis, solventogenesis, and spore morphogenesis. *Journal of Bacteriology* **193**, 1414–1426.

Mixed acid fermentation

Bizzini, A., Zhao, C., Budin-Verneuil, A., Sauvageot, N., Giard, J.-C., Auffray, Y. & Hartke, A. (2010). Glycerol is metabolized in a complex and strain-dependent manner in *Enterococcus faecalis*. *Journal of Bacteriology* **192**, 779–785.

Laurinavichene, T. V., Zorin, N. A. & Tsygankov, A. A. (2002). Effect of redox potential on activity of hydrogenase 1 and hydrogenase 2 in *Escherichia coli*. *Archives of Microbiology* **178**, 437–442.

Moons, P., Van Houdt, R., Vivijs, B., Michiels, C. M. & Aertsen, A. (2011). Integrated regulation of acetoin fermentation by quorum sensing and pH in *Serratia plymuthica* RVH1. *Applied and Environmental Microbiology* **77**, 3422–3427.

Scheu, P. D., Witan, J., Rauschmeier, M., Graf, S., Liao, Y.-F., Ebert-Jung, A., Basché, T., Erker, W. & Unden, G. (2012). CitA/CitB two-component system regulating citrate fermentation in *Escherichia coli* and its relation to the DcuS/DcuR system in vivo. *Journal of Bacteriology* **194**, 636–645.

van Houdt, R., Moons, P., Hueso Buj, M. & Michiels, C. W. (2006). *N*-acyl-L-homoserine lactone quorum sensing controls butanediol fermentation in *Serratia plymuthica* RVH1 and *Serratia marcescens* MG1. *Journal of Bacteriology* **188**, 4570–4572.

Propionate fermentation

Koussemon, M., Combet-Blanc, Y. & Ollivier, B. (2003). Glucose fermentation by *Propionibacterium microaerophilum*: effect of pH on metabolism and bioenergetics. *Current Microbiology* **46**, 141–145.

Reichardt, N., Duncan, S. H., Young, P., Belenguer, A., McWilliam Leitch, C., Scott, K. P., Flint, H. J. & Louis, P. (2014). Phylogenetic distribution of three pathways for propionate production within the human gut microbiota. *ISME J* **8**, 1323–1335.

Scott, K. P., Martin, J. C., Campbell, G., Mayer, C.-D. & Flint, H. J. (2006). Whole-genome transcription profiling reveals genes up-regulated by growth on fucose in the human gut bacterium *Roseburia inulinivorans*. *Journal of Bacteriology* **188**, 4340–4349.

Seeliger, S., Janssen, P. H. & Schink, B. (2002). Energetics and kinetics of lactate fermentation to acetate and propionate via methylmalonyl-CoA or acrylyl-CoA. *FEMS Microbiology Letters* **211**, 65–70.

Fermentation of amino acids

Ato, M., Ishii, M. & Igarashi, Y. (2014). Enrichment of amino acid-oxidizing, acetate-reducing bacteria. *Journal of Bioscience & Bioengineering* **118**, 160–165.

Buckel, W. (2001). Unusual enzymes involved in five pathways of glutamate fermentation. *Applied Microbiology and Biotechnology* **57**, 263–273.

Debnar-Daumler, C., Seubert, A., Schmitt, G. & Heider, J. (2014). Simultaneous involvement of a tungsten-containing aldehyde: ferredoxin oxidoreductase and a phenylacetaldehyde dehydrogenase in anaerobic phenylalanine metabolism. *Journal of Bacteriology* **196**, 483–492.

Lan, J. & Newman, E. B. (2003). A requirement for anaerobically induced redox functions during aerobic growth of *Escherichia coli* with serine, glycine and leucine as carbon source. *Research in Microbiology* **154**, 191–197.

Fermentation of dicarboxylic acids

Abratt, V. R. & Reid, S. J. (2010). Oxalate-degrading bacteria of the human gut as probiotics in the management of kidney stone disease. *Advances in Applied Microbiology*. **72**, 63–87.

Janssen, P. H. & Hugenholtz, P. (2003). Fermentation of glycolate by a pure culture of a strictly anaerobic Gram-positive bacterium belonging to the family *Lachnospiraceae*. *Archives of Microbiology* **179**, 321–328.

Ye, L., Jia, Z., Jung, T. & Maloney, P. C. (2001). Topology of OxlT, the oxalate transporter of *Oxalobacter formigenes*, determined by site-directed fluorescence labeling. *Journal of Bacteriology* **183**, 2490–2496.

Hyperthermophilic archaeal fermentation

Kanai, T., Matsuoka, R., Beppu, H., Nakajima, A., Okada, Y., Atomi, H. & Imanaka, T. (2011). Distinct physiological roles of the three [NiFe]-hydrogenase orthologs in the hyperthermophilic archaeon *Thermococcus kodakarensis*. *Journal of Bacteriology* **193**, 3109–3116.

Kim, Y. J., Lee, H. S., Kim, E. S., Bae, S. S., Lim, J. K., Matsumi, R., Lebedinsky, A. V., Sokolova, T. G., Kozhevnikova, D. A., Cha, S.-S., Kim, S.-J., Kwon, K. K., Imanaka, T., Atomi, H., Bonch-Osmolovskaya, E. A., Lee, J.-H. & Kang, S. G. (2010). Formate-driven growth coupled with H_2 production. *Nature* **467**, 352–355.

Schut, G. J., Boyd, E. S., Peters, J. W. & Adams, M. W. W. (2013). The modular respiratory complexes involved in hydrogen and sulfur metabolism by heterotrophic hyperthermophilic archaea and their evolutionary implications. *FEMS Microbiology Reviews* **37**, 182–203.

Yang, H., Lipscomb, G. L., Keese, A. M., Schut, G. J., Thomm, M., Adams, M. W. W., Wang, B. C. & Scott, R. A. (2010). SurR regulates hydrogen production in *Pyrococcus furiosus* by a sulfur-dependent redox switch. *Molecular Microbiology* **77**, 1111–1122.

Chapter 9

Anaerobic respiration

In the previous chapter, respiration was defined as an energy conservation process achieved through electron transport phosphorylation (ETP) using externally supplied electron acceptors. Electron acceptors used in anaerobic respiration include oxidized sulfur and nitrogen compounds, metal ions, organic halogens and carbon dioxide. Other oxidized compounds reduced under anaerobic conditions include iodate, (per)chlorate, and phosphate. There is evidence to suggest that these compounds are used as electron acceptors in anaerobic ecosystems, but there are some exceptions. ATP synthesis mechanisms dependent on a proton motive force are known in some fermentative bacteria. These include Na^+-dependent decarboxylation, fumarate reduction, product/proton symport and energy-converting hydrogenase, as described earlier (Section 5.8.6). Sulfidogenesis and methanogenesis are described as fermentations in some cases, since a small amount of energy is conserved in these anaerobic processes. However, in these processes ATP is generated mainly through the proton motive force and they can therefore be classified as anaerobic respiration.

Many ecosystems become anaerobic when oxygen consumption is greater than its supply. Even under anaerobic conditions, natural organic compounds are continuously recycled. Anaerobic respiratory microbes convert organic materials to carbon dioxide and methane under anaerobic conditions, in conjunction with fermentative microbes.

Energy is required for all forms of life. At any given conditions, those organisms utilizing energy sources more efficiently will become dominant over the others. Among the anaerobic respiratory prokaryotes, denitrifiers conserve more energy than other groups. For this reason, sulfidogenesis and methanogenesis are inhibited in the presence of nitrate, and sulfate inhibits methanogenesis. Ferric iron is ubiquitous on Earth, and has a redox potential higher than sulfate (Table 9.1). Because of its availability, ferric iron is a more important electron acceptor than nitrate. It has been estimated that more than half of the degradation of organic compounds under anaerobic conditions is coupled to the reduction of ferric iron. Halogenated hydrocarbons are generally toxic and recalcitrant to degradation under aerobic conditions, but can serve as electron acceptors under anaerobic conditions.

9.1 Denitrification

Denitrification is an economically and environmentally important microbial process. Nitrogen fertilizer is lost from farmland through this process, which is also exploited to remove nitrogen from wastewater treatment plants before discharge to prevent eutrophication. Many facultative anaerobes use nitrate and nitrite as electron acceptors under oxygen-limited conditions (Table 9.2). Strains of *Paracoccus*, *Ralstonia* (*Alcaligenes*) and *Pseudomonas* are the best-known denitrifiers. These metabolize carbohydrates through glycolysis and the TCA cycle, as do the aerobes. The electrons from these metabolic pathways are consumed, reducing nitrate and nitrite.

Table 9.1 Free energy from NADH oxidation coupled to electron acceptors used by prokaryotes.

Reduction half reaction	$\Delta G^{0\prime}$ (kJ/2e$^-$)
$\frac{1}{2}O_2 \rightarrow 2H_2O$	−219.07
$2NO_3^- \rightarrow N_2$	−206.12
$Fe^{3+} \rightarrow Fe^{2+}$	−209.46
$CH_3Cl \rightarrow CH_4+HCl$	−135.08
$MnO_2 \rightarrow Mn^{2+}$	−134.52
Se(VI) → Se(IV)	−129.96
Cr(VI) → Cr(III)	−90.04
As(V) → As(0)	−46.11
$SO_4^{2-} \rightarrow HS^-$	−20.24
$CO_2 \rightarrow CH_4$	−14.58

Source: FEMS Microbiol. Rev. **23**, 615–627, 1999, and other sources.

In addition to the facultative anaerobic denitrifiers, various aerobes and strict anaerobes can respire nitrate. These include fungi, chemolithotrophs and bacteria performing anaerobic ammonia oxidation coupled to nitrite reduction (anammox) and anaerobic nitrate ammonification. These are described later (Section 9.1.4). Methane is oxidized under anaerobic conditions, coupled with nitrate or sulfate reduction (Section 9.9.2). Suprisingly, *Streptomyces coelicolor* and mycobacteria can use nitrate as their electron acceptor under anaerobic conditions. Methylotrophic denitrifiers play an important role in nitrate removal from wastewaters when methanol is used as the electron donor.

9.1.1 Biochemistry of denitrification

Denitrification is defined as a microbial process reducing nitrate (NO_3^-) or nitrite (NO_2^-) to generate a proton motive force under anaerobic conditions. Gaseous nitrogen (N_2) is the main product, with small amounts of NO (nitric oxide) and N_2O (nitrous oxide). The denitrifiers have a similar electron transport system to aerobic respiratory organisms. Reduced coenzyme Q provides electrons for nitrate reductase, and the other enzymes oxidize reduced cytochrome *c* (Figure 9.1).

Table 9.2 Typical denitrifying prokaryotes.

Organism	Electron donor	Electron acceptor
Alcaligenes cycloclastes	organics	NO_2^- [a]
Alcaligenes faecalis	organics	NO_3^-
Ralstonia eutropha (Alcaligenes eutrophus)	H_2	NO_3^-
Bacillus licheniformis	organics	NO_3^-
Bacillus azotoformans	organics	NO_2^-
Hyphomicrobium vulgare	CH_4	NO_3^-
Methylophaga sp.	methanol	NO_3^-- NO_2^-
Paracoccus denitrificans	organics, H_2	NO_3^-
Propionibacterium pentosaceum	organics	NO_3^-
Pseudomonas fluorescens	organics	NO_3^-
Thiobacillus denitrificans	S^{2-}	NO_3^-
Pyrobaculum aerophilum [b]	organics, H_2	NO_3^-

[a] NO_3^- is not reduced.
[b] archaeon.

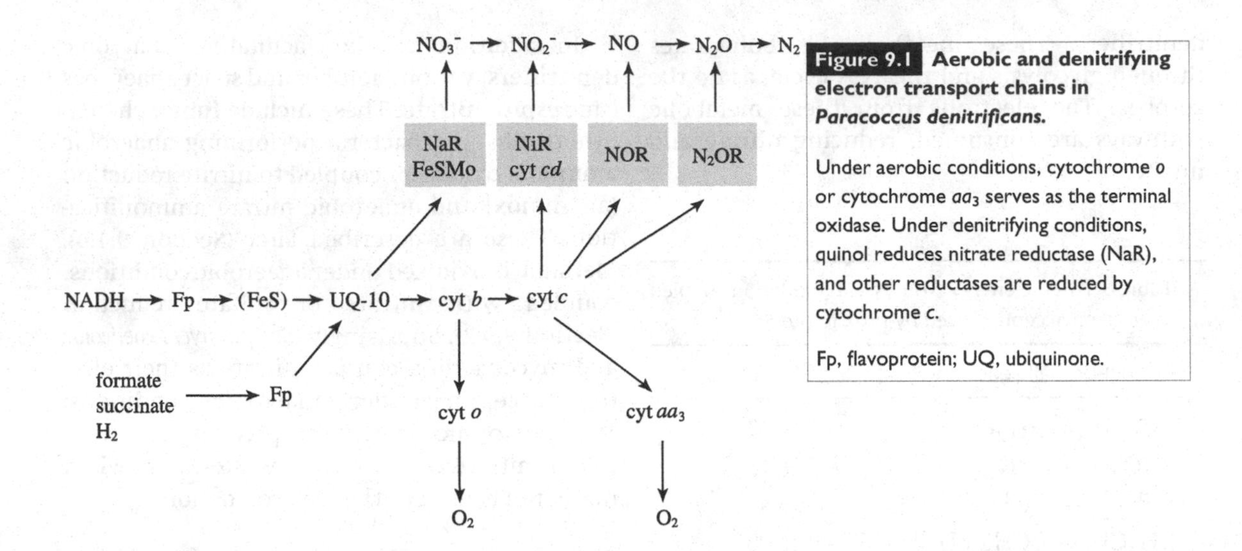

Figure 9.1 Aerobic and denitrifying electron transport chains in *Paracoccus denitrificans*.

Under aerobic conditions, cytochrome *o* or cytochrome *aa*$_3$ serves as the terminal oxidase. Under denitrifying conditions, quinol reduces nitrate reductase (NaR), and other reductases are reduced by cytochrome *c*.

Fp, flavoprotein; UQ, ubiquinone.

Nitrate reductase (NaR) reduces NO_3^- to NO_2^- coupled to the oxidation of quinol. Subsequently nitrite reductase (NiR) oxidizes reduced cytochrome *c* to reduce NO_2^- to nitric oxide (NO). Two molecules of NO are reduced further to nitrous oxide (N_2O) by nitric oxide reductase (Figure 9.2). Finally, nitrous oxide reductase reduces N_2O to N_2. In *Paracoccus denitrificans*, nitrate reductase and nitric oxide reductase are cytoplasmic membrane proteins, while nitrite reductase is located in the periplasm.

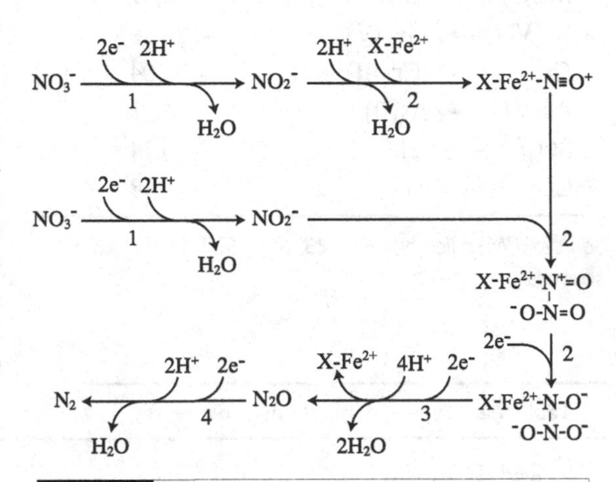

Figure 9.2 Dissimilatory nitrate reduction.

(Gottschalk, G. 1986, *Bacterial Metabolism*, 2nd edn., Figure 5.18. Springer, New York)

Nitrate is reduced to nitrite by nitrate reductase (1) that oxidizes quinol. Nitrite reductase (2) reduces nitrite to nitric oxide. Two molecules of nitric oxide are reductively condensed to nitrous oxide by nitric oxide reductase (3). Nitrous oxide reductase (4) reduces its substrate to gaseous nitrogen. Reduced cytochrome *c* supplies electrons for the last three reducing steps. These enzymes function either at the cytoplasmic membrane or in the periplasm.

9.1.1.1 Nitrate reductase

Nitrate reductase is a complex protein consisting of α, β and γ subunits. This enzyme is an [Fe–S] protein containing molybdenum in addition to iron. The [Fe–S] centres of the β subunit participate in electron transfer within the molecule, and the *b*-type cytochromes of the γ subunit are involved in electron transfer from quinol to the [Fe–S] clusters (Figure 9.3). The α subunit containing molybdopterin and a [4Fe–4S] cluster catalyses the reductive reaction:

$$NO_3^- + 2H^+ + 2e^- \longrightarrow NO_2^- + H_2O$$

The β subunit connects the α subunit at the cytoplasmic side and the γ subunit embedded in the membrane. The γ subunit takes electrons using *b*-type cytochromes from the reduced quinone and transfers them to the α subunit through the β subunit, involving a [3Fe–4S] cluster and three [4Fe–4S] clusters. The *b*-type cytochromes export protons during the electron transfer reactions. *Paracoccus denitrificans* has a separate nitrate reductase in the periplasm, of unknown function.

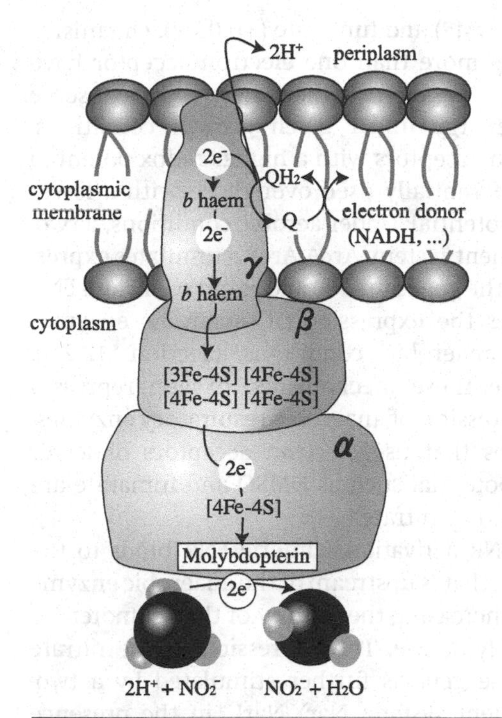

Figure 9.3 Model of the dissimilatory nitrate reductase of *Escherichia coli*.

(*Biochim. Biophys. Acta.* **1446**: 1–23, 1999)

Nitrate reductase is a complex protein consisting of α, β and γ subunits. The *b*-type cytochrome associated with the γ subunit is reduced, oxidizing quinol and liberating protons. The [Fe–S] cluster of the β subunit transfers electrons from the γ subunit to the α subunit, where nitrate is reduced to nitrite.

The nitrate reductase that initiates denitrification is different from the assimilatory nitrate reductase that is involved in the use of nitrate as a nitrogen source (Section 6.2.2). This enzyme is referred to as the dissimilatory nitrate reductase and is inhibited by oxygen but not by NH_3. On the other hand, NH_3 inhibits the assimilatory enzyme but O_2 does not. Energy in the form of the proton motive force is not conserved by the assimilatory enzyme.

9.1.1.2 Nitrite reductase

Nitrite produced by nitrate reductase is excreted to the periplasmic region by a specific transporter, to be reduced by nitrite reductase. Two different nitrite reductases are known, one containing *c*-type and *d*-type cytochromes (cdNiR) and the other a copper protein (CuNiR). CuNiR is widely distributed in prokaryotes including Gram-positive denitrifiers, such as species of the genus *Bacillus*, Gram-negative bacteria such as *Pseudomonas aureofaciens* and in archaea (species of the genus *Haloarcula*); cdNiR is found only in Gram-negative bacteria.

Cytochrome *c* or a Cu-containing small protein, pseudoazurin, provides electrons to the homodimeric cdNiR. Cytochrome *c* of the cdNiR takes electrons and nitrite is reduced to nitric oxide (NO) at cytochrome *d*. cdNiR reduces not only nitrite but also nitric oxide. The gene for nitric oxide reductase has been identified and its mutant cannot grow under denitrifying conditions. The function of nitric oxide reducing activity in cdNiR is not known.

This dissimilatory nitrite reductase is different from the assimilatory nitrite reductase, which is a cytoplasmic enzyme catalysing the reduction of NO_2^- directly to NH_4^+ (Section 6.2.2):

$$NO_2^- + 3NADH + 5H^+ \longrightarrow NH_4^+ + 3NAD^+ + 2H_2O$$

9.1.1.3 Nitric oxide reductase and nitrous oxide reductase

Nitric oxide produced by nitrite reductase is toxic (Section 12.2.5.2) and is reduced by nitric oxide reductase to nitrous oxide as soon as it is produced. Nitric oxide reductase is a complex enzyme consisting of a small subunit containing *c*-type cytochromes and two large subunits containing *b*-type cytochromes. This enzyme reduces two molecules of NO to N_2O using the electrons available from reduced cytochrome *c*. Protons are not translocated during this reaction.

N_2O is reduced to N_2 by nitrous oxide reductase, generating a proton motive force through consuming $2H^+$ in the cytoplasm. Reduced cytochrome *c* provides electrons for this reaction.

9.1.2 ATP synthesis in denitrification

During aerobic respiration, the H^+/O ratio is approximately 10 in prokaryotes, depending on the electron carriers involved in ETP (Section 5.8.4). The redox potentials of the denitrification reactions are shown in Table 9.3. The redox

Table 9.3 | Redox potentials and ATP yield of denitrification reactions.

Half reaction	$E^{0'}$ (mV)	P/2e$^-$
$\frac{1}{2}O_2 / H_2O$	+815	3
NO_3^- / NO_2^-	+421	2
NO_2^- / NO	+337	2
$NO / \frac{1}{2}N_2O$	+1180	3
N_2O / N_2	+1350	3

potentials of the first two reactions are lower but those of the other two reactions are much higher than that of O_2. Assuming that the P/2e^- (ATP synthesized/2e^- transported) ratio is 3 in aerobic respiration (Section 5.8.4), the ratio can be 2 in the first two reactions and 3 in the other reactions.

9.1.3 Regulation of denitrification

Denitrification is an alternative respiratory metabolism in facultative anaerobes under O_2-limited conditions. Since less energy is conserved in denitrification than in aerobic respiration, denitrification is strongly inhibited by O_2, with few exceptions. The expression of the genes for denitrification is regulated, and so are the enzyme activities after they are expressed. The enzyme activities appear 4–120 minutes after the culture becomes anaerobic, and their expression is stimulated by nitrate, indicating that their expression is repressed by O_2. When the culture is transferred from anaerobic to aerobic conditions, the enzymes are slowly irreversibly inactivated. The enzymes become inactive under aerobic conditions because their affinity for reduced coenzyme Q and cytochrome *c* is lower than that of aerobic respiratory enzymes. Of the four enzymes in denitrification, nitrous oxide reductase (N_2OR) is most sensitive to O_2, potentially leading to increased N_2O production under suboxic or fluctuating oxygen conditions. N_2O has a strong greenhouse gas effect.

Many denitrifiers can use various other electron acceptors, including oxygen, nitrate, dimethyl sulfoxide (DMSO, +0.15 V), dimethyl sulfide (DMS) and fumarate (+0.03 V). Organisms utilizing more than one electron acceptor have elaborate regulatory mechanisms to conserve more energy under given growth conditions. Electron acceptors with a higher redox potential are preferentially used over those with a lower redox potential. Under aerobic conditions, a two-component system, ArcA/ArcB, stimulates expression of the genes for aerobic respiration, and FNR activates the expression of anaerobic enzymes under anaerobic conditions (Section 12.2.4). Through these mechanisms, oxygen represses the expression of anaerobic respiratory enzymes. Enzymes that use electron acceptors of lower redox potential such as DMSO and fumarate are repressed by nitrate.

In FNR activation, dimeric FNR binds to the FNR box that is upstream of the anaerobic enzyme genes, increasing the affinity of the promoter for RNA polymerase. The expression of the nitrate reductase gene is further stimulated by a two-component system, NarX/NarL, in the presence of nitrate (Section 12.2.10). The membrane sensor protein NarX is phosphorylated, consuming ATP in the presence of nitrate, and transfers phosphate to the regulatory protein NarL. The phosphorylated NarL is an activator of nitrate reductase gene expression. Another two-component system, NarQ/NarP, controls expression of the other enzymes of denitrification, including nitrite reductase, nitric oxide reductase and nitrous oxide reductase. A similar two-component system, YhcS/YhcR, is involved in regulation of the nitrate respiratory pathway of *Staphylococcus aureus*. Regulation by two-component systems is discussed later (Section 12.2.10).

9.1.4 Denitrifiers other than facultatively anaerobic chemoorganotrophs

Denitrification has been regarded as a purely prokaryotic metabolism occurring mainly in facultative anaerobes and using organic electron donors. However, this process has now been identified in many other organisms including fungi such as *Fusarium oxysporum* and other species including *Cylindrocarpon tonkinense*, *Fusarium solani*, *Gibberella fujikuroi*, *Talaromyces flavus*, *Trichoderma hamatum* and *Trichosporon cutaneum*. These reduce nitrate and nitrite to N_2 or N_2O.

Some chemolithotrophs can use nitrate as their electron acceptor. *Thiobacillus denitrificans* and *Thiomicrospira denitrificans* use sulfide as the electron donor and nitrate as the electron acceptor. This anaerobic chemolithotrophic metabolism is ubiquitous in freshwater and marine environments. *Thiobacillus denitrificans* can couple uranium(IV) oxidation to nitrate reduction. Some alkalophilic bacteria can also use reduced sulfur as the electron donor and nitrate or nitrite as electron acceptors. *Thioalkalivibrio denitrificans* and *Thioalkalivibrio nitratireducens* oxidize thiosulfate, reducing nitrate, and *Thioalkalivibrio thiocyanodenitrificans* grows chemolithotrophically, oxidizing thiocyanate and thiosulfate to sulfate coupled to the reduction of nitrate. These are facultative anaerobes (Section 10.3.1).

A carboxydobacterium (Section 10.6), *Pseudomonas carboxydoflava*, uses nitrate under anaerobic conditions with carbon monoxide as its electron donor. Physiologically diverse H_2-oxidizing bacteria and archaea can also use nitrate as their electron acceptor (Section 10.5). A thermophilic chemolithotrophic hydrogen bacterium, *Hydrogenobacter thermophilus*, uses O_2 as well as nitrate as electron acceptors (Section 10.5.1). A group of thermophilic bacteria isolated from hydrothermal vents grow chemolithotrophically using H_2 as the electron donor, reducing nitrate to ammonia. These include *Caminibacter* spp., *Desulfurobacterium crinifex*, *Thermovibrio ruber*, *Thermovibrio ammonificans* and *Hydrogenomonas thermophila*. They reduce elemental sulfur to hydrogen sulfide as an alternative electron acceptor (Section 9.3). Species of *Caminibacter* and *Hydrogenomonas thermophila* are microaerophiles, and therefore use O_2 as their electron acceptor. *Ferroglobus placidus*, a strictly anaerobic archaeon, grows chemoautotrophically, oxidizing hydrogen coupled to nitrate reduction. Many bacteria oxidize Fe(II) coupled to nitrate reduction. These include *Citrobacter freundii*, *Paracoccus denitrificans*, *Pseudomonas stutzeri* and *Sphaerotilus natans*.

Another group of chemolithotrophs use NH_4^+ as an electron donor reducing nitrite:

$$NH_4^+ + NO_2^- \longrightarrow N_2 + 2H_2O \qquad (\Delta G^{0\prime} = -358\ \text{kJ/mol}\ NH_4^+)$$

These have not been isolated in pure culture but have been enriched in a reactor treating wastewater containing a high concentration of NH_4^+ in a process known as the anaerobic ammonia oxidation (ANAMMOX) process. Based on 16S ribosomal RNA gene sequences, these organisms have been named as *Brocadia anammoxidans*, *Kuenenia stuttgartiensis*, *Scalindua brodae* and *Scalindua wagneri*. Similar clones are widely distributed in natural and artificial ecosystems. A nitrifier, *Nitromonas europaea*, oxidizes ammonia under anaerobic conditions using nitrogen dioxide (NO_2) as the electron acceptor.

Some strictly anaerobic bacteria grow chemoheterotrophically, reducing nitrate to NH_4^+. These include *Denitrovibrio acetiphilus*, *Thauera selenatis*, *Wolinella succinogenes* and *Desulfovibrio gigas*. Since this process generates a proton motive force and is not inhibited by NH_4^+, the reduction of nitrate to NH_4^+ is regarded as dissimilatory nitrate reduction. This process may also be referred to as nitrate ammonification. The thermophilic anaerobic bacteria *Thermosulfurimonas dismutans* and *Dissulfuribacter thermophilus*, isolated from deep-sea hydrothermal vents, grow chemolithototrophically with elemental sulfur as an electron donor and nitrate as an electron acceptor, producing sulfate and ammonium.

A similar reaction is found in *Bacillus subtilis* and *Moorella thermoacetica* (formerly *Clostridium thermoaceticum*). Homoacetogenic (Section 9.5) *Moorella thermoacetica* ferments glucose to three molecules of acetate, two arising from pyruvate from glycolysis and the third from CO_2 produced by pyruvate: ferredoxin oxidoreductase. A culture grown with nitrate produces only two molecules of acetate since the electrons used for the synthesis of the third acetate on glucose are consumed to reduce nitrate. *Shewanella loihica* produces ammonia when nitrite is used as the electron acceptor while nitrate is reduced to N_2 in this bacterium. The electron transport system in nitrate ammonification has not yet been established.

Filamentous sulfur bacteria of the genera *Thioploca* and *Beggiatoa* and a related bacterium, *Thiomargarita namibiensis*, reduce nitrate to ammonia (Section 10.3.1). They accumulate nitrate within intracellular vacuoles. In these bacteria, nitrate is used as an electron acceptor and is reduced to ammonia with sulfide or sulfur as the electron donors when the oxygen supply is limited.

Clostridium perfringens and propionate-producing anaerobes reduce nitrate to NH_4^+ as in nitrate ammonification, but this reaction in these bacteria is not coupled to the generation of a proton motive force. Since electrons are disposed of in nitrate reduction, the electron acceptor, acetyl-CoA, is converted to acetate, synthesizing ATP. The fermentative propionate producers do not produce propionate in the presence of nitrate, but produce only acetate from lactate (Section 8.7). This process is referred to as fermentative nitrate reduction, and nitrate in this case is termed an electron sink.

Though it is well known that O_2 represses expression of denitrification enzyme genes, aerobic nitrate reduction has also been known for a long time. A facultative anaerobic bacterium isolated from an anaerobic digester, *Microvigula aerodenitrificans*, can reduce nitrate under air-saturated conditions. This bacterium can simultaneously use O_2 and nitrate as electron acceptors. A similar property is found in some other bacteria, including *Alcaligenes faecalis*, *Citrobacter diversus*, *Marinobacter* sp., *Pseudomonas nautica* and *Thiospaera pantotropha*. Some of these organisms use ammonia as the electron donor while growing heterotrophically (Section 10.2.1) and are referred to as 'heterotrophic nitrification–aerobic denitrification' bacteria.

9.1.5 Oxidation of xenobiotics under denitrifying conditions

Many xenobiotics can be oxidized under denitrifying conditions, including aromatic compounds such as benzene, toluene, alkylbenzene, xylene, phenol and resorcinol, and chlorinated hydrocarbons such as atrazine and 3-chlorobenzoate (Section 9.9).

9.2 Metal reduction

Some chemolithotrophs oxidize various metal ions acting as electron donors (Sections 10.4 and 10.7), while oxidized metal ions can serve as electron acceptors in anaerobic respiration. Microbes play important roles in the cycling of metals and metalloids, such as iron, manganese, selenium, and arsenic (as well as all the elements comprising cellular constituents, such as carbon, nitrogen, sulfur, phosphorus, etc.). Microbes reduce metal ions as electron acceptors (dissimilatory metal reduction) and for biosynthetic purposes (assimilatory metal reduction), both of which may reduce metal toxicity. During the reduction of the metals and metalloids listed in Table 9.4, free energy is conserved in the form of a proton motive force. Some metal reductions, for example Au(III) to Au(0), are not, however, coupled to energy conservation. Table 9.4 shows a partial list of metal-reducing bacteria.

Before a Fe(III)-reducing bacterium, *Shewanella oneidensis*, was characterized, metal reduction in anaerobic ecosystems was regarded as a chemical process coupled to the oxidation of sulfide (HS^-) produced through sulfidogenesis (Section 9.3). Fe(III) is virtually insoluble in water and this electron acceptor is reduced on the cell surface. Cr(VI), As(V), U(VI) and Tc(VII) are water soluble and toxic. Dissimilatory reducers of these metal species can reduce toxicity and remove them from the aqueous phase.

9.2.1 Fe(III) and Mn(IV) reduction

Most Fe(III) and Mn(IV) reducers use fermentation products such as acetate and lactate as their electron donors, though some use carbohydrates and glycerol (Table 9.4). Carbohydrate and glycerol metabolism in these organisms is not fully elucidated.

Shewanella alga and *Shewanella oneidensis* oxidize lactate and pyruvate to acetate, and the resulting electrons are consumed in reducing the metal ions. They cannot oxidize acetate further and are referred to as incomplete oxidizers, being similar to some strains of the sulfate-reducing *Desulfovibrio* genus:

$$\text{lactate}^- + 4Fe^{3+}(2Mn^{4+}) + 2H_2O \longrightarrow \text{acetate}^- + 4Fe^{2+}(2Mn^{2+}) + HCO_3^- + 5H^+$$

$$\text{pyruvate}^- + 2Fe^{3+}(Mn^{4+}) + 2H_2O \longrightarrow \text{acetate}^- + 2Fe^{2+}(Mn^{2+}) + HCO_3^- + 3H^+$$

Table 9.4 Anaerobes capable of using oxidized metal ions as electron acceptors.

Electron acceptor	Organism	Electron donor
Fe(III)	*Acidiphilum cryptum*	carbohydrate
	Aeromonas hydrophila	glycerol
	Deferribacter thermophilus	acetate, amino acids
	Desulfuromonas acetoxidans	acetate
	Ferroglobus placidus[a]	acetate
	Ferroplasma acidarmanus[a]	carbohydrate
	Geoalkalibacter ferrihydriticus	acetate
	Geobacter chapellei	acetate
	Geobacter hydrogenophilus	acetate, propionate, H_2
	Geobacter metallireducens	acetate
	Geobacter sulfurreducens	acetate
	Geoglobus ahangari	acetate
	Geothrix fermentans	acetate
	Geovibrio ferrireducens	acetate
	Pantoea agglomerans	acetate
	Pelobacter carbinolicus	ethanol
	Pyrobaculum islandicum[a]	H_2
	Shewanella alga	lactate
	Shewanella frigidimarina	carbohydrate
	Shewanella saccharophilia	carbohydrate
	Shewanella oneidensis	lactate
	Thermoterrabacterium ferrireducens	carbohydrate, glycerol, amino acids
Mn(IV)	*Desulfurimonas acetoxidans*	acetate
	Geobacter metallireducens	acetate
	Geobacter sulfurreducens	acetate
	Pyrobaculum islandicum[a]	H_2
	Shewanella oneidensis	lactate
U(VI)	*Desulfovibrio desulfuricans*	lactate
	Geobacter metallireducens	acetate
	Pyrobaculum islandicum[a]	H_2
	Shewanella alga	lactate
Selenate [Se(VI)]	*Aeromonas hydrophila*	glycerol
	Bacillus arsenicoselenatis	lactate
	Bacillus selenitireducens	amino acids
	Desulfotomaculum auripigmentum	lactate
	Enterobacter cloacae	amino acids
	Geospirillum barnesii	amino acids
	Pyrobaculum aerophilum[a]	H_2

Table 9.4 | (cont.)

Electron acceptor	Organism	Electron donor
	Selenihalanaerobacter shriftii	acetate, glucose
	Thauera selenatis	acetate
	Wolinella succinogenes	formate, hydrogen
Chromate [Cr(VI)]	*Bacterium dechromaticans*	lactate
	Enterobacter cloacae	amino acids
	Pyrobaculum islandicum	H_2
	Shewanella oneidensis	lactate
Co(III)	*Aeromonas hydrophila*	glycerol
	Pyrobaculum islandicum[a]	H_2
Arsenate [As(V)]	*Bacillus arsenicoselenatis*	lactate
	Chrysiogenes arsenatis	acetate
	Pyrobaculum arsenaticum[a]	H_2
	Sulfurospirillum arsenophilum	lactate
	Sulfurospirillum barnesii	lactate
Pertechnetate [Tc(VII)]	*Geobacter metallireducens*	acetate
	Shewanella oneidensis	lactate
Vanadate [V(V)]	*Shewanella oneidensis*	lactate
Plutonium (V/VI)	*Geobacter metallireducens*	acetate
	Shewanella oneidensis	lactate

[a] Archaea.

Geobacter metallireducens, *Geobacter sulfurreducens* and *Desulfuromonas acetoxidans* oxidize acetate completely to CO_2 through the modified TCA cycle described later (Figure 9.6):

$$CH_3COO^- + 8Fe^{3+}(Mn^{4+}) + 4H_2O \longrightarrow 8Fe^{2+}(4Mn^{2+}) + 2HCO_3^- + 9H^+$$

In addition to acetate, *Geobacter metallireducens* can use aromatic compounds such as phenol, benzoate and toluene as electron donors. Other xenobiotics, including benzene, toluene, ethylbenzene, xylene and trinitrotoluene, can also be removed under Fe(III)-reducing conditions (Section 9.9).

Fe(III) and Mn(IV) are practically insoluble in water under physiological conditions. Bacteria using these insoluble electron acceptors need to export electrons to reduce them extracellularly, as they cannot import these metal species. The electrons are transferred from the bacterial cell to the insoluble electron acceptor either through direct contact between them and a mineral surface, or are facilitated by soluble mediators such as humic acid and flavin. A third mechanism of electron transfer to the insoluble electron acceptor is known in *Shewanella oneidensis* and *Geobacter sulfurreducens*. They possess electrically conductive pilus-like appendages called bacterial nanowires through which electrons are transferred.

The whole genome has been sequenced in *Shewanella oneidensis* and *Geobacter sulfurreducens* and more than 40 genes encoding cytochromes have been identified in these bacteria. More than four *c*-type cytochromes are present on the outer membrane of *Shewanella oneidensis*. Figure 9.4 depicts these *c*-type cytochromes in *Shewanella oneidensis*. Electrons from quinol reduced during carbon metabolism are transferred to the periplasmic *c*-type cytochrome, MtrA, through the

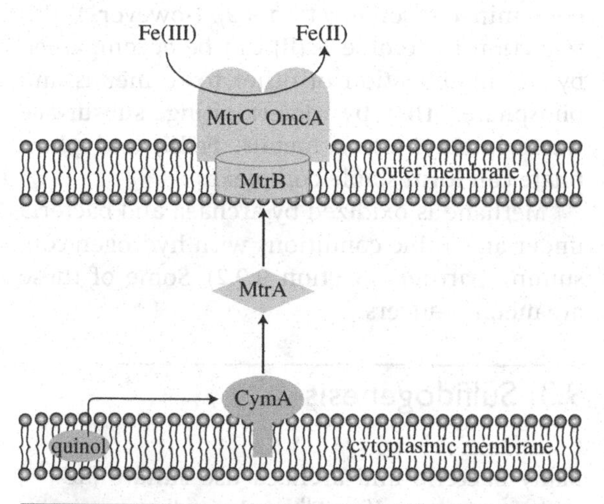

Figure 9.4 Electron transport to water-insoluble Fe(III) in *Shewanella oneidensis*.

The cytoplasmic membrane *c*-type cytochrome, CymA, accepts electrons from quinols and transfers them to the outer membrane MtrB through the periplasmic *c*-type cytochrome, MtrA. The cell surface MtrC–OmcA complex reduces Fe(III) using electrons from MtrB.

cytoplasmic membrane-bound *c*-type cytochrome, CymA. The OmcA–MtrC complex on the outer surface of the outer membrane reduces Fe(III) using electrons from MtrA through the outer membrane *c*-type cytochrome, MtrB. Other Fe(III) reducers are believed to have a similar electron transport system. A Gram-positive Fe(III) reducing bacterium, *Thermincola potens*, expresses a number of mutihaem cytochromes localized in the cell wall or cell surface during growth on Fe(III), as in the Gram-negative bacteria *Shewanella oneidensis* and *Geobacter sulfurreducens*, while the Gram-positive bacteria *Orenia metallireducens* and *Desulfotomaculum reducens* do not have *c*-type cytochromes that are typically associated with iron reduction in *Geobacter* and *Shewanella* spp., suggesting the existence of different Fe(III) reduction mechanisms.

Many Fe(III)-reducers are electrochemically active and capable of exchanging electrons with an electrode. They metabolize electron donors, with the electrode being used as an electron acceptor or electron sink in a fuel-cell-type electrochemical device. Similar devices are used to enrich microbial consortia oxidizing organic contaminants in wastewater with concomitant electricity generation.

Since Fe(III)-reducers metabolize a wide range of electron donors, including many xenobiotics, and Fe(III) is a major constituent of the Earth's crust, these organisms have considerable potential for the bioremediation of contaminated soil. They are also responsible for reduced methanogenesis in freshwater ecosystems, such as paddy fields, since they can outcompete methanogens.

9.2.2 Microbial reduction of other metal ions

In addition to Fe(III) and Mn(IV), many other metal ions are reduced by prokaryotes for biosynthetic purposes, as electron acceptors or in detoxification mechanisms. Selenium (a metalloid) is a component of some enzymes, including glycine reductase and formate dehydrogenase in the form of selenocysteine. Selenate is reduced in an assimilatory process, and also reduced as an electron acceptor. Other metal ions used as electron acceptors include As(V), Co(III), Cr(VI), Mo(VI), Pu(VI), Sb(V), Se (VI), U(VI) and V(V) (Table 9.4). Many microbes, including fermentative bacteria, reduce toxic As(V), Cr(VI), Se(VI) and V(V) without conserving energy available from the oxidation-reduction. These reactions result in detoxification. It is not known if energy is conserved during the microbial reduction of mercury(II), palladium(II), tellurium(VI), neptunium(V) and technetium(VII).

As(V), Cr(VI), Mo(VI), Se(VI) and U(VI) are rapidly reduced in anaerobic ecosystems, and various bacteria and archaea have been isolated based on their abilities to use these metal(loid)s as their electron acceptor. These organisms are phylogenetically diverse and found widely in nature. Though many prokaryotes have been identified as metal reducers, their carbon metabolism and electron transport processes have been less well studied.

The reduction of metals as electron acceptors is different from their reduction for biosynthetic purposes. As in nitrate reduction and sulfate reduction, metal reduction using the metal as an electron acceptor is referred to as dissimilatory metal reduction, while biosynthetic metal reduction is referred to as assimilatory reduction. The arsenate reductase of *Chrysiogenes arsenatis* is a [Fe–S] protein containing molybdenum and zinc. Various respiratory inhibitors (Section 5.8.2.4) inhibit Cr(VI) and V(V) reduction by *Shewanella oneidensis*, indicating the involvement of quinone and cytochromes in this electron transport process.

Metal reducers offer potential for treating wastewater containing toxic metals and metalloids, including As(V), Cr(VI), Mo(VI), Se(VI), U(VI), Np(V), Tc(VII) and V(V). For example, a radiation-resistant bacterium, *Deinococcus radiodurans*, reduces water-soluble U(VI) and Tc(VII) to insoluble U(IV) and Tc(IV), respectively, in effluents from nuclear power plants. A marine bacterium, *Marinobacter santoriniensis*, respires As (V) under anaerobic conditions, and oxidizes As (III) aerobically. Co(III) and U(VI) reduction by the Fe(III)-reducing bacterium *Shewanella oneidensis* is catalysed by the same outer membrane *c*-type cytochrome, OmcA.

9.2.3 Metal reduction and the environment

Before molecular oxygen accumulated in the atmosphere during the evolution of the Earth, Fe(III) and Mn(IV) were the most important electron acceptors in the carbon cycle since they are widely distributed in the Earth's crust, and chemolithotrophs used reduced metal ions as electron donors. Through these microbial activities, metal ions were subject to solubilization–immobilization cycles, thereby concentrating metal ores upon which some of the mining industries depend.

Among the electron acceptors used by microbes under anaerobic conditions, Fe(III) and Mn(IV) are most commonly used, since nitrate is not widely distributed. These metal ions have a higher redox potential than other electron acceptors and their reduction conserves more energy than sulfidogenesis and methanogenesis. It is estimated that most organic compound oxidation under anaerobic conditions is coupled to reduction of these metal ions. This has an important environmental impact, as metal reducers can oxidize xenobiotics in contaminated soil (Section 9.9). However, Fe(III) reduction to soluble Fe(II) can be accompanied by the mobilization of other toxic metals and phosphate, thereby deteriorating subsurface water quality. Toxic metals, Fe(III) and phosphate can form stable complexes.

Methane is oxidized by archaea and bacteria under anaerobic conditions with hydrogen-consuming partners (Section 9.9.2). Some of these are metal reducers.

9.3 Sulfidogenesis

Many bacteria and archaea use sulfate (SO_4^{2-}) and elemental sulfur (S^0) as their electron acceptors. Almost all of these organisms are obligate anaerobes (Table 9.5). Sulfidogens are grouped into mesophilic Gram-negative bacteria, spore-forming Gram-positive bacteria, thermophilic bacteria and hyperthermophilic archaea. In Table 9.5 they are listed as sulfate reducers and sulfur reducers, with each of these groups being further divided into complete oxidizers and incomplete oxidizers.

The main habitat of sulfidogens is sediments rich in organic electron donors and sulfate. They cause corrosion of underground and underwater structures and are especially troublesome in petroleum refineries and sewage works, causing great economic loss.

Sulfur can be reduced by some metal reducers, including species of *Wolinella*, *Shewanella*, *Sulfurospirillum* and *Geobacter*. A sulfur-reducing archaeon, *Pyrobaculum yellowstonensis*, reduces arsenate.

Electron donors used by sulfidogens are mainly fatty acids and alcohols produced by fermentation, with a few exceptions. Incomplete oxidizers metabolize ethanol and lactate to acetate, while acetate is completely oxidized by the complete oxidizers. Sulfidogens couple electron transport to sulfate or sulfur, with the generation of a proton motive force in most cases, but incomplete oxidizing sulfur-reducing archaea do not. They dispose of electrons from fermentative metabolism in a process referred to as fermentative sulfidogenesis. A similar metabolism is found among the denitrifiers (Section 9.1.4).

Table 9.5 | Sulfidogens

Organism (genus or species)	Electron donor	Acetate catabolism
Sulfate reducers		
Incomplete oxidizer		
Eubacteria		
Desulfovibrio	lactate, ethanol, malate, H_2, methanol,[a] glycerol[a]	–
Desulfomonas	lactate	–
Desulfomicrobium	lactate, ethanol, malate, succinate, fumarate, H_2	–
Desulfobulbus	lactate, ethanol, H_2, propionate	–
Desulfobotulus	lactate, long-chain fatty acids	–
Desulfotomaculum nigrificans	lactate, ethanol, H_2	–
Thermodesulfobacterium	lactate, H_2	–
Complete oxidizer		
Eubacteria		
Desulfoarculus	acetate, propionate, long-chain fatty acids	CO[b]
Desulfobacca acetoxidans	acetate, propionate, butyrate, ethanol, propanol	CO
Desulfobacter	acetate, H_2,[a] ethanol[a]	TCA[c]
Desulfobacterium	lactate, acetate, ethanol	CO
Desulfococcus	lactate, ethanol, propionate, long-chain fatty acids, benzoate	CO
Desulfosarcina	lactate, ethanol, H_2, acetate, propionate, long-chain fatty acids, benzoate	CO
Desulfonema	lactate, acetate, H_2, succinate, fumarate, malate, propionate, formate, benzoate, long-chain fatty acids	CO
Desulfosarcina	lactate, ethanol, H_2, acetate, propionate, long-chain fatty acids, benzoate	CO
Desulfotomaculum acetoxidans	ethanol, butanol, H_2, acetate, butyrate, long-chain fatty acids	CO
Desulfovibrio baarsii	acetate, propionate, butyrate, long-chain fatty acids	CO
Desulfovirga adipica	acetate, propionate, long-chain fatty acids, alcohols	?[d]
Archaeon		
Archaeoglobus fulgidus	lactate, malate, H_2, succinate,[a] fumarate,[a] glucose[a]	CO
Sulfur reducers		
Incomplete oxidizer		
Eubacteria		
Desulfovibrio gigas	lactate, ethanol, malate	–
Desulfomicrobium	lactate	–
Wolinella	lactate	–

Table 9.5 (cont.)

Organism (genus or species)	Electron donor	Acetate catabolism
Shewanella	lactate	–
Sulfurospirillum arcachonense (microaerophile)	lactate	–
Archaea		
Thermoproteus	H_2, yeast extract	– (fermentative)
Pyrobaculum	H_2, yeast extract	– (fermentative)
Desulfurococcus	H_2, yeast extract	– (fermentative)
Complete oxidizer		
Eubacteria		
Desulfuromonas	lactate,[a] ethanol, acetate, propionate, succinate, glutamate	TCA
Desulfurella	acetate	TCA
Desulfuromusa	acetate, propionate, yeast extract, peptone	?[d]
Geobacter	acetate	TCA

[a] Used by some species.
[b] Carbon monoxide dehydrogenase (acetyl-CoA or Wood–Ljungdahl) pathway.
[c] Modified TCA cycle.
[d] Not known.

As shown in Table 9.5, some sulfidogenic bacteria and archaea use H_2 as their electron donor. Some of them are facultative chemolithotrophs and others are heterotrophs using organic compounds as their major electron donor. Methane is oxidized by archaea and bacteria under anaerobic conditions with hydrogen-consuming partners (Section 9.9.2). Some of these are chemolithotrophic sulfate reducers. A group of thermophilic obligately chemolithotrophic bacteria have been isolated from hydrothermal vents, using H_2 as the electron donor, and reducing elemental sulfur to hydrogen sulfide (Section 10.5.4). These include *Balnearium lithotrophicum*, *Caminibacter* spp., *Desulfurobacterium crinifex*, *Thermovibrio ruber*, *Thermovibrio ammonificans* and *Hydrogenomonas thermophila*. They all reduce nitrate to ammonia, except for *Balnearium lithotrophicum* (Section 9.1.4). *Hydrogenomonas thermophila* and species of *Caminibacter* are microaerophiles while the others are strict anaerobes.

Desulfovibrio sp. can use nitrate as an electron acceptor (Section 9.1.4). It is interesting to note that they use the thermodynamically less favourable sulfate preferentially over nitrate. Their genes for denitrification are induced by nitrate and repressed by sulfate.

9.3.1 Biochemistry of sulfidogenesis

9.3.1.1 Reduction of sulfate and sulfur

Sulfate is activated to adenosine-5′-phosphosulfate (APS), consuming two high energy phosphate bonds of ATP. This reaction is similar to the initial reaction of assimilatory sulfate reduction (Section 6.3), but APS is not activated further to adenosine-3′-phosphate-5′-phosphosulfate (PAPS). The redox potential of the HSO_4^-/HSO_3^- half reaction is –0.454 V, which is much lower than that of any biological electron carriers, including ferredoxin (–0.41 V). The activation process makes the electron transfer to the electron acceptor a downhill reaction:

$$ATP + SO_4^{2-} \rightarrow adenosine\text{-}5'\text{-}phosphosulfate + PP_i$$

Pyrophosphate (PP_i) produced by ATP sulfurylase is hydrolysed to two P_i by the action of inorganic phosphatase in most cases, but strains of *Desulfotomaculum* use PP_i to activate acetate to acetyl-phosphate through the reaction catalysed by acetate-pyrophosphate kinase.

APS reductase liberates AMP on reducing APS to sulfite that is further reduced to sulfide (HS^-) by sulfite reductase consuming $6e^-$ (Figure 9.5).

Membrane-bound sulfur reductase reduces water-insoluble elemental sulfur in sulfur reducers. Electrons are transferred to sulfur reductase from electron donors through menaquinone and cytochrome *c*. Sulfur reducers grow attached to sulfur granules for efficient utilization of this water-insoluble electron acceptor. Sulfide in solution reacts with granular sulfur to form polysulfides such as tetrasulfide and pentasulfide. Polysulfides are highly soluble and might be the actual electron acceptors of sulfur reductase.

9.3.1.2 Carbon metabolism

Incomplete oxidizers provide electrons for the reduction of sulfate and sulfur from the oxidation of lactate and ethanol to acetate. Lactate dehydrogenase oxidizes lactate to pyruvate which is further oxidized to acetyl-CoA through the phosphoroclastic reaction catalysed by pyruvate:ferredoxin oxidoreductase (Section 8.5.1). The membrane-bound flavoprotein, lactate dehydrogenase, transfers electrons to menaquinone (Figure 9.9). Acetyl-CoA is converted to acetate, synthesizing ATP through substrate-level phosphorylation (Figure 9.5). Alcohol dehydrogenase and acetaldehyde dehydrogenase oxidize ethanol to acetyl-CoA:

$$ethanol \xrightarrow[NAD^+ \;\rightarrow\; NADH + H^+]{alcohol\ dehydrogenase} acetaldehyde \xrightarrow[NAD^+ \;\rightarrow\; NADH + H^+]{acetaldehyde\ dehydrogenase} acetyl\text{-}CoA$$

Acetate is metabolized through the acetyl-CoA pathway (Sections 9.5.2 and 10.8.3) in most complete oxidizers, except in strains of *Desulfobacter*, *Desulfuromonas* and *Desulfurella*, that oxidize acetate through a modified TCA cycle (Table 9.5). Each of these organisms has its own form of modified TCA cycle (Figure 9.6), with differences occurring in the reactions of acetyl-CoA and citrate synthesis. Acetate is activated to acetyl-CoA by succinyl-CoA: acetate CoA-SH transferase in strains of *Desulfobacter* and *Desulfuromonas*, while acetate kinase and phosphotransacetylase catalyse the reactions in strains of *Desulfurella*, consuming ATP. ATP: citrate lyase synthesizes citrate and ATP from acetyl-CoA and oxaloacetate in strains of *Desulfobacter*, and citrate synthase condenses acetyl-CoA and oxaloacetate into citrate without ATP synthesis in *Desulfuromonas* and *Desulfurella* strains.

Ferredoxin is reduced coupled to the oxidation of 2-ketoglutarate in a reaction catalysed by 2-ketoglutarate:ferredoxin oxidoreductase; succinate oxidation is coupled to the reduction of menaquinone. Electrons from the reduced menaquinone undergo reverse electron transport to be used in reducing sulfur (Figure 9.11b, and Section 10.1).

All complete oxidizing sulfate reducers, except strains of *Desulfobacter*, oxidize acetate through the acetyl-CoA pathway, which is not known in sulfur reducers (Figure 9.7). Since the key enzyme of this pathway is carbon monoxide

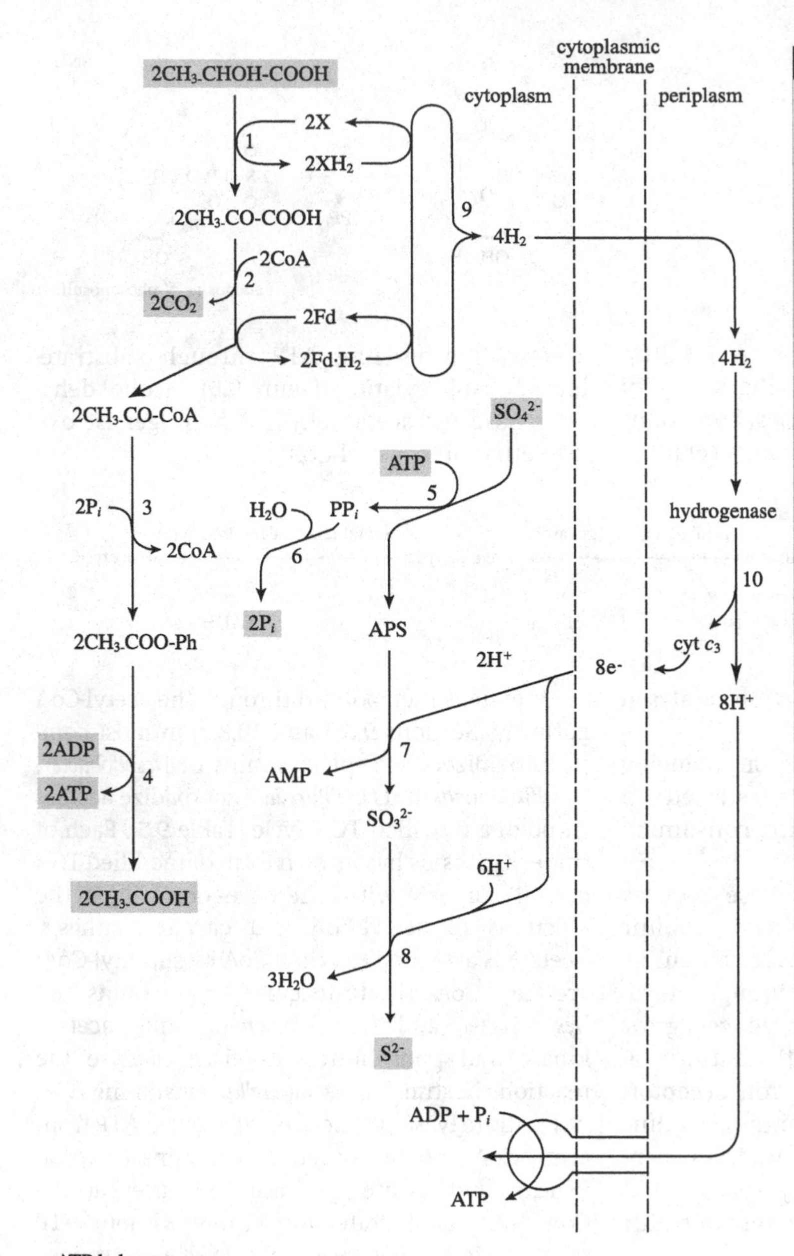

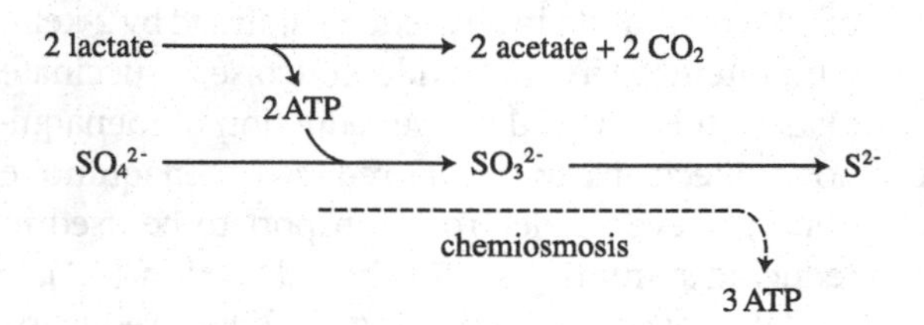

Figure 9.5 Hydrogen cycling in sulfate reduction by *Desulfovibrio* species.

(Gottschalk, G. 1986, *Bacterial Metabolism*, 2nd edn., Figure 8.29. Springer, New York)

Incomplete oxidizing sulfate-reducers, including *Desulfovibrio* species, oxidize two lactate to acetate to reduce one sulfate. Through substrate-level phosphorylation (reaction 4) two ATPs are produced, and two high energy phosphate bonds are consumed (reaction 5) per sulfate reduced. Extra ATP needs to be generated for bacterial growth and this is achieved by hydrogen cycling. Cytoplasmic hydrogenase produces H_2 and oxidizes the reduced electron carriers that result during lactate oxidation. H_2 diffuses to the periplasm where another hydrogenase reduces cytochrome c_3, leaving protons. The reduced cytochrome c_3 supplies electrons for sulfate reduction across the membrane.

1, lactate dehydrogenase; 2, pyruvate: ferredoxin oxidoreductase; 3, phosphotransacetylase; 4, acetate kinase; 5, ATP sulfurylase; 6, pyrophosphatase; 7, APS reductase; 8, sulfite reductase; 9, cytoplasmic hydrogenase; 10, periplasmic hydrogenase.

dehydrogenase (CODH), this pathway is also referred to as the CODH pathway or the Wood-Ljungdahl pathway, to honour the individuals who discovered this metabolism. CODH splits acetyl-CoA to form methyltetrahydrofolate (CH_3-H_4F) and enzyme-bound [CO]. The enzyme-bound [CO] is oxidized to CO_2, reducing ferredoxin. CH_3-H_4F is oxidized to CO_2. In the archaeon

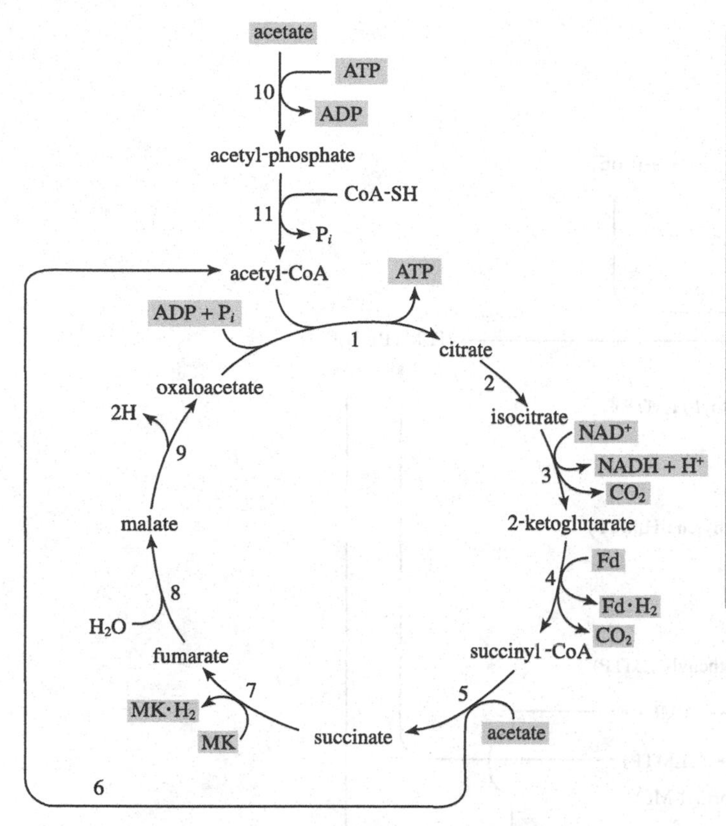

Figure 9.6 Modified TCA cycle in complete oxidizing sulfidogens.

Acetate is activated in a reaction catalysed by succinyl-CoA: acetate CoA-SH transferase (6) in *Desulfobacter* and *Desulfuromonas*, and by acetate kinase (10) and phosphotransacetylase (11) in *Desulfurella*. Citrate is formed by ATP: citrate lyase in *Desulfobacter*, and by citrate synthase in *Desulfuromonas* and *Desulfurella*.

1, ATP: citrate lyase (*Desulfobacter*) or citrate synthase (*Desulfuromonas* and *Desulfurella*); 2, 3, 8, 9 as in the TCA cycle; 4, 2-ketoglutarate: ferredoxin oxidoreductase; 5, succinyl-CoA synthetase (*Desulfurella*); 6, succinyl-CoA: acetate CoA transferase (*Desulfobacter* and *Desulfuromonas*); 7, succinate: menaquinone oxidoreductase; 10, acetate kinase; 11, phosphotransacetylase.

Archaeoglobus fulgidus, the archaeal C1-carriers tetrahydromethanopterin (H_4MPT) and methanofuran (MF) replace tetrahydrofolate.

Since all reactions of the acetyl-CoA pathway are reversible, some chemolithotrophic anaerobes synthesize acetyl-CoA for biosynthetic purposes from CO_2 and H_2 through this pathway (Section 10.8.3). These include methanogens, homoacetogens and chemolithotrophic sulfidogens.

Strains of *Desulfobulbus*, incomplete oxidizers, oxidize propionate to acetate through a similar pathway to the succinate-propionate pathway in strains of *Propionibacterium* (Section 8.7.1). These bacteria oxidize ethanol and lactate to acetate via acetyl-CoA and pyruvate, respectively (Figure 9.8).

9.3.2 Electron transport and ATP yield in sulfidogens

9.3.2.1 Incomplete oxidizers

As shown below, incomplete oxidizing strains of the genus *Desulfovibrio* gain two ATP through substrate-level phosphorylation, and consume the same number of ATP molecules to reduce a molecule of sulfate, oxidizing two molecules of lactate to acetate:

$$SO_4^{2-} + 8H^+ + ATP \longrightarrow S^{2-} + AMP + PP_i + 4H_2O$$
$$PP_i + H_2O \longrightarrow 2P_i$$
$$2\text{lactate} + 2ADP + 2P_i \longrightarrow 2\text{acetate} + 8H^+ + 2CO_2 + 2ATP$$
$$ATP + AMP \longrightarrow 2ADP$$

$$\text{sum)}\ SO_4^{2-} + 2\text{lactate} \longrightarrow S^{2-} + 2\text{acetate} + 2H_2O + 2CO_2$$

These bacteria need other mechanisms to synthesize ATP to grow on lactate using sulfate as the electron acceptor. They synthesize ATP through an electron transport phosphorylation (ETP) mechanism. Hydrogenases play an important role in generating a proton motive force, which is coupled to sulfidogenesis in these bacteria. The cytoplasmic membrane-bound [FeNi] hydrogenase reduces $8H^+$ to produce $4H_2$. The H_2 diffuses into the periplasm where it is oxidized by periplasmic hydrogenase to reduce cytochrome c_3. Electrons are transferred back into the cytoplasm

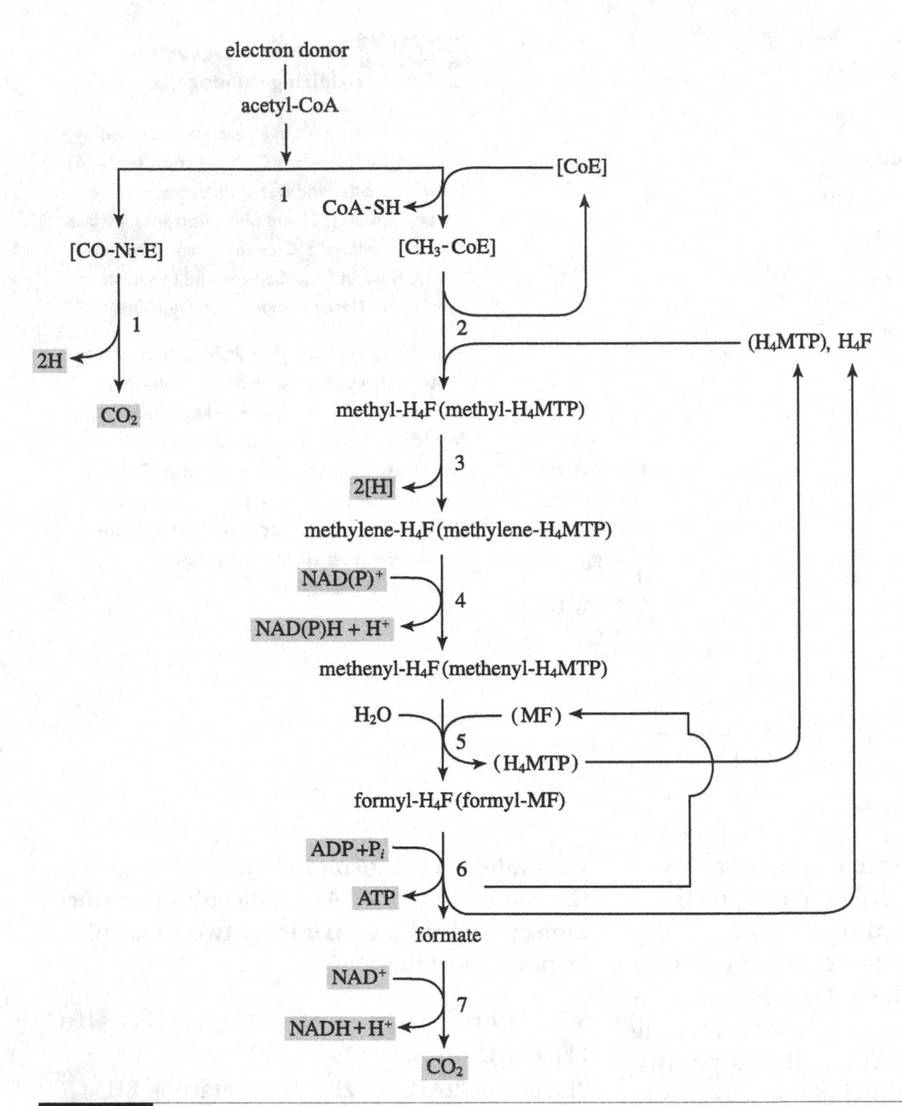

Figure 9.7 Complete oxidation of acetate through the acetyl-CoA pathway in sulfidogens.

All complete oxidizing sulfate-reducers, except *Desulfobacter*, employ the acetyl-CoA pathway, involving carbon monoxide dehydrogenase (1, or acetyl-CoA synthase). This pathway is the reverse of acetate production from $H_2 + CO_2$ in homoacetogens. H_4F (tetrahydrofolate) carries C_1 compounds in bacteria; the archaeon *Archaeoglobus fulgidus* uses tetrahydromethanopterin (H_4MPT) and methanofuran (MF) for this purpose.

1, carbon monoxide dehydrogenase; 2, methyltransferase; 3, methyl-H_4F dehydrogenase; 4, methylene-H_4F dehydrogenase; 5, methenyl-H_4F cyclohydrolase; 6, formyl-H_4F synthetase; 7, formate dehydrogenase.

[CO-Ni-E]: enzyme-bound carbon monoxide; CoE, corrinoid enzyme.

to reduce sulfate, consuming another $8H^+$. These reactions result in the generation of a proton motive force and this process is referred to as the hydrogen cycling mechanism (Figure 9.5).

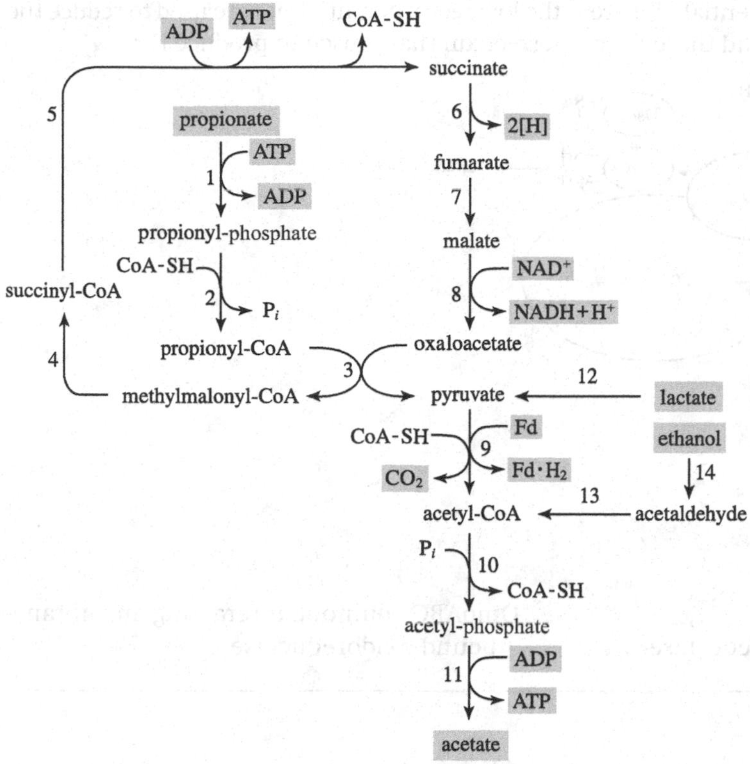

Figure 9.8 Oxidation of propionate, lactate and ethanol to acetate by strains of *Desulfobulbus*.

1, propionate kinase; 2, phosphotransacylase; 3, methylmalonyl-CoA: pyruvate transcarboxylase; 4, methylmalonyl-CoA mutase; 5, succinyl-CoA synthetase; 6, succinate dehydrogenase; 7, fumarase; 8, malate dehydrogenase; 9, pyruvate: ferredoxin oxidoreductase; 10, phosphotransacetylase; 11, acetate kinase; 12, lactate dehydrogenase; 13, aldehyde dehydrogenase; 14, alcohol dehydrogenase.

In the hydrogen cycling mechanism it is assumed that all electrons available from the oxidation of lactate to acetate are used to reduce protons to hydrogen. However, the redox potential of pyruvate/lactate, −190mV, is too high for the electrons from the reaction catalysed by lactate dehydrogenase to be used to reduce protons. Because of this a modified hydrogen cycling mechanism is proposed (Figure 9.9). According to this mechanism, the energy-converting membrane-bound [FeNi] hydrogenase oxidizes a part of the reduced ferredoxin to produce the hydrogen that participates in the hydrogen cycling. The remaining electrons of reduced ferredoxin and from lactate oxidation are transferred to quinone to be used to reduce sulfate. Additionally the energy-converting membrane-bound [FeNi] hydrogenase exports H^+ (Section 5.8.6.5), as shown in Figure 9.9.

APS reductase (AprAB) and sulfite reductase (DsrAB) are membrane-bound protein complexes in *Desulfovibrio* spp. APS reductase (AprBA) reduces APS to sulfite, liberating AMP with electrons supplied from the QmoABC (quinone-interacting membrane-bound oxidoreductase) complex and reduced ferredoxin, in an electron confurcation reaction (Figure 9.10a), since the redox potential of menaquinol ($E^{0\prime}$ = −75 mV) is not low enough to allow reduction of APS to sulfite ($E^{0\prime}$ = −60 mV), or sulfite to sulfide ($E^{0\prime}$ = −116 mV). The AprA contains a FAD group and the AprB contains two [4Fe–4S] clusters. The dissimilatory sulfite reductase DsrAB is a sirohaem protein, reducing sulfite to sulfide, consuming six electrons supplied from the membrane protein complex DsrMKJOP, mediated by a soluble protein, DsrC (Figure 9.10b).

When *Desulfovibrio vulgaris* is cultivated on ethanol/sulfate, the NADH: flavin oxidoreductase (FloxABCD) oxidizes NADH, reducing a flavoprotein. The reduced flavoprotein is oxidized in an electron bifurcation reaction (Section 5.8.6.4).

Electrons from the high redox potential site are used in sulfate reduction directly and those from the low redox potential site are used to reduce the ferredoxin that is used to produce H_2:

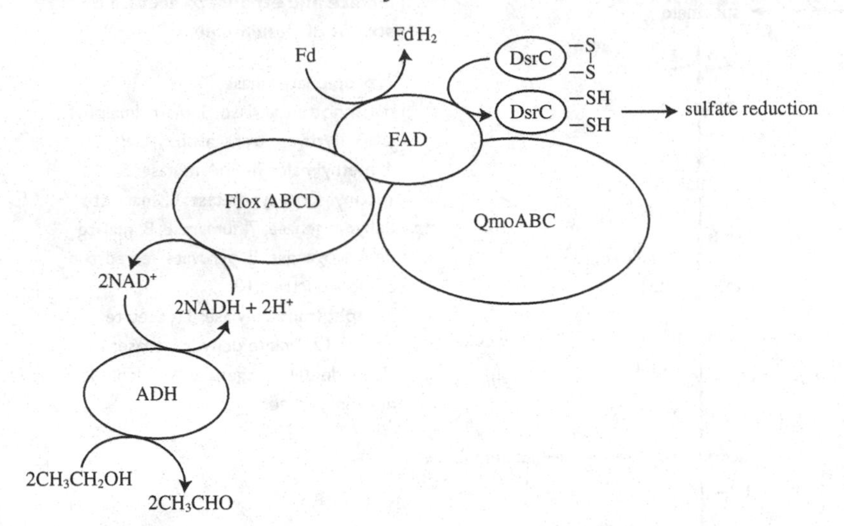

ADH: alcohol dehydrogenase
FloxABCD: NADH:flavin oxidoreductase
QmoABC: quinone-interacting membrane-bound oxidoreductase.

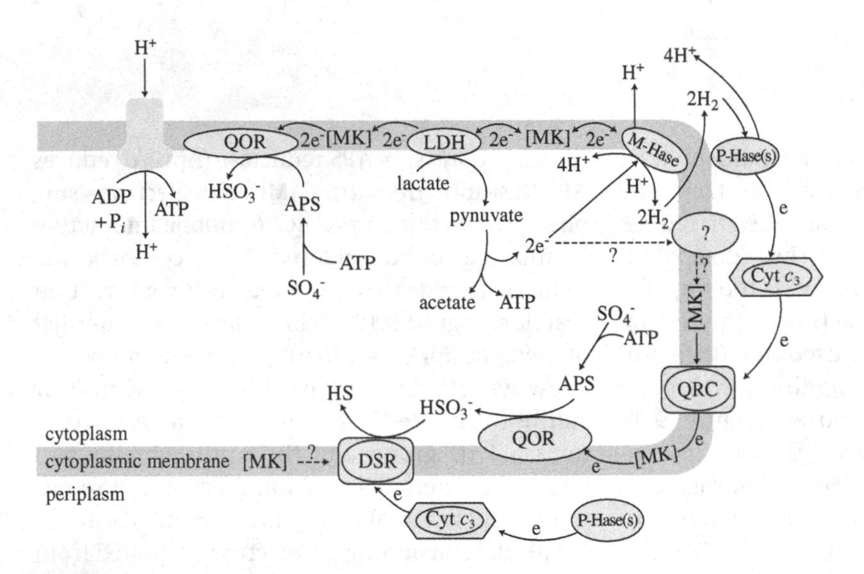

Figure 9.9 The modified hydrogen cycling mechanism.

(Modified from *Front. Microbiol.* **2**: 135, 2011)

Electrons from lactate oxidation to pyruvate cannot be used to reduce protons to hydrogen since the redox potential of pyruvate/lactate is −190mV, which is much higher than that of H^+/H_2, −0.41mV. In this modified hydrogen cycling mechanism, the energy-converting membrane-bound [FeNi] hydrogenase oxidizes a part of reduced ferredoxin to produce hydrogen that participates in the hydrogen cycling mechanism. The remaining electrons from lactate oxidation to acetate are directly transferred to quinone to be used in sulfate reduction. Additionally the energy-converting hydrogenase translocates H^+.

Cyt c_3, cytochrome c_3; DSR, dissimilatory sulfite reductase; LDH, lactate dehydrogenase; M-Hase, membrane-bound [FeNi] hydrogenase; MK, menaquinone; P-Hase, periplasmic hydrogenase; QOR, quinone oxidoreductase; QRC, cytochrome c_3: menaquinone oxidoreductase

(a)

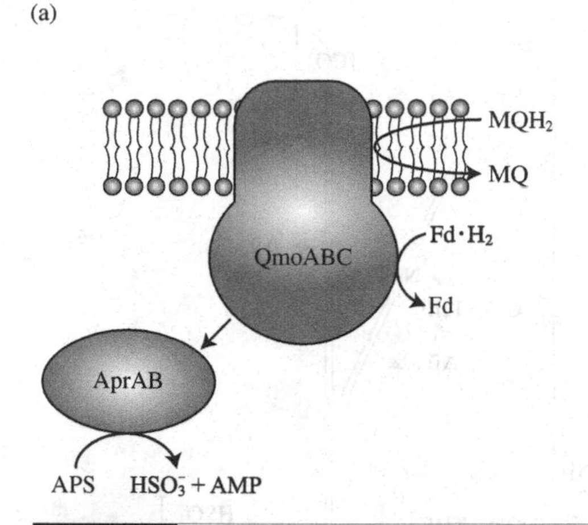

(b)

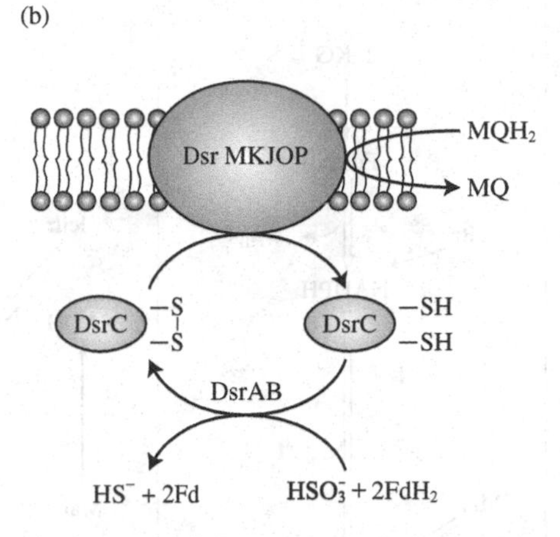

Figure 9.10 Structures of membrane-bound APS reductase (a) and sulfite reductase (b) complexes in *Desulfovibrio desulfuricans*.

(Modified from *Biochim. Biophys. Acta.* 1827: 145–160, 2013)

APS reductase (AprAB) reduces APS to sulfite, liberating AMP, with electrons supplied from the quinone-interacting membrane-bound oxidoreductase (QmoABC) complex and a low redox potential electron donor, in an electron confurcation reaction (a), since the redox potential of menaquinol ($E^{0'}$ = –75 mV) is not low enough to allow reduction of APS to sulfite ($E^{0'}$= –60 mV) or sulfite to sulfide ($E^{0'}$= –116 mV). The dissimilatory sulfite reductase (DsrAB) is a sirohaem protein, reducing sulfite to sulfide, consuming electrons supplied from membrane protein complex DsrMKJOP and from reduced ferredoxin (b). Quinone is reduced, coupled to oxidation of either lactate or cytochrome c_3.

The hydrogen-cycling mechanism is not known in strains of *Desulfotomaculum*. They use PP_i produced by ATP sulfurylase to phosphorylate acetate through the action of acetate-pyrophosphate kinase. These bacteria synthesize 2 ATP from lactate oxidation to acetate and invest 1 ATP in sulfate reduction. Pyrophosphatase activity is low in these bacteria, and PP_i is not hydrolysed to 2 P_i. In many bacteria, polyphosphate is used to store energy (Section 13.2.4):

$$ATP + SO_4^{2-} \xrightarrow{\text{ATP sulfurylase}} PP_i + \text{adenosine-5}'\text{-phosphosulfate}$$

$$PP_i + \text{acetate} \xrightarrow{\text{acetate–pyrophosphate kinase}} \text{acetyl-phosphate} + P_i$$

$$\text{acetyl-phosphate} + ADP \xrightarrow{\text{acetate kinase}} \text{acetate} + ATP$$

9.3.2.2 Complete oxidizers

Though ATP is synthesized through SLP in complete oxidizers (reaction 1 in Figure 9.6 and reaction 6 in Figure 9.7), the number of ATP molecules is less than that required for the activation of acetate and sulfate. As shown below, enough free energy is generated from acetate oxidation coupled to the reduction of sulfate or sulfur. This free energy is conserved in the form of a proton motive force:

$$\text{acetate}^- + SO_4^{2-} + 3H^+ \longrightarrow 2CO_2 + H_2S + 2H_2O \quad (\Delta G^{0'} = -63\ \text{kJ/mol acetate})$$

$$\text{acetate}^- + H^+ + 4S^0 + 2H_2O \longrightarrow 2CO_2 + 4H_2S \quad (\Delta G^{0'} = -39\ \text{kJ/mol acetate})$$

Acetate oxidation is coupled to the reduction of $NAD(P)^+$ and ferredoxin. Electrons from NAD(P)H and reduced ferredoxin are transported to menaquinone, exporting H^+ to generate a proton motive force for ATP synthesis (Figure 9.11). Since the redox potential of S^0/H_2S is lower than that of menaquinone, electrons of the reduced menaquinone with succinate oxidation (reaction 6 in Figure 9.8) undergo reverse electron transport (Section 10.1) to reduce S^0 in

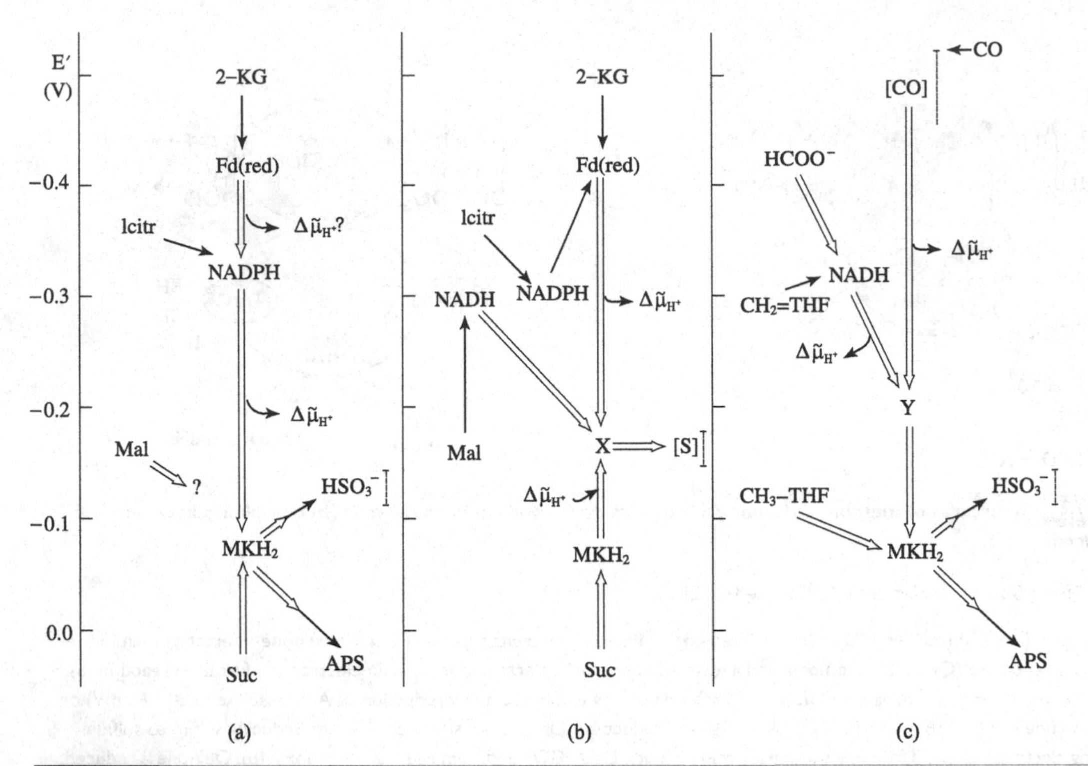

Figure 9.11 Electron metabolism in complete oxidizing sulfidogens.

(Widdel, F. & Bak, F. 1992. In *The Prokaryotes*, 2nd edn., pp. 3352–3378. Springer, New York)

Acetate oxidation is mediated through a modified TCA cycle to reduce sulfate in *Desulfobacter postgatei* (a) and to reduce sulfur in *Desulfuromonas acetoxidans* (b). *Desulfotomaculum acetoxidans* (c) oxidizes acetate through the acetyl-CoA pathway to reduce sulfate. The TCA cycle intermediate 2-ketoglutarate (2-KG) reduces ferredoxin (Fd), and isocitrate (Icitr) reduces $NADP^+$. Reduced ferredoxin and NADPH transfer electrons to menaquinone (MKH_2) in *Desulfobacter postgatei* (a) to an unknown electron carrier (X) in *Desulfuromonas acetoxidans* (b), generating a proton motive force. The electrons are finally consumed to reduce adenosine-5′-phosphosulfate (APS), sulfite (HSO_3^-) or sulfur ([S]). Another TCA cycle intermediate, malate (Mal), reduces NAD^+ in *Desulfuromonas acetoxidans* (b), but the electron carrier in *Desulfobacter postgatei* (a) is not known. Succinate (Suc) reduces menaquinone. Electrons from menaquinone are transferred to the unknown electron carrier (X) to reduce sulfur through a reverse electron transport process in *Desulfuromonas acetoxidans* (b). Electrons from the acetyl-CoA pathway are transferred to an unknown electron carrier (Y), with generation of a proton motive force. Reactions indicated with single arrows take place in the cytoplasm, and those with double arrows occur in the cytoplasmic membrane. Electrons from acetyl-CoA pathway intermediates, formate ($HCOO^-$) and methylene-tetrahydrofolate (CH_2=THF) are transferred to NAD^+, and those from methyl-tetrahydrofolate (CH_3–THF) to menaquinone.

sulfur reducers, employing a modified TCA cycle (Figure 9.11b).

The ATP yield of the complete oxidizers is 0.6 ATP/acetate in a sulfur reducer (*Desulfuromonas acetoxidans*) and 0.8 ATP/acetate in a sulfate reducer (*Desulfotomaculum acetoxidans*). These figures are calculated from cell yield, Y_{ATP} and the maintenance energy (Section 6.14.3).

9.3.3 Carbon skeleton supply in sulfidogens

Some complete oxidizing sulfate reducers can grow chemolithotrophically using H_2 as the electron donor (Table 9.5). Among these organisms, *Desulfobacter hydrogenophilus* fixes CO_2 through a reductive TCA cycle (Section 10.8.2) and the acetyl-CoA pathway

(Section 10.8.3) is employed in others, including *Desulfosarcina variabilis*, *Desulfonema limicola*, and *Desulfotomaculum orientis*.

Acetyl-CoA synthesized from $H_2 + CO_2$ or ethanol is reduced to pyruvate by pyruvate:ferredoxin oxidoreductase. Pyruvate is metabolized through the incomplete TCA fork (Section 5.4.1) and gluconeogenesis (Section 4.2) to supply carbon skeletons for biosynthesis.

9.3.4 Oxidation of xenobiotics under sulfidogenic conditions

Lactate and acetate are common electron donors for most sulfidogens. However, it has been found that many xenobiotics can be oxidized under sulfidogenic conditions. These include aliphatic, aromatic and halogenated hydrocarbons. Microbial consortia that include sulfidogens are generally responsible for xenobiotic oxidation. Methane is also oxidized under these conditions. Sulfate reducers have been isolated from soil contaminated with petroleum and related substances. Xenobiotics that can be used as electron donors in anaerobic respiratory processes are discussed later (Section 9.9).

9.4 Methanogenesis

The conversion of organic materials to methane (CH_4) has been known for a long time in anaerobic ecosystems. This microbial process has been applied to the treatment of wastewaters rich in organic content. Methanogens are strictly anaerobic archaea. They flourish in anaerobic organic-rich ecosystems with low sulfate concentrations, but they are one of the most tedious groups of microbes to cultivate in pure culture. CO_2 is the electron acceptor in methanogens. The common electron donors used by methanogens to reduce CO_2 are formate, methanol, acetate, methylamines, carbon monoxide and H_2. Some methanogens can use ethanol, 2-propanol, 2-butanol and ketones.

9.4.1 Methanogens

Methanogens occupy seven phylogenetic orders in the archaea (Table 9.6). They are grouped according to the electron donors they use and comprise hydrogenotrophic, methylotrophic and aceticlastic methanogens.

Table 9.6 Methanogens

Methanogen	Electron donor	Characteristics
Order ***Methanobacteriales***		
Methanobacterium	$H_2 + CO_2$, formate,[a] CO,[a] 2° alcohol + CO_2[a]	pseudopeptidoglycan
Methanobrevibacter	$H_2 + CO_2$, formate[a]	pseudopeptidoglycan, S-layer
Methanosphaera	H_2 + methanol	pseudopeptidoglycan, acetate or CO_2 as carbon source in mammalian intestine
Methanothermobacter	$H_2 + CO_2$, formate	sulfur reducer
Methanothermus	$H_2 + CO_2$	pseudopeptidoglycan, S-layer, thermophile, hot springs
Order ***Methanococcales***		
Methanococcus	$H_2 + CO_2$, formate[a]	S-layer
Methanothermococcus	$H_2 + CO_2$, formate[a]	S-layer
Methanocaldococcus	$H_2 + CO_2$	S-layer, fast growth
Methanotorris	$H_2 + CO_2$	S-layer

Table 9.6 (cont.)

Methanogen	Electron donor	Characteristics
Order ***Methanomicrobiales***		
Methanomicrobium	$H_2 + CO_2$, formate[a]	acetate as carbon source
Methanoculleus	$H_2 + CO_2$, formate,[a] 2° alcohol + CO_2[a]	acetate as carbon source
Methanofollis	$H_2 + CO_2$, formate,[a] 2° alcohol + CO_2[a]	S-layer, acetate as carbon source
Methanogenium	$H_2 + CO_2$, formate,[a] 2° alcohol + CO_2[a]	S-layer, acetate as carbon source
Methanolacinia	$H_2 + CO_2$, 2° alcohol	acetate as carbon source
Methanoplanus	$H_2 + CO_2$, formate	sulfur reducer
Methanocorpusculum	$H_2 + CO_2$, formate, 2° alcohol + CO_2[a]	yeast extract or peptone as carbon source
Methanospirillum	$H_2 + CO_2$, formate, 2° alcohol + CO_2[a]	grows in chains
Methanocalculus	$H_2 + CO_2$	halophilic (12% NaCl), disc-shaped, S-layer
Order ***Methanosarcinales***		
Methanosarcina	$H_2 + CO_2$,[a] acetate,[a] CO,[a] methanol, methylamines	forms clumps
Methanococcoides	methanol, methylamines	halophilic (0.2M NaCl), requires Mg^{2+}
Methanohalobium	methanol, methylamines	halophilic (2.6–5.1M NaCl)
Methanohalophilus	methanol, methylamines	halophilic (1.0–2.5M NaCl)
Methanolobus	methanol, methylamines, dimethylsulfide[a]	S-layer
Methanosalsum	methanol, methylamines, dimethylsulfide	halophilic and alkalophilic, S-layer
Methanosaeta	acetate	grows in chains, halophilic (2.5M NaCl)
Order ***Methanopyrales***		
Methanopyrus	$H_2 + CO_2$	hyperthermophilic (110°C), sulfur reducer, pseudopeptidoglycan
Order ***Methanocellaes***		
Methanocella	$H_2 + CO_2$	paddy field
Order ***Methanomassiliicoccales***		
Methanomassiliicoccus	H_2 + methanol, H_2 + methylamines	

[a] Dependent on strain.
2°, secondary.

9.4.1.1 Hydrogenotrophic methanogens

The majority of known methanogens grow on the free energy available from the reduction of CO_2 to CH_4 using H_2 as the electron donor. These are referred to as chemolithotrophic methanogens:

$CO_2 + 4H_2 \longrightarrow CH_4 + 2H_2O$
$(\Delta G^{0'} = -136\,kJ/mol\,CH_4)$

They use formate and CO in addition to CO_2 and H_2. These are oxidized to CO_2 before being reduced to CH_4:

$4HCOOH \longrightarrow CH_4 + 3CO_2 + 2H_2O$
$(\Delta G^{0'} = -144\,kJ/mol\,CH_4)$
$4CO + 2H_2O \longrightarrow CH_4 + 3CO_2$
$(\Delta G^{0'} = -211\,kJ/mol\,CH_4)$

Some of the hydrogenotrophic methanogens oxidize ethanol, 2-propanol and 2-butanol completely or partially using CO_2 as the electron acceptor. These include strains of *Methanobacterium*, *Methanoculleus*, *Methanogenium*, *Methanolacinia*, *Methanospirillum* and *Methanocorpusculum*.

9.4.1.2 Methylotrophic methanogens

Methyl compounds such as methanol and methylamines are used as the substrate by methylotrophic methanogens. These include strains of the order *Methanosarcinales*, except those of the genus *Methanosaeta*. Strains of *Methanolobus* can use methyl sulfide:

$4CH_3OH \longrightarrow 3CH_4 + CO_2 + 2H_2O$
$(\Delta G^{0'} = -106\,kJ/mol\,CH_4)$

Methylotrophic methanogens without cytochromes (*Methanosphaera* spp. and *Methanomassiliicoccus* spp.) are obligately H_2 dependent and unable to oxidize methyl groups, while those that possess cytochromes (members of the order *Methanosarcinales*) oxidize methyl groups to CO_2 via a membrane-bound electron transport chain, to supply electrons to reduce methyl-coenzyme M (methyl-CoM) to CH_4.

9.4.1.3 Aceticlastic methanogens

Limited numbers of methanogens can use acetate as their substrate. These include strains of *Methanosaeta* and *Methanosarcina*. The former use only acetate and grow in chains. Strains of *Methanosarcina* are most versatile in terms of their substrate. They can use all known methanogenic substrates except secondary alcohols:

$CH_3COOH \longrightarrow CH_4 + CO_2$
$(\Delta G^{0'} = -37\,kJ/mol\,CH_4)$

9.4.2 Coenzymes in methanogens

Methanogenic archaea employ some unique coenzymes. These coenzymes are known in some hyperthermophilic archaea and in a limited number of the eubacteria (Sections 4.3.4, 7.7 and 7.10.2). The presence of unique coenzymes in methanogenic archaea is one of the bases for the hypothesis of separate archaeal evolution. The presence of archaeal coenzymes in eubacteria is taken as evidence of recent lateral gene transfer.

Tetrahydrofolate (H_4F) and *S*-adenosylmethionine (SAM) are C1-carriers in eubacteria and eukaryotes (Section 6.6.2). Methanofuran (MF), 5,6,7,8-tetrahydromethanopterin (H_4MPT) and coenzyme M replace them in methanogenic archaea. MF is a formyl-carrier, H_4MPT is the archaeal analogue of tetrahydrofolate in eubacteria carrying formyl, methenyl, methylene and methyl groups, while coenzyme M is a methyl-group carrier. In addition to these C1-carriers, methanogenic archaea use coenzyme F_{420}, coenzyme F_{430}, 7-mercaptoheptanoylthreonine phosphate and methanophenazine (Figure 9.12).

Coenzyme F_{420} is reduced by hydrogenase, formate dehydrogenase and carbon monoxide dehydrogenase, and oxidized in reducing CO_2 to CH_4. Coenzyme F_{420} is involved in the reactions catalysed by $NADP^+$ reductase, pyruvate synthase and 2-ketoglutarate synthase, in methanogenic archaea. Since coenzyme F_{420} is fluorescent, methanogenic archaea containing this coenzyme can be distinguished from eubacteria using fluorescence microscopy. The structure of F_{420} shown in Figure 9.12 is found in *Methanobacterium thermoautotrophicum* and is referred to as F_{420}-2 because it has a side chain consisting of two glutamates. In other strains, 4–7 glutamate residues form the F_{420} side chain.

Coenzyme F_{430} is another methanogenic electron carrier involved in the reaction catalysed by methyl-coenzyme M methylreductase. Structural variants of F_{430}, F_{430}-2 and F_{430}-3 are known in methanogens belonging to *Methanococcales*, and there are other variants found in anaerobic methanotrophic archaea. These variants are

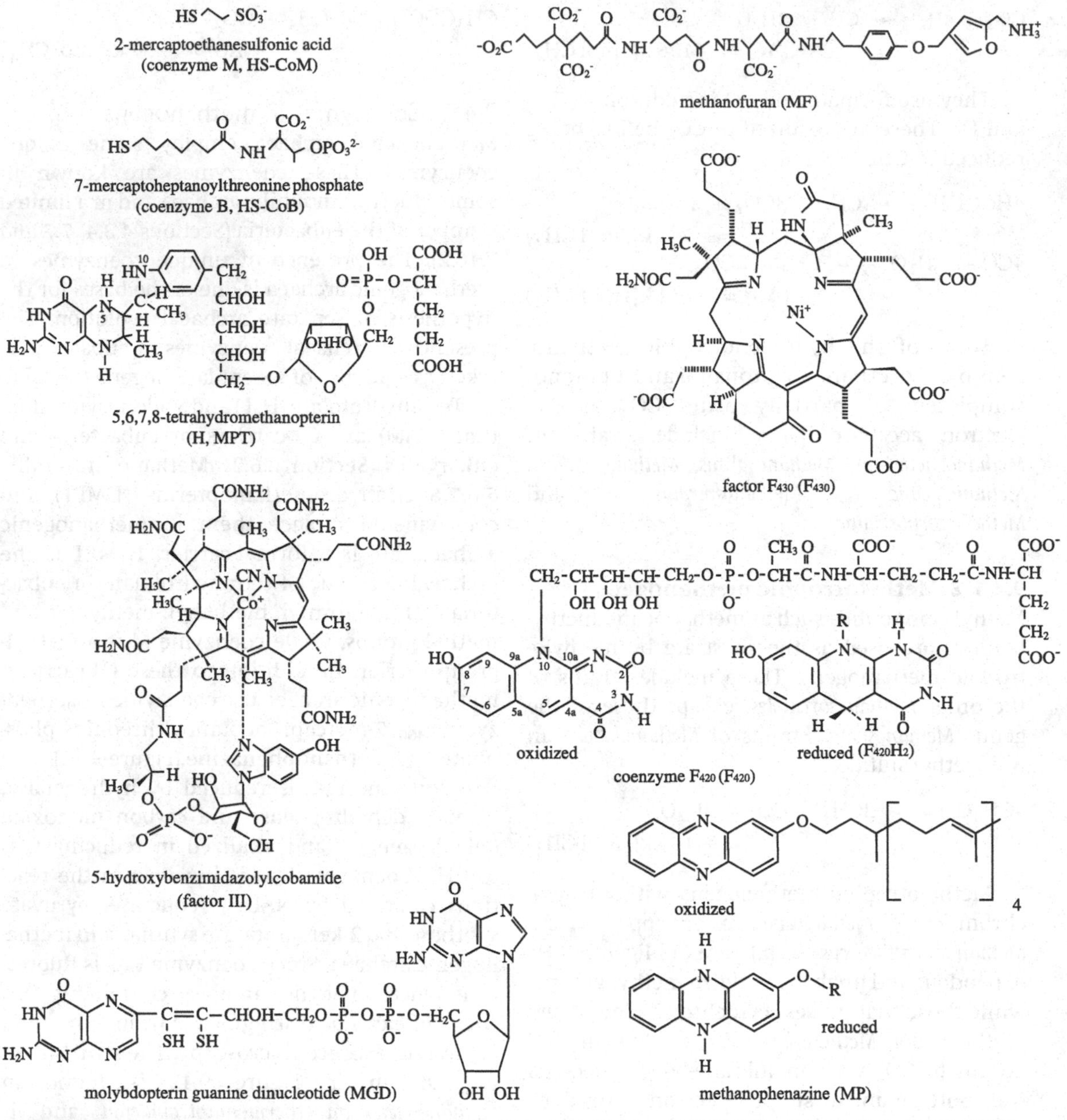

Figure 9.12 Methanogenic cofactors and their structures.

(Modified from *Microbiol. Mol. Biol. Rev.* **63**: 570–620, 1999 and *FEMS Microbiol. Rev.* **23**: 13–38, 1999)

believed to be involved in methane oxidation under anaerobic conditions (Section 9.9.2).

MF, H_4MPT and F_{420} are known in some thermophilic archaea, e.g. *Archaeoglobus fulgidus* and *Solfolobus acidocaldarius*. Methanogenic coenzymes are also known in eubacteria. Coenzyme F_{420} is used in strains of *Streptomyces*, in reactions synthesizing a range of antibiotics including tetracycline and lincomycin. F_{420}-dependent glucose-6-phosphate dehydrogenase catalyses the first reaction of the HMP pathway in eubacteria, *Mycobacterium smegmatis* and some species of

Nocardia (Section 4.3.4). F_{420} contributes to the survival of mycobacteria in challenging environments. This coenzyme is involved in the synthesis of mycolic acid, the major component of mycobacterial membranes (Section 2.3.3.2), and in detoxification of antibiotics and agents of oxidative and nitrosative stress (Section 12.2.5). F_{420} is also known in other actinobacterial genera, including *Rhodococcus*, *Nocardia* and *Nocardioides*, although its roles in these bacteria are not yet well established.

Coenzyme M is also known in eubacteria. *Rhodococcus rhodochrous* uses Co-M in the oxidative metabolism of propylene (Section 7.7). A methylotroph, *Methylobacterium extorquens*, metabolizes C1-compounds, bound not only to tetrahydrofolate but also to H_4MPT (Section 7.10.2).

9.4.3 Methanogenic pathways

Molecular oxygen not only inhibits the growth of methanogens, but also irreversibly inactivates many of the methanogenic enzymes. Because of such oxygen sensitivity, it can be difficult to study the reactions of methanogenesis. However, most of the enzymes of methanogenesis have been characterized and further information is now available from genomic analysis.

9.4.3.1 Hydrogenotrophic methanogenesis

The membrane-bound formyl-MF dehydrogenase reduces CO_2 to formate, oxidizing H_2 or reducing ferredoxin. Formate binds to methanofuran (MF). Ferredoxin is reduced through the electron bifurcation reaction by the heterodisulfide reductase described below. H_2 oxidation is coupled to consumption of the proton motive force, whereas the proton motive force is not consumed when reduced ferredoxin is used as the electron donor. In the next reaction, the formyl group is transferred from formyl-MF to H_4MPT before being reduced to the methyl group. The reduced F_{420} provides electrons for these reduction reactions. The cytoplasmic F_{420}-reducing hydrogenase reduces F_{420}, oxidizing H_2. Some methanogens belonging to the order *Methanomicrobiales* reduce F_{420}, which is coupled to the oxidation of secondary alcohols to ketones (Table 9.6). Methyl-H_4MPT:CoM-transferase forms methyl-CoM by transferring the methyl group. Energy is conserved in this reaction by translocation of Na^+ (Figure 9.14). Methyl-CoM is finally reduced to CH_4 in a reaction catalysed by methyl-CoM reductase, forming a heterodisulfide CoM-S-S-CoB. Methyl-CoM reductase contains coenzyme F_{430} as a prosthetic group. The heterodisulfide is reduced to CoM-SH and CoB-SH (Figure 9.13). The heterodisulfide reductase in hydrogenotrophic methanogens is a soluble flavoprotein complex (HdrABC) that oxidizes H_2 as the electron donor. This enzyme is membrane-bound in other methanogens (see below). The heterodisulfide reductase is an electron bifurcation enzyme oxidizing two H_2 to reduce ferredoxin and heterodisulfide. Electrons from the high redox potential site are used to reduce heterodisulfide and those from the low redox potential site to reduce ferredoxin in an electron bifurcation reaction (Reaction 7 in Figure 9.13, Section 5.8.6.4).

Hydrogenotrophic methanogens synthesize carbon skeletons for biosynthesis from methyl-CoM. Carbon monoxide dehydrogenase synthesizes acetyl-CoA from methyl-CoM and enzyme-bound [CO] (Figure 9.18).

Methanogens require Na^+ for growth, since this cation is used in an energy conservation process in a reaction catalysed by the membrane-bound enzyme methyl-H_4MPT:CoM-methyltransferase. This enzyme exports Na^+ in the methyl-CoM-forming reaction to generate a sodium motive force (Figure 9.14). Methyl-H_4MPT:CoM-methyltransferase consists of 6–8 subunits depending on the strain, and contains a cobalamin (vitamin B_{12}) derivative, 5-hydroxybenzimidazolyl cobamide and [4S-4Fe] clusters.

Methyl-CoM reductase catalyses the last reaction of methanogenesis and forms CH_4 and CoM-S-S-CoB heterodisulfide from methyl-CoM and reduced 7-mercaptoheptanoylthreonine phosphate (HS-CoB). This enzyme contains coenzyme F_{430} as a prosthetic group. Heterodisulfide reductase reduces CoM-S-S-CoB heterodisulfide to HS-CoM and HS-CoB in an electron bifurcation reaction, as mentioned above. Energy is conserved in this reaction, indirectly providing reduced ferredoxin that is used by formyl-MF dehydrogenase without consuming the proton motive force.

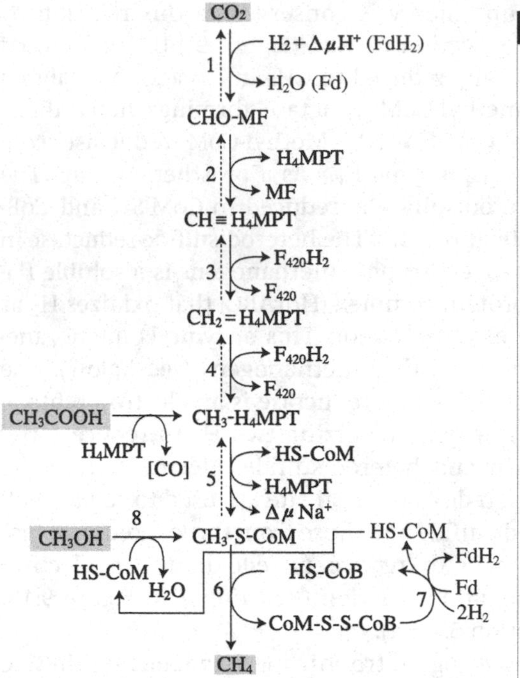

Figure 9.13 Reduction of methanol and carbon dioxide to methane.

(Modified from *Microbiol. Mol. Biol. Rev.* **63**: 570–620, 1999)

Formyl-MF dehydrogenase (1) reduces carbon dioxide to form formyl-MF. The formyl group is transferred from formyl-MF to tetrahydromethanopterin (H_4MPT) to be reduced to methyl-H_4MPT. The methyl group is transferred to coenzyme M (CoM-SH) before being reduced to methane with the electrons supplied from reduced forms of coenzyme F_{420} or H_2 indirectly. Energy is conserved in the reactions catalysed by methyl-H_4MPT:CoM-methyltransferase (5), in the form of a sodium motive force. The hydrogenotrophic methanogens have a soluble enzyme, heterodisulfide reductase (HdrABC), that reduces ferredoxin and heterodisulfide with hydrogen as the electron donor in an electron bifurcation reaction, while the methylotrophic enzyme (HdrED) is membrane-bound and uses reduced F_{420} (Figure 9.15). A proton motive force is generated during the reaction catalysed by HdrED, but not HdrABC. The ferredoxin reduced by HdrABC is oxidized by formyl-MF dehydrogenase without consuming the proton motive force. The proton motive force is consumed in the reaction catalysed by formyl-MF dehydrogenase consuming H_2 (1). Methanol is converted to methyl-CoM, part of which is oxidized (dotted line) to supply reducing equivalents for the energy-conserving reaction catalyzed by heterodisulfide reductase (7).

1, formyl-MF dehydrogenase; 2, formyl-MF:H_4MPT formyltransferase and methenyl-H_4MPT cyclohydrolase; 3, F_{420}-dependent methylene-H_4MPT dehydrogenase; 4, F_{420}-dependent methylene-H_4MPT reductase; 5, methyl-H_4MPT:CoM-methyltransferase; 6, methyl-CoM reductase; 7, heterodisulfide reductase; 8, methyltransferase.

MF, methanofuran; H_4MPT, tetrahydromethanopterin; HS-CoB, 7-mercaptoheptanoylthreonine phosphate; Fd H_2, reduced ferredoxin (Figure 9.13).

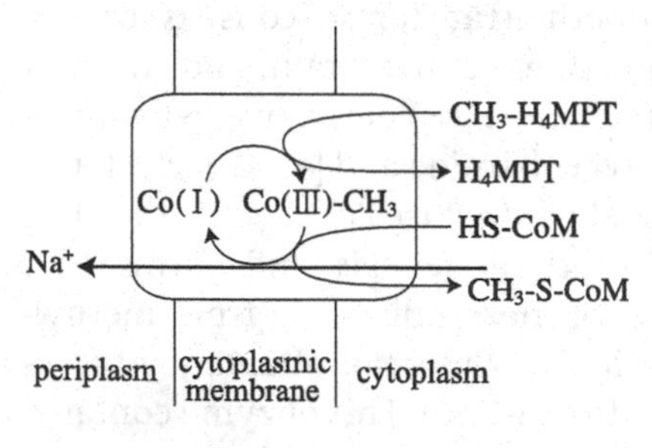

Figure 9.14 Sodium motive force generation by the methyl-H_4MPT: CoM-methyltransferase complex.

(Modified from *Microbiol. Mol. Biol. Rev.* **63**: 570–620, 1999)

The methyl-H_4MPT:CoM-methyltransferase complex is a Na^+-dependent enzyme present in the cytoplasmic membrane. The prosthetic group cobamide mediates methyl group transfer from methyl-H_4MPT to coenzyme M. During this process, Na^+ is exported, generating a sodium motive force. When methanol is oxidized to CO_2, the sodium motive force is consumed in the reverse reaction.

9.4.3.2 Methylotrophic methanogenesis

Methyltransferase forms methyl-CoM from methanol and methylamine (Figure 9.13). As in hydrogenotrophic methanogens, methyl-CoM is reduced to CH_4, forming heterodisulfide. To supply electrons for the reaction catalysed by heterodisulfide reductase, a part of methyl-CoM is oxidized to CO_2 through the reverse reactions of CO_2 reduction to methyl-CoM (Figure 9.13). This occurs in members of the order *Methanosarcinales* that possess cytochromes, while H_2 is consumed in *Methanosphaera* spp.

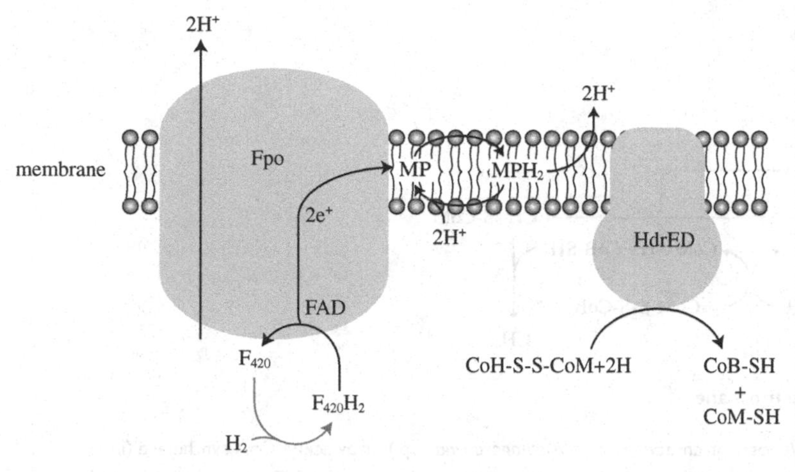

Figure 9.15 Proton motive force generation by methylotrophic methanogens that possess heterodisulfide reductase.

(Modified from *Mol. Microbiol.* **75**: 843–853. 2010)

Methylotrophic heterodisulfide reductase (HdrED) reduces CoB-S-S-CoM using electrons from reduced methanophenazine (MPH2) in the membrane. Protons are translocated during the reaction. Methanophenazine (MP) is reduced by the proton-translocating F_{420}: methanophenazine oxidoreductase (Fpo).

and *Methanomassiliicoccus* spp. that do not possess cytochromes and cannot oxidize methyl-CoM. During the oxidation of methyl-CoM, the sodium motive force is consumed in the reaction catalysed by methyl-H_4MPT:CoM-methyltransferase (Figure 9.14). It is not known what electron carrier formyl-MF dehydrogenase reduces or if the enzyme reaction is coupled to the generation of a proton motive force. The reaction catalysed by heterodisulfide reductase in methylotrophic methanogens is different from that in hydrogenotrophic methanogens. The latter (HdrABC), a soluble enzyme, uses hydrogen as the electron donor for the reaction, while the former (HdrED) is membrane bound and uses reduced methanophenazine (MP, Figure 9.15). Ferredoxin:methanophenazine oxidoreductase (Rnf complex, Section 5.8.6.6) reduces MP, oxidizing F_{420}, coupled to proton motive force generation. A proton motive force is generated during the reaction catalysed by HdrED, but not HdrABC. Many methylotrophic methanogens carry genes for the primary enzyme complex HdrEd, as well as HdrABC.

9.4.3.3 Aceticlastic methanogenesis

Strains of *Methanosarcina* and *Methanosaeta* use acetate to produce CH_4. Species of the genus *Methanosaeta* have a higher affinity for acetate and activate it to acetyl-CoA in a one-step reaction catalysed by acetyl-CoA synthetase:

$$\text{acetate} + \text{CoA-SH} + \text{ATP} \xrightarrow{\text{acetyl-CoA synthetase}} \text{acetyl-CoA} + \text{AMP} + \text{PP}_i \quad (\Delta G^{0'} = -6\,\text{kJ/mol acetate})$$

However, in species of *Methanosarcina* that have a low affinity for acetate, acetyl-CoA is formed through a two-step process catalysed by acetate kinase and phosphotransacetylase:

$$\text{acetate} + \text{ATP} \xrightarrow{\text{acetyl-CoA synthetase}} \text{acetyl-phosphate} + \text{ADP} \quad (\Delta G^{0'} = -13\,\text{kJ/mol acetate})$$

$$\text{acetyl-phosphate} + \text{CoA-SH} \xrightarrow{\text{phosphotransacetylase}} \text{acetyl-CoA} + \text{P}_i \quad (\Delta G^{0'} = -9\,\text{kJ/mol acetyl-phosphate})$$

Carbon monoxide dehydrogenase splits acetyl-CoA to form CH_3-H_4MPT and enzyme-bound [CO]. The methyl group of CH_3-H_4MPT is transferred to methyl-CoM, and [CO] is oxidized to CO_2, reducing ferredoxin (Figure 9.16). In *Methanosarcina barkeri*, H_4MPT is replaced by tetrahydrosarcinapterin which has a similar structure to H_4MPT. Electrons from the reduced ferredoxin are transferred to heterodisulfide reductase through ferredoxin:methanophenazine oxidoreductase which contains cytochrome *b*. This reaction is similar to that occurring in the methylotrophic methanogens. It is believed that ferredoxin:methanophenazine oxidoreductase

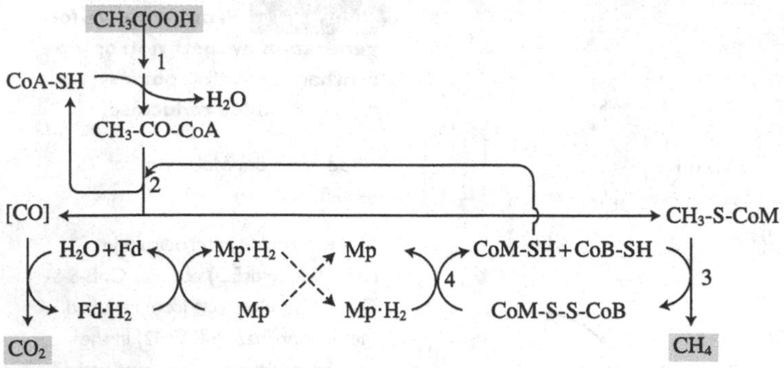

Figure 9.16 Acetate degradation to methane.

Acetate is activated (1) by acetate kinase/phosphotransacetylase (in *Methanosarcina* spp.) or by acetyl-CoA synthetase (in *Methanosaeta* spp.). Carbon monoxide dehydrogenase (2) transfers methyl groups to form methyl-S-CoM, oxidizing the carbonyl group to carbon dioxide to reduce ferredoxin. Methyl-CoM methylreductase (3) reduces methyl-CoM to methane, oxidizing HS-CoB to form CoM-S-S-CoB heterodisulfide. The membrane-bound heterodisulfide reductase (HdrED, 4) reduces the heterodisulfide to CoM-SH and Co-B, oxidizing methanophenazine (MP). Methanophenazine oxidoreductase reduces MP, oxidizing reduced ferredoxin.

(Rnf complex, Section 5.8.6.6) and heterosulfide reductase generate a proton motive force, as shown in Figure 9.15.

9.4.4 Energy conservation in methanogenesis

The free energy change in each step of methanogenesis (Table 9.7) is less than the phosphorylation potential (−51.8 kJ/mol, Section 5.6.3) needed for ATP synthesis through substrate-level phosphorylation (SLP). The enzymes of SLP are not known in methanogenesis. The methanogenic archaea synthesize ATP through electron transport phosphorylation (ETP) with the generation of a proton (or sodium) motive force. Experiments using uncouplers and ATPase inhibitors support this hypothesis (Figure 9.17).

With the addition of methanol and H_2, cell suspensions of *Methanosarcina barkeri* produce CH_4, with concomitant generation of a proton motive force (Δp) and an increase in ATP concentration (Figure 9.17). Δp increases with the decrease in ATP concentration when an ATPase inhibitor, *N,N′*-dicyclohexylcarbodiimide (DCCD), is added. DCCD inhibits ATP synthesis and Δp consumption. The high level of Δp inhibits methanogenesis, the Δp generation process. A rapid decrease in Δp and ATP concentration is observed on addition of uncouplers such as carbonylcyanide *m*-chlorophenylhydrazone (CCCP) and carbonylcyanide *p*-trifluoromethoxy-phenylhydrazone (FCCP). Uncouplers dissipate Δp. At a low Δp level, ATP cannot be synthesized, and methanogenesis is accelerated.

As described previously, a proton motive force is generated by heterodisulfide reductase (HdrED) and a sodium motive force is generated in the methyl-CoM-forming reaction catalysed by methyl-H_4MPT:CoM-methyltransferase. The sodium motive force is used in various energy consuming reactions, including transport, motility and ATP synthesis. The sodium motive force is used for ATP synthesis either through the proton motive force by the action of a Na^+/H^+ antiport, or directly with the Na^+-ATPase. A Na^+-ATPase is known in many prokaryotes and has a V-type ATPase structure (Section 5.8.4.1).

Table 9.7 | Changes in free energy in each step of methanogenesis.

Reaction	$\Delta G^{0'}$ = (kJ/mol reactants)
$CO_2 + H_2 + MF \longrightarrow HCO\text{-}MF + H_2O$	+16
$HCO\text{-}MF + H_4MPT \longrightarrow HCO\text{-}H_4MPT + MF$	−5
$HCO\text{-}H_4MPT + H^+ \longrightarrow CH{\equiv}H_4MPT^+ + H_2O$	−2
$CH{\equiv}H_4MPT^+ + F_{420}H_2 \longrightarrow CH{=}H_4MPT + F_{420} + H^+$	+6.5
$CH_2{=}H_4MPT + F_{420}H_2 \longrightarrow CH_3\text{-}H_4MPT + F_{420}$	−5
$CH_3\text{-}H_4MPT + HS\text{-}CoM \longrightarrow CH_3\text{-}S\text{-}CoM + H_4MPT$	−29
$CH_3\text{-}S\text{-}CoM + HS\text{-}CoB \longrightarrow CH_4 + CoM\text{-}S\text{-}S\text{-}CoB$	−43
$CoM\text{-}S\text{-}S\text{-}CoB + H_2 \longrightarrow CH_3\text{-}S\text{-}CoM + HS\text{-}CoB$	−42
$CH_3OH + HS\text{-}CoM \longrightarrow CH_3S\text{-}CoM + H_2O$	−27.5
$CH_3\text{-}CO\text{-}S\text{-}CoA + H_4MPT \longrightarrow CH_3\text{-}H_4MPT + [CO] + HS\text{-}CoA$	+62
$CoM\text{-}S\text{-}S\text{-}CoB + F_{420}H_2 \longrightarrow CH_3\text{-}S\text{-}CoM + HS\text{-}CoB + F_{420}$	−29

Source: *Antonie van Leeuwenhoek*, **66**, 187–208, 1994.

Energy conservation in methanogenesis can be summarized as below:

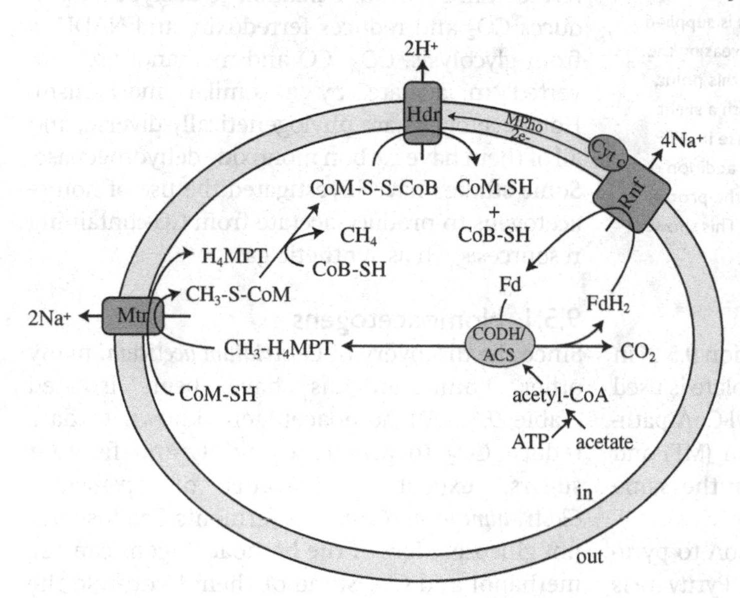

Figure 9.16. Other methanogens use carbon monoxide dehydrogenase in the synthesis of acetyl-CoA from methyl-CoM and enzyme-bound [CO], as in the acetyl-CoA pathway in reaction 2 in Figure 9.16. The acetyl-CoA pathway is the reverse of the acetate oxidation pathway (Figure 9.7) and this pathway is also known as the carbon monoxide dehydrogenase pathway or the Wood-Ljungdahl pathway (Section 9.3.1.2). This pathway is known in some complete oxidizing sulfidogens

9.4.5 Biosynthesis in methanogens

The hydrogenotrophic methanogens use CO_2 as their carbon source, and simple carbon compounds such as formate, methanol, methylamines and acetate are used by others. They metabolize these carbon sources to acetyl-CoA to supply carbon skeletons for biosynthesis. Aceticlastic methanogens activate acetate as in reaction 1 in

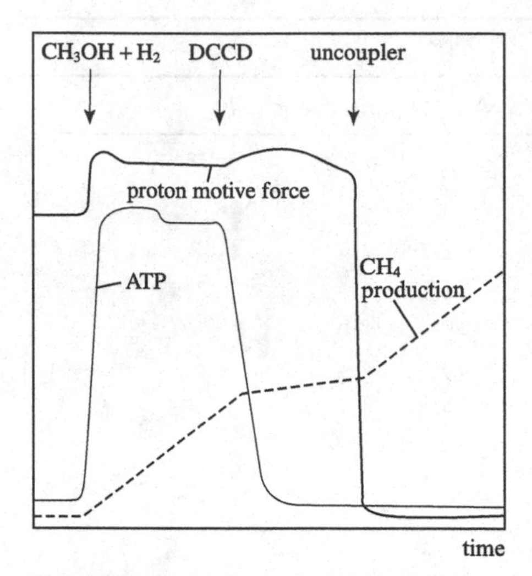

Figure 9.17 Proton motive force generation during methanol reduction to methane by *Methanosarcina barkeri*.

(Gottschalk, G. 1986, *Bacterial Metabolism*, 2nd edn., Figure 8.26. Springer, New York)

When a cell suspension of *Methanosarcina barkeri* is supplied with methanol and H_2, methane is generated, increasing the proton motive force and ATP concentration. At this point, ATPase inhibitors inhibit methane generation, with a slight increase in the proton motive force and a decrease in ATP concentration to the background level. With the addition of an uncoupler, methane generation resumes, but the proton motive force and ATP concentration remain low. This shows that the proton motive force is generated during methanogenesis before ATP is synthesized.

(Table 9.5) and in homoacetogens (Section 9.5.2) in addition to methanogens. Tetrahydrofolate is used as the C1-carrier in the bacterial acetyl-CoA pathway, while archaea use methanofuran (MF) and tetrahydromethanopterin (H_4MPT) for the same purpose.

Pyruvate synthase reduces acetyl-CoA to pyruvate, oxidizing reduced coenzyme F_{420}. Pyruvate is carboxylated to oxaloacetate (OAA), either directly or through phosphoenolpyruvate (PEP). OAA is metabolized to 2-ketoglutarate through an incomplete TCA fork (Section 5.4.1). Species of *Methanobacterium* and *Methanococcus* that lack citrate synthase employ the reductive branch of the incomplete TCA fork, while species of *Methanosarcina* metabolize PEP via the oxidative branch (Figure 9.18).

Gluconeogenesis has not been firmly established in methanogens, but some EMP pathway enzymes, including fructose-1,6-diphosphate aldolase, have been identified, suggesting that normal gluconeogenesis (Section 4.2) occurs in methanogens (Figure 9.18).

9.5 Homoacetogenesis

Homoacetogens are a group of strictly anaerobic bacteria growing on the free energy available from the reduction of CO_2 to acetate. In addition to CO_2 + H_2, they metabolize sugars and methanol to acetate. Saccharolytic clostridia and enteric bacteria ferment carbohydrates to acetate with other fermentation products, such as butyrate, lactate, succinate and others. Homoacetogens ferment a molecule of glucose to three molecules of acetate, two via pyruvate and acetyl-CoA. The third acetate is produced from two CO_2, NADH and reduced ferredoxin. Pyruvate oxidation to acetyl-CoA produces CO_2 and reduces ferredoxin, and NADH is from glycolysis. CO_2, CO and methanol are converted to acetate by a similar mechanism. Homoacetogens are phylogenetically diverse, and all of them have carbon monoxide dehydrogenase. Some studies have investigated the use of homoacetogens to produce acetate from CO-containing resources such as synthetic gas.

9.5.1 Homoacetogens

Since the discovery of *Clostridium aceticum*, many other homoacetogens have been isolated (Table 9.8). All homoacetogens known to date reduce CO_2 to acetate using H_2 and ferment sugars, except for strains of *Sporomusa*. *Clostridium formicoaceticum* ferments fructose but not glucose. Most of the homoacetogens can use methanol and CO. Some of them hydrolyse the methyl esters in lignin and its degradation products to produce acetate. Many aromatic compounds are metabolized by homoacetogens and halogenated hydrocarbons such as CH_2Cl_2 are dehalogenated.

Homoacetogens are involved in microbially influenced corrosion (MIC), consuming electrons from metals to produce acetate that can be used by *Desulfovibrio* spp. as a carbon source.

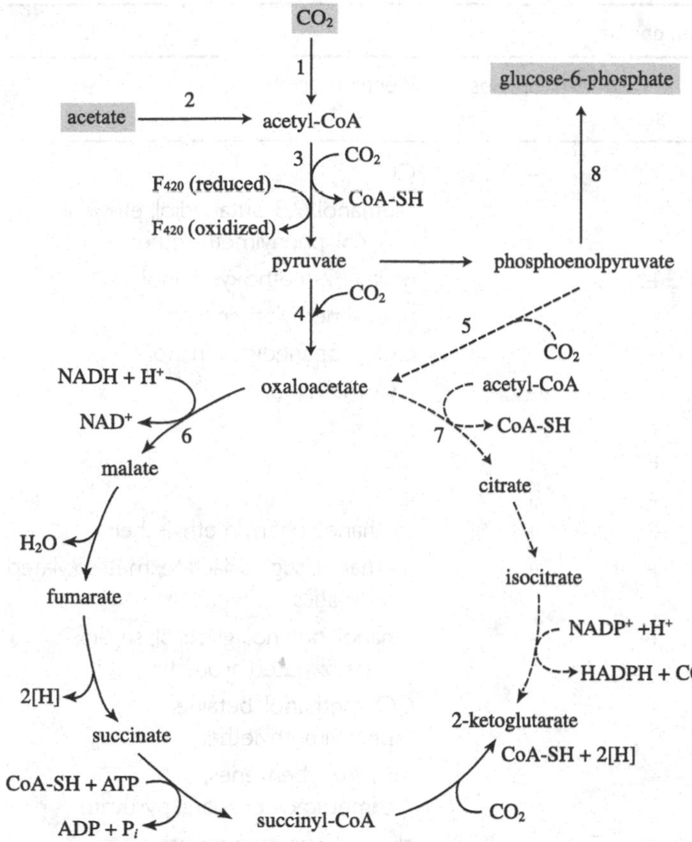

Figure 9.18 Supply of carbon skeletons for biosynthesis in methanogens.

(*Microbiol. Rev.* **51**: 135–177, 1987)

Pyruvate synthase (3) produces pyruvate from the reduction of acetyl-CoA that is produced either from the acetyl-CoA pathway (1) or through acetate activation (2). PEP carboxylase (5) starts an oxidative TCA fork (7, dotted line) in species of *Methanosarcina*, while species of *Methanobacterium* and *Methanococcus* produce oxaloacetate, catalysed by pyruvate carboxylase (4) for the reductive TCA fork (6). Gluconeogenesis (8) is used to provide glucose-6-phosphate.

9.5.2 Carbon metabolism in homoacetogens

9.5.2.1 Sugar metabolism

Homoacetogens metabolize hexoses through the EMP pathway in most cases, while the modified ED pathway (Section 4.4.3) is the glycolytic pathway in *Clostridium aceticum*. The resulting pyruvate is oxidized to acetyl-CoA through the phosphoroclastic reaction. Phosphotransacetylase and acetate kinase cleave acetyl-CoA to acetate with ATP synthesis. NADH and reduced ferredoxin are oxidized coupled with the reduction of CO_2 to acetate (Figure 9.19).

Formate dehydrogenase reduces some of the CO_2 produced from the phosphoroclastic reaction to formate, before it is bound to tetrahydrofolate (H_4F) to be reduced to methyl-H_4F. The remaining CO_2 is reduced to an enzyme-bound form of [CO] by carbon monoxide dehydrogenase (CODH). CODH catalyses acetyl-CoA synthesis from methyl-H_4F and [CO]. CODH is a dual-function enzyme catalysing the reversible reaction of CO oxidation and acetyl-CoA synthesis. This enzyme is referred to as either CO dehydrogenase or acetyl-CoA synthase.

This CO_2-fixation acetyl-CoA pathway or CO dehydrogenase pathway is found in autotrophic methanogens (Section 9.4.5) and sulfidogens (Section 9.3.3) for the supply of carbon skeletons for biosynthesis, as well as in homoacetogens (Section 10.8.3).

A tungsten-containing aldehyde oxidoreductase (AOR) is known in *Clostridium formicoaceticum* and *Moorela thermoacetica* (*Clostridium thermoaceticum*). This enzyme oxidizes aldehyde directly to the corresponding acid. Formate dehydrogenase is another enzyme containing tungsten. *Clostridium formicoaceticum* has a molybdenum-containing AOR in addition to the enzyme containing tungsten. A similar enzyme, glyceraldehyde: ferredoxin oxidoreductase, is known in the

Table 9.8 Homoacetogens and their electron donors.

Organism	Gram stain	Spores	Electron donor[a]
Acetitomaculum ruminis	+	–	CO
Acetobacterium carbinolicum	+	–	methanol, 2,3-butanediol, ethylene glycol, phenylmethylether
Acetobacterium malicum	+	–	malate, 2-methoxyethanol
Acetobacterium wieringae	+	–	phenylmethylether
Acetobacterium woodii	+	–	CO, propanediol, ethanol
Butyribacterium methylotrophicum[b]	+	–	CO, methanol
Clostridium aceticum	+	+	sugars
Clostridium autoethanogenum[c]	+	+	CO
Clostridium carboxidivorans[d]	+	+	CO
Clostridium formicoaceticum	+	+	methanol, phenylmethylether
Clostridium methoxybenzovorans	+	+	methanol, sugars, lactate, methoxylated aromatics
Clostridium scatologenes	+	+	ethanol, butanol, glycerol, sugars, methoxylated aromatics
Eubacterium limosum[b]	+	–	CO, methanol, betaine, phenylmethylether
Halophaga foetida	–	–	trihydroxybenzenes, trimethoxybenzoate, pyruvate
Moorella glycerini	+	+	glycerol, sugars, pyruvate
Moorella thermoacetica[e]	+	+	CO, methanol, phenylmethylether
Moorella thermoautotrophica[f]	+	+	CO, phenylmethylether
Ruminococcus hydrogenotrophicus	+	–	
Ruminococcus productus[g]	+	–	CO, phenylmethylether
Sporomusa malonica[h]	+	+	malonate
Sporomusa silvacetica[h]	+	+	vanillate, sugars, fumarate, ethanol
Sporomusa sphaeroides[h]	+	+	
Sporomusa termitida[g]	+	+	CO, phenylmethylether
Syntrophococcus sucromutans[b]	+	–	phenylmethylether
Thermoanaerobacter kivui[i]	+	–	

[a] All grow on $CO_2 + H_2$, and sugars are used except for species of Sporomusa that do not use sugars as their electron donor.
[b] Butyrate is produced in addition to acetate by *Butyribacterium methylotrophicum*, *Eubacterium limosum* and *Syntrophococcus sucromutans*.
[c] Lactate and butanediol are produced.
[d] Butyrate, ethanol and butanol are produced.
[e] Formerly *Clostridium thermoaceticum*.
[f] Formerly *Clostridium thermoautotrophicum*.
[g] Formerly *Peptostreptococcus productus*.
[h] Formerly considered to be Gram-negative.
[i] Formerly known as *Acetogenium kivui*.

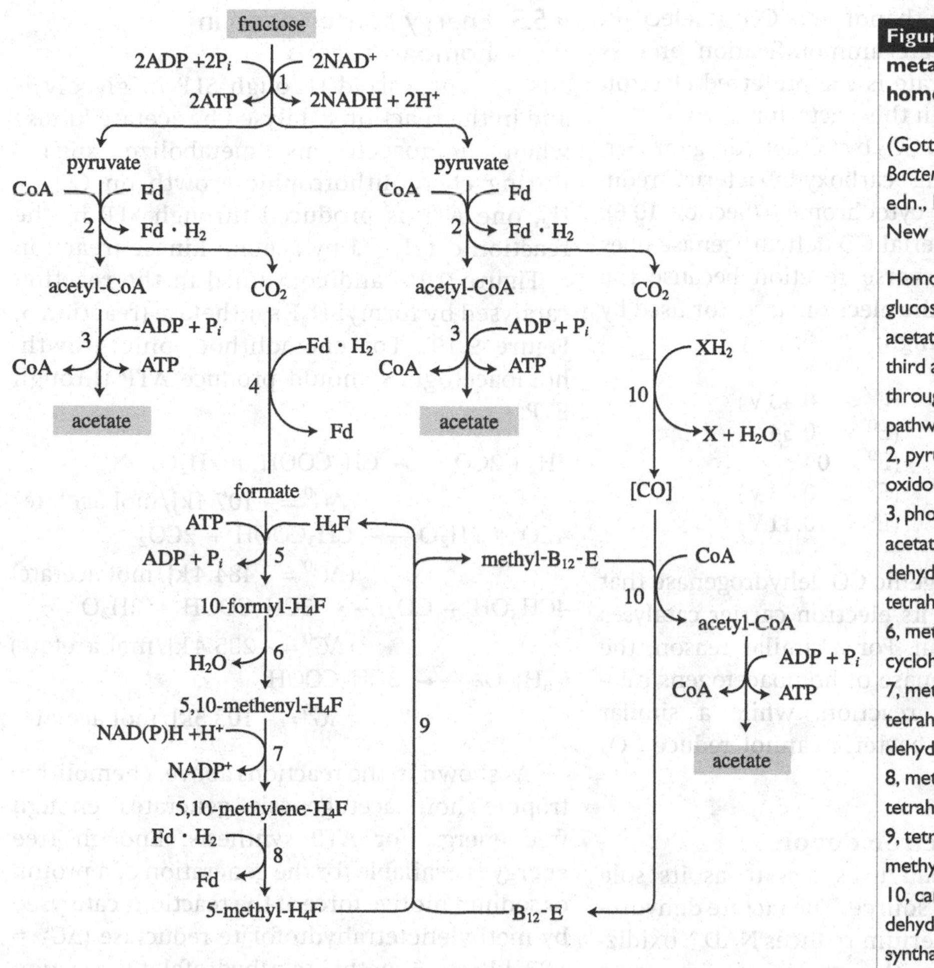

Figure 9.19 Hexose metabolism by homoacetogens.

(Gottschalk, G. 1986, *Bacterial Metabolism*, 2nd edn., Figure 8.23. Springer, New York)

Homoacetogens ferment glucose to produce two acetate and synthesize the third acetate from CO_2 through the acetyl-CoA pathway. 1, glycolysis; 2, pyruvate: ferredoxin oxidoreductase; 3, phosphotransacetylase and acetate kinase; 4, formate dehydrogenase; 5, formyl tetrahydrofolate synthetase; 6, methenyl-tetrahydrofolate cyclohydrolase; 7, methylene-tetrahydrofolate dehydrogenase; 8, methylene-tetrahydrofolate reductase; 9, tetrahydrofolate: B_{12} methyltransferase; 10, carbon monoxide dehydrogenase (acetyl-CoA synthase).

[CO], enzyme-bound carbon monoxide.

hyperthermophilic archaea *Sulfolobus acidocaldarius* and *Thermoproteus tenax* that metabolize sugars without phosphorylation (Section 4.4.3.2).

It is interesting to note that AOR in homoacetogens oxidizes aldehyde without conserving free energy from the aldehyde oxidation to the corresponding acids. Aldehydes are oxidized through acyl-CoA and acyl-phosphate in most anaerobic metabolic pathways, such as sulfidogenesis, synthesizing ATP through the action of kinases (Section 9.3.1). AOR oxidizes various aldehydes, including aromatic aldehydes derived from lignin, to corresponding acids to provide electrons for the reduction of CO_2 to acetate.

All reactions of the acetyl-CoA pathway are reversible. Acetate is used as the electron donor in aceticlastic methanogens and in some of the complete oxidizing sulfidogens, and anaerobic chemolithotrophs fix CO_2 through the same pathway. The syntrophic bacteria *Clostridium ultunense* and *Thermoacetogenium phaeum* oxidize acetate to CO_2 and H_2, in association with H_2-consuming sulfidogens or methanogens. The oxidation of acetate to CO_2 and H_2 is an endergonic reaction ($\Delta G^{0'} = +107.1$ kJ/mol acetate) and possible under very low hydrogen partial pressures (Section 9.8.2). These organisms grow chemolithotrophically, reducing CO_2 to acetate using H_2 under high H_2 partial pressures. They are referred to as reversibacteria.

Moorella thermoacetica produces two acetate from one hexose in the presence of nitrate. This bacterium uses methanol and CO as electron donors in the nitrate ammonification process (Section 9.1.4). Nitrate is the preferred electron acceptor over CO_2 in this bacterium.

CO is oxidized to CO_2 by CO dehydrogenase in aerobic CO-utilizing carboxydobacteria, reducing ubiquinone or cytochrome *b* (Section 10.6). The carboxydobacterial CO dehydrogenase does not catalyse the reverse reaction because the redox potential of the electron acceptor used by this enzyme is too high:

$HCO_3^-/HCOO^-$	$(E^{0\prime} = -0.43\,V)$
CO_2/CO	$(E^{0\prime} = -0.54\,V)$
Q/QH_2	$(E^{0\prime} = 0\,V)$
$NAD^+/NADH + H^+$	$(E^{0\prime} = -0.32\,V)$
$Fd/Fd{\cdot}H_2$	$(E^{0\prime} = -0.41\,V)$

The homoacetogenic CO dehydrogenase that uses ferredoxin as its electron carrier catalyses the reverse reaction. For a similar reason, the formate dehydrogenase of homoacetogens catalyses the reverse reaction, while a similar enzyme in enteric bacteria cannot reduce CO_2 to formate.

9.5.2.2 Other electron donors

Acetobacterium woodii uses lactate as its sole energy and carbon source. The lactate dehydrogenase of this bacterium reduces NAD^+, oxidizing lactate only in the presence of reduced ferredoxin, which is oxidized in the reaction. Lactate oxidation ($E^{0\prime}$ = −190 mV) coupled to NAD^+ reduction ($E^{0\prime}$ = −320 mV) is thermodynamically unfavourable. To overcome this thermodynamic barrier, reduced ferredoxin is oxidized simultaneously. This is an example of an electron confurcation reaction (Section 8.1.2). Other electron donors used by homoacetogens include CO, methanol, ethanol, propanediol and butanediol among others (Table 9.8).

9.5.2.3 Synthesis of carbon skeletons for biosynthesis in homoacetogens

Homoacetogens synthesize acetyl-CoA from CO_2 + H_2. Acetyl-CoA is metabolized through gluconeogenesis to supply carbon skeletons for biosynthesis, as in methanogens (Section 9.4.5).

9.5.3 Energy conservation in homoacetogens

ATP is synthesized through SLP in glycolysis and in the reaction catalysed by acetate kinase when homoacetogens metabolize sugars. During chemolithotrophic growth on CO_2 + H_2, one ATP is produced through SLP in the reaction catalysed by acetate kinase (reaction 3, Figure 9.19) and consumed in the reaction catalysed by formyl-H_4F synthetase (reaction 5, Figure 9.19). For chemolithotrophic growth, homoacetogens should produce ATP through ETP:

$$4H_2 + 2CO_2 \longrightarrow CH_3COOH + 2H_2O \quad (\Delta G^{0\prime} = -107.1\,kJ/mol\ acetate)$$

$$4CO + 2H_2O \longrightarrow CH_3COOH + 2CO_2 \quad (\Delta G^{0\prime} = -484.4\,kJ/mol\ acetate)$$

$$4CH_3OH + CO_2 \longrightarrow 3CH_3COOH + 2H_2O \quad (\Delta G^{0\prime} = -235.4\,kJ/mol\ acetate)$$

$$C_6H_{12}O_6 \longrightarrow 3CH_3COOH \quad (\Delta G^{0\prime} = -103.5\,kJ/mol\ acetate)$$

As shown in the reactions above, chemolithotrophic homoacetogenesis generates enough free energy for ATP synthesis. Enough free energy is available for the generation of a proton or sodium motive force at the reactions catalysed by methylenetetrahydrofolate reductase ($\Delta G^{0\prime}$ = −22 kJ/mol 5-methyl-tetrahydrofolate, reaction 8, Figure 9.19) and tetrahydrofolate:B_{12} methyltransferase ($\Delta G^{0\prime}$ = −38 kJ/mol 5-methyl-tetrahydrofolate, reaction 9, Figure 9.19). However, these enzymes are not membrane-bound and do not generate the proton or sodium motive force. Energy is conserved by different mechanisms, depending on the strains, through an electron bifurcation reaction, the proton-translocating ferredoxin:NAD^+ oxidoreductase (Rnf complex) or the energy-converting hydrogenase (Section 5.8.6.5).

In the bacteria known as H^+-dependent acetogens (*Clostridium ljungdahlii*, *Moorella thermoacetica* and *Moorella thermoautotrophica*), a proton motive force is generated, while a sodium motive force is generated in Na^+-dependent acetogens (*Acetobacterium woodii*, *Ruminococcus productus* and *Thermoanaerobacter kivui*).

Acetobacterium woodii does not possess cytochromes or coenzyme Q and energy is conserved by the membrane-bound proton-translocating ferredoxin:NAD^+ oxidoreductase (Rnf complex, Section 5.8.6.6), exporting Na^+. This bacterium oxidizes H_2 to reduce ferredoxin and NAD^+ in an electron bifurcation reaction:

$$2H_2 + Fd + NAD^+ \longrightarrow Fd{\cdot}H_2 + NADH + H^+$$

CO_2 is reduced to [CO] using H_2 as the electron donor by carbon monoxide dehydrogenase in this bacterium.

Clostridium ljungdahlii conserves energy through an electron bifurcation reaction and the Rnf complex, as in *A. woodii*, but exports H^+ not Na^+.

Moorella thermoacetica and *Thermoanaerobacter kivui* have an energy-converting hydrogenase (Ech) exporting H^+ with hydrogen production coupled to oxidation of ferredoxin (Section 5.8.6.5).

9.6 Organohalide respiration

Marine algae synthesize various halogenated compounds, and synthetic halogen compounds are widely used as solvents, agricultural chemicals and for other industrial purposes. Some of them are toxic, carcinogenic and recalcitrant. They are dehalogenated at a higher rate under anaerobic conditions than under aerobic conditions. Enzyme cofactors found in homoacetogens, methanogens and other anaerobic bacteria remove halogens reductively in aqueous solution. These include corrinoids, iron porphyrins and F_{430}. These non-specific reactions are slower than those that occur when halogenated compounds are used as electron acceptors. This metabolism is referred to as dehalorespiration or organohalide respiration. The redox potential ($E^{0\prime}$) of halogenated compounds ranges between +250 and −600 mV:

$$CH_3Cl + NADH + H^+ \longrightarrow CH_4 + HCl + NAD^+ \quad (\Delta G^{0\prime} = -135.08\ \text{kJ/mol}\ CH_3Cl)$$

Organohalide respiratory organisms convert chlorinated compounds to non- or lesser-halogenated compounds that are mostly less toxic to the environment or more easily degraded. The reductive dehalogenases are membrane-bound enzymes that conserve energy using halogenated hydrocarbons as their electron acceptors.

9.6.1 Organohalide respiratory organisms

A few pure cultures have been identified that use chlorinated benzoate, phenol and ethane as electron acceptors (Table 9.9), and enrichment cultures have been made which dehalogenate polychlorinated biphenyls (PCBs), chlorinated benzene derivatives and others. A pure culture of *Dehalococcoides mccartyi* was isolated based on its ability to dechlorinate PCB.

An electrochemically active Fe(III)-reducing bacterium, *Geobacter lovleyi*, reduces tetrachloroethene (PCE) to cis-dichloroethene, with a poised electrode at −300 mV as the sole electron donor, as well as with acetate.

Species of *Desulfomonile* and *Desulfitobacterium* use not only halogenated compounds but also sulfite and thiosulfate as their electron acceptors. Dehalogenating enzymes are induced in these bacteria by these electron acceptors. The transcription and activity of dehalogenating enzymes is repressed by sulfite and thiosulfate in *Desulfomonile tiedjei*. Halogenated compounds are used as electron sinks in sulfur-reducing *Desulfuromonas chloroethenica* coupled to the oxidation of acetate through the modified TCA cycle.

Sulfurospirillum multivorans (formerly *Dehalospirillum multivorans*) is closely related to the metal-reducing bacterium *Geospirillum barnesii* and uses arsenate and selenate as well as halogenated compounds as electron acceptors. The metalloid oxyanions inhibit transcription and activity of the dehalogenating enzymes in this bacterium. *Dehalococcoides ethenogenes* is the only known pure culture capable of complete dehalogenation of tetrachloroethene to ethane. *Enterobacter agglomerans* is the only known dehalogenating facultative anaerobe to date.

9.6.2 Energy conservation in organohalide respiration

Electron transport and energy conservation are not fully understood in organohalide respiration.

Table 9.9 | Organohalide respiratory bacteria.

Strain	Electron acceptor	Electron donor	Remark[a]
Gram-positive			
Desulfitobacterium chlororespirans	2,4,6-TCP	hydrogen, formate	ortho
Desulfitobacterium dehalogenans	PCE, 2,4,6-TCP	hydrogen, formate	ortho
Desulfitobacterium frappieri	2,4,6-TCP	pyruvate	ortho meta para
Desulfitobacterium hafniense	PCP	pyruvate	ortho meta
Dehalobacter restrictus	PCE, TCE	hydrogen	–
Gram-negative			
Desulfomonile tiedjei	PCE, TCE, 3-CB	hydrogen, formate	meta
Desulfomonile limimaris	3-CB, sulfate, nitrate	lactate	meta
Sulfurospirillum multivorans[b]	PCE, TCE	hydrogen, formate	–
Desulfuromonas chloroethenica	PCE, TCE	pyruvate, acetate	–
Dehalococcoides ethenogenes	PCE, TCE	hydrogen	–
Enterobacter agglomerans	PCE, TCE, DCE, chloroethene	hydrogen	–
Desulfovibrio dechloracetivorans[c]	2-CP, 2,6-DCP	acetate	ortho
Geobacter lovleyi	PCE	electrode,[d] acetate	–
Chloroflexi			
Dehalococcoides mccartyi	CBs, dioxin, PCB	Hydrogen	para, meta

2,4,6-TCP, 2,4,6-trichlorophenol; 2-CP, 2-chlorophenol; 2,6-DCP, 2,6-dichlorophenol; 3-CB, 3-chlorobenzoate; PCE, tetrachloroethene; TCE, trichloroethene; PCP, pentachlorophenol; DCE, dichloroethene; CBs, chlorobenzenes.

[a] Dehalogenation position.

[b] Formerly *Dehalospirillum multivorans.*

[c] Non-sulfidogenic.

[d] Cathode of an electrochemical device.

Most of the organisms listed in Table 9.9 can grow with hydrogen as the sole electron donor, which shows that dehalogenation with hydrogen is coupled to ATP generation. Energy is conserved through different mechanisms with different electron carriers, depending on the organism. Under dehalogenating conditions, a *c*-type cytochrome is induced with the dehalogenating enzymes in *Desulfomonile tiedjei*. This bacterium has hydrogenase and formate dehydrogenase in the periplasmic region. Inhibitors of quinone function eliminate the dehalogenation capacity of this bacterium. Menaquinone and cytochromes *b* and *c* are known in some dehalogenating organisms, including *Desulfitobacterium*, *Dehalobacter restrictus* and *Sulfurospirillum multivorans*, but are not found in *Dehalococcoides mccartyi*.

Sulfurospirillum multivorans is one of the best studied organisms capable of organohalide respiration. In addition to halogenated hydrocarbons such as tetrachloroethene and trichloroethene, this bacterium uses a wide range of electron acceptors, including fumarate, nitrate, thiosulfate, selenate and arsenate for anaerobic respiration, with various electron donors such as H_2, formate, pyruvate and lactate. The tetrachloroethene reductive dehalogenase (PceA) in this bacterium is a corrinoid and iron–sulfur

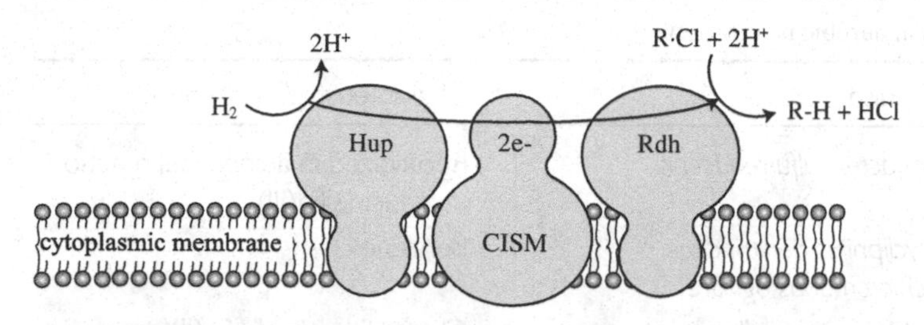

Figure 9.20 The reductive dehalogenase complex of *Dehalococcoides mccartyi*.

(Modified from *Environ. Microbiol.* **18**: 3044–3056, 2016)

In this bacterium, quinones and cytochromes are not known. Electrons from the hydrogenase (Hup) are transferred to dehalogenase (Rdh) through a formate dehydrogenase-like Fe–S molybdoenzyme (CISM). The reductive dehalogenase is a membrane-bound outwards-oriented iron–sulfur protein containing a corrinoid cofactor in the active centre.

Hup, uptake hydrogenase; CISM, complex iron–sulfur molybdoenzyme; Rdh, reductive dehalogenase.

cluster containing membrane-bound enzyme, like nearly all described reductive dehalogenases. The corrinoid of this enzyme is a structurally unique cobamide, norpseudo-B_{12}, that is not found in other organohalide respiratory bacteria. Electrons from the electron donors are transferred to menaquinone by appropriate dehydrogenases and quinone oxidoreductases. Protons are translocated through the membrane during the electron transfer by quinone oxidoreductase and dehalogenase. The Co(II)/Co(I) redox potential of the norpseudo-B_{12} in the dehalogenase is −350 to −380 mV, much lower than that for menaquinone (0 mV). It is not known how the corrinoid reduction is coupled to quinol oxidation.

The reductive dehalogenase in *Dehalococcoides mccartyi* is a membrane-bound outwards-oriented [Fe–S] protein containing a corrinoid cofactor in the active centre. This enzyme forms a complex with a [NiFe] uptake hydrogenase and a formate dehydrogenase-like [Fe–S] molybdoenzyme. Quinone and cytochromes are not known in this bacterium. Electrons from the hydrogenase are transferred to the dehalogenase through the complex iron–sulfur molybdoenzyme (CISM, Figure 9.20). It has not been elucidated what the function of CISM is or how protons are translocated during electron transport.

Dehalogenase in *Desulfitobacterium dehalogenans* is a corrinoid-containing [Fe–S] protein as in other organohalide respiratory bacteria. Menaquinone and cytochromes are present in this bacterium, but the electron metabolism has not been fully elucidated. A periplasmic flavoprotein mediates electron transfer from menaquinone to dehalogenase.

9.7 Miscellaneous electron acceptors

In addition to the electron acceptors discussed above, various other organic and inorganic compounds can be used as electron acceptors by anaerobic bacteria (Table 9.10). Humic acid, derived from lignin, is used as an electron acceptor by a metal-reducing species of *Geobacter*. Reduced humic acid chemically reduces Fe(III) and is diffusible through the membrane; this compound may mediate electron transfer from the bacterial cells to the insoluble electron acceptor.

Perchlorate and chlorate are strong oxidizing agents and toxic, but some bacteria use them as electron acceptors. Perchlorate is reduced to chlorite in two steps by a single enzyme, (per)chlorate reductase that contains

Table 9.10 | Electron acceptors used in anaerobic respiration.

Electron acceptor	Organism	Characteristics
Humic acid	*Geobacter sulfurreducens*	Reoxidized in a chemical reaction reducing Fe(III)
(Per)chlorate	*Alicycliphilus denitrificans*	Benzene
	Dechloromonas agitate	
	Dechlorosoma suillum	Chemolithotroph, Fe(II) as electron donor
	Dechloromonas sp.	Chemolithotroph, H_2 as electron donor
	Haloferax mediterranei	
	Ideonella dechloratans	
	Moorella perchloratireducens	
	Pseudomonas chloritidismutans	
	Sedimenticola selenatireducens	
	Sporomusa ovata	
Iodate	*Shewanella putrefaciens*	
	Desulfovibrio desulfuricans	
Organic sulfonate	*Bilophila wadsworthia*	Electron sink (?)
Organic nitro compounds	*Denitrobacterium detoxificans*	

molybdenum and [Fe–S] clusters as cofactors. A haem protein, chlorite dismutase, cleaves chlorite to Cl^- and O_2. There are two different chlorate reductases, one reduces perchlorate and chlorate and the other only chlorate, depending on the organism:

$$ClO_4^- \xrightarrow{\text{perchlorate reductase}} ClO_3^- \xrightarrow{\text{perchlorate reductase}} ClO_2^- \xrightarrow{\text{chlorite dismutase}} Cl^- + O_2$$

These two enzymes are located in the periplasm. Electrons are transferred to (per)chlorate reductase through the membrane-bound cytochrome *bc* complex and a periplasmic soluble cytochrome *c* coupled to proton translocation across the membrane. (Per)chlorate reductase reactions are not coupled to proton motive force generation. Oxygen produced by the dismutase is used as the terminal electron acceptor, as in aerobes.

A similar oxygenated halogen, iodate, is present in seawater and can be used as an electron acceptor by some bacteria, including *Shewanella putrefaciens*.

Taurine, a natural sulfonate compound, is reduced to sulfide as an electron acceptor by bacteria. Nitropropanol and nitropropionate are toxic compounds found in some plants. They are reduced as electron acceptors in the rumen, thus protecting the host animals.

Some anaerobic bacteria, including *Lactobacillus casei, Streptococcus lactis, Clostridium acetobutylicum, Aerobacter polymyxa* and sulfidogens, are known to reduce phosphate to phosphine. However, the redox potential of these reactions is too low for them to be used as electron acceptors:

$$H_2PO_4^- \xrightarrow[E^{0\prime}=-0.65V]{} HPO_3^{2-} \xrightarrow[-0.74V]{} H_2PO_2^- \xrightarrow[-0.66V]{} PH_3$$

9.8 Syntrophic associations

organic materials introduced into the anaerobic environment are fermented to various alcohols and acids before being completely oxidized through anaerobic respiration, coupled to the reduction of nitrate, metal ions, sulfate and others. Under anaerobic conditions, where

Table 9.11 Syntrophic bacteria.

Strains	Gram stain	Substrate used
Clostridium ultunense[a]	+	acetate
Smithella propionica[b]	−	propionate
Syntrophaceticus schinkii	+	acetate
Syntrophobacter fumaroxidans[c]	−	propionate
Syntrophobacter wolinii	−	propionate
Syntrophobotulus glycolicus	+	glycolate
Syntrophomonas sapovorans	−	butyrate and fatty acids (up to C_{18})
Syntrophomonas wolfei	−	butyrate and fatty acids (up to C_8)
Syntrophospora bryantii	+	butyrate and fatty acids (up to C_{11})
Syntrophothermus lipocalidus[d]	−	butyrate and fatty acids (up to C_{10})
Syntrophus aciditrophicus[e]	−	benzoate, cyclohexane carboxylate
Syntrophus buswellii[e]	−	benzoate, 3-phenylbenzoate
Syntrophus gentianae	−	benzoate
Thermoacetogenium phaeum[a]	+	acetate
Thermosyntropha lipolytica[b]	−	butyrate and fatty acids (up to C_{18})

[a] Pure culture on crotonate fermentation.
[b] Pure culture on fumarate fermentation and on anaerobic respiration using sulfate or fumarate.
[c] Pure culture on fumarate fermentation and on anaerobic respiration using nitrate, sulfate or fumarate.
[d] Pure culture on crotonate fermentation or on benzoate + crotonate.
[e] Reversibacteria growing on $H_2 + CO_2$, producing acetate and acetate oxidation under low partial pressure of H_2.

these electron acceptors are not available, methanogens are the final scavenger. Methanogens convert a limited range of substrates to methane (Table 9.6). The common fermentation products that cannot be utilized directly by methanogens are not accumulated under methanogenic conditions. These include ethanol, propionate and butyrate. These are oxidized to acetate and carbon dioxide by certain bacteria in a syntrophic association with methanogens. These bacteria are referred to as syntrophic bacteria or obligate proton-reducing acetogens (Table 9.11).

9.8.1 Syntrophic bacteria

Syntrophic bacteria catalyse the following reactions in syntrophic association with methanogens or sulfidogens:

$$\text{ethanol} + H_2O \longrightarrow \text{acetate} + H^+ + 2H_2 \quad (\Delta G^{0'} = +9.6\ \text{kJ/mol ethanol})$$

$$\text{butyrate} + H_2O \longrightarrow 2\text{acetate} + H^+ + 2H_2 \quad (\Delta G^{0'} = +48.1\ \text{kJ/mol ethanol})$$

$$\text{propionate} + H_2O \longrightarrow \text{acetate} + HCO_3^- + H^+ + 3H_2 \quad (\Delta G^{0'} = +76.1\ \text{kJ/mol ethanol})$$

These endergonic reactions cannot take place, and therefore cannot support growth of the syntrophic bacteria, under standard conditions. However, these reactions can be exergonic when the reaction product concentrations are kept very low (Section 5.5.1). Methanogens and sulfidogens remove hydrogen efficiently, which keeps its partial pressure low as these reactions become exergonic. Coculture of syntrophic bacteria and methanogens on butyrate and propionate is well documented. These mixed cultures are referred to as syntrophic associations, and the reactions are described as interspecies hydrogen transfer (Section 9.8.3).

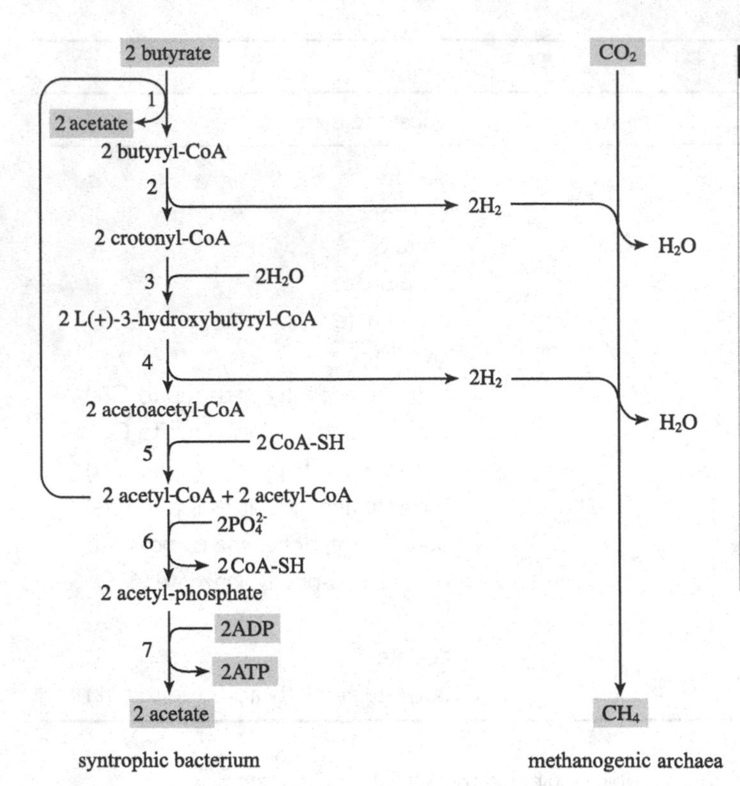

Figure 9.21 Butyrate oxidation by *Syntrophomonas wolfei* in a syntrophic association with *Methanospirillum hungatei*.

Syntrophomonas wolfei is an obligately syntrophic bacterium capable of oxidizing butyrate to acetate in a syntrophic association with methanogens that keeps the hydrogen partial pressure extremely low. This pathway was predicted from the enzyme activities of cell-free extracts of the bacterium prepared by selective lysis.

1, coenzyme A transferase; 2, butyryl-CoA (acyl-CoA) dehydrogenase; 3, 3-enoyl-CoA hydratase(crotonase); 4, 3-hydroxybutyryl-CoA dehydrogenase; 5, 3-ketoacyl-CoA thiolase; 6, phosphotransacetylase; 7, acetate kinase.

Smithella spp. oxidize alkanes to acetate in syntrophic association with methanogens in crude oil contaminated marine sediments.

9.8.2 Carbon metabolism in syntrophic bacteria

Syntrophobacter wolinii and *Syntrophomonas wolfei* grow fermentatively on propionate and butyrate under low hydrogen partial pressures. These are obligate syntrophs growing in co-culture with hydrogen- and acetate-consuming sulfidogens or methanogens. Since they cannot be cultivated in pure culture, their carbon metabolism has been studied indirectly.

When a co-culture of a Gram-negative syntrophic bacterium, *Syntrophomonas wolfei*, and an archaeon, *Methanospirillum hungatei*, is treated with lysozyme, the bacterial cells are lysed but not the archaeal cells. Enzyme activities were measured using this cell-free extract to establish butyrate catabolism to acetate, as shown in Figure 9.21.

Syntrophomonas wolfei grows on crotonate in pure culture. This bacterium activates crotonate to crotonyl-CoA, and oxidizes half of it to 2 acetate, synthesizing ATP, and reduces the other half to butyrate. Other syntrophic bacteria, including *Smithella propionica*, *Syntrophus aciditrophicus* and *Syntrophus buswellii*, can use crotonate fermentatively. *Syntrophus aciditrophicus* and *Syntrophus buswellii* oxidize benzoate fermentatively, reducing crotonate to butyrate (Table 9.11).

Fumarate is fermented in pure culture to acetate and succinate by *Syntrophobacter wolinii*, that oxidizes propionate in a syntrophic association. In this fermentation, a part of fumarate is oxidized to acetate with ATP synthesis, and the remaining fumarate is reduced as an electron acceptor. From this fumarate fermentation, the propionate oxidation reactions are predicted as shown in Figure 9.22. This metabolism is similar to propionate oxidation by an incomplete oxidizing sulfidogen, *Desulfobulus propionicus* (Section 9.3.1.2). Another propionate oxidizing syntrophic bacterium, *Syntrophobacter fumaroxidans*,

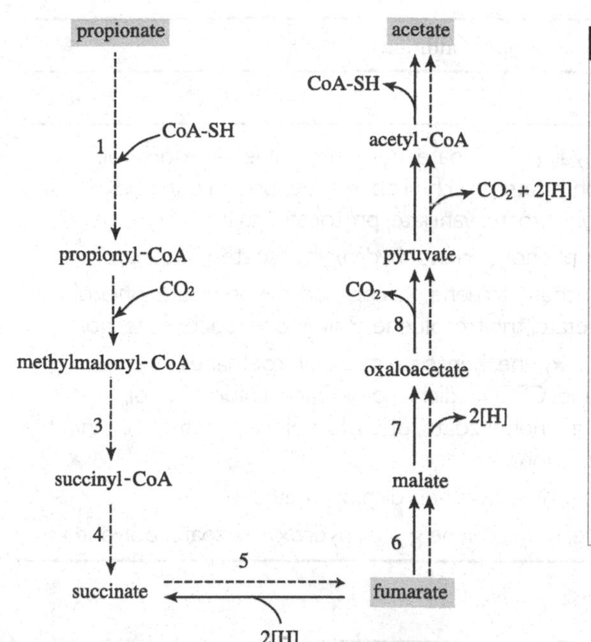

Figure 9.22 Metabolism of fumarate by *Syntrophobacter wolinii* (solid line) and the deduced propionate oxidation pathway (dotted line).

The oxidation of propionate to acetate is an endergonic reaction requiring the consumption of a large amount of free energy under standard conditions. *Syntrophobacter wolinii* disproportionates fumarate to succinate and acetate. A bacterial cell-free extract of the syntrophic consortium showed the enzyme activities as in the figure. It is most probable that *Syntrophobacter wolinii* oxidizes propionate, as shown (dotted lines).

1, propionate kinase/phosphotransacylase; 2, methylmalonyl-CoA:pyruvate transcarboxylase; 3, methylmalonyl-CoA mutase; 4, succinyl-CoA synthetase; 5, succinate dehydrogenase; 6, fumarase; 7, malate dehydrogenase; 8, oxaloacetate decarboxylase.

ferments fumarate in pure culture, and the fatty acid oxidizing *Syntrophothermus lipocalidus* not only ferments fumarate but also uses fumarate, nitrate and sulfate as electron acceptors to oxidize fatty acids to acetate (Table 9.11).

The term disproportionation is used to describe a metabolism where part of the substrate is used as electron donor and the other portion is used as electron acceptor.

Clostridium ultunense, *Syntrophaceticus schinkii* and *Thermoacetogenium phaeum* grow syntrophically with methanogens, oxidizing acetate to CO_2. In pure culture they grow on $H_2 + CO_2$ as homoacetogens, and are referred to as reversibacteria (Section 9.5.2).

Generally, ATP is synthesized from acetyl-CoA by the action of phosphotransacetylase and acetate kinase:

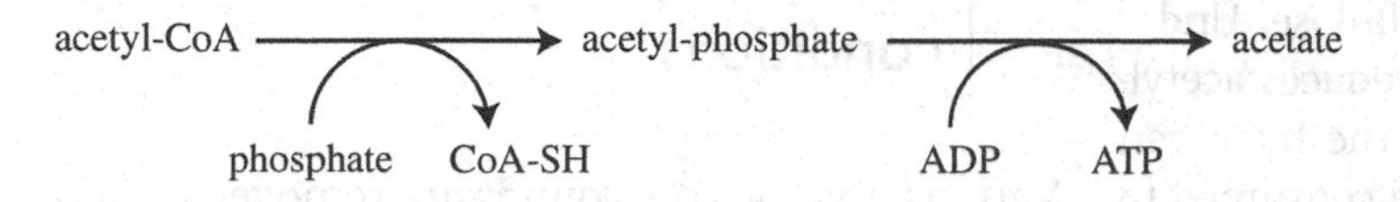

Syntrophus aciditrophicus does not have these enzymes nor their genes. This syntrophic bacterium synthesizes ATP from acetyl-CoA in a one-step reaction catalysed by acetyl-CoA synthetase:

$$\text{acetyl-CoA} + \text{AMP} + \text{PP}_i \longrightarrow \text{acetate} + \text{ATP}$$

9.8.3 Interspecies electron transfer

Electrons are transferred from syntrophic bacteria to methanogens in the form of H_2 or formate. All tested H_2-producing syntrophic bacteria have genes for an electron confurcating hydrogenase (Section 8.1.2):

$$\text{NADH} + \text{H}^+ + \text{ferredoxin}\cdot\text{H}_2 \longrightarrow \text{NAD}^+ + \text{ferredoxin} + 2\text{H}_2$$

This enzyme makes the H_2 production reaction even more favourable at low hydrogen partial pressure condition.

In some cases, electrons are transferred directly through electrically conductive cell surface materials, such as nanowires and the outer membrane MtrC-OmcA complex found in Fe(III) reducers (Section 9.2.1), in a process known as direct interspecies electron transfer (DIET).

Table 9.12 Examples of xenobiotics metabolized under anaerobic conditions.

Reaction	Xenobiotics
Denitrification	phenol, *p*-cresol, hydroxybenzoate, benzoate, toluene, resorcinol, phenylacetate, phthalate polychlorobenzoate, polychlorinated biphenyls, aminobenzoate, vanillate, protocatechuate
Metal reduction	benzoate, toluene, phenol, *p*-cresol hydroxybenzoates, phthalate
Sulfidogenesis	chlorophenols, benzoate, toluene, xylene, hydroxybenzoate, phenol, indole, phenylacetate, trinitrotoluene, 3-aminobenzoate, catechol
Methanogenesis[a]	benzene, toluene, *o*-xylene, benzoate, tetrachloroethanes, 1,2-dichloroethane, CFC-11 dichloromethane, chlorophenol, benzaldehyde 3-aminobenzoate, phenylpropionate, catechol, vanillate, methylbenzoate, chlorobenzene
Homoacetogenesis	ethylene glycol, phenylmethylether, dichloromethane
Fermentation	3-hydroxybenzoate, polyethylene glycol, hydroxybenzoate, polyphenol

[a] Mixed culture.

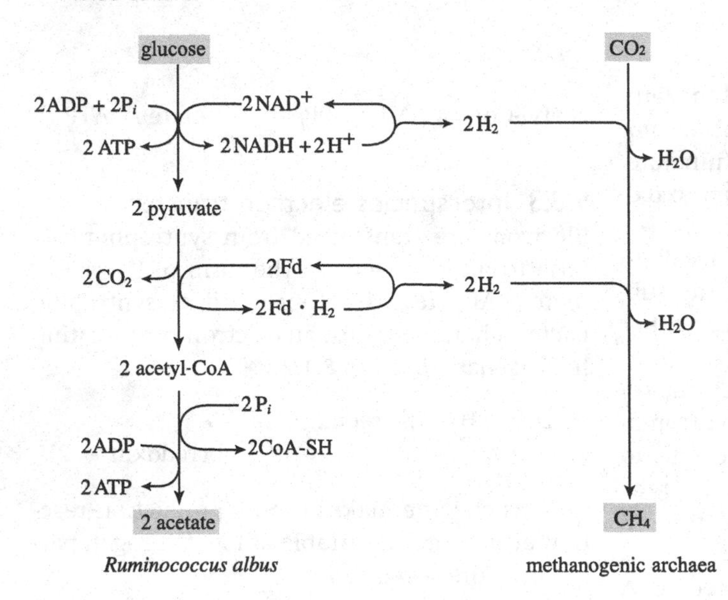

Figure 9.23 Facultative syntrophic association between *Ruminococcus albus* and methanogens through interspecies hydrogen transfer.

(Gottschalk, G. 1986, *Bacterial Metabolism*, 2nd edn., Figure 8.31. Springer, New York)

The cell yield of the rumen bacterium *Ruminococcus albus* increases in association with the hydrogen-consuming methanogens that keep the hydrogen partial pressure low. Under low hydrogen partial pressure, this bacterium synthesizes ATP from acetyl-CoA metabolism to acetate. With high hydrogen partial pressures, acetyl-CoA is reduced to ethanol or butyrate without ATP synthesis. This relationship is referred to as a facultative syntrophic association.

9.8.4 Facultative syntrophic associations

A rumen bacterium, *Ruminococcus albus*, ferments various carbohydrates including cellulose. Under normal conditions, this bacterium reduces acetyl-CoA to ethanol or butyrate. When the hydrogen partial pressure is low, electrons are consumed to reduce protons to hydrogen and acetyl-CoA is metabolized to acetate with ATP synthesis. With extra ATP, the growth yield of the bacterium is higher under low hydrogen partial pressures. Similarly, the bacterium grows better in co-culture with a hydrogen-consuming partner (Figure 9.23). This co-culture is referred to as facultative syntrophy.

9.9 Element cycling under anaerobic conditions

Various organic compounds are removed as CO_2 and CH_4, and various electron acceptors are reduced under anaerobic conditions. Through these oxidation–reduction reactions many compounds are transformed and such processes have been discussed in several previous sections. Many xenobiotics can be degraded under anaerobic conditions (Table 9.12).

9.9.1 Oxidation of hydrocarbons under anaerobic conditions

For a long time, hydrocarbons were believed to be oxidized only under aerobic conditions by the action of oxygenases utilizing molecular oxygen (Section 7.8.3). Hydrocarbons can also be oxidized by different mechanisms under anaerobic conditions using a variety of electron acceptors. These reactions are thermodynamically feasible. The free energy changes for toluene oxidation under anaerobic respiratory conditions are:

Denitrification:

$$C_6H_5(CH_3) + 7.2\,NO_3^- + 0.2H^+ \longrightarrow 7HCO_3^- + 3.6N_2 + 0.6H_2O \quad (\Delta G^{0'} = -493.6\ \mathrm{kJ/mol\ NO_3^-})$$

Nitrate ammonification:

$$C_6H_5(CH_3) + 4.5\,NO_3^- + 0.2H^+ + 7.5H_2O \longrightarrow 7HCO_3^- + 4.5NH_4^+ \quad (\Delta G^{0'} = -493.1\ \mathrm{kJ/mol\ NO_3^-})$$

Ferric iron reduction:

$$C_6H_5(CH_3) + 36Fe(OH)_3 + 29HCO_3^- + 29H^+ \longrightarrow 36FeCO_3 + 87H_2O \quad (\Delta G^{0'} = -39.1\ \mathrm{kJ/mol\ Fe})$$

Sulfate reduction:

$$C_6H_5(CH_3) + 4.5SO_4^{2-} + 2H^+ + 3H_2O \longrightarrow 7HCO_3^- + 4.5H_2S \quad (\Delta G^{0'} = -45.6\ \mathrm{kJ/mol\ SO_4^{2-}})$$

Methanogenesis:

$$C_6H_5(CH_3) + 7.5H_2O \longrightarrow 2.5HCO_3^- + 4.5CH_4 + 2.5H^+ \quad (\Delta G^{0'} = -29.1\ \mathrm{kJ/mol\ CH_4})$$

(*FEMS Microbiol. Rev.* **22**: 459–473, 1998)

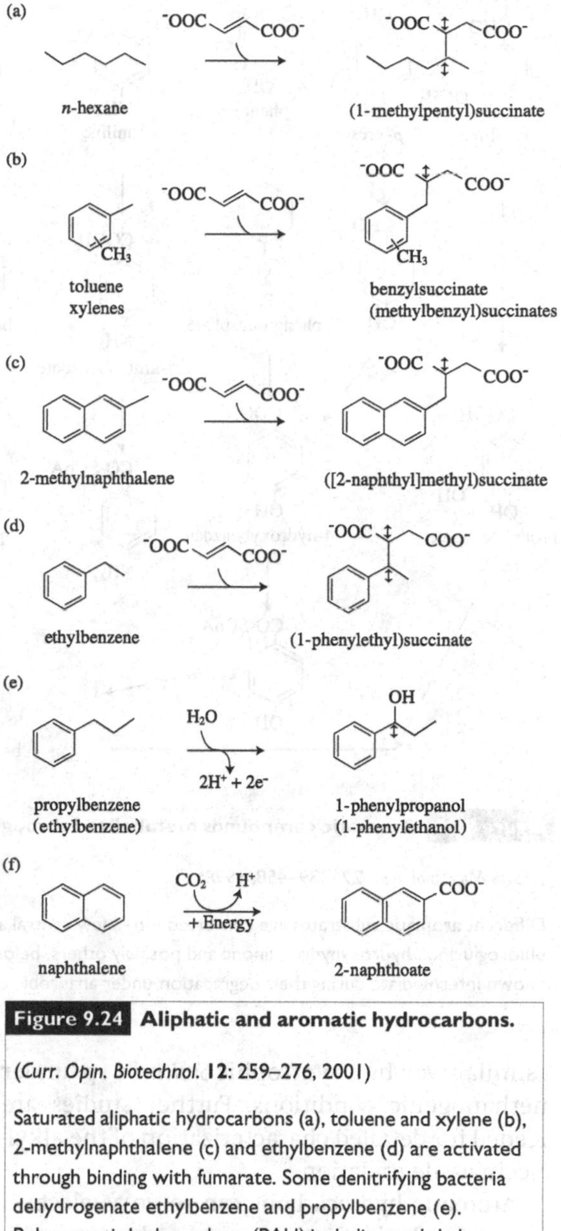

(a) n-hexane → (1-methylpentyl)succinate
(b) toluene, xylenes → benzylsuccinate, (methylbenzyl)succinates
(c) 2-methylnaphthalene → ([2-naphthyl]methyl)succinate
(d) ethylbenzene → (1-phenylethyl)succinate
(e) propylbenzene (ethylbenzene) → 1-phenylpropanol (1-phenylethanol)
(f) naphthalene → 2-naphthoate

Figure 9.24 Aliphatic and aromatic hydrocarbons.

(*Curr. Opin. Biotechnol.* **12**: 259–276, 2001)

Saturated aliphatic hydrocarbons (a), toluene and xylene (b), 2-methylnaphthalene (c) and ethylbenzene (d) are activated through binding with fumarate. Some denitrifying bacteria dehydrogenate ethylbenzene and propylbenzene (e). Polyaromatic hydrocarbons (PAH) including naphthalene are carboxylated (f).

Hydrocarbon oxidation under anaerobic conditions is best known with enriched mixed cultures. Toluene is oxidized under anaerobic conditions by denitrifying *Aromatoleum aromaticum, Thauera aromatica* and *Azoarcus tolulyticus*, metal-reducing *Geobacter metallireducens* and *Geobacter grbiciae*, and sulfidogenic *Desulfobacula toluolica* and *Desulfobacter centonicum*. Benzene is less degradable than toluene and ethylbenzene. Species of *Dechloromonas* oxidize toluene under denitrifying conditions. A member of the *Betaproteobacteria* and a hyperthermophilic sulfate-reducing archaeon, *Archaeoglobus fulgidus*, use *n*-alkanes as the energy and carbon source, anaerobically activating them to (1-methylalkyl) succinates (Figure 9.24a). Paraffin is degraded in

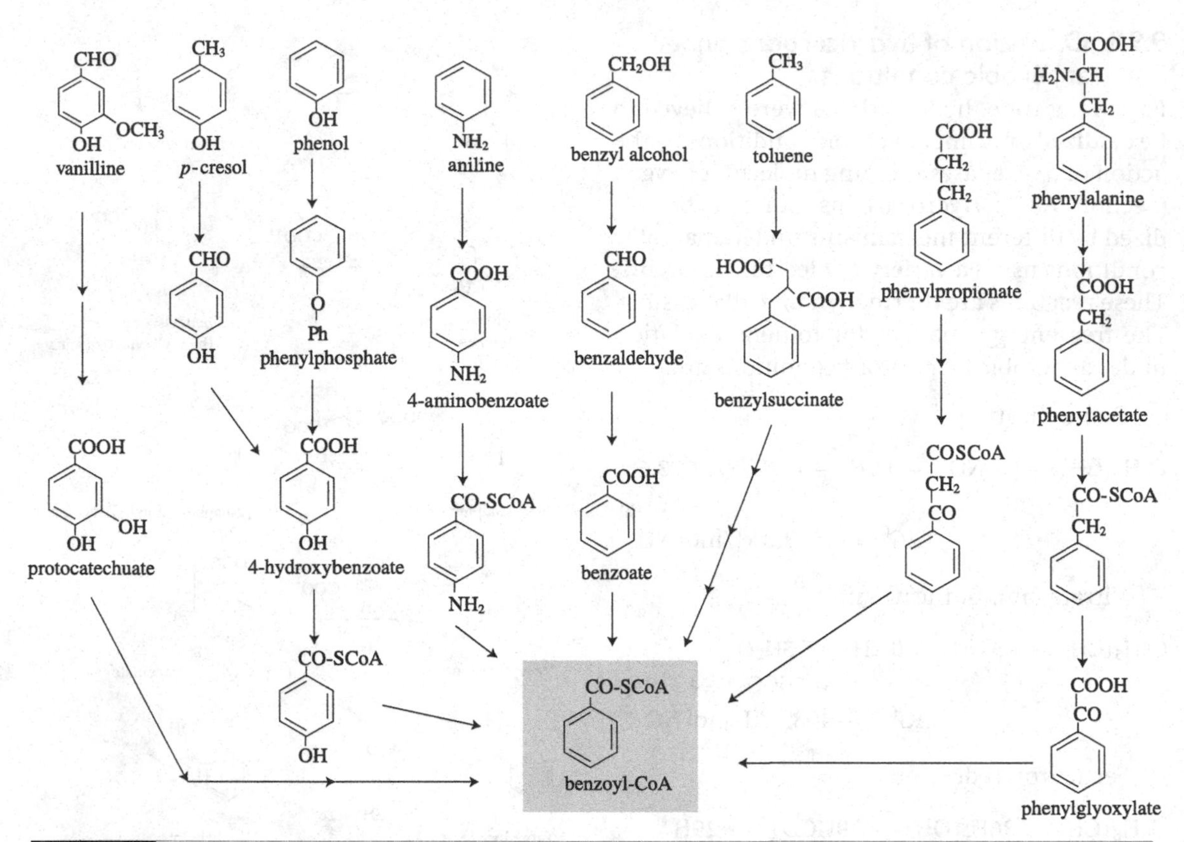

Figure 9.25 **Aromatic compounds metabolized through benzoyl-CoA under anaerobic conditions.**

(*FEMS Microbiol. Rev.* **22**: 439–458, 1998)

Different aromatic substrates are converted into a few central aromatic intermediates, including benzoyl-CoA, resorcinol, phloroglucinol, hydroxyhydroquinone and possibly others, before further degradation. Among them, benzoyl-CoA is the best known intermediate during their degradation under anaerobic conditions.

a similar way by a syntrophic *Smithella* sp. under methanogenic conditions. Further studies are needed for detailed characterization of the alkyl-succinate degradation.

Aromatic hydrocarbons can serve as electron donors for denitrification, metal reduction, sulfidogenesis and methanogenesis. They include toluene, benzene, and aromatic acids such as benzoate and gentisate. The anaerobic degradation is initiated by one of three reactions (Figure 9.24):

(1) binding with fumarate
(2) dehydrogenation
(3) carboxylation.

Fumarate binding is the initial reaction in the anaerobic degradation of saturated aliphatic hydrocarbons and aromatic hydrocarbons such as toluene, xylene, 2-methylnaphthalene and ethylbenzene, under all electron-accepting conditions. Denitrifiers dehydrogenate ethylbenzene and propylbenzene to initiate anaerobic degradation. Polyaromatic hydrocarbons (PAHs) are carboxylated before anaerobic degradation. Benzene is degraded via phenol and benzoate, but the exact mechanisms are not yet fully established.

After the initial activation reactions, aromatic compounds are further metabolized via benzoyl-CoA, resorcinol, phloroglucinol or hydroxyhydroquinone. Benzoyl-CoA is the best known intermediate, as shown in Figure 9.25. Based on enzyme activities, the benzoyl-CoA

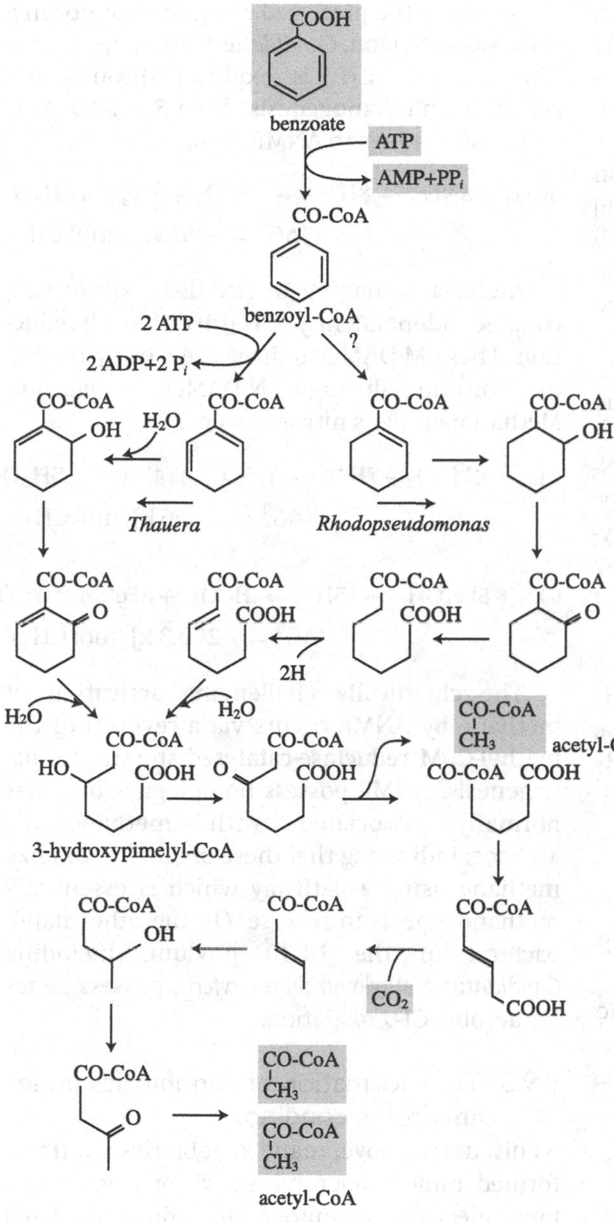

Figure 9.26 Proposed metabolic routes of the benzoyl-CoA pathway.

(*FEMS Microbiol. Rev.* **22**: 439–458, 1998)

Benzoate is activated to benzoylCoA, consuming ATP before being reduced to cyclic dienoyl-CoA by benzoyl-CoA reductase. Since the redox potential ($E^{0'}$) of the benzoylCoA/cyclic dienoyl-CoA half reaction is −1.8 V, much lower than that of ferredoxin, 2ATP is consumed in the reaction. The resulting cyclic dienoyl-CoA is further metabolized to 3-hydroxypimelyl-CoA through different reactions, depending on the strains, and finally to acetyl-CoA.

pathway has been deduced in photosynthetic *Rhodopseudomonas palustris*, denitrifying *Thauera aromatica* and *Azoarcus tolulyticus*, and syntrophic *Syntrophus gentianae* (Figure 9.26). Benzoyl-CoA reductase reduces the benzene ring of benzoyl-CoA. At this reaction 2ATP is consumed to make the substrate reduction ($E^{0'}$ −1.8 V) thermodynamically possible, coupling the oxidation of reduced ferredoxin ($E^{0'}$ −0.41 V). The resulting cyclic dienoyl-CoA is dehydrated and metabolized to acetyl-CoA through 3-hydroxypimelyl-CoA.

9.9.2 Methane oxidation under anaerobic conditions

The anaerobic oxidation of methane (AOM) is a microbially mediated process which consumes a large part of this greenhouse gas in nature. Three different processes of AOM have been recognized, depending on the electron acceptor used:

sulfate-dependent anaerobic methane oxidation (S-DAMO), nitrate/nitrite-dependent anaerobic methane oxidation (N-DAMO) and metal ion [Mn(VI) and Fe(III)]-dependent anaerobic methane oxidation (M-DAMO). The organisms carrying out this difficult biochemical reaction are anaerobic methanotrophic archaea (ANME) in S-DAMO and M-DAMO syntrophically, with sulfate-reducing bacteria and metal-reducing bacteria, respectively. ANME are classified into at least three groups which are phylogenetically related to methanogens: Group 1 ANME (ANME-1) are most closely related to the *Methanomicrobiales*, while ANME-2 and ANME-3 are affiliated with the *Methanosarcinales*. N-DAMO is carried out independently by uncharacterized bacteria belonging to the NC10 phylum using nitrite, or by members of ANME-2 using nitrate. The AOM organisms have not been isolated in pure culture.

In S-DAMO, ANME oxidize CH_4 through the reverse of methanogenesis, producing H_2 that is consumed by sulfate-reducing bacteria, but the exact metabolism is not firmly established:

$$CH_4 + SO_4^{2-} \longrightarrow HCO_3^- + HS^- + H_2O \quad (\Delta G^{0'} = -16.6 \text{ kJ/mol } CH_4)$$

All three groups of ANME are found in the S-DAMO process.

N-DAMO using nitrite is carried out by bacteria, not archaea. An anaerobic nitrite-reducing methanotrophic bacterium, *Candidatus* Methylomirabilis oxyfera has been identified:

$$3CH_4 + 8NO_2^- + 8H^+ \longrightarrow 3CO_2 + 4N_2 + 10H_2O \quad (\Delta G^{0'} = -928 \text{ kJ/mol } CH_4)$$

This bacterium has complete genes for CH_4 oxidation, as in aerobic methanotrophs (Section 7.10.2), including methane monooxygenase (MMO), and genes for denitrification except for nitrous oxide reductase. It is hypothesized that a novel enzyme (nitric oxide dismutase) converts two nitric oxide to N_2 and O_2. O_2 is used as the substrate of MMO and the electron acceptor. A similar oxygenic enzyme, chlorite dismutase, is known in (per)chloride respiration (Section 9.7).

Nitrate is the preferred electron acceptor in a N-DAMO archaeon, *Candidatus* Methanoperedens nitroreducens. CH_4 is oxidized through the reverse of methanogenesis, as in S-DAMO. This archaeon belongs to ANME-2:

$$5CH_4 + 8NO_3^- + 8H^+ \longrightarrow 5CO_2 + 4N_2 + 14H_2O \quad (\Delta G^{0'} = -523 \text{ kJ/mol } CH_4)$$

Archaea belonging to ANME-2 oxidize CH_4 coupled independently to Fe(III) or Mn(IV) reduction. These M-DAMO archaea are closely related to nitrate-reducing N-DAMO *Candidatus* Methanoperedens nitroreducens:

$$CH_4 + 4MnO_2 + 7H^+ \longrightarrow HCO_3^- + 4Mn^{2+} + 5H_2O \quad (\Delta G^{0'} = -556 \text{ kJ/mol } CH_4)$$

$$CH_4 + 8Fe(OH)_3 + 15H^+ \longrightarrow HCO_3^- + 8Fe^{2+} + 21H_2O \quad (\Delta G^{0'} = -270.3 \text{ kJ/mol } CH_4)$$

The chemically challenging activation of methane by ANME occurs via a reversal of the methyl-CoM reductase-catalysed step in methanogenesis. ANME possess homologues of genes normally associated with methanogenic archaea, indicating that these organisms oxidize methane using a pathway which is essentially methanogenesis in reverse. On the other hand, bacteria of the NC10 phylum, including *Candidatus Methylomirabilis oxyfera*, possess genes for aerobic CH_4 oxidation.

9.9.3 Transformation of xenobiotics under anaerobic conditions

As discussed above, many xenobiotics are transformed under anaerobic conditions using various electron acceptors, including oxidized sulfur and nitrogen compounds and metal ions. These microbial activities play an important role in the cycling of elements, including degradation of natural and anthropogenic xenobiotics. Such properties can be exploited to clean up contaminated water and soil in a process known as bioremediation. Not only are organic xenobiotics oxidized, but toxic metals can also be immobilized, extracted from soil or removed from aqueous solution.

Further Reading

Note this section contains key references only. Additional recommended references are available at www.cambridge.org/ProkaryoticMetabolism.

General

Liebensteiner, M. G., Tsesmetzis, N., Stams, A. J. M. & Lomans, B. P. (2014). Microbial redox processes in deep subsurface environments and the potential application of (per)chlorate in oil reservoirs. *Frontiers in Microbiology* **5**, 428.

Schoepp-Cothenet, B., van Lis, R., Atteia, A., Baymann, F., Capowiez, L., Ducluzeau, A.-L., Duval, S., ten Brink, F., Russell, M. J. & Nitschke W. (2013). On the universal core of bioenergetics. *Biochimica et Biophysica Acta* **1827**, 79–93.

Strous, M. & Jetten, M. S. M. (2004). Anaerobic oxidation of methane and ammonium. *Annual Review of Microbiology* **58**, 99–117.

Teske, A. P. (2005). The deep subsurface biosphere is alive and well. *Trends in Microbiology* **13**, 402–404.

Warren, L. A. & Kauffman, M. E. (2003). Geoscience: microbial geoengineers. *Science* **299**, 1027–1029.

Denitrification

Borrero-de Acuña, J. M., Rohde, M., Wissing, J., Jänsch, L., Schobert, M., Molinari, G., Timmis, K. N., Jahn, M. & Jahn, D. (2016). Protein network of the *Pseudomonas aeruginosa* denitrification apparatus. *Journal of Bacteriology* **198**, 1401–1413.

Cabello, P., Roldan, M. D. & Moreno-Vivian, C. (2004). Nitrate reduction and the nitrogen cycle in archaea. *Microbiology* **150**, 3527–3546.

Dalsgaard, T., Stewart, F. J., Thamdrup, B., De Brabandere, L., Revsbech, N. P., Ulloa, O., Canfield, D. E. & DeLong, E. F. (2014). Oxygen at nanomolar levels reversibly suppresses process rates and gene expression in anammox and denitrification in the oxygen minimum zone off northern Chile. *mBio* **5**, e01966-14.

Khan, A. & Sarkar, D. (2012). Nitrate reduction pathways in mycobacteria and their implications during latency. *Microbiology* **158**, 301–307.

Liu, Y., Ai, G.-M., Miao, L.-L. & Liu, Z.-P. (2016). *Marinobacter* strain NNA5, a newly isolated and highly efficient aerobic denitrifier with zero N_2O emission. *Bioresource Technology* **206**, 9–15.

Mania, D., Heylen, K., van Spanning, R. J. M. & Frostegård, A. (2016). Regulation of nitrogen metabolism in the nitrate-ammonifying soil bacterium *Bacillus vireti* and evidence for its ability to grow using N_2O as electron acceptor. *Environmental Microbiology* **18**, 2937–2950.

Oshiki, M., Shimokawa, M., Fujii, N., Satoh, H. & Okabe, S. (2011). Physiological characteristics of the anaerobic ammonium-oxidizing bacterium '*Candidatus* Brocadia sinica'. *Microbiology* **157**, 1706–1713.

Park, D., Kim, H. & Yoon, S. (2017). Nitrous oxide reduction by an obligate aerobic bacterium, *Gemmatimonas aurantiaca* strain T-27. *Applied and Environmental Microbiology* **83**, e00502-17.

Park, S., Kim, D.-H., Lee, J.-H. & Hur, H.-G. (2014). *Sphaerotilus natans* encrusted with nanoball-shaped Fe(III) oxide minerals formed by nitrate-reducing mixotrophic Fe(II) oxidation. *FEMS Microbiology Ecology* **90**, 68–77.

Philippot, L. (2005). Denitrification in pathogenic bacteria: for better or worse? *Trends in Microbiology* **13**, 191–192.

Sawers, R. G., Falke, D. & Fischer, M. (2016). Oxygen and nitrate respiration in *Streptomyces coelicolor* A3(2). *Advances in Microbial Physiology* **68**, 1–40.

Slobodkina, G. B., Mardanov, A. V., Ravin, N. V., Frolova, A. A., Chernyh, N. A., Bonch-Osmolovskaya, E. A. & Slobodkin, A. I. (2017). Respiratory ammonification of nitrate coupled to anaerobic oxidation of elemental sulfur in deep-sea autotrophic thermophilic bacteria. *Frontiers in Microbiology* **8**, 87.

Srivastava, M., Kaushik, M. S., Singh, A., Singh, D. & Mishra, A. K. (2016). Molecular phylogeny of heterotrophic nitrifiers and aerobic denitrifiers and their potential role in ammonium removal. *Journal of Basic Microbiology* **56**, 907–921.

Torregrosa-Crespo, J., Martínez-Espinosa, R. M., Esclapez, J., Bautista, V., Pire, C., Camacho, M., Richardson, D. J. & Bonete, M. J. (2016). Anaerobic metabolism in *Haloferax* genus: denitrification as case of study. *Advances in Microbial Physiology* **68**, 41–85.

Ward, B. B. & Jensen, M. M. (2014). The microbial nitrogen cycle. *Frontiers in Microbiology* **5**, 553.

Metal reduction

Abin, C. A. & Hollibaugh, J. T. (2014). Dissimilatory antimonate reduction and production of antimony trioxide microcrystals by a novel microorganism. *Environmental Science and Technology* **48**, 681–688.

Badalamenti, J. P., Summers, Z. M., Chan, C. H., Gralnick, J. A. & Bond, D. R. (2016). Isolation and genomic characterization of '*Desulfuromonas soudanensis* WTL', a metal- and electrode-respiring bacterium from anoxic deep subsurface brine. *Frontiers in Microbiology* **7**, 913.

Byrne, J. M., Klueglein, N., Pearce, C., Rosso, K. M., Appel, E. & Kappler, A. (2015). Redox cycling of Fe (II) and Fe(III) in magnetite by Fe-metabolizing bacteria. *Science* **347**, 1473–1476.

Carlson, H. K., Iavarone, A. T., Gorur, A., Yeo, B. S., Tran, R., Melnyk, R. A., Mathies, R. A., Auer, M. & Coates, J. D. (2012). Surface multiheme c-type cytochromes from *Thermincola potens* and implications for respiratory metal reduction by Gram-positive bacteria. *Proceedings of the National Academy of Sciences of the USA* **109**, 1702–1707.

Denton, K., Atkinson, M., Borenstein, S., Carlson, A., Carroll, T., Cullity, K., DeMarsico, C., Ellowitz, D., Gialtouridis, A., Gore, R., Herleikson, A., Ling, A., Martin, R., McMahan, K., Naksukpaiboon, P., Seiz, A., Yearwood, K., O'Neill, J. & Wiatrowski, H. (2013). Identification of a possible respiratory arsenate reductase in *Denitrovibrio acetiphilus*, a member of the phylum *Deferribacteres*. *Archives of Microbiology* **195**, 661–670.

Gadd, G. M. (2010). Metals, minerals and microbes: geomicrobiology and bioremediation. *Microbiology* **156**, 609–643.

Hau, H. H., Gilbert, A., Coursolle, D. & Gralnick, J. A. (2008). Mechanism and consequences of anaerobic respiration of cobalt by *Shewanella oneidensis* strain MR-1. *Applied and Environmental Microbiology* **74**, 6880–6886.

Icopini, G. A., Lack, J. G., Hersman, L. E., Neu, M. P. & Boukhalfa, H. (2009). Plutonium(V/VI) reduction by the metal-reducing bacteria *Geobacter metallireducens* GS-15 and *Shewanella oneidensis* MR-1. *Applied and Environmental Microbiology* **75**, 3641–3647.

Kim, B. H., Kim, H. J., Hyun, M. S. & Park, D. H. (1999). Direct electrode reaction of an Fe(III)-reducing bacterium, *Shewanella putrefaciens*. *Journal of Microbiology and Biotechnology* **9**, 127–131.

Marsili, E., Baron, D. B., Shikhare, I. D., Coursolle, D., Gralnick, J. A. & Bond, D. R. (2008). *Shewanella* secretes flavins that mediate extracellular electron transfer. *Proceedings of the National Academy of Sciences of the USA* **105**, 3968–3973.

Nancharaiah, Y. V. & Lens, P. N. L. (2015). Ecology and biotechnology of selenium-respiring bacteria. *Microbiology and Molecular Biology Reviews* **79**, 61–80.

Oni, O. E. & Friedrich, M. W. (2017). Metal oxide reduction linked to anaerobic methane oxidation. *Trends in Microbiology* **25**, 88–90.

Shi, L., Dong, H., Reguera, G., Beyenal, H., Lu, A., Liu, J., Yu, H.-Q. & Fredrickson, J. K. (2016). Extracellular electron transfer mechanisms between microorganisms and minerals. *Nature Reviews Microbiology* **14**, 651–662.

Snider, R. M., Strycharz-Glaven, S. M., Tsoi, S. D., Erickson, J. S. & Tender, L. M. (2012). Long-range electron transport in *Geobacter sulfurreducens* biofilms is redox gradient-driven. *Proceedings of the National Academy of Sciences of the USA* **109**, 15467–15472.

Sure, S., Ackland, M. L., Torriero, A. A. J., Adholeya, A. & Kochar, M. (2016). Microbial nanowires: an electrifying tale. *Microbiology* **162**, 2017–2028.

Wall, J. D. & Krumholz, L. R. (2006). Uranium reduction. *Annual Review of Microbiology* **60**, 149–166.

Sulfidogenesis

Bradley, A. S., Leavitt, W. D. & Johnston, D. T. (2011). Revisiting the dissimilatory sulfate reduction pathway. *Geobiology* **9**, 446–457.

Enning, D. & Garrelfs, J. (2014). Corrosion of iron by sulfate-reducing bacteria: new views of an old problem. *Applied and Environmental Microbiology* **80**, 1226–1236.

Fauque, G. D. & Barton, L. L. (2012). Hemoproteins in dissimilatory sulfate- and sulfur-reducing prokaryotes. *Advances in Microbial Physiology* **60**, 1–90.

Grein, F., Ramos, A. R., Venceslau, S. S. & Pereira, I. A. C. (2013). Unifying concepts in anaerobic respiration: insights from dissimilatory sulfur metabolism. *Biochimica et Biophysica Acta* **1827**, 145–160.

Hockin, S. and Gadd, G. M. (2006). Removal of selenate from sulphate-containing media by sulphate-reducing bacterial biofilms. *Environmental Microbiology* **8**, 816–826.

Jay, Z. J., Beam, J. P., Dohnalkova, A., Lohmayer, R., Bodle, B., Planer-Friedrich, B., Romine, M. & Inskeep, W. P. (2015). *Pyrobaculum yellowstonensis* strain WP30 respires on elemental sulfur and/or arsenate in circumneutral sulfidic geothermal sediments of Yellowstone National Park. *Applied and Environmental Microbiology* **81**, 5907–5916.

Krumholz, L. R., Wang, L., Beck, D. A. C., Wang, T., Hackett, M., Mooney, B., Juba, T. R., McInerney, M. J., Meyer, B., Wall, J. D. & Stahl, D. A. (2013). Membrane protein complex of APS reductase and Qmo is present in *Desulfovibrio vulgaris* and *Desulfovibrio alaskensis*. *Microbiology* **159**, 2162–2168.

Marietou, A., Griffiths, L. & Cole, J. (2009). Preferential reduction of the thermodynamically less favorable electron acceptor, sulfate, by a nitrate-reducing strain of the sulfate-reducing bacterium *Desulfovibrio desulfuricans* 27774. *Journal of Bacteriology* **191**, 882–889.

Meyer, B., Kuehl, J., Deutschbauer, A. M., Price, M. N., Arkin, A. P. and Stahl, D. A. (2013). Variation among *Desulfovibrio* species in electron transfer systems used for syntrophic growth. *Journal of Bacteriology* **195**, 990–1004.

Rabus, R., Venceslau, S. S., Wöhlbrand, L., Voordouw, G., Wall, J. D. & Pereira, I. A. C. (2015). A post-genomic view of the ecophysiology, catabolism and biotechnological relevance of sulphate-reducing prokaryotes. *Advances in Microbial Physiology* **66**, 55–321.

Ramos, A. R., Grein, F., Oliveira, G. P., Venceslau, S. S., Keller, K. L., Wall, J. D. & Pereira, I. A. C. (2015). The FlxABCD-HdrABC proteins correspond to a novel NADH dehydrogenase/heterodisulfide reductase widespread in anaerobic bacteria and involved in ethanol metabolism in *Desulfovibrio vulgaris* Hildenborough. *Environmental Microbiology* **17**, 2288–2305.

Yan, Z., Wang, M. & Ferry, J. G. (2017). A ferredoxin- and $F_{420}H_2$-dependent, electron-bifurcating, heterodisulfide reductase with homologs in the domains bacteria and archaea. *mBio* **8**, e02285-16.

Methanogenesis

Allen, K. D., Wegener, G. & White, R. H. (2014). Discovery of multiple modified F_{430} coenzymes in methanogens and anaerobic methanotrophic archaea suggests possible new roles for F_{430} in nature. *Applied and Environmental Microbiology* **80**, 6403–6412.

Benedict, M. N., Gonnerman, M. C., Metcalf, W. W. & Price, N. D. (2012). Genome-scale metabolic reconstruction and hypothesis testing in the methanogenic archaeon *Methanosarcina acetivorans* C2A. *Journal of Bacteriology* **194**, 855–865.

Buan, N. R. & Metcalf, W. W. (2010). Methanogenesis by *Methanosarcina acetivorans* involves two structurally and functionally distinct classes of heterodisulfide reductase. *Molecular Microbiology* **75**, 843–853.

Costa, K. C. & Leigh, J. A. (2014). Metabolic versatility in methanogens. *Current Opinion in Biotechnology* **29**, 70–75.

Greening, C., Ahmed, F. H., Mohamed, A. E., Lee, B. M., Pandey, G., Warden, A. C., Scott, C., Oakeshott, J. G., Taylor, M. C. & Jackson, C. J. (2016). Physiology, biochemistry, and applications of F_{420}- and F_0-dependent redox reactions. *Microbiology and Molecular Biology Reviews* **80**, 451–493.

Kaster, A.-K., Moll, J., Parey, K. & Thauer, R. K. (2011). Coupling of ferredoxin and heterodisulfide reduction via electron bifurcation in hydrogenotrophic methanogenic archaea. *Proceedings of the National Academy of Sciences of the USA* **108**, 2981–2986.

Lie, T. J., Costa, K. C., Lupa, B., Korpole, S., Whitman, W. B. & Leigh, J. A. (2012). Essential anaplerotic role for the energy-converting hydrogenase Eha in hydrogenotrophic methanogenesis. *Proceedings of the National Academy of Sciences of the USA* **109**, 15473–15478.

Matschiavelli, N., Oelgeschläger, E., Cocchiararo, B., Finke, J. & Rother, M. (2012). Function and regulation of isoforms of carbon monoxide dehydrogenase/acetyl coenzyme A synthase in *Methanosarcina acetivorans*. *Journal of Bacteriology* **194**, 5377–5387.

Purwantini, E., Daniels, L. & Mukhopadhyay, B. (2016). $F_{420}H_2$ is required for phthiocerol dimycocerosate synthesis in mycobacteria. *Journal of Bacteriology* **198**, 2020–2028.

Thauer, R. K., Kaster, A.-K., Seedorf, H., Buckel, W. & Hedderich, R. (2008). Methanogenic archaea: ecologically relevant differences in energy conservation. *Nature Reviews Microbiology* **6**, 579–591.

Wagner, T., Ermler, U. & Shima, S. (2016). The methanogenic CO_2 reducing-and-fixing enzyme is bifunctional and contains 46 [4Fe-4S] clusters. *Science* **354**, 114–117.

Welte, C. & Deppenmeier, U. (2011). Membrane-bound electron transport in *Methanosaeta thermophila*. *Journal of Bacteriology* **193**, 2868–2870.

Wongnate, T., Sliwa, D., Ginovska, B., Smith, D., Wolf, M. W., Lehnert, N., Raugei, S. & Ragsdale, S. W. (2016). The radical mechanism of biological methane synthesis by methyl-coenzyme M reductase. *Science* **352**, 953–958.

Yan, Z., Wang, M. & Ferry, J. G. (2017). A ferredoxin- and $F_{420}H_2$-dependent, electron-bifurcating, heterodisulfide reductase with homologs in the domains bacteria and archaea. *mBio* **8**, e02285-16.

Zheng, K., Ngo, P. D., Owens, V. L., Yang, X.-p. & Mansoorabadi, S. O. (2016). The biosynthetic pathway of coenzyme F_{430} in methanogenic and methanotrophic archaea. *Science* **354**, 339–342.

Homoacetogenesis

Diender, M., Stams, A. J. M. & Sousa, D. Z. (2015). Pathways and bioenergetics of anaerobic carbon monoxide fermentation. *Frontiers in Microbiology* **6**, 1275.

Hess, V., Poehlein, A., Weghoff, M. C., Daniel, R. & Müller, V. (2014). A genome-guided analysis of energy conservation in the thermophilic, cytochrome-free acetogenic bacterium *Thermoanaerobacter kivui*. *BMC Genomics* **15**, 1139.

Huang, H., Wang, S., Moll, J. & Thauer, R. K. (2012). Electron bifurcation involved in the energy metabolism of the acetogenic bacterium *Moorella thermoacetica* growing on glucose or H_2 plus CO_2. *Journal of Bacteriology* **194**, 3689–3699.

Jeong, J., Bertsch, J., Hess, V., Choi, S., Choi, I.-G., Chang, I. S. & Müller, V. (2015). Energy conservation model based on genomic and experimental analyses of a carbon monoxide-utilizing, butyrate-forming acetogen, *Eubacterium limosum* KIST612. *Applied and Environmental Microbiology* **81**, 4782–4790.

Köpke, M., Held, C., Hujer, S., Liesegang, H., Wiezer, A., Wollherr, A., Ehrenreich, A., Liebl, W., Gottschalk, G. & Dürre, P. (2010). *Clostridium ljungdahlii* represents a microbial production platform based on syngas. *Proceedings of the National Academy of Sciences of the USA* **107**, 13087–13092.

Ljungdahl, L. G. (2009). A life with acetogens, thermophiles, and cellulolytic anaerobes. *Annual Review of Microbiology* **63**, 1–25.

Mock, J., Zheng, Y., Mueller, A. P., Ly, S., Tran, L., Segovia, S., Nagaraju, S., Köpke, M., Dürre, P. & Thauer, R. K. (2015). Energy conservation associated with ethanol formation from H_2 and CO_2 in *Clostridium autoethanogenum* involving electron bifurcation. *Journal of Bacteriology* **197**, 2965–2980.

Schuchmann, K. & Müller, V. (2014). Autotrophy at the thermodynamic limit of life: a model for energy conservation in acetogenic bacteria. *Nature Reviews Microbiology* **12**, 809–821.

Spahn, S., Brandt, K. & Müller, V. (2015). A low phosphorylation potential in the acetogen *Acetobacterium woodii* reflects its lifestyle at the thermodynamic edge of life. *Archives of Microbiology* **197**, 745–751.

Wang, S., Huang, H., Kahnt, J. & Thauer, R. K. (2013). A reversible electron-bifurcating ferredoxin- and NAD-dependent [FeFe]-hydrogenase (HydABC) in *Moorella thermoacetica*. *Journal of Bacteriology* **195**, 1267–1275.

Weghoff, M. C., Bertsch, J. & Müller, V. (2015). A novel mode of lactate metabolism in strictly anaerobic bacteria. *Environmental Microbiology* **17**, 670–677.

Organohalide respiration

Adrian, L., Dudkova, V., Demnerova, K. & Bedard, D. L. (2009). "*Dehalococcoides*" sp. strain CBDB1 extensively dechlorinates the commercial polychlorinated biphenyl mixture aroclor 1260. *Applied and Environmental Microbiology* **75**, 4516–4524.

Bommer, M., Kunze, C., Fesseler, J., Schubert, T., Diekert, G. & Dobbek, H. (2014). Structural basis for organohalide respiration. *Science* **346**, 455–458.

Goris, T., Schubert, T., Gadkari, J., Wubet, T., Tarkka, M., Buscot, F., Adrian, L. & Diekert, G. (2014). Insights into organohalide respiration and the versatile catabolism of *Sulfurospirillum multivorans* gained from comparative genomics and physiological studies. *Environmental Microbiology* **16**, 3562–3580.

Holliger, C., Wohlfarth, G. & Diekert, G. (1999). Reductive dechlorination in the energy metabolism of anaerobic bacteria. *FEMS Microbiology Reviews* **22**, 383–398.

Janssen, D. B. (2004). Evolving haloalkane dehalogenases. *Current Opinion in Chemical Biology* **8**, 150–159.

Kruse, T., van de Pas, B. A., Atteia, A., Krab, K., Hagen, W. R., Goodwin, L., Chain, P., Boeren, S., Maphosa, F., Schraa, G., de Vos, W. M., van der Oost, J., Smidt, H. & Stams, A. J. M. (2015). Genomic, proteomic, and biochemical analysis of the organohalide respiratory pathway in *Desulfitobacterium dehalogenans*. *Journal of Bacteriology* **197**, 893–904.

Kublik, A., Deobald, D., Hartwig, S., Schiffmann, C. L., Andrades, A., von Bergen, M., Sawers, R. G. & Adrian, L. (2016). Identification of a multi-protein reductive dehalogenase complex in *Dehalococcoides mccartyi* strain CBDB1 suggests a protein-dependent respiratory electron transport chain obviating quinone involvement. *Environmental Microbiology* **18**, 3044–3056.

Lorenz, A. & Löffler, F. E. (eds) (2016). *Organohalide-Respiring Bacteria*. Berlin: Springer-Verlag.

Smidt, H. & de Vos, W. M. (2004). Anaerobic microbial dehalogenation. *Annual Review of Microbiology* **58**, 43–73.

Tang, S., Wang, P. H., Higgins, S., Loeffler, F. & Edwards, E. A. (2016). Sister *Dehalobacter* genomes reveal specialization in organohalide respiration and recent strain differentiation likely driven by chlorinated substrates. *Frontiers in Microbiology* **7**, 100.

Anaerobic respiration on miscellaneous electron acceptors

Arkhipova, O. & Akimenko, V. (2005). Unsaturated organic acids as terminal electron acceptors for reductase chains of anaerobic bacteria. *Microbiology-Moscow* **74**, 629–639.

Bardiya, N. & Bae, J.-H. (2011). Dissimilatory perchlorate reduction: a review. *Microbiological Research* **166**, 237–254.

Martínez-Espinosa, R. M., Richardson, D. J. & Bonete, M. J. (2015). Characterisation of chlorate reduction in the haloarchaeon *Haloferax mediterranei*. *Biochimica et Biophysica Acta* **1850**, 587–594.

Syntrophic associations

Cao, X., Liu, X. & Dong, X. (2003). *Alkaliphilus crotonatoxidans* sp. nov., a strictly anaerobic, crotonate-dismutating bacterium isolated from a methanogenic environment. *International Journal of Systematic and Evolutionary Microbiology* **53**, 971–975.

Cheng, Q. & Call, D. F. (2016). Hardwiring microbes via direct interspecies electron transfer: mechanisms and applications. *Environmental Science: Processes and Impacts* **18**, 968–980.

de Bok, F. A. M., Plugge, C. M. & Stams, A. J. M. (2004). Interspecies electron transfer in methanogenic propionate degrading consortia. *Water Research* **38**, 1368–1375.

Gray, N. D., Sherry, A., Grant, R. J., Rowan, A. K., Hubert, C. R. J., Callbeck, C. M., Aitken, C. M., Jones, D. M., Adams, J. J., Larter, S. R. & Head, I. M. (2011). The quantitative significance of *Syntrophaceae* and syntrophic partnerships in methanogenic degradation of crude oil alkanes. *Environmental Microbiology* **13**, 2957–2975.

Kung, J. W., Seifert, J., von Bergen, M. & Boll, M. (2013). Cyclohexanecarboxyl-coenzyme A (CoA) and cyclohex-1-ene-1-carboxyl-CoA dehydrogenases, two enzymes involved in the fermentation of benzoate and crotonate in *Syntrophus aciditrophicus*. *Journal of Bacteriology* **195**, 3193–3200.

Sieber, J. R., McInerney, M. J. & Gunsalus, R. P. (2012). Genomic insights into syntrophy: the paradigm for anaerobic metabolic cooperation. *Annual Review of Microbiology* **66**, 429–452.

Storck, T., Virdis, B. & Batstone, D. J. (2016). Modelling extracellular limitations for mediated versus direct interspecies electron transfer. *ISME Journal* **10**, 621–631.

Summers, Z. M., Fogarty, H. E., Leang, C., Franks, A. E., Malvankar, N. S. & Lovley, D. R. (2010). Direct exchange of electrons within aggregates of an evolved syntrophic coculture of anaerobic bacteria. *Science* **330**, 1413–1415.

Oxidation of hydrocarbons under anaerobic conditions

Abu Laban, N., Selesi, D., Rattei, T., Tischler, P. & Meckenstock, R. U. (2010). Identification of enzymes involved in anaerobic benzene degradation by a strictly anaerobic iron-reducing enrichment culture. *Environmental Microbiology* **12**, 2783–2796.

Bergmann, F., Selesi, D., Weinmaier, T., Tischler, P., Rattei, T. & Meckenstock, R. U. (2011). Genomic insights into the metabolic potential of the polycyclic aromatic hydrocarbon degrading sulfate-reducing *Deltaproteobacterium* N47. *Environmental Microbiology* **13**, 1125–1137.

Carmona, M., Zamarro, M. T., Blazquez, B., Durante-Rodriguez, G., Juarez, J. F., Valderrama, J. A., Barragan, M. J. L., Garcia, J. L. & Diaz, E. (2009). Anaerobic catabolism of aromatic compounds: a genetic and genomic view. *Microbiology and Molecular Biology Reviews* **73**, 71–133.

Eberlein, C., Johannes, J., Mouttaki, H., Sadeghi, M., Golding, B. T., Boll M., & Meckenstock, R. U. (2013). ATP-dependent/-independent enzymatic ring reductions involved in the anaerobic catabolism of naphthalene. *Environmental Microbiology* **15**, 1832–1841.

Jarling, R., Kühner, S., Janke, E. B., Gruner, A., Drozdowska, M., Golding, B. T., Rabus, R. & Wilkes, H. (2015). Versatile transformations of hydrocarbons in anaerobic bacteria: substrate ranges and regio- and stereo-chemistry of activation reactions. *Frontiers in Microbiology* **6**, 880.

Khelifi, N., Amin Ali, O., Roche, P., Grossi, V., Brochier-Armanet, C., Valette, O., Ollivier, B., Dolla, A. & Hirschler-Rea, A. (2014). Anaerobic oxidation of long-chain n-alkanes by the hyperthermophilic sulfate-reducing archaeon, *Archaeoglobus fulgidus*. *ISME Journal* **8**, 2153–2166.

Meckenstock, R. U. & Mouttaki, H. (2011). Anaerobic degradation of non-substituted aromatic hydrocarbons. *Current Opinion in Biotechnology* **22**, 406–414.

Philipp, B. & Schink, B. (2012). Different strategies in anaerobic biodegradation of aromatic compounds: nitrate reducers versus strict anaerobes. *Environmental Microbiology Reports* **4**, 469–478.

Porter, A. W. & Young, L. Y. (2014). Benzoyl-CoA, a universal biomarker for anaerobic degradation of aromatic compounds. *Advances in Applied Microbiology* **88**, 167–203.

Wawrik, B., Marks, C. R., Davidova, I. A., McInerney, M. J., Pruitt, S., Duncan, K., Suflita, J. M. & Callaghan, A. V. (2016). Methanogenic paraffin degradation proceeds via alkane addition to fumarate by "*Smithella*" spp. mediated by a syntrophic coupling with hydrogenotrophic methanogens. *Environmental Microbiology* **18**, 2604–2619.

Methane oxidation under anaerobic conditions

Beal, E. J., House, C. H. & Orphan, V. J. (2009). Manganese- and iron-dependent marine methane oxidation. *Science* **325**, 184–187.

Ettwig, K. F., Zhu, B., Speth, D., Keltjens, J. T., Jetten, M. S. M. & Kartal, B. (2016). Archaea catalyze iron-dependent anaerobic oxidation of methane. *Proceedings of the National Academy of Sciences of the USA* **113**, 12792–12796.

Haroon, M. F., Hu, S., Shi, Y., Imelfort, M., Keller, J., Hugenholtz, P., Yuan, Z. & Tyson, G. W. (2013). Anaerobic oxidation of methane coupled to nitrate reduction in a novel archaeal lineage. *Nature* **500**, 567–570.

Krukenberg, V., Harding, K., Richter, M., Glöckner, F. O., Gruber-Vodicka, H., Adam, B., Berg, J. S., Knittel, K., Tegetmeyer, H. E., Boetius, A. & Wegener, G. (2016). *Candidatus Desulfofervidus auxilii*, a hydrogenotrophic sulfate-reducing bacterium involved in the thermophilic anaerobic oxidation of methane. *Environmental Microbiology* **18**, 3073–3091.

Oni, O. E. & Friedrich, M. W. (2017). Metal oxide reduction linked to anaerobic methane oxidation. *Trends in Microbiology* **25**, 88–90.

Scheller, S., Goenrich, M., Boecher, R., Thauer, R. K. & Jaun, B. (2010). The key nickel enzyme of methanogenesis catalyses the anaerobic oxidation of methane. *Nature* **465**, 606–608.

Shima, S. & Thauer, R. K. (2005). Methyl-coenzyme M reductase and the anaerobic oxidation of methane in methanotrophic Archaea. *Current Opinion in Microbiology* **8**, 643–648.

Degradation of xenobiotics under anaerobic conditions

Esteve-Nunez, A., Caballero, A. & Ramos, J. L. (2001). Biological degradation of 2,4,6-trinitrotoluene. *Microbiology and Molecular Biology Reviews* **65**, 335–352.

Eyers, L., George, I., Schuler, L., Stenuit, B., Agathos, S. N. & El Fantroussi, S. (2004). Environmental genomics: exploring the unmined richness of microbes to degrade xenobiotics. *Applied Microbiology and Biotechnology* **66**, 123–130.

Yu, H.-Y., Bao, L.-J., Liang, Y. & E. Zeng, Y. (2011). Field validation of anaerobic degradation pathways for dichlorodiphenyltrichloroethane (DDT) and 13 metabolites in marine sediment cores from China. *Environmental Science and Technology* **45**, 5245–5252.

Zhang, C. & Bennett, G. N. (2005). Biodegradation of xenobiotics by anaerobic bacteria. *Applied Microbiology and Biotechnology* **67**, 600–618.

Chapter 10

Chemolithotrophy

Some prokaryotes grow by using reduced inorganic compounds as their energy source and CO_2 as the carbon source. These are called chemolithotrophs. The electron donors used by chemolithotrophs include nitrogen and sulfur compounds, Fe(II), H_2 and CO. Certain bacteria and archaea can use metalloid anions such as arsenite [As(III)] as the electron donor. The Calvin cycle is the most common CO_2 fixation mechanism, and the reductive TCA cycle, acetyl-CoA pathway, 3-hydroxypropionate cycle and 4-hydroxybutyrate cycles are found in some chemolithotrophic prokaryotes. Some can use organic compounds as their carbon source while metabolizing an inorganic electron donor. This kind of bacterial metabolism is referred to as mixotrophy.

10.1 Reverse electron transport

As with chemoorganotrophs, the metabolism of chemolithotrophs requires ATP and NAD(P)H for carbon metabolism and biosynthetic processes. Some of the electron donors used by chemolithotrophs have a redox potential higher than that of $NAD(P)^+/NAD(P)H$ (Table 10.1). Electrons from these electron donors are transferred to coenzyme Q or to cytochromes. Some of the electrons are used to generate a proton motive force reducing O_2, while the remaining electrons are used to reduce $NAD(P)^+$ to NAD(P)H through a reverse of the electron transport chain. The latter is an uphill reaction and is coupled with the consumption of the proton motive force (Figure 10.1). This is referred to as reverse electron transport. In most cases, electron donors with a redox potential lower than $NAD(P)^+/NAD(P)H$ are oxidized and this is coupled with the reduction of coenzyme Q or cytochromes for the efficient utilization of the electron donors at low concentrations. The energy consumed in reverse electron transport from cytochrome *c* to $NAD(P)^+$ is about five times the energy generated from the forward electron transport process.

10.2 Nitrification

A group of bacteria oxidize ammonia to nitrite, which is further oxidized to nitrate by another group of bacteria in an energy generating process known as nitrification. With the exception of the heterotrophic nitrifying–aerobic denitrifiers (see below), the ammonia oxidizing bacteria (AOB) are all Gram-negative, mostly obligate chemolithotrophs, and have an extensive membrane structure within the cytoplasm except for *Nitrosospira tenuis*. A separate group of bacteria oxidize nitrite to nitrate. These organisms are referred to as nitrifying bacteria and are widely distributed in soil and water. The nitrogen cycle cannot be completed without denitrifiers. Bacterial ammonia and nitrite oxidizers fix CO_2 through the Calvin cycle except for strains of the *Nitrospira* genus (see below).

A strain of *Nitrospira* (*Candidatus* Nitrospira inopinata) in an enriched culture oxidizes

Table 10.1 Redox potential of inorganic electron donors used by chemolithotrophs.

Electron donating reaction	Redox potential ($E^{0\prime}$, V)
$CO + H_2 \rightarrow CO_2 + 2H^+ + 2e^-$	−0.54
$SO_3^{2-} + H_2O \rightarrow SO_4^{2-} + 2H^+ + 2e^-$	−0.45
$H_2 \rightarrow 2H^+ + 2e^-$	−0.41
$NAD(P)H + H^+ \rightarrow NAD(P)^+ + 2H^+ + 2e^-$	−0.32
$H_2S \rightarrow S^0 + 2H^+ + 2e^-$	−0.25
$S^0 + 3H_2O \rightarrow SO_3^{2-} + 6H^+ + 4e^-$	+0.05
$NO_2^- + H_2O \rightarrow NO_3^- + 2H^+ + 2e^-$	+0.42
$NH_4^+ + 2H_2O \rightarrow NO_2^- + 8H^+ + 6e^-$	+0.44
$Fe^{2+} \rightarrow Fe^{3+} + e^-$	+0.78
$O_2 + 4H^+ + 4e^- \rightarrow 2H_2O$	+0.86

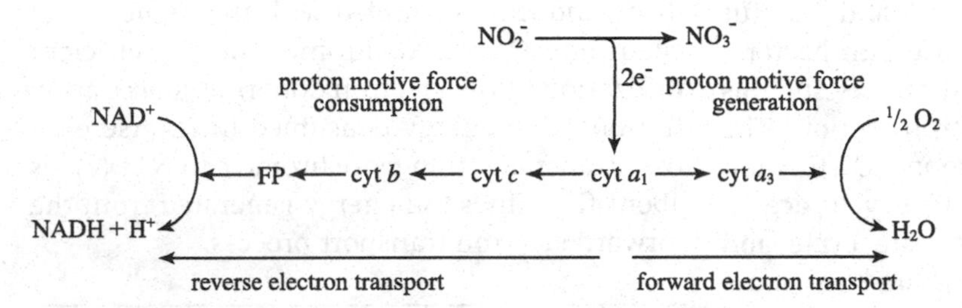

Figure 10.1 Reverse electron transport to reduce NAD(P)+ in a nitrite oxidizer, *Nitrococcus mobilis*.

Nitrococcus mobilis oxidizes nitrite as the electron donor, reducing cytochrome a_1. Some of the electrons are consumed to reduce O_2 and generate a proton motive force (forward electron transport). The remaining electrons are transferred to $NAD(P)^+$ to supply reducing power for biosynthesis. The latter process requires energy in the form of a proton motive force, and is referred to as reverse electron transport.

ammonia completely to nitrate (complete ammonia oxidizer, comammox), but does not grow on nitrite. This bacterium fixes CO_2 through the reductive TCA cycle (Section 10.8.2), as do other members of the same genus. Comammox bacteria have not been isolated in pure culture. This bacterium has urease like other species of the same genus (see below).

A group of aerobic heterotrophic bacteria can simultaneously use ammonia and organic compounds as their electron donors. These include *Thiosphaera pantotropha*, *Alcaligenes faecalis*, *Bacillus* sp., *Providencia rettgeri*, *Anoxybacillus contaminans* and *Paenibacillus uliginis*. Ammonia is oxidized as in AOB. They use nitrate and oxygen simultaneously (Section 9.1.4) and are known as the heterotrophic nitrifying–aerobic denitrifiers.

Bacteria were regarded as the sole nitrifiers until ammonia oxidizing archaea (AOA) were identified. AOA are more important ammonia oxidizers than AOB and are dominant over their bacterial counterparts (AOB) in most natural habitats. AOA actively oxidize ammonia to nitrite where the ammonia concentration is below the affinity threshold for AOB. Known AOA are members of the phylum *Thaumarchaeota*, including *Nitrosopumilus maritimus* and *Nitrososphaera viennensis*. Although they are ubiquitous in large numbers in the environment, they have not been characterized in detail

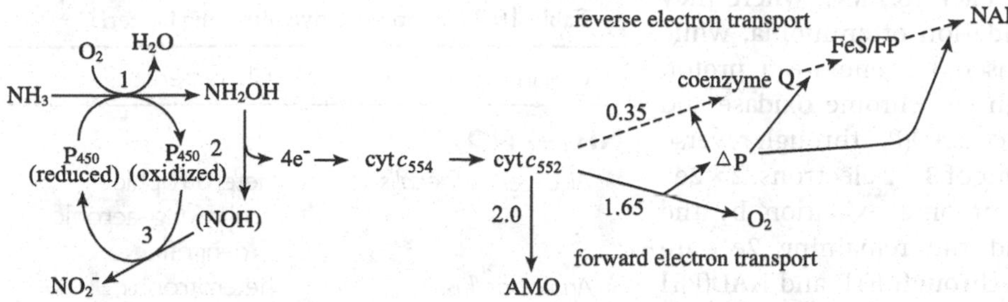

Figure 10.2 Ammonia oxidation to nitrite by denitrifiers.

(Modified from *Arch. Microbiol.* **178**: 250–255, 2002)

Since the redox potential of NH_2OH/NH_3 (+0.899 V) is higher than that of $\frac{1}{2}O_2/H_2O$, ammonia monooxygenase (AMO) (1) oxidizes ammonia, consuming $2e^-$ from the reduced form of cytochrome c_{552} mediated by P_{450}. Hydroxylamine oxidation is coupled to the reduction of cytochrome c_{554}. Out of the four electrons released in the oxidation of NH_2OH by hydroxylamine oxidoreductase (HAO, reaction 2 and 3), two electrons are consumed by AMO where they are used for the oxidation of ammonia. The other 1.65 electrons are routed to cytochrome oxidase to generate a proton motive force and the remaining 0.35 electrons pass to NAD^+ through reverse electron transport to supply reducing power for biosynthesis.

due to difficulties in cultivation of the slow growers. It is generally assumed that ammonia monooxygenase, encoded in all studied AOA, is responsible for the first step of ammonia oxidation, but the second step to nitrite production and the contributing cofactors and electron carriers have still not been identified. AOA possess genes for the key enzymes of the 3-hydroxypropionate–4-hydroxybutyrate cycle, fixing CO_2 (Section 10.8.5.2). Catalase genes are largely absent in AOA, and their growth is stimulated by 2-keto acids, such as pyruvate, that chemically detoxify H_2O_2. An ammonia-oxidizing archaeon *Nitrososphaera gargensis* converts cyanate to NH_3 and CO_2 using cyanase and uses NH_3 as the electron donor.

10.2.1 Ammonia oxidation

NH_3 is oxidized in a two-step reaction via hydroxylamine (NH_2OH) and nitroxyl (NOH) in reactions catalysed by ammonia monooxygenase (AMO) and hydroxylamine oxidoreductase (HAO) (Figure 10.2):

$$2NH_3 + 3O_2 \longrightarrow 2NO_2^- + 2H^+ + 2H_2O \quad (\Delta G^{0'} = -272\ \text{kJ/mol}\ NH_3)$$

NH_3 is oxidized to hydroxylamine by AMO, consuming two electrons available from the oxidation of hydroxylamine, probably through a membrane-bound cytochrome c_{552}. Since the redox potential of NH_2OH/NH_3 (+0.899 V) is higher than that of $\frac{1}{2}O_2/H_2O$, ammonia monooxygenase oxidizes ammonia, consuming $2e^-$ from the reduced form of cytochrome c_{552} mediated by P_{450}, as shown in Figure 10.2. Hydroxylamine oxidation to nitrite is coupled to the reduction of cytochrome c_{554}:

$$2NH_3 + O_2 \xrightarrow{AMO} 2NH_2OH \quad (\Delta G^{0'} = +16\ \text{kJ/mol}\ NH_3)$$

$$\text{hydroxylamine} + H_2O \xrightarrow{HAO} NO_2^- + 5H^+ + 4e^-$$

HAO from *Nitrosomonas europaea* is a homotrimer, with each subunit containing eight *c*-type haems, giving a total of 24 haems. Seven of the haems in each subunit are covalently attached to the protein by two thioester linkages. The eighth haem, designated P-460, is an unusual prosthetic group and has an additional covalent bond through a tyrosine residue. The P-460 haem is located at the active site. The function of the *c*-haems is believed to be the transfer of electrons from the active site of P-460 to cytochrome c_{554}.

The four electrons released from the oxidation of NH_2OH by HAO in *Nitrosomonas europaea* are channelled through cytochrome c_{554} to a membrane-bound cytochrome c_{552}. Two of the electrons are routed back to AMO, where they are used for the oxidation of ammonia, while 1.65 electrons are used to generate a proton motive force through cytochrome oxidase and 0.35 are used to reduce $NAD(P)^+$ through reverse electron transport. Out of 3 × 2 electrons, 2 × $2e^-$ are consumed in ammonia oxidation by the monooxygenase, and the remaining $2e^-$ are used to generate Δp through ETP and NAD(P)H through reverse electron transport.

Although ammonia oxidizers are known as obligate chemolithotrophs, they can utilize a limited number of organic compounds, including amino acids and organic acids. The complete genome sequence of *Nitrosomonas europaea* has revealed a potential fructose permease, and this bacterium metabolizes fructose and pyruvate mixotrophically.

10.2.2 Nitrite oxidation

Nitrite produced from the oxidation of ammonia can be used by a separate group of bacteria as their energy source (Table 10.2):

$$2NO_2^- + O_2 \longrightarrow 2NO_3^- \quad (\Delta G^{0\prime} = -74.8 \text{ kJ/mol } NO_2^-)$$

Nitrite oxidoreductase (NOR) oxidizes nitrite to nitrate, reducing cytochrome a_1. It is a membrane-associated iron–sulfur molybdoprotein, and is part of an electron transfer chain which channels electrons from nitrite to molecular oxygen. NOR in *Nitrobacter hamburgensis* is a heterodimer consisting of α and β subunits. The NOR of *Nitrobacter winogradskyi* is composed of three subunits, as well as haem a_1, haem c, non-haem iron and molybdenum. This enzyme transfers electrons to the cytochrome c oxidase through a membrane-bound cytochrome c (Figure 10.3). Since the free energy change is small, only one H^+ is transported coupled to these reactions. The electron transfer is not well understood. Hydride ion (H^-) is transferred from NOR to cytochrome c, consuming the inside negative membrane potential, because electron transfer from NOR with a redox potential of +420 mV to cytochrome c (+270 V) is an uphill reaction.

NOR in members of the *Nitrospira* genus is a cytoplasmic membrane-bound iron–sulfur

Table 10.2 Representative nitrifying bacteria.

Organism	Characteristics
$NH_3 \rightarrow NO_2^-$	
Alcaligenes faecalis	heterotrophic nitrifying–aerobic denitrifier
Anoxybacillus contaminans	heterotrophic nitrifying–aerobic denitrifier
Nitrosomonas europaea	soil, freshwater, seawater, sewage works
Nitrosospira (Nitrosovibrio) tenuis	soil
Nitrosococcus nitrosus	soil
Nitrosococcus oceanus	seawater
Nitrosospira briensis	soil
Nitrosolobus multiformis	soil
Nitrosopumilus maritimus	seawater, archaeon
Nitrososphaera gargensis	hot spring, archaeon, cyanate
Nitrososphaera viennensis	soil, archaeon
Paenibacillus uliginis	heterotrophic nitrifying–aerobic denitrifier
Providencia rettgeri	heterotrophic nitrifying–aerobic denitrifier
Thiosphaera pantotropha	heterotrophic nitrifying–aerobic denitrifier
$NO_2^- \rightarrow NO_3^-$	
Nitrococcus mobilis	seawater
Nitrobacter winogradskyi	soil, freshwater, seawater, facultative chemolithotroph
Nitrospina gracilis	seawater
Nitrospira marina	seawater
Nitrotoga arctica	cold-adapted
Nitrolancea hollandica	Gram-positive
Complete oxidizer	
Candidatus Nitrospira inopinata	

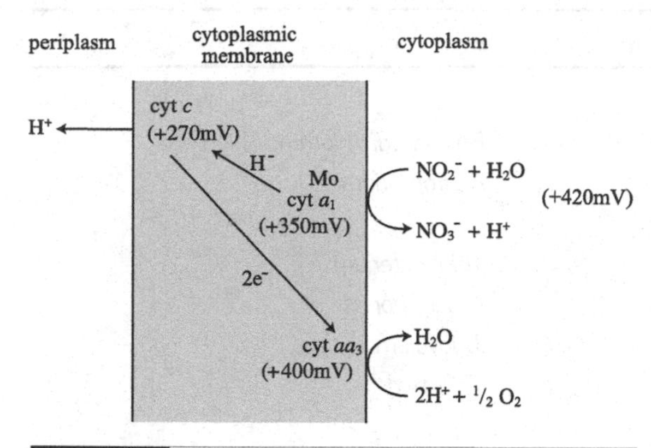

Figure 10.3 The electron transport system in nitrite oxidation by *Nitrobacter winogradskyi*.

(Dawes, E. A. 1986, *Microbial Energetics*, Figure 9.4. Blackie & Son, Glasgow)

The oxidation of nitrite transports 1H$^+$ across the membrane.

molybdoprotein enzyme, facing the periplasm, consisting of three subunits.

Nitrite oxidizers are obligate chemolithotrophs, with the exception of *Nitrobacter winogradskyi*, which is a facultative chemolithotroph.

Urease is active in nitrite-oxidizing bacteria of the *Nitrospira* genus, such as *N. moscoviensis* and *N. lenta*, as in the complete ammonia oxidizing bacterium, *Candidatus* Nitrospira inopinata. This enzyme enables them to supply ammonia oxidizers with ammonia from urea, which is fully nitrified by this consortium through reciprocal feeding. *N. moscoviensis* can use H_2 as the sole energy source in addition to nitrite (Section 10.5.1).

10.2.3 Anaerobic nitrification

As discussed in Chapter 9 (Section 9.1.4), some bacteria of the phylum *Planctomycetes* oxidize ammonia under anaerobic conditions using nitrite as the electron acceptor. These are known as ANAMMOX bacteria. *Nitrosomonas europaea* oxidizes ammonia under anaerobic conditions using nitrogen dioxide (NO_2) as the electron acceptor.

10.3 Sulfur bacteria and the oxidation of sulfur compounds

Certain prokaryotes can use inorganic sulfur compounds, including sulfide (HS$^-$), elemental sulfur (S^0), thiosulfate ($HS_2O_3^-$) and sulfite (SO_3^{2-}) as their energy source. These are known as sulfur bacteria.

10.3.1 Sulfur bacteria

To distinguish them from photolithotrophic sulfur bacteria (Chapter 11), chemolithotrophic sulfur bacteria are referred to as colourless sulfur bacteria. These are phylogenetically diverse and include bacteria and archaea. They are grouped either according to the cellular location of sulfur deposition after sulfide is oxidized (Table 10.3), or by their ability to use polythionate. Many can thrive at the aerobic–anaerobic interface where sulfide, produced by sulfate-reducing bacteria, diffuses from anaerobic regions. These organisms have to compete with molecular oxygen for sulfide, which is rapidly oxidized by molecular oxygen. Some other sulfur bacteria are acidophilic, oxidizing pyrite (FeS_2), and include species of *Thiobacillus*. The genus *Thiobacillus* is designated as small Gram-negative rod-shaped bacteria deriving energy from the oxidation of one or more reduced sulfur compounds including sulfides, thiosulfate, polythionate and thiocyanate. They fix CO_2 through the Calvin cycle. However, they are phylogenetically diverse in terms of 16S ribosomal RNA gene sequences, DNA G + C content and DNA homology, in addition to physiological differences. Many species have been reclassified to *Paracoccus*, *Acidiphilium*, *Thiomonas*, *Thermithiobacillus*, *Acidithiobacillus* and *Halothiobacillus* genera. In addition to species of *Beggiatoa*, *Acidithiobacillus* and *Thiomicrospira* within the sulfur bacteria, many other prokaryotes can oxidize sulfur compounds mixotrophically or chemolithotrophically. These include bacteria such as species of *Aquaspirillum*, *Aquifex*, *Bacillus*, *Paracoccus*, *Pseudomonas*, *Starkeya* and *Xanthobacter*, and archaea such as species of *Sulfolobus* and *Acidianus*. Species of the Gram-negative bacterial genus *Thermothrix* and the archaea *Sulfolobus* and *Acidianus* are thermophiles. *Acidianus brierleyi* uses elemental sulfur not only as an electron donor under aerobic conditions, but also as an electron acceptor to reduce hydrogen under anaerobic conditions.

Species of *Acidithiobacillus* and *Thiomicrospira* can be isolated from diverse ecosystems

Table 10.3 | Sulfur bacteria.

1. Accumulating sulfur intracellularly				
Gliding, filamentous cells			*Beggiatoa, Thiothrix, Thioploca*	
Gliding, very large unicells			*Achromatium*	
Immotile or motile with flagella, coccus or rod-shaped				
immotile, rod			*Thiobacterium*	
motile with flagella, rod			*Macromonas*	
motile with flagella, coccus			*Thiovulum*	
motile with flagella, vibrioid			*Thiospira*	
2. Accumulating sulfur extracellularly				
	Acidithiobacillus	*Thiomicrospira*	*Thioalkalimicrobium*	*Thioalkalivibrio*
Morphology	rod	vibrioid	rod to spirillum	rod to spirillum
Flagellum	+	+	+	+
DNA G + C (%)	34–70	48	61.0–65.6	48.0–51.2
Growth pH	1–8.5	5.0–8.5	7.5–10.6	7.5–10.6
Chemolithotrophy	facultative	obligate	obligate	obligate
3. Thermophilic				
		Thermothrix	*Sulfolobus*	*Acidianus*
Classification		Gram-negative	archaeon	archaeon
Morphology		rod	coccus	coccus
DNA G + C content (%)		?	36–38	~31
Growth temperature (°C)		40–80	50–85	45–95
Chemolithotrophy		facultative	facultative	facultative

Source: *Appl. Environ. Microbiology*, **67**: 2873–2882, 2001.

including soil, freshwater and seawater. *Sulfolobus acidocaldarius* is an archaeon isolated from an acidic hot spring. In addition to reduced sulfur compounds, *Acidithiobacillus ferrooxidans* and *Sulfolobus acidocaldarius* oxidize Fe(II) to Fe (III) as their electron donor. *Acidithiobacillus ferrooxidans* and *Acidithiobacillus thiooxidans* grow optimally at around pH 2.0 and oxidize metal sulfides, solubilizing the metals. Low quality ores are treated with these bacteria to recover metals such as Cu, U and others. This process is referred to as bacterial leaching or bioleaching. This property can also be applied to remove sulfur (in pyrite, FeS_2) from coal. *Thiobacillus denitrificans* is an anaerobe which oxidizes sulfur compounds using nitrate as the electron acceptor.

Filamentous sulfur bacteria of the genera *Thioploca* and *Beggiatoa* accumulate nitrate in intracellular vacuoles. Nitrate, acting as electron acceptor, is reduced to ammonia with sulfide or sulfur as electron donors in these bacteria (Section 9.1.4). Many nitrate-accumulating sulfur bacteria inhabit the sediments of upwelling areas characterized by high sediment concentrations of soluble sulfide, and low levels of dissolved oxygen. The ecological implication of nitrate ammonification is that nitrogen is conserved within the ecosystem. *Thiomargarita namibiensis* is another sulfur bacterium, with vacuoles containing nitrate. This bacterium has not been isolated in pure culture, but is found in sediments in the coastal waters of Namibia, measuring up to 0.75 mm in size, which is about 100 times bigger than a 'normal' sized bacterium. Other giant bacteria have been identified, including *Thioploca araucae*, *Thiomargarita* sp., *Thioploca chileae* and species of *Leucothrix*, *Thiothrix* and *Achromatium*. These are all members of the family *Thiotrichaceae*.

In addition to the acidophilic and neutrophilic sulfur bacteria, alkaliphilic sulfur bacteria thrive in alkaline soda lakes. These are species of *Thioalkalimicrobium* and *Thioalkalivibrio*. They accumulate elemental sulfur extracellularly before oxidizing it when sulfide is depleted. Members of the former genus are obligate aerobes, while some of the latter genus are facultative anaerobes, using nitrate as electron acceptor (Section 9.1.4). An anaerobic Gram-negative bacterium is also known that can grow chemolithotrophically using HS^- as the electron donor and arsenate as the electron acceptor.

Filamentous cable bacteria are found in freshwater and marine sediment environments. These bacteria have micro-cables located in the periplasm that transfer electrons, enabling sulfide oxidation in the sulfide-rich anaerobic zone coupled to oxygen reduction in the aerobic zone (Section 2.3.3.3). They belong to the family *Desulfobulbaceae* and transfer electrons through centimetre-long filaments spanning the aerobic surface to the anaerobic sulfide-rich zone of marine sediments. These have not yet been isolated in pure culture. Based on 16S ribosomal RNA sequences, two candidate genera have been proposed. These are the mostly marine '*Candidatus* Electrothrix', with four candidate species, and the mostly freshwater '*Candidatus* Electronema', with two candidate species. They have the genes for dissimilatory sulfite reductase as in the phylogenetically close sulfate-reducing bacterial genus, *Desulfobulbus*, but their function is not yet known.

A thermophilic, anaerobic, chemolithoautotrophic bacterium, *Thermosulfurimonas dismutans*, isolated from a deep-sea hydrothermal vent disproportionates elemental sulfur, thiosulfate and sulfite under anaerobic conditions. Sulfate was not used as an electron acceptor. The dissimilatory perchlorate-reducing bacterium (Section 9.7) *Azospira suillum* oxidizes H_2S to S^0. H_2S is preferentially utilized in this bacterium over physiological electron donors such as lactate or acetate.

Culture-independent analyses have revealed many uncharacterized chemolithotrophic sulfur bacteria. Many double-stranded DNA viruses that putatively infect sulfur-oxidizing bacteria contain genes for reverse dissimilatory sulfite reductase. These viruses are a reservoir of genetic diversity for bacterial sulfur oxidation.

10.3.2 Biochemistry of sulfur compound oxidation

Sulfur compound oxidation mechanisms have not been clearly established, partly because sulfur chemistry is complicated and also because different organisms have different oxidative mechanisms that use different enzymes and coenzymes. Figure 10.4 outlines the sulfur compound oxidation pathways in bacteria and archaea. These oxidize HS^-, S^0, $HS_2O_3^-$ and HSO_3^- through a common pathway, transferring electrons to the electron transport chain to generate a proton motive force.

Inorganic sulfur oxidation enzyme systems are best known in *Paracoccus pantotrophus*, a facultative chemolithotrophic Gram-negative bacterium. The genes for sulfur oxidation (Sox) are encoded in the *sox* gene cluster, consisting of 15 genes. The Sox system, reconstituted from SoxB, SoxCD, SoxXA and SoxYZ, oxidizes HS^-, S^0, $HS_2O_3^-$ and HSO_3^-, reducing cytochrome *c*, but each of the proteins is catalytically inactive *in vitro*. These form the periplasmic Sox system that oxidizes the substrates to sulfate without accumulation of any intermediates. Other gene products include *c*-type cytochromes, regulator proteins and others.

The *sox* gene cluster or its component parts have been identified in various sulfur bacteria, but not in all. The *sox* genes are absent in acidophilic *Acidithiobacillus ferrooxidans* and related strains, and in archaea. These have genes for sulfur oxidation, such as sulfur oxygenase reductase (SOR), sulfide dehydrogenase (*sorAB*), sulfide:quinone reductase (SQR), reverse dissimilatory sulfite reductase (*dsr*), sulfite dehydrogenase, adenosine 5′-phosphosulfate (APS) reductase, ATP sulfurylase, thiosulfate:quinol oxidoreductase (*doxDA*) tetrathionate hydrolase (*tetH*) and others. Heterodisulfide reductase (Hdr, Section 9.4.3) is involved in S^0

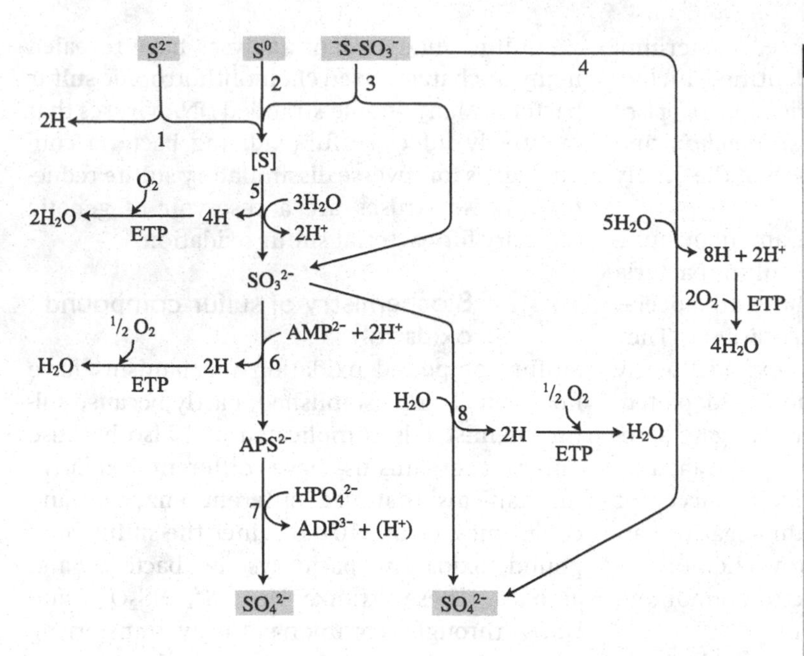

Figure 10.4 Oxidation of sulfur compounds by *Thiobacillus ferrooxidans.*

(Dawes, E. A. 1986, *Microbial Energetics*, Figure 9.5. Blackie & Son, Glasgow)

Sulfide and sulfur are converted to polysulfide before being oxidized further. Thiosulfate is metabolized to sulfite and polysulfide by rhodanese (3), or directly oxidized to sulfate by the thiosulfate-oxidizing multienzyme complex (4). Sulfur compound oxidation is coupled to the reduction of cytochrome *b* or *c*.

1, sulfide-oxidizing enzyme; 2, conversion to polysulfide; 3, rhodanese; 4, thiosulfate-oxidizing multienzyme complex; 5, sulfur-oxidizing enzyme; 6, APS reductase; 7, ADP sulfurylase; 8, sulfite cytochrome *c* reductase.

oxidation by *Acidithiobacillus ferrooxidans*. These enzymes are located in the membrane, periplasm or cytoplasm, depending on the strain. This emphasises the diversity of sulfur oxidation metabolism in bacteria. These enzymes are believed to catalyse the reactions shown in Figure 10.4.

SOR has two catalytic sites, each catalysing sulfur oxidation and disproportionation in the presence of O_2:

Oxidation	$S^0 + O_2 + H_2O \longrightarrow H_2SO_3$
Disproportionation	$3S^0 + 3H_2O \longrightarrow 2H_2S + H_2SO_3$
Sum	$4S^0 + O_2 + 4H_2O \longrightarrow 2H_2S + 2H_2SO_3$

HS^- and S^0 are converted to polysulfide [S], before being oxidized to sulfate. The reduced form of glutathione (GSH) is involved in the oxidation of sulfide to [S]:

$$nS^{2-} + GSH \longrightarrow GS_nSH + 2ne^-$$

Sulfur oxidase catalyses [S] oxidation to sulfite. Sulfite is oxidized to sulfate, either through the direct reaction catalysed by sulfite cytochrome reductase (SCR), or through the reactions catalysed by adenosine-5′-phosphosulfate (APS) reductase and ADP sulfurylase. Direct oxidation appears to be far more widespread than the APS reductase pathway. More energy is conserved in the latter reactions than in the former. The nature of SCR is different among the sulfur oxidizers. The SCR of *Thiobacillus thioparus* is a cytoplasmic soluble enzyme containing molybdenum, and that of *Paracoccus pantotrophicus* is a periplasmic enzyme containing Mo and *c*-type haem. It is not known if Mo is contained in the membrane-bound SCR of *Thiobacillus thiooxidans*. The soluble and periplasmic SCRs oxidize sulfite coupled to the reduction of *c*-type cytochromes, while the membrane-bound enzyme is coupled to the reduction of Fe(III). The electron transport chain is not known in detail, though genes of sulfur-oxidizing enzymes have been characterized in many organisms. The mid-point redox potentials of the sulfur compounds are:

$$S^0/H_2S = -0.25\ V$$
$$SO_3^{2-}/S = +0.5\ V$$
$$SO_4^{2-}/SO_3^{2-} = -0.454\ V$$

The cell yield on sulfide is higher in the anaerobic denitrifier *Thiobacillus denitrificans* than in *Thiobacillus thiooxidans*. This shows that some sulfur compounds are oxidized by oxidases, directly reducing molecular oxygen in aerobic *T. thiooxidans* without energy conservation, while the oxidative reactions are coupled to denitrification with energy conservation in *T. denitrificans*. A sulfur bacterium *Paracoccus denitrificans* (*Thiosphaera pantotropha*) can grow chemolithotrophically, oxidizing carbon disulfide (CS_2) or carbonyl sulfide (COS).

Acidithiobacillus ferrooxidans and *Sulfolobus acidocaldarius* can use Fe(III) as their electron acceptor and S^0 as the electron donor:

$$S^0 + 6Fe^{3+} + 4H_2O \longrightarrow HSO_4^- + 6Fe^{2+} + 7H^+$$
$$(\Delta G^{0\prime} \text{ at pH } 2.0 = -314 \text{ kJ/mol } S^0)$$

10.3.3 Carbon metabolism in colourless sulfur bacteria

All species of the *Thiomicrospira* genus and some species of *Thiobacillus* and *Sulfolobus* are obligate chemolithotrophs. Other species are either facultative chemolithotrophs or mixotrophs capable of using organic compounds. Obligately chemolithotrophic colourless sulfur bacteria fix CO_2 through the Calvin cycle, while the reductive TCA cycle (Section 10.8.2) and the 3-hydroxypropionate/4-hydroxybutyrate cycle (Section 10.8.5.2) are used in the archaea *Sulfolobus acidocaldarius* and *Acidianus brierleyi*, respectively.

10.4 Iron bacteria: ferrous iron oxidation

Many microorganisms can oxidize Fe(II) to Fe(III). These are divided into four main physiological groups: (1) acidophilic aerobes, (2) neutrophilic aerobes, (3) neutrophilic anaerobes (nitrate-dependent) and (4) anaerobic phototrophs (Section 11.1.2). Some of these are known as iron bacteria and use the free energy generated from the oxidation. Many heterotrophic bacteria also oxidize Fe(II), but the function of such ferrous iron oxidation is not known and they do not conserve the free energy.

The acidophiles are the largest Fe(II) oxidizing group, including Gram-negative bacteria, e.g. *Acidithiobacillus ferrooxidans*, *Acidithiobacillus ferrivorans* and species of *Leptospirillum*, Gram-positive bacteria, e.g. species of *Sulfobacillus*, *Alicyclobacillus*, *Acidimicrobium* and *Acidibacillus*, and archaea, e.g. *Sulfolobus acidocaldarius*, *Acidianus brierleyi* and species of *Acidiplasma* and *Ferroplasma*. The Gram-negative bacteria *Ferriphaselus amnicola*, *Mariprofundus ferrooxydans* and *Sideroxydans lithotrophicus* are neutrophiles. Nitrate-dependent and phototrophic Fe(II) oxidizers grow at neutral pH.

Acidithiobacillus ferrooxidans, *Sulfolobus acidocaldarius* and *Acidianus brierleyi* use sulfur compounds as their electron donors, but the others do not use them. Some of the Fe(II) oxidizers are facultative chemolithotrophs and the others obligate chemolithotrophs. Some facultative lithotrophs oxidize Fe(II) mixotrophically.

Since Fe(II) is chemically oxidized easily at neutral pH, iron bacteria growing at neutral pH are microaerophilic and prefer a medium redox potential of around 200–320 mV at a slightly acidic pH of 6.0. They flourish at redox boundaries where opposing gradients of Fe(II) and O_2 develop in natural habitats. Acidophilic Fe(II) oxidizers grow optimally at around pH 2.0 where the chemical oxidation of Fe(II) is slow.

A water-insoluble mineral, pyrite (FeS_2), is the natural electron donor used by acidophilic iron bacteria including *Acidithiobacillus ferrooxidans*. These oxidize the electron donor at the cell surface, reducing the outer membrane cytochrome c_2 which transfers electrons to the terminal oxidase located at the inner face of the cytoplasmic membrane through the periplasmic proteins, rusticyanin and cytochrome c_1 (Figure 10.5). The redox potential of cytochrome c_2 is 560 mV at pH 4.8. The copper-containing blue protein, rusticyanin, is not involved in the electron transport chain that oxidizes sulfur compounds in this bacterium. These proteins form a complex spanning the cytoplasmic membrane and outer membrane through the periplasm. Cytochrome c_4, which transfers electrons to the cytochrome *bc* complex for reverse electron transport (Section 10.1), is associated with the complex. Electron carriers involved in Fe(II) oxidation differ depending on the strain.

The free energy change in Fe(II) oxidation is small since the redox potential of Fe(III)/Fe(II) is +0.78 V, which is very similar to the

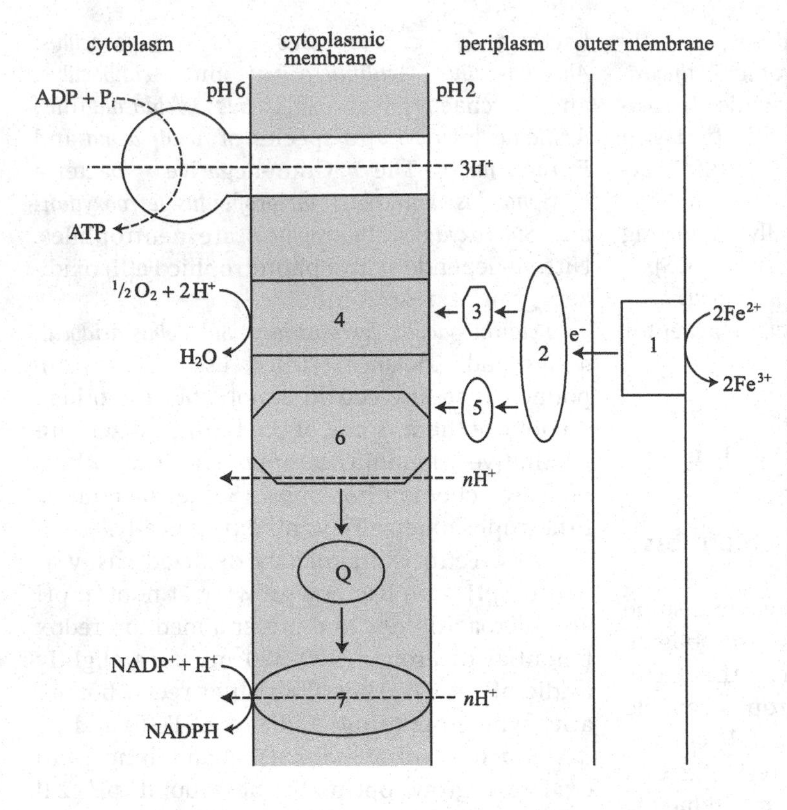

Figure 10.5 ATP synthesis and reverse electron transport coupled to the oxidation of Fe^{2+} in *Acidithiobacillus ferrooxidans*.

(Modified from *J. Biol. Chem.* **283**: 25803–25811, 2008; *Trends Microbiol.* **19**: 330–340, 2011)

Water-insoluble ferrous iron, such as located in pyrite, is oxidized at the cell surface reducing the outer membrane cytochrome c_2. Electrons from reduced cytochrome c_2 are transferred to the cytoplasmic membrane cytochrome *c* oxidase through periplasmic rusticyanin and cytochrome c_1. These form a complex with cytochrome c_4 that transfers electrons to the cytochrome *bc* complex for reverse electron transport to reduce $NADP^+$ for anabolism.

1, cytochrome c_2; 2, rusticyanin; 3, cytochrome c_1; 4, cytochrome *c* oxidase; 5, cytochrome c_4; 6, cytochrome *bc* complex; 7, NADH dehydrogenase complex; Q, coenzyme Q.

+0.86 V of O_2/H_2O (+1.12 V at pH 2.0). The acidophilic Fe(II) oxidizers maintain their internal pH around neutrality with a H^+ gradient of 10^3–10^6 (Section 5.7.3). They maintain a low or inside positive membrane potential to compensate for the large potential generated by the H^+ gradient. Electron transfer from the periplasmic region to cytochrome oxidase contributes to the membrane potential, and proton consumption by the oxidase contributes to the proton gradient part of the proton motive force.

A few neutrophilic Fe(II) oxidizers grow under microaerophilic conditions, including *Gallionella ferruginea*, *Gallionella capsiferriformans*, *Sideroxydans lithotrophicus*, *Sideroxydans paludicola*, *Ferritrophicum radicicola* and *Mariprofundus ferrooxidans*. Less is known about the neutral Fe(II) oxidizers than the acidophiles. Rusticyanin is not known in these bacteria. Electrons available from Fe(II) oxidation at the cell surface are transferred to a cytoplasmic membrane terminal oxidase for energy generation, and to the cytochrome *bc* complex for reverse electron transport via periplasmic cytochrome *c* and other proteins, in a similar way as in the acidophilic Fe(II) reducers. A cell-surface multihaem cytochrome *c* (MtoAB) complex is responsible for Fe(II) oxidation in *Sideroxydans lithotrophicus*. Some of these organisms, including *Sideroxydans lithotrophicus*, can use reduced sulfur compounds such as thiosulfate and can fix N_2.

As the Fe(III) reducers that reduce the electron acceptor at the cell surface are electrochemically active (Section 9.2.1), a strain of the neutral Fe(II) oxidizer *Mariprofundus ferrooxydans* can use electrons provided by an electrode in a fuel cell as the energy source replacing Fe(II).

The hyperthermophilic archaeon *Sulfolobus acidocaldarius* grows chemolithotrophically using sulfide and Fe(II) as its electron donors, like *Acidithiobacillus ferrooxidans*. The former is a facultative chemolithotroph while the latter is an obligate chemolithotroph. Species of *Leptospirillum*, *Gallionella ferruginea* and *Acidithiobacillus ferrooxidans* fix CO_2 through the Calvin cycle (Section 10.8.1), while the 4-hydroxybutyrate cycles (Section 10.8.5) are employed by archaea such as *Sulfolobus acidocaldarius*, *Acidianus brierleyi* and *Metallosphaera yellowstonensis*.

One of the problems aerobic Fe(II) oxidizing microbes must overcome is the presence of highly reactive oxygen species (ROS, Section 12.2.5) that are produced via Fenton chemistry in the presence of oxygen and iron. Catalase, peroxidase and superoxide dismutase are the most common defence mechanisms against ROS and, surprisingly, a minimal set of catalase, peroxidase and superoxide dismutase genes are present in the neutrophilic Fe(II) oxidizers *Gallionella capsiferriformans* and *Sideroxydans lithotrophicus*. These have multiple genes for haemerythrins and globins. Their functions include oxygen sensing and oxygen binding. They are therefore involved in oxygen storage and detoxification, and even the binding of iron. These haemoglobins function as an intracellular oxygen buffer that reduces ROS production by binding oxygen and Fe(II).

Another problem associated with Fe(II) oxidizers is the uncontrolled precipitation of Fe(III) around the cell blocking cellular access to substrates. To overcome this, they produce extracellular polymers which control the precipitation of the iron oxides and prevent the cells from becoming encrusted.

Thiobacillus denitrificans and an archaeon, *Ferroglobus placidus*, oxidize Fe(II) under anaerobic conditions using nitrate as the electron acceptor. This is unusual since the redox potential of the electron donor ($E^{0'}$, Fe(III)/Fe(II) = +0.78 V) is higher than that of the electron acceptor ($E^{0'}$, NO_3^-/NO_2^- = +0.42 V). The $E^{0'}$ of Fe(III)/Fe(II) becomes less positive in the presence of chelating agents and in various salt forms with certain anions. For example, the $E^{0'}$ of Fe_3O_4 (magnetite)/Fe(II)aq is −0.314 V and becomes +0.385 V when Fe(II) and Fe(III) are chelated to citrate. A strain of the family *Gallionellaceae* oxidizes Fe(II) in collaboration with a denitrifying bacterium in a process known as nitrate-dependent Fe(II) oxidation. A few photosynthetic bacteria use Fe(II) as the electron donor for photolithotrophic or photoorganotrophic growth (Section 11.1.2). Several dissimilatory perchlorate-reducing bacteria, including *Dechlorosoma suillum*, use Fe(II) as their electron donor (Section 9.7).

In addition to Fe(II), As(III), Sb(III) and Mn(II) can be used as electron donors by various prokaryotes. These are discussed later (Section 10.7).

10.5 Hydrogen oxidation

10.5.1 Hydrogen-oxidizing bacteria

Various bacteria grow chemolithotrophically on a $H_2 + CO_2$ mixture. With a few exceptions (e.g. *Hydrogenobacter thermophilus* and *Hydrogenovibrio marinus*), these organisms are facultative chemolithotrophs (Table 10.4). They are phylogenetically diverse, and are grouped not on their chemolithotrophy but on their heterotrophic characteristics. Hydrogen is used as an alternative energy source by *Helicobacter pylori*. This provides the pathogen with metabolic flexibility, which can aid persistence (Section 13.3.4). The nitrite oxidizer *Nitrospira moscoviensis* (Section 10. 2. 2) can also use H_2 as the sole energy source.

Most of the organisms listed in Table 10.4 have a hydrogenase with a low affinity for the substrate. For example, the K_m of the enzyme in *Paracoccus denitrificans* is 1.1 μM. On the other hand, many soil samples actively consume hydrogen at nM concentrations. This is due to soil bacteria, such as streptomycetes and mycobacteria, that possess high affinity hydrogenases. Soil isolates such as *Streptomyces avermitilis*, *Mycobacterium smegmatis* and *Rhodococcus equi* can consume H_2 at concentrations as low as 50 pM with a K_s value of 50 nM. These have a special group of enzymes, the Group 5 [NiFe]-hydrogenases, that have a high affinity for the substrate (Section 10.5.2). These enzymes are unusually O_2 insensitive while others are O_2 sensitive or tolerant. Dormant cells of these bacteria have higher H_2 consuming activities than their vegetative cells, suggesting that these enzymes supply energy to the dormant cells for survival, consuming atmospheric H_2 at low concentration. They are not chemolithotrophs and consume H_2 mixotrophically.

Table 10.4 | Hydrogen-oxidizing bacteria and their hydrogenase enzymes.

Organism	Hydrogenase	
	Soluble (NAD^+ dependent)	Particulate (cytochrome dependent)
Facultative chemolithotroph		
Gram-negative		
Alcaligenes denitrificans	+	–
Ralstonia eutropha (Alcaligenes eutrophus)	+	+
Alcaligenes latus	–	+
Alcaligenes ruhlandii	+	–
Aquaspirillum autotrophicum	–	+
Azospirillum lipoferum	–	+
Derxia gummosa	–	+
Flavobacterium autothermophilum	?	+
Helicobacter pylori	–	+
Microcyclus aquaticus	–	+
Paracoccus denitrificans	–	+
Pseudomonas facilis	–	+
Pseudomonas hydrogenovara	–	+
Cupriavidus necator (Ralstonia eutropha)	+	+
Renobacter vacuolatum	–	+
Rhizobium japonicum	–	+
Xanthobacter flavus	–	+
Gram-positive		
Arthrobacter sp.	–	+
Bacillus schlegelii	–	+
Mycobacterium gordonae	?	+
Nocardia autotrophica	+	–
Rhodococcus opacus	+	–
Obligate chemolithotroph		
Hydrogenobacter thermophilus	–	+
Hydrogenovibrio marinus	–	+[a]

[a] The particulate hydrogenase in this bacterium is NAD^+ dependent and catalyses the reverse reaction.

10.5.2 Hydrogenase

The bacteria listed in Table 10.4 use hydrogen as their electron donor to grow chemolithotrophically with the aid of hydrogenase. Most of these have a cytochrome-dependent particulate hydrogenase on their cytoplasmic membrane and *Cupriavidus necator* (*Ralstonia eutropha*) and *Nocardia autotrophica* possess a NAD^+-dependent soluble hydrogenase in addition to the particulate enzyme (Figure 10.6). Only the soluble enzyme is found in the third group, which includes *Alcaligenes denitrificans*, *Alcaligenes ruhlandii* and *Rhodococcus opacus*. The soluble hydrogenase reduces NAD^+, and the particulate enzyme transfers electrons from H_2 to

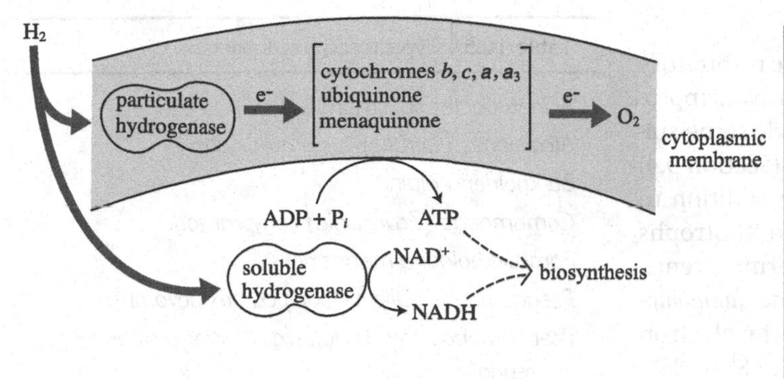

Figure 10.6 Hydrogen utilization by *Cupriavidus necator*.

(Gottschalk, G. 1986, *Bacterial Metabolism*, 2nd edn., Figure 9.1. Springer, New York)

This bacterium has a NAD^+-dependent soluble hydrogenase in addition to the particulate enzyme. The soluble enzyme reduces pyridine nucleotide for biosynthesis, and the particulate hydrogenase channels electrons from hydrogen directly to the electron transport chain to generate a proton motive force.

coenzyme Q of the electron transport chain. The soluble hydrogenase gene in *Cupriavidus necator* (*Ralstonia eutropha*) is plasmid encoded. Since the affinity of the soluble enzyme for the substrate is low, this enzyme cannot oxidize the substrate at low concentrations.

Hydrogenases are active in anaerobic bacteria and archaea in addition to the aerobic hydrogen-oxidizing bacteria. Hydrogenases can be grouped into three classes according to the active centre metal: [NiFe]-, [FeFe]- and [Fe]-hydrogenase. [Fe]-hydrogenase produces H_2 coupled to the oxidation of methylene-tetrahydromethanopterin (methylene-H_4MPT) to methenyl-H_4MPT in some methanogens (Section 9.4), and is known as H_2-forming methylene-H_4MPT dehydrogenase (Hmd). [FeFe]-hydrogenases are bidirectional in many anaerobic prokaryotes, including clostridia (Section 8.5) and sulfidogens (Section 9.3). [NiFe]-hydrogenases are the most numerous hydrogenases known in bacteria. This enzyme consists of an α-subunit with a bimetallic active centre and a β-subunit with [Fe–S] clusters. The [NiFeSe]-hydrogenase known in species of *Desulfovibrio* is a member of the [NiFe]-hydrogenases. [NiFe]-hydrogenases are further divided into five groups according to their gene sequence and function. The membrane-bound uptake hydrogenases of aerobic bacteria are Group 1, and Group 2 includes the cyanobacterial uptake hydrogenase. Bidirectional hydrogenases coupled to oxidation/reduction of NAD(P)H or ferredoxin are in Group 3, and some sulfidogens (Section 9.3) and methanogens (Section 9.4) have an energy-converting hydrogenase that belongs to Group 4. The high affinity O_2-insensitive hydrogenases in the soil bacteria mentioned previously are Group 5 (Section 10.5.1).

The particulate Group 1 [NiFe]-hydrogenase has a high affinity for the substrate, enabling the bacterium to use hydrogen at low concentrations. Organisms with only the particulate hydrogenase employ reverse electron transport to reduce $NAD(P)^+$. In most cases, the hydrogenase in hydrogen-oxidizing bacteria cannot produce hydrogen. These enzymes are referred to as uptake hydrogenases, to differentiate them from those of anaerobic bacteria. The anaerobic bacterial [FeFe]-hydrogenases are called evolution (or production) hydrogenases. The function of the evolution hydrogenase is to dispose of electrons generated from fermentative metabolism by reducing protons:

$$2H^+ + 2e^-(\text{reduced ferredoxin}) \xrightarrow{\text{evolution hydrogenase}} H_2$$

$$H_2 + \text{electron carrier (oxidized)} \xrightarrow{\text{uptake hydrogenase}} \text{electron carrier (reduced)}$$

The particulate hydrogenase of the thermophilic chemolithotroph *Hydrogenovibrio marinus* is NAD^+ dependent and catalyses the reverse reaction.

10.5.3 CO_2 fixation in H_2-oxidizers

The reductive TCA cycle (Section 10.8.2) is used to fix CO_2 in the obligate chemolithotroph *Hydrogenobacter thermophilus*, while all the other hydrogen bacteria tested to date fix CO_2 through the Calvin cycle (Section 10.8.1).

10.5.4 Anaerobic H_2-oxidizers

It has been stated that some anaerobic respiratory prokaryotes can grow on $H_2 + CO_2$ with an appropriate electron acceptor. These include some sulfidogens (Section 9.3), methanogens (Section 9.4) and homoacetogens (Section 9.5). In addition to these, H_2-oxidizing anaerobic chemolithotrophs have been isolated from hydrothermal vents. *Desulfurobacterium crinifex*, *Thermovibrio ammonificans* and *Thermovibrio ruber* use H_2 as the electron donor, reducing nitrate to ammonia, or S^0 to HS^-. The archaeon *Ignicoccus hospitalis* reduces S^0 to HS^- with H_2 as the electron donor, fixing CO_2 through the dicarboxylate/4-hydroxybutyrate pathway (Section 10.8.5.1). These are strict anaerobes. Species of *Caminibacter* and *Hydrogenomonas thermophila* are microaerophilic H_2-oxidizers using O_2, nitrate or elemental sulfur as electron acceptors (Sections 9.1.4 and 9.3). Another strict anaerobe, *Balnearium lithotrophicum*, oxidizes H_2, reducing nitrate to ammonia, but cannot reduce sulfur. A *Dechloromonas* sp. isolated from a sewage works grows chemolithotrophically on H_2 and perchlorate. *Ferroglobus placidus*, a strictly anaerobic archaeon, grows chemoautotrophically, oxidizing hydrogen coupled to nitrate reduction.

10.6 Carbon monoxide oxidation: carboxydobacteria

Annually, about 1×10^9 tons of carbon monoxide (CO) are produced industrially, and biologically as a by-product in photosynthesis, while a similar amount is oxidized so that the concentration remains at around 0.1 ppm in the atmosphere. The concentration can, however, reach up to 100 ppm in industrial areas. CO is oxidized photochemically in the atmosphere, while bacterial oxidation is predominant in soil. Aerobic CO oxidizers are referred to as carboxydobacteria (Table 10.5). They are facultative chemolithotrophs like the hydrogen-oxidizing bacteria.

Table 10.5 Typical carboxydobacteria.

Acinetobacter sp. strain IC-1
Alcaligenes (Carbophilus) carboxidus
Burkholderia alpina
Comamonas (Zavarzinia) compransoris
Paraburkholderia metrosider
Pseudomonas (Oligotropha) carboxidovorans
Pseudomonas carboxidoflava (Hydrogenovibrio pseudoflava)
Pseudomonas carboxidohydrogena
Pseudomonas gazotropha
Terrabacter carboxydivorans

Most of the carboxydobacteria known to date are Gram-negative bacteria and, except for *Alcaligenes carboxidus*, can use H_2, but not all H_2-oxidizing bacteria can use CO. H_2-oxidizing carboxydobacteria oxidize CO and H_2 simultaneously. *Burkholderia alpina* and *Paraburkholderia metrosider* have been isolated from vocanoic soils. Unexpectedly, many species of *Mycobacterium*, including *Mycobacterium tuberculosis*, can grow chemolithotrophically using CO as their sole carbon and energy source. *Pseudomonas carboxydoflava* uses nitrate as electron acceptor under anaerobic conditions, using carbon monoxide as electron donor (Section 9.1.4).

Carboxydobacteria possess carbon monoxide dehydrogenase. This aerobic enzyme is different from the anaerobic CO dehydrogenases of methanogens and homoacetogens. The latter are dual-function enzymes catalysing CO oxidation and acetyl-CoA synthesis in both directions (Sections 9.4.3 and 9.5.2). They are soluble enzymes and use low redox potential electron carriers such as F_{420} and ferredoxin in methanogens and homoacetogens, respectively. The membrane-bound aerobic enzyme in the carboxydobacteria catalyses CO oxidation only, reducing coenzyme Q:

$$CO + H_2O \xrightarrow{\text{CO dehydrogenase}} CO_2 + 2H^+ + 2e^-$$

As stated above, carboxydobacteria can use CO efficiently at low concentrations, employing CO dehydrogenase to reduce coenzyme Q with a redox potential of around zero V, much higher than −0.54 V, the redox potential of CO_2/CO. Carboxydobacteria reduce $NAD(P)^+$ through a reverse electron transport chain, not directly coupled to CO oxidation. The Calvin cycle is the CO_2-fixing mechanism in carboxydobacteria.

Many methanogens and homoacetogens can use CO as their electron donor (Sections 9.4 and 9.5).

10.7 Chemolithotrophs using other electron donors

In addition to the electron donors discussed previously, other inorganic compounds can serve as an energy source in chemolithotrophic metabolism. *Pseudomonas arsenitoxidans* was isolated from a gold mine based on its ability to use arsenite as its electron donor:

$$2H_3AsO_3 + O_2 \longrightarrow HAsO_4^{2-} + H_2AsO_4^{-} + 3H^+$$
$$(\Delta G^{0'} = -128\ kJ/mol\ H_3AsO_3)$$

Similar isolates include *Agrobacterium albertimagni* and *A. tumefaciens* from freshwater, *Thiomonas* sp. from acid mine drainage, *Hydrogenobaculum* sp. from geothermal springs, *Pseudomonas arsenicoxydans* from a desert and several strains of α-proteobacteria isolated from a gold mine. *Ancylobacter dichloromethanicus* is a facultative chemolithotroph reducing arsenite to fix CO_2 through the Calvin cycle. A thermophilic facultative chemolithotroph, *Anoxybacillus flavithermus*, and the alkaliphile *Alkalilimnicola ehrlichii* use arsenite as the electron donor with CO_2 as the carbon source. A *Thermus* strain isolated from an arsenic-rich terrestrial geothermal environment rapidly oxidized inorganic As(III) to As(V) under aerobic conditions, but energy was not conserved. The same strain could use As(VI) as an electron acceptor in the absence of oxygen. The arsenite oxidase large subunit gene, *aoxB*, is found in many bacterial genera including *Achromobacter*, *Acidovorax*, *Acinetobacter*, *Agromyces*, *Albidiferax*, *Bacillus*, *Bosea*, *Flavobacterium*, *Hydrogenophaga*, *Polymorphum*, *Pseudomonas* and *Rhodococcus*.

Various bacteria oxidize antimonite, Sb(III), to antimonite, Sb(V), aerobically. Among these, *Variovorax paradoxus* conserves energy from the oxidation reaction to fix CO_2. Another chemolithotrophic bacterium, *Hydrogenophaga taeniospiralis*, oxidizes Sb(III) coupled to nitrate reduction for chemolithotrophic growth. Although antimonite oxidase has been identified, arsenite oxidase can also oxidize antimonite in some bacteria including *Agrobacterium tumefaciens*.

Mn(II) is oxidized by many bacterial species inhabiting a wide variety of environments, including marine and freshwater habitats, soil, sediments, Mn nodules, hydrothermal vents and industrial locations, e.g. water pipes. Phylogenetically, Mn(II)-oxidizing bacteria appear to be quite diverse, falling within the low G + C Gram-positive bacteria, the Actinobacteria or the Proteobacteria. The most well-characterized Mn(II)-oxidizing bacteria are *Aurantimonas manganoxydans*, a *Bacillus* sp., *Leptothrix discophora*, *Pedomicrobium manganicum* and *Pseudomonas putida*. The oxidation of Mn(II) is thermodynamically favourable and therefore bacteria may derive energy from this reaction, although this has never been unequivocally proven. Mn(II) oxidation is catalysed by a member of the multicopper oxidase family that utilizes multiple copper ions as cofactors. Quinones play an integral role in bacterial Mn(II) oxidation. An aerobic anoxygenic phototrophic bacterium (Section 11.1.3), *Erythrobacter* sp., grows better with Mn (II) oxidation and probably conserves some energy from the oxidation. *Thiobacillus denitrificans* conserves energy from uranium(IV) oxidation coupled to nitrate reduction.

Thioalkalivibrio spp. isolated from soda lakes convert thiocyanate (SCN^-) to H_2S and use this as an electron donor in chemolithotrophic metabolism.

10.8 CO_2 fixation pathways in chemolithotrophs

The Calvin cycle is the most common CO_2 fixation pathway in aerobic chemolithotrophs and in photolithotrophs, and some fix CO_2 through the reductive TCA cycle. The anaerobic chemolithotrophs, including methanogens, homoacetogens and sulfidogens, employ the acetyl-CoA pathway to fix CO_2. A fourth CO_2 fixation pathway, the 3-hydroxypropionate cycle, is known in

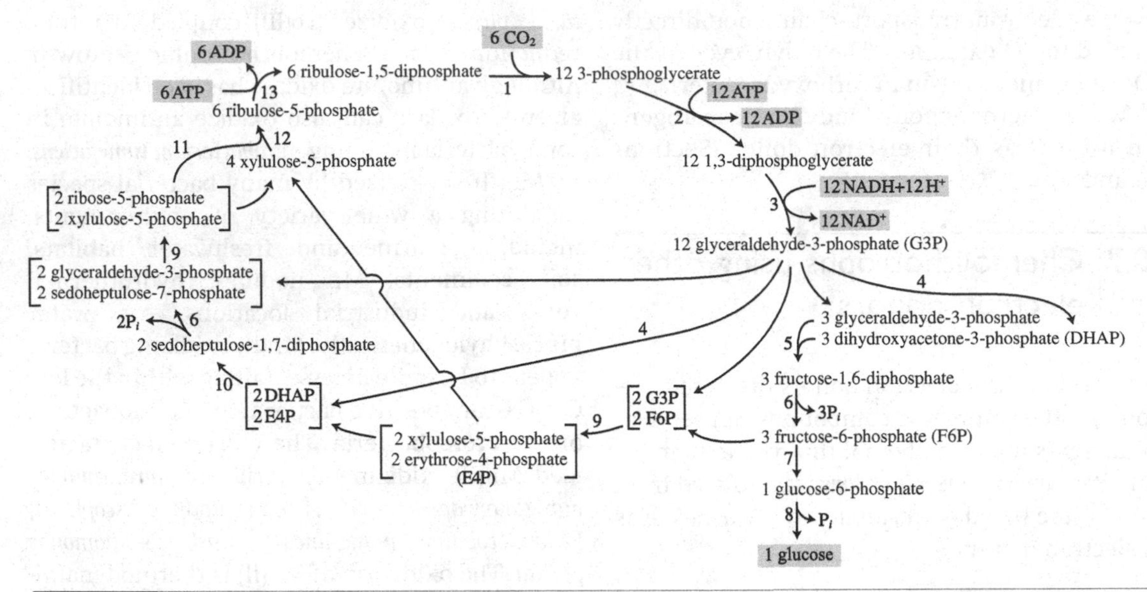

Figure 10.7 CO_2 **fixation through the Calvin cycle.**

1, ribulose-1,5-bisphosphate carboxylase; 2, 3-phosphoglycerate kinase; 3, glyceraldehyde-3-phosphate dehydrogenase; 4, triose phosphate isomerase; 5, fructose-1,6-diphosphate aldolase; 6, fructose-1,6-diphosphatase; 7, glucose-6-phosphate isomerase; 8, glucose-6-phosphatase; 9, transketolase; 10, sedoheptulose-1,7-diphosphate aldolase (the same as number 5); 11, ribose-5-phosphate isomerase; 12, ribose-5-phosphate-3-epimerase; 13, phosphoribulokinase.

some chemo- and photolithotrophs. In thermophilic archaea, 4-hydroxybutyrate cycles are known. The Calvin cycle is the only known CO_2-fixing metabolism in eukaryotes.

10.8.1 Calvin cycle

This is the commonest CO_2 fixation pathway, but is not known in hyperthermophiles that grow at temperatures higher than 75°C, probably because some intermediates of the cycle are unstable at high temperature. All enzymes of the cycle are stable under aerobic conditions. As shown in Figure 10.7, CO_2 is condensed to ribulose-1,5-bisphosphate to produce two molecules of 3-phosphoglycerate, which is reduced to glyceraldehyde-3-phosphate:

$$\begin{array}{c} CH_2\text{-}O\text{-}P \\ | \\ C{=}O \\ | \\ HCOH \\ | \\ HCOH \\ | \\ CH_2\text{-}O\text{-}P \end{array} \xrightarrow[CO_2]{\text{ribulose-1,5-diphosphate carboxylase}} \begin{array}{c} CH_2\text{-}O\text{-}P \\ | \\ HOCH \\ | \\ COOH \\ + \\ COOH \\ | \\ HCOH \\ | \\ CH_2\text{-}O\text{-}P \end{array}$$

ribulose-1,5-diphosphate → 3-phosphoglycerate

$$\begin{array}{c} COOH \\ | \\ HCOH \\ | \\ CH_2\text{-}O\text{-}P \end{array} \xrightarrow[ATP \rightarrow ADP]{\text{3-phosphoglycerate kinase}} \begin{array}{c} COO\text{-}P \\ | \\ HCOH \\ | \\ CH_2\text{-}O\text{-}P \end{array} \xrightarrow[NADH+H^+ \rightarrow NAD^+,\ P_i]{\text{glyceraldehyde-3-phosphate dehydrogenase}} \begin{array}{c} CHO \\ | \\ HCOH \\ | \\ CH_2\text{-}O\text{-}P \end{array}$$

3-phosphoglycerate → 1,3-diphosphoglycerate → glyceraldehyde-3-phosphate

A molecule of glyceraldehyde-3-phosphate is isomerized to dihydroxyacetone phosphate before being condensed to fructose-1,6-diphosphate with the second glyceraldehyde-3-phosphate molecule through the reverse reactions of the EMP pathway. Fructose-1,6-diphosphate is dephosphorylated to fructose-6-phosphate by the action of fructose-1,6-diphosphatase:

dihydroxyacetone phosphate + glyceraldehyde-3-phosphate
↓
fructose-1,6-diphosphate
↓ (H_2O → P_i)
fructose-6-phosphate

Through carbon rearrangement reactions similar to those of the HMP pathway, two molecules of fructose-6-phosphate (2 × C6) and six molecules of glyceraldehyde-3-phosphate (6 × C3) are converted to six molecules of ribulose-5-phosphate (6 × C5), before being phosphorylated to ribulose-1,5-bisphosphate to begin the next round of reactions. The net result of these complex reactions is the synthesis of fructose-6-phosphate from $6CO_2$, consuming 18 ATP and 12 NAD(P)H:

$$6CO_2 + 18ATP + 12NAD(P)H + 12H^+ \longrightarrow \text{fructose-6-phosphate} + 18ADP + 18P_i + 12NAD(P)^+$$

ATP is consumed in the reactions catalysed by 3-phosphoglycerate kinase and phosphoribulokinase, and glyceraldehyde-3-phosphate dehydrogenase oxidizes NAD(P)H.

10.8.1.1 Key enzymes of the Calvin cycle

Ribulose-1,5-bisphosphate carboxylase and phosphoribulokinase are key enzymes of the Calvin cycle, and are present only in the organisms fixing CO_2 through this highly energy-demanding pathway, with a few exceptions (see below). Their activities are controlled at the transcriptional level and also after they are expressed. The enzymes are encoded by *cbb* genes, organized in *cbb* operons differing in size and composition depending on the organism. In a facultative chemolithotroph, *Ralstonia eutropha*, the transcription of the operons, which may form regulons, is strictly controlled, being induced during chemolithotrophic growth but repressed to varying extents during heterotrophic growth. CbbR is a transcriptional regulator and the key activator protein of *cbb* operons. The *cbbR* gene is located adjacent to its cognate operon. The activating function of CbbR is modulated by metabolites which signal the nutritional state of the cell to the *cbb* system. Phosphoenolpyruvate is a negative effector of CbbR, whereas NADPH is a coactivator of the protein. In organisms with more than one copy of the *cbb* operon, each operon is controlled by a separate CbbR. In the photolithotrophs *Rhodobacter capsulatus* and *Rhodobacter sphaeroides*, a global two-component signal transduction system, RegBA, serves this function. Different *cbb* control systems have evolved in diverse chemolithotrophs with different metabolic capabilities.

Phosphoribulokinase activity is regulated by similar physiological signals. NADH activates enzyme activity, while AMP and phosphoenolpyruvate (PEP) are inhibitory. The increase in NADH concentration means the cells are ready to grow, activating the Calvin cycle. On the other hand, the biosynthetic pathway cannot be operated under a poor energy state with an increased AMP concentration. When a facultative chemolithotroph is provided with organic carbon, enzyme activity is inhibited by PEP. Similarly, 6-phosphogluconate inhibits ribulose-1,5-bisphosphate carboxylase activity:

ribulose-5-phosphate
↓ phosphoribulokinase
(+) ← NADH (activates)
(−) ← AMP, PEP (inhibits)
ribulose-1,5-diphosphate

Ribulose-1,5-bisphosphate carboxylase is the most abundant single protein on Earth and is synthesized by all organisms fixing CO_2 through the Calvin cycle, including plants. This enzyme is typically categorized into two forms. Type I, the most common form, consists of eight large and eight small subunits in a hexadecameric (L_8S_8) structure. This type is widely distributed in CO_2-fixing organisms, including all higher plants, algae, cyanobacteria and many chemolithotrophic bacteria. The type II enzyme, on the other hand, consists of only large subunits (L*x*), the number of which may be 2, 4 or 8 in different organisms. Type II ribulose-1,5-bisphosphate carboxylase is found in anaerobic purple photosynthetic bacteria and in some chemolithotrophs. Both types are found in some bacteria, especially in sulfur bacteria such as *Halothiobacillus neapolitanus* (formerly *Thiobacillus neapolitanus*), *Thiomonas intermedia* (formerly *Thiobacillus intermedius*) and *Thiobacillus denitrificans*, in photolithotrophs such as *Rhodobacter sphaeroides* and *Rhodobacter capsulatus*, and in the obligately chemolithotrophic hydrogen bacterium *Hydrogenovibrio marinus*. These have genes for both type I and type II enzymes.

In addition to type I and type II enzymes, two novel type III and type IV enzymes have been revealed by the complete genome sequences of some archaea and bacteria in which the Calvin cycle is not yet known. The type III enzyme is found in many thermophilic archaea, including *Thermococcus* (formerly *Pyrococcus*) *kodakaraensis*, *Methanococcus jannaschii* and *Archaeoglobus fulgidus*. The gene for the enzyme from *Thermococcus kodakaraensis* has been cloned and expressed in *Escherichia coli*. The recombinant enzyme shows carboxylase activity with a decameric structure. The existence of the type III enzyme is not limited to thermophilic archaea. Culture-dependent and -independent analyses have revealed that type III enzyme genes are found in archaea as well as bacteria. The complete genomes of the mesophilic heterotrophic methanogens *Methanosarcina acetivorans*, *Methanosarcina mazei* and *Methanosarcina barkeri* were also found to contain genes encoding putative ribulose-1,5-bisphosphate carboxylase proteins, but these were not found in *Methanobacterium thermoautotrophicum* and *Methanococcus maripaludis*. These methanogenic archaea fix CO_2 through the acetyl-CoA pathway. The type III ribulose-1,5-bisphosphate carboxylase in *Thermococcus kodakarensis* is involved in an AMP metabolism with AMP phosphorylase (AMPpase) and ribose-1,5-diphosphate isomerase. AMPpase converts AMP to adenine and ribose-1,5-diphosphate, which is isomerized to ribulose-1,5-diphosphate to be carboxylated to two 3-phosphoglycerate. The AMP concentration controls this metabolism. AMP activates ribose-1,5-diphosphate isomerase activity while low AMP reduces the activity, preventing excess degradation of AMP:

$$\text{AMP} \xrightarrow[\text{AMP phosphorylase}]{P_i \;\to\; \text{base}} \text{ribose-1,5-diphosphate} \xrightarrow{\text{ribose-1,5-diphosphate isomerase}} \text{ribulose-1,5-diphosphate} \xrightarrow[\text{RuBisCO}]{CO_3 + H_2O} 2\ \text{3-phosphoglycerate}$$

Microbial genome sequences have revealed open reading frames with a similar sequence to that of ribulose-1,5-bisphosphate carboxylase in *Bacillus subtilis* and green sulfur photosynthetic bacteria. This protein is referred to as type IV and is involved in the oxidative stress response or in sulfur metabolism. Photosynthetic green sulfur bacteria fix CO_2 through the reductive TCA cycle.

Ribulose-1,5-bisphosphate carboxylase is located in the carboxysome, a proteinaceous microcompartment in many, but not all, CO_2-fixing bacteria. This microcompartment contains up to 2000 molecules of RuBisCO. The shell acts as a semipermeable barrier, allowing the passive import of the negatively charged reactants, HCO_3^- and ribulose-1,5-bisphosphate, and excluding the competing substrate O_2 (Section 10.8.1.2). In the carboxysome, carbonic anhydrase converts HCO_3^- to CO_2. The cyanobacterium *Synechococcus elongatus* contains around four carboxysomes per cell that segregate with the cytoskeletal protein ParA during cell division. The rate of CO_2 fixation is much higher than expected from the HCO_3^- concentration in the cytoplasm and the low affinity of the enzyme for CO_2. The carboxysomes are believed to have a function in concentrating HCO_3^- to achieve a higher rate of CO_2 fixation.

10.8.1.2 Photorespiration

In addition to carboxylase activity, ribulose-1,5-bisphosphate carboxylase has oxygenase activity under CO_2-limited conditions and a high O_2 concentration. Under these conditions the enzyme oxidizes ribulose-1,5-bisphosphate to 3-phosphoglycerate and phosphoglycolate. For this reason this enzyme is referred to as ribulose-1,5-bisphosphate carboxylase/oxygenase (RuBisCO):

CH_2-O-P
C=O
HCOH ribulose-1,5-diphosphate
HCOH
CH_2-OH-P

+ CO_2 → 2 × [COOH / HCOH / CH_2-O-P] 3-phosphoglycerate

+ O_2 → [CH_2-O-P / COOH] phosphoglycollate + [COOH / HCOH / CH_2-O-P] 3-phosphoglycerate

The enzyme efficiency is usually measured by the specificity factor (τ), which is the ratio of the rate constants for both carboxylase and oxygenase activities. The higher the RuBisCO τ value, the better the RuBisCO can discern CO_2 from O_2. The τ value is generally over 80 for type I RuBisCO in higher plants, between 25 and 75 for type I RuBisCO in bacteria and under 20 for type II RuBisCO. The expression of the type I enzyme gene is activated at low CO_2 concentrations, and a high CO_2 concentration results in increased type II enzyme activity in organisms with both enzyme types, including *Halothiobacillus neapolitanus*, *Hydrogenovibrio marinus* and *Rhodobacter sphaeroides*. It is hypothesized that photosynthetic bacteria cannot grow under aerobic conditions due to the low τ value of their RuBisCO (Section 11.1.2).

10.8.2 Reductive TCA cycle

The green sulfur bacterium *Chlorobium limicola* (*Chlorobi*) can grow photolithotrophically, but does not have enzymes of the Calvin cycle. This bacterium and other chemolithotrophic bacteria

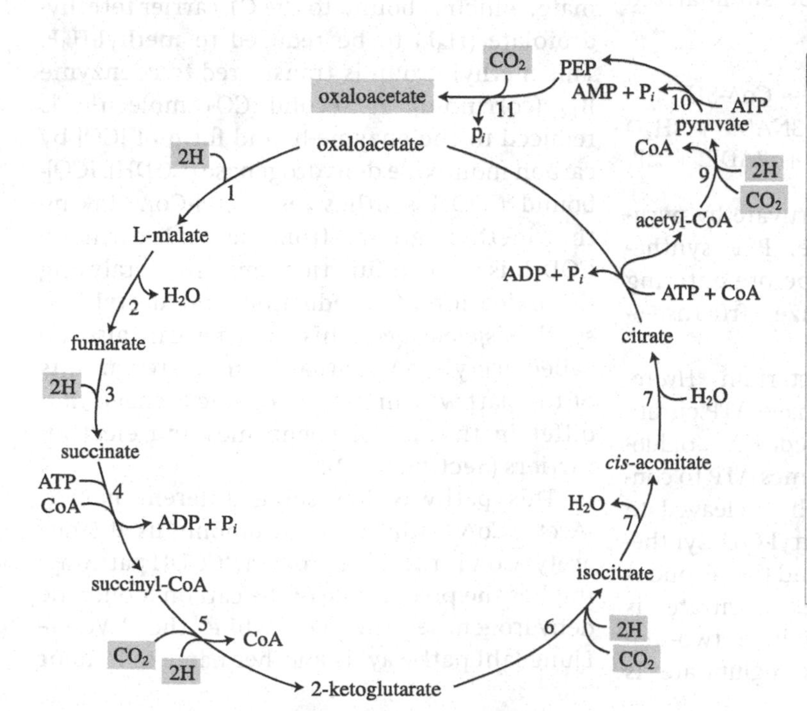

Figure 10.8 CO_2 fixation through the reductive TCA cycle.

Some bacteria and archaea reduce CO_2 to acetyl-CoA in a reverse direction of the TCA cycle. The TCA cycle enzymes that are unable to catalyse the reverse reaction, i.e. citrate synthase, 2-ketoglutarate dehydrogenase and succinate dehydrogenase, are replaced by ATP: citrate lyase (8), 2-ketoglutarate synthase (5) and fumarate reductase (3).

1, malate dehydrogenase; 2, fumarase; 3, fumarate reductase; 4, succinyl-CoA synthetase; 5, 2-ketoglutarate synthase; 6, isocitrate dehydrogenase; 7, aconitase; 8, ATP:citrate lyase; 9, pyruvate synthase; 10, PEP synthetase; 11, PEP carboxylase.

fix CO_2 through the reductive TCA cycle (Figure 10.8), but this metabolism is not widely found in archaea. This cyclic pathway of CO_2 fixation occurs in some archaea including species of *Thermoproteus* and *Sulfolobus*, and in photosynthetic green sulfur bacteria. This CO_2-fixing metabolism shares TCA cycle enzymes that can catalyse the reverse reactions. ATP:citrate lyase, fumarate reductase and 2-ketoglutarate:ferredoxin oxidoreductase replace the TCA cycle enzymes that are irreversible, i.e. citrate synthase, succinate dehydrogenase and 2-ketoglutarate dehydrogenase. The reductive TCA cycle can be summarized as:

$$2CO_2 + 3NADH + 3H^+ + Fd{\cdot}H_2 + CoA\text{-}SH + 2ATP \longrightarrow CH_3CO\text{-}CoA + 3NAD^+ + 3H_2O + Fd + 2ADP + 2P_i$$

Acetyl-CoA is reduced to pyruvate by pyruvate:ferredoxin oxidoreductase. PEP synthetase converts pyruvate to PEP before entering gluconeogenesis, to synthesize fructose-6-phosphate.

The hydrogenotrophic bacterium *Hydrogenobacter thermophilus* does not have ATP:citrate lyase or 2-ketoglutarate:ferredoxin oxidoreductase. This bacterium consumes ATP to convert citrate to citryl-CoA, which is cleaved to oxaloacetate and acetyl-CoA. Citryl-CoA synthetase catalyses the first reaction and the second is catalysed by citryl-CoA lyase. Isocitrate is produced from 2-ketoglutarate in a two-step reaction. Carboxylation of 2-ketoglutarate is catalysed by 2-ketoglutarate carboxylase, consuming ATP. The resulting oxalosuccinate is reduced to isocitrate by oxalosuccinate reductase:

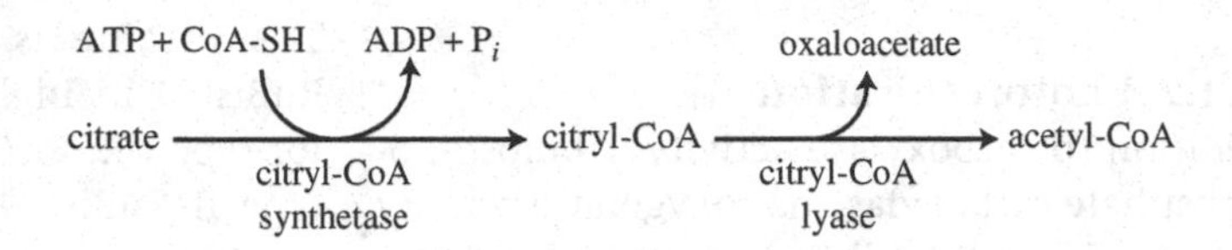

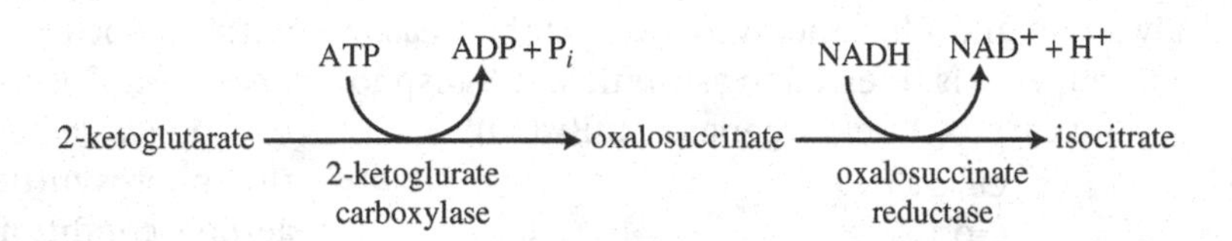

10.8.3 Anaerobic CO_2 fixation through the acetyl-CoA pathway

The acetyl-CoA pathway is employed for CO_2 fixation by anaerobic chemolithotrophs, including sulfidogens, methanogens and homoacetogens.

Figure 10.9 shows the acetyl-CoA pathway that occurs in homoacetogens (Section 9.5). Formate dehydrogenase reduces CO_2 to formate, which is bound to the C1-carrier tetrahydrofolate (H_4F) to be reduced to methyl-H_4F. This methyl group is transferred to coenzyme B_{12} (corrinoid). A second CO_2 molecule is reduced to the enzyme-bound form of [CO] by carbon monoxide dehydrogenase (CODH). [CO]-bound CODH synthesizes acetyl-CoA, taking the methyl group from methyl-corrinoid. CODH is a dual-function enzyme catalysing CO oxidation/CO_2 reduction and acetyl-CoA synthesis/cleavage. This enzyme can also be called acetyl-CoA synthase. There are variants of the pathway in methanogenic archaea that differ in the use of coenzymes and electron carriers (Section 9.4.3).

This pathway has some different names. 'Acetyl-CoA pathway' is commonly used since acetyl-CoA is the final product. 'CODH pathway' implies the pivotal role of the carbon monoxide dehydrogenase enzyme, while the 'Wood-Ljungdahl pathway' is another name to honour

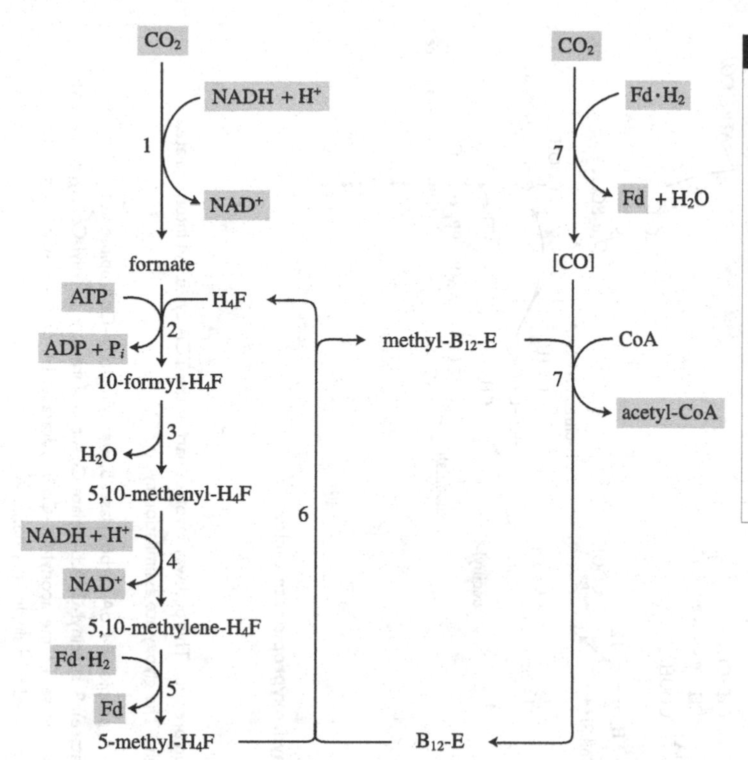

Figure 10.9 The acetyl-CoA pathway – an anaerobic CO_2 fixation mechanism.

The [CO]-bound carbon monoxide (CO) dehydrogenase (7) synthesizes acetyl-CoA with methyl-corrinoid. This pathway is also called the CO dehydrogenase (CODH) pathway, or the Wood–Ljungdahl pathway.

1, formate dehydrogenase; 2, formyl-tetrahydrofolate (H_4F) synthetase; 3, methenyl-H_4F cyclohydrolase; 4, methylene-H_4F dehydrogenase; 5, methylene-H_4F reductase; 6, H_4F:B_{12} methyltransferase; 7, carbon monoxide dehydrogenase. [CO], enzyme-bound carbon monoxide; methyl-B_{12}-E, methylcorrinoid.

those who elucidated the pathway. The pathway can be summarized as:

$$2CO_2 + 2NADH + 2H^+ + 2Fd{\cdot}H_2 + CoA\text{-}SH + ATP \longrightarrow CH_3CO\text{-}CoA + 2NAD^+ + 3H_2O + 2Fd + ADP + P_i$$

In comparison with the reductive TCA cycle, this pathway consumes one ATP less, and 2 NADH and 2 Fd·H_2 are oxidized while the reductive TCA cycle oxidizes 3 NADH and 1 Fd·H_2. Additional energy is conserved in the form of a sodium motive force at the reaction catalysed by methylene-tetrahydrofolate reductase (Section 9.5.3). This comparison shows that the anaerobic process is more efficient than the reductive TCA cycle.

10.8.4 CO_2 fixation through the 3-hydroxypropionate cycle

CO_2 is reduced to glyoxylate in a green gliding bacterium, *Chloroflexus aurantiacus*, through a prokaryote-specific 3-hydroxypropionate cycle (Figure 10.10). Acetyl-CoA is carboxylated to malonyl-CoA by ATP-dependent acetyl-CoA carboxylase before being reduced to 3-hydroxypropionate. A bifunctional enzyme, malonyl-CoA reductase, catalyses this two-step reductive reaction. 3-hydroxypropionate is reduced to propionyl-CoA. A single enzyme, propionyl-CoA synthase, catalyses the three reactions from 3-hydroxypropionate to propionyl-CoA via 3-hydroxypropionyl-CoA and acrylyl-CoA, consuming NADPH and ATP (AMP + PP_i). Propionyl-CoA is carboxylated to methylmalonyl-CoA followed by isomerization of methylmalonyl-CoA to succinyl-CoA. Succinyl-CoA is used for malate activation by CoA transfer, forming succinate and malyl-CoA; succinate in turn is oxidized to malate by the TCA cycle enzymes. Malyl-CoA is cleaved by malyl-CoA lyase with regeneration of the starting acetyl-CoA molecule and production of the first net CO_2 fixation product, glyoxylate.

In the second part of the cycle, pyruvate is synthesized from glyoxylate, fixing another molecule of CO_2. Acetyl-CoA condenses with

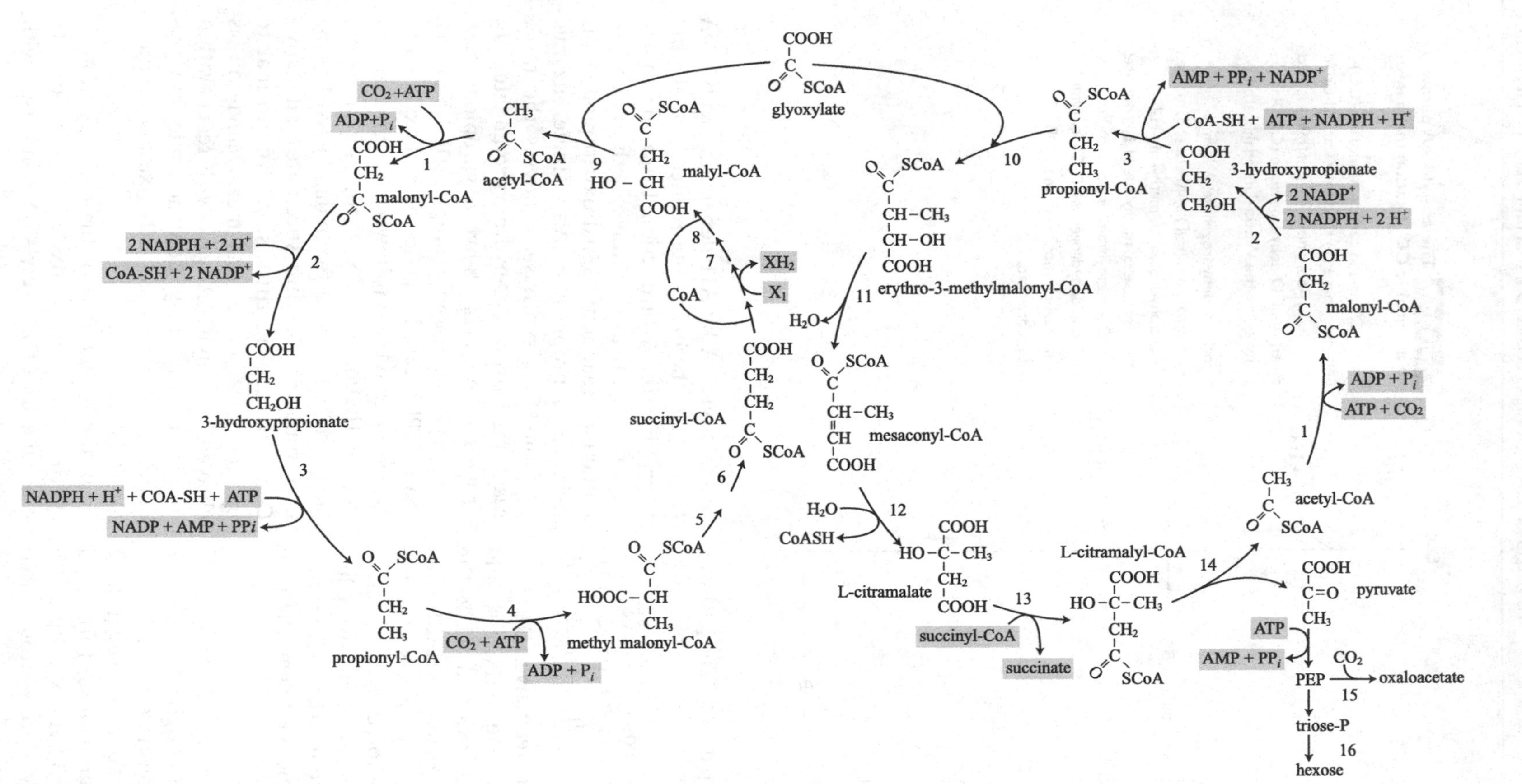

Figure 10.10 CO_2 **fixation in photosynthetic *Chloroflexus aurantiacus* through the 3-hydroxypropionate cycle.**

(Modified from *J. Bacteriol.* **184**: 5999–6006, 2002)

(Left) The 3-hydroxypropionate cycle for CO_2 fixation with glyoxylate as the first net CO_2 fixation product. This pathway involves part of the TCA cycle. An intermediate, 3-hydroxypropionate, was isolated and the cyclic pathway was named after this intermediate. (Right) The glyoxylate assimilation cycle.

1, acetyl-CoA carboxylase; 2, malonyl-CoA reductase (bifunctional); 3, propionyl-CoA synthetase; 4, propionyl-CoA carboxylase; 5, methylmalonyl-CoA epimerase; 6, methylmalonyl-CoA mutase; 7, citrate cycle enzymes (succinate dehydrogenase, fumarate hydratase); 8, succinyl-CoA:L-malate CoA transferase; 9, L-malyl-CoA lyase; 10, *erythro*-3-methylmalyl-CoA lyase; 11, mesaconyl-C1-CoA hydratase; 12, mesaconyl-CoA C2–C4 CoA transferase; 13, mesaconyl-C4-CoA hydratase; 14, citramalyl-CoA lyase; 15, phosphoenolpyruvate carboxylase; 16, gluconeogenesis enzymes. Reactions 9, 10 and 14 are catalysed by a single enzyme.

CO_2 and is converted to propionyl-CoA as in the first part of the cycle. Glyoxylate and propionyl-CoA are condensed to *erythro*-3-methylmalyl-CoA before being cleaved to acetyl-CoA and pyruvate via *erythro*-3-methylmalyl-CoA, mesaconyl-CoA and citramalate. Pyruvate is the product of the second part of the cycle through the reduction of glyoxylate and CO_2, with regeneration of acetyl-CoA, the primary CO_2 acceptor molecule.

There are three unique processes in this CO_2 fixation pathway involving multifunctional enzymes that are not present in other chemolithotrophs. A bifunctional enzyme, malonyl-CoA reductase, catalyses the two-step reduction of malonyl-CoA to 3-hydroxypropionate (reaction 2 in Figure 10.10). 3-hydroxypropionate is further metabolized to propionyl-CoA, catalysed by a trifunctional enzyme, propionyl-CoA synthase (reaction 3 in Figure 10.10). Another bifunctional enzyme, malyl-CoA lyase/*erythro*-3-methylmalyl-CoA lyase, cleaves malyl-CoA to acetyl-CoA and glyoxylate (reaction 9 in Figure 10.10), and condenses glyoxylate and propionyl-CoA to *erythro*-3-methylmalyl-CoA (reaction 10 in Figure 10.10).

The 3-hydroxypropionate cycle can be summarized as:

$$3CO_2 + 6NADPH + 6H^+ + X + 5ATP \longrightarrow \text{pyruvate} + 6NADP^+ + XH_2 + 3ADP + 2AMP + 2PP_i + 3P_i$$

X is an unknown electron carrier reduced by succinate dehydrogenase.

10.8.5 CO_2 fixation through the 4-hydroxybutyrate cycles

Chemolithotrophic members of the phylum *Crenarchaeota* employ yet more CO_2 fixing 4-hydroxybutyrate cycles. These are the 3-hydroxypropionate/4-hydroxybutyrate (HP/HB) cycle in aerobes and the dicarboxylate/4-hydroxybutyrate (DC/HB) cycle in anaerobes. These share some common enzymes and intermediates, but differ in their sensitivity to oxygen. The enzymes of the HP/HB cycle tolerate oxygen, while some of the enzymes and electron carriers of the DC/HB cycle are sensitive to oxygen. Consequently, the HP/HB cycle functions in (micro)aerobic strains of the order *Sulfolobales*, and the DC/HB cycle is present in mostly anaerobic chemolithotrophic representatives of the orders *Thermoproteales* and *Desulfurococcales*. A strictly anaerobic member of the *Sulfolobales*, *Stygiolobus azoricus*, operates the aerobic HP/HB cycle.

10.8.5.1 Dicarboxylate/4-hydroxybutyrate (DC/HB) cycle

In this anaerobic process, CO_2 is fixed by pyruvate:ferredoxin oxidoreductase, and PEP carboxylase incorporates carbonate into oxaloacetate which is reduced to acetoacetyl-CoA via 4-hydroxybutyrate. Acetoacetyl-CoA is cleaved to two acetyl-CoA (Figure 10.11). Acetyl-CoA is used for biosynthesis and as the CO_2 acceptor in the next round of the carbon fixation cycle.

The DC/HB cycle can be summarized as below to show that five high-energy phosphate bonds are invested to synthesize acetyl-CoA:

$$CO_2 + HCO_3^+ + Fd{\cdot}H_2 + NAD(P)H + 2XH_2 + CoA\text{-}SH + 3ATP \longrightarrow \text{acetyl-CoA} + Fd + NAD(P)^+ + 2X + ADP + 2AMP + PP_i + 3P_i$$

10.8.5.2 3-hydroxypropionate/4-hydroxybutyrate (HP/HB) cycle

In this cycle, acetyl-CoA and propionyl-CoA function as the carbonate acceptors. Acetyl-CoA is carboxylated to malonyl-CoA which is reduced to propionyl-CoA that is then carboxylated to methylmalonyl-CoA, fixing the second carbonate, before being reduced to acetoacetyl-CoA via 4-hydroxybutyrate (Figure 10.12). In this cycle, one more high-energy phosphate bond is consumed than in the anaerobic DC/HB cycle, as shown below:

$$2HCO_3^+ + 4NADPH + CoA\text{-}SH + 4ATP \longrightarrow \text{acetyl-CoA} + 4NADP^+ + 2ADP + 2AMP + 2PP_i + 2P_i$$

10.8.6 Energy expenditure in CO_2 fixation

Assuming that the electrons consumed in the different CO_2 fixation processes are in the same

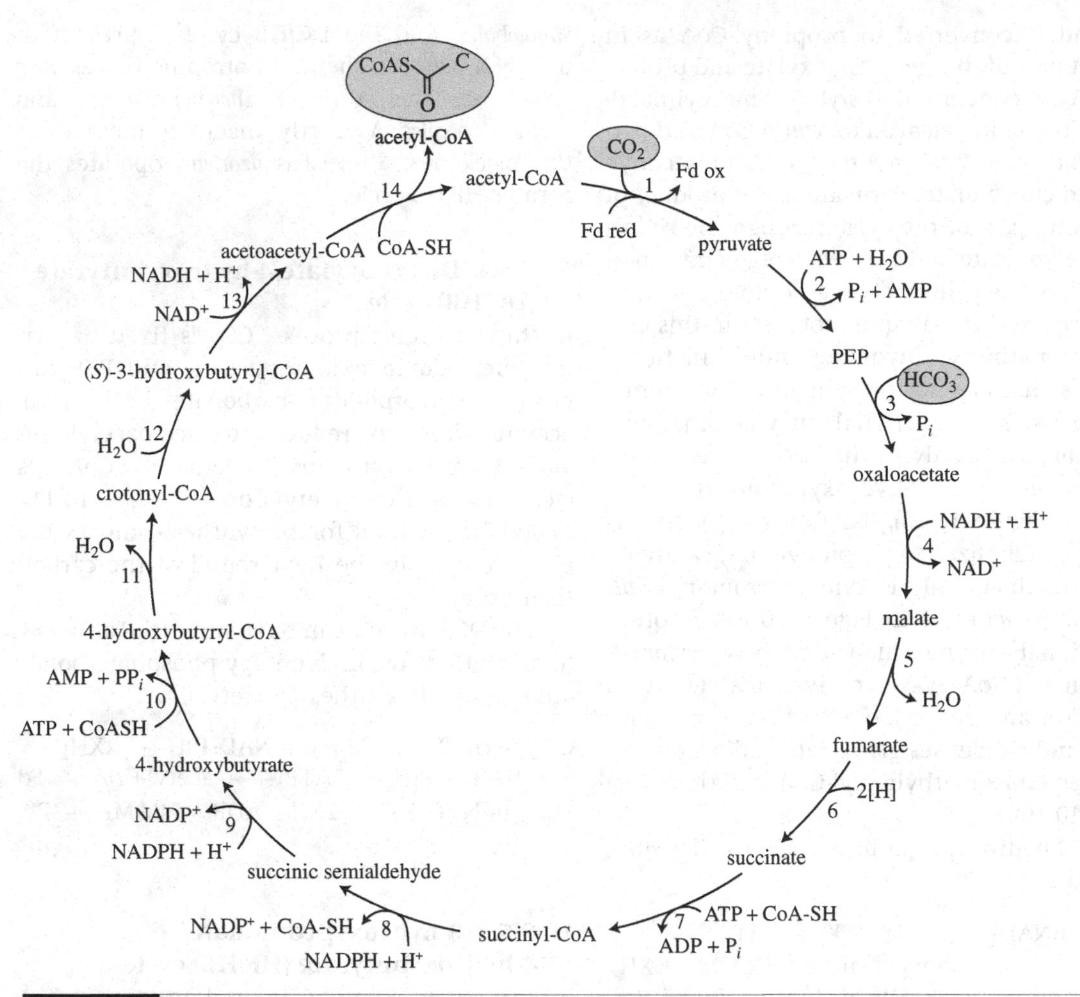

Figure 10.11 **CO_2 fixation through the dicarboxylate/4-hydroxybutyrate (DC/HB) cycle in an anaerobic member of the *Crenarchaeota*, *Ignicoccus hospitalis*.**

(Modified from *Microbiol.* **156**: 256–269, 2010)

1, pyruvate synthase; 2, pyruvate:water dikinase; 3, PEP carboxylase; 4, malate dehydrogenase; 5, fumarate hydratase; 6, fumarate reductase; 7, succinyl-CoA synthetase; 8, succinyl-CoA reductase; 9, succinic semialdehyde reductase; 10, 4-hydroxybutyrate-CoA ligase; 11, 4-hydroxybutyryl-CoA dehydratase; 12, crotonyl-CoA hydratase; 13, 3-hydroxybutyryl-CoA dehydrogenase; 14, acetoacetyl-CoA-β-ketothiolase.

Fd, ferredoxin; PEP, phosphoenolpyruvate; MV, methyl viologen.

energy state, efficiencies can be compared in terms of ATP required for the synthesis of acetyl-CoA after normalization of the different products. Fructose-1,6-phosphate is produced in the Calvin cycle, consuming 18 ATP. When fructose-1,6-phosphate is metabolized through the EMP pathway to 2 pyruvate and then to 2 acetyl-CoA, 4ATP are generated. This gives a net 7 ATP/acetylCoA from the Calvin cycle. The same parameters give 2 ATP and 1 ATP for the reductive TCA cycle and acetyl-CoA pathways, respectively. Pyruvate is produced consuming 5 ATP to

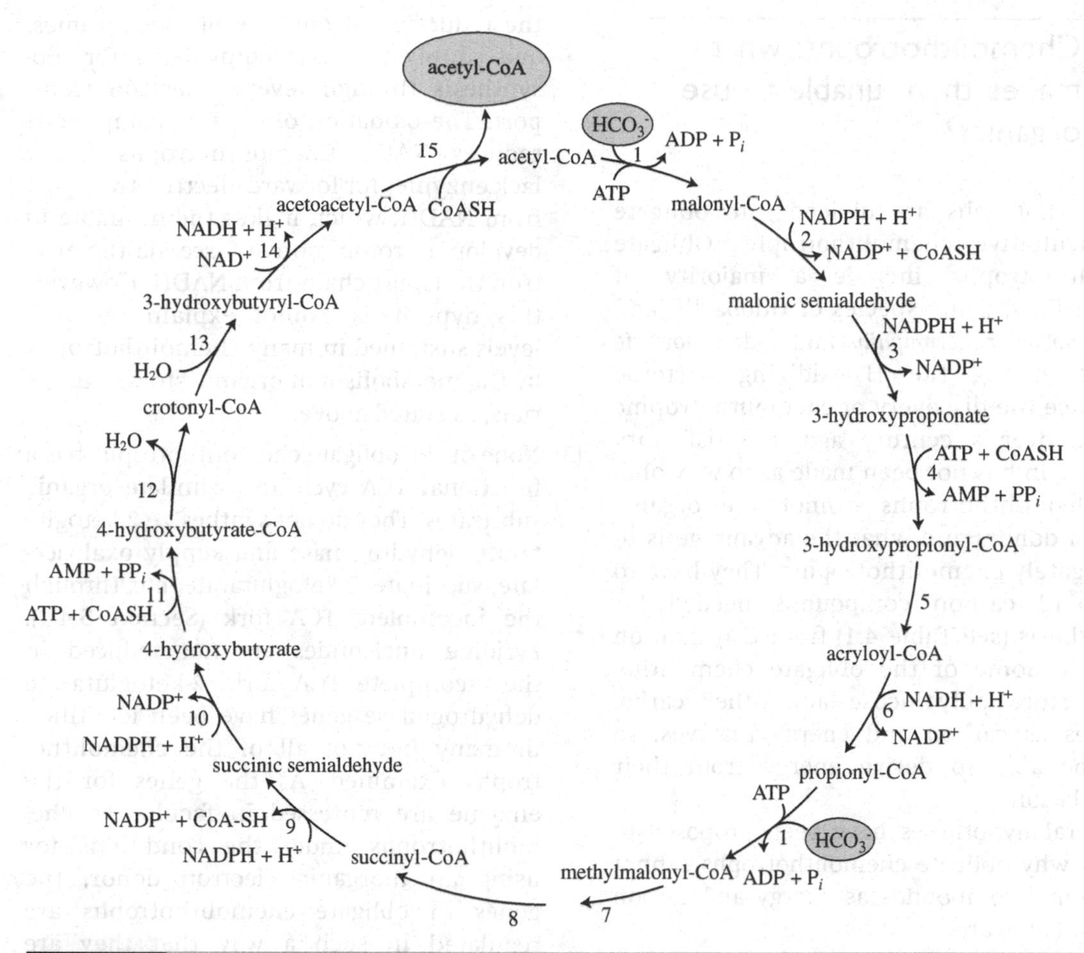

Figure 10.12 **CO_2 fixation through the 3-hydroxypropionate/4-hydroxybutyrate (HP/HB) cycle in aerobic *Crenarchaeota*.**

(Modified from *Microbiol.* **156**: 256–269, 2010)

1, acetyl-CoA/propionyl-CoA carboxylase; 2, malonyl-CoA reductase; 3, malonic semialdehyde reductase; 4, 3-hydroxypropionate-CoA ligase; 5, 3-hydroxypropionyl-CoA dehydratase; 6, acryloyl-CoA reductase; 7, methylmalonyl-CoA epimerase; 8, methylmalonyl-CoA mutase; 9, succinyl-CoA reductase; 10, succinic semialdehyde reductase; 11, 4-hydroxybutyrate-CoA ligase; 12, 4-hydroxybutyryl-CoA dehydratase; 13, crotonyl-CoA hydratase; 14, 3-hydroxybutyryl-CoA dehydrogenase; 15, acetoacetyl-CoA β-ketothiolase.

3 ADP and 2 AMP through the 3-hydroxypropionate cycle. Assuming that PP_i is hydrolysed to $2P_i$ without conserving energy, the ratio of ATP consumed/acetyl-CoA (pyruvate) is 7. In the anaerobic DC/HB cycle, 3 ATP are hydrolysed to ADP and 2 ADP, consuming five high energy phosphate bonds, while the aerobic HP/HB cycle hydrolyses 4 ATP to 2 ADP and 2 AMP, consuming six high energy phosphate bonds. From this comparison, it can be seen that the energy efficiency varies from 1 to 7 ATP/acetyl-CoA in the different CO_2 fixation mechanisms. Generally the anaerobic CO_2 fixation processes consume less energy than the aerobic processes.

10.9 Chemolithotrophs: what makes them unable to use organics?

Chemolithotrophs are divided into obligate and facultative chemolithotrophs. Obligate chemolithotrophs include a majority of the nitrifiers, some species of thiobacilli and *Hydrogenobacter thermophilus* and *Hydrogenovibrio marinus* among the H_2-oxidizing bacteria. Ever since the discovery of chemolithotrophic bacteria over a century ago, a satisfactory explanation has not been made as to why obligate chemolithotrophs cannot use organic electron donors and what the advantage is to be obligately chemolithotrophic. They have to supply 12 carbon compounds needed for biosynthesis (see Table 4.1) from CO_2 fixation products. Some of the obligate chemolithotrophs store polyglucose and other carbohydrates as carbon and energy reserves, so must be able to derive energy from their dissimilation.

Several hypotheses have been proposed to explain why obligate chemolithotrophs cannot use organic compounds as energy and carbon sources. These are:

(1) Obligate chemolithotrophs cannot use organic carbon compounds because the organisms are unable to transport them into cells. As discussed above (Section 10.2.1), a potential fructose permease has been revealed in *Nitrosomonas europaea* through complete sequence determination of its genome, and this organism, as an obligate chemolithotroph, metabolizes fructose mixotrophically. However, amino acids and other monomers can diffuse into cells and are incorporated into biosynthesis in many chemolithotrophs.

(2) Chemolithotrophs are unable to synthesize ATP from NADH oxidation. With a few exceptions, obligate chemolithotrophs couple the oxidation of their electron donors to the reduction of quinone or cytochromes, and supply reducing equivalents for biosynthesis through reverse electron transport. The oxidation of organic compounds reduces NAD^+. Chemolithotrophs might lack enzymes for forward electron transport from NADH, which makes them unable to develop a proton motive force via the electron transport chain from NADH. However, this hypothesis cannot explain the ATP levels sustained in many chemolithotrophs by the metabolism of organic storage materials, as stated above.

(3) None of the obligate chemolithotrophs has a functional TCA cycle to assimilate organic substrates. They do not synthesize 2-ketoglutarate dehydrogenase and supply oxaloacetate, succinate, 2-ketoglutarate, etc. through the incomplete TCA fork (Section 5.4.1). Pyridine nucleotides are not reduced in the incomplete TCA fork. 2-ketoglutarate dehydrogenase genes have been identified in many but not all of the chemolithotrophs examined. As the genes for the enzyme are repressed in facultative chemolithotrophs under the conditions for using an inorganic electron donor, the genes in obligate chemolithotrophs are regulated in such a way that they are repressed permanently.

(4) Obligate chemolithotrophs might not be able to use organic electron donors due to metabolic control. The facultative chemolithotroph *Ralstonia eutropha* (*Alcaligenes eutrophus*) metabolizes fructose through the ED pathway. When H_2 is supplied to the fructose culture, the bacterium stops growing because the enzymes for fructose metabolism are inhibited by H_2. When a mixture of $H_2/CO_2/O_2$ is supplied, growth is resumed.

These hypotheses do not provide completely plausible reasons for the inability of obligate chemolithotrophs to grow on organic electron donors. The mechanisms might be much more complex than expected.

Further Reading

Note this section contains key references only. Additional recommended references are available at www.cambridge.org/ProkaryoticMetabolism.

General

Akerman, N. H., Price, R. E., Pichler, T. & Amend, J. P. (2011). Energy sources for chemolithotrophs in an arsenic- and iron-rich shallow-sea hydrothermal system. *Geobiology* **9**, 436–445.

Claassens, N. J., Sousa, D. Z., dos Santos, V. A. P. M., de Vos, W. M. & van der Oost, J. (2016). Harnessing the power of microbial autotrophy. *Nature Reviews Microbiology* **14**, 692–706.

Gadd, G. M., Semple, K. T. & Lappin-Scott, H. M. (2005). *Micro-organisms and Earth Systems: Advances in Geomicrobiology*. Cambridge: Cambridge University Press.

Maden, B. E. H. (1995). No soup for starters? Autotrophy and the origins of metabolism. *Trends in Biochemical Sciences* **20**, 337–341.

Srinivasan, V., Morowitz, H. & Huber, H. (2012). What is an autotroph?*Archives of Microbiology* **194**, 135–140.

Stevens, T. O. (1997). Lithoautotrophy in the subsurface. *FEMS Microbiology Reviews* **20**, 327–337.

Wood, A. P., Aurikko, J. P. & Kelly, D. P. (2004). A challenge for 21st century molecular biology and biochemistry: what are the causes of obligate autotrophy and methanotrophy? *FEMS Microbiology Reviews* **28**, 335–352.

Reverse electron transport

Elbehti, A., Brasseur, G. & Lemesle-Meunier, D. (2000). First evidence for existence of an uphill electron transfer through the bc_1 and NADH-Q oxidoreductase complexes of the acidophilic obligate chemolithotrophic ferrous ion-oxidizing bacterium *Thiobacillus ferrooxidans*. *Journal of Bacteriology* **182**, 3602–3606.

Jin, Q. & Bethke, C. M. (2003). A new rate law describing microbial respiration. *Applied and Environmental Microbiology* **69**, 2340–2348.

Nitrification

Chen, J., Zheng, J., Li, Y., Hao, H.-h. & Chen, J.-m. (2015). Characteristics of a novel thermophilic heterotrophic bacterium, *Anoxybacillus contaminans* HA, for nitrification–aerobic denitrification. *Applied Microbiology and Biotechnology* **99**, 10695–10702.

Costa, E., Perez, J. & Kreft, J. U. (2006). Why is metabolic labour divided in nitrification? *Trends in Microbiology* **14**, 213–219.

Daims, H., Lebedeva, E. V., Pjevac, P., Han, P., Herbold, C., Albertsen, M., Jehmlich, N., Palatinszky, M., Vierheilig, J., Bulaev, A., Kirkegaard, R. H., von Bergen, M., Rattei, T., Bendinger, B., Nielsen, P. H. & Wagner, M. (2015). Complete nitrification by *Nitrospira* bacteria. *Nature* **528**, 504–509.

Kerou, M., Offre, P., Valledor, L., Abby, S. S., Melcher, M., Nagler, M., Weckwerth, W. & Schleper, C. (2016). Proteomics and comparative genomics of *Nitrososphaera viennensis* reveal the core genome and adaptations of archaeal ammonia oxidizers. *Proceedings of the National Academy of Sciences of the USA* **113**, 7937–7946.

Kim, J.-G., Park, S.-J., Sinninghe Damsté, J. S., Schouten, S., Rijpstra, W. I. C., Jung, M.-Y., Kim, S.-J., Gwak, J.-H., Hong, H., Si, O.-J., Lee, S., Madsen, E. L. & Rhee, S.-K. (2016). Hydrogen peroxide detoxification is a key mechanism for growth of ammonia-oxidizing archaea. *Proceedings of the National Academy of Sciences of the USA* **113**, 7888–7893.

Koch, H., Galushko, A., Albertsen, M., Schintlmeister, A., Gruber-Dorninger, C., Lücker, S., Pelletier, E., Le Paslier, D., Spieck, E., Richter, A., Nielsen, P. H., Wagner, M. & Daims, H. (2014). Growth of nitrite-oxidizing bacteria by aerobic hydrogen oxidation. *Science* **345**, 1052–1054.

Nicol, G. W. & Schleper, C. (2006). Ammonia-oxidising Crenarchaeota: important players in the nitrogen cycle? *Trends in Microbiology* **14**, 207–212.

Prosser, J. I. & Nicol, G. W. (2012). Archaeal and bacterial ammonia-oxidisers in soil: the quest for niche specialisation and differentiation. *Trends in Microbiology* **20**, 523–531.

van Kessel, M. A. H. J., Speth, D. R., Albertsen, M., Nielsen, P. H., Op den Camp, H. J. M., Kartal, B., Jetten, M. S. M. & Lücker, S. (2015). Complete nitrification by a single microorganism. *Nature* **528**, 555–559.

Ye, R. W. & Thomas, S. M. (2001). Microbial nitrogen cycles: physiology, genomics and applications. *Current Opinion in Microbiology* **4**, 307–312.

Colourless sulfur bacteria

Dopson, M. & Johnson, D. B. (2012). Biodiversity, metabolism and applications of acidophilic sulfur-metabolizing microorganisms. *Environmental Microbiology* **14**, 2620–2631.

Friedrich, C. G. (1998). Physiology and genetics of sulfur-oxidizing bacteria. *Advances in Microbial Physiology*, **39**, 235–289.

Ghosh, W. & Dam, B. (2009). Biochemistry and molecular biology of lithotrophic sulfur oxidation by taxonomically and ecologically diverse bacteria and archaea. *FEMS Microbiology Reviews* **33**, 999–1043.

Han, Y. & Perner, M. (2016). Sulfide consumption in *Sulfurimonas denitrificans* and heterologous expression of its three sulfide-quinone reductase homologs. *Journal of Bacteriology* **198**, 1260–1267.

Kelly, D. P. (1999). Thermodynamic aspects of energy conservation by chemolithotrophic sulfur bacteria in relation to the sulfur oxidation pathways. *Archives of Microbiology* **171**, 219–229.

Liu, Y., Beer, L. L. & Whitman, W. B. (2012). Sulfur metabolism in archaea reveals novel processes. *Environmental Microbiology* **14**, 2632–2644.

Pfeffer, C., Larsen, S., Song, J., Dong, M., Besenbacher, F., Meyer, R. L., Kjeldsen, K. U., Schreiber, L., Gorby, Y. A., El-Naggar, M. Y., Leung, K. M., Schramm, A., Risgaard-Petersen, N. & Nielsen, L. P. (2012). Filamentous bacteria transport electrons over centimetre distances. *Nature* **491**, 218–221.

Rother, D., Ringk, J. & Friedrich, C. G. (2008). Sulfur oxidation of *Paracoccus pantotrophus*: the sulfur-binding protein SoxYZ is the target of the periplasmic thiol-disulfide oxidoreductase SoxS. *Microbiology* **154**, 1980–1988.

Salman, V., Bailey, J. & Teske, A. (2013). Phylogenetic and morphologic complexity of giant sulphur bacteria. *Antonie van Leeuwenhoek* **104**, 169–186.

Ferrous iron and other metal oxides

Amouric, A., Brochier-Armanet, C., Johnson, D. B., Bonnefoy, V. & Hallberg, K. B. (2011). Phylogenetic and genetic variation among Fe(II)-oxidizing acidithiobacilli supports the view that these comprise multiple species with different ferrous iron oxidation pathways. *Microbiology* **157**, 111–122.

Bird, L. J., Bonnefoy, V. & Newman, D. K. (2011). Bioenergetic challenges of microbial iron metabolisms. *Trends in Microbiology* **19**, 330–340.

Bonnefoy, V. & Holmes, D. S. (2012). Genomic insights into microbial iron oxidation and iron uptake strategies in extremely acidic environments. *Environmental Microbiology* **14**, 1597–1611.

Castelle, C., Guiral, M., Malarte, G., Ledgham, F., Leroy, G., Brugna, M. & Giudici-Orticoni, M.-T. (2008). A new iron-oxidizing/O_2-reducing supercomplex spanning both inner and outer membranes, isolated from the extreme acidophile *Acidithiobacillus ferrooxidans*. *Journal of Biological Chemistry* **283**, 25803–25811.

Emerson, D., Fleming, E. J. & McBeth, J. M. (2010). Iron-oxidizing bacteria: an environmental and genomic perspective. *Annual Review of Microbiology* **64**, 561–583.

Hedrich, S., Schlömann, M. & Johnson, D. B. (2011). The iron-oxidizing proteobacteria. *Microbiology* **157**, 1551–1564.

Summers, Z. M., Gralnick, J. A. & Bond, D. R. (2013). Cultivation of an obligate Fe(II)-oxidizing lithoautotrophic bacterium using electrodes. *mBio* **4**, 00420–12.

Hydrogen oxidizers and carboxydobacteria

Anantharaman, K., Breier, J. A., Sheik, C. S. & Dick, G. J. (2013). Evidence for hydrogen oxidation and metabolic plasticity in widespread deep-sea sulfur-oxidizing bacteria. *Proceedings of the National Academy of Sciences of the USA* **110**, 330–335.

Fritsch, J., Lenz, O. & Friedrich, B. (2013). Structure, function and biosynthesis of O_2-tolerant hydrogenases. *Nature Reviews Microbiology* **11**, 106–114.

Greening, C., Constant, P., Hards, K., Morales, S. E., Oakeshott, J. G., Russell, R. J., Taylor, M. C., Berney, M., Conrad, R. & Cook, G. M. (2015). Atmospheric hydrogen scavenging: from enzymes to ecosystems. *Applied and Environmental Microbiology* **81**, 1190–1199.

Kim, Y. & Park, S. (2012). Microbiology and genetics of CO utilization in mycobacteria. *Antonie van Leeuwenhoek* **101**, 685–700.

Koch, H., Galushko, A., Albertsen, M., Schintlmeister, A., Gruber-Dorninger, C., Lücker, S., Pelletier, E., Le Paslier, D., Spieck, E., Richter, A., Nielsen, P. H., Wagner, M. & Daims, H. (2014). Growth of nitrite-oxidizing bacteria by aerobic hydrogen oxidation. *Science* **345**, 1052–1054.

Kuhns, L. G., Benoit, S. L., Bayyareddy, K., Johnson, D., Orlando, R., Evans, A. L., Waldrop, G. L. & Maier, R. J. (2016). Carbon fixation driven by molecular hydrogen results in chemolithoautotrophically enhanced growth of *Helicobacter pylori*. *Journal of Bacteriology* **198**, 1423–1428.

Oh, J.-I., Park, S.-J., Shin, S.-J., Ko, I.-J., Han, S. J., Park, S. W., Song, T. & Kim, Y. M. (2010). Identification of *trans*- and *cis*-control elements involved in regulation of the carbon monoxide dehydrogenase genes in *Mycobacterium* sp. strain JC1 DSM 3803. *Journal of Bacteriology* **192**, 3925–3933.

Parkin, A. & Sargent, F. (2012). The hows and whys of aerobic H_2 metabolism. *Current Opinion in Chemical Biology* **16**, 26–34.

Vignais, P. M. & Billoud, B. (2007). Occurrence, classification, and biological function of hydrogenases: an overview. *Chemical Reviews* **107**, 4206–4272.

Other inorganic electron donors

Anderson, C. R., Johnson, H. A., Caputo, N., Davis, R. E., Torpey, J. W. & Tebo, B. M. (2009). Mn(II) oxidation is catalyzed by heme peroxidases in "*Aurantimonas manganoxydans*" strain SI85-9A1 and *Erythrobacter* sp. strain SD-21. *Applied and Environmental Microbiology* **75**, 4130–4138.

Gadd, G. M. (2010). Metals, minerals and microbes: geomicrobiology and bioremediation. *Microbiology* **156**, 609–643.

Heinrich-Salmeron, A., Cordi, A., Brochier-Armanet, C., Halter, D., Pagnout, C., Abbaszadeh-fard, E., Montaut, D., Seby, F., Bertin, P. N., Bauda, P. & Arsene-Ploetze, F. (2011). Unsuspected diversity of arsenite-oxidizing bacteria as revealed by widespread distribution of the *aoxB* gene in prokaryotes. *Applied and Environmental Microbiology* **77**, 4685–4692.

Johnson, H. & Tebo, B. (2008). In vitro studies indicate a quinone is involved in bacterial Mn(II) oxidation. *Archives of Microbiology* **189**, 59–69.

Li, J., Wang, Q., Oremland, R. S., Kulp, T. R., Rensing, C. & Wang, G. (2016). Microbial antimony biogeochemistry: enzymes, regulation, and related metabolic pathways. *Applied and Environmental Microbiology* **82**, 5482–5495.

Wang, Q., Warelow, T. P., Kang, Y.-S., Romano, C., Osborne, T. H., Lehr, C. R., Bothner, B., McDermott, T. R., Santini, J. M. & Wang, G. (2015). Arsenite oxidase also functions as an antimonite oxidase. *Applied and Environmental Microbiology* **81**, 1959–1965.

Zargar, K., Conrad, A., Bernick, D. L., Lowe, T. M., Stolc, V., Hoeft, S., Oremland, R. S., Stolz, J. & Saltikov, C. W. (2012). ArxA, a new clade of arsenite oxidase within the DMSO reductase family of molybdenum oxidoreductases. *Environmental Microbiology* **14**, 1635–1645.

CO_2 fixation

Berg, I. A. (2011). Ecological aspects of the distribution of different autotrophic CO_2 fixation pathways. *Applied and Environmental Microbiology* **77**, 1925–1936.

Huegler, M., Huber, H., Stetter, K. O. & Fuchs, G. (2003). Autotrophic CO_2 fixation pathways in archaea (Crenarchaeota). *Archives of Microbiology* **179**, 160–173.

Jennings, R. d. M., Moran, J. J., Jay, Z. J., Beam, J. P., Whitmore, L. M., Kozubal, M. A., Kreuzer, H. W. & Inskeep, W. P. (2017). Integration of metagenomic and stable carbon isotope evidence reveals the extent and mechanisms of carbon dioxide fixation in high-temperature microbial communities. *Frontiers in Microbiology* **8**, 88.

Montoya, L., Celis, L., Razo-Flores, E. & Alpuche-Solís, Á. (2012). Distribution of CO_2 fixation and acetate mineralization pathways in microorganisms from extremophilic anaerobic biotopes. *Extremophiles* **16**, 805–817.

Calvin cycle

Cannon, G. C., Baker, S. H., Soyer, F., Johnson, D. R., Bradburne, C. E., Mehlman, J. L., Davies, P. S., Jiang, Q. L., Heinhorst, S. & Shively, J. M. (2003). Organization of carboxysome genes in the thiobacilli. *Current Microbiology* **46**, 115–119.

Dangel, A. W. & Tabita, F. R. (2015). CbbR, the master regulator for microbial carbon dioxide fixation. *Journal of Bacteriology* **197**, 3488–3498.

Finn, M. W. & Tabita, F. R. (2004). Modified pathway to synthesize ribulose 1,5-bisphosphate in methanogenic Archaea. *Journal of Bacteriology* **186**, 6360–6366.

Savage, D. F., Afonso, B., Chen, A. H. & Silver, P. A. (2010). Spatially ordered dynamics of the bacterial carbon fixation machinery. *Science* **327**, 1258–1261.

Tabita, F. R., Hanson, T. E., Li, H., Satagopan, S., Singh, J. & Chan, S. (2007). Function, structure, and evolution of the RubisCO-like proteins and their RubisCO homologs. *Microbiology and Molecular Biology Reviews* **71**, 576–599.

Witte, B., John, D., Wawrik, B., Paul, J. H., Dayan, D. & Tabita, F. R. (2010). Functional prokaryotic RubisCO from an oceanic metagenomic library. *Applied and Environmental Microbiology* **76**, 2997–3003.

Reductive TCA cycle

Hugler, M., Huber, H., Molyneaux, S. J., Vetriani, C. & Sievert, S. M. (2007). Autotrophic CO_2 fixation via the reductive tricarboxylic acid cycle in different lineages within the phylum *Aquificae* : evidence for two ways of citrate cleavage. *Environmental Microbiology* **9**, 81–92.

Hugler, M., Wirsen, C. O., Fuchs, G., Taylor, C. D. & Sievert, S. M. (2005). Evidence for autotrophic CO_2 fixation via the reductive tricarboxylic acid cycle by members of the ε-subdivision of Proteobacteria. *Journal of Bacteriology* **187**, 3020–3027.

Miura, A., Kameya, M., Arai, H., Ishii, M. & Igarashi, Y. (2008). A soluble NADH-dependent fumarate reductase in the reductive tricarboxylic acid cycle of *Hydrogenobacter thermophilus* TK-6. *Journal of Bacteriology* **190**, 7170–7177.

Acetyl-CoA pathway

Liew, F., Henstra, A. M., Winzer, K., Köpke, M., Simpson, S. D. & Minton, N. P. (2016). Insights into CO_2 fixation pathway of *Clostridium autoethanogenum* by targeted mutagenesis. *mBio* **7**, 00427-16.

Russell, M. J. & Martin, W. (2004). The rocky roots of the acetyl-CoA pathway. *Trends in Biochemical Sciences* **29**, 358–363.

3-hydroxypropionate cycle

Friedmann, S., Alber, B. E. & Fuchs, G. (2006). Properties of succinyl-coenzyme A:D-citramalate coenzyme A transferase and its role in the autotrophic 3-hydroxypropionate cycle of *Chloroflexus aurantiacus*. *Journal of Bacteriology* **188**, 6460–6468.

Herter, S., Fuchs, G., Bacher, A. & Eisenreich, W. (2002). A bicyclic autotrophic CO_2 fixation pathway in *Chloroflexus aurantiacus*. *Journal of Biological Chemistry* **277**, 20277–20283.

Zarzycki, J., Brecht, V., Mueller, M. & Fuchs, G. (2009). Identifying the missing steps of the autotrophic 3-hydroxypropionate CO_2 fixation cycle in *Chloroflexus aurantiacus*. *Proceedings of the National Academy of Sciences of the USA* **106**, 21317–21322.

4-hydroxybutyrate cycles

Berg, I. A., Kockelkorn, D., Buckel, W. & Fuchs, G. (2008). A 3-hydroxypropionate/4-hydroxybutyrate autotrophic carbon dioxide assimilation pathway in Archaea. *Science* **318**, 1732–1733.

Berg, I. A., Ramos-Vera, W. H., Petri, A., Huber, H. & Fuchs, G. (2010). Study of the distribution of autotrophic CO_2 fixation cycles in *Crenarchaeota*. *Microbiology* **156**, 256–269.

Jennings, R. d. M., Moran, J. J., Jay, Z. J., Beam, J. P., Whitmore, L. M., Kozubal, M. A., Kreuzer, H. W. & Inskeep, W. P. (2017). Integration of metagenomic and stable carbon isotope evidence reveals the extent and mechanisms of carbon dioxide fixation in high-temperature microbial communities. *Frontiers in Microbiology* **8**, 88.

Ramos-Vera, W. H., Labonte, V., Weiss, M., Pauly, J. & Fuchs, G. (2010). Regulation of autotrophic CO_2 fixation in the archaeon *Thermoproteus neutrophilus*. *Journal of Bacteriology* **192**, 5329–5340.

Chapter 11

Photosynthesis

Photosynthetic organisms use light energy to fuel their biosynthetic processes. Oxygen is generated in oxygenic photosynthesis where water is used as the electron donor. In anoxygenic photosynthesis, organic or sulfur compounds are used as electron donors. Plants, algae and cyanobacteria carry out oxygenic photosynthesis, whereas photosynthetic bacteria obtain energy from anoxygenic photosynthesis. Aerobic anoxygenic phototrophic bacteria use light energy in a similar way to the purple bacteria, and are a group of photosynthetic bacteria that grow under aerobic conditions.

Phototrophic organisms have a photosynthetic apparatus consisting of a reaction centre intimately associated with antenna molecules or a light-harvesting complex. The antenna molecules and the reaction centre absorb light energy. The energy is concentrated at the reaction centre which is activated and initiates light-driven electron transport. Halophilic archaea convert light energy through a photophosphorylation process.

11.1 Photosynthetic microorganisms

Microorganisms utilizing light energy include eukaryotic algae and cyanobacteria, photosynthetic bacteria and aerobic anoxygenic phototrophic bacteria among the prokaryotes. The halophilic archaea synthesize ATP through photophosphorylation, but they are not considered to be photosynthetic organisms since they lack photosynthetic pigments.

Algae and cyanobacteria possess chlorophyll and have similar photosynthetic processes to plants. However, cyanobacteria are members of the proteobacteria, according to their cell structure and ribosomal RNA sequences. Photosynthetic bacteria are different from other photosynthetic organisms. They have different photosynthetic pigments and do not use water as their electron donor. Some of them can grow chemoorganotrophically in the dark.

11.1.1 Cyanobacteria

Cyanobacteria (also known as blue-green algae or sometimes blue-green bacteria) grow photolithotrophically and fix CO_2 through the Calvin cycle. They do not generally require growth factors except for some that require vitamin B_{12}. They are classified into four groups according to their morphology. These are a unicellular group, the *Pleurocapsa* group, the *Oscillatoria* group and a heterocystous group (Table 11.1). N_2 is fixed by some members of the unicellular cyanobacteria and the *Oscillatoria* group, and by all members of the heterocystous group. N_2-fixing cyanobacteria play a major role in nitrogen influx to the global marine ecosystem. Under N_2-fixing conditions, some of the cells within the filaments of heterocystous group cyanobacteria transform into heterocysts that lack photosystem II and this protects the nitrogenase from O_2 (Section 6.2.1.4).

Certain cyanobacteria, such as *Pseudoanabaena* spp., can simultaneously use

Table 11.1 Cyanobacteria.

	Morphology	Cell division[a]	Heterocysts	N_2 fixation
Unicellular group	single cells	single	−	+[b]
Pleurocapsa group	single cells	multiple	−	+[b]
Oscillatoria group	filaments	single	−	−
Heterocystous group	filaments	single	+	+

[a] Cell divides into two daughter cells (single) or into more than two daughter cells (multiple).
[b] Several species do not fix N_2.

Table 11.2 Photosynthetic bacteria.

Character	Purple bacteria		Green bacteria		Heliobacteria	AAPB
	Non-sulfur	Sulfur	Sulfur	FAPB		
BCHL	*a, b*	*a, b*	*a, c, d, e*	*a, c, d*	*g*	*a*
H_2S as e^- donor	±[a]	+	+	+	−	−
S accumulation	−	intracellular[b]	extracellular	−	−	−
H_2 as e^- donor	+	+	+	+	−	−
Organics as e^- donor	+	+	−	+	+	+
Carbon source	CO_2, organics	CO_2, organics	CO_2, organics	CO_2, organics	organics	organics
Aerobic respiration	+	−	−	+	−	+
CO_2 fixation	Calvin cycle	Calvin cycle	reductive TCA cycle	Calvin cycle[c]	−	−

FAPB, filamentous anoxygenic phototrophic bacteria; AAPB, aerobic anoxygenic phototrophic bacteria; BCHL, bacteriochlorophyll.
[a] Depending on the strain.
[b] Members of the family *Ectothiorhodospiraceae* accumulate sulfur granules extracellularly.
[c] Species of the genus *Chloroflexus* employ the 3-hydroxypropionate cycle.
Note: Halophilic archaea use light energy through photophosphorylation.

H_2S and water as electron donors. These are known as 'versatile cyanobacteria'. Sulfide:quinone reductase transfers electrons from H_2S to plastocyanin.

11.1.2 Anaerobic photosynthetic bacteria

Photosynthetic bacteria are grouped according to their photosynthetic pigments and the electron donors used. These are purple bacteria, green bacteria and heliobacteria. Purple and green bacteria are further divided into purple non-sulfur and purple sulfur bacteria, and green sulfur and filamentous anoxygenic phototrophic bacteria (Table 11.2). In addition to sulfur and organic compounds, Fe(II) is used as the electron donor by some photosynthetic bacteria. These include the green sulfur bacterium, *Chlorobium ferrooxidans*, and the purple non-sulfur bacterium, *Rhodobacter capsulatus* (Section 10.4).

The purple sulfur bacteria include members of the *Chromatiaceae* and *Ectothiorhodospiraceae* within the γ-proteobacteria. The former accumulate sulfur granules intracellularly and the latter extracellularly. The purple non-sulfur bacteria are more diverse, belonging to α- and β-proteobacteria. They grow photosynthetically under anaerobic conditions, and many of them can grow chemoorganotrophically under aerobic conditions. The purple non-sulfur bacteria grow under all electron-accepting conditions (aerobic respiration, anaerobic respiration and fermentative conditions) in addition to anaerobic photosynthesis. The purple sulfur and non-sulfur bacteria have phaeophytin-quinone-type reaction centres (Section 11.3).

The photosynthetic green bacteria include two physiologically and phylogenetically distinct groups. These are the strictly anaerobic and obligately photolithotrophic green sulfur bacteria, and the filamentous anoxygenic photolithotrophic bacteria that are facultatively anaerobic. These have different reaction centres. The latter have the phaeophytin–quinone type, while the former have the iron–sulfur type. The green sulfur bacteria cannot grow heterotrophically, while the filamentous anoxygenic phototrophic bacteria can grow heterotrophically under aerobic dark conditions. The latter, members of the *Chloroflexaceae*, belong to a deep-branching lineage of bacteria. These are also called photosynthetic flexibacteria. They stain Gram-negative but lack lipopolysaccharide.

The photoheterotrophic heliobacteria include three genera: *Heliobacterium, Heliobacillus* and *Heliophilum*. They do not grow aerobically in the dark, and can fix N_2. They do not have photosynthetic organelles and the photosynthetic pigments, including the unique bacteriochlorophyll g, are located in the cytoplasmic membrane. They have an iron–sulfur-type reaction centre. Heliobacterial cells have several unusual features. They are extremely fragile and lyse when approaching the stationary phase. They stain Gram-negative but lack lipopolysaccharide, like the filamentous anoxygenic phototrophic bacteria, and do not fix CO_2. Phylogenetically they belong to the low G + C Gram-positive bacteria.

11.1.3 Aerobic anoxygenic phototrophic bacteria

Photosynthetic bacteria utilize light energy under anaerobic conditions and their photosynthetic pigments are not synthesized under aerobic conditions. However, many bacteria are known to synthesize bacteriochlorophyll (BCHL) *a* and carotenoids under aerobic conditions. These can harvest light energy while they respire oxygen. These are referred to as quasi-photosynthetic bacteria or aerobic anoxygenic phototrophic bacteria (AAPB). They inhabit a variety of locations, including the extreme environments of acidic mine drainage waters, hot springs, polar ice, deep-sea sediments and hydrothermal vent plumes. They are metabolically versatile, using a multitude of organic compounds, sulfur and carbon monoxide as their electron donors in addition to aerobic anoxygenic photosynthesis. AAPB can comprise 11–25 per cent of the total prokaryotic community in their habitats, and up to 65 per cent in lakes.

AAPB found in seawater include species of *Erythrobacter*, *Roseibium*, *Roseivivax*, *Roseobacter*, *Roseovarius* and *Rubrimonas*. Freshwater is the habitat of other aerobic anoxygenic phototrophic bacteria, including species of *Sandaracinobacter*, *Erythromonas*, *Erythromicrobium*, *Roseococcus*, *Porphyrobacter* and *Acidiphilium*. These have the phaeophytin–quinone-type reaction centre. These genera include not only aerobic phototrophs but also species unable to synthesize BCHL *a*. It is not clear if the ability to synthesize the photosynthetic pigment was lost or transferred through lateral gene transfer during their evolution. Aerobic anoxygenic photosynthesis is also known in species of *Bradyrhizobium*, syntrophically growing on the stems of tropical legume plants. Although these are obligate aerobes with a high carotenoid and low BCHL content, they are closely related to purple photosynthetic bacteria in several aspects.

Under light conditions with photophosphorylation in the AAPB, aerobic respiration decreases by approximately 25 per cent from its dark value. The additional energy from light allows the AAPB to assimilate the organic carbon which would otherwise be respired. The higher efficiency of organic carbon utilization may provide an important competitive advantage during growth under carbon-limited conditions.

Some of the AAPB divide in an unusual manner. Budding in addition to binary division occurs in *Porphyrobacter neustonensis* and

Table 11.3 | Photosynthetic pigments.

Pigment	Cyanobacteria	Purple bacteria	Green bacteria	Heliobacteria	Aerobic anoxygenic phototrophic bacteria
Reaction centre pigment	CHL *a*	BCHL *a*, *b*	BCHL *a*	BCHL *g*	BCHL *a*[a]
Antenna pigment	phycobiliprotein CHL *a*	BCHL *a* or *b*	BCHL *c*, *d* or *e*	BCHL *g*	BCHL *a*
Main carotenoid	dicyclic	aliphatic	aryl	aliphatic	dicyclic, aliphatic[b]

[a] BCHL *a* with Zn^{2+} or Mg^{2+}.
[b] Various carotenoids depending on the strain.

Erythromonas ursincola (Section 6.14.1.4). Ternary fission and branching are exhibited by *Erythromicrobium ramosum* and *Erythromicrobium hydrolyticum* (Section 6.14.1.3).

11.2 Photosynthetic pigments

Photosynthetic pigments include chlorophylls, carotenoids and phycobiliproteins in plants and cyanobacteria; phycobiliproteins are not found in photosynthetic bacteria (Table 11.3). Bacteria use bacteriochlorophylls in place of chlorophyll, and bacteriophaeophytin to replace phaeophytin. Since these pigments generate reactive oxygen species (ROS, Section 12.2.5), genes for their synthesis and their binding proteins are repressed under aerobic conditions in facultative anaerobic purple bacteria. In the purple non-sulfur bacterium *Rhodobacter sphaeroides*, the repressor protein PpsR binds to the upstream region of the photosynthetic genes to repress their expression under aerobic conditions. When the oxygen tension decreases, the anti-repressor protein AppA binds to PpsR and releases it from the DNA, allowing transcription. AppA synthesis is activated by the PrrB/PrrA (RegB/RegA) two-component system (Section 12.2.4.1). PrrB is phosphorylated at low redox potential, due to the accumulation of a reduced quinone pool as the Arc protein, and transfers phosphate to PrrA. The phosphorylated PrrA activates genes of AppA and the small RNA PcrZ (photosynthesis control RNA Z). The latter negatively controls the translation of the photosynthetic mRNAs. In addition, a separate redox sensing protein, FnrL, activates genes for AppA and photosynthesis. AppA senses light to activate photosynthesis genes.

Over 20 small regulatory RNAs (sRNAs, Section 12.1.9.4) are induced in *Rhodobacter sphaeroides* when this bacterium is treated with ROS. Some of their expression is controlled by an extracytoplasmic sigma factor (Section 12.1.1). These are involved in control of various functions, including activation of ROS detoxification and aerobic respiration, and repression of photosynthesis. The RNA chaperone, Hfq (Section 12.1.9.4), participates in these control mechanisms. ROS shows similar effects on the AAPB (Section 11.1.3).

In addition to transcriptional regulation of the photosynthetic genes, photosynthesis is regulated according to the Earth's rotation through a process known as the circadian clock (Section 11.4.3.1).

11.2.1 Chlorophylls

Chlorophylls have a general structure of four pyrrole derivatives with covalently bound Mg^{2+} (Figure 11.1). They form a complex with proteins embedded in the membranes of the

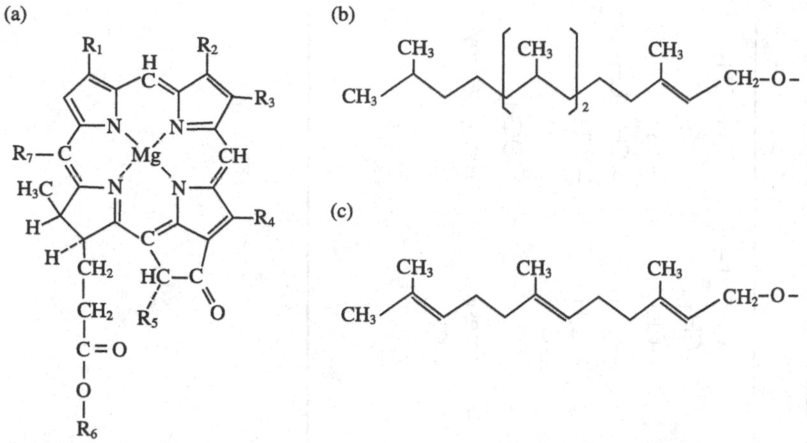

Figure 11.1 The structure of chlorophyll (see Table 11.4 for the side chains R1–R7).

(a) Chlorophyll; (b) phytyl group; (c) farnesyl group.

photosynthetic apparatus. There are several structurally different chlorophylls that have different side chains from the pyrrole rings (Table 11.4). These include chlorophyll *a*, and bacteriochlorophyll (BCHL) *a*, *b*, *c*, *d*, *e* and g. Bacteriochlorophyll *a* with Zn^{2+} in place of Mg^{2+} occurs in aerobic anoxygenic phototrophic bacteria.

As listed in Table 11.3, cyanobacteria possess chlorophyll *a*, as do plants. Green bacteria have reaction centres with BCHL *a* and antenna molecules with BCHL *c*, *d* or *e*. In contrast, purple bacterial reaction centres and antenna molecules contain BCHL *a* or *b*. Aerobic anoxygenic phototrophic bacteria contain BCHL *a* with Zn^{2+} or Mg^{2+} in both the reaction centre and antenna molecules. Cyanobacteria with CHL *a* absorb relatively short wavelength light while light of wavelengths over 700 nm is absorbed by photosynthetic bacteria containing BCHL *a* or *b*. The purple bacteria absorb near infrared light of wavelength around 800 nm (see Figure 11.4).

11.2.2 Carotenoids

In addition to chlorophylls, photosynthetic organisms have carotenoids. These absorb light over wavelengths of 400–600 nm and transfer the energy to chlorophylls, and also protect biological materials from photooxidation caused by reactive oxygen derivatives. The common carotenoid in cyanobacteria is *β*-carotene. Over 30 different carotenoids are known in purple bacteria and in aerobic anoxygenic phototrophic bacteria, and spirilloxanthin (without a benzene ring in its structure) is the most common among them. The typical carotenoid of green bacteria is isorenieratene (Figure 11.2). The major carotenoid in heliobacteria is neurosporene which is similar in structure to spirilloxanthin.

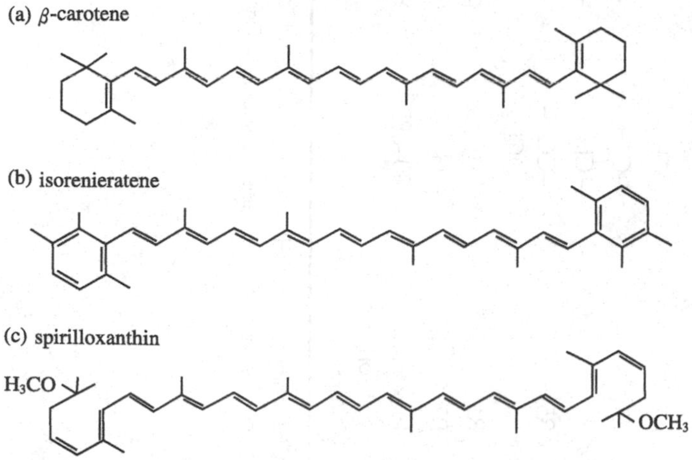

Figure 11.2 Common carotenoids in cyanobacteria and photosynthetic bacteria.

Main carotenoids in (a) cyanobacteria, (b) green bacteria and (c) purple bacteria.

Table 11.4 The structure and maximum absorbance wavelengths of chlorophylls.

Side chain[a]	CHL *a*	BCHL *a*	BCHL *b*	BCHL *C*	BCHL *d*	BCHL e	BCHL *g*
R1	$—CH{=}CH_2$	$—CO—CH_3$	$—CO—CH_3$	$—CHOH—CH_3$	$—CHOH—CH_3$	$—CHOH—CH_3$	$—CH{=}CH_2$
R2	$—CH_3$	$—CH_3$	$—CH_3$	$—CH_3$	$—CH_3$	—CHO	$—CH_3$
R3	$—C_2H_5$	$—C_2H_5$	$=CH—CH_3$	$—C_2H_5$	$—C_2H_5$	$—C_2H_5$	$=CH—CH_3$
R4	$—CH_3$	$—CH_3$	$—CH_3$	$—C_2H_5$	$—C_2H_5$	$—C_2H_5$	$—CH_3$
R5	$—CO—OCH_3$	$—CO—OCH_3$	$—CO—OCH_3$	—H	—H	—H	$—CO—OCH_3$
R6	phytyl	phytyl	phytyl	Farnesyl	farnesyl	farnesyl	geranyl-geranyl
R7	—H	—H	—H	$—CH_3$	—H	$—CH_3$	—H
Maximum absorbance (nm)	680–685	850–910	1020–1035	745–760	725–745	715–725	788, 670

CHL, chlorophyll; BCHL, bacteriochlorophyll.
[a] The positions of side chains are shown in Figure 11.1.

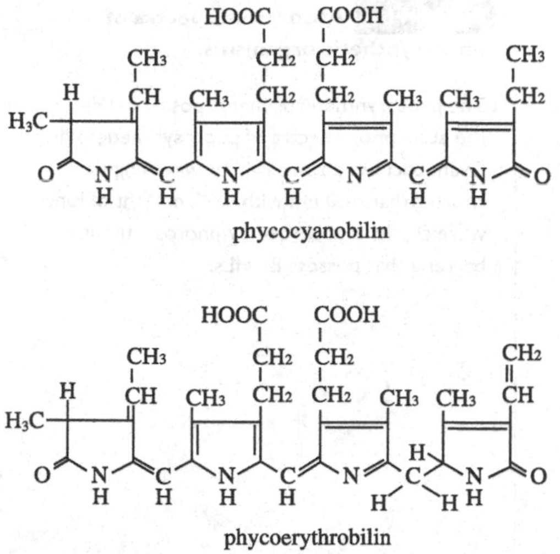

phycocyanobilin

phycoerythrobilin

Figure 11.3 The structure of bilins in phycobiliproteins.

11.2.3 Phycobiliproteins

Phycobiliproteins are soluble proteins containing bilin, which has a structure of four pyrroles in a linear form and is found in photosynthetic eukaryotes and cyanobacteria (Figure 11.3). Cyanobacteria with a blue-green colour contain allophycocyanin and phycocyanin, while phycoerythrin is the phycobiliprotein in red-coloured cyanobacteria. This protein is not found in photosynthetic bacteria.

11.2.4 Phaeophytin

This pigment has the structure of chlorophyll but without Mg^{2+} or Zn^{2+}. Cyanobacteria and photosynthetic eukaryotes possess phaeophytin, while bacteriophaeophytin is found in photosynthetic bacteria with the phaeophytin-quinone-type reaction centre serving as an electron carrier. This pigment is not found in green sulfur bacteria and heliobacteria that have the iron-sulfur-type reaction centre. Before its function was established, phaeophytin was regarded as a chlorophyll degradation product.

11.2.5 Absorption spectra of photosynthetic cells

Each photosynthetic organism has specific photosynthetic pigments in different ratios, and absorbs light of a specific wavelength depending on such pigments (Figure 11.4). Cyanobacteria with CHL *a*, carotenoids and phycobiliproteins absorb light at wavelengths shorter than 700 nm. On the other hand, photosynthetic bacteria absorb light at wavelengths shorter than 600 nm, with carotenoids and the BCHLs absorbing light at wavelengths above 700 nm. Green bacterial cells absorb light at wavelengths between 700 and 800 nm, and also below 600 nm, using BCHL *a*, *b* and *e* and carotenoids in their antenna molecules. The heliobacteria contain BCHL *g* which has maximum light absorption at 788 nm, and shows a similar absorption spectrum to green bacteria. BCHL *a* and *b* absorb long wavelength light over 800 nm in purple bacteria. Aerobic anoxygenic phototrophic bacteria absorb light at 450 nm with carotenoids and at 750 nm with BCHL *a*.

The absorption spectrum of cyanobacteria changes according to the light wavelength. *Oscillatoria sancta* is reddish-brown when grown in green light but blue-green when grown in red light. These colour differences are due to altered pigment synthesis. When grown in red light, cyanobacteria synthesize CHL *b*, *d* or *f* in addition to CHL *a*, many functionally distinct carotenoids and spectroscopically diverse phycobiliproteins. These changes are known as complementary chromatic acclimation. Cyanobacteria adjust the total cellular CHL content and the ratio of photosystem II to photosystem I at different light intensities. These changes enable them to maximize their photosynthetic efficiency in response to the incident irradiation.

11.3 Photosynthetic apparatus

Photosynthetic organisms utilize light energy to reduce $NADP^+$ and to synthesize ATP through the proton motive force. This energy transduction is facilitated by the photosynthetic pigments and electron carriers arranged in the photosynthetic apparatus. Separate photo-

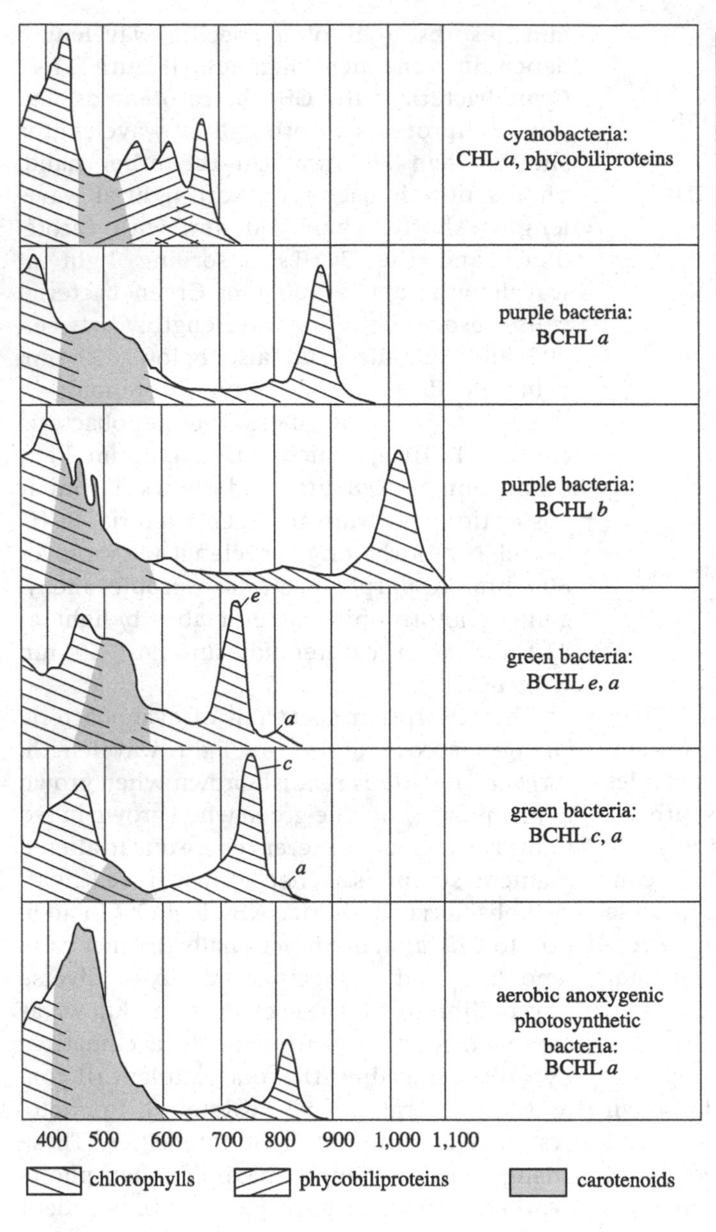

Figure 11.4 Absorption spectra of photosynthetic organisms.

The photosynthetic pigments possessed determine the absorption spectra of photosynthetic cells. Cyanobacteria absorb light at wavelengths shorter than 700 nm with CHL *a*. Light of longer wavelengths is absorbed by photosynthetic bacteria that possess BCHLs.

synthetic structures are not found in heliobacteria and aerobic anoxygenic phototrophic bacteria. Their reaction centres and antenna molecules are located at the cytoplasmic membrane. The reaction centre is the key component for the primary events in the photochemical conversion of light into biological energy. Coupling to secondary electron donors and acceptors allows the electrons and accompanying protons to be transferred to other components of the photosynthetic apparatus, synthesizing ATP or reducing $NADP^+$.

Photosynthetic reaction centres can be classed in two categories based on the nature of the electron acceptors. Those of the purple bacteria, filamentous anoxygenic phototrophic bacteria and photosystem II of cyanobacteria belong to the phaeophytin–quinone type, while the iron–sulfur

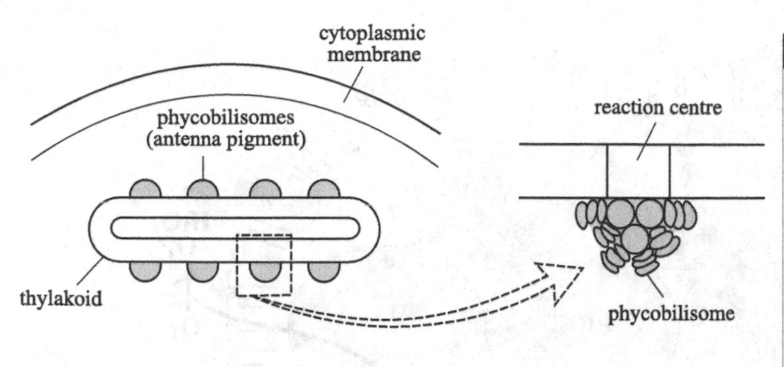

Figure 11.5 The thylakoid, the photosynthetic organelle in cyanobacteria.

The thylakoid with a galactosyl diglyceride bilayer membrane structure is the intracellular photosynthetic organelle in cyanobacteria. The reaction centres are spread on the membrane, and phycobilisomes are attached to the reaction centres, which are associated with the electron transport chains to use light energy for proton motive force generation and for the reduction of $NAD(P)^{+}$.

type is found in green sulfur bacteria, heliobacteria and photosystem I of cyanobacteria. Phaeophytin (or bacteriophaeophytin) and iron-sulfur centres participate in the electron transfer reactions of each type of reaction centre.

11.3.1 Thylakoids of cyanobacteria

Cyanobacteria have thylakoids with the photosynthetic pigments arranged similarly to those of the chloroplasts in photosynthetic eukaryotic cells. The thylakoid has a bilayer membrane structure, consisting of galactosyl diglyceride containing one or two galactose molecules in place of the phosphate of phospholipids (Figure 11.5). Thylakoids convert light energy into biological energy in the cytoplasm of the cyanobacteria.

Chlorophyll *a* and proteins form the reaction centre on the thylakoid membrane. A small fraction of chlorophyll *a* is found in the reaction centre, while the majority forms antenna molecules with phycobiliproteins and carotenoids in a structure known as the phycobilisome. In addition to chlorophyll *a*, reaction centres have various electron carriers that convert the light energy into a proton motive force. Photosystem I contains [Fe–S] proteins and quinones, and photosystem II contains phaeophytin and quinones. The phycobilisomes occupy the cytoplasmic side of the reaction centre (Figure 11.5).

11.3.2 Green bacteria

Green bacteria have a photosynthetic apparatus called the chlorosome on their cytoplasmic membrane. Chlorosomes contain antenna molecules, and their baseplates are bound to the reaction centre which is a part of the cytoplasmic membrane (Figure 11.6). The chlorosome has the structure of a galactosyl diglyceride monolayer membrane filled with rod-shaped antenna molecules. Bacteriochlorophyll (BCHL) *c*, *d* or *e* constitute the antenna molecules together with carotenoids. The baseplate contains BCHL *a* which transfers photons to the reaction centre. The reaction centre contains BCHL *a* and the electron transport chains. In addition to BCHLs, [Fe–S] proteins are found in the reaction centres of green sulfur bacteria, and bacteriophaeophytin (BPHE) in those of filamentous anoxygenic phototrophic bacteria.

The obligate anaerobe *Chlorobium tepidum* thrives in anaerobic aquatic environments where sulfide is available with very dim light. To capture light efficiently this bacterium contains about 200–250 chlorosomes per cell with more than 200×10^3 BCHL *c* molecules per chlorosome. Thus, this bacterium contains up to 50×10^6 BCHL *c* molecules per cell.

11.3.3 Purple bacteria

The purple bacteria have a less well-developed photosynthetic structure than cyanobacteria and the green bacteria. An intracellular membrane structure contains antenna molecules and reaction centres (Figure 11.7). This is a phospholipid bilayer membrane continuous with the cytoplasmic membrane. The shape of the intracellular membrane structure varies depending on the strain. The intracellular membrane structure is developed when the photosynthetic proteins and pigments are

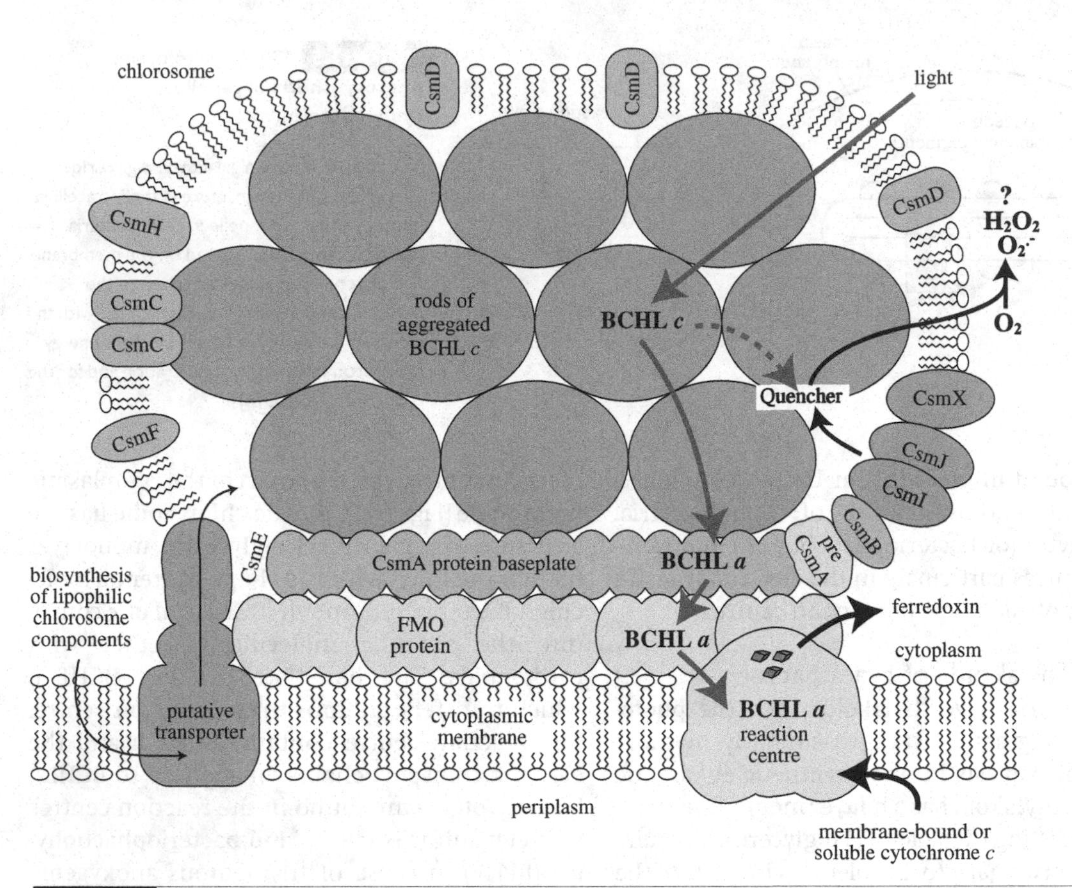

Figure 11.6 Chlorosome bound to the cytoplasmic membrane in green bacteria.

(Modified from *Arch. Microbiol.* **182**: 265–276, 2004)

The monolayered chlorosome contains antenna molecules, and its baseplate is bound to a reaction centre in the cytoplasmic membrane.

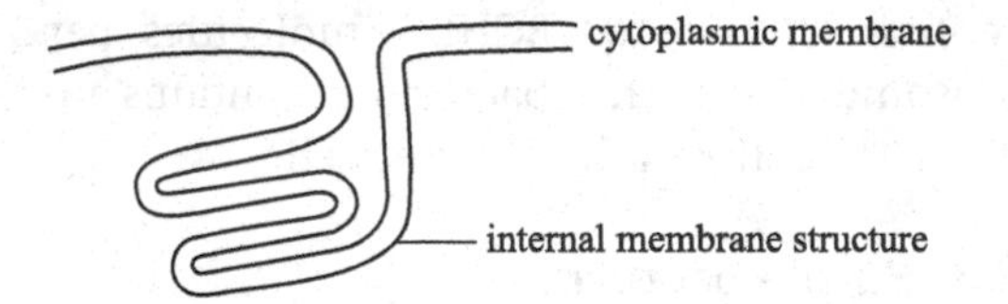

Figure 11.7 The photosynthetic apparatus in purple bacteria.

The intracellular membrane structure, continuous with the cytoplasmic membrane, contains antenna molecules and the reaction centres.

produced under anaerobic light conditions and are controlled by AppA, PrrB/PrrA (RegB/RegA) and PcrZ (Section 11.2).

11.3.4 Heliobacteria and aerobic anoxygenic phototrophic bacteria

These organisms are the least well-developed in terms of the photosynthetic apparatus and lack differentiated structures such as chlorosomes or intracytoplasmic membranes. The antenna molecules and reaction centres reside within the cytoplasmic membrane.

11.4 Light reactions

Photosynthesis is a process utilizing light energy for biosynthesis. It can be divided into light

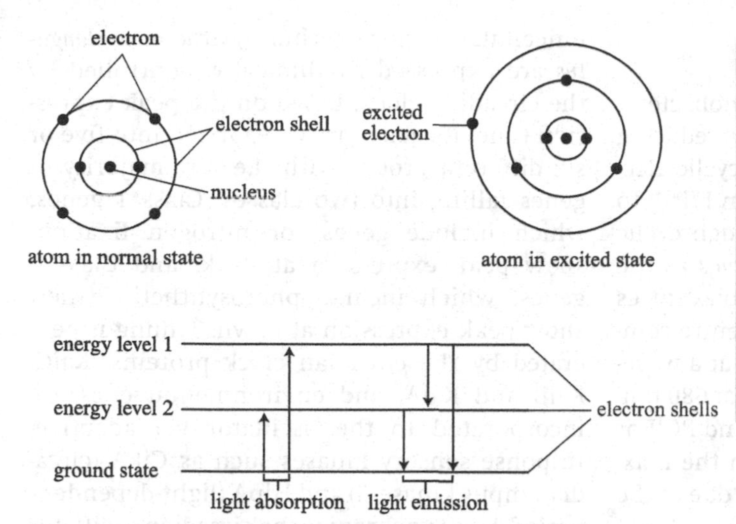

Figure 11.8 Photon exciting an electron of a light-absorbing pigment molecule.

The excitation takes place when the energy of the photon is similar to the energy needed to excite the electron. The excited electron returns to the original unexcited state coupled to one of three reactions. These reactions are (1) transfer of energy to the adjacent molecule exciting its electron (resonance transfer), (2) oxidation of the excited pigment transferring an electron to a second compound or (3) emission of fluorescent light.

reactions that convert light energy into biological energy and dark reactions that utilize the biological energy in biosynthesis.

11.4.1 Properties of light

Light is a form of electromagnetic radiation that travels in rhythmic waves transported in discrete particle units called photons. The number of photons per unit time is the intensity of the light. The energy carried by the photons is related to the frequency of the wave. The higher the frequency, the higher the energy carried.

When a photon is radiated to a surface, it may be reflected, transmitted or absorbed. Different pigments absorb photons of different wavelengths depending on the nature of the absorbing pigment. When a photon is absorbed by a pigment, the light energy is converted to kinetic energy in the form of an excited electron. When a molecule (or atom) absorbs a photon, an electron is boosted to a higher energy level by transferring an electron from the normal shell to the outer shell (Figure 11.8).

11.4.2 Excitation of antenna molecules and resonance transfer

An excited electron is very unstable and returns to the original unexcited state coupled to one of three reactions. These reactions are (1) transfer of energy to an adjacent molecule exciting its electron (resonance transfer), (2) reduction of a second compound (photo-induced charge separation) or (3) emission of fluorescent light.

The antenna molecules are excited on absorbing photons. The energy gained by the pigments through such excitation is referred to as the exciton. The exciton is transferred from the antenna molecules to the reaction centre through resonance transfer, and this reaction takes about 0.1 picosecond. This energy is referred to as the resonance energy. Electron transport reaction at the reaction centre is initiated by oxidation, with the consumption of resonance energy.

Resonance transfer takes place from molecules of a higher exciton to those with a lower exciton. For this reason, photosynthetic pigments are arranged in such a way in the phycobilisome of the cyanobacterial thylakoid and in the bacterial photosynthetic apparatus to facilitate such resonance transfer.

11.4.3 Electron transport

When the resonance energy excites chlorophyll in the reaction centre, its redox potential becomes very low, enough to transfer electrons to a lower potential electron carrier. The oxidized chlorophylls are reduced again, either oxidizing externally supplied electron donor(s) in a process known as noncyclic electron transport, or by taking the original electrons through the cyclic electron transport system.

11.4.3.1 Photosystem I and II in cyanobacteria

As in photosynthetic eukaryotes, cyanobacteria have photosystem II (PSII) to supply reducing power, oxidizing water through non-cyclic electron transport, as well as photosystem I (PSI) to generate a proton motive force through cyclic electron transport. Chlorophyll *a* serves as the photosynthetic pigment in the reaction centres of both photosystems. The reaction centre complex of PSI has maximum absorption at a wavelength of 700 nm, while that of PSII is at 680 nm. These are referred to as RCI or P700 and RCII or P680, respectively. The differences in the maximum absorption of chlorophyll *a* are due to the proteins incorporated in the complex.

When excited by photons, the redox potential of RCI (P700) decreases to −1.0 V from +0.5 V. The excited P700 reduces the primary acceptor (A_0, CHL *a*). Electrons are transferred from A_0 to ferredoxin through phylloquinone and several [Fe–S] centres. The electrons from the reduced ferredoxin are transferred back to P700 through the cytochrome *bf* complex and plastocyanin in a process known as cyclic electron transport (Figure 11.9). The free energy changes in cyclic electron transport are conserved as a proton motive force transporting protons into the thylakoid.

Alternatively, electrons are transferred from the reduced ferredoxin to $NADP^+$. PSII replaces the electrons used to reduce $NADP^+$ in PSI. When RCII absorbs light, the redox potential decreases from +1.0 V to −0.8 V. Electrons move from the excited RCII to phaeophytin before being transferred to PSI via various electron carriers (Figure 11.9).

The P680 (RCII) of PSII has a redox potential of +1.0 V which is higher than that of O_2/H_2O. Oxidized P680 is reduced, oxidizing water to molecular oxygen. This reaction is catalysed by a manganoprotein. Since O_2 is evolved, this is referred to as oxygenic photosynthesis.

The expression of photosynthetic genes is not only activated by light (Section 11.2), but also controlled by the circadian clock which is a timing system that induces rhythms of biological activity in synchrony with day and night. Nearly all genes in the genome of the unicellular cyanobacterium *Synechococcus elongatus* are expressed rhythmically, controlled by the circadian clock. Based on the peak expression time, its genes are categorized into five or six different groups with the vast majority of genes falling into two classes. Class 1 genes, which include genes for nitrogen fixation, show peak expression at dusk, and Class 2 genes, which include photosynthetic genes, show peak expression at dawn. Timing is generated by the circadian clock proteins, KaiC, KaiB and KaiA, and environmental cues are incorporated to the oscillator via adaptive-response sensory kinases such as CikA (circadian input kinase A) and LdpA (light-dependent period A) to synchronize the circadian oscillator with the external environment. Information from the oscillator is transmitted via an output pathway consisting of a two-component system, comprising SasA (*Synechococcus* adaptive sensor A) and RpaA (regulator of phycobilisome association A).

Oscillations in metabolic processes persist under constant conditions. Thus, circadian rhythms control the timing of physiological processes for adaptation to daily environmental changes such as light–dark or temperature cycles. There are several hypotheses explaining the advantage of the timing mechanism. The most common hypothesis is that the timekeeper enables the cell to prepare for the daily routines of environmental change. Such preparation might be most easily discerned in the case of cyanobacteria. It is a disadvantage to maintain energetically expensive photosynthetic pathways during the night. Likewise, it is advantageous to prepare protective responses against damage caused by sunlight (especially UV irradiation). Unicellular cyanobacteria rhythmically fix nitrogen during the night, approximately twelve hours out of phase from photosynthesis, which peaks at midday. This is adaptive biochemistry, as nitrogenase is inactivated by oxygen and its activity is therefore incompatible with oxygenic photosynthesis which releases oxygen (Section 6.2.1.4). This is one simple example of why cyanobacteria have evolved timekeeping systems to coordinate metabolic events during the day, to optimize performance and to temporally separate potentially

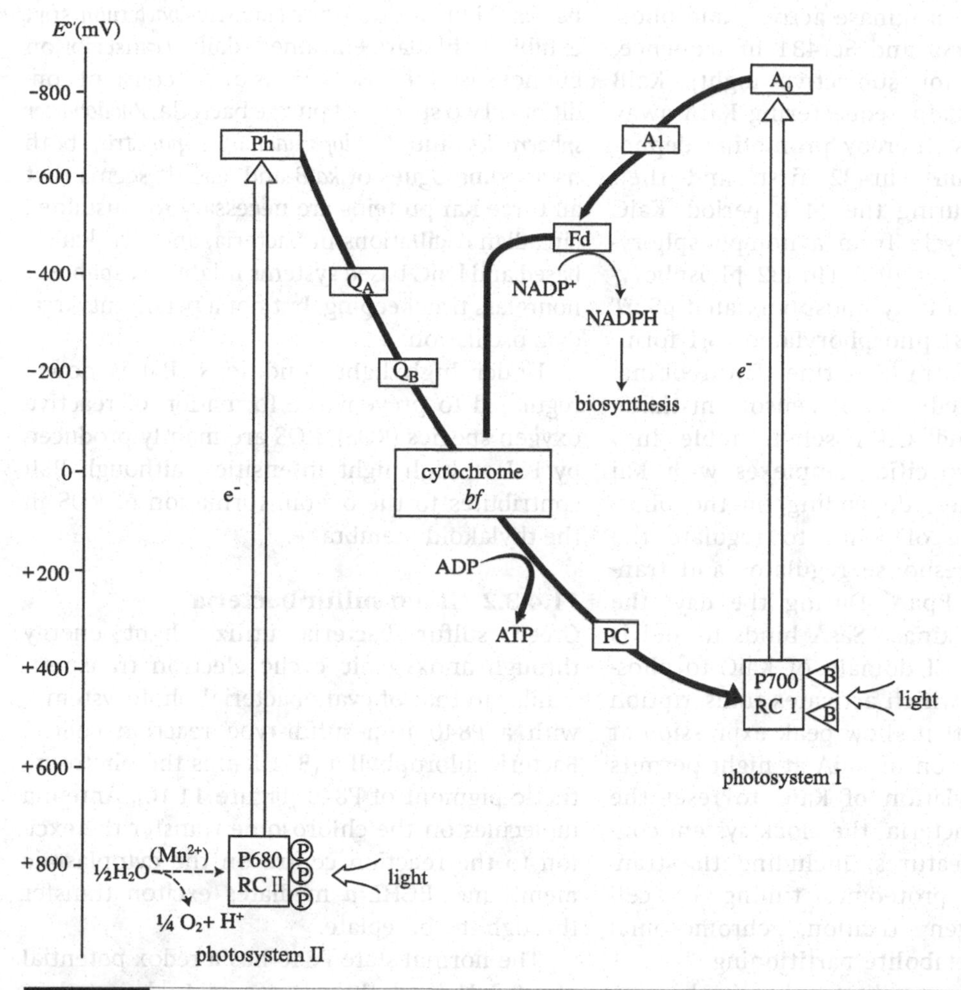

Figure 11.9 Photosynthetic electron transport chains of cyanobacteria.

Cyanobacteria have two different reaction centres, RCI or P700 and RCII or P680. These consist of chlorophyll *a* and a protein complex. When RCI absorbs photons, chlorophyll *a* is excited and transfers electrons to the primary acceptor (A_0, CHL *a*). Electrons are transferred from A_0 to ferredoxin through phylloquinone (A_1) and several [Fe–S] centres. The electrons from the reduced ferredoxin are transferred back to P700 through the cytochrome *bf* complex and plastocyanin (PC) in a process known as cyclic electron transport. The free energy changes in cyclic electron transport are conserved as a proton motive force transporting protons into the thylakoid. $NADP^+$ is reduced, taking electrons from the ferredoxin (Fd). To replace the electrons channelled to NADPH, RCII oxidizes water, reducing phaeophytin (Ph) and transferring the electrons to the cyclic electron transport chain. Electron transport at RCII is referred to as non-cyclic electron transport.

Ph, phaeophytin; Q_A and Q_B, plastoquinone A and B, respectively; PC, plastocyanin; A_0, primary electron acceptor (CHL *a*); A_1, phylloquinone; Fd, ferredoxin.

conflicting metabolic processes. The circadian oscillation is independent of cell cycle or growth rate.

KaiC with autokinase and autophosphatase activities contains CI, CII and C-terminal A-loop domains. KaiC forms a homohexamer with two rings formed by CI and CII domains and KaiA binding the A-loop tail. The CII domain has two phosphorylation sites, Ser431 and Thr432. During the illuminated portion of the day (or the subjective day) KaiA binds to the A-loop segment of KaiC, promoting KaiC autokinase activity and phosphorylating Thr432 and Ser431 in sequence. Over the night (or subjective night), KaiB binds the CI domain sequestering KaiA away from the A-loops, thereby promoting dephosphorylation from Thr432 first, and then from Ser 431. During the 24 h period, KaiC is subject to a cycle from a non-phosphorylated S/T form through a Thr432 phosphorylated S/pT form, a fully phosphorylated pS/pT form and a Ser431 phosphorylated pS/T form, back to the S/T form (S, serine; T, threonine; p, phosphorylated). Two sensor histidine kinases SasA and CikA self-assemble into day- and night-specific complexes with Kai protein complexes, depending on the phosphorylation state of KaiC, to regulate the activity of the response regulator and transcription factor RpaA. During the day, the sensor histidine kinase SasA binds to and is activated by the CI domain of KaiC to phosphorylate RpaA, which activates transcription of class 1 genes that show peak expression at dusk. The inhibition of KaiA at night permits autodephosphorylation of KaiC to reset the cycle. In cyanobacteria, the clock system controls multiple features, including the transcriptome, the proteome, timing of cell division, nitrogen fixation, chromosomal topology and metabolite partitioning.

While the cyanobacterial timekeeping mechanism is the best understood system to date, it is proposed that there is widespread daily timekeeping among eubacteria and archaea through distinct mechanisms that share some common elements with the cyanobacterial clock. Other bacteria may have evolved alternative timekeeping systems, such as an hourglass timer, which once set in motion, keeps track of time linearly and does not self-sustain a cycle. Homologues of *kaiB* and *kaiC* are distributed widely in both photosynthetic and non-photosynthetic members of the eubacteria and the archaea, but *kaiA* is only found in cyanobacteria. In the cyanobacterial genus *Prochlorococcus*, *kaiA* is not present and a simpler hourglass timer operates. Another potential hourglass timer is found in the halophilic archaeal *Halobacterium* genus, which has *kaiC* but not *kaiA* nor *kaiB*. *Halobacterium* spp. exhibit light–dark-entrained daily transcription but not sustained oscillations under constant conditions. Two species of purple bacteria, *Rhodobacter sphaeroides* and *Rhodopseudomonas palustris*, both have homologues of *kaiB* and *kaiC*. It seems that all three Kai proteins are necessary for sustained circadian oscillations in bacteria, and that kaiBC-based and kaiC-based systems might be capable of hourglass timekeeping, but not a persistent circadian oscillation.

Under high light conditions, PSI is down-regulated to prevent the formation of reactive oxygen species (ROS). ROS are mainly produced by PSI at high light intensities, although PSII contributes to the overall formation of ROS in the thylakoid membrane.

11.4.3.2 Green sulfur bacteria

Green sulfur bacteria utilize light energy through anoxygenic cyclic electron transport, similar to that of cyanobacterial photosystem I, with a P840 iron–sulfur-type reaction centre. Bacteriochlorophyll *a* (BCHL *a*) is the photosynthetic pigment of P840 (Figure 11.10). Antenna molecules on the chlorosome transfer the exciton to the reaction centre on the cytoplasmic membrane. BCHL *a* mediates exciton transfer through the baseplate.

The normal state P840 has a redox potential of +0.3 V, and this decreases to lower than −1.0 V, low enough to reduce the primary acceptor (A_0, BCHL *a*) when excited by photons. Electrons either flow back to P840 generating a proton motive force, or are transferred to $NAD(P)^+$. Electrons that are consumed to reduce $NAD(P)^+$ are replaced by oxidizing electron donors such as sulfide. These organisms cannot grow chemoorganotrophically under aerobic conditions. On the other hand, the other members of the green sulfur bacteria, the filamentous anoxygenic phototrophic bacteria, can grow chemoorganotrophically under aerobic conditions. These latter organisms have a different photosynthetic electron transport system from that of the green sulfur bacteria, which is similar to that of the purple bacteria (see Figure 11.11), with a phaeophytin–quinone-type reaction centre.

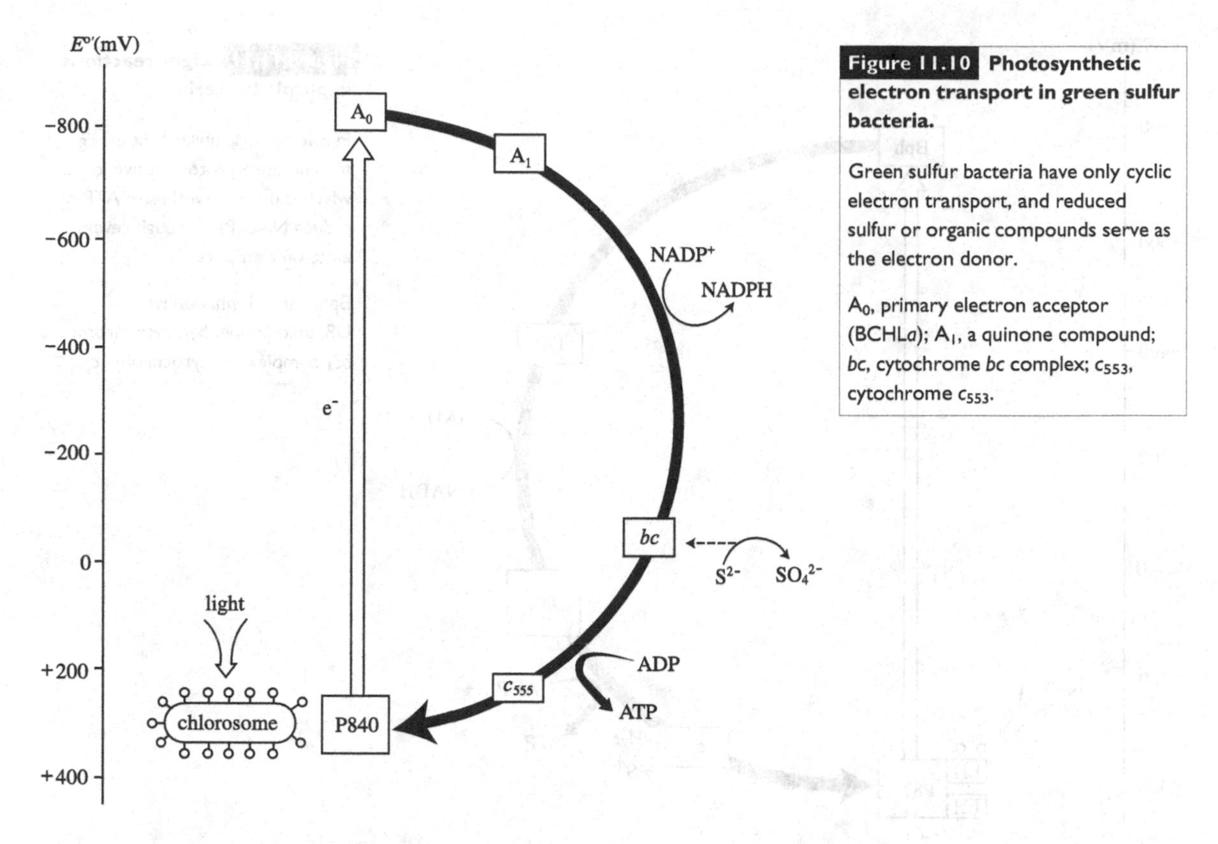

Figure 11.10 Photosynthetic electron transport in green sulfur bacteria.

Green sulfur bacteria have only cyclic electron transport, and reduced sulfur or organic compounds serve as the electron donor.

A_0, primary electron acceptor (BCHL*a*); A_1, a quinone compound; *bc*, cytochrome *bc* complex; c_{553}, cytochrome c_{553}.

Heliobacteria have P788 as the reaction centre, which is different from the green sulfur bacteria although the photosynthetic mechanisms are similar.

11.4.3.3 Purple bacteria

Purple bacteria have a reaction centre with maximum absorption at 870 nm, which is excited by photons, with a decrease in redox potential. The excited P870 reduces bacteriophaeophytin to begin the cyclic electron transport for the generation of a proton motive force. $NAD(P)^+$ is reduced through reverse electron transport oxidizing quinol. Sulfur and organic compounds are used as electron donors, reducing cytochrome c_2 (Figure 11.11). Purple sulfur bacteria cannot grow under aerobic conditions, while purple non-sulfur bacteria can grow chemoorganotrophically under aerobic conditions.

11.4.3.4 Aerobic anoxygenic photosynthetic bacteria (AAPB)

These bacteria only synthesize bacteriochlorophyll *a* and carotenoids under aerobic conditions. They therefore do not use light energy under anaerobic conditions. They have a phaeophytin-quinone-type reaction centre, with the cytoplasmic membrane accommodating the reaction centres and antenna molecules. They possess a similar cyclic electron transport system to the purple bacteria, as depicted in Figure 11.11, but NAD(P)H is supplied from the metabolism of organic compounds.

Since a photosynthetic apparatus is absent in AAPB, they have fewer photosynthetic reaction centres and a lower light-harvesting capacity than true phototrophic bacteria. Even with low numbers of photosynthetic reaction centres, photophosphorylation provides up to three times higher electron fluxes than aerobic respiration in AAPB.

11.5 Carbon metabolism in phototrophs

According to their carbon sources, phototrophs are classified into photoorganotrophs and

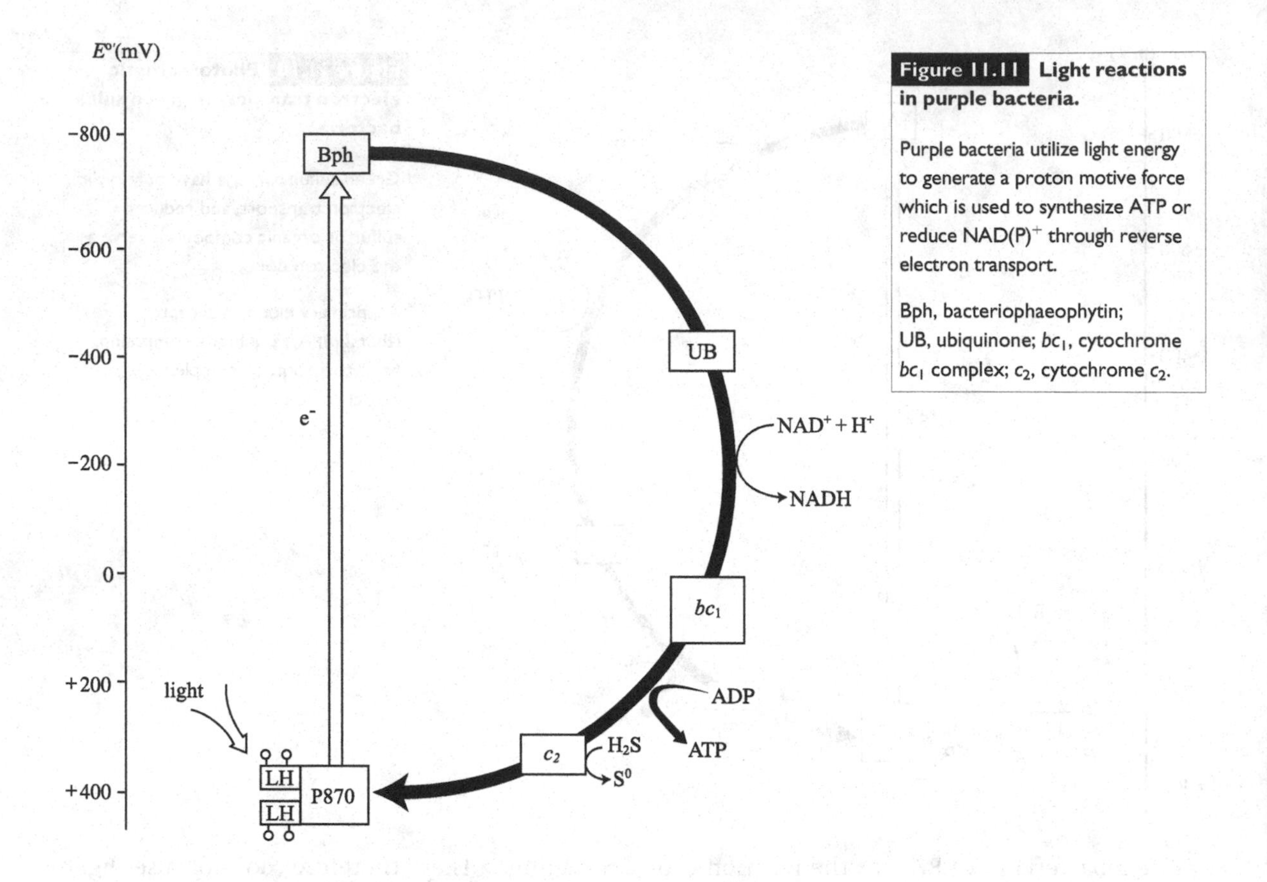

Figure 11.11 Light reactions in purple bacteria.

Purple bacteria utilize light energy to generate a proton motive force which is used to synthesize ATP or reduce $NAD(P)^+$ through reverse electron transport.

Bph, bacteriophaeophytin; UB, ubiquinone; bc_1, cytochrome bc_1 complex; c_2, cytochrome c_2.

photolithotrophs. With a few exceptions, cyanobacteria are photolithotrophs, and the green bacteria and purple bacteria can grow photolithotrophically, fixing CO_2, as well as photoorganotrophically (Table 11.2). Heliobacteria and aerobic anoxygenic photosynthetic bacteria do not fix CO_2. A rhodopsin-containing *Dokdonia* sp. fixes CO_2 through various anaplerotic reactions (Section 11.6).

11.5.1 CO_2 fixation

The Calvin cycle is the most common CO_2 fixing mechanism in phototrophs. The key enzymes of the Calvin cycle, phosphoribulokinase and ribulose-1,5-bisphosphate carboxylase, are not found in green sulfur bacteria or in the filamentous anoxygenic phototrophic bacterium, *Chloroflexus aurantiacus*. Green sulfur bacteria fix CO_2 into acetyl-CoA through the reductive TCA cycle (Section 10.8.2), while the less common 3-hydroxypropionate pathway is employed by *Chloroflexus aurantiacus* (Section 10.8.4). CO_2 is fixed through the Calvin cycle in cyanobacteria, purple bacteria and filamentous anoxygenic phototrophic bacteria (Table 11.2).

11.5.2 Carbon metabolism in photoorganotrophs

Most photosynthetic bacteria use simple organic compounds as their carbon source and electron donor, with ATP and NAD(P)H being generated from the light reactions.

11.5.2.1 Purple bacteria, heliobacteria and aerobic anoxygenic photosynthetic bacteria

As photoorganotrophs, purple non-sulfur bacteria preferentially use organic compounds as their carbon source. Sugars are metabolized through the EMP or ED pathway depending on the organism. CO_2 from glycolysis is fixed under photosynthetic conditions (Figure 11.12). Under dark conditions, purple non-sulfur bacteria can grow with or without molecular oxygen.

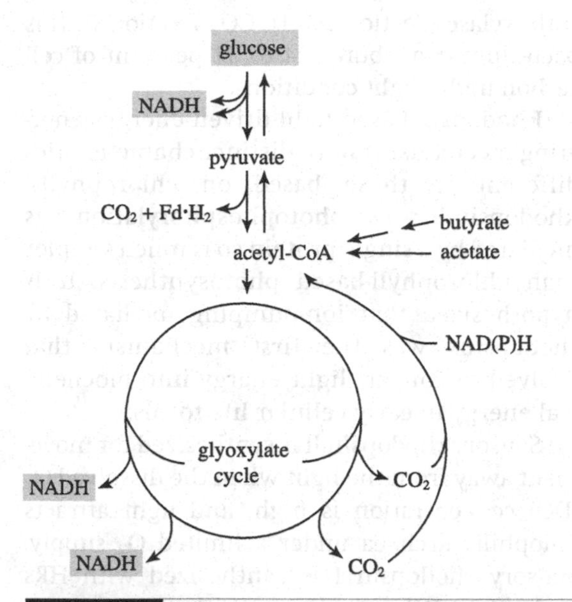

Figure 11.12 Photoorganotrophic metabolism in purple bacteria.

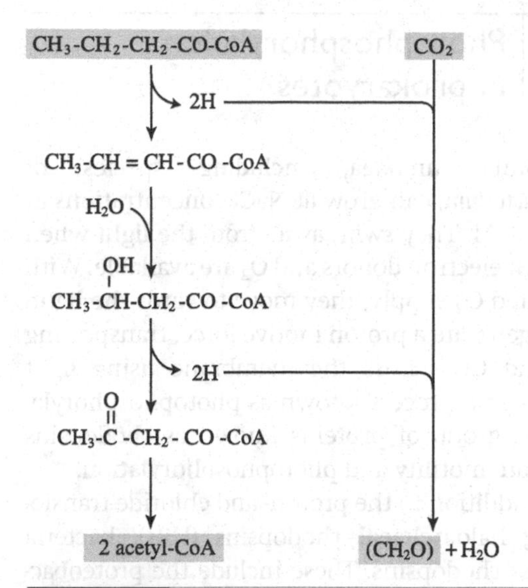

Figure 11.13 Butyrate metabolism by purple bacteria.

Heliobacteria and aerobic anoxygenic photosynthetic bacteria supplement their energy requirements through light reactions while growing as chemoorganotrophs.

Acetate is converted to acetyl-CoA before being metabolized through the TCA cycle and glyoxylate cycle, or by pyruvate:ferredoxin oxidoreductase to pyruvate. CO_2 is fixed to consume the excess reducing equivalents generated from the metabolism of organic compounds under photosynthetic conditions. For this reason, purple bacteria require CO_2 for photoorganotrophic growth on compounds more reduced than acetate. Butyrate metabolism is a good example of this (Figure 11.13).

The purple non-sulfur bacterium *Rhodobacter sphaeroides* does not possess isocitrate lyase, the key enzyme of the glyoxylate cycle, and metabolizes acetate through the TCA cycle and the ethylmalonyl-CoA pathway (Section 5.3.3). Calvin cycle enzymes are induced by the regulator protein CbbR and a two-component system (CbbRRS) when non-sulfur purple bacteria encounter photolithotrophic conditions.

11.5.2.2 Green sulfur bacteria

Green sulfur bacteria obtain energy required for their growth from light reactions, and can use simple organic carbon sources such as acetate, but not as the electron donor. Acetate is assimilated only when CO_2 and sulfur compounds are available as electron donors. On the other hand, filamentous anoxygenic phototrophic bacteria can use organic electron donors in phototrophic metabolism and can grow chemoorganotrophically like purple non-sulfur bacteria.

11.5.2.3 Cyanobacteria

The majority of cyanobacteria grow photolithotrophically, and a few grow photoorganotrophically. Obligately photolithotrophic cyanobacteria cannot use glucose, although they metabolize glycogen as storage materials with glycolytic enzymes. It is likely that they do not possess sugar transport systems.

The photoorganotrophic cyanobacteria use glucose and other organic compounds through glycolysis and the modified TCA cycle under photoorganotrophic conditions. Cyanobacteria lack 2-ketoglutarate dehydrogenase of the TCA cycle, and many cyanobacteria have genes encoding a novel 2-ketoglutarate decarboxylase and succinic semialdehyde dehydrogenase. These two enzymes catalyse the oxidation of 2-ketoglutarate to succinate in a modified TCA cycle (Section 5.2.2).

11.6 Photophosphorylation in prokaryotes

Halophilic archaea, including species of *Halobacterium*, can grow at NaCl concentrations of over 2.5 M. They swim away from the light when enough electron donors and O_2 are available. With a limited O_2 supply, they move towards the light. They generate a proton motive force, transporting H^+ and Cl^- across the membrane using light energy in a process known as photophosphorylation. A group of proteins known as rhodopsins facilitate motility and photophosphorylation.

In addition to the proton- and chloride-translocating haloarchaeal rhodopsins (HRs), bacteria possess rhodopsins. These include the proteobacterial proton pump rhodopsin (PR), an inward chloride pump (ClR) and sodium pump rhodopsin (NaR). The marine flavobacterium *Dokdonia eikastus* possesses two rhodopsins, a PR and a NaR. This NaR translocates H^+, Li^+ and Na^+, but not larger cations such as K^+. *Nonlabens marinus* possesses all three rhodopsins, PR, ClR and NaR. The extremely halophilic eubacterium, *Salinibacter ruber*, has a carotenoid that transfers light energy to its rhodopsin. Two variants are known among PR. The blue PR is prevalent in bacteria in deeper regions of the photic zone where blue light can still penetrate, while the green PR occurs in bacteria found in surface regions. A small proton pump rhodopsin is known in the Gram-positive bacterium *Exiguobacterium sibiricum*. Photosynthetic bacteria, including cyanobacteria, possess PR genes although their role is obscure. Some marine viruses contain PR genes and may be gene carriers for lateral gene transfer.

It is estimated that 50 per cent of microorganisms in the aquatic photic zone possess PR genes. Rhodopsin-containing bacteria use light to enhance growth or to promote survival during starvation. A *Dokdonia* sp. produces over four-fold-higher cell yields in the light, compared to the dark, on low concentrations of substrate. CO_2 is fixed in this bacterium through anaplerotic reactions involving pyruvate (or PEP) carboxylase (Section 5.3.1). CO_2 fixation in this bacterium contributes 24 to 31 per cent of cell carbon under light conditions.

Rhodopsin-based light-driven energy-generating mechanisms have distinct characteristics different to those based on chlorophylls. Rhodopsin-based photophosphorylation is mediated by a single protein so is much simpler than chlorophyll-based photosynthesis. It is hypothesized that ion-pumping mediated by rhodopsins was the first mechanism that evolved to convert light energy into biochemical energy in early cellular life forms.

Sensory rhodopsin II is synthesized for movement away from the light when the dissolved O_2 (DO) concentration is high, and light attracts halophilic archaea under a limited O_2 supply. Sensory rhodopsin I is synthesized with HRs under low DO conditions. With light energy, bacteriorhodopsin exports H^+ and halorhodopsin imports Cl^- to generate a proton motive force. These rhodopsins are purple proteins bound with retinal (Figure 11.14).

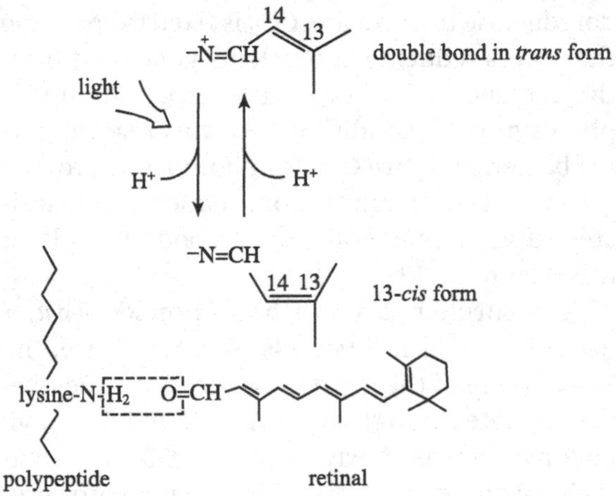

Figure 11.14 The structure of retinal bound to rhodopsins in species of *Halobacterium*.

(Gottschalk, G. 1986, *Bacterial Metabolism*, 2nd edn., Figure 9.20. Springer, New York)

Rhodopsin has the structure of a Schiff base between the chromophore retinal and a lysine residue of the apoprotein. The Schiff base exports H^+ with light energy.

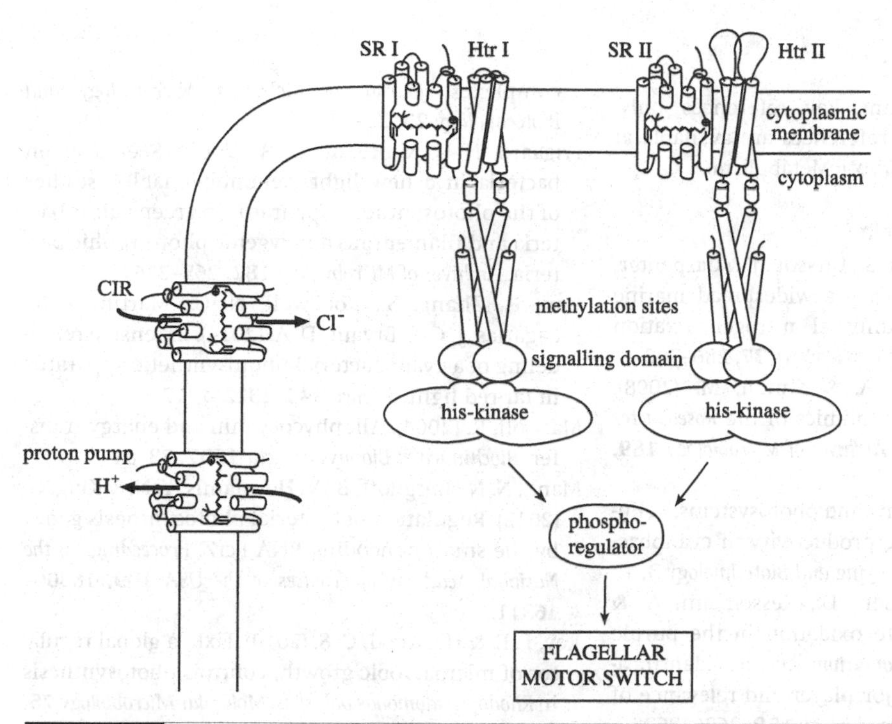

Figure 11.15 Rhodopsins and their functions in halophilic archaea.

(Modified from *Mol. Microbiol.* **28**: 1051–1058, 1998)

Halophilic archaea have sensory rhodopsin I and II to control phototaxis, and bacteriorhodopsin and halorhodopsin to generate a proton motive force utilizing light energy. Proton pump rhodopsin exports H^+ and CIR imports Cl^-, utilizing light energy. Sensory rhodopsins with a similar structure pass the information to the transducer proteins to control phototaxis. At low dissolved O_2 concentrations, the organisms move towards the light through the actions of sensory rhodopsin I and transducer I. Proton pump rhodopsin and CIR convert light energy into a proton motive force. Sensory rhodopsin II is synthesized in order for them to move away from the light at high dissolved O_2 concentrations.

Proton pump, proton pump rhodopsin; CIR, inward chloride pump; SRI, sensory rhodopsin I; HtrI, transducer I; SRII, sensory rhodopsin II; HtrII, transducer II.

Archeal rhodopsins have a similar structure, with retinal forming a Schiff base with a lysine residue of the peptide. The Schiff base releases H^+ when it receives light. Proton pump rhodopsin exports H^+ and CIR imports Cl^-, and the sensory rhodopsins pass the information to transducer proteins to control phototaxis (Figure 11.15). Bacterial rhodopsins have different amino acid sequences depending on the strain.

In *Dokdonia* sp., PR gene expression levels are substantially higher on growth in the light compared with growth in darkness. Little is known about how rhodopsin synthesis is regulated. A strong promoter sequence, that is recognized by the housekeeping sigma factor (σ^{70}, Section 12.1.1), is present in the upstream region of the PR gene in a *Dokdonia* sp. This bacterium has a considerable number of genes for light-sensing proteins, and some small RNAs are highly expressed in the light, together with the PR genes, but their exact role has not been elucidated.

Further Reading

Note this section contains key references only. Additional recommended references are available at www.cambridge.org/ProkaryoticMetabolism.

Photosynthetic bacteria

Bergman, B., Sandh, G., Lin, S., Larsson, J. & Carpenter, E. J. (2013). *Trichodesmium* – a widespread marine cyanobacterium with unusual nitrogen fixation properties. *FEMS Microbiology Reviews* **37**, 286–302.

Brinkhoff, T., Giebel, H. A. & Simon, M. (2008). Diversity, ecology, and genomics of the *Roseobacter* clade: a short overview. *Archives of Microbiology* **189**, 531–539.

Burnap, R. L. (2015). Systems and photosystems: cellular limits of autotrophic productivity in cyanobacteria. *Frontiers in Bioengineering and Biotechnology* **3**, 1.

Dahl, C., Franz, B., Hensen, D., Kesselheim, A. & Zigann, R. (2013). Sulfite oxidation in the purple sulfur bacterium *Allochromatium vinosum*: identification of SoeABC as a major player and relevance of SoxYZ in the process. *Microbiology* **159**, 2626–2638.

Fleischman, D. & Kramerb, D. (1998). Photosynthetic rhizobia. *Biochimica et Biophysica Acta* **1364**, 17–36.

Hauruseu, D. & Koblížek, M. (2012). Influence of light on carbon utilization in aerobic anoxygenic phototrophs. *Applied and Environmental Microbiology* **78**, 7414–7419.

Morgan-Kiss, R. M., Priscu, J. C., Pocock, T., Gudynaite-Savitch, L. & Huner, N.P.A. (2006). Adaptation and acclimation of photosynthetic microorganisms to permanently cold environments. *Microbiology and Molecular Biology Reviews* **70**, 222–252.

Zhang, C. C., Laurent, S., Sakr, S., Peng, L. & Bedu, S. (2006). Heterocyst differentiation and pattern formation in cyanobacteria: a chorus of signals. *Molecular Microbiology* **59**, 367–375.

Photosynthetic apparatus and pigments

Barber, J. (2002). Photosystem II: a multisubunit membrane protein that oxidises water. *Current Opinion in Structural Biology* **12**, 523–530.

Berghoff, B. A., Glaeser, J,. Nuss, A. M., Zobawa, M., Lottspeich, F. & Klug, G. (2011). Anoxygenic photosynthesis and photooxidative stress: a particular challenge for *Roseobacter*. *Environmental Microbiology* **13**, 775–791.

Drews, G. (2013). The intracytoplasmic membranes of purple bacteria – assembly of energy-transducing complexes. *Journal of Molecular Microbiology and Biotechnology* **23**, 35–47.

Frigaard, N.-U. & Bryant, D. A. (2004). Seeing green bacteria in a new light: genomics-enabled studies of the photosynthetic apparatus in green sulfur bacteria and filamentous anoxygenic phototrophic bacteria. *Archives of Microbiology* **182**, 265–276.

Gan, F., Zhang, S., Rockwell, N. C., Martin, S. S., Lagarias, J. C. & Bryant, D. A. (2014). Extensive remodeling of a cyanobacterial photosynthetic apparatus in far-red light. *Science* **345**, 1312–1317.

MacColl, R. (2004). Allophycocyanin and energy transfer. *Biochimica et Biophysica Acta* **1657**, 73–81.

Mank, N. N., Berghoff, B. A., Hermanns, Y. N. & Klug, G. (2012). Regulation of bacterial photosynthesis genes by the small noncoding RNA PcrZ. *Proceedings of the National Academy of Sciences of the USA* **109**, 16306–16311.

Rey, F. E. & Harwood, C. S. (2010). FixK, a global regulator of microaerobic growth, controls photosynthesis in *Rhodopseudomonas palustris*. *Molecular Microbiology* **75**, 1007–1020.

Samsonoff, W. A. & MacColl, R. (2001). Biliproteins and phycobilisomes from cyanobacteria and red algae at the extremes of habitat. *Archives of Microbiology* **176**, 400–405.

Umeno, D., Tobias, A. V. & Arnold, F. H. (2005). Diversifying carotenoid biosynthetic pathways by directed evolution. *Microbiology and Molecular Biology Reviews* **69**, 51–78.

Light reactions

Bryant, D. A. & Frigaard, N. U. (2006). Prokaryotic photosynthesis and phototrophy illuminated. *Trends in Microbiology* **14**, 488–496.

Chan, L.-K., Morgan-Kiss, R. M. & Hanson, T. E. (2009). Functional analysis of three sulfide:quinone oxidoreductase homologs in *Chlorobaculum tepidum*. *Journal of Bacteriology* **191**, 1026–1034.

González, A., Sevilla, E., Bes, M. T., Peleato, M. L. & Fillat, M. F. (2016). Pivotal role of iron in the regulation of cyanobacterial electron transport. *Advances in Microbial Physiology* **68**, 169–217.

Gregersen, L. H., Bryant, D. A. & Frigaard, N.-U. (2011). Mechanisms and evolution of oxidative sulfur metabolism in green sulfur bacteria. *Frontiers in Microbiology* **2**, 116.

Hanson, T. E., Bonsu, E., Tuerk, A., Marnocha, C. L., Powell, D. H. & Chan, C. S. (2016). *Chlorobaculum tepidum* growth on biogenic S(0) as the sole photosynthetic electron donor. *Environmental Microbiology* **18**, 2856–2867.

Koblížek, M., Mlčoušková, J., Kolber, Z. & Kopecký, J. (2010). On the photosynthetic properties of marine bacterium COL2P belonging to *Roseobacter* clade. *Archives of Microbiology* **192**, 41–49.

Rodriguez, J., Hiras, J. & Hanson, T. E. (2011). Sulfite oxidation in *Chlorobaculum tepidum*. *Frontiers in Microbiology* **2**, 112.

Vinyard, D. J., Ananyev, G. M. and Dismukes, G. C. (2013). Photosystem II: The reaction center of oxygenic photosynthesis. *Annual Review of Biochemistry* **82**, 577–606.

Circadian clock

Cohen, S. E. & Golden, S. S. (2015). Circadian rhythms in cyanobacteria. *Microbiology and Molecular Biology Reviews* **79**, 373–385.

Johnson, C. H., Zhao, C., Xu, Y. & Mori, T. (2017). Timing the day: what makes bacterial clocks tick? *Nature Reviews Microbiology* **15**, 232–242.

Rust, M. J., Golden, S. S. & O'Shea, E. K. (2011). Light-driven changes in energy metabolism directly entrain the cyanobacterial circadian oscillator. *Science* **331**, 220–223.

Snijder, J., Schuller, J. M., Wiegard, A., Lössl, P., Schmelling, N., Axmann, I. M., Plitzko, J. M. Förster, F. & Heck, A. J. R. (2017). Structures of the cyanobacterial circadian oscillator frozen in a fully assembled state. *Science* **355**, 1181–1184.

Tseng, R., Goularte, N. F., Chavan, A., Luu, J., Cohen, S. E., Chang, Y-G., Heisler, J., Li, S., Michael, A. K., Tripathi, S., Golden, S. S., LiWang, A. & Partch, C. L. (2017). Structural basis of the day-night transition in a bacterial circadian clock. *Science* **355**, 1174–1180.

Carbon metabolism

Joshi, G. S., Bobst, C. E. & Tabita, F. R. (2011). Unravelling the regulatory twist – regulation of CO_2 fixation in *Rhodopseudomonas palustris* CGA010 mediated by atypical response regulator(s). *Molecular Microbiology* **80**, 756–771.

Leroy, B., De Meur, Q., Moulin, C., Wegria, G. & Wattiez, R. (2015). New insight into the photoheterotrophic growth of the isocitrate lyase-lacking purple bacterium *Rhodospirillum rubrum* on acetate. *Microbiology* **161**, 1061–1072.

Rae, B. D., Long, B. M., Badger, M. R. & Price, G. D. (2013). Functions, compositions, and evolution of the two types of carboxysomes: Polyhedral microcompartments that facilitate CO_2 fixation in cyanobacteria and some proteobacteria. *Microbiology and Molecular Biology Reviews* **77**, 357–379.

Tang, K.-H., Tang, Y. J. & Blankenship, R. E. (2011). Carbon metabolic pathways in phototrophic bacteria and their broader evolutionary implications. *Frontiers in Microbiology* **2**, 00165.

Zhang, S. & Bryant, D. A. (2011). The tricarboxylic acid cycle in cyanobacteria. *Science* **334**, 1551–1553.

Photophosphorylation

Bamann, C., Bamberg, E., Wachtveitl, J. & Glaubitz, C. (2014). Proteorhodopsin. *Biochimica et Biophysica Acta* **1837**, 614–625.

Béjà, O., Aravind, L., Koonin, E. V., Suzuki, M. T., Hadd, A., Nguyen, L. P., Jovanovich, S. B., Gates, C. M., Feldman, R. A., Spudich, J. L., Spudich, E. N. & DeLong, E. F. (2000). Bacterial rhodopsin: evidence for a new type of phototrophy in the sea. *Science* **289**, 1902–1906.

Dutta, S., Weiner, L. & Sheves, M. (2015). Cation binding to halorhodopsin. *Biochemistry* **54**, 3164–3172.

Grote, M., Engelhard, M. & Hegemann, P. (2014). Of ion pumps, sensors and channels – perspectives on microbial rhodopsins between science and history. *Biochimica et Biophysica Acta* **1837**, 533–545.

Pinhassi, J., DeLong, E. F., Béjà, O., González, J. M. & Pedrós-Alió, C. (2016). Marine bacterial and archaeal ion-pumping rhodopsins: genetic diversity, physiology, and ecology. *Microbiology and Molecular Biology Reviews* **80**, 929–954.

Schertler, G. F. (2005). Structure of rhodopsin and the metarhodopsin I photointermediate. *Current Opinion in Structural Biology* **15**, 408–415.

Spudich, J. L. (2006). The multitalented microbial sensory rhodopsins. *Trends in Microbiology* **14**, 480–487.

Chapter 12

Metabolic regulation

Life processes transform materials available from the environment into cell components. Organic materials are converted to carbon skeletons for monomer and polymer synthesis, as well as being used to supply energy. Microbes synthesize monomers in the proportions needed for growth. This is possible through regulation of the reactions of anabolism and catabolism. With a few exceptions, microbial ecosystems are oligotrophic with a limited availability of nutrients, the raw materials used for biosynthesis. Furthermore, nutrients are not usually found in balanced concentrations while the organisms have to compete with each other for available nutrients.

Unlike animals and plants, unicellular microbial cells are more directly coupled to their environment, which changes continuously. Many of these changes are stressful so organisms have evolved to cope with this situation. They regulate their metabolism to adapt to the ever-changing environment.

Since almost all biological reactions are catalysed by enzymes, metabolism is regulated by controlling the synthesis of enzymes and their activity (Table 12.1). Metabolic regulation through the dynamic interactions between DNA or RNA and the regulatory apparatus employed determines major characteristics of organisms. In this chapter, the different mechanisms of metabolic regulation are discussed in terms of enzyme synthesis through transcription and translation and enzyme activity modulation.

Table 12.1 Regulatory mechanisms that control the synthesis and activity of enzymes.

		Mechanism
Enzyme synthesis		
	Transcription	promoter structure and sigma (σ) factors
		activator – positive control
		repressor – negative control
		termination – antitermination
	Translation	attenuation
		autogenous translational repression
		mRNA stability
Enzyme activity		feedback inhibition
		feedforward activation
		chemical modification
		physical modification
		degradation

12.1 Mechanisms regulating enzyme synthesis

The rate of biological reactions catalysed by enzymes is determined by the concentration and activity of the enzymes. Various mechanisms regulating the synthesis of individual enzymes are discussed here before multigene regulation is considered.

12.1.1 Regulation of transcription by promoter structure and sigma (σ) factor activity

RNA polymerase synthesizes messenger RNA (mRNA) that is needed for protein synthesis according to the DNA template. RNA synthesis is initiated with the recognition of the promoter by the σ-factor of the RNA polymerase (Section 6.11) in bacteria. Several σ-factors have been found in all the bacteria tested. They are grouped in two main families: the σ-70 and σ-54 families. The latter is composed exclusively of σ^{54}, while the σ-70 family includes all others. The σ-70 family is further divided into four groups (Table 12.2). Group 1 σ^{D} is essential, and responsible for the transcription of most genes in exponentially growing cells. A Group 2 σ-factor, σ^{S}, participates in the expression of stationary phase proteins. Group 3 σ-factors include σ^{F}, responsible for flagella synthesis, and σ^{H}, the heat shock response σ-factor in *Escherichia coli*. Sigma factors involved in spore formation in *Bacillus subtilis* belong to Group 3. σ^{E} is a member of Group 4 σ-factors. This group includes the extracytoplasmic function (ECF) σ-factor subfamily, the largest and most divergent group.

Each σ-factor recognizes a different promoter. The promoter region in *Escherichia coli* consists of six bases, each at −35 and −10 bases upstream of the transcription start site. Genes for proteins needed in large quantities have strong promoters with a high affinity for the σ-factors. For example, the housekeeping σ^{D} (σ^{70}) recognizes the promoter region consisting of TTGACA at −35 and TATAAT at −10 most efficiently, but is much less effective for those with substituted base(s) in *Escherichia coli*. It should be emphasized that promoter activity is regulated by other transcription factors such as a repressor and an activator, as well as others that interact with DNA around the promoters, as discussed later.

During active growth, *Escherichia coli* mainly uses the housekeeping sigma factor (σ^{D}), and its association with core RNA polymerase (RNAP) is generally favoured because of its higher

Table 12.2 RNA polymerase σ-factors in *Escherichia coli*.

σ-factor	Gene	Function	Promoter consensus sequence −35	−10	Anti-σ factor
σ–70 family					
Group 1					
σ^{D} (σ^{70})	*rpoD*	housekeeping	TTGACA	TATAAT	AsiA[a]
Group 2					
σ^{S} (σ^{38})	*rpoS, katF*	stringent response, general stress response	–	CTATACT	RssB
Group 3					
σ^{F} (σ^{28})	*rpoF, fliA*	chemotaxis	TAAA	GCCGATAA	FlgM
σ^{H} (σ^{32})	*rpoH, htpR*	heat shock protein	CTTGAAA	CCCATnT	DnaK
Group 4					
σ^{E} (σ^{24})	*rpoE*	stress response	GAACTT	TCTRA	RseA
σ–54 family					
σ^{N} (σ^{54})	*rpoN, ntrA*	nitrogen metabolism	TGGCAC	TTGCW	

R: A or G; n: A, T, G, or C; W: A or T.
[a]Bacteriophage T4.

intracellular level and higher affinity for core RNAP. When alternative sigma factors are needed for environmental stresses, σ^D is reversibly separated from core RNAP for an alternative σ factor to form complete RNAP. Various facilitators are involved in this shifting process. These are the alarmone (p)ppGpp produced by the stringent response (Section 12.2.1), and proteins (DksA, RNA polymerase-binding transcription factor; Rsd, σ^D-binding protein; and sigma factor-binding Crl protein) and a small RNA (6S RNA).

While the essential housekeeping σ^D functions in exponentially growing cells, other σ-factors are activated according to growth conditions, including σ^E, σ^F, σ^H, σ^N and σ^S. The promoters recognized by σ^S (σ^{38}) do not have a −35 region but have a longer −10 region, known as the extended −10 region. As cells enter the stationary phase, *Escherichia coli* accumulates σ^S and becomes resistant to general stress. Different stress conditions lead to induction of specific sRNAs that stimulate σ^S and anti-adaptor synthesis. Anti-adaptors protect the sigma factor from proteolytic degradation. Disruption of central metabolism induces similar effects.

ECF sigma factors are generally kept inactive by specific anti-σ factors that are separated from their cognate σ factors to cope with environmental stimuli. Free σ factors trigger adaptive responses for survival of the cell. *Escherichia coli* has five ECFs. These are σ^E, Cpx, Bae, Rcs, and the phage shock protein response. σ^E is indispensable in *E. coli*. Single σ^E mutants have not been obtained, while double mutants of σ^E and antitoxin (HicB) can grow. The toxin/antitoxin system (Section 13.4.2) is related to the function of σ^E.

Sigma factors with similar functions have different names in different organisms. The housekeeping σ-factor is called σ^D in Gram-negative bacteria and σ^A in Gram-positive bacteria. HrdB is the housekeeping σ-factor in *Streptomyces* spp. Consequently, σ-factors with the same name have different functions in different bacteria: σ^D is the housekeeping σ-factor in *Escherichia coli* but is the σ-factor for flagella formation in *Bacillus subtilis*. In a related terminological problem, the extracytoplasmic function σ-factor, σ^E, should not be regarded as functioning extracytoplasmically. Instead, σ^E participates in the expression of genes needed to repair denatured proteins of extracytoplasmic location.

Since a σ-factor recognizes multiple promoters throughout the chromosome, a specific σ-factor participates in the transcription of functionally unrelated genes. Regulation by σ-factor–promoter interaction is a type of global regulation system, discussed later (Section 12.2).

Proteins known as anti-sigma (anti-σ) factors bind σ-factors and inhibit their activity. When *Escherichia coli* is infected with bacteriophage T4, a phage-originating anti-σ factor, AsiA, inhibits the activity of the bacterial housekeeping σ^D to produce more phage proteins. An anti-σ factor for σ^E, RseA, is a membrane protein. A periplasmic protein, RseB, binds the anti-σ factor, stabilizing its complex with σ^E under normal conditions. When *Escherichia coli* cells are stressed, outer membrane proteins (OMPs) become misfolded. These misfolded OMPs remove RseB from RseA. A membrane protease, DegS, senses misfolded OMPs and becomes active to degrade the periplasmic domain of RseA free of RseB. Two consecutive proteolytic cleavages of RseA are required to release σ^E. Following degradation by DegS, the remaining RseA protein becomes sensitive to proteolysis by an intramembrane protease, RseP and by a cytoplasmic protease, ClpXP, subsequently liberating σ^E (Figure 12.1). Previously, RseB was regarded as an anti-anti-sigma factor. The activity of σ^E is further increased by the alarmone (p)ppGpp and DksA (RNA polymerase-binding transcription factor) which inhibits the housekeeping σ^D activity. In the case of σ^F, which is responsible for flagellin synthesis, the anti-σ factor, FlgM, is exported into the environment through the flagellin export mechanism when more flagellin is needed (Section 12.2.11). The stationary phase σ-factor is inactivated by the anti-σ factor RssB under normal growth conditions. RssB is hydrolysed by an ATP-dependent protease (ClpXP) when σ^S is needed (Section 12.2.1).

In alphaproteobacteria, the general stress σ^S is not known; an alternative sigma factor EcfG is the general stress sigma factor. EcfG activity is controlled by NepR-like anti-σ factors and PhyR-

(a)

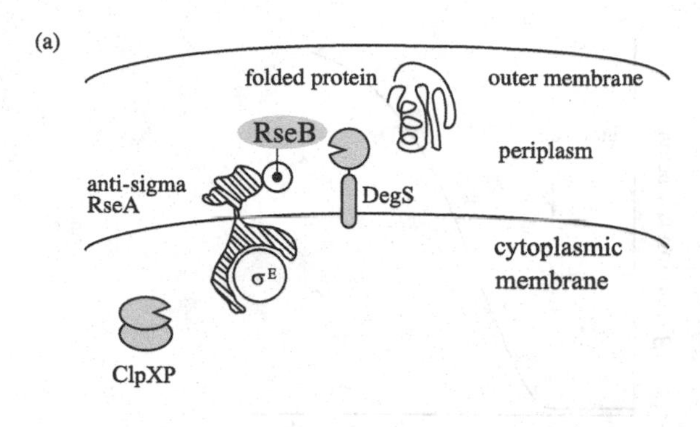

(b)

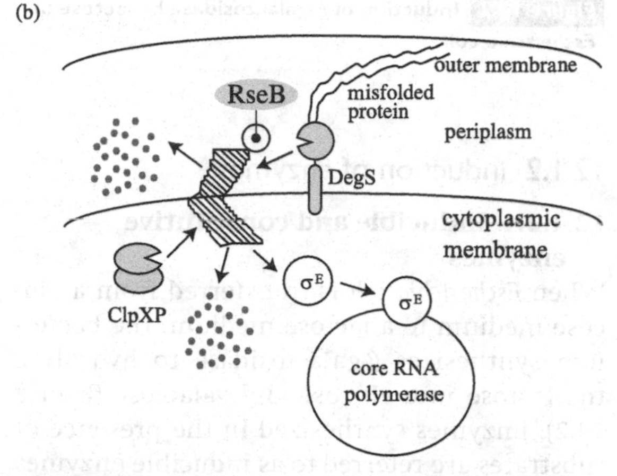

Figure 12.1 Regulation of extracytoplasmic function sigma factor σ^E activity by anti-σ factor RseA and RseB.

(Modified from *Trends Biochem. Sci.* **42**: 232–242, 2017)

The σ^E response is activated by a series of proteolysis steps that degrade the anti-σ factor RseA and liberate σ^E into the cytosol, where it can increase target gene transcription. The RseA periplasmic domain is first cleaved by DegS. RseB binds RseA and inhibits DegS activity. DegS senses unfolded outer membrane proteins. RseB is displaced from RseA in response to lipopolysaccharide binding. RseB displacement allows DegS to cleave RseA and initiate the response. RseA is subsequently cleaved by the intramembrane protease RseP. The cytoplasmic domain of RseA is then degraded by the ClpXP protease to release σ^E.

like proteins that act as anti-anti-σ factors by a similar mechanism.

Gram-positive bacteria, including species of *Bacillus*, *Listeria* and *Staphylococcus*, possess an alternative sigma factor σ^B that is activated under various environmental conditions such as nutrient limitation, and stresses due to antibiotics, elevated osmotic pressure, heat or cold shock and acidic pH values. This sigma factor is also involved in virulence and host invasion in pathogenic *Listeria* spp. and *Staphylococcus* spp. In spite of obvious and significant differences between these stresses, the result is the same, i.e. the activation of σ^B. Rsb (regulator of sigma B) proteins regulate its activity. During exponential growth, the anti-sigma factor RsbW binds σ^B, inactivating it. Under stressful conditions, the anti-anti-sigma factor RsbV is dephosphorylated to the active form which releases RsbW from σ^B. Under energy stress, RsbP dephosphorylates RsbV and the environmental stresses enable RsbU to dephosphorylate RsbV. A 1.8 MDa protein complex called the stressosome orchestrates activation of σ^B by environmental stress. The stressosome comprises multiple copies of RsbS and RsbR with the switch kinase RsbT. With environmental stress, RsbT phosphorylates conserved serine and threonine residues in RsbS and RsbR. Phosphorylated RsbS and RsbR release RsbT that activates RsbU. RsbX dephosphorylates the stressosome when σ^B activity is not needed. Over 130 genes are recognized by σ^B.

Seven extracytoplasmic function σ-actors are recognized in *Bacillus subtilis*. These are σ^M, σ^W, σ^X, σ^V, σ^Y, σ^Z and σ^{YlaC}, and have roles in cell envelope homeostasis. Less well known are the functions of the latter three ECFs. When *B. subtilis* cells are challenged by lysozyme, σ^V activates genes for enzymes modifying peptidoglycan and teichoic acid. σ^V has a similar function in *Enterococcus faecalis*. Four sigma factors are involved in sporulation of *Bacillus subtilis*. They are σ^F and σ^G in the forespore and σ^E and σ^K in the parental cell which act sequentially. An anti-σ^F (Fin) inactivates σ^F allowing σ^G to function. Similarly activities of σ^E and σ^K are controlled in the parental cell.

Clostridium thermocellum possesses an alternative sigma factor, σ^I, and its membrane-bound anti-sigma factor, RsgI. RsgI consists of an extracellular carbohydrate-active module and cytoplasmic anti-σ peptide domain. When RsgI binds cellulose or xylan, σ^I is released from RsgI to transcribe cellulosome protein genes.

An ECF, σ^C, activates cytochrome *bd*-type quinol oxidase in *Corynebacterium glutamicum* under conditions where the main terminal oxidase, aa_3-type cytochrome *c* oxidase, is not functional.

Pseudomonas aeruginosa possesses several ECF sigma factors. An ECF, SbrI, activates the gene *muiA* whose product inhibits swarming motility and promotes biofilm formation. Its activity is controlled by the anti-sigma factor, SbrR. Another ECF, SigX is involved in its pathogenicity by altering fatty acid composition of the cell membrane, increasing swarming activity to reduce biofilm formation, and reducing c-di-GMP levels.

The highly aerotolerant anaerobic *Bacteroides fragilis* has about 14 ECFs including EcfO, an oxidative stress response sigma factor. Its activity is controlled by the anti-sigma factor Reo.

ECFs control transcription of various genes according to environmental signals. ECF-dependent signalling is versatile and diverse, and the majority of mechanisms that regulate ECF activity remain to be characterized.

Sigma factors are unknown in archaea. Transcription factors, repressors and activators recognize the promoter region in archaea in a similar mechanism to that occurring in eukaryotes (Section 6.11.1).

Promoter activity is also regulated by the superhelix DNA structure and various general DNA-binding proteins such as H-NS, Fis and StpA, in addition to the specific activators and repressors known as transcription factors (TFs). Promoters associated with the genes involved in stress responses are under the control of multiple TFs, each monitoring one specific environmental condition or factor. In *Escherichia coli*, the *sdiA* gene encoding the master regulator of cell division and quorum sensing is controlled by at least 15 TFs, including five two-component system regulators. Under normal growth conditions, all these TFs repress *sdiA* expression. In the stationary phase, these TFs repress transcription of the *sdiA* gene, ultimately leading to suppression of cell division.

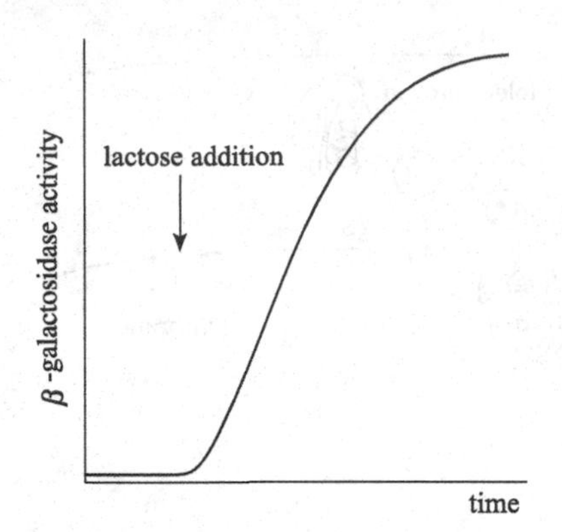

Figure 12.2 **Induction of β-galactosidase by lactose in *Escherichia coli*.**

12.1.2 Induction of enzymes

12.1.2.1 Inducible and constitutive enzymes

When *Escherichia coli* is transferred from a glucose medium to a lactose medium, the bacterium synthesizes β-galactosidase to hydrolyse the lactose into glucose and galactose (Figure 12.2). Enzymes synthesized in the presence of substrates are referred to as inducible enzymes and the substrate is termed the inducer. Constitutive enzymes are those enzymes that are produced under all growth conditions. Inducible enzymes are generally those used in the catabolism of carbohydrates such as polysaccharides (cellulose, starch, etc.), oligosaccharides (lactose, trehalose, raffinose, etc.), minor sugars (arabinose and rhamnose) and aromatic compounds.

When a single inducer induces more than two enzymes, they are produced either simultaneously or sequentially. The former is referred to as coordinate induction, and the latter as sequential induction (Figure 12.3). Genes of coordinate induction are in the same operon, and genes from separate operons are induced sequentially.

12.1.2.2 Enzyme induction

Enzyme induction is regulated at the level of transcription. Lactose induces the production of

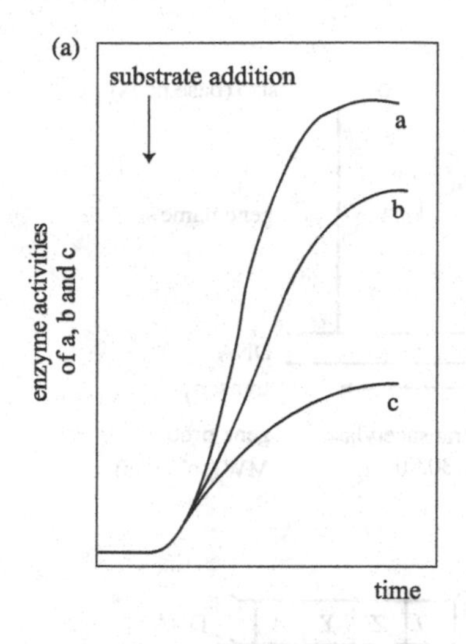

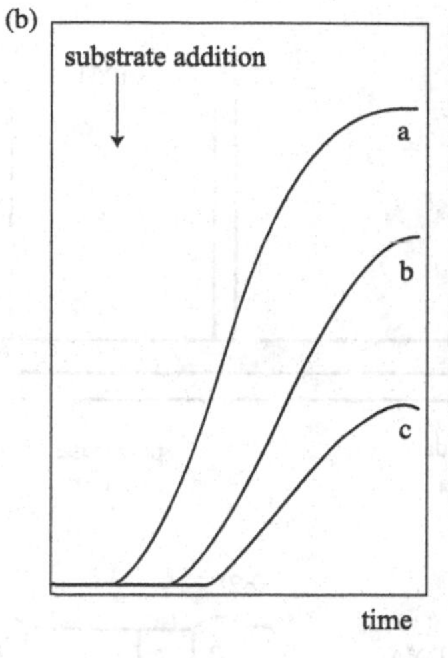

Figure 12.3 Coordinate induction and sequential induction of multiple enzymes by a single inducer.

(Gottschalk, G. 1986, *Bacterial Metabolism*, 2nd edn., Figure 7.2. Springer, New York)

Enzymes from genes of the same operon are induced simultaneously in coordinate induction (a) and genes from different operons are induced sequentially (b). The product of the first enzyme reaction is the inducer of the second enzyme in sequential induction.

β-galactosidase, permease and transacetylase. Their structural genes form an operon (*lac* operon) with a promoter and operator (Figure 12.4a). The regulatory gene (*lacI*) next to the 5′ end of the operon is expressed constitutively with its own promoter. In the absence of the inducer, the repressor protein binds the operator region of the *lac* operon, inhibiting RNA polymerase from binding the promoter region. Consequently, the structural genes are not transcribed (Figure 12.4b1). On the other hand, when the inducer is available, it binds the repressor protein, removing it from the operator region (Figure 12.4b2). The repressor protein is not produced in a *lacI* mutant. This mutant transcribes the structural genes as constitutive enzymes in the absence of the inducer. In this sense, the regulation by repressor proteins is referred to as negative control.

12.1.2.3 Positive and negative control

As stated previously, the *lac* operon is regulated by a negative control mechanism. Activator proteins are involved in the regulation of catabolic genes for arabinose, rhamnose, maltose and others. Genes for arabinose catabolism consist of *araA*, *B*, *C*, *D*, *E* and *F*; *araC* is a regulatory gene encoding an activator protein. *araC* mutants are unable to use arabinose, since an AraC complex with the inducer activates transcription of the structural genes (Figure 12.5). *araA*, *B*, *C* and *D* form an operon, and *araE* and *F* occupy other parts of the chromosome. The term regulon is used to define genes of the same metabolism controlled by the same effectors scattered around the chromosome, such as *ara* genes. Regulation by an activator, as in the *ara* regulon, is referred to as positive control.

12.1.3 Catabolite repression

When *Escherichia coli* or *Bacillus subtilis* is cultivated in a medium containing glucose and lactose, they grow in a distinct two-phase pattern, as shown in Figure 12.6. This is called diauxic growth or diauxie. This is due to the fact that the readily utilizable glucose and its metabolites repress the utilization of lactose. The term catabolite repression is used to describe this regulatory mechanism, which regulates many catabolic genes and operons in bacteria.

Gram-negative bacteria have two separate catabolite repression mechanisms, one involving a cAMP-CRP (cAMP receptor protein) complex and the other a catabolite repressor/activator (Cra) protein. In Gram-positive bacteria, the catabolite control protein A (CcpA) has a similar function.

It is interesting to note that the hyperthermophilic archaeon *Sulfolobus acidocaldarius* utilizes xylose and glucose simultaneously without diauxie. Gene expression in cells grown on xylose alone is very similar to that in cells grown on a mixture of xylose and glucose, and substantially different from that in cells grown on glucose

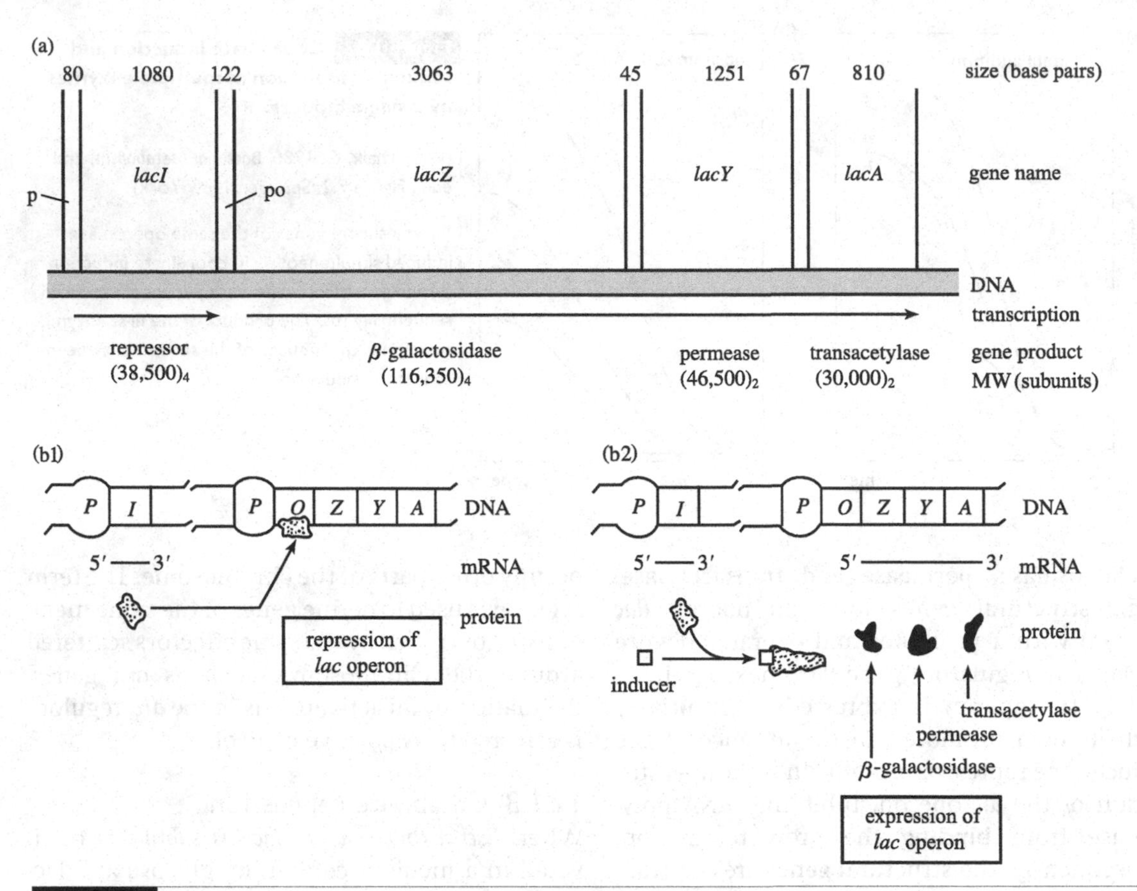

Figure 12.4 Induction mechanism of the *lac* operon in *Escherichia coli*.

(Gottschalk, G. 1986, *Bacterial Metabolism*, 2nd edn., Figure 7.4. Springer, New York)

The *lac* operon consists of a promoter, operator and structural genes, and a repressor protein is produced from the *lacI* gene that has its own promoter (a). In the absence of inducer, the repressor protein binds the operator region of the operons, preventing transcription of the structural genes (b1). The inducer forms a complex with the repressor protein. The complex cannot bind the operator region, and the structural genes are transcribed (b2). The structural genes are transcribed as constitutive enzymes in the *lacI* mutant. The regulation by repressor proteins is referred to as negative control.

alone. The mechanism by which the organism utilizes a mixture of sugars has yet to be elucidated.

12.1.3.1 Carbon catabolite repression by the cAMP-CRP complex

The primary carbon catabolite repression (CCR) that occurs in many Gram-negative bacteria is related to the intracellular cyclic AMP (cAMP) concentration. When the readily utilizable substrate is exhausted, the cAMP concentration increases, and this cyclic nucleotide forms a complex with the cAMP receptor protein (CRP or catabolite activator protein, CAP). This complex controls many operons (Section 12.2).

Glucose is transported through the phosphotransferase (PT) system (group translocation mechanism) in many bacteria. When the glucose concentration is low, enzyme II_A of the PT system remains phosphorylated (Section 3.5). The phosphorylated enzyme II_A activates the activity of adenylate cyclase, a cytoplasmic membrane enzyme. This enzyme converts ATP to cAMP:

$$\text{ATP} \xrightarrow{\text{adenylate cyclase}} \text{adenosine-3}',5'\text{-monophosphate (cyclic AMP)} + PP_i$$

A cytoplasmic enzyme, phosphodiesterase, hydrolyses cAMP but the activity is very low:

$$\text{cAMP} + H_2O \xrightarrow{\text{phosphodiesterase}} \text{AMP}$$

When adenylate cyclase activity is low due to the low level of phosphorylated enzyme II_A with active glucose transport, the cAMP concentration is kept low, and CRP cannot form the complex. It is known that cAMP is actively exported to keep the intracellular concentration low with sufficient glucose. When glucose is exhausted, the rate of cAMP formation is higher than that of hydrolysis and export, facilitating cAMP–CRP complex formation. This complex activates the transcription of many operons including the *lac* operon.

When lactose is provided with a low level of glucose, lactose binds the repressor protein, as shown in Figure 12.4, and the cAMP–CRP complex binds the CRP site of the promoter region, activating transcription of the structural genes (Figure 12.7). Studies with CRP and adenylate cyclase mutants have shown that the cAMP–CRP complex regulates the expression of over

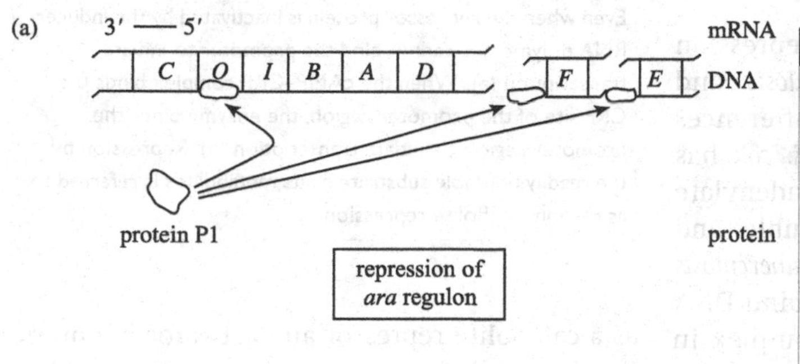

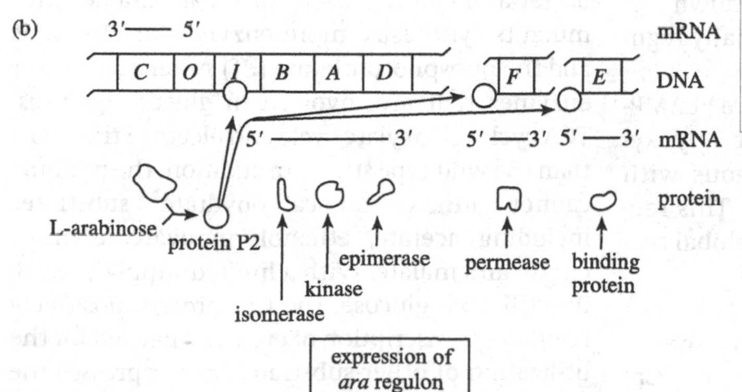

Figure 12.5 Induction of enzymes of arabinose metabolism in *Escherichia coli*.

(Gottschalk, G. 1986, *Bacterial Metabolism*, 2nd edn., Figure 7.5. Springer, New York)

The regulator protein AraC is in the protein P1 form to bind the operator region, inhibiting transcription of the structural genes in the absence of arabinose (a). When the inducer arabinose binds, protein P1 is converted to protein P2. Protein P2 activates transcription of the structural genes. *araC* mutants cannot use arabinose. The term positive control is used to describe metabolic regulation by activators.

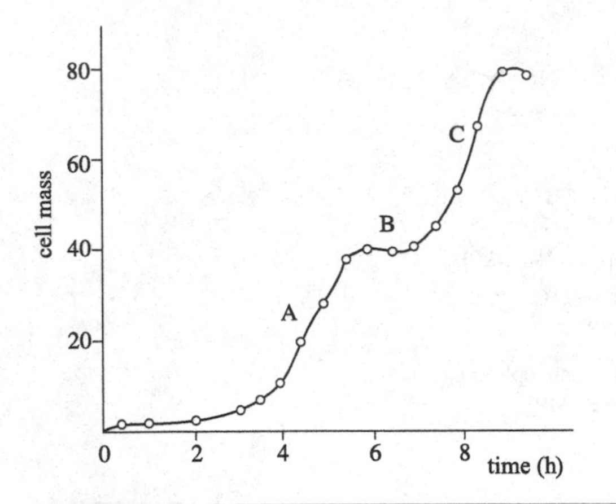

Figure 12.6 Diauxic growth of *Escherichia coli* on glucose and lactose.

When *Escherichia coli* is cultivated on a mixture of glucose and lactose, the bacterium grows on glucose, repressing lactose utilization (A) at the beginning. When glucose is exhausted the bacterium grows again on lactose (C) after a lag period (B). This is referred to as diauxic growth or diauxie. Diauxic growth is the result of a regulatory mechanism known as catabolite repression.

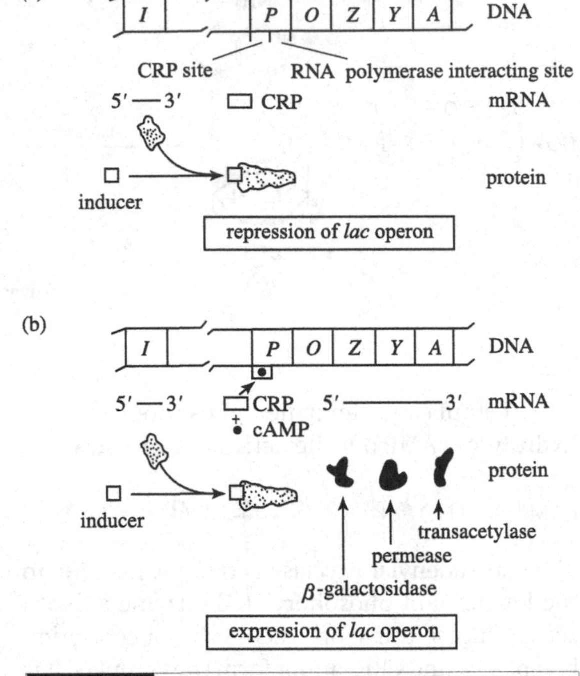

Figure 12.7 Activation of the *lac* operon by the inducer and the cAMP–CRP complex.

(Gottschalk, G. 1986, *Bacterial Metabolism*, 2nd edn., Figure 7.7. Springer, New York)

Even when the repressor protein is inactivated by the inducer, RNA polymerase cannot bind the promoter to initiate transcription (a). When the cAMP–CRP complex binds the CRP site of the promoter region, the enzyme binds the promoter to initiate transcription (b). Repression by the readily utilizable substrate or its metabolites is referred to as carbon catabolite repression.

200 proteins. Carbon catabolite repression (CCR) by the cAMP–CRP complex is known only in Gram-negative bacteria.

Cyclic-AMP dependent catabolite repression is known in *Mycobacterium tuberculosis* and *Pseudomonas putida*, with some differences from that of *Escherichia coli*. While *E. coli* has only one adenylate cyclase, at least 16 adenylate cyclase-like proteins, including soluble and membrane proteins, are known in *M. tuberculosis* and two in *P. putida*. CRP and cAMP bind DNA independently without forming a complex in these bacteria. Two CRPs are known in *Mycobacterium smegmatis* that differentially regulate genes.

Corynebacterium glutamicum has a cAMP-dependent transcriptional regulator (glyoxylate cycle regulator, GxlR) homologous with the CRP and with a similar function. This regulatory mechanism is an example of global regulation (Section 12.2).

12.1.3.2 Catabolite repressor/activator

In addition to the cAMP–CRP complex, the Cra (catabolite repressor/activator) protein functions as a catabolite repressor and activator in enteric bacteria including *Escherichia coli*. Cra-negative mutants synthesize more enzymes of glycolysis and the phosphotranferase (PT) system and fewer enzymes that are involved in gluconeogenesis, TCA cycle, glyoxylate cycle and electron transport than the wild-type strain. In addition, the mutants cannot utilize non-carbohydrate substrates including acetate, ethanol, pyruvate, alanine, citrate and malate. With a limited supply of readily utilizable glucose, the Cra protein positively regulates transcription of enzymes needed for the utilization of other substrates, and represses the genes for glycolytic enzymes (Figure 12.8). When

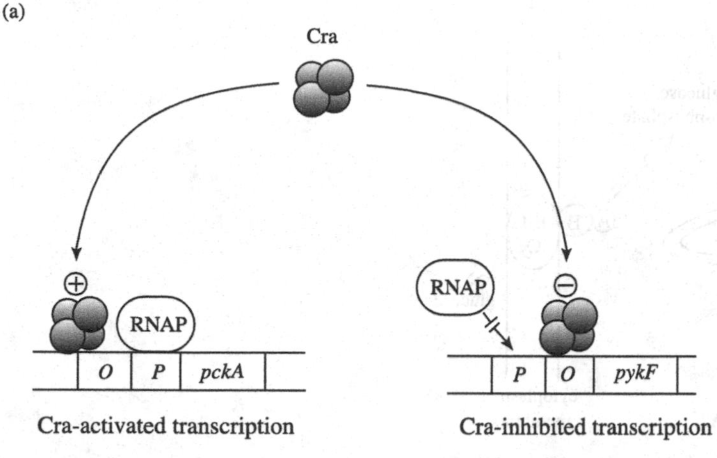

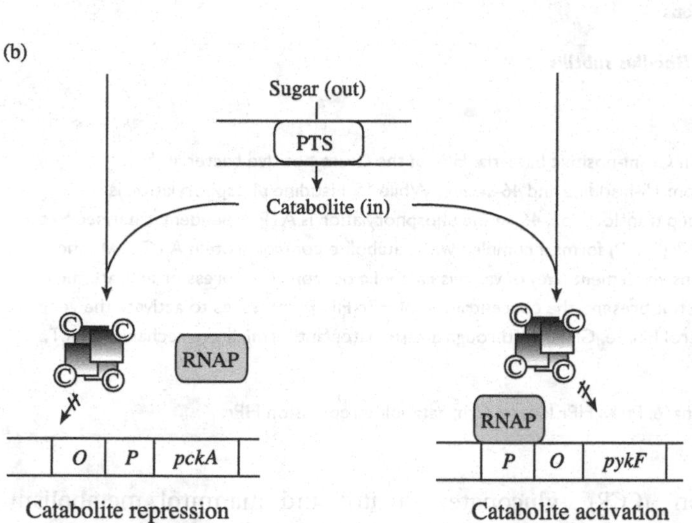

Figure 12.8 cAMP–CRP independent catabolite repression by Cra protein in *Escherichia coli*.

(*J. Bacteriol.* **178**: 3411–3417, 1996)

Cra (catabolite repressor/activator) protein functions as a catabolite repressor and activator for enzymes related to sugar metabolism. When a non-carbohydrate substrate is used, the Cra protein binds operator regions of the genes for PEP carboxykinase (*pckA*) and pyruvate kinase (*pykF*), activating the former and repressing the latter (a). When glucose is used as the substrate (b), the Cra protein is inactivated through complex formation with the catabolite and loses its function as an activator or repressor.

acetate or ethanol is used as the substrate, the Cra protein activates the expression of genes for the enzymes of gluconeogenesis (fructose-1,6-diphosphatase, Section 4.2.4) and the anaplerotic sequence (PEP carboxykinase, Section 5.3.1), and represses the transcription of genes for glycolytic enzymes (phosphofructokinase). Genes activated by the Cra protein are those that have a Cra-binding site upstream of the promoter region (left-hand side of Figure 12.8), and the genes are repressed where the Cra-binding site overlaps the promoter region or occupies it downstream (right-hand side of Figure 12.8).

In species of *Pseudomonas* and *Acinetobacter*, succinate and acetate repress catabolism of aromatic compounds through the protein Crc (catabolite repression control). Crc proteins bind mRNAs of the target genes to inhibit translation. Crc protein activity in *Pseudomonas* spp. is modulated by small non-coding RNAs, CrcZ and CrcY, as in the CsrA/CsrB system (Section 12.1.9.3), in concert with catabolites.

Sinorhizobium meliloti uses succinate preferentially over less favourable carbon sources such as lactose. Succinate-mediated catabolite repression in this bacterium is controlled by a two-component regulatory system (Section 12.1.7) that senses succinate.

12.1.3.3 Carbon catabolite repression in Gram-positive bacteria with a low G + C content

The cAMP–CRP complex is not known in Gram-positive bacteria, and these bacteria have a

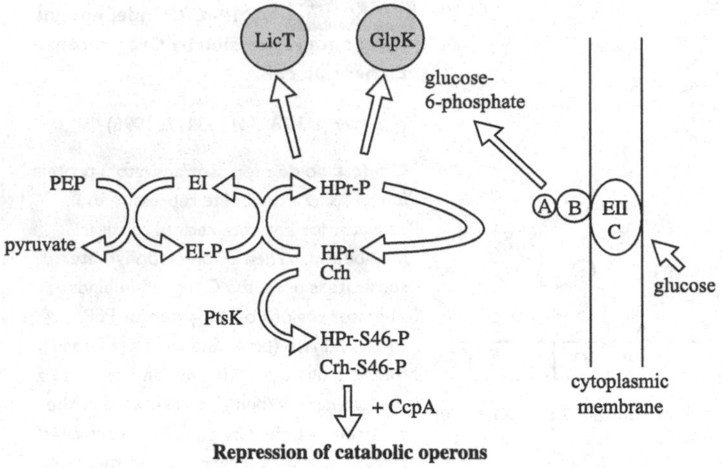

LicT GlpK glucose-6-phosphate PEP EI HPr-P pyruvate EI-P HPr Crh A B EII C glucose PtsK HPr-S46-P Crh-S46-P + CcpA cytoplasmic membrane Repression of catabolic operons

Figure 12.9 Carbon catabolite repression in *Bacillus subtilis*.

(*Curr. Opin. Microbiol.* **2**: 195–201, 1999)

Catabolite repression by cAMP–CRP is not known in Gram-positive bacteria. HPr of the Gram-positive bacterial phosphotransferase has two sites for phosphorylation: 15-histidine and 46-serine. While 15-histidine phosphorylation is phosphoenolpyruvate (PEP)-dependent for sugar group translocation, 46-serine phosphorylation is ATP dependent, catalysed by the HPr kinase activated by fructose-1,6-diphosphate. HPr(Ser-P) forms a complex with catabolite control protein A (CcpA). The CcpA–HPr(Ser-P) complex binds a catabolite-responsive element (*cre*) of various catabolic operons to repress or activate their transcription. When readily utilizable substrates are not present, the concentration of HPr(His-P) increases to activate the genes for alternative carbon sources, either directly (glycerol kinase, GlpK) or through a terminator/antiterminator mechanism (LicT, Section 12.1.6).

LicT, transcription antiterminator; GlpK, glycerol kinase; PtsK, HPr kinase; Crh, catabolite repression HPr.

different carbon catabolite repression (CCR) mechanism. The HPr protein of the phosphotransferase system in Gram-positive bacteria with a low G + C content is phosphorylated at the histidine-15 residue by being coupled to the conversion of PEP to pyruvate as in Gram-negative bacteria. HPr in these bacteria has a second phosphorylation site at serine-46, in addition to histidine-15. PEP phosphorylates the histidine-15 for the group translocation while HPr(ser) kinase (PstK) phosphorylates the serine-46, consuming ATP. When glucose is available, the cell maintains high levels of glucose metabolites (fructose-1,6-diphosphate) and non-phosphorylated HPr. Fructose-1,6-diphosphate activates HPr(ser) kinase to phosphorylate the serine residue of the non-phosphorylated HPr protein. This phosphorylated HPr protein [HPr(Ser-P)] forms a complex with CcpA (catabolite control protein A). The CcpA–HPr(Ser-P) complex represses the expression of genes for enzymes of gluconate, glucitol and mannitol metabolism, and activates others including the *ilv-leu* operon for branched-chain amino acid synthesis through binding to the *cre* (catabolite responsive element) that is located either upstream or within their promoters (Figure 12.9). ATP-dependent HPr(ser) phosphorylation is not known in Gram-negative bacteria.

The CcpA–HPr(Ser-P) complex is an important regulator in *Bacillus subtilis*. More than 28 genes or operons have been experimentally demonstrated to be directly regulated by this complex. Moreover, recent global gene expression analyses combined with analyses of *cre* occurrences in the chromosome sequence suggest that approximately 10 per cent of the genes might be directly regulated by this regulator in *Bacillus subtilis*.

Control by the CcpA–HPr(Ser-P) complex is known in low G + C content Gram-positive bacteria, including the genera *Bacillus*, *Staphylococcus*,

Streptococcus, *Lactococcus*, *Enterococcus*, *Mycoplasma*, *Clostridium* and *Listeria*. This complex regulates many genes analogous to the cAMP–CRP complex in Gram-negative bacteria. The expression of virulence factors is controlled by the CcpA–HPr(Ser-P) complex in pathogenic low G + C content Gram-positive bacteria. In addition to catabolite repression, HPr(Ser-P) is responsible for the repression of gene expression through the inducer exclusion/expulsion mechanism. When glucose is present, the expression of proteins for lactose utilization is repressed (inducer exclusion), and lactose is exported (inducer expulsion). HPr(Ser-P) facilitates this inducer exclusion/expulsion.

When the glucose supply is limited, the expression of some hydrolases, such as levanase and β-glucosidase, is activated in *Bacillus subtilis*. The CcpA–HPr(Ser-P) complex is not involved in this regulation (upper part of Figure 12.9). With a limited supply of readily utilizable substrate, HPr is phosphorylated at histidine-15. HPr(His-P) phosphorylates the transcription regulatory protein (LevR) or transcription antiterminator (LicT). Phosphorylated LevR and LicT activate expression of the hydrolases. This regulatory mechanism is referred to as a terminator/antiterminator mechanism (Section 12.1.6).

A difference between CRP-dependent catabolite repression found in Gram-negative bacteria and CcpA-dependent catabolite repression found in Gram-positive bacteria is the tightness of the coupling between the PTS (phosphotransferase system), transport and regulatory functions. In the CRP-dependent mechanism, PTS enzyme II_A activates the primary sensor adenylate cyclase, while HPr(ser) kinase, the primary sensor in the CcpA-dependent mechanism, is activated by glycolytic intermediates. A search for HPr(ser) kinase gene homologues showed that this gene is widespread not only in Gram-positive bacteria, but also in Gram-negative bacteria and mycoplasmas.

Metabolic control by the effectors of catabolite repression, cAMP–CRP and CcpA–HPr (Ser-P) complexes, and the Cra protein, are examples of global regulation mechanisms (Section 12.2).

Similar regulators to CcpA have been identified in *Bacillus subtilis*. Crh (catabolite repression HPr) protein has a similar function in catabolite repression in *Bacillus subtilis*. Crh has a similar structure to HPr, but does not participate in group translocation. There is evidence to suggest that Crh is involved in the repression of catabolic genes by non-sugar electron donors. CcpB is the regulator for catabolite repression of gluconate (*gnt*) and xylose (*xyl*) operons by glucose, mannitol and sucrose. CcpC represses transcription of genes that encode enzymes of the TCA cycle and the CcpN genes of gluconeogenesis.

A regulator catabolite control protein E (CcpE) is known in *Staphylococcus aureus*. CcpE functions as an activator of the TCA cycle, increasing the transcription of genes for aconitase and citrate synthase. In addition, the GTP-sensing transcriptional pleiotropic repressor (CodY) and pentose phosphate pathway regulator (RpiRc) are involved in CCR in the low G + C Gram-positive bacteria. HPr (Ser-P) regulates CcpA-independent carbon catabolite repression in conjunction with sugar-specific EII permeases in *Streptococcus mutans*.

As stated earlier (Section 4.1.5.4), CggR (central glycolytic gene regulator protein) represses the genes encoding enzymes catalysing the glycolytic steps from glyceraldehyde-3-phosphate to PEP in *Bacillus subtilis*, when the bacterium is in a gluconeogenic state, while fructose-1,6-diphosphate inhibits CggR activity to activate glycolysis.

The multiple CCR systems enable bacteria to monitor intracellular energy and extracellular pools of carbohydrate to optimize expression of genes for carbon catabolism.

12.1.4 Repression and attenuation by final metabolic products

Under any given conditions, microbes reproduce themselves in the most efficient way and have to synthesize cell constituents in the right proportions. For this reason, anabolism should be regulated. Just as the enzymes of catabolism are induced by inducers, the final products repress the anabolic enzymes. When *Escherichia coli* grows on a glucose-mineral salts medium, glucose flux is tightly controlled to produce the cell constituents in

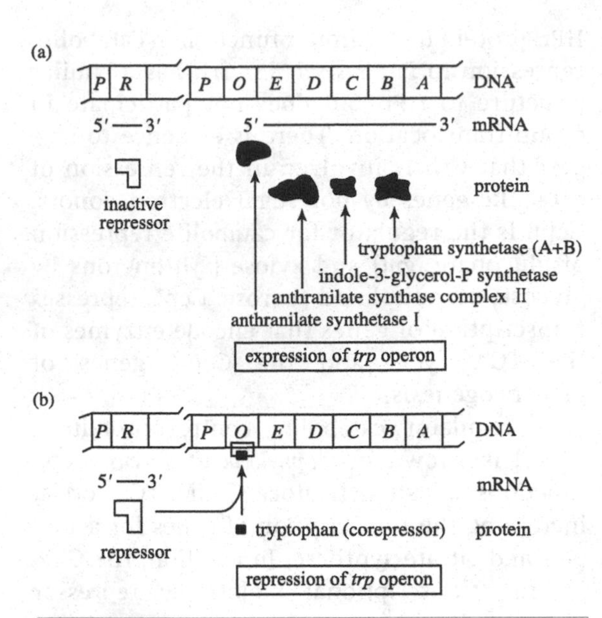

Figure 12.10 Transcriptional regulation of the tryptophan operon through repression.

(Gottschalk, G. 1986, *Bacterial Metabolism*, 2nd edn., Figure 7.8. Springer, New York)

When the intracellular tryptophan concentration is low, the repressor protein is inactive and the operon is transcribed (a). When the tryptophan concentration increases, it binds to the repressor protein to make the active form. The active repressor binds the operator region of the operon to inhibit transcription (b).

the right proportions. If a monomer of a cell constituent is available, e.g. an amino acid, the bacterium shuts down production of this amino acid. Repression and attenuation by the final anabolic product(s) are the regulatory mechanisms involved. Regulation by repression depends on repressor proteins, while proteins are not directly involved in the attenuation mechanism.

12.1.4.1 Repression

The expression of genes for the enzymes of amino acid biosynthesis is regulated through repression or attenuation in *Escherichia coli*. Figure 12.10 shows the repression of enzymes for tryptophan biosynthesis. A repressor protein is synthesized from a separate operon. When the tryptophan concentration is low, the repressor protein is inactive, and the *trp* operon is transcribed. When the tryptophan concentration increases, the amino acid binds the repressor protein to make the repressor active. In this case tryptophan is referred to as a corepressor. The active repressor binds to the operator region of the operon to repress its transcription.

12.1.4.2 Attenuation

In bacteria, gene expression is commonly regulated at the level of the initiation of transcription in response to external signals. Because of the low stability of most bacterial mRNA (Section 12.1.9), this provides an efficient way to control the synthesis of any given protein. Most frequently, transcription initiation is tightly regulated by σ-factors, repressors or activators, as discussed above. In several operons, however, transcription is constitutively initiated, but elongation can only proceed under specific conditions. Transcription elongation is regulated in response to external signals by the nascent mRNA with or without the aid of other effectors. In both cases the mRNA forms alternative hairpin structures known as a terminator and an antiterminator. For convenience, the discussion here is on the attenuation process where mRNA senses the physiological conditions directly without the aid of other effector molecules. In some cases the formation of the terminator/antiterminator is aided by small proteins, metabolites, tRNAs and antisense RNAs, as discussed later (Section 12.1.6).

The repressor protein for the histidine operon has not been identified in *Escherichia coli*. The histidine operon is regulated through a mechanism known as attenuation. Control by attenuation is possible only in prokaryotic cells, where mRNA is translated while it is being synthesized.

The tryptophan operon is under dual control, primarily through repression and attenuation. Even when the *trp* operon is fully derepressed, about 90 per cent of the RNA polymerase transcribing the operon does not reach the structural genes due to the attenuation mechanism. About 160 base pairs, known as the leader sequence, occupy a region between the promoter and the start of the structural genes of the *trp* operon. The transcript of the leader sequence functions as the regulator (Figure 12.11).

(a)

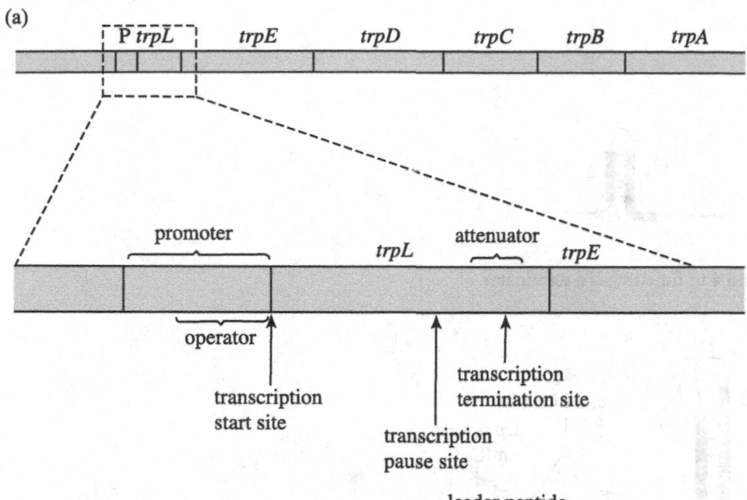

leader peptide

Met Lys Ala Ile Phe Val Leu Lys Gly Trp Trp Arg Thr Ser

(b)

1 20 40 60

pppAAGUUCACGUAAAAAGGGUAUCGACAAUGAAAGCAAUUUUCGUACUGAAAGGUUGGUGGCGCACUUCC

80 100 120

UGAAACGGGCAGUGUAUUCACCAUGCGUAAAGCAAUCAGAUACCCAGCCCGCCUAAUGAGCGGGCUUUU

140 160 180

UUUUGAACAAAAUUAGAGAAUAACAAUGCAAACACAAAAAACCGACUCUCGAACUGCU

Met Gln Thr Gln Lys Pro Thr Leu Glu Leu Leu

trpE polypeptide

Figure 12.11 The structure of the *trp* operon with the leader sequence for transcriptional control through attenuation.

The distance is about 160 base pairs between the promoter and the start of the structural genes in the *trp* operon. This segment is the leader sequence (*trpL*) that controls the transcription through attenuation. The transcript from the leader sequence has four regions that can make hairpin structures (see Figure 12.11).(a) *trp* operon; (b) base sequence of the leader sequence.

As stated above, attenuation is known only in prokaryotes, where translation starts while mRNA is synthesized. The transcript of the leader sequence (leader mRNA) contains a relatively high proportion of codons for the amino acid whose biosynthesis is encoded by the operon, and four regions that can make hairpin (stem-and-loop) structures (Figure 12.12).

When tryptophan is available at adequate levels to meet cellular requirements, the ribosomes move to the end of the second sequence of the four hairpin-making regions on the mRNA during its translation. At this point the third and the fourth regions form a hairpin structure. This hairpin removes the RNA polymerase from the DNA before it reaches the start of the structural genes. This hairpin is referred to as an attenuator or terminator. However, when the tryptophan level is low, the tryptophanyl-tRNA becomes limiting and the ribosome translating the leader sequence stalls at a trytophan codon before it reaches the second hairpin-forming region. At this point the leader mRNA forms a hairpin structure between the second and the third region, preventing the formation of the terminator between the third and the fourth segments, and the transcription continues to the structural genes. This hairpin between the second and the third region is referred to as an antiterminator.

(a) tryptophan in adequate level

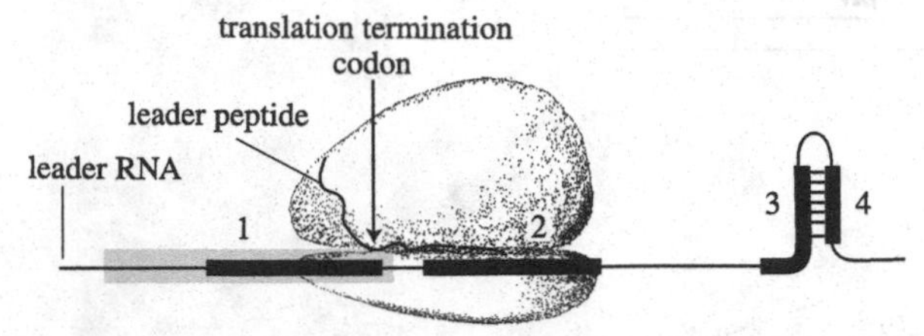

Hairpin formation between sequences 3 and 4 to function as a terminator.

(b) tryptophan in limited conditions

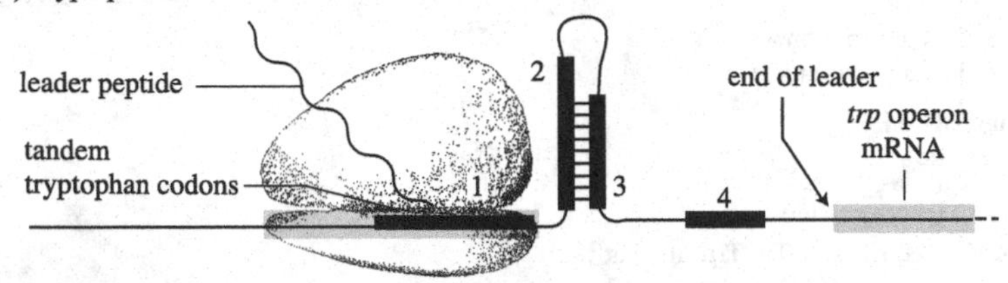

Hairpin formation between sequences 2 and 3 when the ribosome stops at sequence 1 due to lack of activated tryptophan.

(c) all amino acids in low concentration

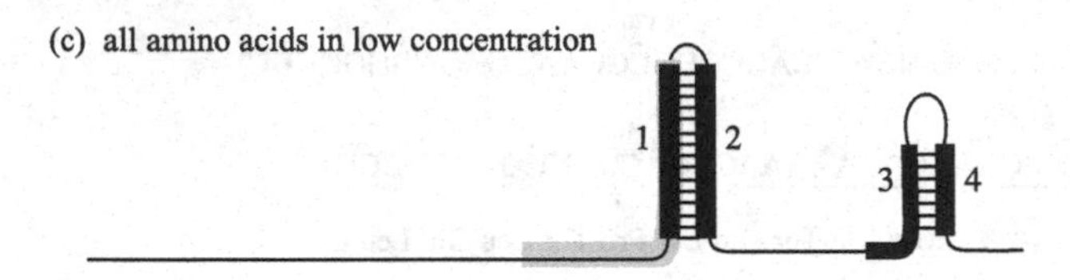

When amino acids are in low concentration, translation is not initiated to form a hairpin between sequences 1 and 2, and 3 and 4.

Figure 12.12 Transcriptional control of the *trp* operon through attenuation.

The leader sequence of the messenger RNA transcribed from the *trp* operon contains codons (UGG) for tryptophan and four segments (thick lines) that can form hairpin structures. The hairpin structure formed between the 3rd and 4th segment functions as a terminator separating transcribing RNA polymerase from DNA before it reaches the structural gene. (a) With adequate tryptophan levels to meet cell requirements, ribosomes move to the end of the second hairpin-forming region that contains the tryptophan codons. At this point the 3rd and 4th segments form the terminator hairpin to remove RNA polymerase from the DNA before it reaches the start of the structural gene. The structural genes are not transcribed. (b) When the tryptophan concentration is low, the ribosomes stop at the tryptophan codon within the leader sequence before they reach the second segment of the hairpin-forming regions. At this point the 2nd and the 3rd segments form a hairpin structure preventing the formation of the terminator hairpin structure between the 3rd and the 4th segments. The transcription continues through the structural genes. The hairpin structure between the 2nd and the 3rd segments is referred to as the antiterminator. (c) Under general starvation conditions, translation cannot proceed and hairpin structures are formed between the 1st and 2nd, and the 3rd and 4th segments: the latter one functions as the terminator.

When the overall amino acid availability is low, the ribosome cannot start translation, and the leader mRNA adopts a conformation with two hairpin structures between the first and the second region, and the third and the fourth region (terminator).

The rate of transcription elongation in vivo is about 40 to 50 nucleotides/second. The decision between termination and antitermination is therefore made within a very short timeframe. Termination should be prevented by sequestration of the third region of the hairpin-forming sequences by pairing with the second region to form the antiterminator element before synthesis of the fourth region. To synchronize translation with transcription of the leader sequence, transcription pauses at a transcription pause site within the leader sequence.

The synthesis of many amino acids and pyridine nucleotides is controlled by this mechanism alone or by a dual mechanism involving repression. Attenuation is not known in archaea and eukaryotes.

Attenuation is a regulatory mechanism in that leader mRNAs directly sense physiological signals by binding an effector molecule without the involvement of regulatory proteins or ribosomes. This is an example of a 'riboswitch'. The term riboswitch is defined as structured domains that usually reside in the non-coding regions of mRNAs, where they bind metabolites and control gene expression (Section 12.1.6).

A similar regulatory mechanism is known to control the expression of structural proteins. The mRNA of *agn43*, encoding the biofilm initiating outer membrane protein Ag43 of *Escherichia coli*, has a long leader region that has two translation start codons and a sequence to form a hairpin secondary structure between the start codons. Translation from the upstream start codon leads to increased downstream *agn43* expression by preventing the formation of the terminator hairpin structure.

A similar premature termination of transcription is mediated through a different mechanism known as the roadblock mechanism in *Bacillus subtilis*. A GTP-sensing transcriptional pleiotropic repressor CodY binds downstream of the promoters of target genes such as *ybgE* (a branched-chain amino acid transaminase that is involved in phenylalanine and tyrosine conversions), blocking RNA polymerase.

12.1.5 Regulation of gene expression by multiple end products

In an anabolic process where more than two products are formed from a common precursor, an excess of one product should not interfere with the synthesis of the other product. To ensure this, the first enzyme of the common pathway has one of the following properties:

(1) Multiple enzymes (isoenzymes) catalyse the same reaction, but their activities and synthesis are regulated by individual products.
(2) If a single enzyme catalyses the reaction, gene expression and its activity are regulated collectively by the products.

The expression of genes for isoenzymes and their activities is regulated separately by individual products, although they catalyse the same reaction. Aspartate is the precursor for the synthesis of lysine, methionine, threonine and isoleucine. The first reaction of this anabolic metabolism is catalysed by aspartate kinase (Figure 12.13). This is an example of an isoenzyme. The activity of an isoenzyme as well as its expression are controlled by its own final product.

One of the three aspartate kinase isoenzymes is regulated collectively by threonine and isoleucine. Expression of the gene encoding this enzyme is regulated through an attenuation mechanism involving threonine and isoleucine. This regulatory process is referred to as divalent attenuation. Figure 12.13 shows some other examples of divalent attenuation. They include one of two homoserine dehydrogenases, homoserine kinase and threonine synthase. Their gene expression and activity are regulated collectively by threonine and isoleucine (Figure 12.13).

Pyruvate is the precursor for the synthesis of valine and leucine, and 2-ketobutyrate the precursor for isoleucine. The same enzymes catalyse each of the first four reactions, synthesizing valine, leucine and isoleucine. Expression of their genes is regulated through an attenuation mechanism mediated collectively by these three amino acids (Figure 12.14). Multivalent attenuation is the term used to describe attenuation systems mediated by more than two effectors.

12.1.6 Termination and antitermination

Since elucidation of the attenuation process (Section 12.1.4.2) in amino acid synthesis by *Escherichia coli*, similar transcription regulatory mechanisms have been discovered in several systems. These include terminator/antiterminator formation aided by small proteins, metabolites, tRNAs and antisense RNAs. These regulatory

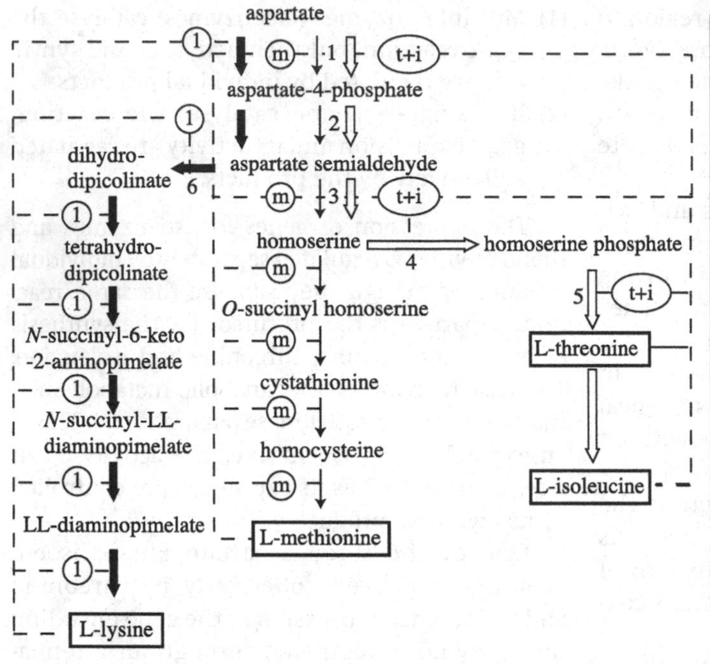

Figure 12.13 Transcription of genes for synthesis of aspartate series amino acids is regulated through a dual repression–attenuation mechanism.

(Gottschalk, G. 1986, *Bacterial Metabolism*, 2nd edn., Figure 7.10. Springer, New York)

Aspartate series amino acids include lysine, methionine, threonine and isoleucine. Isoleucine is synthesized through threonine. Aspartate kinase, the enzyme catalysing the first reaction, occurs as three isoenzymes. The final products control their transcription separately. Lysine and methionine regulate transcription of the specific isoenzymes through repression, while the third isoenzyme is regulated by threonine and isoleucine through attenuation. This regulation is referred to as divalent attenuation. Since other reactions branch off from aspartate semialdehyde and homoserine, the enzymes catalysing the first reaction of each branch are subject to a similar regulatory control.

1, aspartate kinase; 2, aspartate semialdehyde dehydrogenase; 3, homoserine dehydrogenase; 4, homoserine kinase; 5, threonine synthase; 6, dihydrodipicolinate cyclohydrolase.

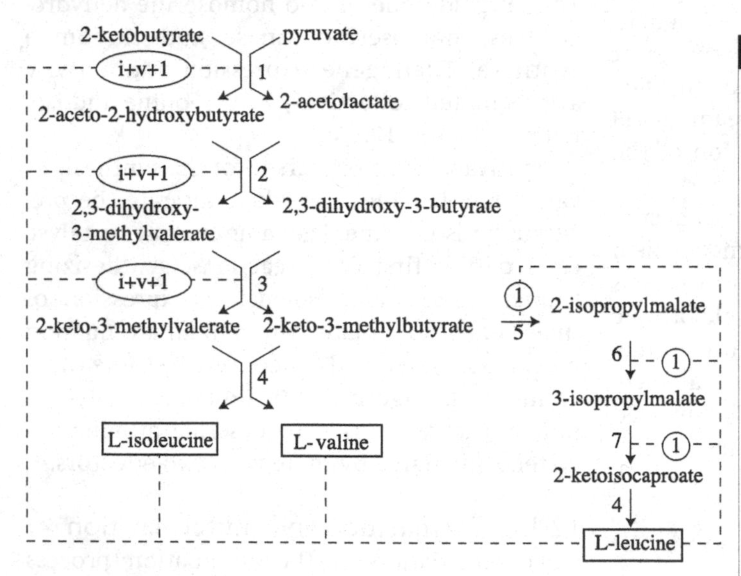

Figure 12.14 Transcriptional control of genes for the enzymes synthesizing isoleucine, valine and leucine by attenuation.

(Gottschalk, G. 1986, *Bacterial Metabolism*, 2nd edn., Figure 7.11. Springer, New York)

Repressors are not known in regulation of transcription of genes for isoleucine and valine synthesis in enteric bacteria, including *Salmonella typhimurium*. The same series of enzymes catalyse the synthesis of isoleucine and valine, and the early stages of leucine synthesis. These are regulated through multivalent attenuation by isoleucine, valine and leucine, as indicated by i + v + l. Genes for leucine synthesis are regulated through repression with leucine as the corepressor.

1, acetohydroxy acid synthase; 2, acetohydroxy acid isomeroreductase; 3, dihydroxy acid dehydratase; 4, transaminase; 5, 2-isopropylmalate synthase; 6, isopropylmalate isomerase; 7, 3-isopropylmalate dehydrogenase.

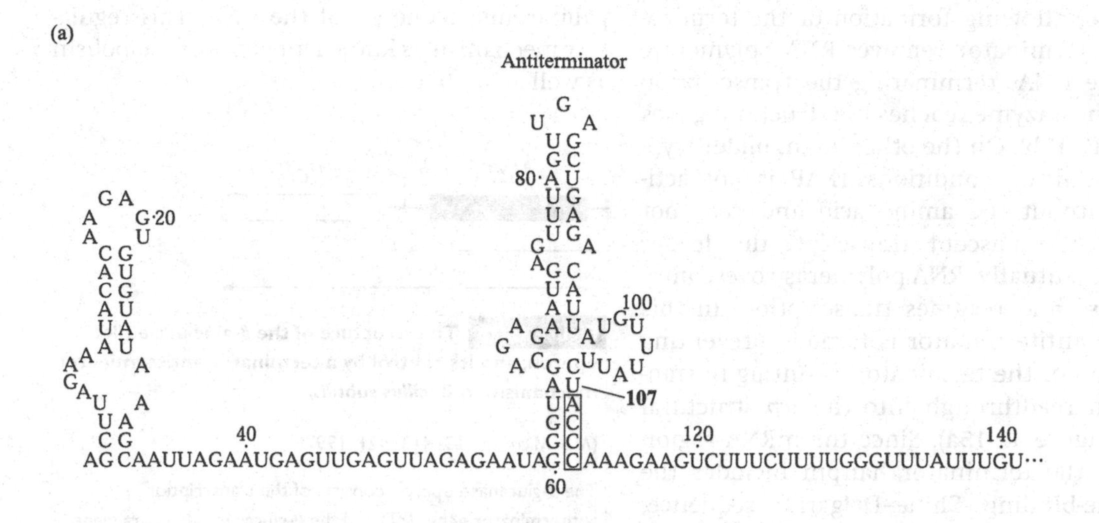

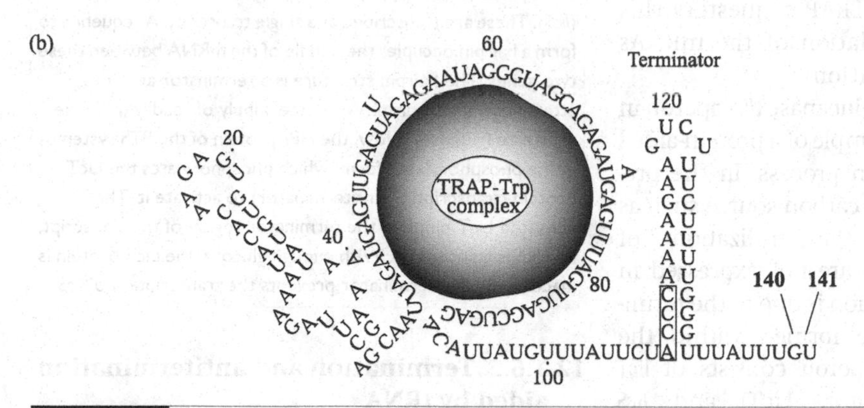

Figure 12.15 Model of *trp* operon attenuation in *Bacillus subtilis*.

(*Curr. Opin. Microbiol.* **7**: 132–139, 2004)

(a) In tryptophan-limiting conditions, TRAP is not activated and does not bind to the nascent transcript. Eventually, RNA polymerase overcomes the pause and resumes transcription. In this case, formation of the antiterminator prevents formation of the terminator, resulting in transcription readthrough into the *trp* structural genes. (b) In tryptophan-excess conditions, TRAP is activated and binds to the nascent mRNAs soon after they are synthesized. Pausing during the transcription of the leader sequence allows additional time for TRAP to bind. TRAP binding releases paused RNA polymerase and transcription resumes. Bound TRAP prevents formation of the antiterminator, thereby allowing formation of the terminator. Boxed nucleotides represent the overlap between the antiterminator and terminator structures.

processes are referred to as 'riboswitch' mechanisms (Section 12.1.6.3).

12.1.6.1 Termination and antitermination aided by protein

The best characterized example of a small protein involved in this mechanism is TRAP (*trp* RNA binding attenuation protein) in *Bacillus subtilis*. When tryptophan is available at adequate levels to meet cellular requirements, the amino acid binds TRAP to activate the protein. The activated TRAP binds to the leader mRNAs soon after they are synthesized, preventing the formation of antiterminator. RNA polymerase pauses at the pause site allowing additional time for the activated TRAP to bind. TRAP binding releases paused RNA polymerase and transcription resumes to the end of the leader sequence, allowing formation of the terminator. The terminator removes RNA polymerase from the DNA, terminating the transcription before the enzyme reaches the structural genes (Figure 12.15b). On the other hand, under tryptophan-limiting conditions TRAP is not activated without the amino acid and does not bind to the nascent transcript, the leader mRNA. Eventually, RNA polymerase overcomes the pause and resumes transcription. In this case the antiterminator is formed, preventing formation of the terminator, resulting in transcription readthrough into the *trp* structural genes (Figure 12.15a). Since the mRNA region forming the terminator hairpin includes the ribosome-binding Shine–Dalgarno sequence, the binding of activated TRAP sequesters this sequence, blocking translation of the mRNAs that have escaped termination.

The expression of the β-glucanase (*lic*) operon in *Bacillus subtilis* is another example of a protein-aided termination/antitermination process. In the presence of a readily utilizable carbon source such as glucose, enzymes for the utilization of β-glucans such as lichenan are not expressed in this bacterium. This regulation is due to the terminator, a hairpin structure formed within the mRNA. The β-glucanase operon consists of *licT* (transcription antiterminator, LicT) and *licS* (β-glucanase) in *Bacillus subtilis* (Figure 12.16). The antiterminator is activated by phosphorylated HPr (see Figure 12.9). When glucose is available, phosphorylated HPr(His-P) participates in glucose transport, and the HPr(His-P) concentration is low. In this case the antiterminator (LicT) is not activated, and the terminator is formed within the β-glucanase operon transcript, inducing a premature transcription termination. In the absence of glucose, HPr(His-P) activates LicT, and the activated LicT prevents formation of the terminator. Consequently *licS* (β-glucanase) is transcribed. The antiterminator-binding site within the mRNA is referred to as the ribonucleic antiterminator (RAT).

Termination/antitermination is responsible for the regulation of 6-phospho-β-glucosidase in *Escherichia coli*, and operons for the utilization of sucrose, glycerol and histidine in addition to β-glucan in *Bacillus subtilis*. The RNA sequence of the terminator hairpin structure is encoded in the palindromic sequence of the DNA. This regulatory mechanism is known for genes of catabolism as well as anabolism.

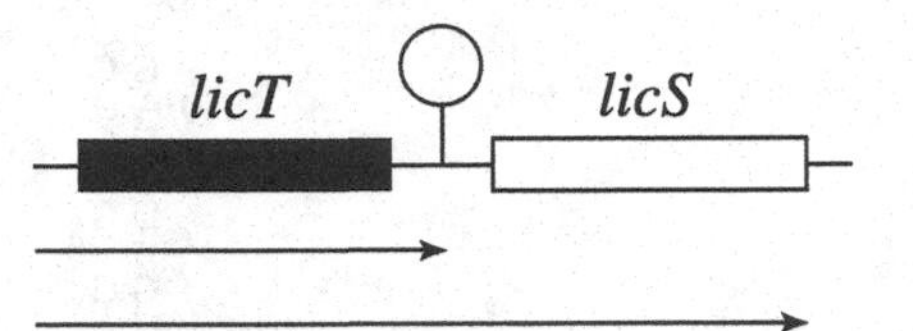

Figure 12.16 The structure of the β-glucanase (*lic*) operon, and its control by a terminator/antiterminator mechanism in *Bacillus subtilis*.

(*Mol. Microbiol.* **23**: 413–421, 1997)

The β-glucanase operon consists of the transcription antiterminator gene (*licT*) and the β-glucanase structural gene (*licS*). These are transcribed as a single transcript. A sequence to form a hairpin occupies the middle of the mRNA between these two genes. The hairpin structure is a terminator as in the attenuation mechanism. When the supply of readily utilizable substrate (glucose) is low, the HPr protein of the PTS system is in the phosphorylated form, which phosphorylates the LicT protein (transcription antiterminator) to activate it. The activated LicT binds to the terminator region of the transcript, and *licS* is transcribed. With enough glucose, the LicT protein is inactive and the terminator prevents the transcription of *licS*.

12.1.6.2 Termination and antitermination aided by tRNA

In *Bacillus subtilis*, transcription of aminoacyl-tRNA synthetases and enzymes for the biosynthesis of certain amino acids are regulated by yet another mechanism known as tRNA-dependent transcription antitermination. Uncharged tRNA functions as the effector. The leader region of tyrosyl-tRNA synthetase mRNA contains a codon for tyrosine and a binding sequence for the 3′ acceptor end of uncharged $tRNA^{Tyr}$. Under tyrosine-excess conditions with a low uncharged $tRNA^{Tyr}$ (effector) concentration, the anticodon of charged tyrosyl$tRNA^{Tyr}$ binds the tyrosine codon of the leader mRNA, but the acceptor end of the tyrosyl-$tRNA^{Tyr}$ does not participate in mRNA binding, preventing antitermination formation (left side, Figure 12.17a). Under these conditions tyrosyl-tRNA synthetase is not expressed. When tyrosine is limiting, the effector, uncharged $tRNA^{Tyr}$, binds the leader mRNA through the acceptor end as well as the anticodon,

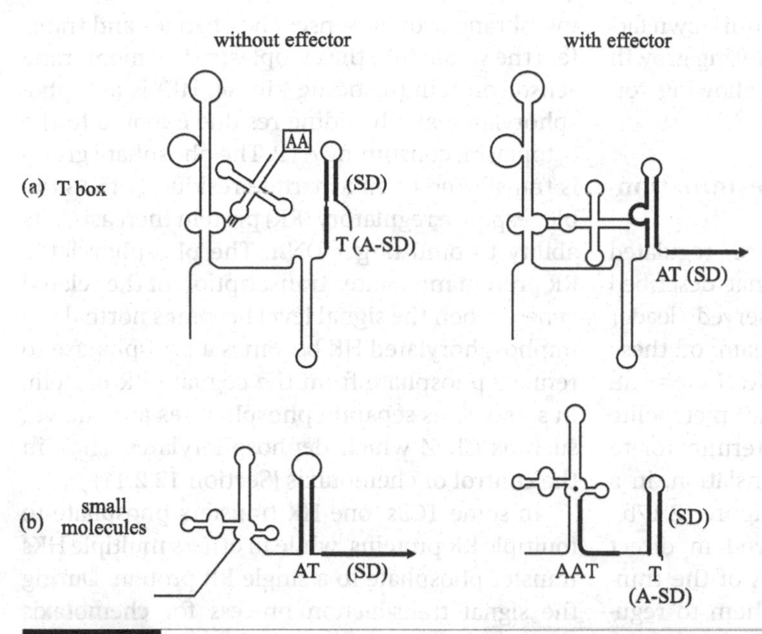

Figure 12.17 RNA elements that directly bind effector molecules in tRNA-dependent transcription antitermination (a) and the termination and antitermination mechanism aided by metabolites (b).

(*Curr. Opin. Microbiol.* **7**: 126–131, 2004)

(a) For regulation by tRNA (T-box system), the uncharged tRNA (shown in cloverleaf form) serves as the effector. In the absence of the effector (or in the presence of charged tRNA, indicated by the boxed AA), the leader RNA folds into the terminator form indicated as T (left-hand side). In the presence of the effector (uncharged tRNA), the antiterminator (AT) is stabilized, preventing formation of the terminator hairpin and allowing transcription of the downstream coding regions indicated by the arrow (right-hand side). This arrangement can also be used to regulate gene expression at the level of translation initiation, in which case the 'terminator' helix (A-SD) sequesters the ribosome-binding Shine–Dalgarno sequence (SD) of the downstream gene. (b) For the termination and antitermination mechanism aided by metabolites, the antiterminator structure forms in the absence of effector (left-hand side), allowing expression of the downstream coding regions (arrow); addition of the effector (•, right-hand side) promotes stabilization of an anti-antiterminator element (AAT) that sequesters sequences required for the formation of the antiterminator (AT), permitting formation of the less stable terminator helix (T). As described above, the terminator hairpin can be replaced with a structure that occludes the Shine–Dalgarno sequence (SD) of the downstream gene to permit regulation at the level of translation initiation. Sequences involved in the formation of terminator as well as antiterminator are shown as bold lines.

leading to the formation and stabilization of the antiterminator to prevent transcription termination. In this case the terminator is not formed, and the enzyme is expressed. The uncharged tRNA binding site of the leader mRNA is referred to as the T-box, and the mechanism as the T-box transcription termination control system. As in the TRAP-mediated attenuation mechanism, translation is blocked since the ribosome-binding Shine–Dalgarno sequence participates in formation of the terminator hairpin.

The T-box transcription termination control system is widely used for the control of gene expression in Gram-positive bacteria, but is rare in Gram-negative organisms. Genomic data analyses reveal high conservation of primary sequence and structural elements of the system. The T-box system regulates a variety of amino acid-related genes.

A different mechanism, known as the S-box system, regulates genes involved in methionine metabolism. While both systems involve gene regulation at the level of premature termination of transcription, the molecular mechanisms employed are very different. In the T-box system, expression is induced by stabilization of an antiterminator structure in the leader by interaction with the uncharged tRNA, while an anti-antiterminator, stabilized by an unknown factor, is involved in the S-box system during growth under methionine-rich conditions, allowing formation of the terminator.

12.1.6.3 Termination and antitermination aided by metabolites

Genes for riboflavin synthesis are regulated through a similar mechanism to that described above for *Bacillus subtilis*. A conserved leader sequence occupies a region upstream of these genes which is designated as the RFN element. When riboflavin is in excess, the metabolite binds the RFN element forming a terminator to inhibit transcription, as well as translation, in a similar manner to that depicted in Figure 12.17b.

Various metabolites are involved in direct interactions with the leader mRNA of the transcripts of enzymes, synthesizing them to regulate gene expression and/or translation. These include cobalamin in *Escherichia coli* (B_{12}-box), thiamine in *Rhizobium etli* (THI-box), guanine in *Bacillus subtilis* (G-box), S-adenosylmethionine (SMK-box) in the Lactobacillales (lactic acid bacteria) and lysine in *B. subtilis* (L-box). Some of these regulate translation while others regulate transcription depending on the system and the bacterium. The metabolite-binding region of the mRNA is referred to as the riboswitch.

In addition to the metabolite-aided riboswitches, a manganese-chelated M-box RNA complex is known in *Bacillus subtilis*. This riboswitch controls magnesium homeostasis in the cell through modulation of gene expression.

12.1.7 Two-component systems with sensor-regulator proteins

A two-component system (TCS) consists of signal transduction proteins and is known in all domains of life. In bacteria, TCSs constitute a dominant form of gene control in response to changes in environmental conditions so that they can regulate their metabolism accordingly. This system is responsible for adaptation to a variety of stress conditions, pathogenesis and symbiotic interactions with eukaryotic hosts, and essential cellular pathways. This signal transduction system is not known in *Mycoplasma* spp. while over 100 occur in certain cyanobacteria. A membrane protein senses the changes and transfers the signal into the cytoplasm. The membrane sensor protein (histidine kinase, HK) is autophosphorylated at a histidine residue exposed to the cytoplasm, consuming ATP. The phosphate group is transferred to an aspartate residue of the soluble response regulatory (RR) protein increasing its ability to bind target DNA. The phosphorylated RR protein modulates transcription of the related genes. When the signal level becomes normal the unphosphorylated HK becomes a phosphatase to remove phosphate from the cognate RR protein. In some cases separate phosphatases are known, such as CheZ which dephosphorylates CheY in the control of chemotaxis (Section 12.2.11).

In some TCSs, one HK transfers phosphate to multiple RR proteins, while in others multiple HKs transfer phosphate to a single RR protein. During the signal transduction process for chemotaxis (Section 12.2.11), CheA transfers phosphate from CheB and CheY. An example of multiple HKs and a single RR protein is the quorum-sensing network of *Vibrio harveyi*, where three different HKs sensing different autoinducers converge on the same RR protein, LuxU (Section 12.2.8).

In some cases, auxiliary regulators modulate the phosphorylating activity of the HK connecting a given two-component system to other regulatory networks. For example, a group of small-sized proteins, referred to as connectors, are known to link TCSs. The EvgS/EvgA system induces acid resistance in *E. coli*. The phosphorylated RR protein, EvgA, transcribes genes for the regulatory proteins YdeO (a transcriptional regulator) and SafA (sensor associating factor A). The former activates transcription of acid resistance genes and the latter phosphorylates PhoP to further increase acid resistance. Other connectors are MzrA (modulator of the EnvZ/OmpR regulon) connecting the CpxA/CpxR (envelope stress two-component system) and EnvZ/OmpR systems, and IraM (RpoS stabilizer during Mg starvation, anti-RssB factor) connecting PhoQ/PhoP and the response regulator RssB in *E. coli*. PmrD in *Salmonella* spp. connects the PhoQ/PhoP and PmrB/PmrA (sensing external stimuli of high Fe^{3+} and mild acidic conditions) systems.

A large number of TCSs contain sRNAs in their regulons. The expression of sRNAs is controlled by the RR protein of a TCS, or sRNAs control activity of the TCS. Expression of these sRNAs is either activated or repressed by binding of the corresponding RR protein to the sRNA promoter. The sRNAs in turn repress or activate targets at the post-transcriptional level, which can either be mRNA or protein. In *Pseudomonas aeruginosa*, the utilization of secondary carbon sources is repressed in the presence of preferred carbon sources by the catabolite repression control (Crc) protein that binds and thereby blocks translation of mRNAs (Section 12.1.3). The phosphorylated RR protein, CbrA of the TCS CbrA/CbrB, activates a small RNA, CrcZ, which is expressed in the absence of the preferred carbon sources. CrcZ contains five of the Crc binding motifs and sequesters the Crc protein, which consequently causes derepression of the Crc mRNA targets. Many TCSs are known to control sRNA synthesis. In some cases sRNAs control TCSs. A sRNA, MicA, transcribed by the extracellular function sigma factor, downregulates genes of the PhoQ/PhoP and OmpA/LamB systems.

Certain regulators lack DNA-binding domains and exert their regulatory effects by establishing direct interactions with protein or RNA targets.

Two-component systems are involved in the regulation of more than one operon, and this is discussed in detail later (Figures 12.24 and 12.25, Section 12.2).

It is interesting to note that some unphosphorylated regulatory proteins of two-component systems regulate expression of genes with functions not related to responses against the signal carried by the cognate sensor protein. In *Pseudomanas aeruginosa*, the phosphorylated regulatory protein AlgR, with the sensor protein AlgZ, activates the *hcn* genes for hydrogen cyanide production. Unphosphorylated AlgR activates genes for alginate production. AlgZ is also a protein involved in alginate production.

In addition to HK, bacteria possess various kinases known as eukaryote-like serine/threonine kinases (eSTK). These are involved in diverse biological processes, including development, cell competence, cell division, cell wall synthesis, central and secondary metabolism, biofilm formation, stress responses and virulence (Section 12.3.2.1).

12.1.8 Autogenous regulation

Some proteins are able to control the transcription or translation of their own genes. These include proline dehydrogenase controlling its transcription (Figure 12.18) and ribosomal proteins regulating their own translation.

A ribosome consists of three ribosomal RNAs (rRNA) and up to 56 ribosomal proteins (r-proteins). Elaborate regulations, including one at the translational level, ensure synthesis of these proteins in the correct proportions. The r-proteins bind rRNAs. When an r-protein is present in excess to bind the rRNAs, this binds its own mRNA to inhibit the initiation of translation. This process is referred to as autogenous regulation. The synthesis of many proteins associated with DNA and RNA is regulated through this regulatory mechanism.

Translation of other proteins is regulated in a similar manner. The S-layer constitutes the cell surface of many prokaryotes, and the S-layer protein is one of the most abundant single proteins (Section 2.3.3). This protein is encoded by a single gene and its mRNA is stable with a half-life of 10–22 minutes producing the protein efficiently. The mRNA has an untranslated region (5′UTR) of 30–358 bases that forms hairpin structures to protect the coding region from ribonuclease (Section 12.1.9.1). When excess S-layer protein is produced, this protein binds the 5′ UTR of its own mRNA, inhibiting translation. These are examples of negative regulation of translation by a protein. Apo-aconitase binds the 3′ UTR of its own mRNA to stabilize it in *Escherichia coli*. This is positive regulation.

When enteric bacteria, including *Escherichia coli*, use proline as a sole carbon source, the transcription of the *put* operon is regulated through autogenous regulation. Proline dehydrogenase and proline permease comprise the *put* operon. When proline is available as an energy source, proline dehydrogenase binds to the cytoplasmic membrane, catalysing a two-step oxidation of proline to glutamate (Section 7.5.6). When the substrate concentration is low, proline dehydrogenase binds the

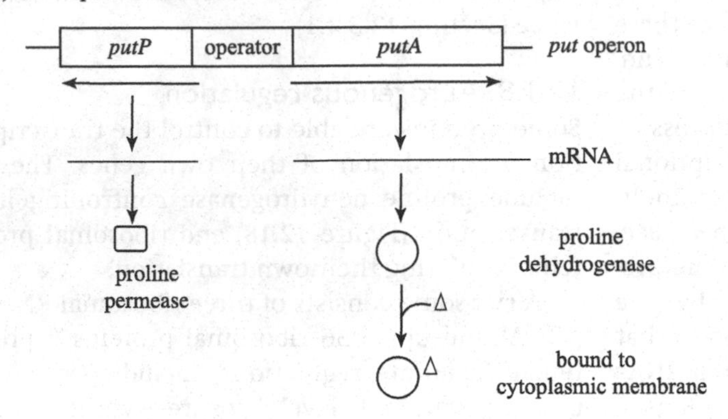

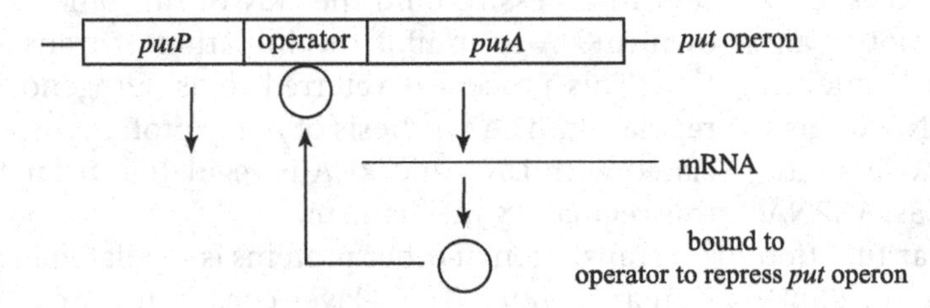

Figure 12.18 Transcriptional control of the proline dehydrogenase gene through autogenous regulation.

Proline dehydrogenase has a DNA-binding site as well as the active site. When proline is supplied as the substrate, the enzyme binds to the cytoplasmic membrane and oxidizes proline (a). When the substrate concentration becomes low, the enzyme binds the operator region of the *put* operon to prevent its transcription (b).

operator region of the *put* operon, inhibiting transcription (Figure 12.18). This is another example of autogenous regulation. When the enzyme concentration is higher than that of the binding site on the membrane, the free enzyme binds DNA, inhibiting the initiation of transcription even when proline is available as an energy source.

12.1.9 Post-transcriptional regulation of gene expression

Gene expression is regulated not only at the transcriptional level, as discussed above, but also at the post-transcriptional level, and these mechanisms are integrated and coordinated to implement the information in the genome according to cellular needs. Post-transcriptional levels of regulation, such as transcript turnover and translational control, are an integral part of gene expression and approach the complexity and importance of transcriptional control. Post-transcriptional control is mediated by various combinations of RNA-binding proteins (RBPs) and small non-coding RNAs (sRNAs) that determine the fate of the tagged transcripts. Without regulation at the post-transcriptional level, regulation at the transcriptional level would be meaningless. Autogenous regulation of ribosomal protein synthesis (discussed above) is an example of post-transcriptional regulation.

12.1.9.1 RNA stability

Generally the half-life of a bacterial mRNA is in the range of several minutes and much shorter than those of rRNA and tRNA. The latter are referred to as stable RNAs. The stability of mRNA is determined by its intrinsic structure as well as by the proteins and RNAs that bind it, and is related to its translation efficiency. Regulation at transcription becomes more meaningful with rapid mRNA turnover. *Escherichia coli* has both endo- and exo-type RNases. Endonuclease RNase E is the main enzyme degrading mRNA. This enzyme recognizes and hydrolyses adenine- and uracil-rich sequences in the RNA. RNase E

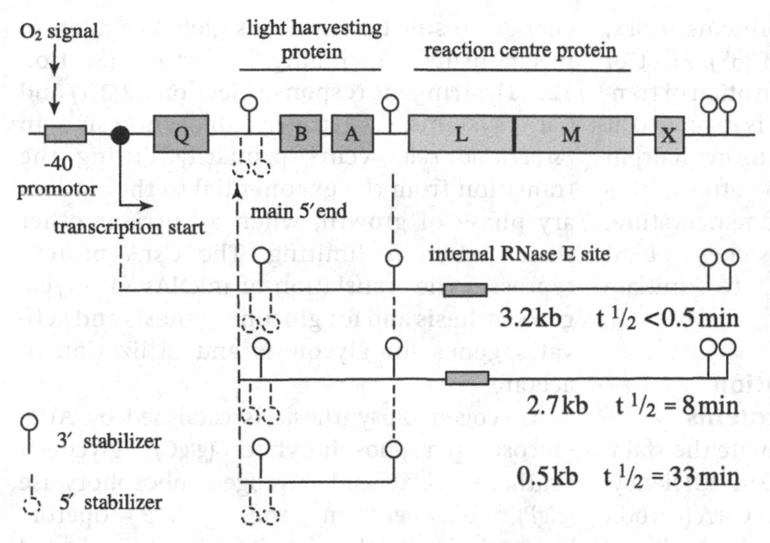

Figure 12.19 The stability of mRNA of *puf* genes in photosynthetic *Rhodobacter capsulatus*.

(*FEMS Microbiol. Rev.* **23**: 353–370, 1999)

The photosynthetic structure contains about 15 antenna molecules per reaction centre. Proteins of the antenna molecule and the reaction centre are encoded by a single *puf* operon that is transcribed into one transcript. Proteins of the antenna molecules are produced 15 times greater than the reaction centre proteins. The difference in the efficiency of translation is due to differences in the stability of their mRNAs. The hairpin in the transcript prevents RNA degradation by endonuclease RNase E. The region with the base sequence for the endonuclease RNase E reaction is more prone to degradation.

interacts with RNA helicase, the glycolytic enzyme enolase and the exoribonuclease PNPase to form a multi-protein complex called the RNA degradosome which is localized at the cytoplasmic membrane in *E. coli*.

A photosynthetic bacterium, *Rhodobacter capsulatus*, produces antenna proteins and reaction centre proteins from the *puf* operon. The photosynthetic apparatus consists of a reaction centre with about 15 antennae in this bacterium. A single transcript is produced from the *puf* operon, and 15-fold more antenna proteins are produced than the reaction centre proteins from this single transcript. This is possible because the mRNA for the antenna proteins is much more stable than that of the reaction centre proteins (Figure 12.19).

The transcript has an untranslated region of about 500 bases (5′-untranslated region, 5′ UTR) that forms three hairpin structures, and two sites for endonuclease RNase E, one each within the 5′UTR and reaction centre proteins encoding region. Additionally, sequences between the coding regions and the 3′-end of this transcript can form hairpin structures. These structural features determine the stability of different translational units within the transcript.

The entire transcription has a half-life of about 30 seconds, but this increases to eight minutes after RNase E cleaves a part of 5′UTR. When the transcript is processed to the translational unit for only the antenna proteins, the half-life is longer than 30 minutes (Figure 12.19).

Among many functions of the bacterial toxin/antitoxin system (Section 13.4.2), toxins degrade mRNAs under starvation conditions when the toxins are released from the antitoxin. Toxin-degrading mRNAs are referred to as mRNA interferases.

NAD-modified RNAs at the 5′-end are found in bacteria. This modification protects RNA from degradation, as in eukaryotes. Modification of the 5′-end of the RNA with 5′,5′-triphosphate-linked 7-methylguanosine protects the messenger RNA from degradation. NAD modification is found mainly in regulatory sRNAs and certain mRNAs with sRNA-like 5′-terminal fragments. Analogous to a eukaryotic cap, 5′-NAD modification stabilizes RNA against RNases. NADH pyrophosphatase decaps NAD-RNA and thereby triggers RNA decay.

12.1.9.2 mRNA structure and translational efficiency

When the growth temperature is shifted from 30°C to 42°C, *Escherichia coli* produces many proteins known as heat shock proteins (HSPs, Section 12.2.6), including sigma H (σ^H). Most of the HSPs are expressed through activation of transcription, and the expression of σ^H is regulated at the translational level. σ^H mRNA forms hairpin structures at physiological temperatures, preventing translation. At an elevated temperature, the structure melts, allowing ribosomes to bind the Shine–Dalgarno sequence to initiate translation.

12.1.9.3 Modulation of translation and stability of mRNA by proteins

Many proteins are known to modulate the stability of mRNAs and their translation efficiency. Examples of these proteins include CsrA (carbon storage regulator) protein in *Escherichia coli* and RsmA (regulator of secondary metabolism) protein in *Erwinia carotovora*.

During the transition from exponential growth to stationary phase, non-sporulating bacteria, including *Escherichia coli*, readjust their physiological status from one that allows robust growth and metabolism to one that provides for greater stress resistance and an enhanced ability to scavenge remaining substrates from the medium. These phenotypic changes result from various global regulatory mechanisms, including σ-factor (Section 12.1.1), stringent response (Section 12.2.1) and Csr systems. Glycogen biosynthesis in *Escherichia coli* occurs primarily during the transition from the exponential to the stationary phase of growth, when a nutrient other than carbon is limiting. The CsrA protein represses the translation of mRNAs for glycogen synthesis and for gluconeogenesis and activates genes for glycolysis and utilization of acetate.

Glycogen biosynthesis is catalysed by ADP-glucose pyrophosphorylase (*glgC*), glycogen synthase (*glgA*) and glycogen phosphorylase (*glgP*), encoded in the *glgCAP* operon. Transcription of the *glgCAP* operon is activated by the cAMP–CRP complex (Section 12.1.3.1) and through the stringent response mediated by ppGpp (Section 12.2.1). CsrA protein modulates the translation and stability of the *glgCAP* transcript by binding to the ribosome-binding site (Shine–Dalgarno sequence, Section 6.12.2.2) of the target mRNAs, including the *glgCAP* transcript, inhibiting translation and promoting rapid degradation of the transcript by RNA degradosome (Figure 12.20). A small non-coding RNA

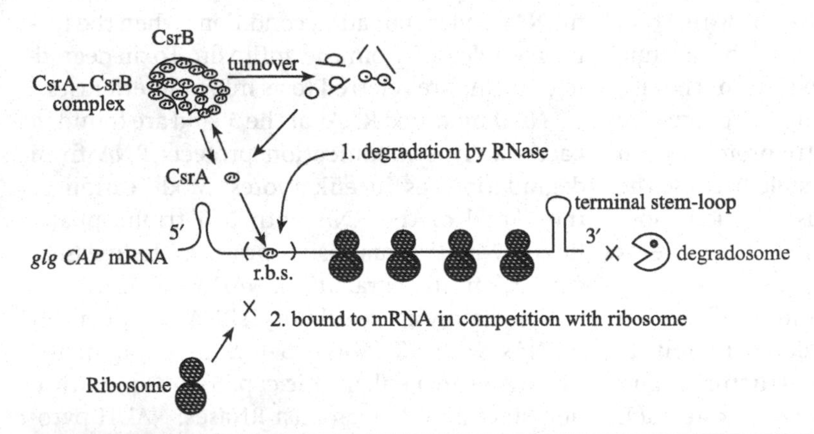

Figure 12.20 The CsrA/CsrB system controls the translation and stability of mRNA encoding enzymes of glycogen metabolism.

(*Mol. Microbiol.* **29**: 1321–1330, 1998)

CrsA (carbon storage regulator) protein binds the ribosome binding site of mRNAs encoding ADP-glucose synthetase (GlgC), glycogen synthase (GlgA) and glycogen phosphorylase (GlgP), inhibiting their translation. CraA-bound mRNAs are prone to degradation by RNase. A small RNA (sRNA), CsrB, binds CsrA protein in the ratio of 1 to 18 to inhibit its function. Another sRNA, CsrC, has a similar function.

(sRNA), CsrB, antagonizes CsrA activity by sequestering this protein. Approximately 18 molecules of CsrA bind one molecule of CsrB. Another sRNA, CsrC, has a similar function. Over 50 sRNAs are known in *Escherichia coli* and are discussed later (Section 12.1.9.3).

The Csr system represses genes for gluconeogenesis and glycogen biosynthesis, and activates those for glycolysis, cell motility and acetate metabolism. In addition, CsrA represses extracellular polysaccharide production that promotes biofilm formation, and flagellin synthesis. Another protein, FliW, antagonizes CsrA-dependent repression of flagellin translation. According to genome analysis, CsrA as a global regulator not only stabilizes a large number of mRNAs, but also positively or negatively regulates the transcription of many genes. These include those for secretion systems, surface molecules and biofilm formation, quorum sensing, motility, pigmentation, siderophore production and the virulence networks of animal and plant pathogens.

The BarA/UvrY two-component system (Section 12.1.7) of *Escherichia coli* consists of a membrane-bound sensor kinase BarA and its cognate response regulator UvrY. BarA senses and responds to the presence of short-chain fatty acids. This BarA/UvrY forms a feedback loop mechanism with the Csr system. CsrA indirectly activates *uvrY* expression at both transcriptional and translational levels, and is required for switching BarA from phosphatase to kinase activity. The phosphorylated UvrY in turn activates expression of CsrB and CsrC. The major catabolite repressor, cAMP-CRP, inhibits transcription of *csrB* and *csrC*, thus shaping the dynamics of global signalling in response to the nutritional environment by poising CsrB/C sRNA levels for a rapid response.

A similar protein, RsmA, in *Erwinia carotovora* destabilizes mRNA for pectinase and represses its translation. This enzyme is a virulence factor in plants, and its transcription is regulated by quorum sensing (Section 12.2.8). An sRNA, AepH (exoenzyme regulatory sRNA), antagonizes RsmA activity.

Escherichia coli produces various cold shock proteins (CSPs) including CspA at low temperature. CspA is another example of a protein modulating translation. This protein prevents mRNAs from forming hairpin structures at low temperatures, facilitating translation by ribosomes. This will be discussed later (Section 12.2.7).

12.1.9.4 Modulation of translation and stability of mRNA by small RNA and small RNA-protein complexes: riboregulation

More than 50 small non-coding RNAs have been identified in *Escherichia coli*. Their sizes range between 40 and 400 nucleotides. They are referred to as small RNAs (sRNAs) or non-coding RNAs (ncRNAs). These have been found in archaea, bacteria and eukaryota. Their functions are not fully understood, but some of them modulate translation and stability of mRNA. Their activities comprise three general mechanisms. They can (1) act as a member in RNA-protein complexes such as 4.5S RNA of the signal recognition particle (SRP, Section 3.10.1.1), (2) mimic the structure of other nucleic acids such as CsrB (Figure 12.20) and (3) base pair with other RNAs. Many of the regulatory sRNAs, such as the DsrA, MicF, OxyS, Spot42 and RyhB RNAs, act by base pairing to activate or repress translation, or to destabilize mRNAs (Figure 12.21).

When an *Escherichia coli* culture is shifted from 37°C to 25°C, an sRNA, DrsA, increases in concentration more than 25 times. This sRNA forms a base pair with the 5′ UTR region of *rpoS* mRNA, which encodes the stationary phase sigma factor σ^S, and leads to its increased translation. The DsrA sRNA promotes translation by preventing the formation of an inhibitory secondary structure that normally occludes the ribosome-binding site of the *rpoS* transcript. This sRNA downregulates the translation of the DNA-binding protein H-NS, promoting degradation of its mRNA. At low temperature, *Escherichia coli* produces many proteins. These are referred to as cold-shock proteins (CSP), and are discussed later (Section 12.2.7).

Similar sRNAs are known in *Escherichia coli*, including MicF and OxyS, all induced under stress

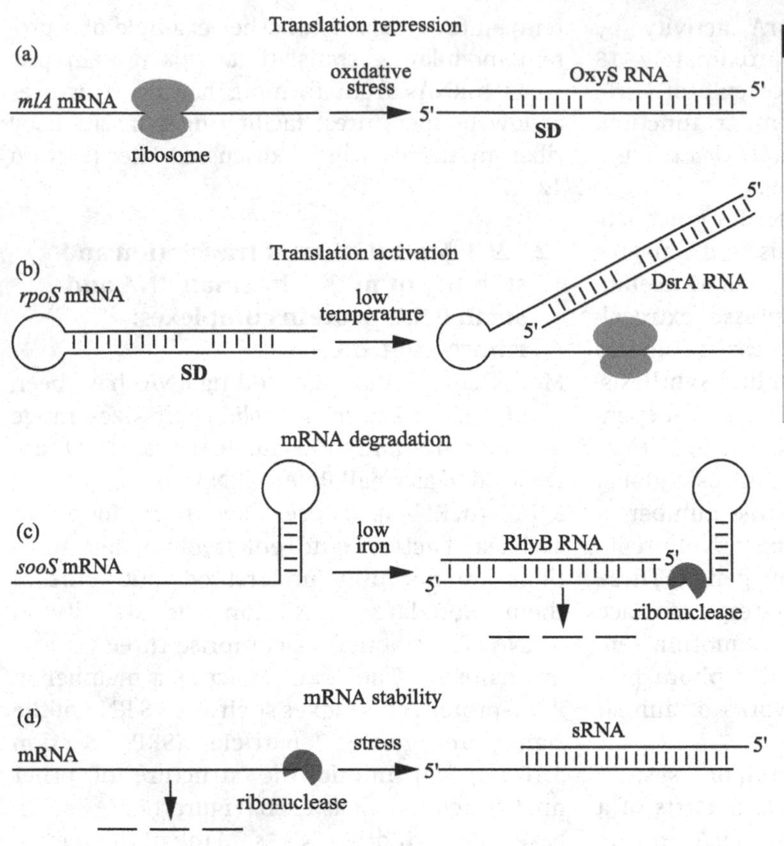

Figure 12.21 Different regulatory outcomes brought about by sRNA base pairing with mRNAs.

(*Curr. Opin. Microbiol.* **7**: 140–144, 2004)

sRNAs can repress or activate translation by blocking or promoting ribosome binding to the Shine–Dalgarno sequence (SD) of mRNAs. sRNAs can also destabilize or possibly stabilize mRNAs by increasing or decreasing accessibility to the RNA degradosome (see text for details).

conditions (Table 12.3). Expression of the OxyS sRNA is strongly induced under oxidative stress imposed by hydrogen peroxide and other substances (Section 12.2.5.1), and this forms a base pair with *fhlA* mRNA and represses its translation. FhlA is the transcriptional activator for the formate hydrogen-lyase operon that is an anaerobic enzyme (Section 8.6.4). The MicF sRNA blocks translation of the OmpF porin by base pairing with the *ompF* mRNA under high osmotic pressure. Another sRNA, MicC, regulates the other major porin, OmpC, under complementary conditions. These sRNAs presumably help the cell to respond to environmental conditions beyond those that are sensed by the phosphorelay system that regulates both the *ompF* and *ompC* genes (Section 12.2.9). Changing the ratio of OmpF to OmpC in the cell envelope modulates the entry of small molecules into the cell (Section 2.3.3). The build-up of phosphorylated sugars inhibits growth at high concentrations. Many enteric bacteria cope with this sugar-phosphate stress by producing a small RNA (sRNA) regulator, SgrS, that represses translation of sugar transporter mRNAs and enhances translation of a sugar phosphatase mRNA.

In the phototrophic bacterium *Rhodobacter sphaeroides*, the PrrB/PrrA two-component system (Section 12.2.4.1) senses low oxygen and the phosphorylated response regulator PrrA activates photosynthetic genes and an sRNA, PcrZ (photosynthesis control RNA Z). The latter represses translation of the former. This is one of the rare cases of an incoherent feed-forward loop involving an sRNA.

Genes for ethanolamine utilization are induced by the substrate through a two-component system, *eutVW* and vitamin B_{12}, aided by an

Table 12.3 | Base pairing sRNAs of known function in *Escherichia coli.*

sRNA	Numbers of nucleotides	Target transcript(s)	Effect
MicC	109	OmpC	translation repression
DicF	53	FtsZ	translation repression
RprA	105	RpoS	translation activation
DsrA	85	RpoS	translation activation
		Hns, RbsD	translation repression
MicF	93	OmpF	translation repression
GcvB	204	OppA, DppA	translation repression
RyhB	90	SodB	translation repression
			mRNA degradation
		SdhCDAB	mRNA degradation
Spot42	109	GalETKM	translation repression
OxyS	109	FhlA, RpoS	translation repression

OmpC, outer membrane protein; FtsZ, GTPase involved in cell division; RpoS, σ^S; Hns, histone-like DNA-binding protein; RbsD, D-ribose high-affinity transport system; OmpF, outer membrane protein F; OppA, oligopeptide transport, periplasmic binding protein; DppA, dipeptide binding protein; SodB, iron superoxide dismutase; SdhCDAB, succinate dehydrogenase; GalFTKM, galactose operon; FhlA, transcriptional activator for formate hydrogen-lyase operon.
(*Curr. Opin. Microbiol.* **7**: 140–144, 2004)

sRNA, Rli55, in *Listeria monocytogenes*. A B_{12}-box riboswitch (Section 12.1.6.3) activates the sRNA gene in the presence of the vitamin.

Spot42 is an sRNA with an interesting function. The galactose operon, *galETKM*, encodes UDP-galactose epimerase, galactose-1-phosphate uridylyltransferase, galactokinase and aldose-1-epimerase. When *Escherichia coli* uses galactose as the substrate, this bacterium needs these enzyme activities. UDP-galactose epimerase (GalE) has a second role in the synthesis of UDP-galactose, a building block for the cell wall and capsule. When cells are growing on glucose, GalE is needed but not the others, including galactokinase (GalK). The expression of Spot42 is under negative regulation by the CRP-cAMP complex, leading to higher levels of sRNA synthesis when cells are growing on glucose, and lower levels when cells are growing on galactose. Spot42 pairs with, and negatively regulates, translation of GalK without perturbing GalE translation. This sRNA is unique in its role in regulating polarity within an operon.

The majority of sRNAs in *Escherichia coli* require a protein named Hfq for their function. Hfq was identified as an *Escherichia coli* host factor (also known as HF-I) required for initiation of plus-strand synthesis by the replicase of the Q RNA bacteriophage. This protein functions as an RNA chaperone, preventing formation of secondary structures in sRNAs for efficient base pairing with the target mRNAs. When sRNAs are overexpressed, Hfq becomes limiting and, consequently, sRNAs are cleaved and less functional. Hfq is widely distributed in bacteria, but not found in some, including cyanobacteria, actinobacteria (actinomycetes) and certain *Chlamydia* and *Deinococcus* spp.

The functions of sRNAs requiring Hfq protein differ depending on their characteristics and on the binding site on the target mRNA, as for other sRNAs. Some repress translation by base pairing with the ribosome-binding Shine–Dalgarno (SD) sequence (Figure 12.21a), and the others activate by preventing secondary structures involving the SD sequence (Figure 12.21b). Some mRNAs become susceptible to RNase E (Figure 12.21c) and others are stabilized (Figure 12.21d) through base pairing with sRNAs. In contrast with Gram-negative bacteria, Hfq does not seem to be required for the stability and function of sRNAs in Gram-positive bacteria.

While most regulatory sRNAs participate in post-transcriptional regulation, 6S RNA acts on the transcription process. This sRNA is made in increased amounts when an *Escherichia coli* culture enters stationary phase to inhibit RNA polymerase containing the housekeeping σ^D (σ^{70}), but to activate that containing σ^S (σ^{38}).

The sRNAs discussed above are *trans*-acting, regulating transcripts from other operons. They are global regulators. A few of the regulatory sRNAs functioning by base pairing are encoded on the opposite strand of the DNA from which the target transcript is encoded. These are *cis*-acting sRNAs. An example is glutamine synthetase (GlnA) in *Clostridium acetobutylicum*. When the ammonia supply is limited, *glnA* is transcribed and an sRNA regulates translation (Figure 12.22). Classically, *cis*-acting sRNA was referred to as 'antisense RNA', but this term is now used as a synonym for sRNA.

Small proteins are produced through the translation of some regulatory RNAs that contain open reading frames, such as SgrS in enteric bacteria, SR1 in *Bacillus subtilis* and RNAIII in *Staphylococcus aureus*. In addition, some protein coding mRNAs contain 50 or 30 untranslated regions (UTRs) at the 5′ or 3′ end that have regulatory functions. A stress-induced protein-encoding mRNA (*irvA*) of *Streptococcus mutans* directly modulates the stability of a target mRNA (*gbpC*, glucan-binding protein) through interaction between the 5′ untranslated region of *irvA* mRNA and the coding region of *gbpC* mRNA. Even RNA fragments cleaved from certain larger coding or non-coding RNAs have regulatory functions. This suggests that the duality of 'protein encoding and regulation' in RNA is not uncommon.

Other kinds of sRNA are known which modulate the classical regulatory sRNAs. These are referred to as sRNA sponges or competing endogenous RNA (ceRNA). A sponge sRNA, MgrR, interacts with an sRNA, SorC, that represses *eptB* encoding a LPS-modifying enzyme responsible for resistance to the antimicrobial peptide polymyxin B in *Salmonella typhimurium*. MgrR is induced by surface stresses including the antimicrobial activity of polymyxin B.

12.1.9.5 Cyclic dimeric (c-di-GMP) riboswitch

Just as cAMP (Section 12.1.3.1) and ppGpp (Section 12.2.1) play important roles in bacterial metabolic regulation, 3′-5′ cyclic diguanylate monophosphate (c-di-GMP) is another nucelotide with a regulatory function. These are referred to collectively as second messengers. The latter nucleotide promotes biofilm formation and represses flagellum-driven swarming motility in many bacteria. Other functions include involvement in cell–cell signalling, responses to blue light, oxygen, nitric oxide and several other environmental challenges, and host colonization. These regulations are the result of control of transcription, translation of mRNAs with a riboswitch, enzyme activities and larger cellular structures modulated by cyclic di-GMP (c-di-GMP):

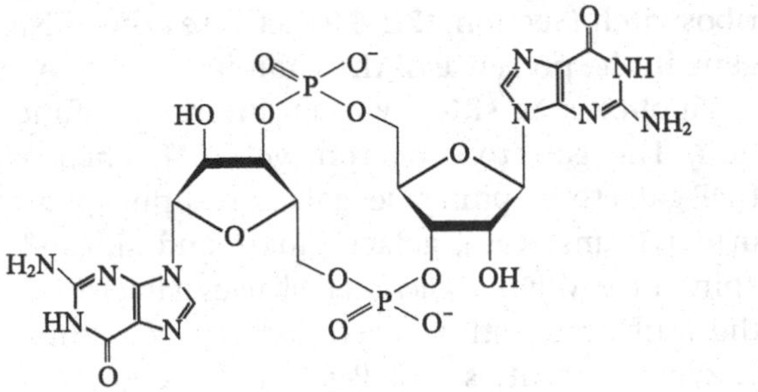

cyclic di-GMP

In natural environments, most microbes exist as surface-associated microbial communities, protected by an extracellular matrix, known as biofilms. Bacteria forming biofilms are physiologically different from genetically identical free-living planktonic cells. When surface-associated growth becomes disadvantageous, bacteria dissociate from the community and migrate towards a more favourable location. When a planktonic bacterial cell

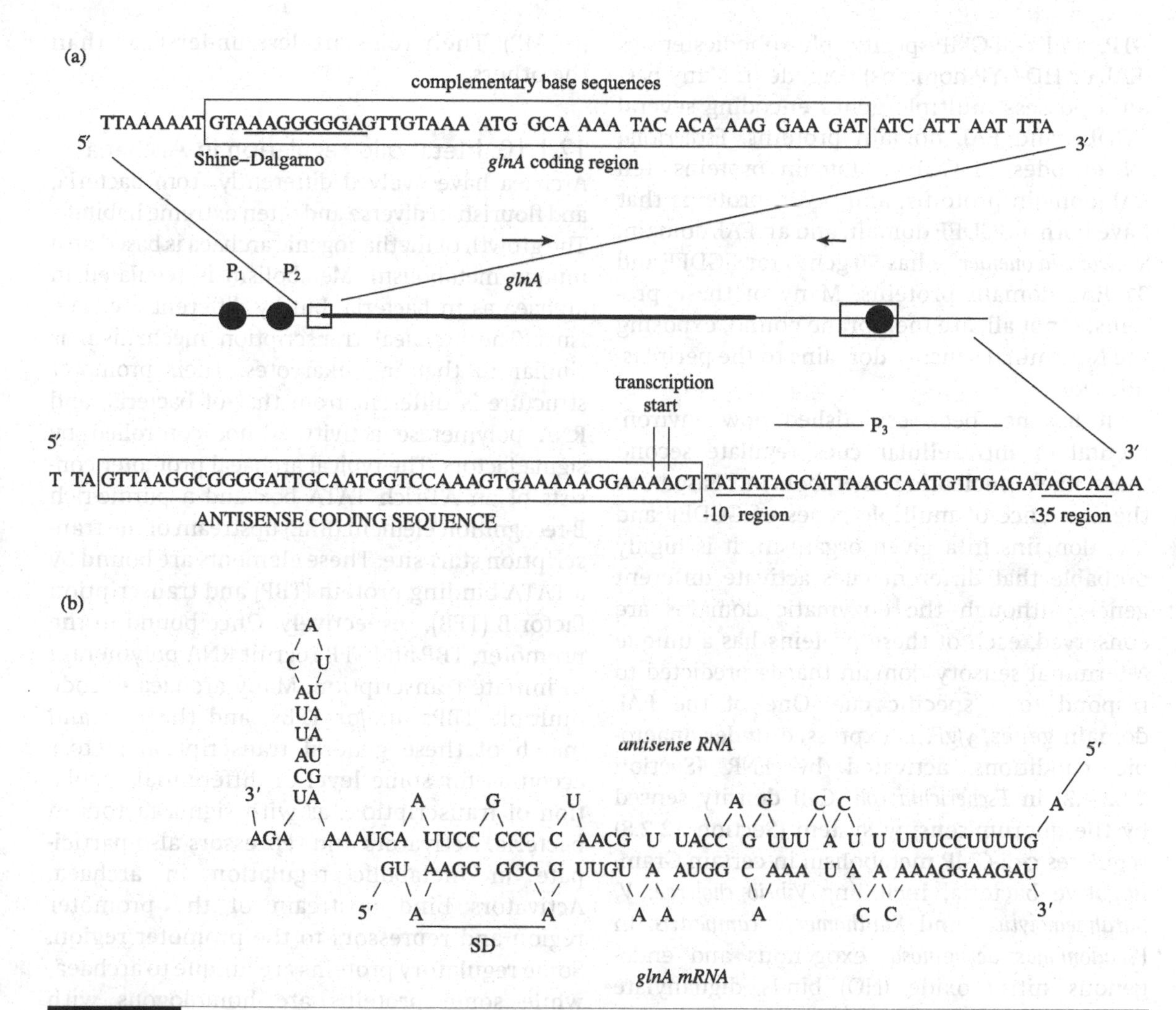

Figure 12.22 **Translational regulation of *glnA* (glutamine synthetase gene) by antisense RNA in *Clostridium acetobutylicum*.**

(*J. Bacteriol.* **174**: 7642–7647, 1992)

Some small RNAs are *cis*-acting regulators. They are encoded on the opposite strand of the DNA from which the target transcript is encoded. (a) In this fermentative bacterium, *glnA* (glutamine synthetase gene) is transcribed under ammonia-limited conditions and the translation is regulated by an sRNA. (b) Classically this sRNA was referred to as antisense RNA. Under conditions that repress *glnA* expression, the amount of antisense RNA exceeds the *glnA* transcript 1.6-fold, while five-fold more *glnA* transcript than the antisense RNA results under the opposite conditions.

recognizes a suitable surface through cell surface structures such as flagella and pili, the cells lose motility due to accumulation of c-di-GMP in the cell and form a new biofilm.

The spores of *Streptomyces* spp. germinate and grow out and produce a new vegetative mycelium when they encounter suitable conditions. The vegetative mycelium undergoes major developmental changes to the aerial hyphae, which are septated with chromosome segregation, eventually producing spores. This developmental change is controlled by a master regulator Bdl (bald without hairy aerial hyphae), the production of which is activated by c-di-GMP.

Diguanylate cyclases (GGDEF domains) synthesize c-di-GMP from two molecules of

GTP, and c-di-GMP-specific phosphodiesterases (EAL or HD-GYP domains) degrade it. Many bacteria possess multiple genes encoding several GGDEF and EAL domain proteins. *Escherichia coli* encodes 12 GGDEF domain proteins, ten EAL domain proteins, and seven proteins that have both a GGDEF domain and an EAL domain. *Shewanella oneidensis* has 50 genes for GGDEF and 31 EAL domain proteins. Many of these proteins, if not all, are membrane bound, exposing the *N*-terminal sensory domains to the periplasmic side.

It has not been established how environmental or intracellular cues regulate second messenger nucleotide metabolism. Based on the existence of multiple genes of GGDEF and EAL domains in a given organism, it is highly probable that different cues activate different genes. Although the enzymatic domains are conserved, each of these proteins has a unique *N*-terminal sensory domain that is predicted to respond to a specific cue. One of the EAL domain genes, *yfgF*, is expressed under anaerobic conditions activated by FNR (Section 12.2.4.2) in *Escherichia coli*. Cell density sensed by the quorum-sensing system (Section 12.2.8) regulates c-di-GMP metabolism in certain Gram-negative bacteria, including *Vibrio cholerae*, *V. parahaemolyticus* and *Xanthomonas campestris*. In *Pseudomonas aeruginosa*, exogenous and endogenous nitric oxide (NO) binds diguanylate cyclases and phosphodiesterases, inhibiting the former and activating the latter. With NO, the c-di-GMP concentration becomes low, and cells disperse from the biofilm. One of the GGDEF-EAL proteins in *E. coli* is inactive due to an intra-molecular disulfide bond under aerobic conditions, and becomes active when the disulfide bond is reduced under anaerobic conditions. Solar UV radiation damages cellular components. Bacteria produce protective pigments and move away from light. Photoreceptors sense the light and pass the signal to GGDEF-EAL proteins to modulate c-di-GMP concentration for the light-driven changes.

Other second messenger nucleotides are also known. They are cyclic di-AMP (c-di-AMP), cyclic AMP–GMP (c-GAMP) and cyclic GMP (c-GMP). Their roles are less understood than the others.

12.1.10 Metabolic regulation in Archaea

Archaea have evolved differently from bacteria, and flourish in diverse and often extreme habitats. The growth of methanogenic archaea is based on a unique metabolism. Metabolism is regulated in archaea as in bacteria, but by different mechanisms. The archaeal transcription mechanism is similar to that in eukaryotes. Their promoter structure is different from that of bacteria, and RNA polymerase activity is not controlled by sigma factors. The typical archaeal promoter consists of an AT-rich TATA box and a purine-rich B-recognition element (BRE) upstream of the transcription start site. These elements are bound by a TATA binding protein (TBP) and transcription factor B (TFB), respectively. Once bound to the promoter, TBP and TFB recruit RNA polymerase to initiate transcription. Many archaea encode multiple TBPs and/or TFBs, and the mix and match of these general transcription factors accounts for some level of differential regulation of transcription, as with sigma factors in bacteria. Activators and repressors also participate in metabolic regulation in archaea. Activators bind upstream of the promoter region and repressors to the promoter region. Some regulatory proteins are unique to archaea, while some proteins are homologous with others found in bacteria.

12.2 Global regulation: responses to environmental stress

In the previous section, various regulatory mechanisms of gene expression were discussed. Though some processes regulate a single operon or regulon, most of them regulate more than one operon or regulon. For example, the cAMP–CRP complex regulates the enzymes of more than 200 metabolic processes. These regulatory processes are referred to as global regulation, a multigene system or pleiotropic control.

Genes modulated by global regulation are encoded in operons and regulons scattered throughout the chromosome, the expression of which is regulated by a common effector. They are involved in many metabolic processes. In this sense they are different from a regulon which is a collection of genes involved in a single metabolic process, such as nitrogen fixation (*nif*) or arabinose catabolism (*ara*). The term modulon is used to define a collection of operons and regulons regulated by a common effector. Those regulated by the cAMP–CRP complex are referred to as *crp* modulons. As the *lac* operon is regulated by the cAMP–CRP complex as well as by the *lac* repressor, each operon and regulon of a modulon is regulated by the common activator or repressor and by effectors specific for each of them.

Stresses such as heat shock and oxidative stress result in regulation of many operons and regulons. It is not known if this regulation is modulated by a common effector. In this case, the term stimulon is used to define a collection of modulons, regulons and operons regulated by a common environmental stimulation. Various global regulation mechanisms are known in prokaryotic organisms, especially in bacteria (Table 12.4).

12.2.1 Stringent response

At the beginning of the stationary phase, most bacteria, including *Escherichia coli*, synthesize at least 50 new proteins, change their cell shape and size and exhibit increased resistance to stresses such as oxidative stress, near-UV irradiation, potentially lethal heat shocks, hyperosmolarity, acidic pH, ethanol and probably others yet to be identified. These phenotypic changes are a part of their survival strategy (Section 13.3). Complex regulatory networks known as the stringent response are involved in these phenotypic changes. Under nutrient-rich conditions, *Escherichia coli* grows at a high growth rate, synthesizing proteins at high speed. When a culture is transferred from nutrient-rich conditions to nutrient-poor conditions, the growth rate is reduced with less protein synthesis. Two different mechanisms are known in the stringent response. These are the ribosomal pathway caused by a limitation in the amino acid supply, and the carbon starvation pathway under carbon- and energy source-limited conditions. In both pathways, nutritional stress results in the production of guanosine-3′-diphosphate-5′-triphosphate (guanosine pentaphosphate, pppGpp) or guanosine-3′-diphosphate-5′-diphosphate (guanosine tetraphosphate, ppGpp), which is known as alarmone.

When the bacterial cell faces amino acid limitation, where the amino acid supply cannot meet the demand for peptide synthesis by the ribosomes, uncharged tRNA enters the peptide-synthesizing ribosome. This activates the ribosome-associated (p)ppGpp synthetase I to produce (p)ppGpp, transferring PP_i from ATP to GTP (GDP). This ribosome-associated enzyme is referred to as the RelA protein. RelA is derived from the fact that a mutant of this protein maintains high-rate peptide synthesis (relaxed mutant) even under amino acid-limited conditions. This RelA-mediated stringent response is referred to as the ribosomal pathway.

In the carbon starvation pathway, (p)ppGpp is produced by a RelA independent mechanism under carbon- and energy source-limited conditions by ppGpp synthetase II. Another enzyme, SpoT, either synthesizes or hydrolyses ppGpp to maintain a proper concentration of this regulatory nucleotide according to environmental conditions. The alarmone concentration increases rapidly upon the stringent response to a peak and then decreases to a steady-state concentration, which is still higher than normal values. The decrease in concentration is due to degradation of mRNAs coding the alarmone by toxins of the toxin/antitoxin system with mRNA interferase activity (Section 13.4.2) and SpoT activity hydrolysing the alarmone. SpoT is a dual-functional enzyme with strong ppGpp hydrolase and weak synthetase activities, while ppGpp synthetase is a mono-functional alarmone synthetase in *E. coli*. In Gram-positive bacteria, ppGpp metabolism is orchestrated by a dual-functional long enzyme and two mono-functional short enzymes, one small alarmone synthetase and a small alarmone hydrolase.

In general, (p)ppGpp downregulates transcription of genes encoding translational machinery and factors required for growth and division, and upregulates stress response genes.

Table 12.4 Global regulation in bacteria.

Environment	Regulation	Bacteria	Control gene	Controlled gene(s)	Mechanism
Substrate availability					
Carbon limitation	catabolite repression	enteric bacteria	*crp* (activator)	lactose, maltose catabolism	expression by cAMP–CRP complex (see Figure 12.7)
Amino acid and energy limitation	stringent response	enteric bacteria	*relA*, *spoT*, *katF* (*rpoS*)	enzymes in stationary phase	expression of σ^S by ppGpp
Substrate limitation	spore	*Bacillus* spp.	*spoOA* (activator) *spoOF* (modulator)	spore related	alternative σ-factors
Nutrient gradient	chemotaxis	enteric bacteria	*rpoF* (*F*)	flagellar and other motility- related proteins	alternative σ-factor
Ammonia limitation	Ntr system	enteric bacteria	*glnB*, *D*, *G*, *L*	glutamine synthetase	activation by phosphorylated protein (see Figure 12.23)
Ammonia limitation	Nif system	nitrogen fixers	*ntrA*	*nif* regulon	alternative σ-factor
Phosphate limitation	Pho system	enteric bacteria	*phoB*, *R*, *U*	alkaline phosphatase	two-component system (see Figure 12.24)
Electron transport chain					
Aerobic	Arc system	*E. coli*	*arcA*, *B*	aerobic enzymes	two-component system (see Figure 12.25)
Anaerobic	anaerobic metabolism	*E. coli*	*fnr*	anaerobic enzymes	activation by reduced FNR (see Figure 12.26)
Deteriorating environment					
UV	SOS response	*E. coli*	*lexA*, *recA*	DNA repair system	repressed by LexA
Heat	heat shock	*E. coli*	*rpoH*	molecular chaperone	alternative σ-factors
Low temperature	cold shock	*E. coli*	CSP	many proteins	RNA chaperone (see Figure 12.29)
H_2O_2 etc.	oxidative stress	many bacteria	*oxyR*	catalase, SOD	activated by OxyR
Cell density					
High cell density	quorum sensing	many bacteria	*luxI*, *luxR*	*luxC*, *D*, *A*, *B*, *E*	activated by LuxR–autoinducer complex (see Figure 12.30)

In addition to transcription, (p)ppGpp can also directly affect chromosome replication and alter the activity of certain enzymes that are involved in stress responses. When the ppGpp concentration increases, RpoS (σ^S) synthesis is activated, and the synthesis of stable tRNA and rRNA and the activity of housekeeping σ^D are inhibited. ppGpp activates the expression of σ^S at the level of translation not transcription. The exact mechanism of this translational activation is not fully understood, but directly and indirectly involves various proteins and sRNAs. σ^S activates the transcription of genes of stationary phase proteins that facilitate the phenotypic changes. In addition to σ^S synthesis, ppGpp is responsible for various physiological changes including virulence, persister cell formation through the induction of the toxin/antitoxin system (Section 13.4.2) and antibiotic resistance. In Gram-positive bacteria, including *Staphylococcus aureus*, *Bacillus subtilis* and *Enterococcus faecalis*, alarmone inhibits ribosomes thereby reducing protein synthesis. (p)ppGpp is not known in those archaea that have been examined.

σ^S- and σ^S-dependent genes are induced not only in stationary phase but also under various stress conditions. Therefore, σ^S is now seen as the master regulator of the general stress response that is triggered by many different stress signals. Its production is often accompanied by a reduction or cessation of growth, and this provides the cells with the ability to survive against the actual stress as well as additional stresses not yet encountered (cross-protection). It should be mentioned that an individual stress signal activates a specific stress response which results in the induction of proteins that allow cells to cope with this specific stress situation only. While specific stress responses tend to eliminate the stress agent and/or mediate repair of cellular damage that has already occurred, the general stress response renders cells broadly stress resistant in such a way that damage is avoided rather than needing to be repaired.

σ^S-controlled gene products generate changes in the cell envelope and overall morphology. Stressed *Escherichia coli* cells tend to become smaller and ovoid. Metabolism is also affected by σ^S-controlled genes, consistent with σ^S being important under conditions where cells switch from a metabolism directed toward maximal growth to a maintenance metabolism. σ^S also controls genes mediating programmed cell death (Section 13.4.1) in stationary phase, which may increase the chances of survival for a bacterial population under extreme stress by sacrificing a fraction of the population in order to provide nutrients for the remaining surviving cells. A number of virulence genes in pathogenic enteric bacteria have been found to be under σ^S control, consistent with the notion that host organisms provide stressful environments for invading pathogens.

Gram-positive bacteria, including *Bacillus subtilis*, have σ^B with the function of σ^S, but the amino acid sequences are different. In spore-forming bacteria, specific σ-factors are expressed in sequence to produce proteins for the transformation of vegetative cells to spores under nutrient-limited conditions (Section 13.3.1).

In addition to the stringent response, under nutrient-limited conditions the Lrp (leucine responsive regulatory protein) controls the expression of as many as 10 per cent of genes in *Escherichia coli*, including amino acid metabolism, transport of small molecules, pili synthesis and other regulatory proteins. In general, Lrp activates genes where the products are required in nutrient-poor conditions, and represses those with functions needed in nutrient-rich conditions.

12.2.2 Response to ammonia limitation

Nitrogen is essential as an element of amino acids and nucleic acid bases. Ammonia is assimilated into organic nitrogenous compounds by the action of glutamate dehydrogenase and glutamine synthetase. Since the affinity for ammonia is low, glutamate dehydrogenase has low activity under low ammonia concentrations. Under these conditions, ammonia is assimilated by glutamine synthetase which has a higher affinity for ammonia than glutamate dehydrogenase (Section 6.2.3). Ammonia inactivates the expression of the *nif* regulon in nitrogen-fixing organisms (Section 6.2.1). This regulation is mediated by an

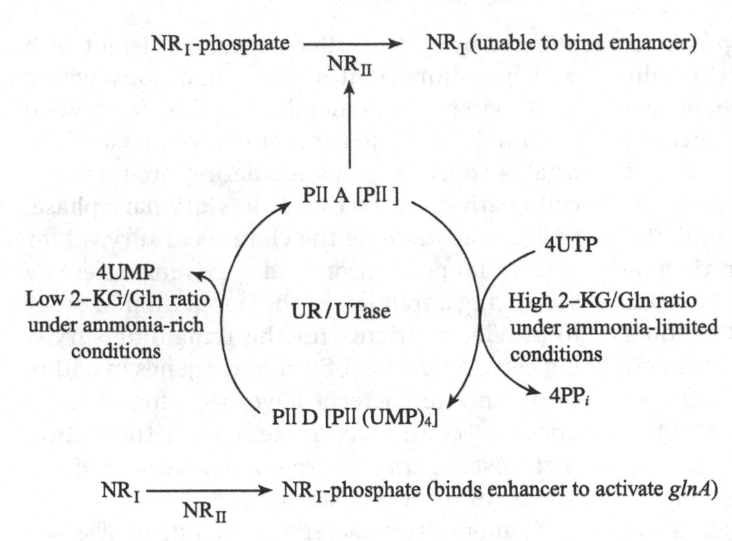

Figure 12.23 Signal transduction cascade of ammonia availability regulating the transcription of *glnA* (glutamine synthetase gene).

The 2-ketoglutarate/glutamine (2-KG/Gln) ratio is dependent on the availability of ammonia. An enzyme, uridylylase/deuridylylase, of PII protein (GlnB), uridylylates PII protein in the PIIA–NR$_{II}$ complex form to the PIID form, liberating NR$_{II}$ under high 2-KG/Gln ratio conditions with limited ammonia supply. The free form of NR$_{II}$ phosphorylates NR$_I$ protein (GlnG, NtrC). The phosphorylated NR$_I$ binds the enhancer region of the glutamine synthetase gene (*glnA*), activating transcription. PIIA under ammonia-excess conditions adenylylates glutamine synthetase to a less active form, and PIID under the reverse conditions removes the adenylyl group from the enzyme to keep the activity high.

alternative σ-factor that affects many regulons. In addition, individual regulons are under complex and separate regulatory mechanisms.

Escherichia coli transcribes the glutamine synthetase gene (*glnA*) at a low level to provide glutamine as the amine donor under ammonia-rich conditions. This is recognized by the house-keeping sigma factor σ^D (σ^{70}), but the same gene is transcribed with the aid of σ^N (σ^{54}) under ammonia-limited conditions. Transcription of *glnA* aided by σ^N requires the binding of phosphorylated NR$_I$ protein (NtrC) to the enhancer region, upstream of the gene. GlnG and NtrC are synonyms for the NR$_I$ protein. NR$_I$ protein is phosphorylated through a complex cascade of reactions to transduce the signal of low ammonia supply and control transcription of the *glnA* gene (Figure 12.23).

The availability of ammonia determines the 2-ketoglutarate/glutamine (2-KG/Gln) ratio. When the ammonia supply is limited, the 2 KG/Gln ratio becomes high. This 2-KG/Gln ratio regulates the activity of an enzyme uridylylase/deuridylylase (GlnD) of a PII protein (GlnB). When the ammonia supply is limited, this enzyme uridylylates the PII protein, transforming the PIIA form to the PIID form [PII(UMP)$_4$], consuming 4UTP. PIIA forms a complex with another protein, NR$_{II}$, when the ammonia concentration is high, and the NR$_{II}$ protein is liberated when PIIA is uridylylated to PIID, with a limited ammonia supply. The PIIA–NR$_{II}$ complex has a phosphatase activity on phosphorylated NR$_I$, while free NR$_{II}$ has kinase activity on NR$_I$. As a consequence of these reactions, NR$_I$ protein is phosphorylated to activate transcription of the *glnA* gene, binding the enhancer region when the ammonia supply is limited. The phosphorylated NR$_I$ under ammonia-limited conditions activates the NAC (nitrogen assimilation control) protein gene. NAC activates genes for the utilization of organic nitrogen compounds such as that for cytosine deaminase (*codBA*). PII proteins in Gram-negative bacteria are subject to post-translational modification, not only by uridylylation, but also by adenylylation, while in some cyanobacteria the PII protein is modified by phosphorylation.

PIIA/IID regulates not only the transcription of glutamine synthetase but also the activity of this enzyme. PIIA under ammonia-rich conditions adenylylates glutamine synthetase to a less active form, and PIID under the reverse conditions removes the adenylyl group from the enzyme to keep the activity high (Figure 12.37, Section 12.3.2). PII or similar proteins have been identified in bacteria, archaea and eukaryota. They are believed to function as transducers for a variety of signals.

The transcription of the *nif* regulon is regulated in a similar manner. Under ammonia-limited conditions, σ^{54} (σ^N, NtrA) initiates transcription of this regulon with binding of an enhancer-binding protein, NR_1 (NtrC) to the enhancer region, as discussed earlier (Section 6.2.1.5). Nitrogenase activity is also regulated by the availability of ammonia.

In addition to *gln* and *nif* operons, σ^N participates in the regulation of many other operons and regulons, including amino acid transporters such as arginine (*argT*), histidine (*hisJQMP*) and glutamine (*glnHPQ*), amino acid metabolism such as arginine (*astCADBE*), carbon metabolism such as acetoacetate (*atoDAEB*) and propionate (prpBCDE), and hydrogenase (*hycABCDEFGHI*, *hydN-hycF* and *hypABCDE-fhlA*). Each of these has its own regulator.

The low G + C Gram-positive bacterium *Bacillus subtilis* does not have an anabolic glutamate dehydrogenase and glutamine synthetase (GS) is the only enzyme used for ammonium assimilation in this bacterium. The expression of this enzyme (*gltAB* operon) is regulated by GlnR (transcriptional regulator) and TnrA (transcriptional regulator). When organic nitrogen compounds are in excess, GlnR and TnrA repress expression of the *gltAB* operon. When the organic nitrogen supply is limited, TnrA activates various genes for nitrogen metabolism, including those for urease, asparagine degradation, nitrate and nitrite assimilation, and ammonia uptake. Other Gram-positive bacteria have a similar regulatory mechanism for nitrogen metabolism as in *B. subtilis*, but the details are different depending on species.

In addition to the phosphotransferase (PT) system that catalyses group translocation of sugars, many Gram-negative bacteria possess the analogous nitrogen PT system (PTS^{Ntr}, Section 3.5). In analogy with the sugar PTS, EI^{Ntr} (PtsP) and NPr (PtsO) catalyse the PEP-dependent phosphorylation of protein $EIIA^{Ntr}$ (PtsN). PTS^{Ntr} exclusively serves regulatory functions using $EIIA^{Ntr}$ as an output domain in the regulation of nitrogen and carbon metabolism. Two of the PTS^{Ntr} genes, *ptsO* (NPr) and *ptsN* ($EIIA^{Ntr}$), map in the same operon of the gene (*rpoN*) encoding the sigma factor σ^N. NPr as well as HPr transfers phosphoryl groups to $EIIA^{Ntr}$, and NPr can be phosphorylated by EI and EI^{Ntr} proteins. However, EI^{ntr} and NPr do not phosphorylate HPr and EI. Phosporylated $EIIA^{Ntr}$ is detectable at all growth stages and is accumulated in the stationary phase, whereas the non-phosphorylated form exclusively appears during rapid growth. The time point of disappearance of non-phosphorylated $EIIA^{Ntr}$ during growth depends on the carbon as well as the nitrogen source, suggesting that phosphorylation of $EIIA^{Ntr}$ is influenced by the physiological status of the cell. A *ptsN* ($EIIA^{Ntr}$) mutant of *Klebsiella pneumonia* exhibited elevated levels of nitrogen fixation. The non-phosphorylated form of $EIIA^{Ntr}$ represses transcription of the *nif* genes and *glnA*. A GAP domain is found at the *N*-terminal of EI^{Ntr} but not in EI. A GAP domain is bound with small signalling molecules, such as cyclic nucleotides, nitric oxide, 2-ketoglutarate, formate and sodium ions. The signalling molecules modulate the activity of the GAP domain protein they bind. It is assumed that glutamine inactivates EI^{Ntr} to reduce phosphorylation of $EIIA^{Ntr}$, connecting carbon (PEP) and nitrogen(glutamine) metabolism.

Polyhydroxybutyrate (PHB) accumulation is controlled by the stringent response and induced under conditions of nitrogen deprivation (Section 13.2.2.1) in *Ralstonia cutropha*. Knockout of $EIIA^{Ntr}$ increases the PHB content. The non-phosphorylated $EIIA^{Ntr}$ inhibits its formation through interaction with a stringent response enzyme, ppGpp synthase/hydrolase (SpoT1).

PTS^{Ntr} is not known in Gram-positive bacteria.

12.2.3 Response to phosphate limitation: the *pho* system

Phosphate is abundant in nature, but it is not always available to microbes due to its low solubility. Under phosphate-limited conditions, enteric bacteria synthesize around 100 new proteins, including alkaline phosphatase (PhoA). The expression of genes for these proteins is regulated through the *pho* system. The *pho* system is a two-component system involving the sensor proteins, PhoU and PhoR, and a modulator protein, PhoB. Since a signal is transferred from a sensor protein to a modulator protein in a

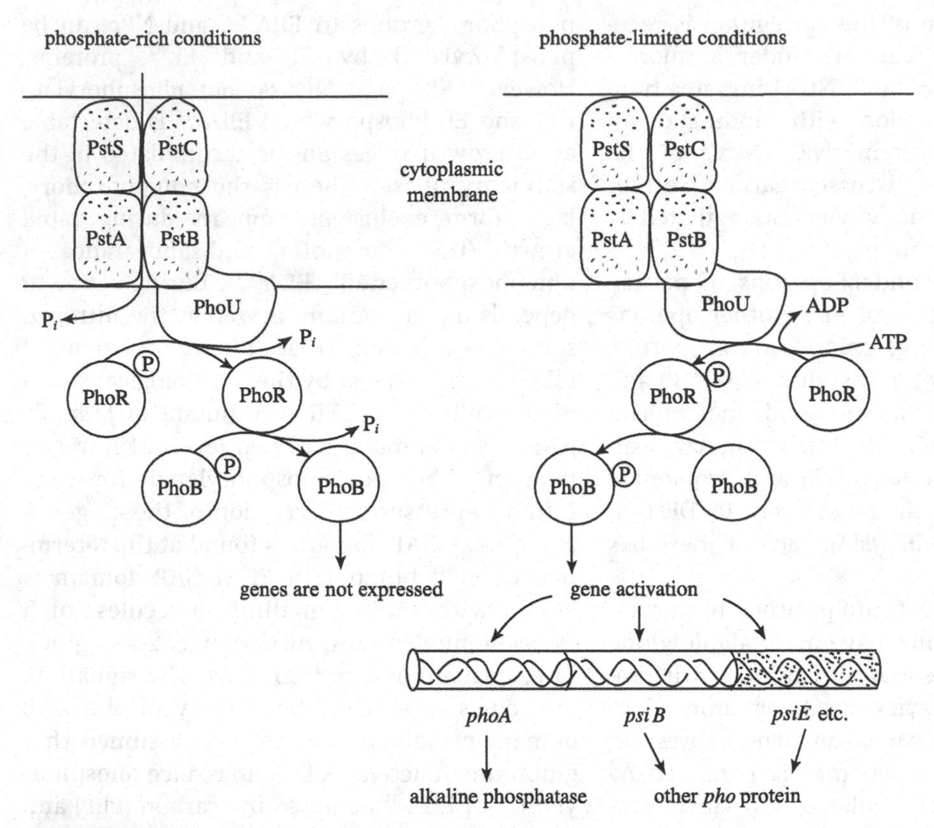

Figure 12.24 Regulation of the *pho* system under phosphate-limited conditions.

The inorganic phosphate specific transporter (PST) senses a limited phosphate supply and passes this information to PhoR through PhoU. PhoR is phosphorylated, consuming ATP, and transfers phosphate to a soluble regulator protein, PhoB. Phosphorylated PhoB activates the transcription of many operons. When phosphate is available, PhoB is dephosphorylated to an inactive form. This is an example of a two-component system. PhoU and PhoR are the sensor proteins and PhoB is the modulator protein.

two-component system, this regulation can also be referred to as a signal-transducing system. In a two-component system, the histidine residue of the membrane-bound sensor protein is phosphorylated before the phosphate group is transferred to the aspartate residue of the soluble modulator protein.

When the phosphate supply is limited, the inorganic phosphate specific transporter (PST), an ABC transporter, senses it and passes this information to PhoR through PhoU. PhoR is phosphorylated, consuming ATP. Phosphorylated PhoR transfers phosphate to PhoB. Phosphorylated PhoB is an activator (Figure 12.24). PST consists of PstS, PstC, PstA and PstB. PhoU and PhoR are membrane proteins, and PhoB is a soluble protein.

PhoU is a peripheral membrane protein that is a negative regulator of the signalling pathway. Mutations in the *phoU* gene of *Escherichia coli* result in PhoR becoming a constitutive PhoB kinase, leading to high expression of the Pho regulon genes, causing the accumulation of high levels of polyphosphate and increased sensitivity to environmental and oxidative stress, high temperature, low pH and antibiotics.

In *Pseudomonas aeruginosa*, the PhoB/PhoR two-component system has a similar function to cope with phosphate-limited conditions. In addition, phosphorylated PhoB activates an extracytoplasmic sigma factor σ^{VreI} that transcribes virulence genes. A similar two-component system, SenX3/RegX3 in *Mycobacterium tuberculosis*, is also involved in virulence enabling the bacterium to survive phosphate limitation during infection. The PhoR sensor kinase and regulatory PhoP two-component system repress genes for teichoic acid synthesis and activate teichuronic acid genes (Section 2.3.3.3) in *Bacillus subtilis*. The synthesis of many classes of secondary metabolites, including clinically important antibiotics, is negatively regulated by high P_i concentrations in culture media (Section 12.5). Evidence is emerging to show that the PhoR/PhoP two-component system regulates secondary metabolism in several *Streptomyces* spp.

It is interesting to note that histidine autokinase (HK), CreC of the CreC/CreB (carbon source responsive) two-component system (TCS), phosphorylates PhoB. This cross-regulation is common between different TCSs and may play a role in the integration of multiple signals (Section 12.4.2). In addition to CreC, other non-partner HKs such as ArcB (Section 12.2.4.1), KdpD (potassium uptake system), QseC (catecholamine sensor), BaeS (drug resistance) and EnvZ (Section 12.2.10) have similar functions.

12.2.4 Regulation by molecular oxygen in facultative anaerobes

Facultative anaerobes including *Escherichia coli* metabolize glucose through the central metabolic pathway to generate ATP, NADPH and the carbon skeletons required for biosynthesis. Under anaerobic conditions, they operate an incomplete TCA fork (Section 5.4.1) instead of the TCA cycle. This metabolic shift means that the activities of TCA cycle enzymes should be regulated (Table 12.5). In addition to the TCA cycle enzymes, transcription of other genes is regulated when the culture becomes anaerobic. Pyruvate is oxidized by pyruvate dehydrogenase under aerobic conditions, while pyruvate:formate lyase catalyses a similar reaction under anaerobic conditions (Section 8.6.4). Genes for other anaerobic enzymes are activated under anaerobic conditions, including fumarate reductase, nitrate reductase and hydrogenase in facultative anaerobes. The products of *arc* (aerobic respiration control) and *fnr* (fumarate nitrate reductase) mediate this regulation.

Table 12.5 Activities of TCA cycle enzymes in *Escherichia coli* growing under aerobic and anaerobic conditions.

Enzyme	Activity (U/mg protein)	
	Aerobic growth	Anaerobic growth
Citrate synthase	51.5	10.5
Aconitase	317	16.1
Isocitrate dehydrogenase	1416	138
2-ketoglutarate dehydrogenase	17.4	0

12.2.4.1 ArcB/ArcA and PrrB/PrrA systems

When the oxygen supply is limited, electrons cannot be transferred to oxygen, and coenzyme Q of the electron transport system is predominantly in the reduced form. This signal is sensed by a membrane protein, ArcB, that is autophosphorylated, consuming ATP. This phosphate is transferred to a soluble protein, ArcA. The phosphorylated ArcA inhibits the transcription of the enzymes of aerobic respiration and activates transcription of the cytochrome *d* gene. This has a higher affinity for oxygen than cytochrome *o* (Figure 12.25). The ArcB/ArcA system modulates the expression of at least 30 regulons, including cytochrome oxidase, enzymes of the TCA and glyoxylate cycles, and fatty acid metabolism. The modulation of carbon metabolism is under dual control by ArcB/ArcA and the CreB/CreC (carbon source responsive) systems. This is another example of a two-component system. The ArcB/ArcA system also activates the gene for the F-pilus.

In another facultative anaerobe, *Shewanella oneidensis*, the sensor kinase ArcS does not transfer phosphate directly to the response regulator ArcA. A separate protein HptA functions as a phosphotransferase from ArcS to ArcA.

A similar redox-sensing two-component system, PrrB/PrrA, is known in other bacteria. Purple non-sulfur bacteria grow photosynthetically under

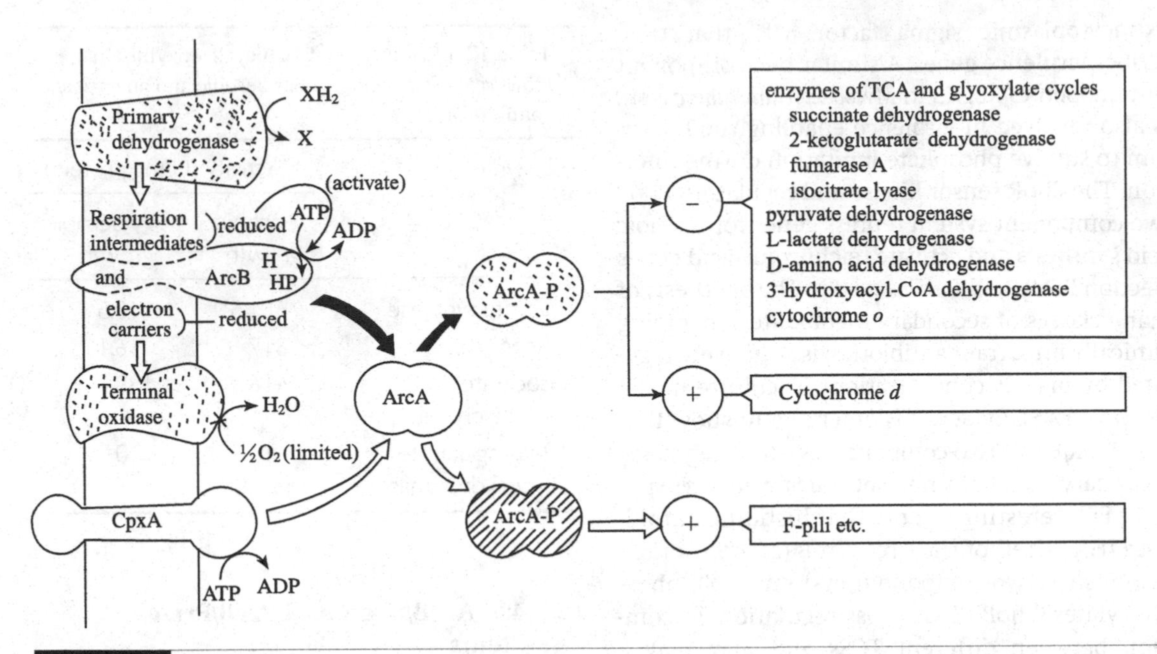

Figure 12.25 Regulation by the *arc* system.

Under oxygen-limited conditions, electron transport chain components are reduced and with this signal a membrane sensor protein, ArcB, is autophosphorylated, consuming ATP. Phosphate is transferred from the histidine residue of ArcB to the aspartate residue of the modulator protein, ArcA. The active ArcA-P modulates the expression of many operons including those for aerobic respiration and F-pili.

anaerobic conditions, and many of them grow chemoorganotrophically under aerobic conditions (Section 11.1.2). The photosynthetic apparatus is synthesized only under anaerobic conditions. The PrrB/PrrA (formerly RegB/RegA) two-component system modulates this regulation in *Rhodobacter capsulatus* and *Rhodobacter sphaeroides*. This system regulates not only photosynthesis but also numerous energy-generating and energy-utilizing processes, such as carbon fixation, nitrogen fixation, hydrogen utilization, cellular lipid accumulation, aerobic and anaerobic respiration, denitrification, electron transport and aerotaxis. Both phosphorylated and unphosphorylated forms of the response regulator PrrA are capable of activating or repressing a variety of genes in the regulon. Highly conserved homologues of PrrB/PrrA have been found in a wide number of photosynthetic and non-photosynthetic bacteria.

In *Brucella abortus*, PrrB/PrrA responds to the redox status and acts as a global regulator, controlling the expression of denitrification and high-affinity cytochrome oxidase genes which are involved in adaptation to the harsh environmental conditions found in mammalian hosts. The nitrogen regulation two-component system, NtrY/NtrX, participates coordinately in this regulation.

12.2.4.2 *fnr* system

A *fnr* (fumarate and nitrate reductase) mutant of *Escherichia coli* cannot use nitrate or fumarate as electron acceptor. The product of the *fnr* gene (FNR) modulates the expression of many operons. FNR is produced independently of oxygen availability. Under anaerobic conditions, the FNR protein with a [4Fe–4S] structure forms a homodimer that can bind DNA, but under aerobic conditions the [4Fe–4S] structure decomposes to a [2Fe–2S] structure that cannot form the dimer and does not bind DNA. When FNR with the [2Fe–2S] structure is exposed to oxygen, [2Fe–2S] is further decomposed. These decomposition processes are reversible (Figure 12.26). Similar reactions are employed for the detection of the reactive products of oxidative or nitrosative stresses (Section 12.2.5).

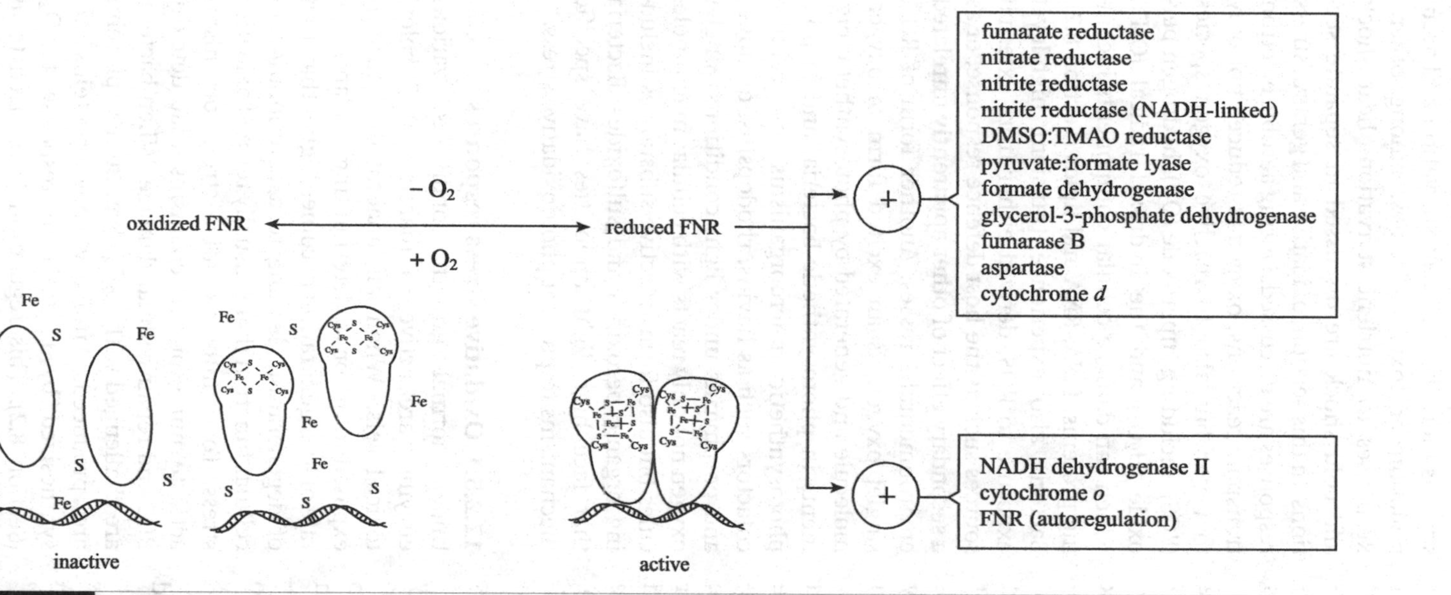

Figure 12.26 Modulation of gene expression by the FNR protein in *Escherichia coli*.

(*Ann. Rev. Microbiol.* **53**: 495–523, 1999)

FNR is produced independently of oxygen availability. FNR with a [4Fe–4S] structure forms a homodimer that binds DNA and modulates the expression of many genes under anaerobic conditions. Under aerobic conditions, [4Fe–4S] decomposes to [2Fe–2S] and decomposes further in the presence of molecular oxygen. FNR with the [2Fe–2S] structure does not bind DNA.

Pseudomonas aeruginosa has a similar regulatory protein, Anr, that is involved in biofilm formation and virulence, and oxidative stress. The activity of the FNR homologous FnrL protein in *Rhodobacter sphaeroides* is inhibited by acetylation (Section 12.3.2.2).

Under anaerobic conditions, nitrate activates genes for enzymes needed for denitrification, and inhibits the expression of genes necessary to utilize fumarate, dimethylsulfoxide (DMSO) and trimethylamine-*N*-oxide (TMAO), as electron acceptors. Through such regulation, enteric bacteria can preferentially use electron acceptors of a higher redox potential over those of lower potential to conserve maximum energy under given conditions. For this regulation, each operon and regulon of a modulon is under the control of separate effector(s). Enteric bacteria, including *Escherichia coli*, can grow fermentatively under anaerobic conditions without electron acceptors (Section 8.6).

The availability of nitrate is sensed and this information is used to modulate nitrate metabolism in a two-component system similar to the *arc* system. When nitrate is available as an electron acceptor, a membrane sensor protein, NarX, is autophosphorylated and transfers the phosphate group to a soluble modulator protein, NarL (Section 9.1.3). Phosphorylated NarL activates genes for denitrification and inhibits genes for enzymes of alternative electron acceptors such as fumarate and DMSO.

In addition to transcriptional regulation by oxygen, an sRNA regulates translation of various mRNAs upon a shift from aerobic to anaerobic conditions in enteric bacteria. An Hfq-binding sRNA (FnrS) induced under anaerobic conditions binds a variety of mRNAs to downregulate their translation. The majority of the target genes are downregulated by FNR. The FnrS extends the FNR regulon and increases the efficiency of anaerobic metabolism by repressing the synthesis of enzymes that are not needed under these conditions. A similar sRNA, AniS, is known in *Neisseria meningitidis*.

12.2.5 Oxidative and nitrosative stress responses

Bacterial cells are exposed to various potential stresses including nutrient starvation and accumulation of toxic molecules, among others. Responses to nutrient starvation, heat shock and cold shock are discussed in separate sections. In this section, oxidative and general stress responses are discussed. During aerobic electron transport reactions, oxygen is reduced not only to water but also to reactive oxygen species (ROS), including superoxide ($O_2^{\bullet}$), hydrogen peroxide (H_2O_2) and the hydroxyl radical ($OH^{\bullet}$), which can damage cellular components including proteins, DNA, RNA and lipids (Section 8.2). Bacteria also experience transient high-level exposure to ROS, derived either from external sources such as the host defence response, or as a secondary effect of other apparently unrelated environmental stresses. Another form of ROS, singlet oxygen, is an excited form of oxygen molecule and generated by photosynthetic pigments in photosynthetic bacteria, and in non-photosynthetic microorganisms by cellular cofactors such as flavins, rhodopsins, quinones and porphyrins under light conditions. Singlet oxygen not only reacts with cellular macromolecules but also forms reactive substances including organic peroxides and sulfoxides. Bacteria that face high light intensities have specific mechanisms to prevent photooxidative stress.

12.2.5.1 Oxidative stress responses

Under normal growth conditions, various enzymes are active to maintain ROS below toxic levels. When an anaerobic culture is exposed to air or to deteriorating oxidants, the rate of ROS generation becomes higher than that of degeneration, leading to a deterioration of cellular function that is usually termed oxidative stress. To counter oxidative stress, organisms activate numerous mechanisms that detoxify ROS and repair cellular damage. When bacteria are challenged with H_2O_2, over 30 new proteins are produced. Similarly, new proteins are synthesized by bacteria in response to $O_2^{\bullet}$ (Section 8.2). This regulation is referred to as the oxidative stress response, and also protects the cell from other deteriorative oxidants such as hypochlorous acid (HOCl) and nitric oxide ($NO^•$), as well as others.

In *Escherichia coli*, a H_2O_2-sensing regulator, OxyR (hydrogen peroxide-inducible genes activator), senses H_2O_2 and activates related operons. OxyR has two cysteine residues that can form a disulfide bond when oxidized. OxyR with a mid-potential of around −185 mV is in the reduced form in the cytoplasm that has a mid-potential of around −280 mV under normal growth conditions. OxyR is oxidized, forming an intramolecular disulfide bond when the intracellular H_2O_2 increases with the increase in cytoplasmic redox potential. The oxidized OxyR binds and activates many operons, including *katG* (catalase), *ahpCF* (alkylhydroperoxide reductase), *gorA* (glutathione reductase) and *grxA* (glutaredoxin 1). When the H_2O_2 concentration reaches a normal level, the oxidized OxyR is reduced by glutathione and glutaredoxin 1. The regulators PerR and OhrR, with a similar function, are known in *Bacillus subtilis*. A redox-sensing protein, CyeR, is known in *Corynebacterium glutamicum*. OxyR and PerR are primarily sensors of H_2O_2, while OhrR senses organic peroxide (ROOH) and HOCl. OhrR senses oxidants by means of the reversible oxidation of cysteine residues, as in OxyR. In contrast, PerR senses H_2O_2 via the Fe-catalysed oxidation of histidine residues. The IprA (inhibitor of hydrogen peroxide resistance) protein in *Salmonella enterica* inhibits expression of genes for catalase and the sigma factor σ^S.

H_2O_2 can be detoxified as a part of the general stress response. HOCl activates the expression of *cpdA*, encoding the cAMP phosphodiesterase that hydrolyses cAMP, leading to dramatically increased resistance to HOCl and H_2O_2 in *Escherichia coli*. The decrease in cellular cAMP levels derepresses *rpoS*, encoding σ^S, which activates the general stress response including the DNA starvation/stationary phase protection protein Dps, the catalase KatE and exonuclease III XthA. H_2O_2 also induces Dps production which promotes survival under stressful conditions, including carbon or nitrogen starvation, oxidative stress, toxic metal exposure and UV irradiation. In addition to activation of the *cdpA* gene, HOCl reduces the intracellular iron concentration. This regulation is mediated by a regulatory protein, HypT (hypochlorite-responsive transcription factor). HOCl oxidizes methionine residues of this protein to activate it. Methionine sulfoxide reductase reduces the oxidized HypT to inactivate it.

In addition to the reactions of H_2O_2 damaging cellular components, this ROS generates more reactive hydroxyl radicals in the presence of iron in a reaction known as the Fenton reaction. Lactobacilli have a protein, HprA1 (hydrogen peroxide resistance), that binds iron to prevent the formation of hydroxyl radicals. Polyphosphate (Section 13.2.4) protects bacterial cells against general stresses, including oxidative stress, as a protein-protective chaperone and preventing the Fenton reaction through reducing the Fe^{2+} concentration. A group of proteins, ferritins, bind and oxidize Fe^{2+} to Fe^{3+}, and thereby can store this metal in a non-reactive form. In addition, Dps binds DNA to protect it in the presence of Fe^{2+} or H_2O_2. Iron exporter genes are activated by OxyR in *Escherichia coli*.

The SoxRS system regulates many operons in response to the concentration of cytoplasmic $O_2^•$ (Section 8.2). SoxR is a redox-sensitive transcriptional activator and SoxS is a regulatory protein. A constitutive protein, SoxR, with two [2Fe–2S] clusters, is oxidized to activate the transcription of *soxS*. SoxS in turn activates the *sox* operon to increase the SoxS concentration further. SoxS activates transcription of many oxidative stress response proteins, including manganese superoxide dismutase (*sodA*), glucose-6-phosphate dehydrogenase (*zwf*), ferredoxin reductase (*fpr*) and endonuclease IV (*nfo*). SodA detoxifies $O_2^•$ (Section 8.2), Zwf and Fpr provide reducing power for SodA and Nfo repairs damaged DNA. This is the only example known so far where a simple change in the [Fe–S] cluster redox state is sufficient to modulate transcriptional activity. SoxR does not sense superoxide ($O_2^•$) directly, but indirectly by means of redox-cycling molecules such as quinones and phenazines. SoxR can be oxidized by redox-cycling molecules (RCMs), such as paraquats and phenazines, under anaerobic conditions. SoxRs among bacteria have different sensitivities to the RCMs depending on their redox potential. *Escherichia coli* SoxR with a lower potential responds to a broader range of RCMs than the *Streptomyces coelicolor* SoxR. In *S. coelicolor*, RCMs that do not activate SoxR do not inhibit growth, suggesting that SoxR is tuned to respond to growth-inhibitory RCMs.

The lipopolysaccharide (LPS) core biosynthetic function operon *waaYZ* is activated by SoxS, but not OxyR in *Escherichia coli*. Cultures treated with superoxide generators are more resistant to superoxide, as well as various drugs including antibiotics that modify LPS. This operon is also activated by MarA (multiple-antibiotic resistance regulator), which is activated by non-oxidative compounds and recognizes the same binding site on the DNA as SoxS.

Mammalian macrophages produce ROS ($O_2^{\bullet}$ and H_2O_2) and NO, from arginine, when infected by bacteria such as *Helicobacter* and *Campylobacter* spp. NO is a reactive nitrogen species (RNS). The bacteria possess the usual detoxifying enzymes, including superoxide dismutase, catalase and Dps for ROS and NO reductase, nitrite reductase and haemoglobin for RNS. Superoxide dismutase is localized on the cell surface in *Helicobacter* spp. Genes for these enzymes are regulated by various regulatory proteins, but not by FNR or the Arc system. Fur (ferric uptake regulator), AcnB (aconitate hydratase) and CrsA (carbon storage regulator) are common in both genera and each genus has its own regulators. The microaerophilic foodborne pathogen *Campylobacter jejuni* does not possess *oxyR* and *soxRS* genes but has *perR*, whose product negatively controls genes for catalase as well as superoxide dismutase.

Bacteriochlorophyll *a* of photosynthetic bacteria generates highly reactive singlet oxygen under aerobic light conditions. Carotenoids quench either excited photosensitizer molecules such as chlorophylls or singlet oxygen itself, and glutathione scavenges this ROS. In *Rhodobacter sphaeroides*, singlet oxygen that escapes from the protective mechanisms is sensed by a factor(s) yet to be identified, which releases the extracytoplasmic sigma factor σ^E from its anti-sigma factor ChrR. The σ^E transcribes another alternative σ, $RpoH_{II}$, which then promotes transcription of photooxidative stress-related genes, including at least 25 sRNAs and protective proteins against singlet oxygen. The sRNAs modulate translation of genes for production of pigments, cell division proteins and ribosomal proteins, aided by the RNA chaperone, Hfq.

It should be noted that activation of the stationary phase sigma factor, σ^S, and the extracytoplasmic sigma factor, σ^E, is a response to general stress. While expression of the σ^s gene is controlled by the alarmone and other mechanisms, the constitutive protein σ^E is controlled by the anti-sigma factor (Section 12.1.1).

The extracytoplasmic sigma factor, σ^B, is known in some Gram-positive bacteria to be regulated post-translationally by the the anti-σ^B-factor, RbsW, and the anti-anti-σ^B-factor, RsbV. Acute environmental stress signals, including heat shock and low pH, and internal signals tied to the metabolic/energetic state of the cell, coordinately regulate RsbV phosphorylation, which in turn controls σ^B activity. *Bacillus subtilis* has a constitutive DNA glycosylase (MutM) that prevents ROS-induced-DNA damage. MutM reduces the adaptive mutation (Section 12.2.12). On the other hand, the alphaproteobacterial extracytoplasmic function (ECF) σ-factor, EcfG (σ^{EcfG}), is activated by a two-component system. The anti-anti-σ^{EcfG}, PhyR, is a dual function protein that is also a regulatory response protein of a two-component system. Under stress conditions, membrane-bound sensor proteins phsophorylate PhyR to increase its affinity for the anti-σ-factor, NepR. Phosphorylated PhyR binds NepR releasing σ^{EcfG}.

Mycobacterium tuberculosis protects itself from oxidative and nitrosative stresses caused by host defence mechanisms and mycobactericidal agents, such as isoniazid, moxifloxacin and clofazimine, through $F_{420}H_2$-dependent quinone reductase (Fqr) and deazaflavin ($F_{420}H_2$)-dependent nitroreductase (Section 9.4.2). Fqr enzymes catalyse an $F_{420}H_2$-specific two-electron reduction of endogenous quinones and thereby prevent the formation of cytotoxic semiquinones.

In addition to activation of genes, redox-regulated chaperones (Section 12.2.6) protect the cell from oxidative stress in an immediate response by preventing protein unfolding and aggregation that leads to cell death. Cysteine residues of the redox-regulated chaperone, Hsp33, in bacteria, are oxidized to form disulfide on oxidative stress, such as exposure to HOCl. Oxidation of the heat-shock protein increases its affinity for unfolding proteins to function as a chaperone.

12.2.5.2 Nitrosative stress responses

Nitric oxide (NO) and related nitrogen oxides, collectively known as reactive nitrogen species (RNS), are reactive and toxic to cells. NO is produced during denitrification by nitrite reductase (Section 9.1.1.2) and by highly regulated multidomain NO synthases of the innate immune system of mammalian cells, consuming L-arginine to defend against bacterial infection. NO is converted to other forms of RNS by enzymatic or chemical reactions. RNS inactivate haem proteins by iron nitrosylation and by forming *S*-nitrosothiol with cysteine residues in proteins.

NO dramatically upregulates expression of the *Escherichia coli* flavohaemoglobin (Hmp), that has enzyme activity to detoxify RNS. Hmp has an *N*-terminal globin domain and a *C*-terminal domain with binding sites for FAD and NAD(P)H. One electron is transferred from NAD(P)H to the *N*-terminal haem domain via a non-covalently bound FAD. This reduced haem oxidizes NO to nitrate, consuming O_2 as a NO-detoxifying enzyme. Many bacteria have Hmp as a NO-detoxifying enzyme. In some bacteria, single-domain haem proteins replace Hmp. Hmp in *E. coli*, in the absence of NO, reduces O_2 to produce superoxide, and the expression of *hmp* is tightly regulated by NO.

The cytochrome *bd*, a quinol oxidase with a high affinity for O_2 in *Salmonella* spp., is resistant against RNS. The induction of cytochrome *bd* helps this bacterium to grow and respire in the presence of inhibitory NO. The antinitrosative defences of cytochrome *bd* and Hmp synergize to promote growth of *Salmonella* spp. in mammalian tissues.

In the case of the anaerobic respiratory bacterium *Wolinella succinogenes*, which lacks both catalase and haemoglobins, a periplasmic multifunctional multihaem cytochrome *c* nitrite reductase (NrfA) mediates NO detoxification.

The RNA polymerase regulatory protein DksA senses RNS and ROS to modulate the expression of up to 427 genes related to NO detoxification and damage repair in *Salmonella* spp., promoting antinitrosative and antioxidative responses. Cysteine biosynthesis genes are among them, showing that cysteine plays an important role in defence against RNS. DksA mediates global adaptation to nitrosative stress in *Salmonella* spp.

In addition, RNS activate a regulator and a NO-related stress sensor protein, NorR, to activate genes involved in RNS detoxification, including *norVW* which encodes a flavorubredoxin with NO reductase activity and an associated flavoprotein with NAD(P)H flavorubredoxin oxidoreductase activity. The NorW reduces NorV. Several proteins are upregulated under nitrosative stress conditions, but how their functions are regulated is still obscure. NorR protein is not activated under nitrosative and oxidative double stress conditions. The sensor kinase, SsrB, of the pathogenicity-related two-component system SsrA/SsrB senses NO in *Salmonella* spp. to modulate related gene expression.

12.2.6 Heat shock response

When a culture of *Escherichia coli* is transferred from the physiological temperature of 37°C to 42°C, a large number of proteins are synthesized. This change to cope with the elevated temperature is referred to as the heat shock response (HSR), and the proteins produced under this condition are called heat shock proteins (HSP). On the increase in temperature, a heat shock σ-factor (RpoH, σ^H or σ^{32}) is synthesized, and genes for HSP are transcribed by the RNA polymerase with σ^H (Figure 12.27).

HSP prevent heat denaturation of cell constituents and repair the denatured constituents. The expression of σ^H is regulated at the translational level. At a normal growth temperature, the *rpoH* transcript (mRNA) is not translated due to a hairpin formed within the molecule preventing ribosome binding. The hairpin structure is melted at an elevated temperature

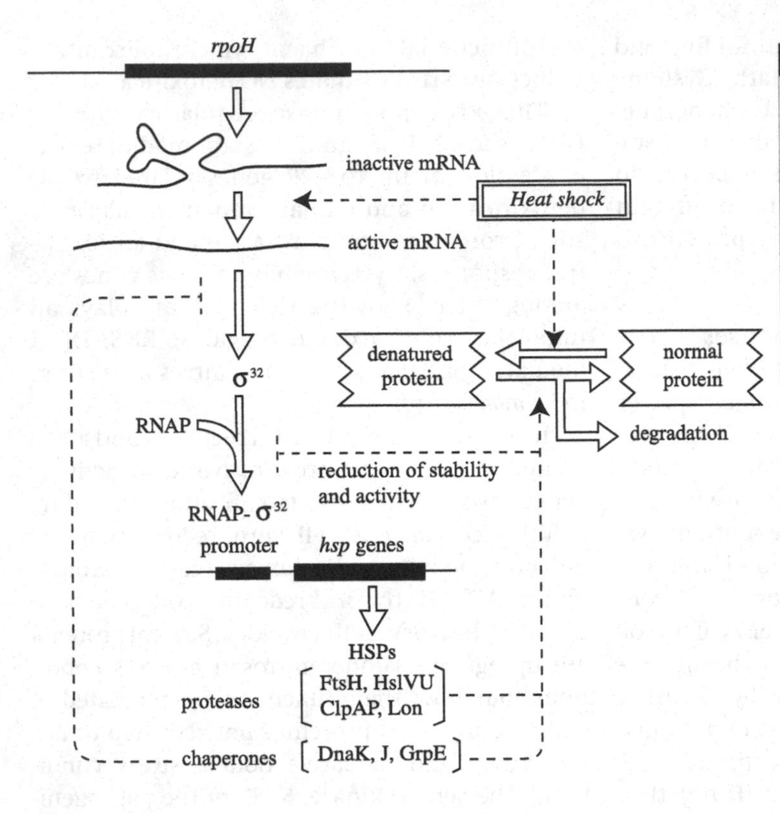

Figure 12.27 Transcriptional regulation of heat shock protein genes in *Escherichia coli* at elevated temperature.

(*Curr. Opin. Microbiol.* **2**: 153–158, 1999)

Under physiological temperature conditions, the *rpoH* transcript forms a hairpin structure preventing ribosome binding, while the hairpin structure is melted at an elevated temperature allowing ribosomes to translate the mRNA. The heat shock protein genes are recognized by RNA polymerase with σ^H and transcribed. HSP prevent denaturation of cellular constituents, and repair or degrade the denatured proteins. When the concentration of HSP reaches a certain level, they separate σ^H from the RNA polymerase and degrade it to prevent overproduction of HSP. HSP interact with σ^H mRNA inhibiting its translation. Through these mechanisms the concentration of σ^H and HSP decreases to a stable level after a sharp rise after the increase in temperature (see Figure 12.28).

allowing ribosomes to translate the mRNA (Section 12.1.9.2).

HSP increase sharply to a maximum within 4–5 minutes after the temperature increase before they decrease to a new steady-state concentration (Figure 12.28). HSP regulate the translation and the activity of σ^H that activates their transcription. This is an example of feedback regulation.

Gram-positive bacteria and archaea have different mechanisms from Gram-negative bacteria in the regulation of HSP production. HSP genes are repressed in Gram-positive bacteria, including *Bacillus subtilis*, at the physiological temperature, by a heat-inducible transcription repressor, HrcA, that binds the upstream region of the genes. The repressor protein is released at high temperature. In the hyperthermophile *Pyrococcus furiosus*, a similar transcriptional regulator of HSP, Phr, binds to promoters of HSP genes and to its own gene at the physiological growth temperature of 95°C, repressing their expression. This releases when the temperature is increased to 107°C to derepress their expression.

In *Synechococcus elongatus*, a two-component system, Hik34–Rre1, activates the heat-shock protein genes. This two-component system is known in all sequenced cyanobacterial genomes.

HSP are synthesized at a lower level at physiological temperature. They have a crucial role in all forms of life. They participate in protein folding into the correct form (Section 6.12.3), and keep the nascent peptide in a translocation-compatible state during its translocation through a membrane (Section 3.10.1). For these reasons HSP are referred to as molecular chaperones. They are grouped as chaperones and chaperonins, according to their functions (Table 12.6). They have homologous amino acid sequences, not only between prokaryotes, but also between prokaryotes and eukaryotes. Some of them that require energy for their activity consume ATP with their ATPase activity.

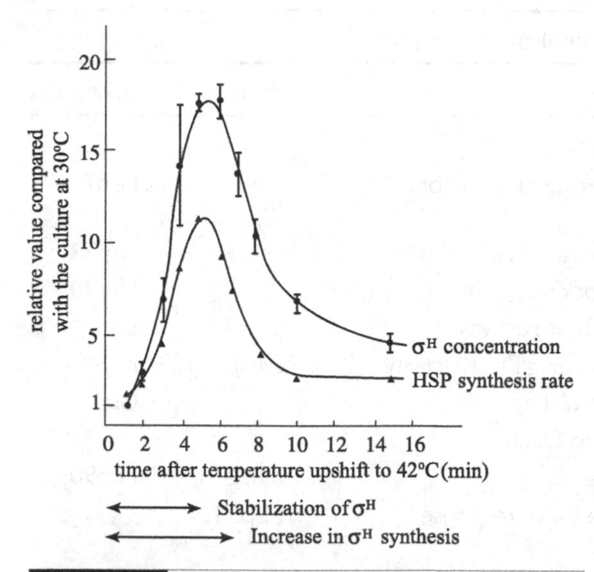

Figure 12.28 Transcriptional regulation of HSP in *Escherichia coli*.

HSP are synthesized at a low level at physiological temperatures for protein folding and translocation through the membrane. When their concentration reaches a certain level after a temperature increase, HSP inhibit the translation and activity of σ^H, preventing the overproduction of HSP. Through these mechanisms, the concentration of HSP is maintained at an elevated level after a sharp rise with the increase in temperature.

Chaperonins are divided into Groups I and II. Bacteria possess Group I chaperonins and Group II chaperonins are found in archaeal and eukaryotic cells. Unusually some methanogenic archaea have both groups of chaperonins.

Escherichia coli has an extracytoplasmic function sigma factor (σ^E) that has a similar function to σ^H. This σ-factor is activated when periplasmic proteins are denatured by heat or chemicals such as ethanol. Mutants defective in σ^E have not been isolated in *Escherichia coli*, suggesting that this σ-factor is an essential protein. σ^E recognizes at least ten genes in this bacterium but their functions have not been fully characterized. The activity of σ^E is regulated by anti-σ-factors RseA and RseB (Section 12.1.1).

Aminoglycoside antibiotics, such as streptomycin, promote translational misreading and misfolding of newly synthesized protein. In many bacteria, chaperone expression increases when they are exposed to these antibiotics leading to resistance. Non-native disulfide bonds in the cytoplasm (disulfide stress) formed during oxidative stress stabilize σ^H in *E. coli*, activating heat shock protein genes.

12.2.7 Cold shock response

When an exponentially growing culture of *Escherichia coli* is shifted from 37°C to 15°C, growth resumes after an acclimation phase of 3–6 h, characterized by the transient dramatic induction of a group of proteins known as cold induced proteins (CIP), with various functions. This regulation is referred to as the cold shock response (CSR). Under low temperature conditions, general protein synthesis is severely inhibited due to the formation of hairpin structures within mRNAs and the double-stranded DNA structure becomes more stable, hindering transcription and replication. The membrane does not function properly due to reduced fluidity, and the reduced flexibility of proteins causes improper function. CIP enable the cell to cope with these problems under low temperature conditions. Among CIP, a group of low molecular weight proteins are involved in the regulation of CIP expression. These are referred to as cold shock proteins (CSP). In most of the literature, CSP is used interchangeably with CIP to describe all the proteins induced at low temperature.

Some CSP are produced at physiological temperatures and these are involved in various cellular processes to promote normal growth and stress adaptation responses. CSP increase tolerance to starvation, and osmotic, oxidative, pH and ethanol stress, as well as assisting host cell invasion.

CIP are grouped into two classes. Class I CIP are those produced only under low temperature conditions or which increase amounts to more than ten times that produced at normal growth temperatures; those that increase less than ten times are termed class II CIP. The latter include RecA (recombinase A), initiation factor IF-2, DNA-binding protein H-NS and RNase R, that are essential for replication, transcription and translation. Class I CIP include CspA (cold shock protein A), CspB, CdsA (cysteine sulfinate desulfinase), NusA (N utilization substance protein A

Table 12.6 Bacterial molecular chaperones and their equivalents in eukaryotes.

Family	Protein	Function	Bacteria	Eukaryotes
Chaperone				
Hsp70	Hsp70	Preventing the aggregation of unfolded peptide	DnaK	Hsp72
		Involved in protein translocation		Hsp73
		Regulating heat shock response		Hsp70
		Associated with ATPase activity		BiP
	Hsp40	Co-chaperone regulating DnaK activity	DnaJ	Hsp40
	GrpE	Co-chaperone modulating ATP/ADP binding to DnaK		GrpE
Hsp90		General chaperone	HtpG	Hsp90
Hsp33		Heat and oxidative shock response	Hsp33	
Chaperonin				
Hsp60	Chaperonin 60	ATP-dependent protein folding	GroEL	Hsp60
	Chaperonin 10	Co-chaperonin to chaperonin 60	GroES	Hsp10
TRiC		Similar function to GroEL	TriC (TF55)	TriC (TCP1)

that participates in the termination and antitermination process), PNP (polynucleotide phosphorylase) and RbfA (ribosome-binding factor A).

The induction of CSP is due to increased translation as well as transcription. The CSP genes, including *cspA*, contain an AT-rich UP region upstream of the promoter region that increases transcription at low temperature. In addition, *csp* transcripts, including *cspA*, contain unique sequences that enhance their translation and stability at low temperature. These are the cold shock (CS) box of about 25 bases in the 5′-untranslated region (5′UTR) of about 150 bases and the downstream box (DB) in the coding region downstream of the ribosome-binding (RB) site (Figure 12.29(1)). CSP have a high affinity to bind to the CS box. The *csp* transcripts are unstable at normal growth temperature, but stable at low temperature. Secondary structures in the unusually long 5′UTR are responsible for their stability at low temperature. When CSP are in excess they bind to the CS box, decreasing the secondary structure. The transcripts become unstable when CSP bind them. The DB has a high affinity for the 16S RNA of the ribosome, increasing translation at low temperature. CspE produced in the early stages of cold adaptation requires the cAPM–CRP complex (Section 12.1.3.1) for gene expression in *E. coli*.

At low temperature, the transcription of genes with a UP region is enhanced, and the *csp* transcripts are stabilized through the secondary structure at the CS box. The stabilized transcript is efficiently translated through the high affinity of the DB with the ribosome. CSP bind mRNAs, reducing the secondary structure and restoring translational efficiency at low temperature (Figure 12.29(2)). For this reason CSP are referred to as RNA chaperones. CSP genes have not been identified in the complete genome sequences of some bacteria pathogenic to warm-blooded animals, including *Helicobacter pylori*, *Campylobacter jejuni* and *Mycobacterium genitalium*.

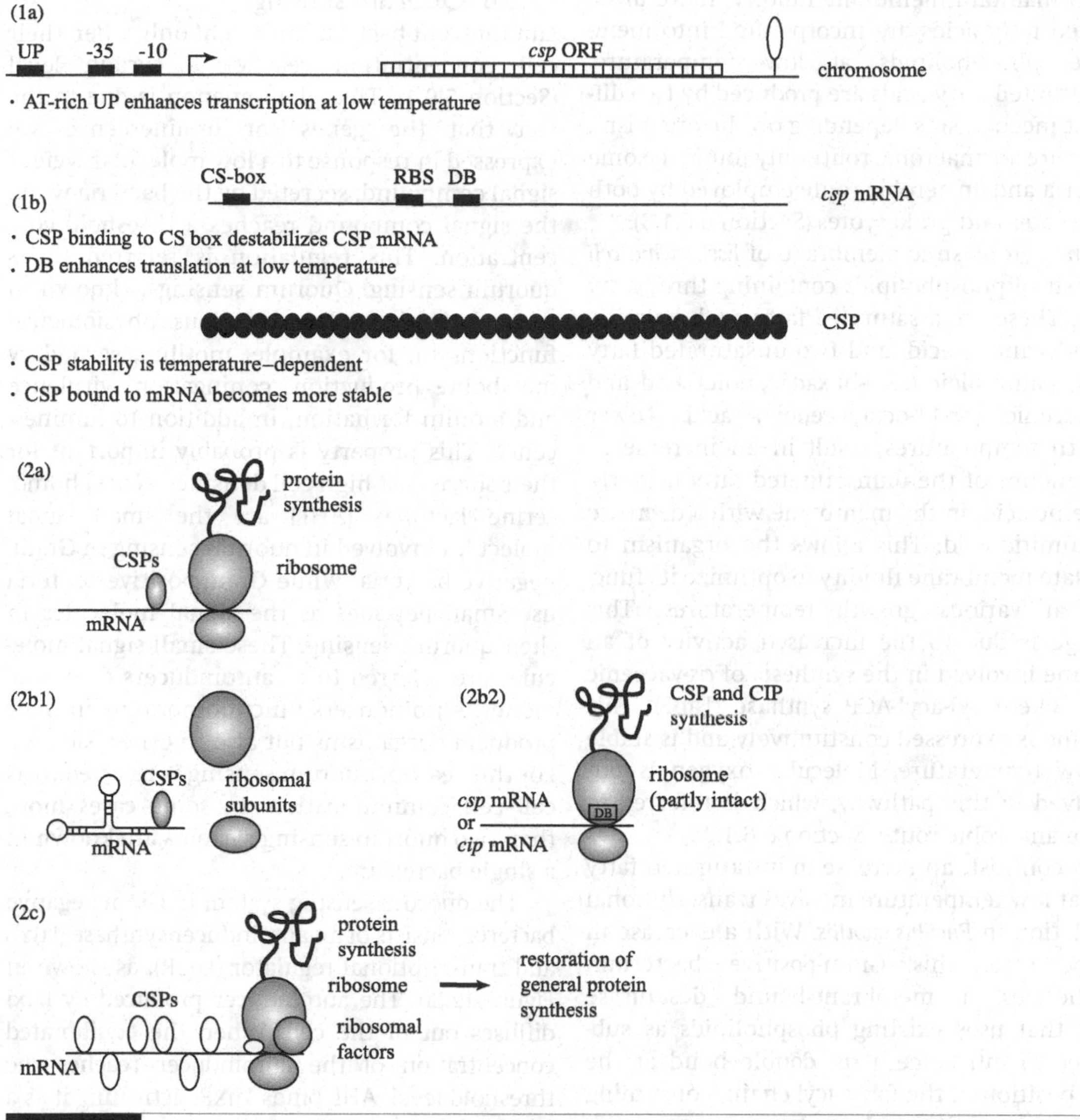

Figure 12.29 Structure of *cspA* (1), and translation of cold induced protein (CIP) and cold shock protein (CSP), and the function of CSP in the restoration of translational efficiency (2).

(*System. Appl. Microbiol.* **23**: 165–173, 2000)

Upstream of the *cspA* promoter is an AT-rich UP region (1a) that enhances transcription of the gene. The *cspA* transcript contains the cold shock (CS) box in the 5′-untranslated region (5′UTR) and downstream box (DB) in the coding region (1b). The 5′UTR forms a secondary structure at low temperature that stabilizes the transcript. At normal growth temperature, the transcript is unstable without the secondary structure. DB increases translation with a high affinity for ribosomes at low temperature. At normal growth temperature, mRNAs are translated (2a), but at low temperature they are not translated due to the secondary structure formed in the mRNAs and due to disruption of ribosomal structure (2b1). Transcripts of *csps* are translated with increased affinity of the DB for ribosomes at low temperature (2b2). CSP bind mRNAs, destabilizing the secondary structure. The CSP bound mRNAs are transcribed normally at low temperature (2c). CSPs that bind mRNA are referred to as RNA chaperones.

To maintain membrane fluidity, more unsaturated fatty acids are incorporated into membrane phospholipids at low temperature. Unsaturated fatty acids are produced by two different mechanisms depending on the organism. These are an anaerobic route only found in some bacteria and an aerobic route employed by both eukaryotes and prokaryotes (Section 6.6.1.3).

The cytoplasmic membrane of *Escherichia coli* consists of phospholipids containing three fatty acids. These are a saturated fatty acid, palmitic (hexadecanoic) acid, and two unsaturated fatty acids, palmitoleic (*cis*-9-hexadecenoic) acid and *cis*-vaccenic (*cis*-11-octadecenoic) acid. Lower growth temperatures result in an increase in the amount of the diunsaturated fatty acid, *cis*-vaccenic acid, in the membrane with a decrease in palmitic acid. This allows the organism to regulate membrane fluidity to optimize its function at various growth temperatures. This change is due to the increased activity of an enzyme involved in the synthesis of *cis*-vaccenic acid, 3-ketoacyl-acyl-ACP synthase (FabF). This enzyme is expressed constitutively and is stable at low temperature. Molecular oxygen is not involved in this pathway, which is referred to as the anaerobic route (Section 6.6.1.3).

In contrast, an increase in unsaturated fatty acid at low temperature involves transcriptional regulation in *Bacillus subtilis*. With a decrease in temperature, this Gram-positive bacterium synthesizes a membrane-bound desaturase (Des) that uses existing phospholipids as substrates to introduce a *cis*- double bond at the fifth position of the fatty acyl chain, consuming O_2. This transcriptional regulation is a two-component system consisting of a membrane-associated kinase, DesK, and a transcriptional regulator, DesR, which stringently controls the transcription of the *des* gene, coding for the desaturase. DesK senses the decrease in membrane fluidity, and autophosphorylates. This reaction is referred to as the aerobic route of unsaturated fatty acid synthesis (Section 6.6.1.3).

Pseudomonas putida produces CrcZ and CrcY sRNAs growing at low temperature, leading to reduced Crc (catabolite repression control, Section 12.1.3)-dependent catabolite repression.

12.2.8 Quorum sensing

Luminescent bacteria emit light only when their cell concentration reaches a certain level (Section 5.9.1). This phenomenon is due to the fact that the genes for luminescence are expressed in response to a low molecular weight signal compound, secreted by the bacteria when the signal compound reaches a threshold concentration. This regulation is referred to as quorum sensing. Quorum sensing is known in many bacteria and has various physiological functions in, for example, motility, secondary metabolite production, conjugation, virulence and biofilm formation, in addition to luminescence. This property is probably important for their survival at high cell densities. *N*-acyl homoserine lactones (AHL) are the small signal molecules involved in quorum sensing in Gram-negative bacteria, while Gram-positive bacteria use small peptides as the signal molecules in their quorum sensing. These small signal molecules are referred to as autoinducers or pheromones. Autoinducers function not only in their producing organisms but also in other bacteria. For this reason, quorum sensing is referred to as cell–cell communication. In some cases more than two quorum-sensing systems are known in a single bacterium.

The quorum-sensing system in Gram-negative bacteria consists of an autoinducer synthase (LuxI) and transcriptional regulator (LuxR), as shown in Figure 12.30. The autoinducer produced by LuxI diffuses out of the cell. When the equilibrated concentration of the autoinducer reaches the threshold level, AHL binds LuxR, activating it as a regulator to activate a number of genes. Alternatively, a membrane-bound sensor/kinase of a two-component system (Section 12.1.7) detects the extracellular AHL concentration to phosphorylate the cognate regulator protein. The phosphorylated regulatory protein modulates various genes.

In addition to two AHL autoinducers (LasI-LasR and RhlI-RhlR systems), *Pseudomonas aeruginosa* uses 4-hydroxy-2-alkylquinolines (HAQ) as quorum-sensing signals. Their production is activated by the stringent response (Section 12.2.1). The plant pathogen *Xanthomonas*

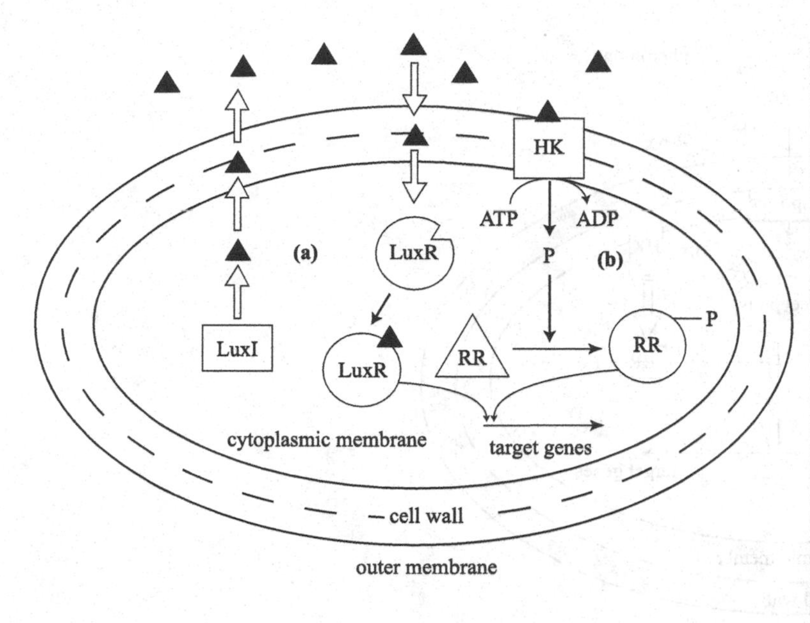

Figure 12.30 Metabolic regulation by quorum sensing in Gram-negative bacteria.

An autoinducer (black triangle) produced by an autoinducer synthase (LuxI) diffuses through the membrane. (a) When its concentration reaches a threshold level the autoinducer binds the transcriptional regulator (LuxR) activating it. The activated LuxR activates the expression of operons related to many physiological functions, depending on the bacterium. (b) Alternatively, a membrane-bound sensor/kinase of a two-component system detects the extracellular AHL concentration to phosphorylate the cognate regulator protein. The phosphorylated regulatory protein modulates various genes.

campestris produces an unsaturated fatty acid, *cis*-11-methyl-dodecenoic acid, that functions similarly as an autoinducer but is structurally not related to AHL. Similar unsaturated fatty acids produced by other bacteria, including *Burkholderia cenocepacia* and *Pseudomonas aeruginosa*, regulate virulence, biofilm formation and antibiotic tolerance. These *cis*-unsaturated fatty acids, known as diffusible signal factors, are also responsible for regulation of the synthesis of extracellular degradative enzymes and extracellular polysaccharide virulence factors. They are also involved in interspecies signalling that modulates bacterial behaviour. A two-component system, RpfC/RpfG (regulation of pathogenicity factors), senses this signal molecule, to modulate gene expression in *Xanthomonas campestris*.

Peptides are used as autoinducers in some Gram-negative bacteria just as in Gram-positive bacteria (see below). These are produced through various mechanisms, including by the toxin of the toxin–antitoxin module (Section 13.4.2). MazF, the toxin of a stress-induced toxin–antitoxin module, *mazEF* in *Escherichia coli*, cleaves the mRNA encoding glucose-6-phosphate dehydrogenase to a short RNA that is translated to a quorum-sensing autoinducer pentapeptide. This peptide is also known as the extracellular death factor (EDF).

There are LuxR homologues without the cognate autoinducer synthase LuxI, such as SdiA in *Salmonella enterica* and *Escherichia coli*, QscR in *Pseudomonas aeruginosa* and PluR in *Photorhabdus luminescens*. These LuxR orphans bind to endogenous AHL. Non-species-specific autoinducers, such as autoinducer 2(AI-2), are used for intra- and interspecies communication even between Gram-negative and Gram-positive bacteria. Intestinal bacteria produce autoinducer 3(AI-3) for pathogen–host interactions

The signal molecules can be inactivated by inhibitors or enzymes. This inactivation is termed quorum quenching. A number of bacteria have been identified which are capable of enzymatic inactivation of AHL. These include Gram-positive bacteria including *Bacillus thuringiensis*, *Bacillus cereus*, *Bacillus mycoides* and *Arthrobacter* sp., as well as Gram-negative bacteria such as *Agrobacterium tumefaciens*, *Pseudomonas aeruginosa* and *Comamonas testosterone*, among others. AHL-inactivating enzymes have been observed not only in the laboratory but also in the soil. These enzymes have been studied as a means to control the virulence of animal and plant pathogens. Quorum quenching enzyme

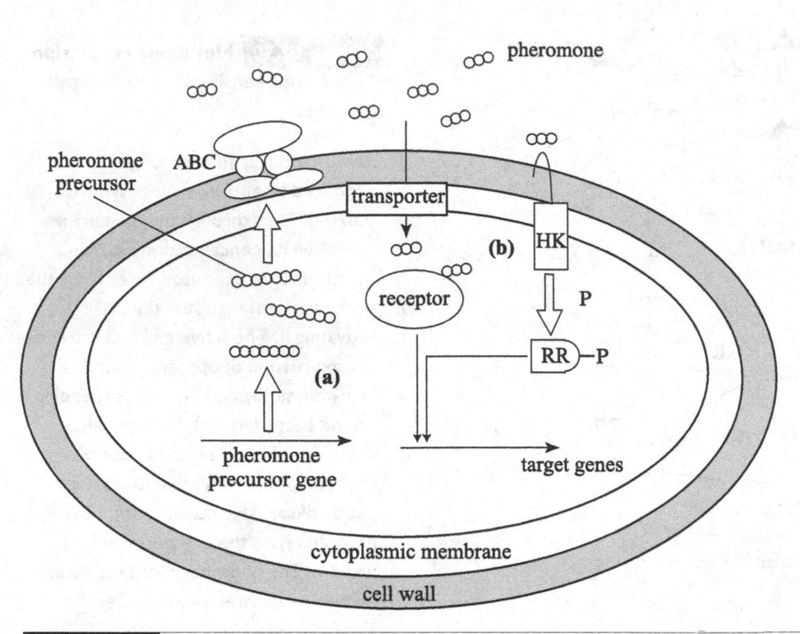

Figure 12.31 Quorum sensing in Gram-positive bacteria.

Peptides or modified peptides are used as the autoinducer (pheromone) of quorum sensing in Gram-positive bacteria. These are not permeable to the membrane and are excreted through the general secretory pathway or ATP-binding cassette mechanism before being modified extracellularly. When their concentration reaches a threshold level, they bind the membrane-bound sensor/kinase protein (H). A histidine residue of H is autophosphorylated when it binds the autoinducer peptide before the phosphate is transferred to the aspartate residue of the cytoplasmic regulator protein (D). The phosphorylated D modulates the expression of various genes (b). In some cases, the autoinducer is imported by a specific transporter into the cell where it binds a receptor to modulate expression of various genes (a).

activity is high in the radiation-resistant *Deinococcus radiodurans*, to keep the AHL level low under non-stress conditions, while the AHL concentration increases when it is exposed to oxidative stress that inactivates the enzyme.

Gram-positive bacteria use post-translationally modified peptides as their autoinducer. These autoinducer peptides are impermeable to the cytoplasmic membrane and are excreted through the general secretory pathway or ATP-binding cassette (ABC) pathways (Section 3.10.1.3) before being modified extracellularly. When their extracellular concentration reaches the threshold level, they are detected by the membrane-bound sensor/kinase protein of a two-component system or imported into the cell, where they bind a receptor to modulate expression of various genes (Figure 12.31).

A-factor is the autoinducer in *Streptomyces griseus*. When its concentration builds up to the threshold level, a two-component system senses and regulates genes to produce aerial mycelium, leading to the formation of spores and secondary metabolites, including streptomycin.

The bacterial immune system CRISPR-Cas targeting of foreign DNA (Section 13.5) in *Pseudomonas aeruginosa* is upregulated by quorum sensing for protection from the high risk of phage infection at high cell densities.

AHL based quorum sensing is also known in halophilic and methanogenic archaea.

12.2.9 Response to changes in osmotic pressure

Gram-negative bacteria, including *Escherichia coli*, have a special class of outer membrane proteins (OMP) called porins, that form water-filled channels through which low molecular weight hydrophilic solutes gain access to the periplasmic region (Section 2.3.3.2). Under normal osmotic pressure,

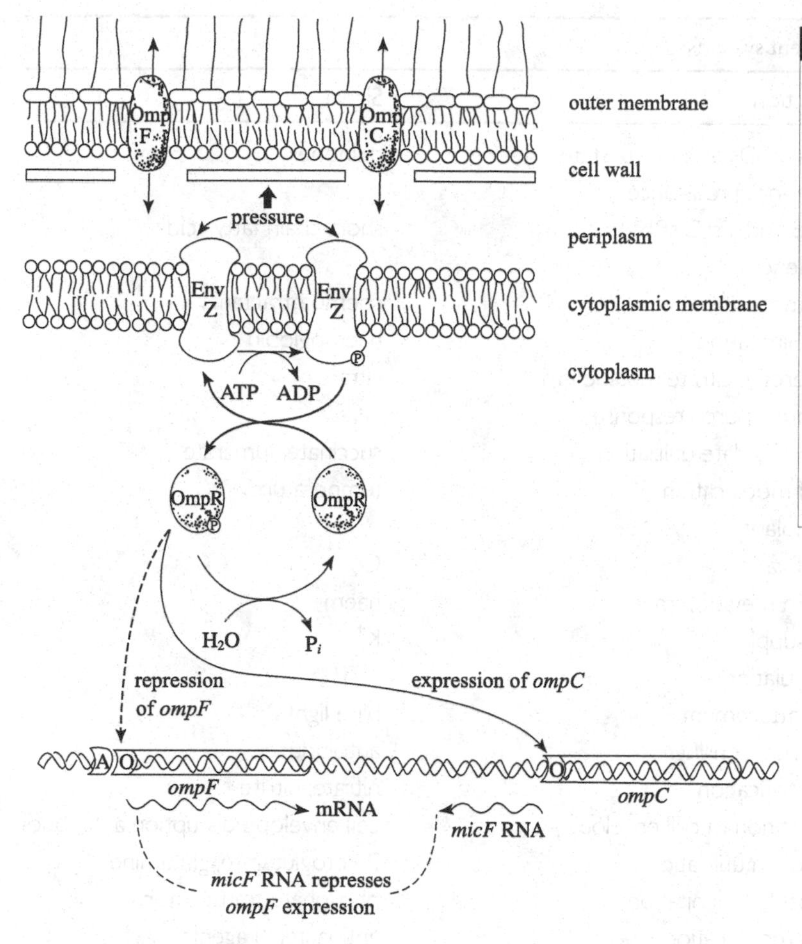

Figure 12.32 The EnvZ/OmpR two-component system, regulating the expression of the outer membrane proteins OmpC and OmpF.

When the environmental osmotic pressure increases, the membrane-bound sensor protein EnvZ is autophosphorylated, consuming ATP at its histidine residue before transferring phosphate to the cytoplasmic modulator protein OmpR. Phosphorylated OmpR activates the expression of OmpC, a porin with a small pore, and represses OmpF, which is another porin expressed at normal osmotic pressure.

the outer membrane protein F (OmpF) is the major porin in *Escherichia coli*. Under increased osmotic pressure, a smaller pore-size OmpC replaces OmpF. A membrane-bound sensor protein, EnvZ, senses the increase in osmotic pressure, to be autophosphorylated at its histidine residue. Subsequently the phosphate is transferred to an aspartate residue of the cytoplasmic modulator protein, OmpR. The phosphorylated OmpR activates the the expression of *ompC* and represses *ompF* (Figure 12.32). The EnvZ/OmpR two-component system modulates various genes in addition to *omp* genes. These include those for compatible solute production and transport. OmpC and OmpF are also subject to post-transcriptional regulation by the small regulatory RNA molecules *micC* and *micF*, respectively (Section 12.1.9.4).

Mutants either in *envZ* or in *ompR* result in a loss of virulence in the opportunistic pathogen *Acinetobacter baumannii*. A regulatory protein, YfeR, is produced by *Salmonella enterica* at low osmotic pressure that modulates expression of various genes.

12.2.10 Two-component systems and cross-regulation

Several two-component systems have already been discussed, including the *pho* system under phosphate-limited conditions, the *arc* system related to the availability of O_2, quorum sensing in Gram-positive bacteria, the EnvZ/OmpR system associated with changes in osmotic pressure and NarC/NarL and NarQ/NarP related to denitrification (Section 9.1.3). Two-component systems

Table 12.7 Selected two-component systems.

System	Function	Signal
ArcB/ArcA	sensing O_2 and redox state	quinones
BaeS/BaeR	multi-drug resistance	
BarA/UvrY	CsrB and CsrC sRNAs	short-chain fatty acids
BvgS/BvgA	virulence	
cheA/CheY	chemotaxis	chemoattractants
ChrS/ChrA	iron limitation	haemoglobin
CitA/CitB	anaerobic citrate metabolism	citrate
CreC/CreB	carbon source response	
DcuS/DcuR	dicarboxylate utilization	succinate, fumarate
DesR/DesK	lipid modification	temperature
EnvZ/OmpR	osmolarity	
FixL/FixJ	N_2 fixation	O_2
HrrS/HrrA	haem metabolism	haem
KdpD/KdpE	K^+ supply	K^+
KinB/SpoOF	sporulation	ATP
LovK/LovR	cell attachment	blue light
LuxQ/LuxO	quorum sensing	autoinducer
NarX/NarL	denitrification	nitrate, nitrite
NsaS/NsaR	disruption of cell envelope	cell envelope disruption antibiotics
NtrB/NtrC	nitrogen utilization	2-ketoglutarate, glutamine
PhoR/PhoB	phosphate limitation	phosphate transporter
PmrB/PmrA	LPS modification	antibacterial agents
QseC/QseB	quorum-sensing sensory protein	autoinducer
SaeS/SaeR	virulence factors	antibiotics, host defence system
TodS/TodT	degradation of benzene	benzene
VanS/VanR	antibiotic responses	vancomycin
YehU/YehT	stationary phase control	cAMP–CRP complex

consist of a membrane-bound sensor protein and a cytoplasmic regulator protein. When the signal recognition domain of the sensor protein at the periplasmic side of the membrane senses the signal, the autokinase domain of the protein at the cytoplasmic side consumes ATP to phosphorylate a histidine residue within the protein. This phosphate is transferred to the aspartate residue of the regulator protein. This phosphorylated regulator protein modulates the expression of the target genes.

Analyses of bacterial genomes have revealed many putative two-component systems with base sequences of sensors/regulators. The function of many of them is not yet known. A few two-component systems are known that have received limited study (Table 12.7). Enzymes of transport and metabolism of 4-carbon dicarboxylates, such as succinate and fumarate, are under the control of a two-component system consisting of a sensor protein, DcuS, and a regulator protein, DcuR, in *Escherichia coli*. Another example of this kind is the utilization of organic sulfur compounds, such as sulfonated detergents, under sulfate-limited conditions in *Escherichia coli*, *Pseudomonas putida* and *Staphylococcus aureus*. This system is believed to be similar to the *pho* system.

The sporulation process in *Bacillus subtilis* is initiated with the phosphorylation of KinA or KinB, consuming ATP in a similar manner to a two-component system. The phosphate is transferred to Spo0A through Spo0F and Spo0B. This is referred to as a phosphorelay (Section 13.3.1). Histidine residues of KinB (or KinA) and Spo0B are phosphorylated as the sensor protein, and aspartate residues of Spo0F and Spo0A are the phosphate receptor.

Some regulatory proteins of two-component systems are phosphorylated by non-cognate sensory proteins. The sensory protein CreC of the carbon source response two-component system, CreC/CreB, transfers phosphate to PhoB. Auxiliary regulators are widespread and influence phosphate transfer, not only to the cognate regulatory protein but also to non-cognate regulatory proteins. This cross-regulation connects a given two-component system to other regulatory networks in response to an expanded range of stimuli (Section 12.4.2).

Less well known are the archaeal two-component systems. Bacterial type two-component systems are known in halophilic and methanogenic archaea (*Euryarchaeota*), while serine, threonine and tyrosine residues of the regulatory proteins are phosphorylated in acidophilic and hyperthermophilic archaea (*Crenarchaeota*).

12.2.11 Chemotaxis

Motile bacteria synthesize proteins for chemotaxis, including those for flagella formation, when the substrate concentration becomes low. Bacterial flagella consist of a basal body, hook and filament (Section 2.3.1). The proteins forming the hook and filament are transported through the basal body. For this reason, the proteins of the basal body are synthesized before those forming the hook and filament.

The synthesis and function of the flagellar and chemotaxis system require the expression of more than 50 genes which are scattered among at least 17 operons that constitute the coordinately regulated flagellar regulon. In *Escherichia coli* and *Salmonella typhimurium*, the flagella and chemotaxis regulons are expressed in three separate steps, by transcriptional as well as post-transcriptional mechanisms, according to substrate concentration and to the morphological development of the flagellar structure itself. With a limited supply of substrate, the cAMP–CRP complex activates the expression of two early genes in the *flhDC* operon. FlhD and FlhC proteins are transcriptional activators that activate the middle gene operons. Various sRNAs bind to the 5′ UTR of the *flhDC* mRNA to regulate their translation according to environmental cues. Four sRNAs, ArcZ, OmrA, OmrB and OxyS, negatively regulate, and one sRNA, McaS, positively regulates translation of *flhDC* by base-pairing with the 5′ UTR of this mRNA. The middle gene operons include *fliA* that encodes σ^{28} (σ^F). RNA polymerase with σ^{28} transcribes the late genes that encode proteins of the hook and filament (Figure 12.33). A similar alternative σ-factor, σ^D, is known in *Bacillus subtilis*.

The anti-σ factor, FlgM, binds to σ^{28} directly to prevent the transcription of the late genes until the formation of the hook-basal body is completed. Once the hook-basal body is completed, FlgM is secreted from the cell through the flagellin export machinery to free σ^{28}. In addition to transcriptional regulation, post-transcriptional control further ensures the efficiency of flagellar assembly. The translation of the anti-σ factor and hook protein is regulated. When the flagella and chemotaxis regulons are expressed, motility is regulated by the gradient of attractant or repellant through the mode of flagellar rotation. A cell typically travels in a three-dimensional random 'walk'. Intervals in which the cells swim in gently curved paths (runs) alternate with briefer periods of chaotic motion (tumbles) that randomly reorientate the next run. Cells run when the left-handed flagellar filaments rotate counterclockwise and coalesce into a bundle to propel the cell. Cells tumble when the flagella turn clockwise and disrupt the bundle. Cells extend runs (suppress tumbles) when they move up

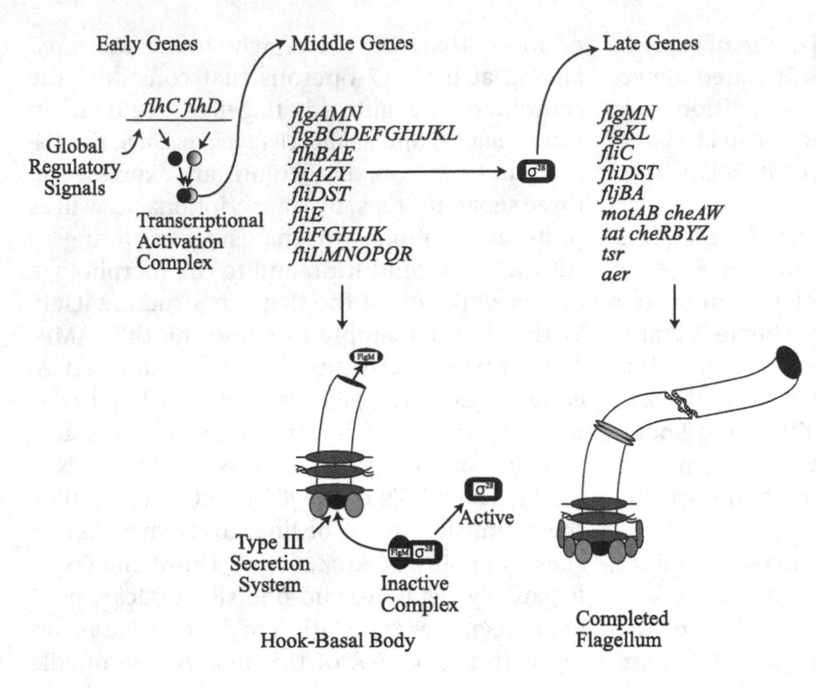

Figure 12.33 The expression of the flagella regulon in three steps.

(*Microbiol. Mol. Biol. Rev.* **64**: 694–708, 2000)

The flagella and chemotaxis regulon includes more than 50 genes in 17 operons. These operons are transcribed in three steps, early, middle and late. The early genes include the master flagellar operon, *flhDC*. FlhC and FlhD proteins are the transcriptional activators that activate expression of the middle genes. The middle genes include those for the basal body and σ^{28} (FliA, σ^F) that activate late genes of proteins for flagellin. The anti-σ factor FlgM, produced in the middle step, binds to σ^{28} directly to prevent the transcription of the late genes until formation of the hook-basal body is completed. Once the hook-basal body is completed, FlgM is secreted from the cell to free σ^{28}.

concentration gradients of attractants or down gradients of repellents.

Three classes of proteins are essential for chemotaxis. These are transmembrane receptors, cytoplasmic signalling components and enzymes for adaptive methylation. Using these proteins, a cell senses the spatial gradient and compares the instantaneous concentration of a compound. The transmembrane receptors consist of an amino-terminal transmembrane helix (TM1), a periplasmic ligand interaction domain, a second transmembrane helix (TM2) and a large cytoplasmic signalling and adaptation domain. The cytoplasmic domain contains five methylatable glutamate residues, and the receptors are therefore also called methyl-accepting chemotaxis proteins (MCP). Many environmental stimuli, such as metabolites and signalling molecules, temperature, light, salinity, oxygen, the magnetic field and pH, induce taxes and the specificity of a tactic response is determined by the MCPs.

The attractant concentration is sensed by the receptor protein, and the signal is transduced to the flagella by four proteins to determine the direction of flagellar rotation. These are a typical two-component system, CheA, the histidine protein kinase and CheY, the response regulator, and CheW, the receptor-coupling factor and CheZ, an enhancer of CheY-P dephosphorylation. When the receptor protein is free from attractant the autophosphorylation activity of CheA is stimulated, which in turn increases phosphotransfer from CheA-P to CheY. CheY-P binds to the flagellar motor-switch complex to cause clockwise rotation. CheZ prevents accumulation of CheY-P by accelerating the decay of its intrinsically unstable aspartyl-phosphate residue. Under steady-state conditions, CheY-P is maintained at a level that generates the random walk. When an attractant (or an attractant–protein complex) binds the periplasmic side of the receptor protein, this protein undergoes a conformational change that suppresses CheA activity. Levels of CheY-P fall, and cells tumble less often. Thus, cells increase their run lengths as they enter areas of higher attractant concentration (Figure 12.34).

This response does not explain, however, how cells sense continually increasing attractant

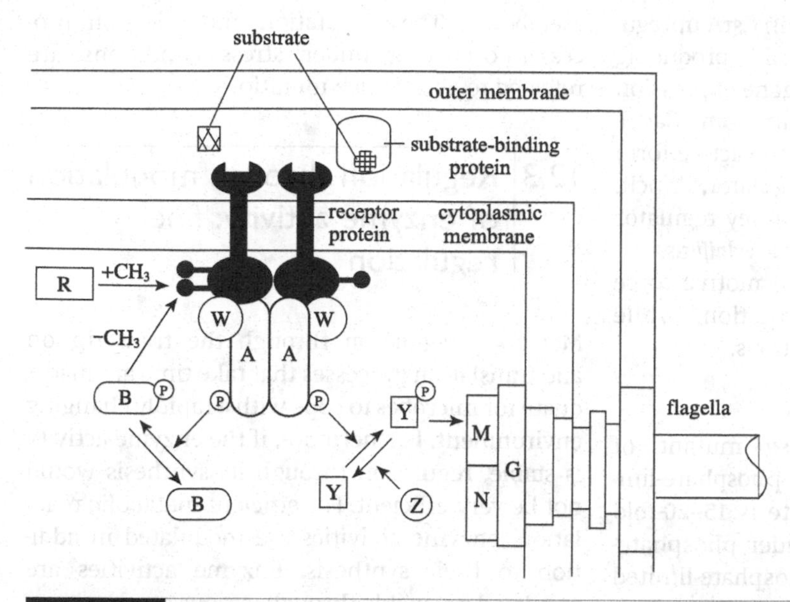

Figure 12.34 **Signal transducing pathway of chemotaxis to swim towards the attractant gradient.**

(*J. Bacteriol.* **180**: 1009–1022, 1998)

The dimeric membrane-spanning chemoreceptors (paired black wrench-like proteins) form a complex with CheA and CheW polypeptides and stimulate the autokinase activity of CheA. CheA-P can transfer the phosphate to CheY. CheY-P interacts with FliM in the motor-switch complex to induce clockwise (CW) flagellar rotation, causing tumbles. The decay of CheY-P is accelerated by CheZ. CheR is a constitutive methyltransferase that methylates certain glutamate residues in the cytoplasmic domains of the receptors. CheB is a methylesterase that is activated by phosphotransfer from CheA-P. CheB-P removes methyl groups from the receptors. When an attractant (cross-hatched square) binds directly or forms a complex with substrate-binding proteins to the periplasmic domain of the receptors, this protein undergoes a conformational change that suppresses CheA activity. The level of CheY-P falls, and the flagella rotate anticlockwise (ACW). Thus, cells tumble less often and increase their run lengths as they enter areas of higher attractant concentration. In the adaptation pathway, reduced CheA activity decreases the CheB-P level, although more slowly than the CheY-P level. As methylesterase activity declines, the receptors become more highly methylated. Increased methylation counteracts the attractant-dependent inhibition of CheA activity. As CheA activity rises, the intracellular CheY-P concentration returns to its pre-stimulus value, and the flagellar motor resumes its pre-stimulus CW-to-ACW switching ratio.

Abbreviations: A, CheA; W, CheW; Y, CheY; Z, CheZ; R, CheR; B, CheB; G, FliG; M, FliM; N, FliN; p, phosphate; CH_3, methyl group (shown as lollipop-like forms on the cytoplasmic domain of the receptors).

concentrations. To accomplish that, adaptation is necessary. Two enzymes, the methyltransferase CheR and the methylesterase CheB, are necessary for adaptive methylation. CheR is a constitutive enzyme that uses S-adenosylmethionine to methylate glutamate residues in the receptor protein. CheB is a target for phosphotransfer from CheA, and CheB-P removes methyl groups from the MCP. In steady state, methyl addition by CheR balances methyl removal by CheB-P to achieve an intermediate level of receptor methylation (0.5 to 1 methyl group per subunit) that maintains run-tumble behaviour. When an attractant binds a receptor and inhibits CheA activity, CheB-P levels fall, although more slowly than CheY-P levels, since CheB-P is not a substrate for CheZ. Increased methylation restores the ability of the receptor to stimulate CheA. Even after basal levels of CheY-P and CheB-P are regained, however, an attractant-bound receptor remains overmethylated, because its properties as a substrate for CheB-P are altered. Through these reactions the cell moves towards the attractant gradient (Figure 12.34).

The second messenger, c-di-GMP, promotes biofilm formation and represses flagellum-driven swarming motility in many bacteria (Section

12.1.9.5). The mRNA binding protein CsrA upregulates the expression of c-di-GMP producing enzymes and downregulates the gene expression of c-di-GMP hydrolysing enzymes and some flagellar proteins in *Salmonella enterica*. The active form of a quorum-sensing master regulator, SmcR, downregulates the expression of a key regulator of flagellar biogenesis, FlhF, in *Vibrio vulnificus*.

In bacteria, a proton (sodium) motive force provides energy for flagellar rotation, while archaea consume ATP for chemotaxis.

12.2.12 Adaptive mutation

When an alkaline phosphatase mutant of *Escherichia coli* is cultivated under phosphate-limited conditions, the reversion rate is 15–20-fold higher than when cultivated under phosphate-excess conditions. Under phosphate-limited conditions many genes, including alkaline phosphatase (*phoA*), are expressed (Section 12.2.3). During the transcription a part of the gene is in an unpaired state (single strand) causing supercoiling of adjacent parts and changes in the DNA secondary stem–loop structures. These DNA structures are vulnerable to mutation. The increased reversion rate is due to an increased mutation rate in the derepression of the *phoA* gene under phosphate-limited conditions.

As discussed previously, bacteria have mechanisms to derepress genes according to various stresses. These stresses are a selection pressure for beneficial mutations. Cells starved after they use up their usual carbon source, in the presence of a carbon source they cannot use, may produce mutants that can use it. The most common mechanism involved is gene derepression, resulting in the constitutive production of a previously inducible enzyme or an enzyme with altered substrate specificity. Many examples exist, such as the use of altrose-galactoside via β-galactosidase, β-glycerol phosphate via alkaline phosphatase and xylitol via ribitol dehydrogenase. The metabolic steps required to metabolize a new related substrate are similar to those for existing pathways. Therefore, relatively minor changes to a duplicate copy of the existing gene may be required for recruitment to a new function. Other frequently observed mutations in response to carbon source starvation confer increased permeability to the limiting metabolite. These mutation and selection processes occurring under stress conditions are referred to as adaptive mutations.

12.3 Regulation through modulation of enzyme activity: fine regulation

Metabolic regulation through the transcription and translation processes that take time are inadequate for microbes to cope with a rapidly changing environment. Furthermore, if the enzyme activity is stable, regulation through its synthesis would not be very efficient. For efficient metabolic regulation, enzyme activities are modulated in addition to their synthesis. Enzyme activities are regulated reversibly through various mechanisms, including feedback inhibition, feedforward (precursor) activation and post-translational modification of the enzyme proteins.

12.3.1 Feedback inhibition and feedforward activation

Amino acids are synthesized through a series of reactions. When an amino acid is available and sufficient to meet the needs of growth, the enzyme catalysing the first reaction is inhibited. This process is referred to as feedback inhibition. For example, threonine dehydratase is inhibited when isoleucine is available (Section 6.4.1), and AMP and GMP inhibit amidophosphoribosyl transferase (Section 6.5.3).

In the case of a single precursor used for multiple product formation, isoenzymes are regulated separately according to the concentration of each product, as in gene expression regulation through repression (Section 12.1.5). Aspartate is used to synthesize lysine, methionine and isoleucine through threonine (Figure 12.35). The first reaction is catalysed by three separate isoenzymes. One is regulated by lysine, the other by both threonine and isoleucine. The third isoenzyme is not regulated, but methionine inhibits homoserine acyltransferase, the first enzyme specifically involved in its synthesis. When methionine is needed with threonine and lysine in excess,

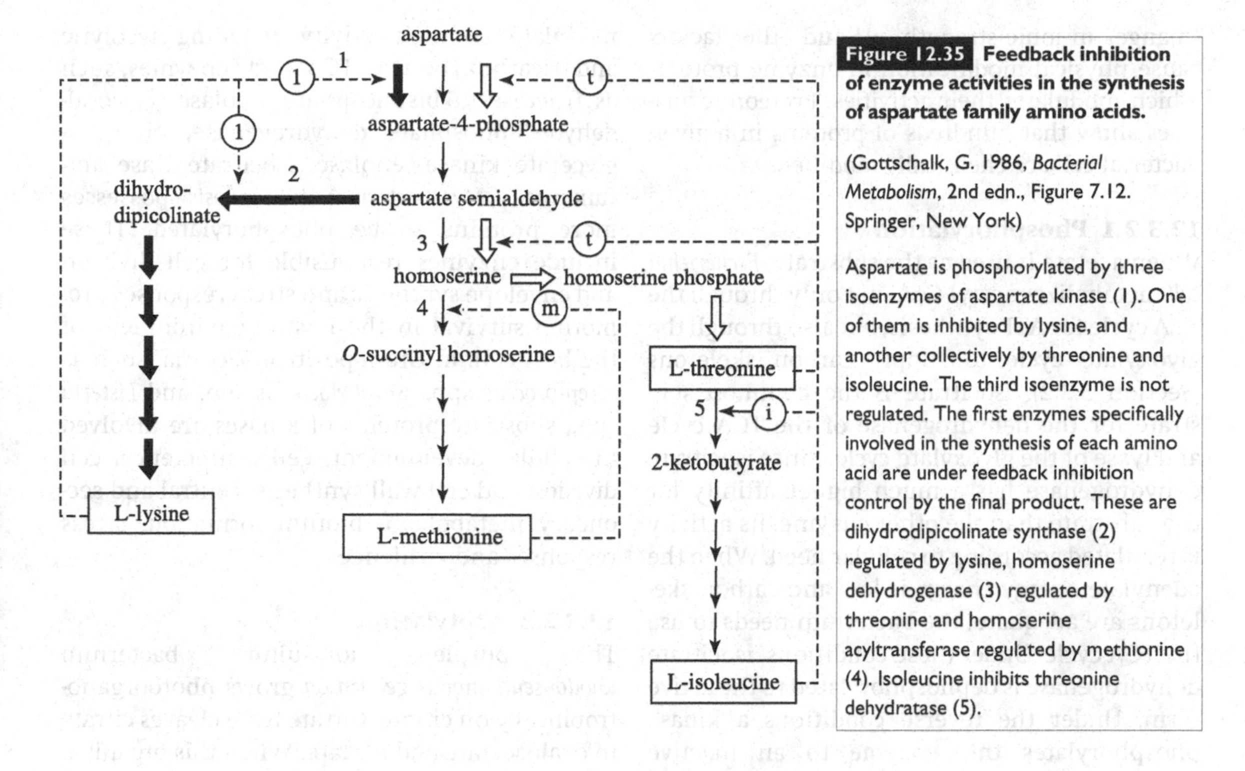

Figure 12.35 Feedback inhibition of enzyme activities in the synthesis of aspartate family amino acids.

(Gottschalk, G. 1986, *Bacterial Metabolism*, 2nd edn., Figure 7.12. Springer, New York)

Aspartate is phosphorylated by three isoenzymes of aspartate kinase (1). One of them is inhibited by lysine, and another collectively by threonine and isoleucine. The third isoenzyme is not regulated. The first enzymes specifically involved in the synthesis of each amino acid are under feedback inhibition control by the final product. These are dihydrodipicolinate synthase (2) regulated by lysine, homoserine dehydrogenase (3) regulated by threonine and homoserine acyltransferase regulated by methionine (4). Isoleucine inhibits threonine dehydratase (5).

the third aspartate kinase isoenzyme is active and the other isoenzymes are inhibited by lysine and threonine. Enzymes using intermediates as common substrates are regulated similarly. These are dihydrodipicolinate synthase regulated by lysine, homoserine dehydrogenase regulated by threonine and homoserine acyltransferase regulated by methionine.

In some cases, a single enzyme catalyses the first reaction that leads to the synthesis of multiple products. For example, *Escherichia coli* has isoenzymes of 3-deoxy-D-arabino-heptulosonate (DAHP) synthase catalysing the first reaction of aromatic amino acid synthesis (Section 6.4.4), but a single enzyme catalyses the reaction in *Ralstonia eutropha* (*Alcaligenes eutrophus*). In the latter case, each product inhibits the enzyme, and the inhibition is stronger in the presence of all the products than with just a single product. This regulatory mechanism is referred to as cumulative inhibition.

Enzymes controlled through feedback inhibition have a unique property. These enzymes have an inhibitor-binding site different from the active site. When the inhibitor binds the enzyme, the structure is changed to an inactive form. The inhibitor-binding site is referred to as an allosteric site, and the enzyme as an allosteric enzyme.

When fructose-1,6-diphosphate (FDP) is accumulated, this EMP pathway intermediate activates pyruvate kinase (Section 4.1.5.2). Such enzyme activity modulation is referred to as feedforward activation or precursor activation. Another example of this regulatory mechanism is found in lactic acid bacteria, where lactate dehydrogenase is activated by FDP.

12.3.2 Enzyme activity modulation through post-translational modification

Inhibitor binding causes structural changes in allosteric enzymes. The activity of many enzymes is modulated through structural modification. Some are modified chemically and others physically. Post-translational modification of proteins contributes significantly to bacterial adaptability.

Chemical modification includes phosphorylation, acetylation, methylation and adenylylation.

Changes in ionic strength, pH and other factors cause physical modification in enzyme proteins which modulates their activities. Proteomic analyses show that hundreds of proteins in a given bacterial cell are chemically modified.

12.3.2.1 Phosphorylation

When acetate is used as the substrate, *Escherichia coli* metabolizes acetyl-CoA, not only through the TCA cycle for ATP synthesis, but also through the glyoxylate cycle to supply carbon skeletons (Section 5.3.2). Isocitrate is the common substrate for the dehydrogenase of the TCA cycle and lyase of the glyoxylate cycle. Since isocitrate dehydrogenase has a much higher affinity for the substrate than the other enzyme, its activity is regulated according to cellular need. When the adenylate energy charge is low and carbon skeletons are adequate, the bacterium needs to use the TCA cycle. Under these conditions, isocitrate dehydrogenase is dephosphorylated to the active form. Under the reverse conditions a kinase phosphorylates this enzyme to an inactive form. Acetylation also inactivates this enzyme, suggesting that the activity of this enzyme is regulated under different conditions. Other enzymes modified by phosphorylation as well as by acetylation include enzymes of glycolysis, the TCA cycle and the metabolism of amino acids and nucleotides.

Phosphorylation is one of the most important of the post-translational modifications that allow proteins to reversibly change their activity for proper function. In addition to phosphorylation of histidine kinase (HK) and response regulator (RR) of the two-component system (TCS), various proteins are phosphorylated at serine, threonine or tyrosine residues through the action of protein kinases. While any given pair of HK and RR protein in a TCS selectively recognize each other to maintain a faithful flow of information with just a few exceptions (Section 12.1.7), protein kinases phosphorylate multiple substrates and are integration nodes in the signalling network. Many enzyme activities are modulated by a phosphorylation mechanism.

In *Escherichia coli*, 342 proteins have tyrosine residues that can be phosphorylated for modulation of their activity, including glycolytic and tricarboxylic acid (TCA) cycle enzymes, such as fructose-1,6-bisphosphate aldolase, glyceraldehyde-3-phosphate dehydrogenase, phosphoglycerate kinase, enolase, isocitrate lyase and fumarase. *Mycobacterium tuberculosis* possesses more proteins to be phosphorylated. These include enzymes responsible for cell division and envelope synthesis and stress responses promoting survival in the hostile environment of the host cell. In Gram-positive bacteria, such as *Streptococcus* spp., *Staphylococcus* spp. and *Listeria* spp., substrate proteins of kinases are involved in cellular development, cell competence, cell division and cell wall synthesis, central and secondary metabolism, biofilm formation, stress responses and virulence.

12.3.2.2 Acetylation

The purple non-sulfur bacterium *Rhodopseudomonas gelatinosa* grows photoorganotrophically on citrate. Citrate lyase cleaves citrate to oxaloacetate and acetate. When this organism grows on carbon sources other than citrate under similar conditions, citrate synthase condenses oxaloacetate and acetyl-CoA into citrate.

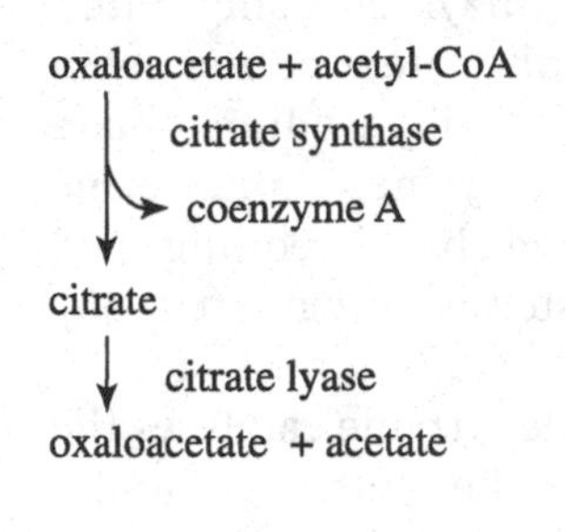

(result) acetyl-CoA ⟶ acetate + CoA-SH

As described above, when citrate synthase and citrate lyase are active at the same time, acetyl-CoA is cleaved to acetate, wasting energy in the form of a high-energy acyl-CoA bond. To avoid this futile cycle, citrate lyase activity is regulated according to the concentration of glutamate. When a metabolite of citrate metabolism, i.e. glutamate, is accumulated, citrate lyase is acetylated to an active form. Citrate lyase is deacetylated to an inactive form under the reverse conditions when citrate synthase activity is needed (Figure 12.36).

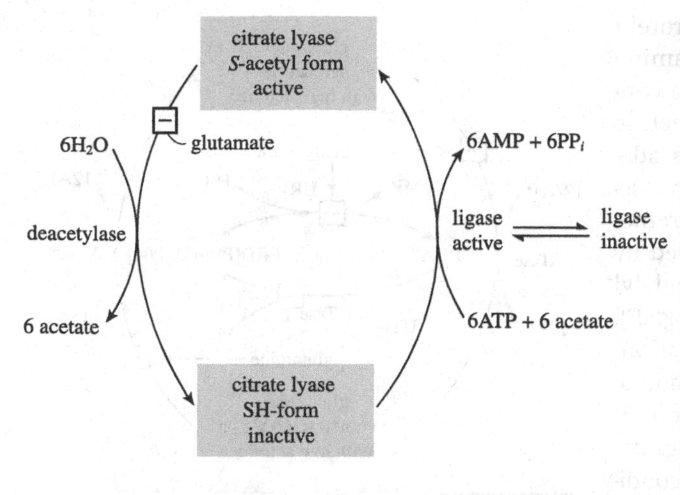

Figure 12.36 Citrate lyase activity is regulated according to the cellular glutamate concentration through acetylation and deacetylation in *Rhodopseudomonas gelatinosa*.

(Gottschalk, G. 1986, *Bacterial Metabolism*, 2nd edn., Figure 12.37. Springer, New York)

When citrate is used as a carbon and energy source, citrate lyase cleaves citrate to oxaloacetate and acetate in *Rhodopseudomonas gelatinosa*. Citrate synthase is active and synthesizes glutamate via 2-ketoglutarate as part of the incomplete TCA fork, when carbon sources other than citrate are used. A futile cycle is possible when citrate lyase and citrate synthase are active at the same time. To avoid this futile cycle, the activity of citrate lyase is regulated. When glutamate is accumulated, a ligase acetylates citrate lyase to an active form. Deacetylase deacetylates this enzyme to an inactive form when citrate synthase activity is needed, due to a low glutamate concentration.

Acetyl-CoA synthetase activates acetate to acetyl-CoA when acetate is used as the carbon and energy source. This enzyme is acetylated by acetyltransferase to the inactive form and reactivated by deacetylase. Deacetylase is constitutively expressed and the expression of acetyltransferase and acetyl-CoA synthetase is upregulated in the stationary phase and in the presence of non-sugar carbon sources, and is positively regulated by cAMP-CRP (Section 12.1.3.1). The concomitant expression of acetyl-CoA synthetase and acetyltransferase seems paradoxical, but this is interpreted as a system for fine-tuning, showing interdependence between transcriptional and post-transcriptional control mechanisms in metabolic pathways. Acetyl-CoA is the acetyl group donor in the enzymatic reaction, and acetyl-phosphate is used to acetylate protein in a non-enzymatic chemical reaction.

Many enzyme activities of transcription, translation, stress response, detoxification, sugar, amino acid and nucleotide metabolism in *Escherichia coli* are modulated through acetylation of lysine residues, and nearly all central metabolic enzymes in *Salmonella enterica* and *Bacillus subtilis* are under a similar control. Serine and threonine residues are also acetylated. Propionyl-CoA, butyryl-CoA and other acyl-CoA are used to acylate proteins to control their activities in similar enzymatic reactions.

12.3.2.3 Adenylylation

Ammonia is used in the synthesis of glutamate and glutamine in the reactions catalysed by glutamate dehydrogenase and glutamine synthetase, respectively. Since glutamate dehydrogenase has a low affinity for ammonia, this enzyme is active only at high ammonia concentrations. When the ammonia concentration is low, glutamate is produced through glutamine catalysed by glutamine synthetase and glutamate synthase, consuming ATP (Section 6.2.3):

$$\text{2-ketoglutarate} + NH_3 + NADPH + H^+ \xrightarrow{\text{glutamate dehydrogenase}} \text{glutamate} + NADP^+$$

$$\text{glutamate} + NH_3 + ATP \xrightarrow{\text{glutamine synthetase}} \text{glutamine} + ADP + P_i$$

$$\text{glutamine} + \text{2-ketoglutarate} + NADPH + H^+ \xrightarrow{\text{glutamate synthase}} \text{glutamate} + NADP^+$$

To prevent ATP consumption under ammonia-rich conditions, the glutamine synthetase gene is expressed at a low level (Section 12.2.2), and its activity is inhibited. Activity is modulated through adenylylation of the enzyme protein according to the 2-ketoglutarate/glutamine ratio which is determined by the ammonia concentration. Regulation of glutamine synthetase by adenylylation/deadenylylation involves adenylyltransferase (ATase, also known as glutamine synthetase adenylyltransferase/removase, encoded by *glnE*), PII protein (encoded by *glnB*) and uridylyltransferase (UTase)/uridylyl removing (UR) enzyme (encoded by *glnD*). UTase/UR senses the ammonia concentration, uridylylating the PII protein under ammonia-limited conditions with the accumulation of 2-ketoglutarate, and removing UMP when glutamine is accumulated under ammonia-rich conditions. PII D-$(UMP)_4$, formed under ammonia-limited conditions, activates ATase to remove AMP from glutamine synthetase, converting the enzyme to the active form. Under ammonia-rich conditions UR removes UMP, converting PII D-$(UMP)_4$ to PIIA. PIIA activates ATase, adenylylating the enzyme into a less active form (Figure 12.37).

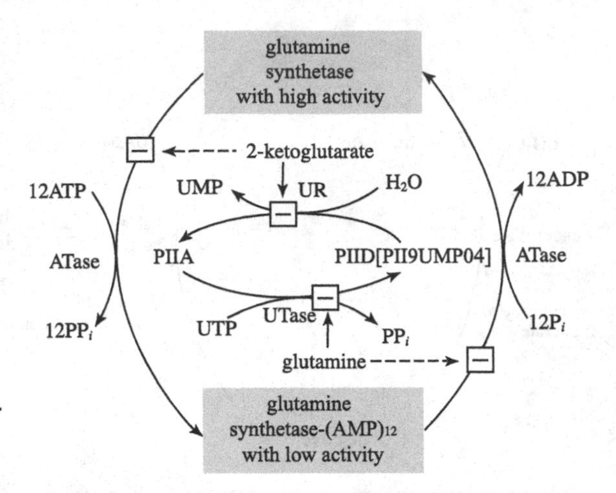

Figure 12.37 Regulation of glutamine synthetase activity in *Escherichia coli* through adenylylation and deadenylylation of the enzyme protein according to the ratio of 2-ketoglutarate/glutamine, determined by ammonia availability.

(Gottschalk, G. 1986, *Bacterial Metabolism*, 2nd edn., Figure 7.17. Springer, New York)

When glutamine accumulates under ammonia-rich conditions, uridylyltransferase/uridylyl-removing enzyme (UTase/UR, encoded by *glnD*) removes UMP from PIID [PII $(UMP)_4$], converting it to PIIA. PIIA directs adenylyltransferase (ATase) adenylylate glutamine synthetase to a less active form. When 2-ketoglutarate is accumulated due to the limited supply of ammonia, PIIA is uridylylated to PIID, which activates ATase to remove AMP from less active adenylylated glutamine synthetase to make the enzyme highly active. PIIA/PIID regulates the expression of the enzyme (Figure 12.23).

It should be mentioned that glutamine synthetase activity is regulated, not only by ammonia availability, but also by the end products of glutamine metabolism, ADP and AMP and other nucleotides, through cumulative feedback inhibition (Section 6.2.3). Enzyme activity is inhibited even under ammonia-limited conditions when the culture is not actively growing.

In Gram-negative bacteria, PII is modified by uridylylation or by adenylylation (Section 12.2.2). Protein modification by uridylylation or adenylylation is not as common as modification by phosphorylation or acetylation.

12.3.2.4 Other chemical modifications

Table 12.8 summarizes enzyme activity modulation through chemical modification of the enzyme protein. As shown in the table, some proteins are modified by means other than those mentioned above. When nitrogenase activity is inhibited through the ammonia switch, this enzyme is modified with ADP-ribose from NAD^+ (Section 6.2.1.5). PII protein (GlnB) participating in the response to ammonia limitation is uridylylated (Figure 12.36). The substrate-receptor protein involved in chemotaxis is methylated for its regulatory function (Section 12.2.11).

In addition to the widely recognized enzyme modifications through phosphorylation and acetylation, enzymes can also be modified by succinylation, methylation, propionylation, butyrylation, glutarylation, lipidation, carboxylation, glycosylation, nitrosylation or thiolation. Over 1000 *E. coli* proteins are modified by succinylation and there are 23 glutarylated proteins. The regulatory mechanisms of succinylation and glutarylation, and their potential effects on metabolism, are yet to be characterized.

Table 12.8 | Chemical modification of enzyme proteins for activity regulation.

Modified with	Source	Example
Phosphate	ATP	isocitrate dehydrogenase
Acetyl groups	Acetate + ATP	citrate lyase
Adenylyl groups	ATP	glutamine synthetase
Uridylyl groups	UTP	PII (regulatory protein in nitrogen metabolism)
ADP-ribose	NAD^+	nitrogenase
Methyl groups	SAM	substrate receptor protein in chemotaxis

12.3.2.5 Regulation through physical modification and dissociation/association

Enzyme activity can be modulated not only through chemical modification but also through physical modification. Enzymes modifiable through this mechanism for regulation of activity consist of multiple subunits. For example, clostridial lactate dehydrogenase is a homotetramer. It is active in the tetramer form, but loses its activity when it is dissociated into dimers. Lactate dehydrogenase is activated by fructose-1,6-diphosphate to accelerate carbohydrate metabolism through the EMP pathway, increasing the rate of NADH oxidation to NAD^+. Under the reverse conditions, clostridia metabolize pyruvate to acetate and butyrate to synthesize more ATP, which inhibits lactate dehydrogenase.

Glutamate dehydrogenase activity is dependent on Mn^{2+}. This enzyme is a hexameric protein and the cation keeps this enzyme in the active taut form. Without the cation, the enzyme becomes an inactive relaxed form. Many proteins are degraded when they are not needed. Degradation is another form of metabolic regulation. Since degradation is an irreversible process, proteolysis needs to be tightly controlled.

12.4 Metabolic regulation and growth

As discussed in Chapter 6, growth conditions determine the growth rate and cell yield for a given organism. This is due to the difference in energy conservation efficiency and maintenance energy when growing on different carbon sources. This difference is the result of elaborate metabolic regulation.

12.4.1 Regulation in central metabolism

For the most efficient growth under given conditions, catabolism and anabolism are coordinately regulated, not only through the expression of the genes, but also through the control of enzyme activities. Glucose is the preferred carbon source in most organisms, and simple organic compounds such as acetate support the growth of many bacteria. These are metabolized through glycolysis (or gluconeogenesis), HMP pathway and TCA cycle with appropriate anaplerotic reactions to supply carbon skeletons, NADPH and ATP in aerobic respiratory organisms. Regulation of gene expression in the control of central metabolism was described in Chapters 4 and 5. Catabolite repression (Section 12.1.3) and regulation of glycogen metabolism by the Csr system (Section 12.1.9.3) have also been described previously, as well as other mechanisms. Most of the regulated genes in central metabolism are controlled through catabolite repression, involving the cAMP receptor protein (Crp) with cAMP and catabolite repressor/activator (Cra) protein with fructose-1,6-bisphosphate and fructose-1-phosphate in bacteria, using glucose through the EMP pathway. In *Pseudomonas* spp., a repressor, HexR, regulates gene expression in the Entner–Doudoroff pathway (Section 4.4.2). In *Rhodobacter sphaeroides*, CceR (carbon and energy metabolism regulator) represses genes

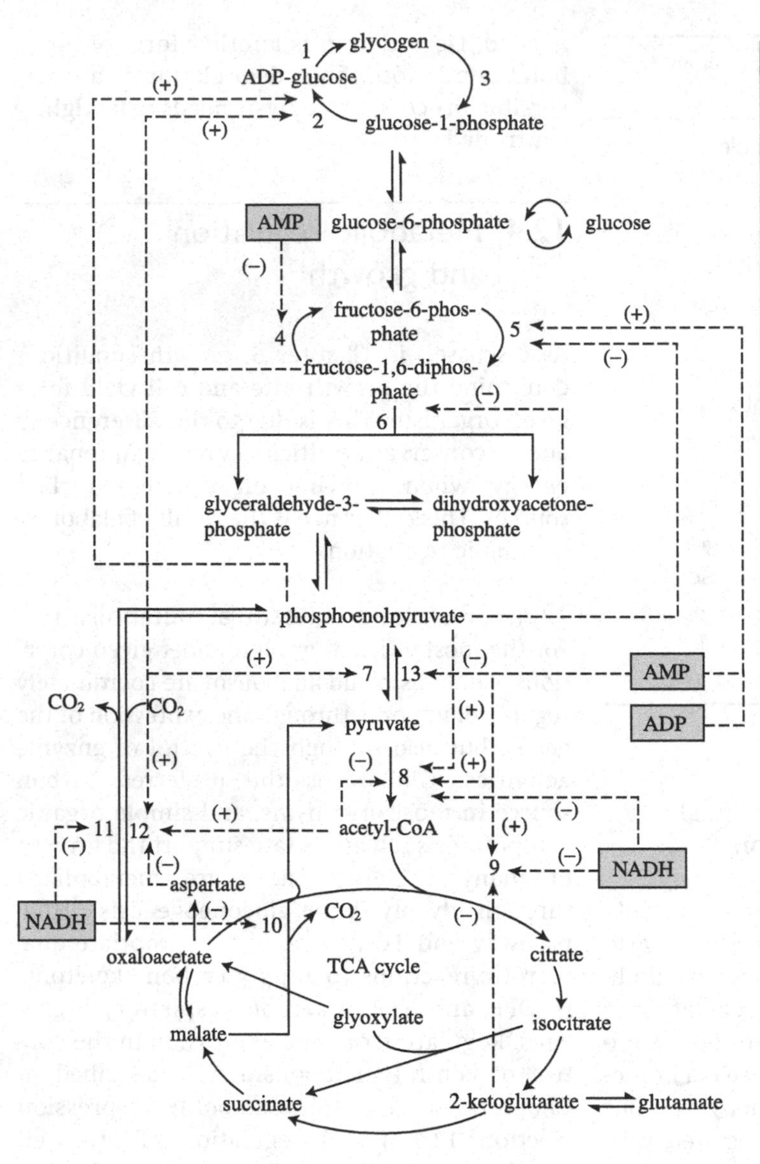

Figure 12.38 Control of central metabolic pathways in bacteria.

The activity of many enzymes in central metabolism is modulated by post-translational modification. In addition their activities are controlled by metabolites. Detailed metabolic control is organism specific. Generally, the adenylate energy charge and the oxidized/reduced ratio of the pyridine nucleotides are the most important effectors, while some metabolic intermediates, such as fructose-1,6-diphosphate and phosphoenolpyruvate, are involved in the regulation as effectors.

1, glycogen synthase; 2, ADP-glucose pyrophosphorylase; 3, phosphorylase; 4, fructose-1,6-diphosphatase; 5, phosphofructokinase; 6, fructose-1,6-diphosphate aldolase; 7, pyruvate kinase; 8, pyruvate dehydrogenase; 9, citrate synthase; 10, malate enzyme; 11, PEP carboxykinase; 12, PEP carboxylase; 13, PEP synthetase. (+), activation; (–), inhibition.

encoding enzymes in the ED pathway and activates those encoding the TCA cycle and gluconeogenesis. This repressor is inactivated by an ED pathway intermediate, 6-phosphogluconate. AkgR (alpha-ketoglutarate regulator) activates genes encoding several TCA cycle enzymes. These proteins cooperatively regulate central metabolism in bacteria with the ED pathway.

Many enzymes of central metabolism are phosphorylated (Section 12.3.2.1) or acetylated (Section 12.3.2.2) for modulation of their activity. In addition, central metabolic pathways are regulated through the control of enzyme activity by inhibitors and activators (Figure 12.38). When the adenylate energy charge (EC, Section 5.6.2) is low with the accumulation of AMP and ADP, catabolic enzymes are activated, while enzymes of gluconeogenesis and glycogen synthesis are inhibited under similar conditions. Reduced pyridine nucleotides inhibit the TCA cycle. Some metabolic intermediates participate in the regulation, such as fructose-1,6-diphosphate, dihydroxyacetone phosphate, phosphoenolpyruvate, acetyl-CoA and 2-ketoglutarate.

12.4.2 Regulatory networks

Control of gene expression is a fundamental process and it pervades most biological processes, from cell proliferation and differentiation to development. As discussed above, gene expression is regulated at transcriptional and post-transcriptional levels, and enzyme activity is regulated for optimum activity under given conditions. Cells need to integrate intrinsic and environmental information and coordinate multiple regulatory mechanisms of gene expression to properly express their biological functions. This can be achieved by the global regulation mechanisms discussed in Section 12.2. Additionally, multiple regulators are involved for the coordinated regulation of genes responsible for related functions.

In this sense, it is interesting to note that nitrogen anabolism is regulated according to carbon catabolism. The CcpA–HPr(Ser-P) complex is an important regulator in *Bacillus subtilis*, regulating approximately 10 per cent of the genes in this bacterium (Section 12.1.3.3). A *ccpA* mutant of this low G + C Gram-positive bacterium cannot grow on minimal media containing glucose and ammonia. *Bacillus subtilis* does not contain an anabolic glutamate dehydrogenase, and glutamine synthetase (GS) is the only enzyme for ammonium assimilation. The expression of this enzyme (*gltAB* operon) is regulated not only by GlnR and TnrA (Section 12.2.2), but also by glucose or other glycolytically catabolized carbon sources. The latter regulation is mediated through CcpA. This is an example of global gene regulation as an orchestrated system. Similarly, the availability of carbon and nitrogen sources coordinately regulates expression of genes in Gram-negative bacteria by cross-talk between PTS and PTS^{Ntr} (Section 12.2.2). It is believed that all catabolic and anabolic processes are regulated coordinately, as exemplified above, for efficient growth or survival in harsh environments.

Generally, any given pair of histidine kinase (HK) and response regulator (RR) proteins in a two-component system (TSC) selectively recognize each other (Section 12.1.7), but cross-talk between TCSs for coordinated metabolic regulation also occurs through integration of multiple signals. The HK CreC of CreC/CreB (carbon source responsive) TCS phosphorylates PhoB (Section 12.3.2.1). Other cross-talk mechanisms include one HK transferring phosphate to multiple RRs, and multiple HKs transferring phosphate to a single RR (Section 12.1.7). In some cases, auxiliary regulators such as sRNA and small-sized proteins, referred to as connectors, link TCSs to other regulatory networks. These include CrcZ (sRNA) between a TCS, CbrA/CbrB and Crc (Section 12.1.3), and YdeO (a transcriptional regulator) and SafA (sensor associating factor A) induced by the EvgS/EvgA system, activate transcription of acid resistance genes and phosphorylate PhoP, respectively (Section 12.1.7).

Although there is scattered evidence that shows coordinated multiple regulatory mechanisms, regulatory networks are still only partially understood.

12.4.3 Growth rate and regulation

The growth rate of microorganisms is dependent on the quality and quantity of their substrate. In a complex medium they grow fast, and need higher enzyme activities. To meet these needs, genes for proteins are activated for higher transcription and more ribosomes are needed to translate the increased transcripts (mRNA). When nutrients are depleted, ribosomal protein (r-proteins) synthesis is inhibited through autogenous regulation (Section 12.1.8), and stationary phase proteins are synthesized through the stringent response (Section 12.2.1). Under starvation conditions, RNAs and proteins are utilized to supply energy needed for survival. This is referred to as endogenous respiration (Section 13.1).

Fast growing cells in a rich medium use electron donors through fermentation and not through respiration, although the latter conserves more energy than the former per electron donor used. This phenomenon is known as overflow metabolism and is due to the fact that the energy cost of protein synthesis for respiratory metabolism is higher than the conserved energy difference between fermentation and respiration. Citrate synthase is inhibited by ATP and

Table 12.9 | Secondary metabolites produced during cell differentiation.

Secondary metabolite	Producing organism	Cellular event
Actinomycin D	*Streptomyces antibioticus*	sporulation
Amylase	*Bacillus subtilis*	sporulation
Butanol	*Clostridium acetobutylicum*	sporulation
Pamamycin	*Streptomyces alboniger*	aerial mycelium formation
Tyrocidin	*Bacillus brevis*	sporulation

NADH that are produced at a rapid rate through fermentation (Section 5.2.3).

12.5 Secondary metabolites

Some microorganisms produce compounds not directly related to their growth when their growth is impaired or during a cell differentiation process such as sporulation. These compounds are referred to as secondary metabolites. Some of them are important industrially (Table 12.9). Antibiotics are the best known secondary metabolites.

A small molecular weight peptide known as A-factor induces the production of streptomycin in *Streptomyces griseus* and *Streptomyces bikiniensis*. Mutants of A-factor do not have the ability to produce the antibiotic compound, and the addition of A-factor to the culture restores this ability. A-factor is an autoinducer of quorum sensing. When its extracellular concentration reaches a certain level, the signal is transduced through a two-component system (Section 12.1.7). Many Gram-positive bacteria produce peptide antibiotics known as bacteriocins. Nisin, a bacteriocin produced by *Lactococcus lactis*, is an autoinducer for quorum sensing in this bacterium. Many of the secondary metabolites are produced during cell differentiation, with a limited supply of nutrients including phosphate.

Genome sequences show that many *Streptomyces* spp. have secondary metabolite biosynthetic pathways some of which are silent under standard laboratory conditions. Each of these pathways is activated by a specific activator. The genome of *Streptomyces coelicolor* has over 700 genes encoding regulatory proteins. Regulation of secondary metabolism is governed by the cooperative or antagonistic action of various global regulators, including catabolite repressors and activators that are controlled by the availability of carbon, nitrogen and phosphate.

12.6 Metabolic regulation and the fermentation industry

'Fermentation' can be defined as 'anaerobic microbial growth through substrate-level phosphorylation under dark conditions' (Chapter 8). This term is also used to describe processes producing useful materials at a large industrial scale using microorganisms. Microorganisms have elaborate regulatory mechanisms for efficient growth but not for the production of specific materials. To improve fermentation efficiency, industry has therefore developed various mutants defective in regulatory mechanisms.

12.6.1 Fermentative production of antibiotics

When Fleming discovered penicillin in 1929, the fungus *Penicillium notatum* produced 1.2 mg l^{-1} penicillin. At present, the fermentation industry uses strains producing over 50 g l^{-1} penicillin. These industrial strains are mutants derived from wild-type strains but with altered regulatory mechanisms. Industrial strains of other antibiotic producers have also been developed through extensive mutation programmes.

12.6.2 Fermentative amino acid production

Bacterial strains have been widely used to produce various amino acids. Coryneform bacteria of the genera *Brevibacterium* and *Corynebacterium* are the most commonly used industrial strains in

Table 12.10 Bacterial strains producing amino acids.

Amino acid	Strain	Genetic trait	Yield (g l^{-1})
DL-alanine	*Microbacterium ammoniaphilum*	ArgHx[a]	60
L-arginine	*Brevibacterium flavum*	Gua[b], TA[a]	35
L-glutamate	*Corynebacterium glutamicum*	wild-type	>100
L-glutamine	*C. glutamicum*	wild-type	40
L-histidine	*B. flavum*	TA,[a] SM,[a] Eth,[a] ABT[a]	10
L-isoleucine	*B. flavum*	OAHV,[a] OMT[a]	15
L-leucine	*Brevibacterium lactofermentum*	Ile,[b] Met,[b] TA[a]	28
L-lysine	*B. flavum*	AEC[a]	57
L-methionine	*C. glutamicum*	Thr,[b] Eth,[a] MetHx[a]	2
L-phenylalanine	*C. glutamicum*	Tyr,[b] PFP,[a] PAP[a]	9
L-threonine	*B. flavum*	Met,[b] AHV[a]	18
L-tryptophan	*C. glutamicum*	Phe,[b] Tyr,[b] 5MT[a]	12
L-valine	*Brevibacterium lactofermentum*	Phe,[b] PFP,[a] PAP,[a] PAT,[a] TyrHx,[a] TA[a]	31

ABT, 2-aminobenzthiazole; AEC, *S*-(3-aminoethyl)-1-cysteine; AHV, 2-amino-3-hydroxdyvalerate; ArgHx, arginine hydroxamate; Eth, ethionine; Gua, guanine; Ile, isoleucine; Met, methionine; MetHx, methionine hydroxamate; 5MT, 5-methyltryptophan; OMT, *o*-methylthreonine; PAP, *p*-aminophenylalanine; PAT, *p*-aminotyrosine; PFP, *p*-fluoro-phenyalanine; Phe, phenylalanine; SM, selenomethionine; TA, 2-thiozolalanine; Thr, threonine; Tyr, tyrosine; TyrHx, tyrosine hydroxamate.
[a] Resistant to the analogue.
[b] Auxotrophic mutant.

amino acid production. These bacteria excrete amino acids into the surrounding environment when the membrane becomes more permeable to amino acids under biotin-limited conditions. Amino acid production is tightly controlled through various mechanisms, as discussed earlier (Section 12.1). The industrial strains used to produce amino acids are mutant strains with defects in regulation. These are selected based on their resistance properties to analogues (Table 12.10). When an amino acid analogue is added to a culture, the wild-type strain cannot grow, since the analogue inhibits expression of the genes for production of the amino acid and the analogue cannot be used for biosynthesis. Auxotrophic mutants are used for the fermentative production of the intermediates. Guanine auxotrophs are used to produce adenine and hypoxanthine. In recent years, whole genome sequences have been determined in various industrially important microorganisms and industrial strains have been developed through molecular biology approaches. These may aim to (1) increase enzyme activities, (2) relieve regulatory mechanisms and (3) improve membrane permeability. Such approaches are referred to as metabolic engineering.

Further Reading

Note this section contains key references only. Additional recommended references are available at www.cambridge.org/ProkaryoticMetabolism.

Promoters and σ-factors

Barnard, A., Wolfe, A. & Busby, S. (2004). Regulation at complex bacterial promoters: how bacteria use different promoter organizations to produce different regulatory outcomes. *Current Opinion in Microbiology* **7**, 102–108.

Battesti, A., Majdalani, N. & Gottesman, S. (2011). The RpoS-mediated general stress response in *Escherichia coli*. *Annual Review of Microbiology* **65**, 189–213.

Fimlaid, K. A. & Shen, A. (2015). Diverse mechanisms regulate sporulation sigma factor activity in the Firmicutes. *Current Opinion in Microbiology* **24**, 88–95.

Grabowicz, M. and Silhavy T. J., (2017). Envelope stress responses: an interconnected safety net. *Trends in Biochemical Sciences* **42**, 232–242.

Gruber, T. M. & Gross, C. A. (2003). Multiple sigma subunits and the partitioning of bacterial transcription space. *Annual Review of Microbiology* **57**, 441–466.

Helmann, J. D. (2016). *Bacillus subtilis* extracytoplasmic function (ECF) sigma factors and defense of the cell envelope. *Current Opinion in Microbiology* **30**, 122–132.

Herrou, J., Foreman, R., Fiebig, A. & Crosson, S. (2010). A structural model of anti-anti-σ inhibition by a two-component receiver domain: the PhyR stress response regulator. *Molecular Microbiology* **78**, 290–304.

Kazmierczak, M. J., Wiedmann, M. & Boor, K. J. (2005). Alternative sigma factors and their roles in bacterial virulence. *Microbiology and Molecular Biology Reviews* **69**, 527–543.

Österberg, S., del Pesos-Santos T. & Shingler, V. (2011). Regulation of alternative sigma factor use. *Annual Review of Microbiology* **65**, 37–55.

Yang, Y., Darbari, V. C., Zhang, N., Lu, D., Glyde, R., Wang, Y.-P., Winkelman, J. T., Gourse, R. L., Murakami, K. S., Buck M., & Zhang, X. (2015). Structures of the RNA polymerase-σ^{54} reveal new and conserved regulatory strategies. *Science* **349**, 882–885.

Enzyme induction – activation and repression

Cai, J., Tong, H., Qi, F. & Dong, X. (2012). CcpA-dependent carbohydrate catabolite repression regulates galactose metabolism in *Streptococcus oligofermentans*. *Journal of Bacteriology* **194**, 3824–3832.

Chavarría, M., Fuhrer, T., Sauer, U., Pflüger-Grau, K. & de Lorenzo, V. (2013). Cra regulates the cross-talk between the two branches of the phosphoenolpyruvate: phosphotransferase system of *Pseudomonas putida*. *Environmental Microbiology* **15**, 121–132.

Fonseca, P., Moreno, R. & Rojo, F. (2013). *Pseudomonas putida* growing at low temperature shows increased levels of CrcZ and CrcY sRNAs, leading to reduced Crc-dependent catabolite repression. *Environmental Microbiology* **15**, 24–35.

Galinier, A. & Deutscher, J. (2017). Sophisticated regulation of transcriptional factors by the bacterial phosphoenolpyruvate: sugar phosphotransferase system. *Journal of Molecular Biology* **429**, 773–789.

Hartmann, T., Zhang, B., Baronian, G., Schulthess, B., Homerova, D., Grubmueller, S., Kutzner, E., Gaupp, R., Bertram, R., Powers, R., Eisenreich, W., Kormanec, J., Herrmann, M., Molle, V., Somerville, G. A. & Bischoff, M. (2013). Catabolite Control Protein E (CcpE) is a LysR-type transcriptional regulator of tricarboxylic acid cycle activity in *Staphylococcus aureus*. *Journal of Biological Chemistry* **288**, 36116–36128.

Joshua, C. J., Dahl, R. Benke, P. I. & Keasling, J. D. (2011). Absence of diauxie during simultaneous utilization of glucose and xylose by *Sulfolobus acidocaldarius*. *Journal of Bacteriology* **193**, 1293–1301.

Attenuation

Babitzke, P. (2004). Regulation of transcription attenuation and translation initiation by allosteric control of an RNA-binding protein: the *Bacillus subtilis* TRAP protein. *Current Opinion in Microbiology* **7**, 132–139.

Gollnick, P., Babitzke, P., Antson, A. & Yanofsky, C. (2005). Complexity in regulation of tryptophan biosynthesis in *Bacillus subtilis*. *Annual Review of Genetics* **39**, 47–68.

Wallecha, A., Oreh, H., van der Woude, M. W. & deHaseth, P. L. (2014). Control of gene expression at a bacterial leader RNA, the *agn43* gene encoding outer membrane protein Ag43 of *Escherichia coli*. *Journal of Bacteriology* **196**, 2728–2735.

Termination/antitermination

Bastet, L., Dubé, A., Massé, E. & Lafontaine, D. A. (2011). New insights into riboswitch regulation mechanisms. *Molecular Microbiology* **80**, 1148–1154.

Dambach, M. D. & Winkler, W. C. (2009). Expanding roles for metabolite-sensing regulatory RNAs. *Current Opinion in Microbiology* **12**, 161–169.

DebRoy, S., Gebbie, M., Ramesh, A., Goodson, J. R., Cruz, M. R., van Hoof, A., Winkler, W. C. & Garsin, D. A. (2014). A riboswitch-containing sRNA controls gene expression by sequestration of a response regulator. *Science* **345**, 937–940.

Fürtig, B., Nozinovic, S., Reining, A. & Schwalbe, H. (2015). Multiple conformational states of riboswitches fine-tune gene regulation. *Current Opinion in Structural Biology* **30**, 112–124.

Garst, A. D., Porter, E. B. & Batey, R. T. (2012). Insights into the regulatory landscape of the lysine riboswitch. *Journal of Molecular Biology* **423**, 17–33.

Johnson Jr, J. E., Reyes, F. E., Polaski, J. T. & Batey, R. T. (2012). B_{12} cofactors directly stabilize an mRNA regulatory switch. *Nature* **49**, 133–137.

Kulshina, N., Baird, N. J. & Ferre-D'Amare, A. R. (2009). Recognition of the bacterial second messenger cyclic diguanylate by its cognate riboswitch. *Nature Structural and Molecular Biology* **16**, 1212–1217.

Serganov, A., Huang, L. & Patel, D. J. (2009). Coenzyme recognition and gene regulation by a flavin mononucleotide riboswitch. *Nature* **458**, 233–237.

Two-component systems

Alvarez, A. F., Barba-Ostria, C., Silva-Jiménez, H. & Georgellis, D. (2016). Organization and mode of action of two component system signaling circuits from the various kingdoms of life. *Environmental Microbiology* **18**, 3210–3226.

Beier, D. & Gross, R. (2006). Regulation of bacterial virulence by two-component systems. *Current Opinion in Microbiology* **9**, 143–152.

Buelow, D. R. & Raivio, T. L. (2010). Three (and more) component regulatory systems - auxiliary regulators of bacterial histidine kinases. *Molecular Microbiology* **75**, 547–566.

Capra, E. J. & Laub, M. T. (2012). Evolution of two-component signal transduction systems. *Annual Review of Microbiology* **66**, 325–347.

Desai, S. K. & Kenney, L. J. (2017). To ~P or Not to ~P? Non-canonical activation by two-component response regulators. *Molecular Microbiology* **103**, 203–213.

Göpel, Y. & Görke, B. (2012). Rewiring two-component signal transduction with small RNAs. *Current Opinion in Microbiology* **15**, 132–139.

Groisman, E. A. (2016). Feedback control of two-component regulatory systems. *Annual Review of Microbiology* **70**, 103–124.

Jung, K., Fried, L., Behr, S. & Heermann, R. (2012). Histidine kinases and response regulators in networks. *Current Opinion in Microbiology* **15**, 118–124.

Podgornaia, A. I. & Laub, M. T. (2013). Determinants of specificity in two-component signal transduction. *Current Opinion in Microbiology* **16**, 156–162.

Salazar, M. E. & Laub, M. T. (2015). Temporal and evolutionary dynamics of two-component signaling pathways. *Current Opinion in Microbiology* **24**, 7–14.

Silversmith, R. E. (2010). Auxiliary phosphatases in two-component signal transduction. *Current Opinion in Microbiology* **13**, 177–183.

Autogenous control

Aseev, L. V., Koledinskaya, L. S. & Boni, I. V. (2016). Regulation of ribosomal protein operons *rplM-rpsI, rpmB-rpmG*, and *rplU-rpmA* at the transcriptional and translational levels. *Journal of Bacteriology* **198**, 2494–2502.

Schneider, D. A., Ross, W. & Gourse, R. L. (2003). Control of rRNA expression in *Escherichia coli*. *Current Opinion in Microbiology* **6**, 151–156.

Post-transcriptional regulation

Mata, J., Marguerat, S. & Bahler, J. (2005). Post-transcriptional control of gene expression: a genome-wide perspective. *Trends in Biochemical Sciences* **30**, 506–514.

Nogueira, T. & Springer, M. (2000). Post-transcriptional control by global regulators of gene expression in bacteria. *Current Opinion in Microbiology* **3**, 154–158.

Romeo, T., Vakulskas, C. A. & Babitzke, P. (2013). Post-transcriptional regulation on a global scale: form and function of Csr/Rsm systems. *Environmental Microbiology* **15**, 313–324.

Stability and translational efficiency of mRNA

Aït-Bara, S. & Carpousis, A. J. (2015). RNA degradosomes in bacteria and chloroplasts: classification, distribution and evolution of RNase E homologs. *Molecular Microbiology* **97**, 1021–1135.

Babitzke, P. (2004). Regulation of transcription attenuation and translation initiation by allosteric control of an RNA-binding protein: the *Bacillus subtilis* TRAP protein. *Current Opinion in Microbiology* **7**, 132–139.

Condon, C. & Bechhofer, D. H. (2011). Regulated RNA stability in the Gram positives. *Current Opinion in Microbiology* **14**, 148–154.

Hui, M. P., Foley, P. L. & Belasco, J. G. (2014). Messenger RNA degradation in bacterial cells. *Annual Review of Genetics* **48**, 537–559.

Kennell, D. (2002). Processing endoribonucleases and mRNA degradation in bacteria. *Journal of Bacteriology* **184**, 4645–4657.

Kushner, S. R. (2002). mRNA decay in *Escherichia coli* comes of age. *Journal of Bacteriology* **184**, 4658–4665.

Mackie, G. A. (2013). RNase E: at the interface of bacterial RNA processing and decay. *Nature Reviews Microbiology* **11**, 45–57.

Wang, Y., Liu, C. L., Storey, J. D., Tibshirani, R. J., Herschlag, D. & Brown, P. O. (2002). Precision and functional specificity in mRNA decay. *Proceedings of the National Academy of Sciences of the USA* **99**, 5860–5865.

Modulation of translation by protein

Altegoer, F., Rensing, S. A. & Bange, G. (2016). Structural basis for the CsrA-dependent modulation of translation initiation by an ancient regulatory protein. *Proceedings of the National Academy of Sciences of the USA* **113**, 10168–10173.

Boni, I. V. (2006). Diverse molecular mechanisms of translation initiation in prokaryotes. *Molecular Biology* **40**, 587–596.

Cahova, H., Winz, M.-L., Hofer, K., Nubel, G. & Jaschke, A. (2015). NAD captureSeq indicates NAD as a bacterial cap for a subset of regulatory RNAs. *Nature* **519**, 374–377.

Schlax, P. J. & Worhunsky, D. J. (2003). Translational repression mechanisms in prokaryotes. *Molecular Microbiology* **48**, 1157–1169.

Modulation of translation by sRNA

Bobrovskyy, M. & Vanderpool, C. K. (2013). Regulation of bacterial metabolism by small RNAs using diverse mechanisms. *Annual Review of Genetics* **47**, 209–232.

Bossi, L. & Figueroa-Bossi, N. (2016). Competing endogenous RNAs: a target-centric view of small RNA regulation in bacteria. *Nature Reviews Microbiology* **14**, 775–784.

Bouloc, P. & Repoila, F. (2016). Fresh layers of RNA-mediated regulation in Gram-positive bacteria. *Current Opinion in Microbiology* **30**, 30–35.

Brantl, S. (2002). Antisense-RNA regulation and RNA interference. *Biochimica et Biophysica Acta* **1575**, 15–25.

DebRoy, S., Gebbie, M., Ramesh, A., Goodson, J. R., Cruz, M. R., van Hoof, A., Winkler, W. C. & Garsin, D. A. (2014). A riboswitch-containing sRNA controls gene expression by sequestration of a response regulator. *Science* **345**, 937–940.

Mars, R. A. T., Nicolas, P., Denham, E. L. & van Dijl, J. M. (2016). Regulatory RNAs in *Bacillus subtilis*: a Gram-positive perspective on bacterial RNA-mediated regulation of gene expression. *Microbiology and Molecular Biology Reviews* **80**, 1029–1057.

Mellin, J. R., Koutero, M., Dar, D., Nahori, M.-A., Sorek, R. & Cossart, P. (2014). Sequestration of a two-component response regulator by a riboswitch-regulated noncoding RNA. *Science* **345**, 940–943.

Sherwood, A. V. & Henkin, T. M. (2016). Riboswitch-mediated gene regulation: novel RNA architectures dictate gene expression responses. *Annual Review of Microbiology* **70**, 361–374.

Storz, G., Opdyke, J. A. & Zhang, A. (2004). Controlling mRNA stability and translation with small, noncoding RNAs. *Current Opinion in Microbiology* **7**, 140–144.

c-di-GMP

Bush, M. J., Tschowri, N., Schlimpert, S., Flardh, K. & Buttner, M. J. (2015). c-di-GMP signalling and the regulation of developmental transitions in streptomycetes. *Nature Reviews Microbiology* **13**, 749–760.

Chou, S.-H. & Galperin, M. Y. (2016). Diversity of cyclic di-GMP-binding proteins and mechanisms. *Journal of Bacteriology* **198**, 32–46.

Gao, J., Tao, J., Liang, W. & Jiang, Z. (2016). Cyclic (di) nucleotides: the common language shared by microbe and host. *Current Opinion in Microbiology* **30**, 79–87.

Hallez, R., Delaby, M., Sanselicio, S. & Viollier, P. H. (2017). Hit the right spots: cell cycle control by phosphorylated guanosines in alphaproteobacteria. *Nature Reviews Microbiology* **15**, 137–148.

Hengge, R., Gründling, A., Jenal, U., Ryan, R. & Yildiz, F. (2016). Bacterial signal transduction by cyclic di-GMP and other nucleotide second messengers. *Journal of Bacteriology* **198**, 15–26.

Jenal, U., Reinders, A. & Lori, C. (2017). Cyclic di-GMP: second messenger extraordinaire. *Nature Reviews Microbiology* **15**, 271–284.

Römling, U., Galperin, M. Y. & Gomelsky, M. (2013). Cyclic di-GMP: the first 25 years of a universal bacterial second messenger. *Microbiology and Molecular Biology Reviews* **77**, 1–52.

Ryan, R. P. (2013). Cyclic di-GMP signalling and the regulation of bacterial virulence. *Microbiology* **159**, 1286–1297.

Metabolic regulation in archaea

Bell, S. D. (2005). Archaeal transcriptional regulation – variation on a bacterial theme? *Trends in Microbiology* **13**, 262–265.

Dennis, P. P., Omer, A. & Lowe, T. (2001). A guided tour: small RNA function in Archaea. *Molecular Microbiology* **40**, 509–519.

Geiduschek, E. P. & Ouhammouch, M. (2005). Archaeal transcription and its regulators. *Molecular Microbiology* **56**, 1397–1407.

Karr, E. A. (2014). Transcription regulation in the third domain. *Advances in Applied Microbiology*. **89**, 101–133.

Marchfelder, A., Fischer, S., Brendel, J., Stoll, B., Maier, L.-K., Jäger, D., Prasse, D., Plagens, A., Schmitz, R. &

Randau, L. (2012). Small RNAs for defence and regulation in archaea. *Extremophiles* **16**, 685–696.

Stringent response

Boutte, C. C. & Crosson, S. (2013). Bacterial lifestyle shapes stringent response activation. *Trends in Microbiology* **21**, 174–180.

Braeken, K., Moris, M., Daniels, R., Vanderleyden, J. & Michiels, J. (2006). New horizons for (p)ppGpp in bacterial and plant physiology. *Trends in Microbiology* **14**, 45–54.

Brown, A., Fernández, I. S., Gordiyenko, Y. & Ramakrishnan, V. (2016). Ribosome-dependent activation of stringent control. *Nature* **534**, 277–280.

Dalebroux, Z. D., Svensson, S. L., Gaynor, E. C. & Swanson, M. S. (2010). ppGpp conjures bacterial virulence. *Microbiology & Molecular Biology Reviews* **74**, 171–199.

Dalebroux, Z. D. & Swanson, M. S. (2012). ppGpp: magic beyond RNA polymerase. *Nature Reviews Microbiology* **10**, 203–212.

Gaca, A. O., Colomer-Winter, C. & Lemos, J. A. (2015). Many means to a common end: the intricacies of (p)ppGpp metabolism and its control of bacterial homeostasis. *Journal of Bacteriology* **197**, 1146–1156.

Hauryliuk, V., Atkinson, G. C., Murakami, K. S., Tenson, T. & Gerdes, K. (2015). Recent functional insights into the role of (p)ppGpp in bacterial physiology. *Nature Reviews Microbiology* **13**, 298–309.

Hengge-Aronis, R. (2002). Signal transduction and regulatory mechanisms involved in control of the σ^S (RpoS) subunit of RNA polymerase. *Microbiology and Molecular Biology Reviews* **66**, 373–395.

Magnusson, L. U., Farewell, A. & Nystrom, T. (2005). ppGpp: a global regulator in *Escherichia coli*. *Trends in Microbiology* **13**, 236–242.

Nitrogen control

Amon, J., Titgemeyer, F. & Burkovski, A. (2010). Common patterns – unique features: nitrogen metabolism and regulation in Gram-positive bacteria. *FEMS Microbiology Reviews* **34**, 588–605.

Arcondeguy, T., Jack, R. & Merrick, M. (2001). P-II signal transduction proteins, pivotal players in microbial nitrogen control. *Microbiology and Molecular Biology Reviews* **65**, 80–105.

Commichau, F. M., Forchhammer, K. & Stulke, J. (2006). Regulatory links between carbon and nitrogen metabolism. *Current Opinion in Microbiology* **9**, 167–172.

Forchhammer, K. (2004). Global carbon/nitrogen control by P_{II} signal transduction in cyanobacteria: from signals to targets. *FEMS Microbiology Reviews* **28**, 319–333.

Huergo, L. F., Chandra, G. & Merrick, M. (2013). P_{II} signal transduction proteins: nitrogen regulation and beyond. *FEMS Microbiology Reviews* **37**, 251–283.

Ninfa, A. J. & Jiang, P. (2005). PII signal transduction proteins: sensors of α-ketoglutarate that regulate nitrogen metabolism. *Current Opinion in Microbiology* **8**, 168–173.

Reitzer, L. (2003). Nitrogen assimilation and global regulation in *Escherichia coli*. *Annual Review of Microbiology* **57**, 155–176.

Pho system

Groisman, E. A. (2001). The pleiotropic two-component regulatory system PhoP-PhoQ. *Journal of Bacteriology* **183**, 1835–1842.

Hsieh, Y.-J. & Wanner, B. L. (2010). Global regulation by the seven-component Pi signaling system. *Current Opinion in Microbiology* **13**, 198–203.

Lamarche, M. G., Wanner, B. L., Crepin, S. & Harel, J. (2008). The phosphate regulon and bacterial virulence: a regulatory network connecting phosphate homeostasis and pathogenesis. *FEMS Microbiology Reviews* **32**, 461–473.

Martin, J. F. (2004). Phosphate control of the biosynthesis of antibiotics and other secondary metabolites is mediated by the PhoR-PhoP system: an unfinished story. *Journal of Bacteriology* **186**, 5197–5201.

Santos-Beneit, F. (2015). The Pho regulon: a huge regulatory network in bacteria. *Frontiers in Microbiology* **6**, 402.

Vershinina, O. A. & Znamenskaya, L. V. (2002). The *pho* regulons of bacteria. *Microbiology-Moscow* **71**, 497–511.

ArcB/ArcA and PrrB/PrrA systems

Alvarez, A. F., Rodriguez, C. & Georgellis, D. (2013). Ubiquinone and menaquinone electron carriers represent the yin and yang in the redox regulation of the ArcB sensor kinase. *Journal of Bacteriology* **195**, 3054–3061.

Bekker, M., Alexeeva, S., Laan, W., Sawers, G., Teixeira de Mattos, J. & Hellingwerf, K. (2010). The ArcBA two-component system of *Escherichia coli* is regulated by the redox state of both the ubiquinone and the menaquinone pool. *Journal of Bacteriology* **192**, 746–754.

Bettenbrock, K., Bai, H., Ederer, M., Green, J., Hellingwerf, K. J., Holcombe, M., Kunz, S., Rolfe, M. D., Sanguinetti, G., Sawodny, O., Sharma, P., Steinsiek, S. & Poole, R. K. (2014). Towards a systems level understanding of the oxygen response of *Escherichia coli*. *Advances in Microbial Physiology* **64**, 65–114.

Elsen, S., Swem, L. R., Swem, D. L. & Bauer, C. E. (2004). RegB/RegA, a highly conserved redox-responding global two-component regulatory system. *Microbiology and Molecular Biology Reviews* **68**, 263–279.

Lemmer, K. C., Dohnalkova, A. C., Noguera, D. R. & Donohue, T. J. (2015). Oxygen-dependent regulation of bacterial lipid production. *Journal of Bacteriology* **197**, 1649–1658.

FNR system

Bauer, C. E., Elsen, S. & Bird, T. H. (1999). Mechanisms for redox control of gene expression. *Annual Review of Microbiology* **53**, 495–523.

Crack, J. C., Green, J., Thomson, A. J. & Le Brun, N. E. (2012). Iron-sulfur cluster sensor-regulators. *Current Opinion in Chemical Biology* **16**, 35–44.

Durand, S. & Storz, G. (2010). Reprogramming of anaerobic metabolism by the FnrS small RNA. *Molecular Microbiology* **75**, 1215–1231.

Green, J. & Paget, M. S. (2004). Bacterial redox sensors. *Nature Reviews Microbiology* **2**, 954–966.

Härtig, E. & Jahn, D. (2012). Regulation of the anaerobic metabolism in *Bacillus subtilis*. *Advances in Microbial Physiology* **61**, 195–216.

Kiley, P. J. & Beinert, H. (2003). The role of Fe–S proteins in sensing and regulation in bacteria. *Current Opinion in Microbiology* **6**, 181–185.

Taylor, B. L., Zhulin, I. B. & Johnson, M. S. (1999). Aerotaxis and other energy-sensing behavior in bacteria. *Annual Review of Microbiology* **53**, 103–128.

Unden, G. & Schirawski, J. (1997). The oxygen-responsive transcriptional regulator FNR of *Escherichia coli*: the search for signals and reactions. *Molecular Microbiology* **25**, 205–210.

Oxidative stress

Dubbs, J. M. & Mongkolsuk, S. (2012). Peroxide-sensing transcriptional regulators in bacteria. *Journal of Bacteriology* **194**, 5495–5503.

Glaeser, J., Nuss, A. M., Berghoff, B. A. & Klug, G. (2011). Singlet oxygen stress in microorganisms. *Advances in Microbial Physiology* **58**, 141–173.

Gray, M. J. & Jakob, U. (2015). Oxidative stress protection by polyphosphate – new roles for an old player. *Current Opinion in Microbiology* **24**, 1–6.

Henningham, A., Döhrmann, S., Nizet, V. & Cole, J. N. (2015). Mechanisms of group A *Streptococcus* resistance to reactive oxygen spp. *FEMS Microbiology Reviews* **39**, 488–508.

Imlay, J. A. (2015). Transcription factors that defend bacteria against reactive oxygen spp. *Annual Review of Microbiology* **69**, 93–108.

Mols, M. & Abee, T. (2011). Primary and secondary oxidative stress in *Bacillus*. *Environmental Microbiology* **13**, 1387–1394.

Thamsen, M. & Jakob, U. (2011). The redoxome: proteomic analysis of cellular redox networks. *Current Opinion in Chemical Biology* **15**, 113–119.

Yesilkaya, H., Andisi, V. F., Andrew, P. W. & Bijlsma, J. J. E. (2013). *Streptococcus pneumoniae* and reactive oxygen spp.: an unusual approach to living with radicals. *Trends in Microbiology* **21**, 187–195.

Zhao, X. & Drlica, K. (2014). Reactive oxygen spp. and the bacterial response to lethal stress. *Current Opinion in Microbiology* **21**, 1–6.

Zuber, P. (2009). Management of oxidative stress in *Bacillus*. *Annual Review of Microbiology* **63**, 575–597.

Nitrosative stress responses

Bowman, L. A. H., McLean, S., Poole, R. K. & Fukuto, J. M. (2011). The diversity of microbial responses to nitric oxide and agents of nitrosative stress: close cousins but not identical twins. *Advances in Microbial Physiology* **59**, 135–219.

Husain, M., Jones-Carson, J., Song, M., McCollister, B. D., Bourret, T. J. & Vázquez-Torres, A. (2010). Redox sensor SsrB Cys203 enhances *Salmonella* fitness against nitric oxide generated in the host immune response to oral infection. *Proceedings of the National Academy of Sciences of the USA* **107**, 14396–14401.

Heat shock

Helmann, J. D., Wu, M. F. W., Kobel, P. A., Gamo, F. J., Wilson, M., Morshedi, M. M., Navre, M. & Paddon, C. (2001). Global transcriptional response of *Bacillus subtilis* to heat shock. *Journal of Bacteriology* **183**, 7318–7328.

Hirtreiter, A. M., Calloni, G., Forner, F., Scheibe, B., Puype, M., Vandekerckhove, J., Mann, M., Hartl, F. U. & Hayer-Hartl, M. (2009). Differential substrate specificity of group I and group II chaperonins in the archaeon *Methanosarcina mazei*. *Molecular Microbiology* **74**, 1152–1168.

Kortmann, J. & Narberhaus, F. (2012). Bacterial RNA thermometers: molecular zippers and switches. *Nature Reviews Microbiology* **10**, 255–265.

Laksanalamai, P., Maeder, D. L. & Robb, F. T. (2001). Regulation and mechanism of action of the small heat shock protein from the hyperthermophilic archaeon *Pyrococcus furiosus*. *Journal of Bacteriology* **183**, 5198–5202.

Lund, P. A. (2009). Multiple chaperonins in bacteria – why so many? *FEMS Microbiology Reviews* **33**, 785–800.

Meyer, A. S. & Baker, T. A. (2011). Proteolysis in the *Escherichia coli* heat shock response: a player at many levels. *Current Opinion in Microbiology* **14**, 194–199.

Schumann, W. (2016). Regulation of bacterial heat shock stimulons. *Cell Stress and Chaperones* **21**, 959–968.

Cold shock

Barria, C., Malecki, M. & Arraiano, C. M. (2013). Bacterial adaptation to cold. *Microbiology* **159**, 2437–2443.

Cavicchioli, R., Thomas, T. & Curmi, P. M. G. (2000). Cold stress response in Archaea. *Extremophiles* **4**, 321–331.

Graumann, P. & Marahiel, M. A. (1996). Some like it cold: response of micro-organisms to cold shock. *Archives of Microbiology* **166**, 293–300.

Keto-Timonen, R., Hietala, N., Palonen, E., Hakakorpi, A., Lindström, M. & Korkeala, H. (2016). Cold shock proteins: a minireview with special emphasis on Csp-family of enteropathogenic *Yersinia*. *Frontiers in Microbiology* **7**, 1151.

Sakamoto, T. & Murata, N. (2002). Regulation of the desaturation of fatty acids and its role in tolerance to cold and salt stress. *Current Opinion in Microbiology* **5**, 206–210.

Shivaji, S. & Prakash, J. (2010). How do bacteria sense and respond to low temperature? *Archives of Microbiology* **192**, 85–95.

Singh, A. K., Sad, K., Singh, S. K. & Shivaji, S. (2014). Regulation of gene expression at low temperature: role of cold-inducible promoters. *Microbiology* **160**, 1291–1296.

Quorum sensing

Antunes, L. C. M., Ferreira, R. B. R., Buckner, M. M. C. & Finlay, B. B. (2010). Quorum sensing in bacterial virulence. *Microbiology* **156**, 2271–2282.

Asfahl, K. L. & Schuster, M. (2017). Social interactions in bacterial cell–cell signaling. *FEMS Microbiology Reviews* **41**, 92–107.

Banerjee, G. & Ray, A. K. (2016). The talking language in some major Gram-negative bacteria. *Archives of Microbiology* **198**, 489–499.

Dandekar, A. A., Chugani, S. & Greenberg, E. P. (2012). Bacterial quorum sensing and metabolic incentives to cooperate. *Science* **338**, 264–266.

Decho, A. W., Norman, R. S. & Visscher, P. T. (2010). Quorum sensing in natural environments: emerging views from microbial mats. *Trends in Microbiology* **18**, 73–80.

Frederix, M. & Downie, A. J. (2011). Quorum sensing: regulating the regulators. *Advances in Microbial Physiology* **58**, 23–80.

Hense, B. A. & Schuster, M. (2015). Core principles of bacterial autoinducer systems. *Microbiology and Molecular Biology Reviews* **79**, 153–169.

Jacob, E. B., Becker, I., Shapira, Y. & Levine, H. (2004). Bacterial linguistic communication and social intelligence. *Trends in Microbiology* **12**, 366–372.

Kalia, V. C. & Purohit, H. J. (2011). Quenching the quorum sensing system: potential antibacterial drug targets. *Critical Reviews in Microbiology* **37**, 121–140.

Monnet, V., Juillard, V. & Gardan, R. (2016). Peptide conversations in Gram-positive bacteria. *Critical Reviews in Microbiology* **42**, 339–351.

Parsek, M. R. & Greenberg, E. P. (2005). Sociomicrobiology: the connections between quorum sensing and biofilms. *Trends in Microbiology* **13**, 27–33.

Rasmussen, T. B. & Givskov, M. (2006). Quorum sensing inhibitors: a bargain of effects. *Microbiology* **152**, 895–904.

Ryan, R. P. & Dow, J. M. (2011). Communication with a growing family: diffusible signal factor (DSF) signaling in bacteria. *Trends in Microbiology* **19**, 145–152.

Schuster, M., Sexton, D. J., Diggle, S. P. & Greenberg, E. P. (2013). Acyl-homoserine lactone quorum sensing: from evolution to application. *Annual Review of Microbiology* **67**, 43–63.

Srivastava, D. & Waters, C. M. (2012). A tangled web: regulatory connections between quorum sensing and cyclic di-GMP. *Journal of Bacteriology* **194**, 4485–4493.

Zhang, G., Zhang, F., Ding, G., Li, J., Guo, X., Zhu, J., Zhou, L., Cai, S., Liu, X., Luo, Y., Zhang, G., Shi, W. & Dong, X. (2012). Acyl homoserine lactone-based quorum sensing in a methanogenic archaeon. *ISME Journal* **6**, 1336–1344.

Osmotic stress

Baños, R. C., Martínez, J., Polo, C., Madrid, C., Prenafeta, A. & Juárez, A. (2011). The *yfeR* gene of *Salmonella enterica* serovar Typhimurium encodes an osmoregulated LysR-type transcriptional regulator. *FEMS Microbiology Letters* **315**, 63–71.

Sleator, R. D. & Hill, C. (2002). Bacterial osmoadaptation: the role of osmolytes in bacterial stress and virulence. *FEMS Microbiology Reviews* **26**, 49–71.

Tipton, K. A. & Rather, P. N. (2017). An ompR-envZ two-component system ortholog regulates phase variation, osmotic tolerance, motility, and virulence in *Acinetobacter baumannii* strain AB5075. *Journal of Bacteriology* **199**, e00705-16.

Wood, J. M. (2011). Bacterial osmoregulation: a paradigm for the study of cellular homeostasis. *Annual Review of Microbiology* **65**, 215–238.

Chemotaxis

Aizawa, S., Harwood, C. S. & Kadner, R. J. (2000). Signaling components in bacterial locomotion and sensory reception. *Journal of Bacteriology* **182**, 1459–1471.

Alexandre, G. (2010). Coupling metabolism and chemotaxis-dependent behaviours by energy taxis receptors. *Microbiology* **156**, 2283–2293.

Brown, M. T., Delalez, N. J. & Armitage, J. P. (2011). Protein dynamics and mechanisms controlling the rotational behaviour of the bacterial flagellar motor. *Current Opinion in Microbiology* **14**, 734–740.

De Lay, N. & Gottesman, S. (2012). A complex network of small non-coding RNAs regulate motility in *Escherichia coli*. *Molecular Microbiology* **86**, 524–538.

Hazelbauer, G. L. (2012). Bacterial chemotaxis: the early years of molecular studies. *Annual Review of Microbiology* **66**, 285–303.

Krell, T., Lacal, J., Muñoz-Martínez, F., Reyes-Darias, J. A., Cadirci, B. H., García-Fontana, C. & Ramos, J. L. (2011). Diversity at its best: bacterial taxis. *Environmental Microbiology* **13**, 1115–1124.

Porter, S. L., Wadhams, G. H. & Armitage, J. P. (2011). Signal processing in complex chemotaxis pathways. *Nature Reviews Microbiology* **9**, 153–165.

Szurmant, H. & Ordal, G. W. (2004). Diversity in chemotaxis mechanisms among the bacteria and archaea. *Microbiology and Molecular Biology Reviews* **68**, 301–319.

Yuan, J., Branch, R. W., Hosu, B. G. & Berg, H. C. (2012). Adaptation at the output of the chemotaxis signalling pathway. *Nature* **484**, 233–236.

Adaptive mutation

Aertsen, A. & Michiels, C. W. (2005). Diversify or die: generation of diversity in response to stress. *Critical Reviews in Microbiology* **31**, 69–78.

Andersson, D. I. & Hughes, D. (2009). Gene amplification and adaptive evolution in bacteria. *Annual Review of Genetics* **43**, 167–195.

Dubnau, D. & Losick, R. (2006). Bistability in bacteria. *Molecular Microbiology* **61**, 564–572.

Foster, P. L. (1993). Adaptive mutation: the uses of adversity. *Annual Review of Microbiology* **47**, 467–504.

Wright, B. E. (2004). Stress-directed adaptive mutations and evolution. *Molecular Microbiology* **52**, 643–650.

Enzyme activity modulation

Dworkin, J. (2015). Ser/Thr phosphorylation as a regulatory mechanism in bacteria. *Current Opinion in Microbiology* **24**, 47–52.

Eoh, H. & Rhee, K. Y. (2014). Allostery and compartmentalization: old but not forgotten. *Current Opinion in Microbiology* **18**, 23–29.

Gur, E., Biran, D. & Ron, E. Z. (2011). Regulated proteolysis in Gram-negative bacteria - how and when? *Nature Reviews Microbiology* **9**, 839–848.

Hu, L. I., Lima, B. P. & Wolfe, A. J. (2010). Bacterial protein acetylation: the dawning of a new age. *Molecular Microbiology* **77**, 15–21.

Itzen, A., Blankenfeldt, W. & Goody, R. S. (2011). Adenylylation: renaissance of a forgotten post-translational modification. *Trends in Biochemical Sciences* **36**, 221–228.

Loi, V. V., Rossius, M. & Antelmann, H. (2015). Redox regulation by reversible protein S-thiolation in bacteria. *Frontiers in Microbiology* **6**, 187.

Mijakovic, I., Grangeasse, C. & Turgay, K. (2016). Exploring the diversity of protein modifications: special bacterial phosphorylation systems. *FEMS Microbiology Reviews* **40**, 398–417.

Pisithkul, T., Patel, N. M. & Amador-Noguez, D. (2015). Post-translational modifications as key regulators of bacterial metabolic fluxes. *Current Opinion in Microbiology* **24**, 29–37.

Soufi, B., Soares, N. C., Ravikumar, V. & Macek, B. (2012). Proteomics reveals evidence of cross-talk between protein modifications in bacteria: focus on acetylation and phosphorylation. *Current Opinion in Microbiology* **15**, 357–363.

Metabolic regulation network

Basan, M., Hui, S., Okano, H., Zhang, Z., Shen, Y., Williamson, J. R. & Hwa, T. (2015). Overflow metabolism in *Escherichia coli* results from efficient proteome allocation. *Nature* **528**, 99–104.

Edwards, J. S., Covert, M. & Palsson, B. (2002). Metabolic modelling of microbes: the flux-balance approach. *Environmental Microbiology* **4**, 133–140.

El-Mansi, M., Cozzone, A. J., Shiloach, J. & Eikmanns, B. J. (2006). Control of carbon flux through enzymes of central and intermediary metabolism during growth of *Escherichia coli* on acetate. *Current Opinion in Microbiology* **9**, 173–179.

Gerosa, L. & Sauer, U. (2011). Regulation and control of metabolic fluxes in microbes. *Current Opinion in Biotechnology* **22**, 566–575.

Hengge, R. & Gourse, R. L. (2004). Cell regulation: tying together the cellular regulatory network. *Current Opinion in Microbiology* **7**, 99–101.

Jung, K., Fried, L., Behr, S. & Heermann, R. (2012). Histidine kinases and response regulators in networks. *Current Opinion in Microbiology* **15**, 118–124.

Leyn, S. A., D. Kazanov, M., Sernova, N. V., Ermakova, E. O., Novichkov, P. S. & Rodionov, D. A. (2013). Genomic reconstruction of the transcriptional regulatory network in *Bacillus subtilis*. *Journal of Bacteriology* **195**, 2463–2473.

Noirot, P. & Noirot-Gros, M. F. (2004). Protein interaction networks in bacteria. *Current Opinion in Microbiology* **7**, 505–512.

Potrykus, K., Murphy, H., Philippe, N. & Cashel, M. (2011). ppGpp is the major source of growth rate control in *E. coli*. *Environmental Microbiology* **13**, 563–575.

van Heeswijk, W. C., Westerhoff, H. V. & Boogerd, F. C. (2013). Nitrogen assimilation in *Escherichia coli*: putting molecular data into a systems perspective. *Microbiology and Molecular Biology Reviews* **77**, 628–695.

Secondary metabolites and fermentation

Liu, G., Chater, K. F., Chandra, G., Niu G. & Tan, H. (2013). Molecular regulation of antibiotic biosynthesis in *Streptomyces*. *Microbiology and Molecular Biology Reviews* **77**, 112–143.

Urem, M., Świątek-Połatyńska, M. A., Rigali, S. & van Wezel, G. P. (2016). Intertwining nutrient-sensory networks and the control of antibiotic production in *Streptomyces*. *Molecular Microbiology* **102**, 183–195.

Chapter 13

Energy, environment and microbial survival

As mentioned repeatedly in this book, the goal of life is preservation of the species through reproduction, but this requires energy. Although there are a few exceptional copiotrophic environments such as foodstuffs and animal guts, most ecosystems where microorganisms are found are oligotrophic. Those organisms that can utilize nutrients efficiently have a better chance of survival in such ecosystems. Further, many microbes synthesize reserve materials when available nutrients are in excess and utilize these under starvation conditions. Various resting cells are produced under conditions where growth is difficult. In this chapter, the main bacterial survival mechanisms are discussed in terms of reserve materials and resting cell types. Also discussed are toxin and antitoxin systems and programmed cell death (apoptosis) as means of population survival, and the bacterial immune system as a protection mechanism against foreign DNA. Another survival mechanism is competence that enables a bacterial cell to repair a damaged genome using DNA available in the environment.

13.1 Survival and energy

As discussed earlier, living microorganisms maintain a certain level of adenylate energy charge (EC) and proton motive force even under starvation conditions (Section 5.6.2). These forms of biological energy are needed for the basic metabolic processes necessary to survive, such as transport and the turnover of macromolecules. Maintenance energy is the term used for this energy. The turnover of proteins and RNA consumes the largest proportion of maintenance energy. Under starvation conditions, cells utilize cellular components, including reserve and non-essential materials for survival. This is referred to as endogenous metabolism. Almost all prokaryotes accumulate at least one type of reserve material under energy-rich conditions. During a period of starvation, the reserve material(s) are consumed through endogenous metabolism before the organism oxidizes other cellular constituents, such as proteins and RNA that are not needed under starvation conditions (Figure 13.1). When a population starves, some individuals die, thereby providing an energy source for other members of the population in a mechanism known as programmed cell death (Section 13.4.1), and in a process mediated by toxin–antitoxin systems (Section 13.4.2).

Most cells respond to energy limitation by reducing cell size and increasing the cell surface to volume ratio within days to weeks, either by shrinking of individual cells or by cell division without growth (fragmentation) to reduce the maintenance energy required. In environmental samples, large numbers of small cells that pass through 0.2 μm filters are common and these have been termed ultramicrobacteria or nanobacteria (Section 13.3.5). These changes are modulated by the stringent response (Section 12.2.1). Many of these cells, including *Sphingopyxis alaskensis*, appear to be starved forms of microbes that grow to a significantly

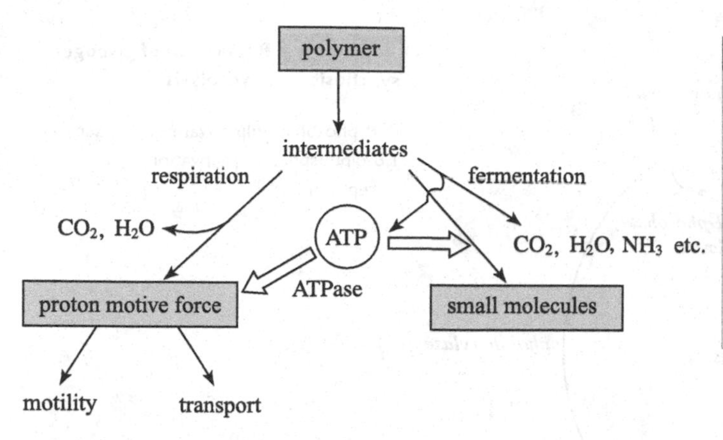

Figure 13.1 Energy sources for survival under starvation conditions.

(Dawes, E. A. 1986, *Microbial Energetics*, Figure 12.1. Blackie & Son, Glasgow)

When nutrients are unavailable in the environment, microbes utilize cellular constituents, including reserve materials and polymers that are not essential for survival. This is referred to as endogenous metabolism.

larger size under energy-rich conditions, but some organisms such as *Candidatus* Pelagibacter ubique and *Candidatus* Nitrosopumilus maritimus are of a small size under energy-rich conditions.

13.2 Reserve materials in bacteria

RNA and proteins used for endogenous metabolism are not referred to as reserve materials since they have specific functions other than as substrates for endogenous metabolism. Reserve materials can be defined as polymers synthesized when an energy source is supplied in excess, and used as a substrate for endogenous metabolism, without any other cellular functions. Some polymers known as reserve materials, such as polyphosphate, have other functions. Almost all prokaryotes accumulate at least one type of reserve material. It is likely that the ability to accumulate reserve materials is advantageous for survival in natural habitats.

Reserve materials in bacteria can be grouped into four categories according to their chemical nature. These are carbohydrates such as glycogen, lipids such as poly-β-hydroxybutyrate (PHB), polypeptides and polyphosphate.

13.2.1 Carbohydrate reserve materials: glycogen

Glycogen is the most common carbohydrate reserve material in prokaryotes, as in mammals. Glycogen is a polysaccharide consisting of glucose with α-1,4-linkages as well as α-1,6-linkages. Many prokaryotes synthesize this polysaccharide when the energy source is in excess and when one or more essential elements are limiting, such as nitrogen.

Like all polysaccharides, glycogen is synthesized from an activated monomer. Glycogen synthase transfers glucose from ADP-glucose to the existing glycogen molecule, forming an α-1,4-linkage (Figure 13.2, Section 6.9.1). Glycogen-synthesizing genes are regulated by the CsrA/CsrB system (Section 12.1.9.3). ADP-glucose pyrophosphorylase (glucose-1-phosphate adenylyltransferase, GlgB) is activated by fructose-6-phosphate, fructose-1,6-diphosphate, phosphoenolpyruvate and pyruvate, and repressed by AMP for regulation of glycogen synthesis.

Mycobacteria including *Mycobacterium tuberculosis* synthesize glycogen from trehalose by a different pathway. Trehalose synthase (TreS) converts this disaccharide to maltose that is phosphorylated to maltose-1-phosphate, consuming ATP. The maltose residue of maltose-1-phosphate is transferred to the existing glycogen molecule by α-1,4-glucan:maltose-1-phosphate maltosyltransferase (GlgE). Genes for this pathway are found in diverse bacteria.

Glycogen is utilized by the action of two enzymes, a debranching enzyme and glycogen phosphorylase. *Escherichia coli* has strong activities of these enzymes and uses glycogen at a high rate at the beginning of starvation; consequently this bacterium does not survive very long under starvation conditions. On the other hand, *Arthrobacter globiformis* stays viable for long time periods under starvation conditions, since this bacterium utilizes the reserve polysaccharide slowly because of low activity of the debranching enzyme.

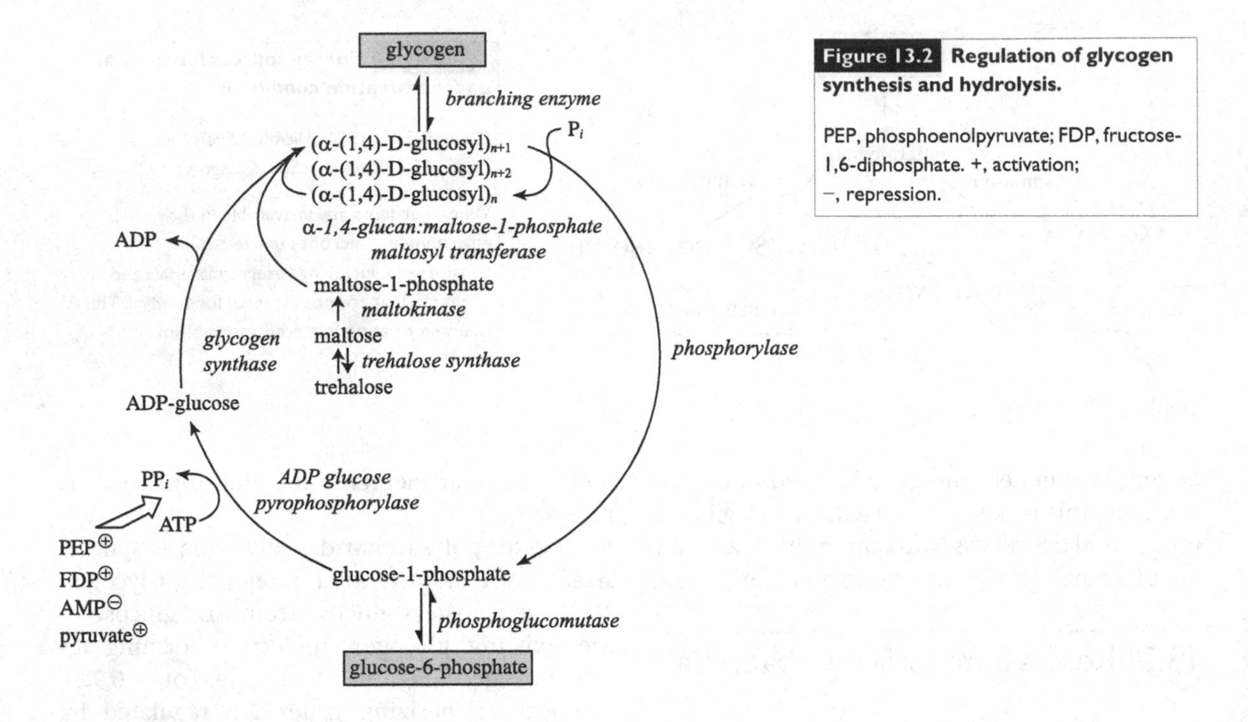

Figure 13.2 Regulation of glycogen synthesis and hydrolysis.

PEP, phosphoenolpyruvate; FDP, fructose-1,6-diphosphate. +, activation; −, repression.

13.2.2 Lipid reserve materials

Poly-*β*-hydroxyalkanoates (PHAs), triacylglycerides (TAG), wax esters and hydrocarbons are synthesized as reserve materials in bacteria. These lipophilic compounds are accumulated as inclusion bodies in the cytoplasm. Many proteins are associated with the inclusion bodies. These are mostly enzymes involved in metabolism of the reserve material. PHAs are the most common reserve material in prokaryotes but are not found in eukaryotes. On the other hand, only a few prokaryotes have the property of accumulating triacylglycerides as a reserve material, a property that is widespread in eukaryotes. Wax esters are accumulated mostly in bacteria.

13.2.2.1 Poly-*β*-hydroxyalkanoate (PHA)

PHA is a reserve material accumulated in bacteria and archaea with the general structure:

$$HO-\underset{\displaystyle R}{\underset{|}{CH}}-CH_2-\underset{\displaystyle O}{\underset{\|}{C}}-\left(-O-\underset{\displaystyle R}{\underset{|}{CH}}-CH_2-\underset{\displaystyle O}{\underset{\|}{C}}-\right)_n-O-\underset{\displaystyle R}{\underset{|}{CH}}-CH_2-COOH$$

The composition of the monomer differs depending on the strain, but 3-hydroxybutyrate (3-HB, R = CH_3) is the most abundant form. For this reason, PHA is usually called poly-*β*-hydroxybutyrate (PHB). Other monomers include 3-hydroxyalkanoate with carbon numbers of 5, 6, 7 and 8. Generally, the PHA in the inclusion body (granule), known as the carbonosome, is a macromolecule with a molecular weight higher than 10^6 daltons. PHA is an ideal reserve material since this water-insoluble polymer does not increase the intracellular osmotic pressure and the energy content is higher than that of carbohydrates. PHA granules are composed of 97.5 per cent PHA, 2 per cent proteins, and there is likely to be some amount of lipid. At least four types of PHA-related proteins are found in bacteria. These are enzymes for PHA synthesis and depolymerization, a regulator (PhaR), and a structural protein called phasin (PhaP). The amphiphilic phasin shields the hydrophobic polymer from the cytoplasm to prevent PHA granules coalescing and ensures granule distribution

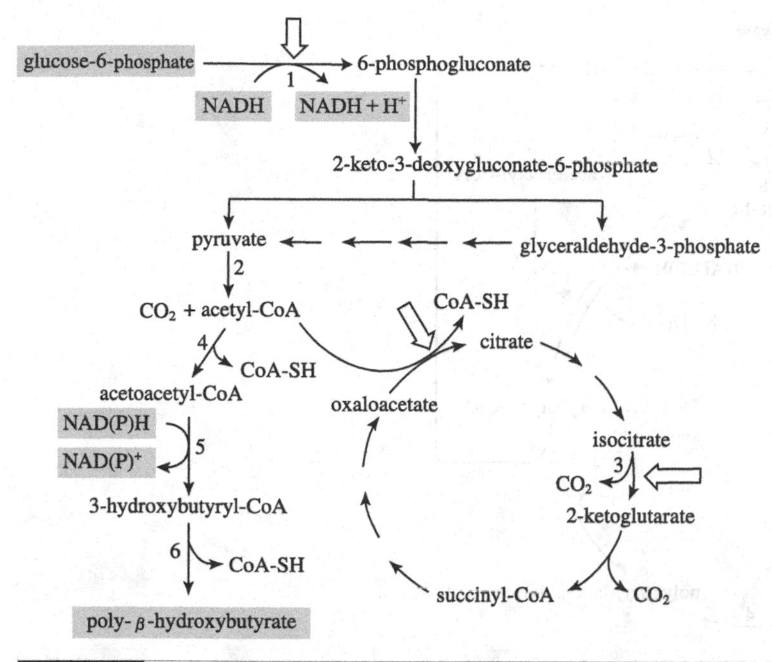

Figure 13.3 Central metabolism and PHB synthesis in *Azotobacter beijerinckii*.

(Dawes, E. A. 1986, *Microbial Energetics*, Figure 11.3. Blackie & Son, Glasgow)

Under oxygen-limited conditions, reduced NADH is accumulated. Under these conditions, many NAD^+-reducing enzymes, including glucose-6-phosphate dehydrogenase (1), pyruvate dehydrogenase (2) and isocitrate dehydrogenase (3), are inhibited, and acetyl-CoA cannot be oxidized through the TCA cycle. Instead, acetyl-CoA is condensed to acetoacetyl-CoA before being reduced to 3-hydroxybutyrate for PHB synthesis. NADH is reoxidized in this process, allowing glycolysis to be possible. PHB is not only a reserve material but also acts as an electron sink, regenerating NAD^+ for metabolism. PHB is synthesized when the reactions marked by block arrows are inhibited by NADH.

1, glucose-6-phosphate dehydrogenase; 2, pyruvate dehydrogenase; 3, isocitrate dehydrogenase; 4, acetyl-CoA acetyltransferase (thiolase); 5, 3-hydroxybutyryl-CoA dehydrogenase; 6, 3-hydroxybutyryl-CoA polymerase.

during cell division. Phasin also controls PHA depolymerization.

In addition to PHB, with over 10^3 3-HB residues, which is found in many prokaryotes, oligo-PHB with 100–200 3-HB residues and conjugated PHB with less than 30 3-HB residues covalently linked to protein are present in all prokaryotes and eukaryotic microbes. Oligo-PHB often forms complexes with polyphosphate and calcium within the cell membrane. This complex mediates the uptake of DNA during transformation (Section 13.6). Several conjugated PHB forms are found in *Escherichia coli* including that in the outer membrane protein OmpA. PHB in OmpA is important for correct integration and orientation of the outer membrane. Under nitrogen- and oxygen-limited conditions, some bacteria, including *Ralstonia eutropha* (*Alcaligenes eutrophus*) and *Azotobacter beijerinckii*, accumulate up to 80 per cent of their dry cell weight as PHA. An industrial process has been developed to produce PHA as a biodegradable 'plastic' using a strain of *Ralstonia eutropha*.

PHB is synthesized through the polymerization of 3-hydroxybutyrate, which is a condensation product of acetyl-CoA, in a series of reactions similar to the clostridial butyrate fermentation (Figure 13.3). In *Ralstonia eutropha*, an operon (*phaCAB*) encodes acetoacetyl-CoA acetyltransferase (β-ketothiolase, PhaA), 3-hydroxybutyryl-CoA dehydrogenase (acetoacetyl-CoA reductase, PhaB) and 3-hydroxybutyryl-CoA polymerase (polyhydroxyalkanoate synthase, PhaC). This bacterium has multiple genes for these enzymes and each gene is expressed under different growth conditions. The regulator protein, PhaR, represses

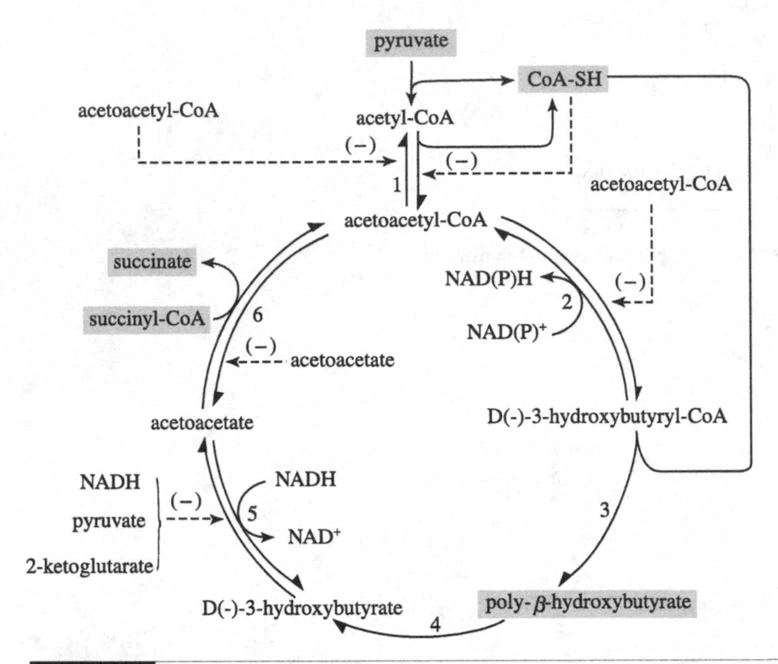

Figure 13.4 **PHB degradation and its regulation in *Azotobacter beijerinckii*.**

(Dawes, E. A. 1986, *Microbial Energetics*, Figure 11.4. Blackie & Son, Glasgow)

PHB degradation is regulated not only by intermediates of central metabolism such as pyruvate and 2-ketoglutarate, but also by NADH, since PHB functions as a reserve material as well as an electron sink. PHB is depolymerized to 3-hydroxybutyrate before being reduced to acetoacetate, which is activated by coenzyme A transferase to acetoacetyl-CoA. Since acetoacetyl-CoA is an intermediate in degradation as well as in the synthesis of this reserve material, reactions involving this intermediate are tightly controlled to avoid a futile cycle which would waste a high energy bond in the form of succinyl-CoA.

1, acetoacetyl-CoA acetyltransferase (β-ketothiolase, PhaA); 2, 3-hydroxybutyryl-CoA dehydrogenase (acetoacetyl-CoA reductase, PhaB); 3, 3-hydroxybutyryl-CoA polymerase (polyhydroxyalkanoate synthase, PhaC); 4, PHB depolymerase (PhaZ); 5, 3-hydroxybutyrate dehydrogenase; 6, coenzyme A transferase. –, inhibition.

transcription of the *phaCAB* operon under normal conditions, and binds to PHA granules under nutrient imbalance conditions, derepressing *phaCAB* transcription. *phaP* and *phaR* form an operon in bacteria and halophilic archaea. PHA production is also controlled by a sRNA MmgR (makes more granules regulator) post-transcriptionally in *Sinorhizobium meliloti*. MmgR inhibits translation of proteins for PHA production under growth conditions, but it is not known how activity of the sRNA is controlled. The reaction catalysed by PhaA is the limiting step.

Azotobacter beijerinckii metabolizes carbohydrates though the ED pathway, and the resulting NADH is reoxidized through the electron transport system, generating a proton motive force when the oxygen supply is not limited. When oxygen is limited, NADH is accumulated to inhibit enzymes reducing NAD^+, including glucose-6-phosphate dehydrogenase, citrate synthase and isocitrate dehydrogenase. When oxygen is limiting, acetyl-CoA is directed toward PHB synthesis, reoxidizing NADH. For this reason, PHB is regarded as a reserve material and as an electron sink allowing the bacterium to continue glycolytic metabolism.

PHB depolymerase removes 3-hydroxybutyrate from PHB, and the product is oxidized to acetoacetate. Coenzyme A transferase activates acetoacetate to acetoacetyl-CoA, consuming succinyl-CoA. Since acetoacetyl-CoA is involved in both synthesis and degradation, the enzymes related to this intermediate need to be strongly regulated to avoid a futile cycle that would waste a high energy bond in the form of succinyl-CoA (Figure 13.4).

PHA has biological functions other than as an energy reserve material. PHA serves as the source of reducing equivalents needed to withstand oxidative stress at low temperature in a psychrotolerant *Pseudomonas* sp.

13.2.2.2 Triacylglyceride (TAG)

TAG can serve as a reserve material in eukaryotes as well as in prokaryotes. Among prokaryotes, TAG is found mainly in the actinomycetes, including species of *Mycobacterium*, *Nocardia*, *Rhodococcus*, *Streptomyces*, *Micromonospora* and *Gordonia*. Among Gram-negative bacteria, species of *Acinetobacter* accumulate small amounts. A strain of *Rhodococcus opacus* accumulates TAG up to 87 per cent of the cell dry weight.

TAG forms lipid droplets (LDs) associated with various proteins. These proteins are related to TAG metabolism. TAG can be used as the energy souce for resuscitation of persister cells of pathogenic mycobacteria (Section 13.3.4). LD-associated proteins are therefore drug targets for treatment of mycobacterial infection.

TAG is synthesized from glycerol-3-phosphate and acyl-ACP. Phosphatidic acid is produced as in phospholipid synthesis (Section 6.6), before phosphatidate phosphatase removes the phosphate group. Finally the third acyl group is added to diacylglycerol from acyl-ACP by the action of diacylglycerol acyltransferase (DGAT). DGAT is believed to be associated with the cytoplasmic membrane. TAG is accumulated with an excess of carbon source but when nutrients other than the carbon source are limited, as with PHA.

13.2.2.3 Wax esters and hydrocarbons

Some bacteria produce wax esters and hydrocarbons intracellularly or extracellularly, but it is not clear if they are reserve materials.

Wax ester (WE) is synthesized by various bacteria, including species of *Acinetobacter*, *Marinobacter*, *Moraxella*, *Micrococcus*, *Corynebacterium* and *Nocardia*. Under nitrogen-limited conditions, a strain of *Acinetobacter calcoaceticus* accumulates WE up to 25 per cent of the cell dry weight. WE is produced through condensation of an alcohol and acyl-CoA catalysed by wax synthase in *Marinobacter aquaeolei*:

$$\underset{\text{alcohol}}{R-CH_2-CH_2-OH} + \underset{\text{acyl-CoA}}{R-CH_2-\overset{O}{\overset{\|}{C}}-S-CoA} \xrightarrow[\text{wax synthase}]{\text{CoA-SH}} \underset{\text{wax ester}}{R-CH_2-\overset{O}{\overset{\|}{C}}-O-CH_2-CH_2-R}$$

Hydrocarbons are accumulated by various microorganisms, including cyanobacteria such as *Synechococcus elongatus*, bacteria such as *Micrococcus luteus*, *Kocuria rhizophila*, *Desulfovibrio* spp. and *Clostridium* spp., and certain fungi. These microbial hydrocarbons are either isoprenoids, alkanes or alkenes. Alkanes are produced from acyl-ACP (Section 6.6.1) catalysed by acyl-ACP reductase and aldehyde deformylating oxygenase:

$$\text{acyl-ACP} \xrightarrow[\text{acyl-ACP reductase}]{\text{NADH} \quad NAD^{+}\text{+ACP}} \text{aldehyde} \xrightarrow[\text{aldehyde deformylating oxygenase}]{O_2 + \text{NADH} \quad \text{HCOOH} + NAD^{+} + H_2O} \text{alkane}$$

Acyl-CoAs are converted to alkenes through head-to-head condensation:

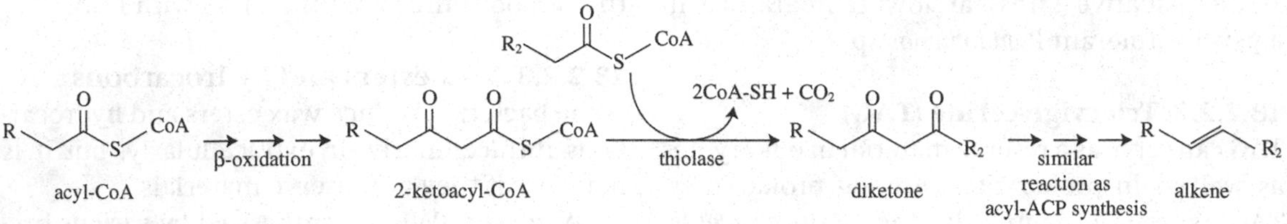

R–CH₂–CO–S-CoA (acyl-CoA) → → → (β-oxidation) R–CH₂–CO–CH₂–CO–S-CoA (2-ketoacyl-CoA) + R₂–CH₂–CO–S-CoA → (thiolase; 2CoA-SH + CO_2) R–CH₂–CO–CH₂–CO–R₂ (diketone) → → → (similar reaction as acyl-ACP synthesis) R–CH=CH–R₂ (alkene)

Since formate or CO_2 is generated, these reactions produce odd-numbered hydrocarbons.

13.2.3 Polypeptides as reserve materials

Cyanobacteria utilize not only the usual reserve materials such as glycogen, poly-*β*-hydroxyalkanoate (PHA) and polyphosphate, but also peptides. Many cyanobacteria accumulate cyanophycin as a reserve material for carbon and nitrogen. Cyanophycin has a structure composed of polyaspartate, each monomer of which is linked with a molecule of arginine (Figure 13.5a). This peptide has a molecular weight of between 25 000 and 125 000 daltons, and is synthesized under phosphate- or sulfate-limited conditions with nitrogen and light in excess. When the synthesis of protein and nucleic acids is inhibited, cyanophycin production is activated. Its synthesis is not inhibited by tetracycline, showing that cyanophycin is synthesized through a non-ribosomal mechanism (Section 6.8.2).

When the nitrogen supply is limited, cyanophycin is degraded, producing ammonia and carbon dioxide. It is not known if energy is conserved during oxidation of the amino acids. In heterocystous cyanobacteria, the heterocysts have a higher enzyme activity for cyanophycin synthesis than the vegetative cells.

Phycocyanin is a pigment peptide in the antenna molecule. When the nitrogen supply is limited, this peptide is also degraded as a nitrogen source. Cyanobacteria have a bluish-green colour under normal conditions, but become yellowish-green under nitrogen-limited conditions because the blue-coloured phycocyanin is degraded. It should be noted that phycocyanin is not a 'true' reserve material according to the definition given earlier.

Gram-positive *Bacillus* spp. produce poly-γ-glutamic acid (γ-PGA) extracellularly through a non-ribosomal mechanism. γ-PGA contains D- and L-glutamate linked by peptide bonds between amino and carboxyl groups at the γ-position (Figure 13.5b). L-glutamate is converted to the D-form by glutamate racemase (RacE) before being transferred to the existing γ-PGA polymer. The γ-PGA synthase complex (PgsBCA) adds glutamate to the carboxyl end of the existing polymer, which is phophorylated. A γ-glutamyl-transpeptidase (PgsS), that possesses exohydrolase activity, releases glutamic acid.

γ-PGA is produced in the early stationary phase and is consumed as a glutamate source in the late stationary phase. In addition, the non-immunogenic γ-PGA protects its producers against

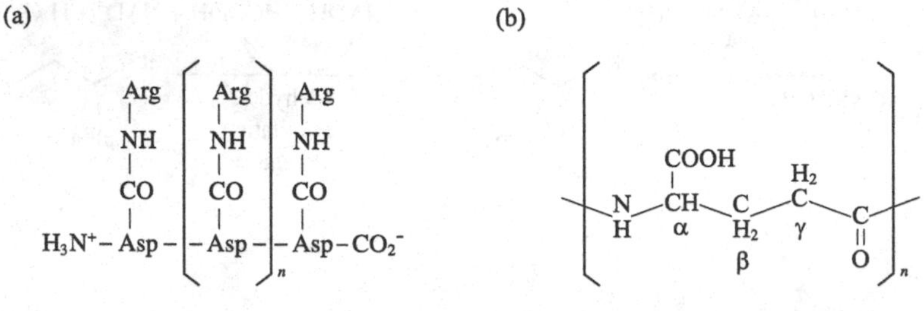

(a)

H_3N^+–Asp–[–Asp–]$_n$–Asp–CO_2^- (each Asp bearing CO–NH–Arg)

(b)

[–NH–CH(COOH)(α)–CH_2(β)–CH_2(γ)–C(=O)–]$_n$

Figure 13.5 Structure of the peptide reserve materials (a) cyanophycin in cyanobacteria and (b) poly-γ-glutamic acid in *Bacillus* spp.

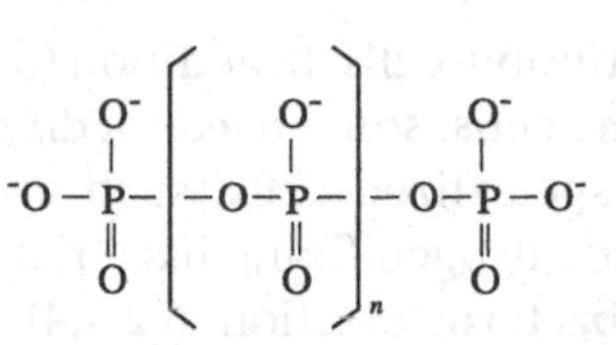

$$^{-}O-\overset{O^-}{\underset{O}{P}}-\left[-O-\overset{O^-}{\underset{O}{P}}-\right]_n-O-\overset{O^-}{\underset{O}{P}}-O^-$$

Figure 13.6 Structure of polyphosphate.

phage infection, antibodies and antimicrobial peptides, and helps pathogenic bacteria to escape phagocytosis, thereby contributing to virulence.

13.2.4 Polyphosphate

It has been known for a long time that many bacteria have cytoplasmic granules that are stainable with basic dyes such as toluidine blue. These granules are composed of polyphosphate and are known in prokaryotes and eukaryotes. The number of phosphate residues ranges between two and over a million (Figure 13.6). Polyphosphate is consumed when the phosphate supply is limited, and in some organisms this can substitute for ATP in energy-requiring reactions. Polyphosphate has functions other than just as a phosphate and energy reserve material. These include regulation of the concentration of cytoplasmic cations because of its strong anionic properties and stabilization of the cytoplasmic membrane. Polyphosphate also regulates gene expression and enzyme activity to increase resistance to environmental stress and antibacterial drugs, and plays a key role in the virulence of various pathogens including *Mycobacterium tuberculosis*. This inorganic polymer protects proteins from reactive oxygen species as a chaperone under conditions of oxidative stress (Section 12.2.5).

Polyphosphate is synthesized through the transfer of phosphate from ATP or from an intermediate of the glycolytic pathway, 1,3-diphosphoglycerate, to the existing polyphosphate. ATP-polyphosphate phosphotransferase (polyphosphate kinase) mediates the reaction with ATP. This reaction is reversible, and phosphate can be transferred, not only to ADP, but also to GDP, UDP and CDP:

$$(P)_n + ATP \xrightleftharpoons{\text{polyphosphate kinase}} (P)_{n+1} + ADP$$

The reaction with 1,3-diphosphoglycerate as the phosphate donor is mediated by 1,3-diphosphoglycerate-polyphosphate phosphotransferase:

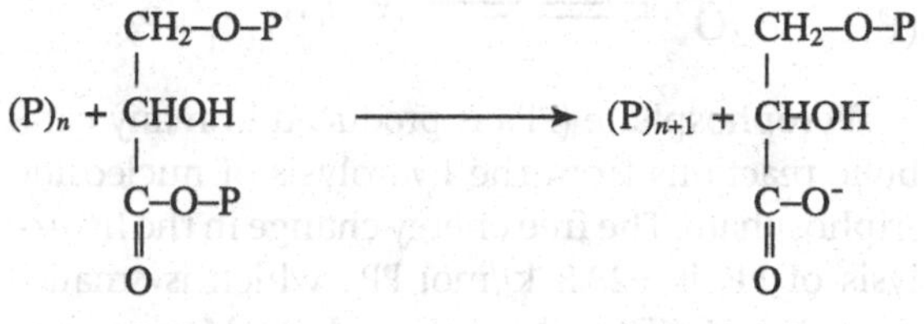

$$(P)_n + \underset{\text{1,3-diphosphoglycerate}}{CH_2\text{–O–P}\,|\,CHOH\,|\,C(=O)\text{–O–P}} \longrightarrow (P)_{n+1} + \underset{\text{3-phosphoglycerate}}{CH_2\text{–O–P}\,|\,CHOH\,|\,C(=O)\text{–O}^-}$$

Polyphosphate synthesis is under elaborate regulation. In *Escherichia coli*, polyphosphate is synthesized during the stringent response (Section 12.2.1) with an increase in (p)ppGpp concentration. Polyphosphate is not synthesized in a PhoB mutant. PhoB is the regulator protein of the two-component *pho* system (Section 12.2.3). Another protein member of the *pho* system, PhoU, negatively controls polyphosphate synthesis.

The free energy change in polyphosphate hydrolysis is −38 kJ/mol phosphate and this is bigger than that of ATP (−30.5 kJ/mol ATP) and smaller than that of 1,3-diphosphoglycerate (−49.4 kJ/mol 1,3-diphosphoglycerate). As mentioned previously, polyphosphate hydrolysis is coupled to the synthesis of nucleotide triphosphate, including ATP, catalysed by ATP polyphosphate phosphotransferase (polyphosphate kinase). In some bacteria, including a strain of *Acinetobacter*, polyphosphate:AMP phosphotransferase converts AMP to ADP coupled to polyphosphate hydrolysis:

$$(P)_n + AMP \xrightleftharpoons{\text{polyphosphate:AMP phosphotransferase}} (P)_{n-1} + ADP$$

Species of the genera *Propionibacterium* and *Micrococcus* possess polyphosphate glucokinase that phosphorylates glucose, consuming polyphosphate:

$$(P)_n + \text{glucose} \xrightleftharpoons{\text{polyphosphate glucokinase}} (P)_{n-1} + \text{glucose-6-phosphate}$$

Klebsiella pneumoniae (*Klebsiella aerogenes*) has a polyphosphatase that hydrolyses polyphosphate without conserving energy. The function of this enzyme in this bacterium is not known:

$$(P)_n + H_2O \xrightleftharpoons{\text{polyphosphatase}} (P)_{n-1} + P_i$$

Pyrophosphate (PP_i) is produced in many catabolic reactions from the hydrolysis of nucleoside triphosphate. The free energy change in the hydrolysis of PP_i is −28.8 kJ/mol PP_i, which is smaller than that of ATP and polyphosphate. Many organisms have phosphatase activity that hydrolyses PP_i without conserving energy. In some bacteria, PP_i functions like ATP. Phosphatase activity is low in species of *Desulfotomaculum*. These have an acetate: pyrophosphate kinase that phosphorylates acetate, consuming PP_i (Section 9.3.2). A purple non-sulfur bacterium, *Rhodospirillum rubrum*, has pyrophosphate phosphohydrolase on the cytoplasmic membrane that synthesizes PP_i using energy available from photosynthetic electron transport, and hydrolyses it to build up a proton motive force. Some prokaryotes have a membrane-bound Na^+-pyrophosphatase and Na^+,H^+-pyrophosphatase that couples Na^+(H^+) export to PP_i hydrolysis.

Propionibacterium shermanii metabolizes glucose through the EMP pathway to pyruvate, which is then fermented to propionate (Section 8.7.1). In this bacterium, PP_i is used as the substrate of pyrophosphate fructokinase (diphosphate - fructose-6-phosphate 1-phosphotransferase) and pyruvate phosphate dikinase. A thermophilic bacterium, *Caldicellulosiruptor saccharolyticus*, also has these enzymes:

$$\text{fructose-6-phosphate} + PP_i \xrightleftharpoons{\text{pyrophosphate fructokinase}} \text{fructose-1,6-diphosphate} + P_i$$

$$\text{PEP} + \text{AMP} + PP_i \xrightleftharpoons{\text{pyruvate phosphate dikinase}} \text{pyruvate} + \text{ATP} + P_i$$

13.3 Resting cells

Many bacteria differentiate into resting cells when the growth environment becomes unfavourable. These include spores in low G + C Gram-positive bacteria and actinomycetes, cysts and viable but non-culturable cells. In addition to the formation of resting cells, some bacteria differentiate into other specialized cells, such as bacteroids in symbiotic nitrogen-fixing bacteria, heterocysts in cyanobacteria (Section 6.2.1.4), swarmer cells in *Caulobacter cereus* and fruiting bodies in myxobacteria.

13.3.1 Sporulation in *Bacillus subtilis*

Some low G + C Gram-positive bacteria, including species of *Bacillus* and *Clostridium*, form spores under nutrient-limited and other adverse growth conditions. *Acetonema longum* is a sporulating Gram-negative bacterium. Spores of this organism can maintain their viability for many years.

As in other cell differentiation processes, sporulation is the result of a complex regulated system which includes signal transduction from environmental and physiological factors. During the spore-forming process the cells cannot propagate. An exception to this is known in the multiple endospore forming *Metabacterium polyspora*. After binary asymmetric cell division, the forespores can undergo division to produce multiple forespores that grow and mature into multiple endospores. Another multiple endospore former, *Anaerobacter polyendosporus*, divides to produce multiple forespores (Section 6.14.1.2).

Predivisional stage sporulation sigma factor σ^H is essential for the transition from vegetative growth to sporulation. σ^H activates the expression of at least 87 genes during the shift from exponential growth to the stationary phase, playing a pivotal role in spore formation, activating the gene for KinA, the first sensor kinase of the phosphorelay. σ^H is also involved in genetic competence.

Sporulation is regulated through a series of phosphate transfers known as a phosphorelay. A sensor kinase of a two-component system (Section 12.1.7), KinA or KinB, autophosphorylates in response to an appropriate stimulus and subsequently transfers the phosphate to Spo0F (a two-component response regulator). The phosphate from Spo0F is transferred to Spo0A (another two-component response regulator) through Spo0B (sporulation initiation phosphotransferase). Phosphorylated Pho0A plays a pivotal role in sporulation, activating σ^F and

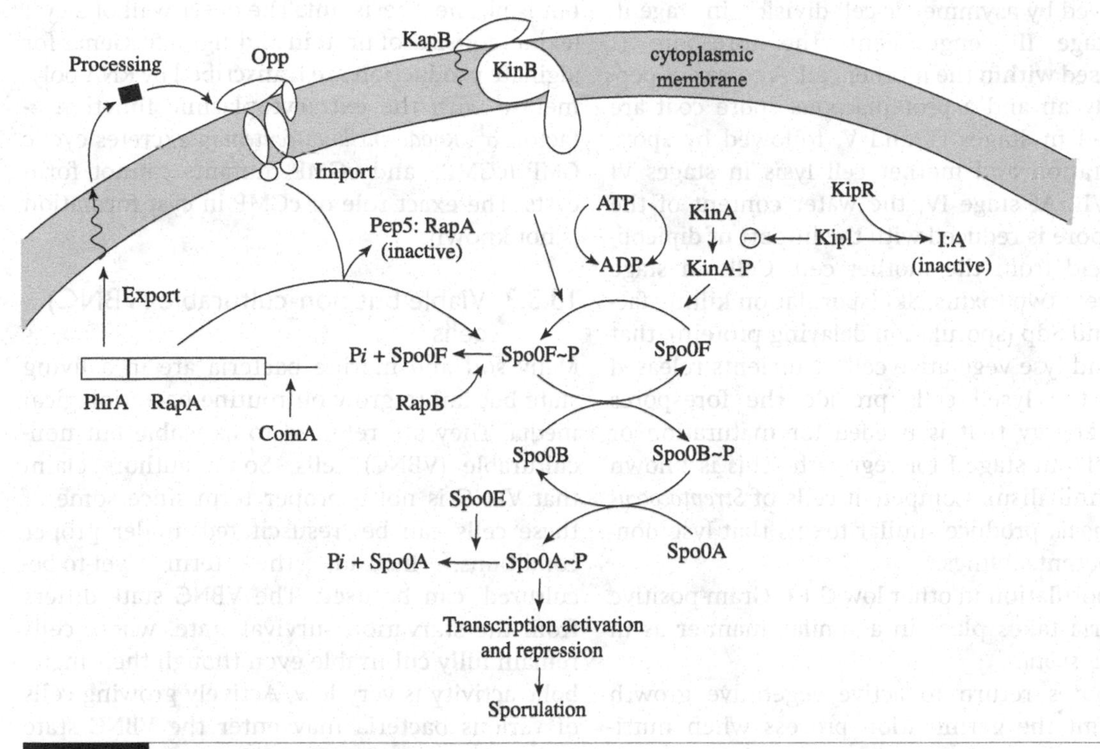

Figure 13.7 Sporulation in *Bacillus subtilis*.

(*Trends Microbiol.* **6**: 366–370, 1998; *Curr. Opin. Microbiol.* **1**: 170–174, 1998)

KinA and KinB consume ATP to phosphorylate Spo0F that transfers phosphate to Spo0A through Spo0B. This phosphate transfer is referred to as a phosphorelay. Phosphorylated Spo0A represses the genes for vegetative cell growth and activates the genes for spore formation. The phosphorelay is under the control of four different steps. KapB regulates the activity of the membrane-bound KinB, and the cytoplasmic enzyme KinA is under the control of antikinase (Kipl) and anti-antikinase (KipR). The phosphorylated Spo0A is dephosphorylated by Spo0E and dephosphorylation of Spo0F-P is under the control of RapA and RapB. It is not fully understood how the signal is transduced to KapB, Kipl, KipR, Spo0E, RapA and RapB. Quorum-sensing systems are involved in control of the phosphorelay.

KapB, kinase-associated protein B; KinA, sporulation kinase A; KinB, sporulation kinase B; Kipl, antikinase (KinA inhibitor); KipR, anti-antikinase (*kip* operon transcriptional regulator); RapA, response regulator aspartate phosphatase A; RapB, response regulator aspartate phosphatase B; PhrA, phosphatase (RapA) regulator; Spo0A, two-component response regulator; Spo0B, sporulation initiation phosphotransferase; Spo0E, negative sporulation regulatory phosphatase; Spo0F, two-component response regulator; ComA, two-component response regulator.

σ^E and the genes for sporulation and repressing the genes for vegetative growth. Various modulators regulate the phosphorelay process (Figure 13.7). In *Bacillus subtilis*, several quorum-sensing systems (Section 12.2.8), consisting of phosphatase regulators (Phr) and response aspartyl phosphatases (Raps), control sporulation phosphorelay signal transduction.

After the forespore is formed through asymmetric cell division, forespore specific σ^F is activated. σ^F activates up to 50 genes, including a signal factor gene *spoIIR* that activates a mother cell specific σ^E. σ^E activates another forespore specific σ^G that in turn activates the final mother cell transcription factor σ^K in sequence. For this signal transduction, the forespore has a channel similar to the Type III secretion system known in Gram-negative bacteria (Section 3.10.2).

The sporulation process can be divided into seven stages. In stage I, Spo0A is phosphorylated followed by asymmetric cell division in stage II. In stage III (engulfment) the forespore is enclosed within the mother cell. A cortex of peptidoglycan and a proteinaceous spore coat are formed in stages IV and V, followed by spore maturation and mother cell lysis in stages VI and VII. At stage IV, the water content of the forespore is reduced with the import of dipicolinic acid from the mother cell. Cells in stage I excrete two toxins, Skf (sporulation killing factor) and Sdp (sporulation delaying protein), that kill and lyse vegetative cells. Nutrients released from the lysed cells provide the forespores with energy that is needed for maturation or for cells in stage I for regrowth. This is known as cannibalism. Competent cells of *Streptococcus pneumonia* produce similar toxins that lyse non-competent siblings.

Sporulation in other low G + C Gram-positive bacteria takes place in a similar manner as in *Bacillus subtilis*.

Spores return to active vegetative growth through the germination process when nutrients are available. Nutrients (germinants) such as amino acids, purine derivatives and sugars are sensed by germinant receptor proteins located in the inner membrane. Specific spore coat proteins facilitate the passage of germinants through the outer layers of spores to the receptors. When germinant receptors sense their cognate germinant, a series of physical events is initiated. These include the release of monovalent cations and calcium dipicolinate. Calcium dipicolinate release triggers hydrolysis of cortex peptidoglycan that allows water uptake into the core to levels found in vegetative cells. Full core hydration then allows enzyme activity and initiation of metabolism, macromolecular synthesis and spore outgrowth.

13.3.2 Cysts

In some bacteria, such as *Azotobacter* and *Cytophaga* spp., resting cells known as cysts are formed. A photosynthetic bacterium, *Rhodospirillum centenum*, and the symbiotic nitrogen-fixing bacterium *Sinorhizobium meliloti* also form cysts. A cyst differs from a vegetative cell in both size and morphology. A cyst is resistant against desiccation and UV radiation like a spore, but is not heat resistant. The outer wall of a cyst (exine) consists of protein and alginate. Genes for alginate production are transcribed by RNA polymerase with the extracytoplasmic function σ-factor, σ^E. *Rhodospirillum centenum* excretes cyclic GMP (cGMP), and cGMP mutants cannot form cysts. The exact role of cGMP in cyst formation is not known.

13.3.3 Viable but non-culturable (VBNC) cells

Many soil and marine bacteria are in a living state but fail to grow on routine bacteriological media. They are referred to as viable but non-culturable (VBNC) cells. Some authors claim that VBNC is not a proper term since some of these cells can be 'resuscitated' under proper conditions. Instead, the term 'yet-to-be-cultured' can be used. The VBNC state differs from the starvation survival state, where cells remain fully culturable even though their metabolic activity is very low. Actively growing cells of various bacteria may enter the VBNC state under diverse environmental stresses, including starvation, non-optimal temperature range for growth, high osmotic pressure and exposure to antibacterial compounds and white light. These stresses might be lethal if the cells did not enter this dormant state. VBNC cells are resistant to such exogenous stresses and confer the properties of long-term survival and of resuscitation.

In addition to the failure to grow on routine bacteriological media, cells in the VBNC state differ from actively growing cells both morphologically and physiologically. VBNC cells are smaller and show increased cell wall cross-linking and extensively modified cytoplasmic membranes with altered fatty acid composition. Expression of starvation and cold shock proteins is increased, and the cells show reduced activities in nutrient uptake, respiration and the synthesis of macromolecules.

The plant pathogenic *Pseudomonas syringae* enters the VBNC state when exposed to the oxidative burst of the plant defence system. Conversion to the VBNC state is accompanied by increased expression of genes for the oxidative stress response, carbohydrate metabolism, energy generation, chemotaxis and peptidoglycan metabolism. These may have critical functions in the transition to and maintenance of the VBNC state. The molecular mechanisms involved in VBNC cell formation are not fully known.

Several hypotheses have been proposed to explain why cells in the VBNC state cannot form colonies on routine media. High concentrations of nutrients may be toxic to cells in the VBNC state because of the formation of free radicals, although this is not known conclusively.

It is noteworthy that a protein produced by *Micrococcus luteus* in the stationary phase promotes resuscitation of VBNC state cells of many high G + C Gram-positive bacteria, including the producing organism. This protein is referred to as a resuscitation-promoting factor (Rpf) or bacterial cytokine. Rpf is known in other bacteria, including *Mycobacterium tuberculosis*, *Corynebacterium glutamicum* and *Streptomyces* spp. Rpf proteins of *M. luteus* and *Streptomyces coelicolor* have cell wall hydrolysing activity, lysing Gram-positive as well as Gram-negative bacterial cells. This suggests that remodelling the cell wall with increased cross-linking is an important part of the resuscitation process. *M. tuberculosis* produces five Rpf proteins, and a mutant strain lacking all of them is viable *in vitro*, but is defective in restoration of growth from the stationary phase. A Rpf-like protein, YeaZ, is known in many Gram-negative bacteria, including *Vibrio harveyi* and *Escherichia coli*, and has a similar function.

VBNC cells of pathogenic *Vibrio cholerae* and *Vibrio vulnificus* are resuscitated by autoinducers of the interspecies quorum-sensing system (Section 12.2.8) that is common in the mammalian intestine, which these bacteria infect.

It has been estimated that less than 1 per cent of microorganisms can be cultivated under laboratory conditions (Chapter 1). Most of the yet-to-be-cultured microbes are not in the VBNC state. Various methods have been tested to improve culturability of the yet-to-be-cultured microbes, but with limited success. It is noteworthy that more diverse microbes grow on agar medium where phosphate is added separately after autoclaving, than on medium where agar and phosphate are autoclaved together.

13.3.4 Persister cells

Persister cells represent a small subpopulation of cells of rapidly growing cultures during the mid-exponential phase, that spontaneously enter a dormant, non-dividing state. They are morphologically similar to actively growing cells but highly tolerant of antibiotics and other stresses. Persister cells are formed at low frequency in actively growing populations, while VBNC cells are formed after longer periods of incubation in environmental conditions that do not sustain growth.

The molecular mechanisms responsible for persister cell formation are not fully elucidated. However, it is known that toxin–antitoxin modules (Section 13.4.2) such as HipA–HipB (high persister), MqsR–MqsA and HokB–SokB, activated by polyphosphate (Section 13.2.4) and (p)ppGpp (Section 12.2.1), are involved.

13.3.5 Nanobacteria

In seawater and soil, bacteria with a size of less than 0.2 μm have been described for many years. Most of these do not form colonies on routine solid media. They are referred to as ultramicrobacteria or nanobacteria. Some isolated nanobacteria including a seawater isolate, *Sphingomonas alaskensis*, grows on standard nutrient media but do not increase in size upon prolonged cultivation in the laboratory, while others grow to a normal bacterial cell size. The latter include bacteria belonging to the *Proteobacteria*, *Actinobacteria* and *Firmicutes* phyla. It is believed that the small-cell-sized bacteria that can increase their cell size are a form of resting cell. However, little is known about how these resting cells are formed.

13.4 Population survival

13.4.1 Programmed cell death (PCD) in bacteria

Programmed cell death, or apoptosis, is a suicide process of active cells in multicellular eukaryotes. Cells no longer needed during a developmental process or those damaged by heat or other lethal agents are destroyed through proteolysis for the preservation of the tissue and/or the individual organism.

Although PCD has been regarded as a eukaryotic phenomenon, similar processes are known in some bacteria. Bacteroids and heterocysts (Section 6.2.1.4) do not divide or return to vegetative cells. When their function is no longer required, they are subject to autolysis, like the mother cells of endospore formers. These are examples of PCD in bacteria that are similar to eukaryotic developmental PCD. Damaged cells are eliminated through autolysis as in a PCD mechanism. Damaged cells that have survived toxic stress induced by various compounds could consume nutrients but would produce few, if any, offspring, which would be a burden for the whole population. PCD of damaged cells therefore benefits the entire population. Bacteria thus appear to be able to exhibit social behaviour. Many live in large, complex and organized communities such as biofilms, while fruiting bodies arise through the initiation of developmental processes that are analogous to those in multicellular eukaryotic organisms. Multicellularity in the bacterial world, in its various forms, might therefore be more prevalent than previously thought. With some of the known similarities to developmental processes in eukaryotes, 'altruistic' behaviour in bacteria might also be a predictable phenomenon.

When *Escherichia coli* cells are stressed to inhibit the transcription or translation of the toxin–antitoxin system *mazEF*, the unstable antitoxin MazE concentration becomes low, activating the cognate toxin MazF that degrades the mRNAs of essential genes leading to cell death. The toxin, usually a RNase, is a stable protein, while the antitoxin is an unstable protein or sRNA (Section 13.4.2). Stresses inhibiting transcription or translation of *mazEF* include ppGpp during the stringent response (Section 12.2.1), antibiotics, high temperatures and others. This toxin–antitoxin system is known in many bacteria.

In *Staphylococcus aureus*, various genes have been identified that are related to the regulation of autolysis and peptidoglycan hydrolase. The genes are believed to code for regulatory proteins involved in PCD. Actively metabolizing bacterial cells maintain a certain level of proton motive force (Section 5.7), with a relatively low pH microenvironment at the cell surface. When a cell is damaged, with dissipation of the proton motive force, changes in the cell surface pH are sensed and transduced by a two-component system consisting of autolysin sensor kinase (LytS) and sensory transduction protein (LytR). It has been hypothesized that this signal represses the activity of antiholin-like protein A (LrgA) and B (LrgB), and activates holin-like protein A (CidA) and B (CidB). LrgA and LrgB inhibit peptidoglycan hydrolase activity, while the enzyme activity is activated by CidA and CidB for PCD. Homologues of *lrgAB* have been identified in many prokaryotes, including archaea and Gram-positive and Gram-negative bacteria, indicating that PCD is probably widespread in prokaryotes.

Sporulating cells of *Bacillus subtilis* secrete two toxins, Skf (sporulation killing factor) and Sdp (sporulation delaying protein), that kill and lyse vegetative cells to provide forespores with the energy needed for maturation (Section 13.3.1).

13.4.2 Toxin–antitoxin systems

In addition to their roles in persister cell formation (Section 13.3.4) and PCD (Section 13.4.1), toxin–antitoxin (TA) systems have other functions, including stress responses, growth control by bacteriostatic toxins, gene regulation by some toxins and an antiphage mechanism through PCD to prevent phage replication.

The stable toxin proteins are responsible for cellular functions and the labile antitoxins neutralize toxin activity through complex formation with the cognate toxin, or regulating the translation of toxin genes. In normally growing cells, toxin activity is kept low by its cognate antitoxin. The labile antitoxins have to be continuously synthesized in order to constantly inhibit stable toxin function. Under stress conditions, stress-induced enzymes digest labile antitoxins,

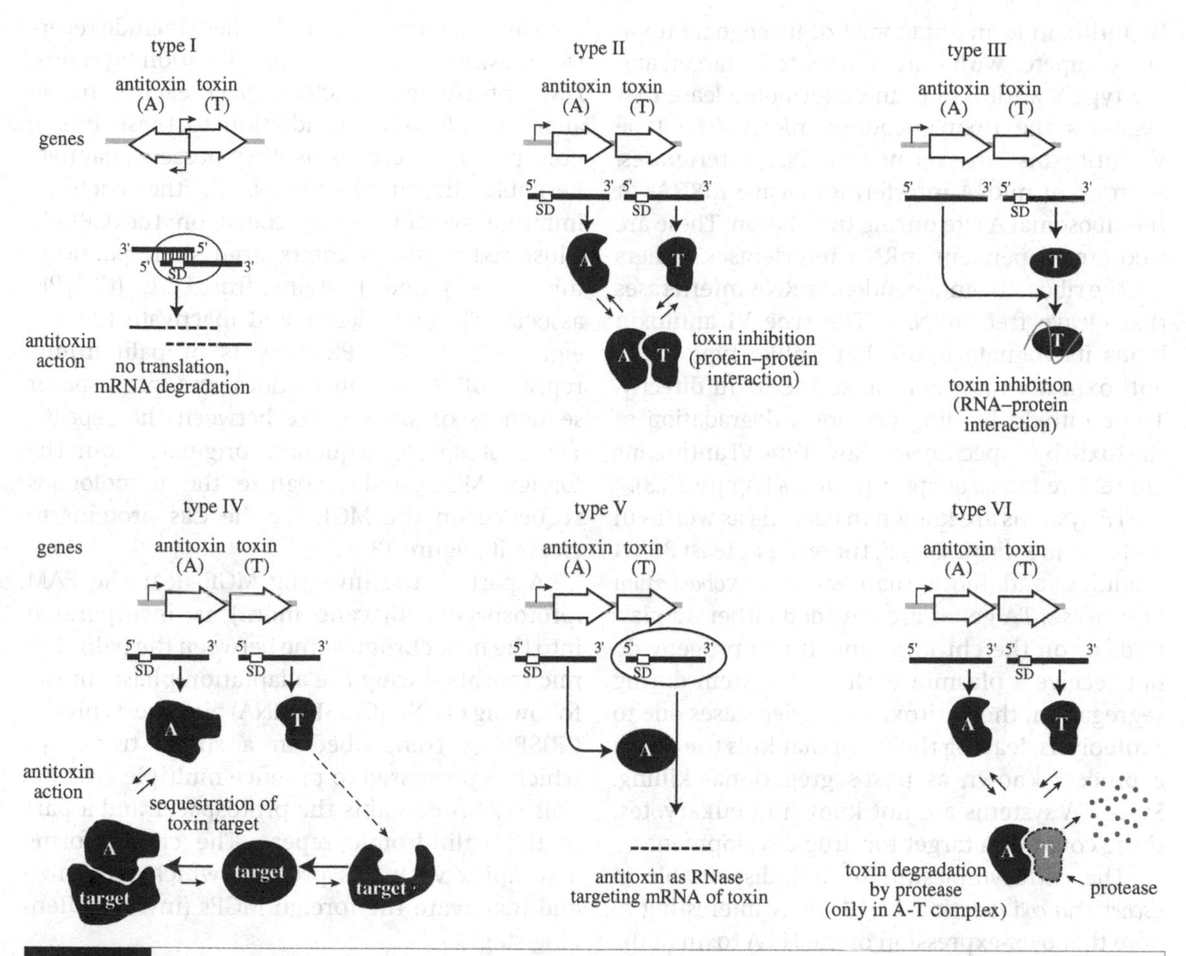

Figure 13.8 Antitoxins of toxin–antitoxin (TA) systems.

(Modified from *FEMS Microbiol. Rev.* **40**: 592–609, 2016)

TA systems are classified according to the nature and activity of the antitoxins. Type I and III antitoxins are small RNAs. Type I antitoxins are antisense RNA that binds toxin mRNA to inhibit translation while type III antitoxins form complexes with toxins neutralizing them.

Antitoxins are proteins in types II, IV, V, and VI. The type II antitoxin forms a complex with its cognate toxin, the type IV antitoxin is an antagonist of its cognate toxin in binding to its target and the type V antitoxin is an endoribonuclease that degrades the toxin-encoding mRNA. Type V antitoxins are termed mRNA interferases. The type VI antitoxin binds its cognate toxin, promoting its degradation by a specific protease.

activating the toxins or their production (Figure 13.8). The toxins interact with various cellular targets to inhibit DNA replication, mRNA stability, protein synthesis, cell wall synthesis and ATP production.

TA systems are currently classified into six types, depending on the nature and activity of the antitoxins. Types I and III antitoxins are small RNAs. Type I antitoxins are antisense RNA binding toxin mRNA to inhibit translation, while type III antitoxins form complexes with toxins neutralizing them. Types II, IV, V and VI, on the other hand, are all proteins. The type II antitoxin forms a complex with its cognate toxin, the type

IV antitoxin is an antagonist of its cognate toxin and competes with it in binding to its target, and the type V antitoxin is an endoribonuclease that degrades the toxin-encoding mRNA. The type V antitoxins are termed mRNA interferases. A group of mRNA interferases cleave mRNAs at the ribosomal A site during translation. These are ribosome-dependent mRNA interferases. Others are the ribosome-independent mRNA interferases that cleave free mRNAs. The type VI antitoxin binds its cognate toxin, but unlike type II the antitoxin does not neutralize the toxin directly. The antitoxin binding promotes degradation of the toxin by a specific protease. Type VI antitoxins are referred to as adaptor proteins (Figure 13.8).

TA systems are known in bacteria as well as in archaea. In *Escherichia coli*, there are at least 36 TA modules, and more than 80 in *Mycobacterium tuberculosis*. TA genes are encoded either on plasmids or on the chromosome. If cell progeny do not receive a plasmid with a TA system during segregation, the antitoxin level decreases due to proteolysis, leaving the toxin that kills the cell in a process known as post-segregational killing. Since TA systems are not known in eukaryotes, these could be a target for drug development.

The extracytoplasmic σ^E is indispensable in *Escherichia coli* (Section 12.1.1). It is interesting to note that over-expression of the HicA toxin of the HicA–HicB TA system rescues growth of a σ^E-less mutant. The toxin MazF of the PCD inducing TA system *mazEF* in *Escherichia coli* is activated by quorum sensing autoinducer peptides (Section 12.2.8) of its own and produced by Gram-positive bacteria such as *Bacillus subtilis*. These peptides are referred to as extracellular death factors.

TA systems in pathogens are related to their virulence. These are involved in survival in the host as persister cells, and toxins of TA systems cause damage to host cells. Toxins ChpI and MazF of the TA systems *chpIK* and *mazEF* in *Leptospira interrogans* are secreted into host cell cytoplasm, causing apoptosis and necrosis.

13.5 Bacterial immune systems

Prokaryotes have multiple defence systems to combat invading mobile genetic elements (MGE) such as viruses and plasmids. These include receptor masking, restriction-modification systems, DNA interference, bacteriophage exclusion and abortive infection. In addition to these innate non-specific mechanisms, they possess adaptive, heritable immune systems. In the bacterial immune system, sRNAs coded on the CRISPR (clustered regularly interspaced short palindromic repeat) and proteins from Cas (CRISPR-associated) genes detect and inactivate the foreign MGE. A CRISPR consists of palindromic repeats of 30–40 nucleotides and protospacer sequences of similar size between the repeats. The protospacer sequences originate from the foreign MGE, and recognize the homologous sequence on the MGE for the Cas proteins to cleave it (Figure 13.9).

A part of the invading MGE near the PAM (protospacer adjacent motif) is incorporated into the host chromosome between the palindromic repeats during the adaptation phase. In the following crRNA (CRISPR RNA) biogenesis phase, CRISPR is transcribed in a single transcript which is processed to produce multiple crRNAs. This crRNA contains the protospacer and a part of the palindromic repeat. The crRNA forms a complex with Cas proteins which recognize and inactivate the foreign MGEs (invader silencing stage).

Although CRISPR–Cas systems share a common molecular principle for genome silencing, the systems are highly variable in their *cas* gene composition. Various Cas proteins are involved in crRNA maturation and the forming of the effector complex. CRISPR–Cas systems are divided into two classes with six types and 17 subtypes. The effector complex of Class 1 consists of the crRNA and multiple proteins, while a single protein forms the effector complex with the crRNA in Class 2. Class 1 includes Types I, III and IV, and Types II, V and VI belong to Class 2. Type I is further divided into Type IA through to Type IF, Type II into IIA–IID and Type III into IIIA–IIID. Types IV, V and VI are single subtype systems. For use in genome editing and other purposes, extensive studies are being made on Type II systems where the effector complex consists of crRNA and a single protein, the Cas nuclease Cas9. Synthetic (reprogrammed) RNA

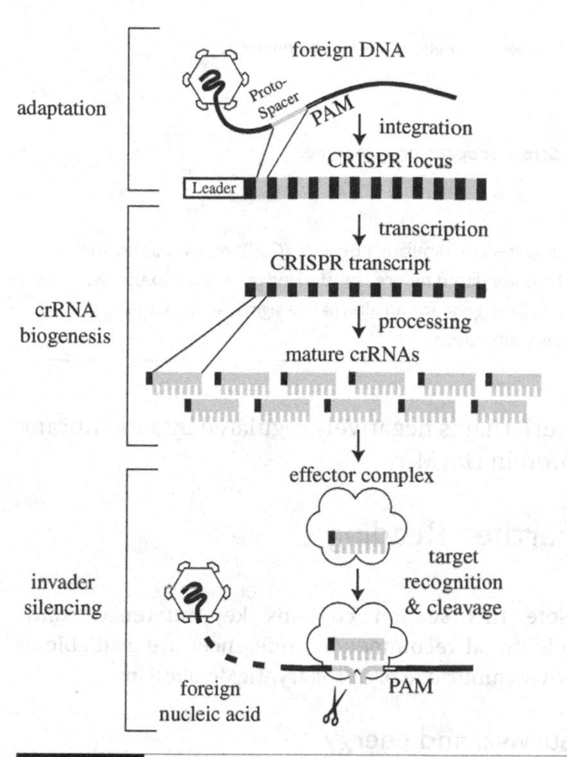

Figure 13.9 Bacterial immune system silencing foreign genetic elements by the action of the CRISPR–Cas complex.

(Modified from *Curr. Opin. Microbiol.* **14**: 321–327, 2011)

During the adaptation phase, a portion of the foreign genetic elements (protospacer) near the protospacer adjacent motifs (PAMs) is incorporated into the CRISPR between the palindromic repeats. The CRISPR transcript is processed to produce crRNAs that contain the protospacer and a part of the palindromic repeat. The crRNAs form complexes with Cas proteins that recognize and cleave the targets.

specific for the target DNA is used in gene editing (see below).

In some bacteria, including *Pseudomonas aeruginosa*, the CRISPR-Cas system is activated by autoinducers of quorum sensing (Section 12.2.8) that are activated at high cell densities when the chance of phage infection is high.

In addition to protection against foreign MGE, there is evidence to show that crRNAs function as sRNA (Section 12.1.9.4) to regulate gene expression and bacterial stress responses in general, and bacterial virulence in particular.

Some lysogenic *Pseudomonas aeruginosa* strains that contain a phage genome in their chromosomes as prophage produce proteins encoded on the phage genome that inhibit the bacterial immune system through blocking DNA-binding activity of the CRISPR-Cas complex or through binding to Cas3, preventing its recruitment to the DNA-bound CRISPR-Cas complex. Anti-CRISPR proteins are known for Class 1 as well as Class 2 systems. Bacteria and phage are constantly coevolving to achieve survival.

CRISPR-Cas systems pose the risk of autoimmunity, inactivating the host DNA in a process known as self-targeting. While many of these protospacers match targets on MGE, noticeable numbers are derived from chromosomal DNA. There are two distinct mechanisms that prevent self-targeting. Types I, II, V and VI CRISPR-Cas systems require a protospacer adjacent motif (PAM) to recognize the target. The host chromosomal DNA does not have PAM. Other types of CRISPR-Cas systems that do not need a PAM require mismatches between the target and a few flanking nucleotides of a protospacer region in the crRNA. If the flanking regions are complementary to each other, the nucleic acid sequence escapes CRISPR-Cas recognition. While self-targeting is lethal to the host in most cases, non-lethal self-targeting has profound effects on the organisms. Self-targeting normally results in cell death, but can also reshape the bacterial genome through deletion of the target region by DNA repair systems. Prophages from lysogenic cells can be removed by this mechanism, thus reshaping the population.

In Type II CRISPR-Cas systems, DNA-cleaving activity is performed by a single enzyme, Cas9, guided by a crRNA and *trans*-activating CRISPR RNA (tracrRNA) duplex. Using a synthetic single-guide RNA chimera (sgRNA), this sytem can be reprogrammed to create specific double-stranded DNA breaks, gene deletions or specific gene insertions in the genomes of a variety of organisms, ranging from bacteria to humans, and thus constitutes a powerful tool for genetic engineering. The technology can be applied in various ways, including alleviation of genetic disorders in animals, crop and livestock breeding, engineering new antimicrobials and controlling

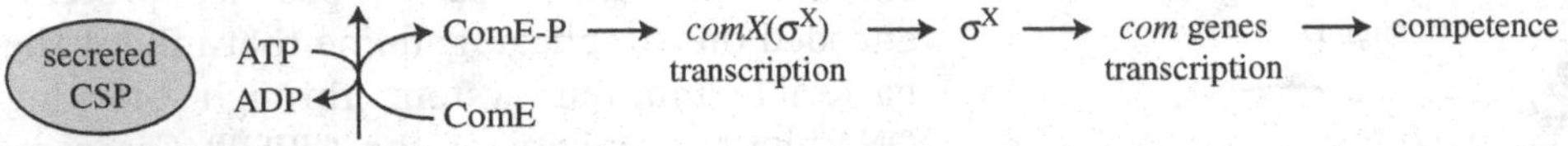

Figure 13.10 Regulation of competence gene expression in *Streptococcus pneumoniae*.

(Modified from *Ann. Rev. Microbiol.* **60**: 451–475, 2006)

When the concentration of the peptide autoinducer, known as the competence-stimulating peptide (CSP), reaches a threshold level, the sensor kinase, ComD, of a two-component system is autophosphorylated before transferring phosphate to the response regulator protein ComE. Phosphorylated ComE activates transcription of the gene for an alternative sigma factor, ComX(σ^X), that transcribes competence genes responsible for DNA uptake and recombination.

disease-carrying insects, among other applications. In addition, nuclease-null Cas9 is used to regulate endogenous gene expression and to label genomic loci in living cells.

13.6 Competence

While foreign mobile genetic elements are inactivated by CRISPR–Cas systems, many bacteria import and incorporate foreign DNA into their chromosome under certain conditions. This property is known as competence. In the Gram-positive bacterium *Streptococcus pneumoniae*, an alternative sigma factor, ComX (σ^X), transcribes *com* genes encoding competence proteins involved in DNA uptake and recombination in the late exponential phase. The membrane-bound sensor kinase of a two-component system (Section 12.1.7), ComD, is phosphorylated when the quorum-sensing peptide autoinducer concentration reaches a threshold level (Section 12.2.8), and transfers phosphate to the response regulator ComE that activates transcription of the ComX(σ^X) gene (Figure 13.10). The peptide autoinducer is referred to as a competence-stimulating peptide (CSP). DNA acquired through competence substitutes for damaged DNA by recombination. In addition to the ComD/E quorum-sensing system, ComX(σ^X) transcription is activated by a regulatory protein HdrR that is negatively regulated by a membrane protein HdrM.

Further Reading

Note this section contains key references only. Additional recommended references are available at www.cambridge.org/ProkaryoticMetabolism.

Survival and energy

Aertsen, A. & Michiels, C. (2004). Stress and how bacteria cope with death and survival. *Critical Reviews in Microbiology* **30**, 263–273.

Errington, J., Daniel, R. A. & Scheffers, D. J. (2003). Cytokinesis in bacteria. *Microbiology and Molecular Biology Reviews* **67**, 52–65.

Ferenci, T. (2001). Hungry bacteria: definition and properties of a nutritional state. *Environmental Microbiology* **3**, 605–611.

Kempes, C. P., van Bodegom, P. M., Wolpert, D., Libby, E., Amend, J. & Hoehler, T. (2017). Drivers of bacterial maintenance and minimal energy requirements. *Frontiers in Microbiology* **8**, 31.

Lever, M. A., Rogers, K. L., Lloyd, K. G., Overmann, J., Schink, B., Thauer, R. K., Hoehler, T. M. & Jørgensen, B. B. (2015). Life under extreme energy limitation: a synthesis of laboratory- and field-based investigations. *FEMS Microbiology Reviews* **39**, 688–728.

Matic, I., Taddei, F. & Radman, M. (2004). Survival versus maintenance of genetic stability: a conflict of priorities during stress. *Research in Microbiology* **155**, 337–341.

Mukamolova, G. V., Kaprelyants, A. S., Kell, D. B. & Young, M. (2003). Adoption of the transiently non-culturable state – a bacterial survival strategy? *Advances in Microbial Physiology* **47**, 65–129.

Nystrom, T. (2004). Growth versus maintenance: a trade-off dictated by RNA polymerase availability and sigma factor competition? *Molecular Microbiology* **54**, 855–862.

Peterson, C. N., Mandel, M. J. & Silhavy, T. J. (2005). *Escherichia coli* starvation diets: essential nutrients weigh in distinctly. *Journal of Bacteriology* **187**, 7549–7553.

Carbohydrate reserve materials

Chandra, G., Chater, K. F. & Bornemann, S. (2011). Unexpected and widespread connections between bacterial glycogen and trehalose metabolism. *Microbiology* **157**, 1565–1572.

Elbein, A. D., Pastuszak, I., Tackett, A. J., Wilson, T. & Pan, Y. T. (2010). Last step in the conversion of trehalose to glycogen: a mycobacterial enzyme that transfers maltose from maltose-1-phosphate to glycogen. *Journal of Biological Chemistry* **285**, 9803–9812.

Wilson, W. A., Roach, P. J., Montero, M., Baroja-Fernández, E., Muñoz, F. J., Eydallin, G., Viale, A. M. & Pozueta-Romero, J. (2010). Regulation of glycogen metabolism in yeast and bacteria. *FEMS Microbiology Reviews* **34**, 952–985.

Lipid reserve materials

Alvarez, H. M. & Steinbuchel, A. (2002). Triacylglycerols in prokaryotic microorganisms. *Applied Microbiology and Biotechnology* **60**, 367–376.

Herman, N. A. & Zhang, W. (2016). Enzymes for fatty acid-based hydrocarbon biosynthesis. *Current Opinion in Chemical Biology* **35**, 22–28.

Jendrossek, D. & Pfeiffer, D. (2014). New insights in the formation of polyhydroxyalkanoate granules (carbonosomes) and novel functions of poly(3–hydroxybutyrate). *Environmental Microbiology* **16**, 2357–2373.

Jiménez-Díaz, L., Caballero, A., Pérez-Hernández, N. & Segura, A. (2017). Microbial alkane production for jet fuel industry: motivation, state of the art and perspectives. *Microbial Biotechnology* **10**, 103–124.

Low, K. L., Shui, G., Natter, K., Yeo, W. K., Kohlwein, S. D., Dick, T., Rao, S. P. S. & Wenk, M. R. (2010). Lipid droplet-associated proteins are involved in the biosynthesis and hydrolysis of triacylglycerol in *Mycobacterium bovis* Bacillus Calmette-Guérin. *Journal of Biological Chemistry* **285**, 21662–21670.

Maestro, B. & Sanz, J. M. (2017). Polyhydroxyalkanoate-associated phasins as phylogenetically heterogeneous, multipurpose proteins. *Microbial Biotechnology* **10**, 1323–1337.

Stubbe, J., Tian, J., He, A., Sinskey, A. J., Lawrence, A. G. & Liu, P. (2005). Nontemplate-dependent polymerization processes: polyhydroxyalkanoate synthases as a paradigm. *Annual Review of Biochemistry* **74**, 433–480.

Waltermann, M. & Steinbuchel, A. (2005). Neutral lipid bodies in prokaryotes: recent insights into structure, formation, and relationship to eukaryotic lipid depots. *Journal of Bacteriology* **187**, 3607–3619.

Polyphosphate

Baykov, A. A., Malinen, A. M., Luoto, H. H. & Lahti, R. (2013). Pyrophosphate-fueled Na^+ and H^+ transport in prokaryotes. *Microbiology and Molecular Biology Reviews* **77**, 267–276.

Brown, M. R. & Kornberg, A. (2004). Inorganic polyphosphate in the origin and survival of species. *Proceedings of the National Academy of Sciences of the USA* **101**, 16085–16087.

Garcia-Contreras, R., Celis, H. & Romero, I. (2004). Importance of *Rhodospirillum rubrum* H^+–pyrophosphatase under low-energy conditions. *Journal of Bacteriology* **186**, 6651–6655.

Gray, M. J. & Jakob, U. (2015). Oxidative stress protection by polyphosphate – new roles for an old player. *Current Opinion in Microbiology* **24**, 1–6.

Kellosalo, J., Kajander, T., Kogan, K., Pokharel, K. & Goldman, A. (2012). The structure and catalytic cycle of a sodium-pumping pyrophosphatase. *Science* **337**, 473–476.

Kulaev, I. & Kulakovskaya, T. (2000). Polyphosphate and phosphate pump. *Annual Review of Microbiology* **54**, 709–734.

Toso, D. B., Henstra, A. M., Gunsalus, R. P. & Zhou, Z. H. (2011). Structural, mass and elemental analyses of storage granules in methanogenic archaeal cells. *Environmental Microbiology* **13**, 2587–2599.

Resting cells

Heinrich, K., Leslie, D. J. & Jonas, K. (2015). Modulation of bacterial proliferation as a survival strategy. *Advances in Applied Microbiology*. **92**, 127–171.

Kaprelyants, A. S., Gottschal, J. C. & Kell, D. B. (1993). Dormancy in nonsporulating bacteria. *FEMS Microbiology Reviews* **10**, 271–286.

Sporulation

Al-Hinai, M. A., Jones, S. W. & Papoutsakis, E. T. (2015). The *Clostridium* sporulation programs: diversity and preservation of endospore differentiation. *Microbiology and Molecular Biology Reviews* **79**, 19–37.

Errington, J. (2001). Septation and chromosome segregation during sporulation in *Bacillus subtilis*. *Current Opinion in Microbiology* **4**, 660–666.

Fimlaid, K. A. & Shen, A. (2015). Diverse mechanisms regulate sporulation sigma factor activity in the Firmicutes. *Current Opinion in Microbiology* **24**, 88–95.

González-Pastor, J. E. (2011). Cannibalism: a social behavior in sporulating *Bacillus subtilis*. *FEMS Microbiology Reviews* **35**, 415–424.

Higgins, D. & Dworkin, J. (2012). Recent progress in *Bacillus subtilis* sporulation. *FEMS Microbiology Reviews* **36**, 131–148.

Hilbert, D. W. & Piggot, P. J. (2004). Compartmentalization of gene expression during *Bacillus subtilis* spore formation. *Microbiology and Molecular Biology Reviews* **68**, 234–262.

McKenney, P. T., Driks, A. & Eichenberger, P. (2013). The *Bacillus subtilis* endospore: assembly and functions of the multilayered coat. *Nature Reviews Microbiology* **11**, 33–44.

Moir, A. (2003). Bacterial spore germination and protein mobility. *Trends in Microbiology* **11**, 452–454.

Paredes-Sabja, D., Setlow, P. & Sarker, M. R. (2011). Germination of spores of *Bacillales* and *Clostridiales* species: mechanisms and proteins involved. *Trends in Microbiology* **19**, 85–94.

Setlow, P. (2014). Germination of spores of *Bacillus* species: what we know and do not know. *Journal of Bacteriology* **196**, 1297–1305.

Stephenson, K. & Hoch, J. A. (2002). Evolution of signalling in the sporulation phosphorelay. *Molecular Microbiology* **46**, 297–304.

Cysts

Loiko, N., Kryazhevskikh, N., Suzina, N., Demkina, E., Muratova, A., Turkovskaya, O., Kozlova, A., Galchenko, V. & El'-Registan, G. (2011). Resting forms of *Sinorhizobium meliloti*. *Microbiology-Moscow* **80**, 472–482.

Marden, J. N., Dong, Q., Roychowdhury, S., Berleman, J. E. & Bauer, C. E. (2011). Cyclic GMP controls *Rhodospirillum centenum* cyst development. *Molecular Microbiology* **79**, 600–615.

Viable but non-culturable (VBNC) cells

Bari, S. M. N., Roky, M. K., Mohiuddin, M., Kamruzzaman, M., Mekalanos, J. J. & Faruque, S. M. (2013). Quorum-sensing autoinducers resuscitate dormant *Vibrio cholerae* in environmental water samples. *Proceedings of the National Academy of Sciences of the USA* **110**, 9926–9931.

Cohen-Gonsaud, M., Keep, N. H., Davies, A. P., Ward, J., Henderson, B. & Labesse, G. (2004). Resuscitation-promoting factors possess a lysozyme-like domain. *Trends in Biochemical Sciences* **29**, 7–10.

Epstein, S. S. (2013). The phenomenon of microbial uncultivability. *Current Opinion in Microbiology* **16**, 636–642.

Kell, D. B. & Young, M. (2000). Bacterial dormancy and culturability: the role of autocrine growth factors. *Current Opinion in Microbiology* **3**, 238–243.

Oliver, J. D. (2005). The viable but nonculturable state in bacteria. *Journal of Microbiology-Seoul* **43**, 93–100.

Pinto, D., Santos, M. A. & Chambel, L. (2015). Thirty years of viable but nonculturable state research: unsolved molecular mechanisms. *Critical Reviews in Microbiology* **41**, 61–76.

Sexton, D. L., St-Onge, R. J., Haiser, H. J., Yousef, M. R., Brady, L., Gao, C., Leonard, J. & Elliot, M. A. (2015). Resuscitation-promoting factors are cell wall-lytic enzymes with important roles in the germination and growth of *Streptomyces coelicolor*. *Journal of Bacteriology* **197**, 848–860.

Persister cells

Gerdes, K. & Maisonneuve, E. (2012). Bacterial persistence and toxin–antitoxin loci. *Annual Review of Microbiology* **66**, 103–123.

Harms, A., Maisonneuve, E. & Gerdes, K. (2016). Mechanisms of bacterial persistence during stress and antibiotic exposure. *Science* **354**, 1390–1399.

Lewis, K. (2010). Persister cells. *Annual Review of Microbiology* **64**, 357–372.

Maisonneuve, E., Castro-Camargo, M. & Gerdes, K. (2013). (p)ppGpp controls bacterial persistence by stochastic induction of toxin–antitoxin activity. *Cell* **154**, 1140–1150.

Nanobacteria

Duda, V., Suzina, N., Polivtseva, V. & Boronin, A. (2012). Ultramicrobacteria: formation of the concept and contribution of ultramicrobacteria to biology. *Microbiology-Moscow* **81**, 379–390.

Silbaq, F. S. (2009). Viable ultramicrocells in drinking water. *Journal of Applied Microbiology* **106**, 106–117.

Vainshtein, M. B. & Kudryashova, E. B. (2000). Nanobacteria. *Microbiology-Moscow* **69**, 129–138.

Programmed cell death

Durand, P. M., Sym, S. & Michod, R. E. (2016). Programmed cell death and complexity in microbial systems. *Current Biology* **26**, R587–R593.

Lewis, K. (2000). Programmed death in bacteria. *Microbiology and Molecular Biology Reviews* **64**, 503–514.

Prozorov, A. & Danilenko, V. (2011). Allolysis in bacteria. *Microbiology-Moscow* **80**, 1–9.

Ramisetty, B. C. M., Natarajan, B. & Santhosh, R. S. (2015). *mazEF*-mediated programmed cell death in bacteria: "What is this?". *Critical Reviews in Microbiology* **41**, 89–100.

Rice, K. C. & Bayles, K. W. (2003). Death's toolbox: examining the molecular components of bacterial programmed cell death. *Molecular Microbiology* **50**, 729–738.

Toxin–antitoxin systems

Brantl, S. & Jahn, N. (2015). sRNAs in bacterial type I and type III toxin–antitoxin systems. *FEMS Microbiology Reviews* **39**, 413–427.

Chan, W. T., Moreno-Córdoba, I., Yeo, C. C. & Espinosa, M. (2012). Toxin-antitoxin genes of the Gram-positive pathogen *Streptococcus pneumoniae*: So few and yet so many. *Microbiology and Molecular Biology Reviews* **76**, 773–791.

Lobato-Márquez, D., Díaz-Orejas, R. & García-del Portillo, F. (2016). Toxin–antitoxins and bacterial virulence. *FEMS Microbiology Reviews* **40**, 592–609.

Yamaguchi, Y., Park, J.-H. & Inouye, M. (2011). Toxin–antitoxin systems in bacteria and archaea. *Annual Review of Genetics* **45**, 61–79.

Bacterial immune systems

Barrangou, R. & Doudna, J. A. (2016). Applications of CRISPR technologies in research and beyond. *Nature Biotechnology* **34**, 933–941.

Bondy-Denomy, J. and Davidson, A.R. (2015). To acquire or resist: the complex biological effects of CRISPR–Cas systems. *Trends in Microbiology* **22**, 218–225.

Bondy-Denomy, J., Pawluk, A., Maxwell, K. L. & Davidson, A. R. (2013). Bacteriophage genes that inactivate the CRISPR/Cas bacterial immune system. *Nature* **493**, 429–432.

Dedrick, R. M., Jacobs-Sera, D., *et al.* (2017). Prophage-mediated defence against viral attack and viral counter-defence. *Nature Microbiology* **2**, 16251.

Doerflinger, M., Forsyth, W., Ebert, G., Pellegrini, M. & Herold, M. J. (2017). CRISPR/Cas9 – the ultimate weapon to battle infectious diseases? *Cellular Microbiology* **19**, e12693.

Doudna, J. A. & Charpentier, E. (2014). The new frontier of genome engineering with CRISPR-Cas9. *Science* **346**, 1258096.

Garneau, J. E., Dupuis, M.-E., Villion, M., Romero, D. A., Barrangou, R., Boyaval, P., Fremaux, C., Horvath, P., Magadan, A. H. & Moineau, S. (2010). The CRISPR/Cas bacterial immune system cleaves bacteriophage and plasmid DNA. *Nature* **468**, 67–71.

Heussler, G. E. & O'Toole, G. A. (2016). Friendly fire: Biological functions and consequences of chromosomal targeting by CRISPR-Cas systems. *Journal of Bacteriology* **198**, 1481–1486.

Jackson, S. A., McKenzie, R. E., Fagerlund, R. D., Kieper, S. N., Fineran, P. C. & Brouns, S. J. J. (2017). CRISPR-Cas: adapting to change. *Science* **356**, eaal5056.

Louwen, R., Staals, R. H. J., Endtz, H. P., van Baarlen, P. & van der Oost, J. (2014). The role of CRISPR–Cas systems in virulence of pathogenic bacteria. *Microbiology and Molecular Biology Reviews* **78**, 74–88.

Makarova, K. S., Wolf, Y. I., Alkhnbashi, O. S., Costa, F., Shah, S. A., Saunders, S. J., Barrangou, R., Brouns, S. J. J., Charpentier, E., Haft, D. H., Horvath, P., Moineau, S., Mojica, F. J. M., Terns, R. M., Terns, M. P., White, M., Yakunin, F. A. F., Garrett, R. A., van der Oost, J., Backofen, R. & Koonin, E. V. (2015). An updated evolutionary classification of CRISPR–Cas systems. *Nature Reviews Microbiology* **13**, 722–736.

Marraffini, L. A. (2015). CRISPR-Cas immunity in prokaryotes. *Nature* **526**, 55–61.

Mojica, F. J. M. & Rodriguez-Valera, F. (2016). The discovery of CRISPR in archaea and bacteria. *FEBS Journal* **283**, 3162–3169.

Peters, J. M., Silvis, M. R., Zhao, D., Hawkins, J. S., Gross, C. A. & Qi, L. S. (2015). Bacterial CRISPR: accomplishments and prospects. *Current Opinion in Microbiology* **27**, 121–126.

Selle, K. & Barrangou, R. (2015). Harnessing CRISPR–Cas systems for bacterial genome editing. *Trends in Microbiology* **23**, 225–232.

Sorek, R., Lawrence, C. M. & Wiedenheft, B. (2013). CRISPR-mediated adaptive immune systems in bacteria and archaea. *Annual Review of Biochemistry* **82**, 237–266.

van Houte, S., Buckling, A. & Westra, E. R. (2016). Evolutionary ecology of prokaryotic immune mechanisms. *Microbiology and Molecular Biology Reviews* **80**, 745–763.

Competence

Claverys, J. P., Prudhomme, M. & Martin, B. (2006). Induction of competence regulons as a general response to stress in Gram-positive bacteria. *Annual Review of Microbiology* **60**, 451–475.

Krüger, N.-J. & Stingl, K. (2011). Two steps away from novelty – principles of bacterial DNA uptake. *Molecular Microbiology* **80**, 860–867.

Martin, B., Quentin, Y., Fichant, G. & Claverys, J. P. (2006). Independent evolution of competence regulatory cascades in streptococci? *Trends in Microbiology* **14**, 339–345.

Mell, J. C. & Redfield, R. J. (2014). Natural competence and the evolution of DNA uptake specificity. *Journal of Bacteriology* **196**, 1471–1483.

Index

Printed in the United States
by Baker & Taylor Publisher Services

Printed in the United States
by Baker & Taylor Publisher Services